W9-CRB-593

NOISE IN PHYSICAL SYSTEMS AND 1/f NOISE – 1985

Organized with the aid of

IIª Universita' 'Tor Vergata', Roma
Consiglio Nazionale delle Ricerche, Roma

Sponsored by

IIª Universita' 'Tor Vergata', Roma
Consiglio Nazionale delle Ricerche, Roma
Balzers SpA, Milano
Istituto Bancario San Paolo di Torino, Torino

NOISE IN PHYSICAL SYSTEMS AND 1/f NOISE – 1985

Proceedings of the 8th International Conference on 'Noise in Physical Systems' and the 4th International Conference on '1/f Noise', Rome, September 9-13, 1985

Edited by

A. D'AMICO
Istituto di Elettronica dello Stato Solido
Roma

and

P. MAZZETTI
Dipartimento di Fisica
Politecnico di Torino
Torino

1986

NORTH-HOLLAND

AMSTERDAM • OXFORD • NEW YORK • TOKYO

ISBN: 0 444 86992 1

Published by:
North-Holland Physics Publishing
a division of
Elsevier Science Publishers B.V.
P.O. Box 103
1000 AC Amsterdam
The Netherlands

Sole distributors for the U.S.A. and Canada:
Elsevier Science Publishing Company, Inc.
52 Vanderbilt Avenue
New York, N.Y. 10017
U.S.A.

Library of Congress Cataloging-in-Publication Data

International Conference on Noise in Physical Systems
(8th : 1985 : Rome, Italy)
Noise in physical systems. And, 1/f noise--1985.

Includes index.
1. Electronic noise--Congresses. I. D'Amico, A.
(Arnaldo), 1940- . II. Mazzetti, P. (Piero),
1934- . III. International Conference on "1/f
Noise" (4th : 1985 : Rome, Italy). 1/f noise.
IV. Il Università degli studi di Roma. V. Title.
VI. Title: 1/f noise.
TK7867.5.I573 1985 621.3815'3 85-32045
ISBN 0-444-86992-1 (U.S.)

Printed in The Netherlands

PREFACE

This book contains the proceedings of the two joint conferences: the "8th International Conference on Noise in Physical Systems" and the "4th International Conference on 1/f noise", held in Rome from September 3 to 13, 1985.

All the manuscripts have been refereed by the International Program Committee before and during the Conferences.

The editors wish to thank all the members of the International Program Committee for their commitment and for their useful suggestions to the Italian Program Committee in the selection procedure of the invited speakers.

It is worthwhile pointing out that during the Conferences the International Program Committee decided to call the next Conference the: "9th International Conference on Noise in Physical Systems", leaving of course all the necessary scientific space for the 1/f noise, and to hold the Conference in Montreal in 1987.

Finally it is a great pleasure to thank the following people who made the Conferences more efficient and enjoyable: Denise Scott of the University of Calabria, Petrocco Giovanni, Alessandro Lucchesini, Massimo Gentili, Andrea Bearzotti, Luigi Maita, Maurizio Salvati of the Istituto di Elettronica dello Stato Solido and Susanna De Panfilis of the Istituto Plasma Spazio Interplanetario.

We also cannot forget the superb work of the North-Holland Physics Publishing organization who helped us a lot for the preparation of the global manuscript; in particular we want to thank Gerda Wolzak and Mary Carpenter.

ORGANIZING COMMITTEE

A. PAOLETTI, Chairman
G. MONTALENTI
I. MODENA
F. BORDONI
A. D'AMICO
G. SPISSU, Secretary

SCIENTIFIC PROGRAM COMMITTEE

P. MAZZETTI, Chairman
E. GATTI
B. PELLEGRINI
G.V. PALLOTTINO
G. BERTOTTI
A. STEPANESCU

INTERNATIONAL PROGRAM COMMITTEE

A. AMBROZY (Hungary)
J. CLARKE (U.S.A.)
E. CONSTANT (France)
L. DE FELICE (U.S.A.)
P.H. HANDEL (U.S.A.)
C. HEIDEN (F.R.G.)
F.N. HOOGE (The Netherlands)
B.K. JONES (U.K.)
P. MAZZETTI (Italy)
T. MUSHA (Japan)
M. SAVELLI (France)
H. THOMAS (Switzerland)
A. VAN DER ZIEL (U.S.A.)
Carolyn M. VAN VLIET (U.S.A.)
D. WOLF (F.R.G.)
R.J. ZIJLSTRA (The Netherlands)

TABLE OF CONTENTS

PART I - NOISE IN PHYSICAL SYSTEMS

INVITED PAPERS

THEORY (GENERAL)

DEVICES

BIOLOGICAL SYSTEMS

VARIOUS PHYSICAL SYSTEMS

PART 2 - 1/F NOISE

THEORY

DEVICES

VARIOUS PHYSICAL SYSTEMS

INVITED PAPERS

NOISE IN PHYSICAL SYSTEMS AND 1/f NOISE - 1985
A. D'Amico and P. Mazzetti (editors)

QUANTUM 1/F NOISE IN SEMICONDUCTORS INVOLVING IONIZED IMPURITY SCATTERING AND VARIOUS TYPES OF ELECTRON-PHONON SCATTERING

Carolyn M. VAN VLIET and Ganesh S. KOUSIK

Department of Electrical Engineering, U. of Florida, Gainesville, FL 32611, USA

The basic results for electron scattering in $\underline{k}$-space, under soft photon emission leading to infrared divergent coupling of the system to the electromagnetic field, are reviewed. The resulting fluctuations in scattering rates lead to mobility fluctuations which satisfy Hooge's formula. Explicit expressions for the Hooge parameter have been derived for scattering with ionized impurities and for acoustical phonon, optical phonon, polar optical phonon, and intervalley scattering. The results are evaluated by computer, and the overall mobility and weighted Hooge parameter are determined for n-type silicon and n-type gallium arsenide from cryogenic temperatures up to 300K. Good agreement with available experimental data mainly based on measurements by Bisschop (Eindhoven), van der Ziel (Minnesota), and Andrian (Florida) is obtained. This is the first detailed computation of the Hooge parameter and derivation of Hooge's law.

1. INTRODUCTION

Noise with a 1/f spectrum has been investigated for almost 60 years. In most cases it is proportional to the square of the current and inversely proportioned to the number of carriers causing the noise. This is expressed by Hooge's law [1]

$$S_I(f) = \alpha_H I^2/fN. \tag{1.1}$$

Though Hooge suggested this law in 1969, with experiments showing that in most cases 1/f noise is a bulk effect, the law (1.1) was basically known in the thirties.

There have been many theories of 1/f noise, some of which are rather universal, and some of which refer to very specific models. A general theory for 1/f noise was developed by Handel [1]. It is based on infrared divergent coupling of the system to the electromagnetic field or other elementary excitations. It uses the fact that any scattering event in $\underline{k}$-space or in real space is accompanied by the emission of soft phonons (frequencies $<10^5$ Hz). This theory derives 1/f noise from the quantum physics of a single charged particle, although the final result depends on the presence of many particles. This is similar to electron diffraction which is a one-particle effect, but can be observed only if many particles are present. Estimates easily show that electromagnetic quantum noise leads to Hooge parameters α_H less than 10^{-6}. Until the last decade it was believed that α_H was a universal constant of about 2×10^{-3}. However, many recent experiments, especially on devices such as mesa structures, microwave transistors, and JFETs have shown that values as low as 10^{-8} may occur. These values come close to the range where quantum 1/f noise can explain the data.

Presently we believe that quantum 1/f noise determines the lower limit of observable 1/f noise. In quartz time standards and in α-particle emission counting experiments (see elsewhere in these conference proceedings), such noise relates to the lowest possible error, present when the so-called flicker floor has been reached. In semiconductors, quantum 1/f noise relates to the slow fluctuations in cross sections for scattering, or emission, or capture. The results for $S_{\Delta\sigma}$, the spectrum of the cross section fluctuations, can in principle be taken from Handel's paper, see section 2. The result is

$$S_{\Delta\sigma}(\Omega) = \langle\sigma(\Omega)\rangle^2(ct)/f, \quad (1.2)$$

where the constant (ct) follows from Handel's paper.

However, (ct) is not Hooge's parameter, as suggested in Handel's papers, since his papers do not consider any specific application to mobility fluctuations in condensed matter. In fact, the latter is the purpose of this paper. We will then show that Hooge's law (1.1) is derivable from (1.2). The factor 1/N of (1.1), disputed and discussed in earlier works by Van Vliet and Zijlstra and by Hooge, will appear automatically, and α_H will be identified. It will be shown that α_H is a very complex function of the material parameters, strongly dependent on the prevailing scattering mechanism (Sections 3 and 4). Explicit curves of α_H versus temperature will be obtained for silicon and gallium arsenide (Section 5).

2. FLUCTUATIONS IN SCATTERING CROSS SECTION

We consider a beam of electrons, scattered by a potential λV, under soft photon emission. Fig. 1 shows the example of Coulomb scattering.

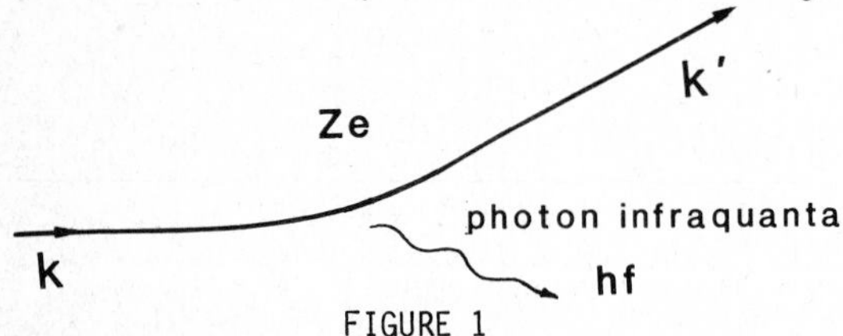

FIGURE 1
Electron-ionized impurity scattering with associated soft photon emission.

According to the scattering theory in the first order Born approximation, the wave function is

$$\psi = C[e^{i(kz-\omega t)} + \psi_1(\underline{r})]. \quad (2.1)$$

The scattered wave is given by

$$\psi_1(\underline{r}) = -\frac{1}{4\pi}\int \frac{1}{|\underline{r}-\underline{r}'|} e^{ik|\underline{r}-\underline{r}'|} e^{i(kz'-\omega t)}$$

$$U(\underline{r}')d^3r' \left[1 + \int_{\varepsilon_0}^{\Lambda} d\varepsilon |b_T(\varepsilon)| e^{-i(\varepsilon t/\hbar+\phi_\varepsilon)}\right]; \quad (2.2)$$

here $U(\underline{r}) = (2m^*/h^2)\lambda V(\underline{r}), |t| \leq T$; the second bracket accounts for soft photon emission, $b_T(\varepsilon)$ being the matrix element and ϕ_ε being a random phase related to the time of photon emission. In the far field approximation we have, if $\underline{k} = (k_z,0,0)$ is the polar axis, the following asymptotic form

$$\psi_1(\underline{r}) = -\frac{1}{4\pi}\frac{e^{ikr}}{r}\int e^{i(\underline{k}\cdot\underline{r}'-\omega t)} U(\underline{r}')$$

$$\times e^{-i\underline{k}'\cdot\underline{r}'} d^3r' \left[1 + \int_{\varepsilon_0}^{\Lambda} d\varepsilon |b_T(\varepsilon)| e^{-i(\varepsilon t/\hbar+\phi_\varepsilon)}\right]. \quad (2.3)$$

The cross section is then found to read

$$\sigma(\theta,\phi,t) = |r\psi_1(\underline{r})|^2$$

$$= \frac{\Omega^2 m^{*2}\lambda^2}{4\pi^2\hbar^4} |\langle\underline{k}'|V|\underline{k}\rangle|^2 |\Phi|_t^2, \quad (2.4)$$

where Ω is the crystal volume, $|k\rangle \to \Omega^{-1/2}e^{i\underline{k}\cdot\underline{r}}$, and Φ is given by

$$\Phi(t) = 1 + \int_{\varepsilon_0}^{\Lambda} d\varepsilon |b_T(\varepsilon)| e^{-i(\varepsilon t/\hbar+\phi_\varepsilon)}. \quad (2.5)$$

For the ensemble average over the random phases, one easily finds, setting

$$\lim_{T\to\infty} \frac{2\pi h}{T} |b_T(\varepsilon)|^2 = |b(\varepsilon)|^2, \quad (2.6)$$

using $\langle \exp i\phi_\varepsilon\rangle = 0$ and

$$\langle e^{i(\phi_\varepsilon-\phi_{\varepsilon'})}\rangle = \Delta\varepsilon\delta(\varepsilon-\varepsilon') = (2\pi\hbar/T)\delta(\varepsilon-\varepsilon') \quad (2.7)$$

($\Delta\varepsilon$ is the uncertainty caused by the observation time T), the result

$$\langle|\Phi|_t^2\rangle = 1 + \int_{\varepsilon_0}^{\Lambda} d\varepsilon |b(\varepsilon)|^2. \quad (2.8)$$

The same result can be obtained from time averaging, see Handel [2]. The matrix element for soft photon emission is found in the literature [3]. The result is $[\Delta k = |\underline{k}-\underline{k}'|]$:

$$|b(\varepsilon)|^2 = \frac{4e^2\hbar^2|\Delta k|^2}{3c^3 m^{*2}\varepsilon h} \equiv \frac{\alpha A}{hf}, \tag{2.9}$$

where α is the fine structure constant, $e^2/\hbar c$, $\varepsilon = hf$, and

$$A = \frac{2}{3\pi}\frac{\hbar^2|\Delta k|^2}{m^{*2}c^2}. \tag{2.10}$$

For (2.8) one finds $1 + \alpha A\ln(\Lambda/\varepsilon_0) \approx 1$ since the coupling constant $\alpha A \ll 1$. Substituting in (2.4) gives $\langle\sigma(\Omega)\rangle$.

Next we need the autocorrelation function and the spectrum. This is found in Handel's paper [2]. The result is

$$S_{\Delta\sigma}(\Omega) = 2\langle\sigma(\Omega)\rangle^2\alpha A/f. \tag{2.11}$$

We now recall that σ is related to the scattering rate $\omega_{\underline{kk}'}$ from $\underline{k}$ to $\underline{k}'$ by

$$\sigma(\Omega')d\Omega' = \frac{\Omega}{v_{\underline{k}}}\sum_{|\underline{k}'|} w_{\underline{kk}'} = \frac{\Omega^2 m^{*2}}{8\pi^3 h^3}\int_0^\infty d\,E' w_{\underline{kk}'}. \tag{2.12}$$

Thus at the origin of the fluctuations in $\sigma(\Omega')$ are fluctuations in the scattering rates. Eq. (2.12) leads to fluctuations consistent with (2.11) if we presume for the cross correlation spectrum of $w_{\underline{k}_1 k'}$ and $w_{\underline{k}_2 k''}$

$$S_{\Delta w}(k_1k';k_2,k'') = \frac{2\alpha}{3\pi}\frac{\hbar^2}{m^{*2}c^2}\bar{w}_{\underline{k}_1\underline{k}'}\bar{w}_{\underline{k}_2\underline{k}''}|\underline{k}_1-\underline{k}'| \times |\underline{k}_2-\underline{k}''|\delta_{\underline{k}_1,\underline{k}_2}/f. \tag{2.13}$$

We used here the overhead bar rather than <> for simplicity. The extension for cross correlations of the scattering rates is similar to the generalization of Handel's result for the cross correlation between screen grid current and anode current in pentodes by van der Ziel et al. (4]. Equation (2.13) is the starting point for mobility fluctuations, since μ is a (complicated) functional of $w_{\underline{kk}'}$.

Our objective is to obtain Hooge's law for the band mobility, in the form

$$S_{\Delta\mu} = \bar{\mu}^2\alpha_H/fN. \tag{2.14}$$

When there are various scattering processes, the overall mobility is

$$1/\mu = \sum_\gamma 1/\mu_\gamma. \tag{2.15}$$

For the fluctuations one then finds an overall Hooge parameter

$$\alpha_H = \sum_\gamma (\bar{\mu}/\bar{\mu}_\gamma)^2\alpha_\gamma. \tag{2.16}$$

In the next two sections we compute α_γ.

3. 1/f NOISE DUE TO ELECTRON-IONIZED IMPURITY SCATTERING

The band mobility due to a given scattering mechanism is expressed by

$$\mu_{band}(t) = \frac{e}{m^*}\frac{\langle\langle v^2\tau(t)\rangle\rangle}{\langle\langle v^2\rangle\rangle}, \tag{3.1}$$

where the double angular brackets mean averaging over $\underline{k}$-space. For the fluctuations we have, therefore,

$$\Delta\mu_{band}(t) = \frac{e}{m^*}\frac{\sum_{\underline{k}} v_{\underline{k}}^2\Delta\tau(t)n_{\underline{k}}}{\langle\langle v^2\rangle\rangle}. \tag{3.2}$$

Here $n_{\underline{k}} = c_{\underline{k}}^\dagger c_{\underline{k}}$ is the fermion occupation number operator which has $|0\rangle$ and $|1\rangle$ as its eigenstates. For the noise spectral density we have

$$\frac{S_{\Delta\mu}(f)}{(\bar{\mu})^2} = \frac{\sum_{\underline{k}_1\underline{k}_2} v_{\underline{k}_1}^2 v_{\underline{k}_2}^2 S_{\Delta\tau}(\underline{k}_1,\underline{k}_2)\langle n_{\underline{k}_1} n_{\underline{k}_2}\rangle}{[\sum v_{\underline{k}}^2\tau(\underline{k})f(E_{\underline{k}})]^2} \tag{3.3}$$

where <> is a canonical nsemble average and where $f(E_{\underline{k}})$ is the Fermi-Dirac distribution function or Boltzmann distribution in nondegenerate materials. Eq. (3.3) holds for any scattering mechanism.

We now consider ionized impurity scattering. If $\theta_{\underline{k}}$ is the angle between $\underline{k}$ and $\underline{k}'$, one finds [5]

$$\frac{1}{\tau(\underline{k})} = \frac{\Omega}{4\pi^2}\iint (1-\cos\theta_{\underline{k}})w_{\underline{kk}'}k'^2dk'\sin\theta_{\underline{k}}d\theta_{\underline{k}}. \tag{3.4}$$

Hence for the spectrum

$$S_{\Delta\tau}(\underline{k}_1,\underline{k}_2) = \tau^2(\underline{k}_1)\tau^2(\underline{k}_2)\left(\frac{\Omega}{4\pi}\right)^2$$

$$\times \iiiint (1-\cos\theta_{\underline{k}_1})(1-\cos\theta_{\underline{k}_2})\, S_{\Delta w}(\underline{k}_1,\underline{k}';\underline{k}_2,\underline{k}'')$$

$$\times \sin\theta_{\underline{k}_1} \sin\theta_{\underline{k}_2} k'^2 k''^2 d\theta_{\underline{k}_1} d\theta_{\underline{k}_2} dk' dk''. \quad (3.5)$$

We now substitute (2.13) and write $|\Delta k_1| = 2k_1 \sin\frac{1}{2}\theta_{\underline{k}_1}$, $|\Delta k_2| = 2k_2 \sin\frac{1}{2}\theta_{\underline{k}_2}$. Further, we know the matrix ele-ment for screened Coulomb scattering

$$w_{\underline{kk}'} = \frac{N_i}{\Omega}\frac{2\pi}{\hbar}\left(\frac{Z^2e^2}{\hat{\varepsilon}}\right)\frac{\delta(\mathcal{E}_{\underline{k}}-\mathcal{E}_{\underline{k}'})}{(\kappa^2+|\underline{k}-\underline{k}'|^2)^2}, \quad (3.6)$$

where N_i is the number of impurities, $\hat{\varepsilon}$ is the dielectric constant, and $1/\kappa$ is the screening length. After much computation, we find after substitution of (3.5) into (3.3) and setting $x = \mathcal{E}_{\underline{k}}/k_BT$,

$$\frac{S_{\Delta\mu}(f)}{\bar{\mu}^2} = \frac{1}{f}\,\frac{1}{\Omega N_c \exp(\mathcal{E}_F/k_BT)}\,\frac{16\sqrt{2}\alpha\kappa^2\hbar^5 N_c}{3m^{*7/2}(k_BT)^{3/2}c^2}$$

$$\times \left\{\int_0^\infty dx\, x^{11/2}e^{-x}[\ln(ax+1) - ax/(ax+1)]^{-4}\right.$$

$$\left.\times [\sqrt{ax} - (3/2)\arctan\sqrt{ax} + \sqrt{ax}/2(ax+1)]^2\right\}$$

$$\times \left\{\int_0^\infty \frac{dx\, x^3e^{-x}}{\ln(ax+1) - ax/(ax+1)}\right\}^{-2}; \quad (3.7)$$

here $a = 8m^*k_BT/\hbar^2\kappa^2$; further, we have multi-plied and divided by N_c, so that the second factor becomes $1/N$. Thus we obtained Hooge's law; all factors but the first two constitute the Hooge parameter α_{impurity}.

4. ELECTRON-PHONON SCATTERING

The starting point is again eq. (3.3). For $1/\tau(\underline{k})$ we have the expression for acoustical phonon scattering

$$\frac{1}{\tau(\underline{k})} = \sum_{\underline{k}'}\left[w^{em}_{\underline{kk}'}\left(1 - \frac{k'\cos\chi'}{k\cos\chi}\right)^{em}\right.$$

$$\left. + w^{abs}_{\underline{kk}'}\left(1 - \frac{k'\cos\chi'}{k\cos\chi}\right)^{abs}\right]. \quad (4.1)$$

The superscripts "em" and "abs" refer to phonon emission and absorption. The angles χ and χ' are shown in Fig. 2.

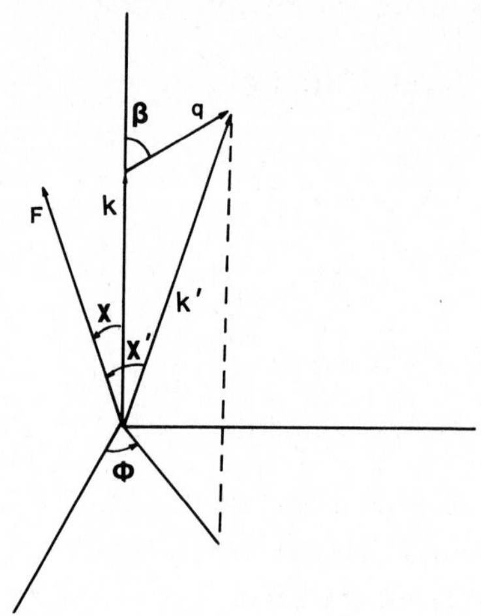

FIGURE 2
Reference coordinate system.

These angles are measured with respect to the applied field. Reference to Fig. 2 shows that

$$k'\cos\chi' = k\cos\chi \mp q\cos\beta\cos\chi$$

$$+ q\sin\beta\cos\phi\sin\chi \quad (4.2)$$

where $\underline{q}$ is the phonon wave vector and the upper sign refers to phonon emission and the lower to absorption. Momentum conservation leads to $\underline{k}' = \underline{k} \mp \underline{q}$ where we note that Umklapp pro-cesses are absent in nondegenerate materials. From (4.2) one has

$$1 - \frac{k'}{k}\frac{\cos\chi'}{\cos\chi} = \pm\frac{q}{k}\cos\beta - \frac{q}{k}\tan\chi\sin\beta\cos\phi. \quad (4.3)$$

Thus $\underline{q}$ is represented by the polar variables $(\underline{q},\beta,\phi)$. From (4.1) $S_{\Delta\tau}$ is expressed in $S_{\Delta w}$. Eq. (2.13) now becomes, using the indicated momentum conservation,

$$S_{\Delta w}(\underline{k}_1,\underline{k}';\underline{k}_2,\underline{k}'') = \frac{4\alpha}{3\pi}\frac{1}{f}\frac{\hbar^2}{m^{*2}c^2}$$

$$q'q''\bar{w}_{\underline{k}_1\underline{k}'}\bar{w}_{\underline{k}_2\underline{k}''}\delta_{\underline{k}_1,\underline{k}_2}. \quad (4.4)$$

For $\bar{w}^{em}$ and $\bar{w}^{abs}$ we take from any standard book on scattering theory, e.g. [5],

$$\bar{w}^{em}_{\underline{kk}'} = \frac{2\pi}{\hbar} \sum_{\underline{q}} |\mathcal{F}(\underline{q})|^2 [F(\underline{q})+1] \delta(\mathcal{E}_{\underline{k}'} - \mathcal{E}_{\underline{k}} + \hbar\omega_{\underline{q}}), \tag{4.5a}$$

$$\bar{w}^{abs}_{\underline{kk}'} = \frac{2\pi}{\hbar} \sum_{\underline{q}} |\mathcal{F}(\underline{q})|^2 F(\underline{q}) \delta(\mathcal{E}_{\underline{k}'} - \mathcal{E}_{\underline{k}} - \hbar\omega_{\underline{q}}), \tag{4.5b}$$

where F(q) is the Bose Einstein distribution,

$$F(\underline{q}) = [e^{\hbar\omega_{\underline{q}}/k_B T} - 1]^{-1} \simeq k_B T/\hbar\omega_{\underline{q}} \tag{4.6}$$

while the matrix element is

$$|\mathcal{F}(\underline{q})|^2 = (\hbar C_1^2/2\rho\,\Omega\,\omega_{\underline{q}})|\underline{q}|^2 \tag{4.7}$$

where C_1 is the deformation potential, $\omega_{\underline{q}}$ the acoustical phonon angular frequency, and ρ is the mass density of the solid.

We now have all of the ingredients to compute (3.3). The computation requires many pages of algebra for which we have no space. Finally, we find again Hooge's law and identify the Hooge parameter as

$$\alpha_{acoustic} = [\frac{32\pi\alpha N_c m^* c_\ell^7 \hbar^3}{3c^2 k_B^4 T^4}] \{ [\int_1^\infty \frac{dx}{x^4} (\frac{2m^* c_\ell^2}{k_B T})^2$$

$$\times [\frac{(x-1)^6}{6} + (1 + \frac{k_B T}{2m^* c_\ell^2}) \frac{(x-1)^5}{5} + \frac{k_B T}{2m^* c_\ell^2} \frac{(x-1)^4}{4}]^2$$

$$\times \exp(-\frac{m^* c_\ell^2}{2k_B T} x^2) + \int_1^\infty \frac{dx}{x^4} [\frac{(x+1)^4}{4} - \frac{(x+1)^5}{5}]^2$$

$$\times \exp(-\frac{m^* c_\ell^2}{2k_B T} x^2) + \int_0^1 \frac{dx}{x^4} [\frac{(x+1)^4}{4} - \frac{(x+1)^5}{5}$$

$$- \frac{(1-x)^4}{4} + \frac{(1-x)^5}{5}]^2 \exp(-\frac{m^* c_\ell^2}{2k_B T} x^2)\} . , \tag{4.8}$$

where c_ℓ is the longitudinal sound velocity.

Similar detailed results have been obtained for optical phonon scattering, polar optical phonon scattering, and intervalley scattering. These results will be published elsewhere (G.S. Kousik, C.M. Van Vliet, G. Bosman, and P.H. Handel, submitted to Adv. in Physics).

5. RESULTS FOR SILICON AND GALLIUM ARSENIDE

The parameters shown in Table 5.1 were used for silicon. Before computing the noise, we notice that as a "byproduct" of our calculations we obtained all of the mobility components. These provide a check on the accuracy of the results and on the reasonableness of the data in Table 5.1. Thus in Figure 3 we computed the

Table 5.1
Parameters for Silicon

Parameter	Magnitude	Units
Density of states effective mass	1.08 m_0	Kg
Conduction band effective mass	0.26 m_0	Kg
Longitudinal acoustic velocity	8.44×10^3	m/sec
Deformation potential energy	9.5	eV
Equivalent intervalley coupling constant	3×10^{10}	eV/m
Intervalley equivalent temperature	700	K

The above parameters are taken from references [5] and [13].

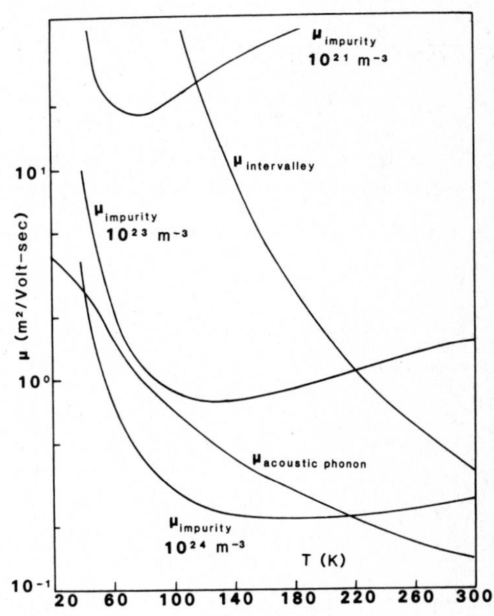

FIGURE 3
Electron mobility due to different mechanisms in silicon.

various mobilities, while in Fig. 4 we give the overall mobility. The results are compared with experimental data. Good agreement is noticed.

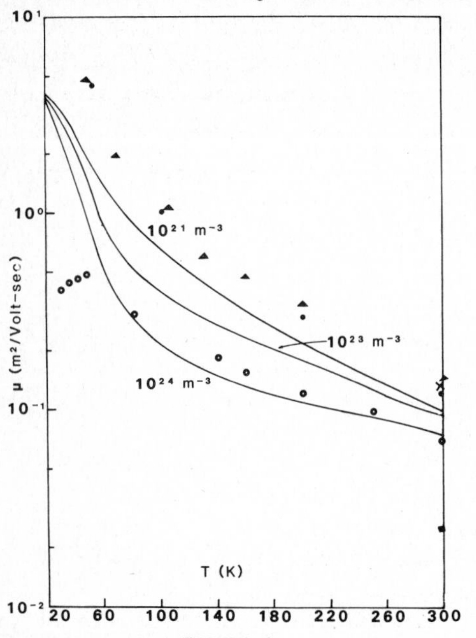

FIGURE 4
Electron mobility vs. T for silicon. Solid curves are theoretical values. Closed circles and triangles are for doping density less than $10^{18}m^{-3}$. Open circles are for $10^{23}m^{-3}$, x and square for 10^{21} and $10^{24}m^{-3}$ respectively [6][7].

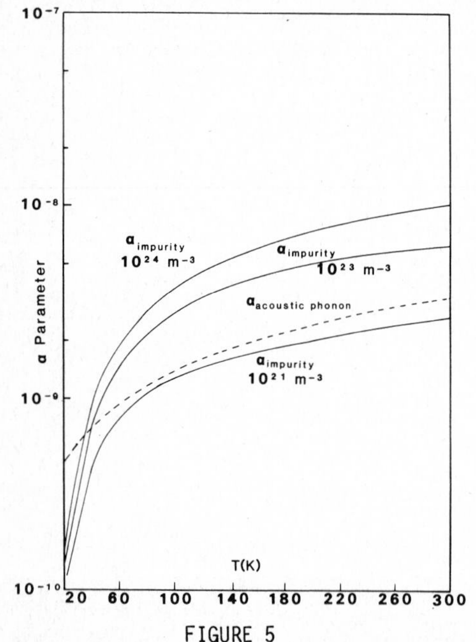

FIGURE 5
$\alpha_{impurity}$ and $\alpha_{acoustic}$ vs. T for silicon.

Next we give the data for α. Figures 5 and 6 show the α-parameter for acoustic, impurity, and intervalley scattering. These are the significant effects in n-type silicon. Finally, employing eq. (2.16) the overall Hooge parameter α_H is shown in Fig. 7. It also contains some

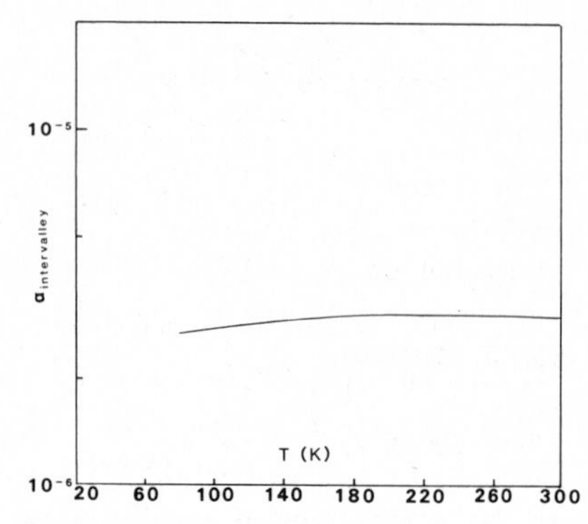

FIGURE 6
$\alpha_{intervalley}$ vs. T for silicon.

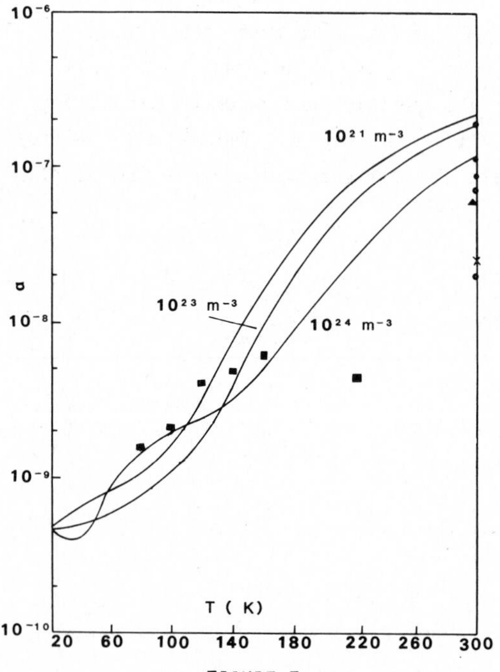

FIGURE 7
α_H vs. T for silicon. Closed circles are obtained from p^+np BJT [8]; closed triangles from BJT [21]; cross is from dual gate n-channel Si JFET [10]; open circles are taken from [11] and the squares are obtained from polysilicon resistors (doping of $1.8\times10^{17}cm^{-3}$) [11].

experimental data. The sources are in the figure caption. In view of the uncertainties in several parameters and the approximation of spherical, rather than ellipsoidal energy surfaces, the agreement is very satisfactory.

For gallium arsenide the data used are listed in Table 5.

TABLE 5.2
Parameters for GaAs

Parameters	Magnitude	Units
Polar optical phonon frequency	5.37×10^{13}	rad/sec
Low-frequency dielectric constant	12.53	
High-frequency dielectric constant	10.82	
Longitudinal acoustic velocity	5.22×10^{3}	m/sec
Deformation potential energy	7.0	eV

The parameters are taken from references [5], [12] and [13].

Again, first we computed the mobilities. The important contributions for n-GaAs are polar optical, acoustic, and impurity scattering. The results are in Fig. 8. The overall mobility is in Fig. 9. Agreement with experiment is clearly very good.

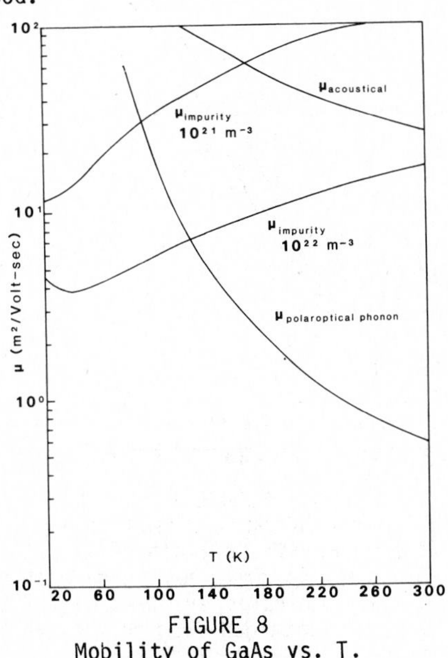

FIGURE 8
Mobility of GaAs vs. T.

Lastly, we give the partial α-parameters and the overall Hooge parameter in Figs. 10 and 11, respectively. It is interesting that, in contrast to silicon, α_H for gallium arsenide is largely temperature independent. The few experimental data available are about one order of magnitude too high. This indicates that in GaAs the quantum 1/f limit has probably not yet been

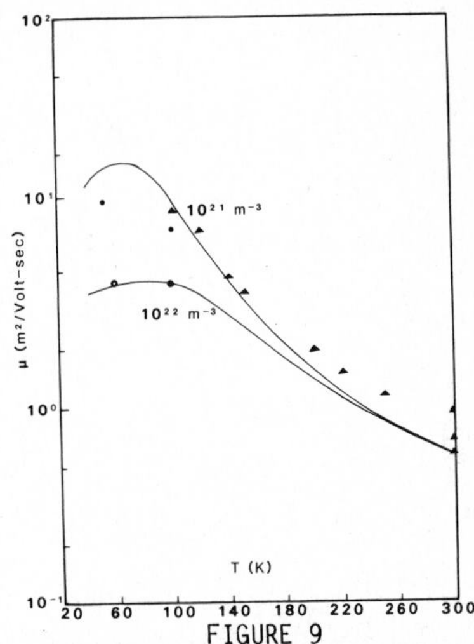

FIGURE 9
Electron mobility vs. temperature for GaAs. Closed circles for 10^{21}m^{-3} and triangles are from various experimental results collected in [6].

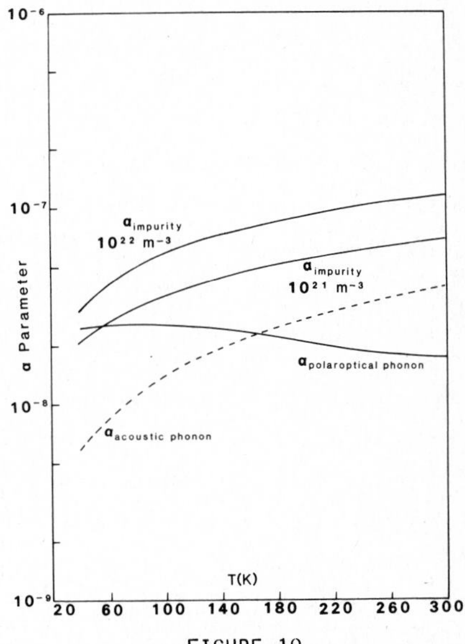

FIGURE 10
α parameters vs. T for GaAs.

reached. Near-ballistic devices, however, show α_H's of order 10^{-8} [15], both at 77 and 300K. Even so, the observed noise in nonballistic devices is coming close to the quantum 1/f noise limit, and the results are encouraging.

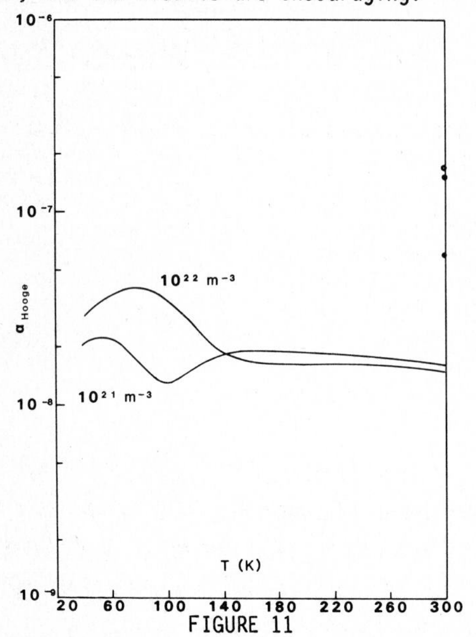

FIGURE 11
α_H vs. T for GaAs. Closed circles from [11] and open circles from measurements on submicron devices [14].

6. CONCLUSIONS

Starting points for the computations were the known solutions of the Boltzmann equation for the relaxation times due to various scattering processes, together with the result (2.13) for the quantum 1/f noise in the scatter-ing rates. The same approach is used to find the various mobilities as well as the various Hooge parameters. In view of the fact that the overall mobility agrees very well with experiment, both for silicon and gallium arsenide, we feel that the computed overall Hooge parameter, which is based on the same data (Tables 5.1 and 5.2) are a reliable result of the theory. A detailed comparison with experiment requires more data on materials and devices which show limiting 1/f noise, as considered here, over a wide temperature range. For silicon, however, comparison with available data is satisfactory and promising.

REFERENCES

[1] F.H. Hooge, Phys. Lett. 29A(3) (1969) 139.

[2] P.H. Handel, Phys. Rev. A22(2) (1980) 745.

[3] D.R. Yennie, S.C. Frautchi and H. Suura, Ann. Phys. (New York) 13 (1961) 379.

[4] A van der Ziel, C.J. Hsieh, P.H. Handel, C.M. Van Vliet and G. Bosman, Phys. B124 (1984) 299.

[5] B.R. Nag, Theory of Electrical Transport in Semiconductors (Pergamon Press, Oxford, 1972).

[6] C. Jacoboni, C. Canali, G.D. Ottoviani and A.A. Quaranta, Solid State Electr. 20 (1977) 77.

[7] M. Castato and L. Reggiani, Phys. Stat. Sol. (b) 38 (1970) 665.

[8] J. Kilmer, A. van der Ziel and G. Bosman, Proc. 7th Inter. Conf. on Noise in Phys. Systems and 3rd Inter. Conf. on 1/f Noise, eds. M. Savelli, G. LeCoy and A. Nougier (North-Holland, Amsterdam, 1984).

[9] X. Zhu and A. van der Ziel, IEEE Trans. Electron Devices ED-32(3) (1985) 658.

[10] K.H. Duh and A. van der Ziel, IEEE Trans. Electron Devices ED-32(3) (1985) 662.

[11] J. Bisschop, Experimental Study of the 1/f Noise Parameter α, Ph.D. dissertation, Technische Hoge School, Eindhoven, The Netherlands (November 1983).

[12] B.R. Nag, Electron Transport in Compound Semiconductors, (Springer-Verlag, Berlin, 1980).

[13] C. Jacoboni and J. Reggiani, Adv. in Physics 28 (1979) 493.

[14] J. Andrian, Noise Properties of Very Short GaAs Devices, Ph.D. dissertation, U. of Florida (April 1985).

[15] R.R. Schmidt, G. Bosman and C.M. Van Vliet, Solid State Electr. 26(5) (1983) 437.

NOISE IN PHYSICAL SYSTEMS AND 1/f NOISE - 1985
A. D'Amico and P. Mazzetti (editors)

THEORY OF AND EXPERIMENTS ON QUANTUM 1/f NOISE

ALDERT VAN DER ZIEL

Electrical Engineering Department, University of Minnesota, Minneapolis, MN 55455 USA

The 1/f noise spectrum of solid state devices is expressed in terms of device parameters such as current, carrier lifetime and Hooge parameter. The Hooge parameter is calculated for collision 1/f noise and Umklapp 1/f noise. The first gives Hooge parameters that are one to several orders of magnitude too small. The expression for Umklapp 1/f noise contains an exponential involving the Debye temperature of the material; this yields often agreement, even for the well-defined temperature dependence. Experiments on Si-JFETs and p-channel silicon MOSFETs imply the validity of the Umklapp theory and preliminary data verify the temperature dependence. The collector current of transistors shows no Umklapp 1/f noise whereas the base current seems to show it. Long $Hg_{1-x}Cd_xTe$ n^+-p diodes seem to show more than Umklapp 1/f noise. At present there is no evidence for injection-extraction and generation-recombination noise, even in the collector current of transistors.

1. INTRODUCTION

We give here a survey of the theory of and several experiments on Handel's theory of quantum 1/f noise[1,2,3], as applied to solid state devices. Recent experimental work[4,5,6] seems to indicate that the theoretical and the experimental values of the Hooge parameter show agreement and that the theory may set a lower limit for all 1/f noise.

Recently van der Ziel et al[7] applied Handel's theory of quantum 1/f noise to partition 1/f noise in pentodes and found a heuristic formula that closely described the experimental data. Later a derivation of this heuristic formula was given from first principles[8]. Gong et al[9] applied Handel's theory to radioactive decay of α-emitters and calculated the 1/f noise associated with this decay; again good agreement between theory and experiment was obtained. This made it interesting to compare theoretical and experimental values of the Hooge parameter α_H for solid state devices. This paper gives the results of such a comparison.

2. THE HOOGE PARAMETER α_H AND 1/f NOISE THEORY

The Hooge parameter α_H is defined as follows. Hooge found[10,11] that the relative 1/f noise of a semiconductor resistor carrying a current I could be represented by the equation

$$S_I(f)/I^2 = \alpha_H/(fN) \qquad (1)$$

where f is the frequency and N the number of carriers in the sample; for non-uniform samples Eq. (1) must be somewhat modified. At first it was thought that α_H was a constant having a value $\simeq 2\times10^{-3}$, but it later appeared that much lower values ($\alpha_H=10^{-6}$-10^{-9}) were possible in small solid state devices[4-6]. Hence it might be convenient to distinguish between "high noise" devices ($\alpha_H=10^{-2}$-10^{-5}) and "low noise" devices ($\alpha_H=10^{-5}$-10^{-9}); the boundary is somewhat fluid. For GaAs MESFETs and MODFETs Eq. (1) may not be satisfied.

In Eq. (1) α_H is an unknown parameter that must be determined experimentally. In order to make the connection between Eq. (1) and the measurements one must express the current spectrum $S_I(f)$ in terms of α_H and the various device parameters such as the current I, the carrier mobility μ the carrier lifetime τ, etc.

For semiconductor resistors of length L this

is simple. If V is the applied voltage, the current I is

$$I = q\mu NV/L^2 \tag{2}$$

where q is the electron charge. Substituting (2) into (1) yields

$$S_I(f) = \alpha_H \frac{q\mu IV}{fL^2} \tag{3}$$

so that α_H can be determined.

For MOSFETs and JFETs one finds in a similar way for all drain voltages V_d smaller than the saturation voltage V_{ds}[12]

$$S_I(f) = \alpha_H \frac{q\mu I_d V_d}{fL^2} \tag{3a}$$

as long as (1) is valid. This is the case for JFETs and MOSFETs but is not true for MESFETs and MODFETs where deep-lying traps seem to play a role in the noise. Here V_d is the drain voltage, I_d the drain current, μ the carrier mobility in the channel and L the channel length.

In early work on n^+-p diodes and p^+-n-p and n^+-p-n some errors were made in the evaluation of $S_I(f)$ that are now being corrected[13]. The errors came about because the devices are non-uniform. Two approaches are then possible. (a) The delta function approach in which (1) is rewritten as[14]

$$S_I(x,x',f) = I^2(x') \frac{\alpha_H \delta(x'-x)}{fN(x')} \tag{4}$$

where N(x') is the carrier density per unit length at x'. By averaging over x and x' one obtains $S_I(f)$. This method turns out to give correct results for short diodes only; for long diodes it gives improper averaging and therefore wrong results[13].
(b) The short section approach[14], in which (4) is applied to a section Δx to evaluate the noise current generator $[S_I(x,f)\Delta f]^{1/2}$ in parallel to the section Δx, resulting in

$$S_I(x,f) = \frac{I^2(x)\alpha_H}{fN(x)\Delta x} \tag{5}$$

One can now evaluate the contribution of $S_I(x,f)$ to the total spectrum $S_I(f)$ of the diode. That contribution is proportional to Δx, so that $S_I(f)$ is found from a simple integration with respect to x. This method is always valid.

There is another possible error for junction devices[14]. In semiconductor resistors and FETs, which are majority carrier devices, $N(x)\Delta x$ is the <u>total</u> number of majority carriers in the section Δx. But for junction devices we are dealing with <u>minority</u> carrier flow and in that case one must reckon with the possibility that only the <u>excess</u> carrier concentration N'(x) contributes to the 1/f noise. That is, one must carry out the calculation <u>twice</u>, once for $N(x)=N'(x)+N_p$ and once for N'(x), where N_p is the equilibrium minority carrier concentration. Experiments must then decide which of the two approaches is valid.

One must here distinguish between <u>long</u> diodes and <u>short</u> diodes. If the p-region of an n^+-p diode has a length w_p, then a short diode has $w_p \ll (D_n\tau_n)^{1/2}$ and a long diode has $w_p > 4(D_n\tau_n)^{1/2}$, where D_n is the carrier diffusion constant and τ_n the carrier lifetime.

The base current I_B of a transistor is due to carrier injection from the base into the emitter; the emitter region can be a "long" diode, since D_n and τ_n (in a p^+-n-p transistor) or D_p and τ_p (in an n^+-p-n transistor) are very small because of the high doping. The base region of a transistor is always a <u>short</u> region, since the base width $w_p \ll (D_n\tau_n)^{1/2}$ (for a n^+-p-n) or $\ll (D_p\tau_p)^{1/2}$ (for a p^+-n-p).

For a short n^+-p diode we have, if $\tau_d = w_p^2/2D_n$ is the diffusion time through the base region[13]

$$S_I(f) = \frac{\alpha_H qI}{2f\tau_d} \ln \left[\frac{N(o)}{N(w_p)}\right] \qquad (6)$$

if $N(x)$ is used, whereas $[N(o)/N(w_p)]$ must be replaced by $[N'(o)/N'(w_p)]$ if $N'(x)$ is applied. Now it is easily shown that[13,14]

$$\frac{N(o)}{N(w_p)} = \frac{D_n/w_p+s_{cn}}{D_n/w_p+s_{cn}\exp(-qV/kT)} ;$$

$$\frac{N'(o)}{N'(w_p)} = \frac{D_n/w_p+s_{cn}}{D_n/w_p} \qquad (6a)$$

where the contact recombination velocity s_{cn} at the p-contact is very poorly known. Hence $S_I(f)/|I|$ has a zero at V=o if $N(x)$ must be used and has a constant value if $N'(x)$ applies.

For the base region Eq. (6) applies, but since usually $N(w_b) \gg N_p$ for forward bias[15]

$$\frac{N(o)}{N(w_b)} \simeq \frac{N'(o)}{N'(w_b)} \simeq 1+(w_b v_c/D_n) \qquad (7)$$

where v_c is the limiting velocity of the carriers at high fields. τ_d is related to the transistor cut-off frequency f_T by the equation $2\Pi f_T \tau_d=1$, so it can be determined accurately. The last half of Eq. (7) is not very accurate, but since Eq. (6) depends only logarithmically on $N(o)/N(w_b)$ the inaccuracy in (7) may not matter very much.

For a long n^+-p diode we have

$$S_I(f) = \frac{\alpha_H qI}{f\tau_n} f(\gamma) \qquad (8)$$

where $\gamma=[\exp(qV/kT)-1]$. If $N(x)$ is used

$$f(\gamma) = [\frac{1}{3} - \frac{1}{2\gamma} + \frac{1}{\gamma^2} - \frac{1}{\gamma^3}(\frac{qV}{kT})] \qquad (8a)$$

whereas if $N'(x)$ applies

$$f(\gamma) = \pm 1/3 \qquad (8b)$$

(plus sign for forward bias, minus sign for back bias).

We note that $S_I(f)/|I|$ has again a zero at V=o if (8a) applies, whereas $S_I(f)/|I|$ is independent of bias if (8b) is valid. Mrs. Wu and van der Ziel[16] plotted $S_I(f)/|I|$ for a long $Hg_{1-x}Cd_xTe$ diode with x=0.30 and found that (8b) was valid. Moreover, (8a) gives a large increase in the noise at back bias, and this was not observed either. We therefore conclude that only the excess minority carriers seem to contribute to the 1/f noise in n^+-p diodes. This was unknown up to now.

According to Eq. (8), $S_I(f)=0$ at I=0, so that when the diode is used as a photovoltaic cell the 1/f noise is zero. This does not mean that the photocurrent has no 1/f noise. Mrs. Wu and van der Ziel[16] found in an irradiated $Hg_{1-x}Cd_xTe$ diode that the photocurrent did show 1/f noise at back bias, as expected from the above derivation.

The final conclusion is therefore that one can always evaluate α_H from the experimental data. For JFETs and MOSFETs this can be done rather accurately, but for the base current I_B of a transistor this is not the case, since the time constant of minority carriers in the emitter region is poorly known.

3. THE HOOGE PARAMETER α_H AND QUANTUM 1/f NOISE

It will now be investigated how the Hooge parameter α_H can be evaluated from Handel's quantum 1/f noise theory. There are two such theories:

(a) Coherent state theory[3], in which

$$\alpha_H = \frac{4\alpha}{3\Pi} \cdot \frac{3}{2} \simeq 4.5\times10^{-3} \qquad (9)$$

where $\alpha=1/(137)$ is the fine structure constant. The theory may, perhaps with some modifications, explain the observed Hooge parameters in high-noise semiconductor resistors ($\alpha_H=10^{-2}-10^{-5}$). We shall not discuss this problem further.

(b) Incoherent state theory[1,2], in which

$$\alpha_H = \frac{4\alpha}{3\pi}\left(\frac{\overline{\Delta v^2}}{c^2}\right) \simeq 3\text{x}10^{-3}\left(\frac{\overline{\Delta v^2}}{c^2}\right) \qquad (10)$$

where Δv is the change in velocity in the collision processes governing the carrier mobility μ and the carrier diffusion constant D and c is the velocity of light. Since $\Delta v^2 \ll c^2$, one would expect $\alpha_H \approx 10^{-5}$-10^{-9}. The theory may therefore explain the low values of α_H found in small solid state devices such as FETs, bipolar transistors and n^+-p junction diodes. We now discuss this problem in detail. The aim is to calculate $\overline{\Delta v^2}$ in each case.

Let us first consider normal collisions in which the carrier interacts with acoustical and optical phonons. Kousik[17] has calculated this accurately. One can find an approximate solution by assuming a Maxwell velocity distribution and elastic scattering over an angle Θ; in that case $\Delta v = 2v \sin\theta/2$, and hence, if m^* is the effective mass

$$\overline{\Delta v^2} = 4\overline{v^2}\,\overline{\sin^2\theta/2} = 2\overline{v^2} = 6kT/m^* \qquad (11)$$

The actual values of $\overline{\Delta v^2}$ turn out to be only a factor 2-3 larger. Consequently

$$\alpha_H \simeq \frac{4\alpha}{3\pi}\cdot\left(\frac{6kT}{m^*c^2}\right) \qquad (11a)$$

For n-type silicon $m^* = m_o$ or $\alpha_{Hn} = 0.94\text{x}10^{-9}$, whereas for p-type silicon $m_1^* = 0.49m_o$, $m_2^* = 0.24m_o$ and $m_3^* = 0.16m_o$. Assuming that these masses occur in equal number we must write $1/m^* = (1/m_1^* + 1/m_2^* + 1/m_3^*)/3$ or $m^* = 0.24m_o$ or $\alpha_{Hp} = 3.9\text{x}10^{-9}$. The measured values of α_H are at least one and sometimes two or three orders of magnitude larger, so that this mechanism does not give enough noise. Note that m^* enters into (11a), so that electrons and holes can have different Hooge parameters.

We must therefore look for mechanisms that give more noise. To that end we consider "Umklapp" processes, in which the lattice takes up an amount of momentum h/a, where a is the lattice spacing. Hence $\Delta v = h/(am^*)$ and

$$\alpha_H = \frac{4\alpha}{3\pi}\left(\frac{h}{m^*ac}\right)^2 \qquad (12)$$

This gives for n-type silicon $\alpha_{Hn} = 6.2\text{x}10^{-8}$ whereas the best experimental value is $2.5\text{x}10^{-8}$. In p-type silicon, if m_1^*, m_2^* and m_3^* again occur with equal probability, $\alpha_{Hp} = 1.13\text{x}10^{-6}$ whereas the best experimental values are $(3\text{-}9)\text{x}10^{-7}$. The theoretical values of α_{Hn} and α_{Hp} are therefore a factor 3 too large. Since the values of α_H for JFETs and MOSFETs can be determined with better than 25% accuracy, this is well outside the limit of experimental error.

A small correction of Eq. (12) that reduces α_H by about a factor 3 for silicon must be introduced. Equation (12) assumes that all processes are of the Umklapp type. Handel and van der Ziel[1,2,6] therefore suggested to multiply the right hand side of (12) by the probability factor $\exp(-\theta_D/2T)$ where θ_D is the Debye temperature of the material. In that case (12) becomes

$$\alpha_H = \frac{4\alpha}{3\pi}\left(\frac{h}{m^*ac}\right)^2 \exp\left(-\frac{\theta_D}{2T}\right) \qquad (12a)$$

Since silicon has $\theta_D = 645°$K, $\exp(-\theta_D/2T) \simeq 1/3$ at 300°K; the correction is of the right order of magnitude, for now we obtain for n type silicon $\alpha_{Hn} = 2.2\text{x}10^{-8}$ at T=300°K and $\alpha_{Hp} = 3.9\text{x}10^{-7}$ for p-type silicon. The best silicon n-channel JFETs gave $\alpha_{Hn} = 2.5\text{x}10^{-8}$ whereas the best silicon p-channel MOSFETs gave $(3\text{-}9)\text{x}10^{-7}$. On the other hand, the best silicon n-channel MOSFETs had $\alpha_H = 1.0\text{x}10^{-6}$. We thus conclude that the n-channel JFET and the p-channel MOSFET practically reach the quantum limit whereas the n-channel MOSFET does not. Moreover, Pawlikiewicz and van der Ziel[18] have measured α_{Hn} in n-channel silicon JFETs as a function of temperature; preliminary data indicate excellent agreement with the factor $\exp(-\theta_D/2T)$.

On the other hand no Umklapp 1/f noise was observed in the collector current of n^+-p-n and p^+-n-p transistors[4,6]. As a matter of fact, the collector current in the best transistors showed no 1/f noise down to 20 Hz. For that reason Zhang and van der Ziel[19] have lowered the limits for α_H in the collector of n^+-p-n and for p^+-n-p transistors to $\alpha_{Hn} \ll 1.6\times10^{-9}$ and to $\alpha_{Hp} \ll 5\times10^{-8}$, where as Zhu and van der Ziel had used the < sign.

The Umklapp noise in the collector of an n^+-p-n transistor would be generated in the base region. Now in modern transistors the base region has a constant gradient in the carrier density. This might be associated with the apparent absence of Umklapp noise in the collector current.

Zhu and van der Ziel[4,6] and Kilmer[20] have found indication of Umklapp noise in the base current of p^+-n-p GE82-185 transistors and in NEC57807 n^+-p-n transistors. Since the base current is due to injection of carriers from the base into the emitter, the diode theory must be applied to the emitter region. It is gratifying that for the GE82 transistor reasonable agreement exists between theory and experiment[6,20].

We now turn to Mrs. Wu and van der Ziel's data on n^+-p $Hg_{1-x}Cd_xTe$ diodes with x=0.30[16]. In that case one assumes that θ_D=250°K and that the gap width, and hence the effective mass, depends on temperature so that the theoretical value of α_{Hn} is practically independent of temperature[21].

Equation (8)-(8b) yield

$$\frac{S_I(f)}{|I|} = \alpha_{Hn} \frac{q}{3f\tau_n} \tag{13}$$

So that measurement of $S_I(f)$ at 20 Hz and near saturation yields α_{Hn}/τ_n. As shown below $\tau_n(T)$ has a maximum at intermediate temperature and is considerably lower at high and at low temperatures. Making fair estimates[16] for $\tau_n(T)$ gives values of $(\alpha_{Hn})_{exp}$ that are about one order of magnitude higher than the theoretical values $(\alpha_{Hn})_{theory}$ found for x = 0.30 and Eq. (12a). Hence the Umklapp noise is masked by a Hooge-type 1/f noise source.

Such a temperature dependence of τ_n is expected for $Hg_{1-x}Cd_xTe$ material with a narrow band gap and band to band transitions. The rate equation is then

$$\frac{dn}{dt} = g_o - \rho(n_a+n)n \tag{14}$$

where n_a is the acceptor concentration. Putting dn/dt = 0 yields the equilibrium electron concentration n_o

$$n_o = -\frac{1}{2} n_a+(n_a^2/4+n_i^2)^{1/2};\quad n_i=(g_o/\rho)^{1/2} \tag{14a}$$

It is easily seen that the life time τ_n is then

$$1/\tau_n = \rho(n_a+2n_o) \tag{14b}$$

It can also be shown that ρ is inversely proportional to T^3. Combining (14a) and (14b) then gives the quoted temperature dependence of τ_n.

4. GENERATION-RECOMBINATION 1/f NOISE IN JUNCTION SPACE CHARGE REGIONS AND INJECTION-EXTRACTION 1/f NOISE ACROSS JUNCTIONS

Van der Ziel and Handel[22] have discussed these two quantum 1/f noise processes in junction devices. They involve Eq. (10) but now Δv^2 is determined by an appropriately chosen energy E such that Δv^2=2E/m*; hence Δv^2 is inversely proportional to m*.

The g-r 1/f noise was first proposed by Jones[23]. It comes about because hole-electron pairs recombine in the space charge region for forward bias and are generated in the space charge region for back bias. The carriers in the channel are decelerated or accelerated and this produces low frequency quanta that interact with the carrier motion and so produce 1/f noise; the mechanism is analogous to the one in quantum partition 1/f noise in pentodes. Applying the Shockley-Read-Hall (S.R.H.) theory, van der Ziel and Handel found[22]

$$S_I(f)_{gr} = \alpha_{Hgr}\, q|I_{gr}|/[f(\tau_{po}+\tau_{no})] \qquad (15)$$

where I_{gr} is the generation-recombination current and τ_{po} and τ_{no} are the S.R.H. carrier lifetimes. Here $E = [q(V_{dif}-V)+3kT]$ for a hole-electron pair, where V_{dif} is the diffusion potential of the junction, but since both electrons and holes are involved one must use an effective mass $m^*_{eff} = [(m^*_n)^{1/2}+(m^*_p)^{1/2}]^2$. Consequently

$$\alpha_{Hgr} = \frac{4\alpha}{3\pi}\left[\frac{2q(V_{dif}-V) + 6kT}{m^*_{eff}\, c^2}\right] \qquad (15a)$$

The injection-extraction 1/f noise comes about because electrons injected across the space charge region or extracted from the p-region are decelerated or accelerated, produce low frequency quanta ... etc, just as before. Van der Ziel and Handel find for a short n^+-p diode

$$S_I(f) = \alpha_{Hi}\, q|I|/(f\tau_d) \qquad (16)$$

where $\tau_d = w_p^2/2D_n$ is the diffusion time through the p-region; for long diodes τ_d must be replaced by the electron life time τ_n. Here $E=[q(V_{dif}-V)+3kT/2]$, so that

$$\alpha_{Hi} = \frac{4\alpha}{3\pi}\left[\frac{2q(V_{dif}-V) + 3KT}{m^*_n c^2}\right] \qquad (16a)$$

In $Hg_{1-x}Cd_xTe$ with x=0.20, m^*_{eff} is about 80 times larger than m^*_n so that the g-r 1/f noise is much smaller than injection-extraction 1/f noise. Because of the $(m^*_n)^{-2}$ term both will be masked by Umklapp 1/f noise, unless it is absent.

Zhang and van der Ziel[24] tried to measure injection-extraction noise in the collector current of transistors, but they were unable to find it. We refer to their paper for details.

5. SUMMARY

In order to test experimentally the validity of Handel's quantum 1/f noise theory, the Hooge relation (1) must be valid, i.e., α_H must be independent of bias. This is the case for n-channel Si-JFETs, n- and p-channel Si-MOSFETs, n^+-p $Hg_{1-x}Cd_xTe$ diodes, and for the base current of p^+-n-p transistors. It does not hold for the base current of n^+-p-n transistors and for 1/f noise in GaAs MESFETs and MODFETs, whereas the collector 1/f noise in the best transistors is extremely small. In junction devices it must be taken into account that only the excess minority carriers give 1/f noise.

Where the Hooge parameter α_H can be defined, there is agreement with a heuristic quantum 1/f noise expression based on Handel's Umklapp 1/f noise mechanism. This is true in n-channel Si-JFETs and p-channel Si-MOSFETs, whereas the noise in n-channel Si-MOSFETs is about a factor 50 too large due to surface 1/f noise. It does not hold for n^+-p $Hg_{1-x}Cd_xTe$ diodes but is true for the base current of p^+-n-p transistors. In the collector 1/f noise of transistors the Umklapp mechanism is definitely absent.

ACKNOWLEDGEMENT

The author is indebted to Dr. P. H. Handel, the originator of the quantum 1/f noise theories, for help, guidance and encouragement. He is indebted to Mr. E. G. Kelso, NVEOL, for support and encouragement. He feels especially indebted to his students and associates.

Parts of this work were supported by an NSF Grant and by Night Vision and Electro-Optics Laboratory through the Army Research Office under contract DAAG-29-81-0-0010.

REFERENCES

1. P. H. Handel, Phys. Rev. Letters, 34, 1492 (1975).

2. P. H. Handel, Phys. Rev., 22, 745 (1980).

3. P. H. Handel, Noise in Physical Systems and 1/f Noise (Eds. M. Savelli, G. Lecoy and J. P. Nougier) Elsevier Science Publishers B.V., 97 (1983).

4. X. C. Zhu and A. van der Ziel, IEEE Trans. El. Dev., ED-32, 658 (1985).

5. K. H. Duh and A. van der Ziel, IEEE Trans. El. Dev., ED-32, 662 (1985).

6. A. van der Ziel, P. H. Handel, X. C. Zhu and K. H. Duh, IEEE Trans. El. Dev. ED-32, 667 (1985).

7. A. van der Ziel, C. J. Hsieh,, P. H. Handel, C. M. van Vliet and G. Bosman, Physica, 124B, 299 (1984).

8. A. van der Ziel and P. H. Handel, Physica, 125B, 293 (1984).

9. J. Gong, C. M. van Vliet, W. H. Ellis, G. Bosman and P. H. Handel, Noise in Physical Systems and 1/f Noise (Eds. M. Savelli, G. Lecoy and J. P. Nougier) Elsevier Science Publishers B.V., 381 (1983).

10. F. N. Hooge, Phys. Lett, A-29, 139 (1969).

11. F. N. Hooge, Physica, 83B, 14 (1976).

12. F. M. Klaassen, IEEE Trans. El. Dev., ED-18, 887 (1971).

13. A. van der Ziel and J. B. Anderson, to be published.

14. T. G. M. Kleinpenning, Physica, 98B, 288 (1980); Proc. U.S. Workshop on HgCdTe, J. Vac. Sci. Tech., A-3, 176 (1985).

15. A. van der Ziel, Solid State Electron., 25, 141 (1982).

16. X. L. Wu and A. van der Ziel, this Conference.

17. G. Kousik, Ph.D. Thesis, U of Florida (1985).

18. A. Pawlikiewicz and A. van der Ziel, to be published.

19. X. N. Zhang and A. van der Ziel, to be published.

20. J. Kilmer, A. van der Ziel and G. Bosman, Solid State Electron., 26, 71 (1983); 28, (1985).

21. X. C. Zhu, X. L. Wu, A. van der Ziel and E. G. Kelso, IEEE Trans. El. Dev., ED-32 (1985).

22. A. van der Ziel and P. H. Handel, IEEE Trans. El. Dev., ED-32 (Oct. 1985).

23. W. A. Radford and C. E. Jones, Proc. U.S. Workshop on HgCdTe, J. Vac. Sci. Techn., A-3, 183 (1985).

NOISE IN PHYSICAL SYSTEMS AND 1/f NOISE - 1985
A. D'Amico and P. Mazzetti (editors)

LOW FREQUENCY NOISE MECHANISMS IN FIELD EFFECT TRANSISTORS

K. KANDIAH*

Department of Electronics, The University of Southampton, Highfield Avenue, Southampton SO9 5NH, U.K.

The paper describes the noise mechanisms attributable to changes of charge state at point defects in silicon and in SiO_2 and the resulting random telegraph signal (RTS) type of current modulation. The dependence of the dwell times in the two charge states on defect characteristics, device parameters and temperature are discussed. The amplitude and dwell times of the RTS current due to individual defects are derived for simple situations in MOSTs and JFETs. The RTS waveform should be easily visible in some situations. New expressions for the equivalent gate noise voltage of MOSTs are derived by summing the modulations produced by all the defects in the oxide using simple approximations for strong and weak inversion. Noise measurements as functions of temperature and bias show good agreement with the calculations.

1. INTRODUCTION

Point defects in the Si and SiO_2 are known to be the main causes of low frequency noise in field effect transistors. Much effort is devoted to eliminating them but the residual defect density in the completed device is significant. This paper is concerned with the noise when defect densities are low. The effects due to correlation between defects are mentioned.

After an introduction to the noise mechanisms in section 2, the conditions governing the changes of charge state at point defects are described in section 3. The random telegraph signal (RTS) type modulation of device current caused by changes of charge state at individual point defects are considered in section 4.

Noise in the channel current of a JFET is discussed in section 5. The noise in surface channel MOSTs is derived in section 6 as the equivalent gate noise voltage for linear and saturation conditions. Simple approximations are used for weak and strong inversion. The validity of this physical model is well demonstrated in the agreement with measurements of noise as functions of temperature and bias. Estimates of the expected low frequency noise in the gate current are made in section 7.

2. GENERAL DESCRIPTION OF MECHANISMS

The drain current of an FET will be modulated by the changes in the field due to charge state transitions at fixed locations (defects or even dopant atoms) in or near the channel. The local loss or gain of a free carrier to a fixed site requires an almost immediate (dielectric relaxation time) change in the current at the device terminal to maintain overall charge neutrality. This results in a RTS type modulation of the drain current whose parameters are the amplitude and the dwell times in the two states.

The amplitude depends on the coupling coefficient M of electric field between the fixed charge site and the carriers in the channel. When the site is inside the channel, at moderate or large channel currents, M=1 and the amplitude is largest. Similarly M=1 for a defect at the oxide interface for a surface channel MOST in strong inversion. When the

* Current address: Building 347.3, AERE, Harwell, Didcot, Oxon, OX11 ORA, UK.

coupling is weak, as for example when the defect is in the depletion region near the gate junction in a JFET, M is nearly zero and the amplitude is very small. M is also small for a MOST in weak inversion.

The dwell times are determined by the processes responsible for the transitions. These are thermal emission of a carrier and capture of a carrier by defects in the Si and tunnelling by charges from the Si through the potential barrier to defects in the oxide nearby.

The current in a diode can also be modulated by defects, particularly in regions of the oxide adjacent to exposed junctions.

3. CHARGE TRANSITIONS AT POINT DEFECTS

3.1 Defects in a semiconductor

A bulk defect behaves as a recombination centre when transitions of charge state are due to the alternate capture of electrons and holes. The mean capture time for electrons is

$$\tau_{cn} = 1/n_n v_{tn} \sigma_n \tag{1}$$

where n_n is the density, v_{tn} the mean thermal velocity and σ_n is the capture cross-section for electrons. A similar expression gives τ_{cp}, the mean capture time for holes.

A generation centre changes charge states by alternate emission of electrons and holes into the conduction and valence bands respectively. The mean emission time for electrons is

$$\tau_{en} = \exp(E/kT)/N_c V_{tn} \sigma_n \tag{2}$$

where N_c is the density of states in the conduction band, E the depth of the defect energy below the edge of the conduction band, k is Boltzmann's constant and T is the temperature. A similar expression gives τ_{ep} for hole emission.

We shall define the behaviour of a defect as a trap if transitions of charge state result from the alternate capture and emission of the same type of carrier.

Sometimes a transition of charge state can be caused either by emission of one type of carrier or the capture of the opposite type. Then the characteristic time for this transition will be the combination of the two times.

3.2 Defects in the oxide

We shall assume that defects in the oxide are electrically active only when they are in the vicinity of the semiconductor and the carriers reach the defect by tunnelling. The mean time for tunnelling of electrons to the defect is

$$\tau_{tn} = \exp(2Ky)/4n_n v_{tn} \sigma_n \tag{3}$$

where y is the distance from the defect to the $Si\text{-}SiO_2$ interface,

$$K = (2mE_0)^{1/2}/h \tag{4}$$

m is the electron mass, E_0 is the height of the potential barrier in the oxide and h is the reduced Planck's constant. The corresponding expression for the tunnelling time from the defect into the Si is related to the hole density and velocity in the Si.

4. AMPLITUDES AND SPECTRA OF RTS DUE TO SINGLE DEFECT

4.1 RTS current amplitude

It can be shown that the amplitude of the current step in a sample such as a linear rod or FET channel(1,2), with a fixed applied voltage, for a change of one electronic charge q at a point inside the sample is

$$I = qv/L \tag{5}$$

where v is the carrier drift velocity at the defect and L is the length of the sample.

Modulation of the channel current in a FET by a charged defect involves a change in local carrier distribution. An acceptor type of defect will reduce the current in an n-channel FET and increase the current in a p-channel FET when it switches to the charged state. The opposite effect will occur for a donor type.

The region affected by a charged site extends to a fraction of the normal mean distance between carriers. For a carrier density of say $10^{16}cm^{-3}$ this is about 46nm. The presence of another defect at a distance much less than this sphere of influence will result in interaction between the carrier capture probabilities of the two defects. This may be the mechanism responsible for the rare phenomena of "intermittent burst" noise[(3)], apparently violating the superposition principle, which has been observed in diodes, transistors and FETs.

RTS currents will be generated even when the defect is outside the channel but the amplitude will be MI where M is the field coupling coefficient of the defect to the channel. M can be defined as

$$M = C_{ch}/C_{tot} \tag{6}$$

where C_{ch} is the capacitance of the defect to the channel and C_{tot} is the total capacitance of the defect.

4.2 RTS Spectrum

If the dwell times of the RTS in the two states are t_1 and t_2 respectively the characteristic time is defined as;

$$t_0 = t_1t_2/(t_1 + t_2) \tag{7}$$

The spectrum of the RTS current is given by

$$i_n^2 = i_{n0}^2/(1 + \omega^2t_0^2) \tag{8}$$

where i_{n0}^2 is the low frequency power density given by

$$i_{n0}^2 = (2MIt_1t_2)^2/(t_1 + t_2)^3 \tag{9}$$

For a symmetric RTS with dwell times τ the spectrum is

$$i_n^2 = (MI)^2(\tau/2)/(1 + [\omega\tau/2]^2) \tag{10}$$

5. NOISE IN THE CHANNEL CURRENT OF JFETs

Generation centres in the gate depletion region[(4)] and majority carrier traps in the channel and in the adjoining Debye region[(5)] will generate RTS drain currents in JFETs. The amplitude of the RTS as given by (5) will be largest for defects near the drain where the drift velocity is highest. Beyond pinch-off the carriers near the drain will drift at approximately the saturation velocity v_s which will give an RTS amplitude

$$I_{max} = qv_s/L \tag{11}$$

If the channel length is 2μm then I_{max} will be 8×10^{-9}A at room temperature. RTS waveforms of even smaller amplitudes due to single defects are readily visible in the drain current of JFETs with narrow gates at low defect densities.

The dwell time in one of the charge states is the carrier emission time given by (2). When the noise is plotted as a function of temperature, peaks as shown in Fig.1 which are characteristic of an activation energy will be seen. The two groups of peaks in this case are due to two classes of defects with energy levels of 0.19eV and 0.22eV. When the energy level is near midband a peak will be seen in the noise at 10Hz at about 260K.

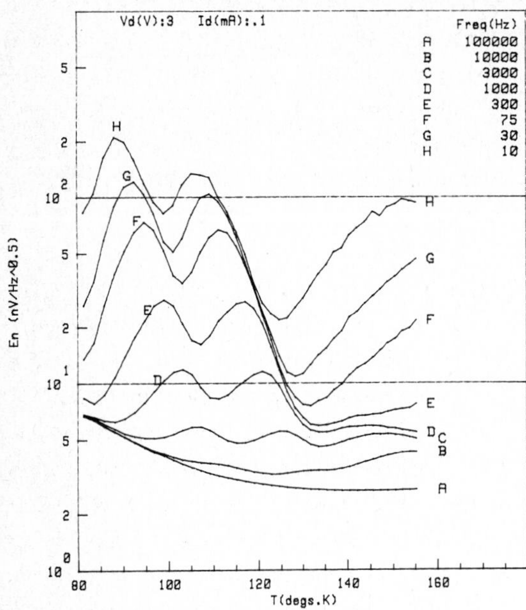

Fig. 1 Noise in a JFET as a function of temperature at 8 frequencies showing peak due to two classes of defects.

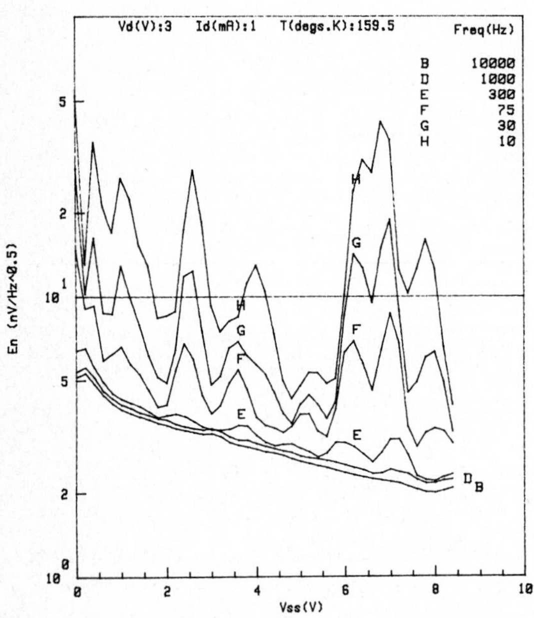

Fig. 2 Noise of 4-terminal JFET as a function of channel position showing peaks due to a number of isolated detects of one class.

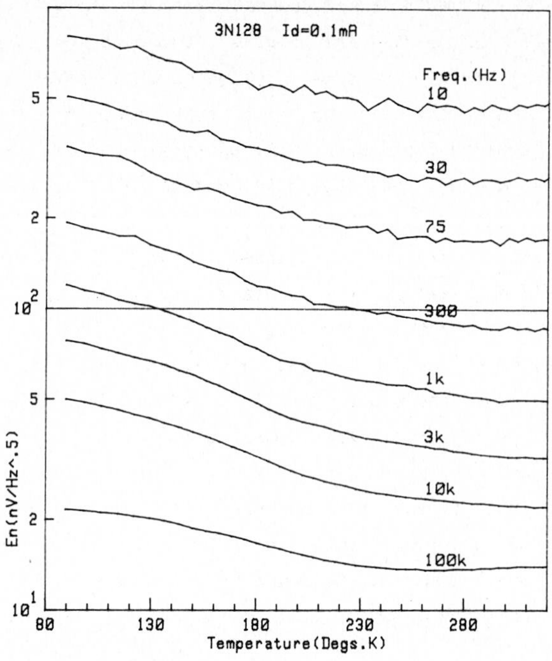

Fig. 3 Noise due to oxide defects in a MOST showing weak dependence on temperature.

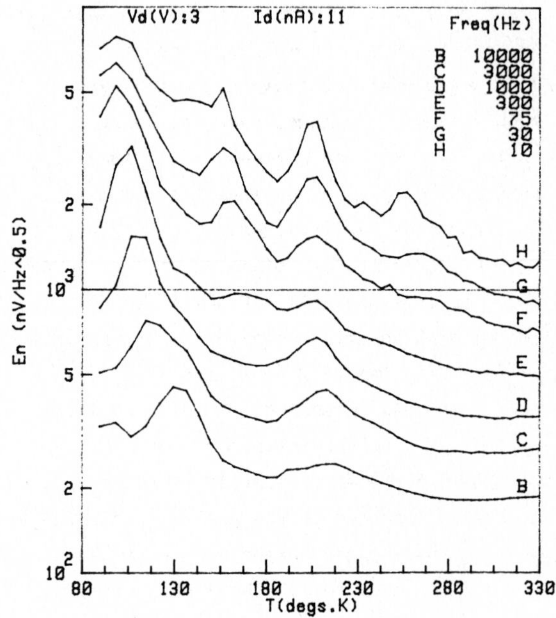

Fig. 4 Noise generated by oxide and bulk defects in a MOST in weak inversion.

At a fixed temperature the emission time τ_{en} is constant for a given type of trap. From (1) the capture time τ_{cn} depends on carrier concentration n_n and hence on the position of the defect in relation to the channel. It can be shown[5] that the noise due to any defect is a maximum when it occupies a position in the Debye region of the channel giving a value of n_n which makes $\tau_{cn} = \tau_{en}$. In a 4-terminal JFET the position of the channel, for a given drain current, can be changed by varying the relative bias on the two gates. The noise at any frequency at a fixed temperature will show peaks as in Fig.2.

For a generation centre in the depletion region the noise will be small unless the energy level is near midband and the temperature is higher than 200K.

6. NOISE IN THE CHANNEL CURRENT OF MOSTs

6.1 Assumptions in the new model

We consider mostly surface channel devices and assume that defects in the oxide near the Si interface act as recombination centres for carriers tunnelling from the Si and cause RTS modulations of drain current. Assuming a barrier height of 3.5ev and maximum dwell times of interest to be a few days we find from (3), assuming reasonable values for n_n and σ_n, that the maximum distance of active defects is 2nm. Owing to this proximity of the defect to the channel and assuming an oxide thickness much greater than 2nm we find from (6) that M = 1 except in weak inversion. The approximations for weak and strong inversion used in this section are chosen to simplify the derivation of the noise referred to the gate. This analysis may perhaps lack the rigor of the usual methods[6] which derive the spectral density of the drain current.

It is clear from (3) that the dwell times are much more sensitive to the distance y from the source of carriers than to the carrier density n_n. It has been shown[7] that the position of the peak of carrier concentration in an inversion layer moves by as much as 20nm with normal variations in surface doping levels and substrate bias.

6.2 The model for strong inversion

We shall assume that the tunnelling times for holes and electrons are both τ, for simplicity. It can be shown however that the results derived in this section are independent of the relative hole and electron tunnelling times. The coordinates for defects in the oxide are x,y,z, the distances from the source, from the interface and the edge of the channel. We make a major simplifying assumption that the carrier drift velocity increases linearly from v_1 at the source to v_2 at the drain in the linear region of the drain I/V characteristic. This is not considered to be a serious distortion in the light of greater uncertainties in the carrier concentration profiles in the inversion layer[7] which strongly affect the times in (3). The drift velocity at x is

$$v = v_1 + \frac{v_2 - v_1}{L} x \tag{12}$$

From (3) we have

$$\delta y = \frac{\delta\tau}{2K\tau} \tag{13}$$

Using this in (10) the spectral density of drain current noise is

$$S_i(\omega) = \int_0^W \int_0^L \int_{\tau_{min}}^{\tau_{max}} \frac{I^2}{1 + (\omega\tau/2)^2} \frac{N_d}{4K} \, d\tau dx dz \tag{14}$$

where N_d is the volumetric defect density in the oxide, W is the channel width, L is the channel length, τ_{min} is the minimum tunnelling time (nearly zero) and τ_{max} is the maximum tunnelling time (almost infinite).

Using (5) and (12) and integrating we have

$$S_i(f) = \frac{WN_d}{24KLf}\,\frac{q^2(v_2^{\,3} - v_1^{\,3})}{v_2 - v_1} \tag{15}$$

For sufficiently large drain voltage $v_2 \gg v_1$ and $v_2 = v_s$ the saturation carrier velocity at the drain.

Therefore

$$S_i(f) = \frac{WN_d}{24KLf}\,q^2 v_s^{\,2} \tag{16}$$

It may be noted that the drain voltage and drain current do not appear explicitly in (16) and the temperature dependence is only due to v_s. The noise power density $S_v(f)$ referred to the gate is obtained by using

$$g_m = WC_{ox}v_s \tag{17}$$

where g_m is the transconductance in saturation[8] and C_{ox} is the gate oxide capacitance per unit area. Therefore

$$S_v(f) = \frac{N_d}{24WKLf}\,\frac{q^2}{C_{ox}^2} \tag{18}$$

N_d can be derived from the areal density D_{it} by using (3) to derive the effective oxide thickness associated with the measurements of D_{it}. Since these cover a frequency (or time) range of about 6 decades we obtain 7Å for the oxide thickness and $N_d = 1.4 \times 10^7 D_{it}$. Using (18) the equivalent noise voltage at 10Hz at the gate of a MOST with W = 200µm, L = 10µm, gate oxide thickness of 100nm, interface state density of $10^9\,cm^{-2}$ and assuming a barrier height of 3.5ev at the oxide is

$$e_n = 810\ nv/\sqrt{Hz} \tag{19}$$

This is in good agreement with the results shown in Fig.3 for a typical enhancement mode surface channel MOST. Fig.3 also demonstrates the weak dependence on temperature. Many measurements confirmed that the gate noise voltage is virtually independent of drain voltage and current in strong inversion.

6.3 Weak inversion

The surface potential in weak inversion is not pinned to a fixed level and the fields due to charges above the interface will penetrate the channel. The value of M will depend on the surface carrier concentration and therefore on gate and drain voltage. The exact profile of the carrier concentration within a few Å of the interface[7] also affects the tunnelling times given by (3) and therefore the noise spectrum. A calculation on the same lines as for strong inversion is not considered feasible.

When the channel is semi-transparent to the field of a charged defect we can assume that the current will only be modulated over a fixed area A_0 of the channel with linear dimensions of the same magnitude as the smaller of the mean distance between dopant atoms or the gate oxide thickness. Then the amplitude of the RTS current due to a single defect will be approximately I_dA_0/WL where I_d is the total drain current. The dwell times given by (3) will be inversely proportional to I_d since n_n in (3) will be proportional to I_d. Therefore

$$S_i(f) = \frac{N_d}{8WKLf}\,A_0^2 I_d^2 \tag{20}$$

The transconductance in weak inversion is

$$g_m = RI_d \tag{21}$$

where R is a the gate control factor which is independent of W or L. Therefore

$$S_v(f) = \frac{N_d}{8WKLf}\,\frac{A_0^2}{R^2} \tag{22}$$

Since the thermal noise depends only on I_d and is independent of gate dimensions it is clear that the relative magnitude of L.F. noise decreases fast with increase in gate dimensions or decrease in oxide thickness.

6.4 Defects in the Si and effects of substrate bias

In ion implanted surface channel devices the noise due to residual defects in the region of falling carrier concentrations at depths of about 100Å to 200Å in the Si will have exactly the same characteristics as the noise in a JFET. An example of noise in a MOST which exhibits noise peaks as a function of temperature is shown in Fig.4. As in JFETs the magnitudes and shapes of these peaks will depend on drain current and substrate bias.

Buried channel operation may reduce the noise due to oxide defects but increase the noise due to bulk defects. The effect will be less important at room temperature since few residual defects are at midband.

7. NOISE IN GATE CURRENT

Gate current in JFETs is mainly due to carrier generation in the depletion region, diffusion from adjoining neutral regions and minority carriers generated by impact ionisation in the channel. The last two components have a white noise spectrum.

The generation current from defects in the depletion region in the Si or interface states exposed to this depletion region arises from the alternate emission of holes and electrons with emission times given by (2). The noise spectrum is related to the energy level of the defects and the temperature. It has been shown[(9)] that the spectral density $S_i(f)$ is $0.67S_0$ at frequencies $f \gg 1/\tau$ where τ is the characteristic time as defined in (7) and S_0 is the expected shot noise for the observed mean gate current. For defects at midband $S_i(f)$ falls at lower frequencies to $0.5S_0$ at $f=0$. Defects with shallower levels (3kT or more from midband) show an increase in $S_i(f)$ at lower frequencies to S_0 at $f=0$.

8. ACKNOWLEDGEMENTS

It is a pleasure to acknowledge the assistance given by Mr. M.O. Deighton with the noise calculations and Mr. F. B. Whiting for the numerous measurements of noise and RTS currents. The work was supported by AERE Harwell and MOD(PE).

9. REFERENCES

(1) A.L.McWhorter, Semiconductor Surface Physics (Ed. R.H.Kingston), p. 219, University of Pensylvania Press (1956)

(2) M.O.Deighton, (Unpublished)

(3) K.F.Knott, Solid-State Electr., 21, 1039 (1978)

(4) C.T.Sah, Proc. IEEE, 52, 795 (1964)

(5) K. Kandiah, M.O.Deighton and F.B.Whiting, Proc. 6th International Conference on Noise in Physical Systems, NBS publication 614, U.S. Department of Commerce (1981)

(6) G.Reimbold, IEE Trans Elec. Dev. ED-31, 1190 (1984)

(7) J.A.Pals, Quantization Effects in Semiconductor Inversion and Accumulation Layers, Thesis Technische Hogeschool Te Eindhoven, Holland (1972)

(8) S.M.Sze, Physics of Semiconductor Devices, p.450, John Wiley & Sons, New York (1981)

(9) M.O.Deighton and K.Kandiah, (To be published)

NOISE IN PHYSICAL SYSTEMS AND 1/f NOISE - 1985
A. D'Amico and P. Mazzetti (editors)

COMPARISON OF 1/f NOISE THEORIES AND EXPERIMENTS

F.N. HOOGE

Eindhoven University of Technology, Eindhoven, The Netherlands

Theoretical models and experimental findings are compared. It is impossible to construct one single picture reconciling them all.

INTRODUCTION

Everyone active in the field of 1/f noise will understand that this survey must be restricted to a few mainstreams in the theory of 1/f noise. I shall not talk about purely mathematical descriptions.

I cannot strive after completeness. Theories and facts already included in review papers[1-5] will be treated superficially. Papers published after the Montpellier Conference[6] have been screened more carefully.

We shall discuss experimental results and theoretical predictions in terms of the noise parameter α. We define α by the phenomenological relation for conductivity fluctuations

$$\frac{S_G}{G^2} = \frac{\alpha}{N} \cdot \frac{1}{f} \tag{1}$$

N is the number of free carriers. Until some five years ago, many experimental results obtained on samples of quite different nature could be summarized by the observation

$$\alpha \simeq 2 \times 10^{-3} \tag{2}$$

Today one can no longer defend this simple statement (2). Nevertheless, the parameter α, as defined by (1), is a good measure of noise intensity.

Δσ: FLUCTUATIONS IN CONDUCTIVITY

One of the very few statements about 1/f noise that has not raised a controversy is that 1/f noise is a fluctuation in the conductivity. A current through the sample does not generate the noise, by turbulence or so, but only serves to measure it. Voss and Clarke[7,8] and others[9] have proved this experimentally by measuring the noise in the thermal noise of a resistor in equilibrium, without a net current flowing through it. There is ample experimental evidence that the conductivity fluctuations have a Gaussian amplitude distribution[10] which is the property easiest to measure when investigating whether the noise has Gaussian statistics.

Recent work of Restle and others[11] has shown that in most cases, the 1/f noise is Gaussian. Non-gaussian effects have been found in silicon-on-sapphire resistors extremely small in area ($A < 1\ \mu m^2$).[12]

An interesting point has been raised by Weissman[13]. Some theoretical models give conductivity fluctuations that are not simply scalars. Experimental investigation of the tensor properties of the 1/f noise can provide arguments for excluding some of the proposed models. Such measurements are difficult and require perfect samples of extreme homogeneity. The 1/f noise in Si samples turned out to be scalar[14,15], which agrees with number fluctuations. Non-scalar fluctuations have been measured in granular-carbon, gold and chrome films[16] and in bismuth[17].

ΔX: THERMALLY ACTIVATED PROCESSES

Metal films have been extensively studied by Dutta, Eberhard and Horn[2]. Values of α obtained at room temperature are not too far from what the temperature fluctuation model predicts, and thus close to the usual empirical value $\sim 10^{-3}$. But a disturbing fact is the temperature dependence of α. The temperature fluctuation model predicts

$$S \propto \frac{1}{R^2}\left(\frac{dR}{dT}\right)^2 \frac{T^2}{C_v} \qquad (3)$$

which for metals, is nearly temperature-independent. C_v is the heat capacity. The metals were investigated between 100 K and 600 K. Towards lower temperature, α usually decreases by a factor of 100. This clearly contradicts the temperature-fluctuations model. Therefore two types of 1/f noise are distinguished in metal films.

A-type: Can be accounted for by the temperature-fluctuation theory. Strongly dependent on the substrate, weakly dependent on temperature.

B-type: Strongly temperature-dependent, independent of the substrate.

Dutta and Horn propose a special model for the B-noise. They assume that the noise-generating process is activated thermally. If there is only one activation energy we find a Lorentzian spectrum $(1+\omega^2\tau^2)^{-1}$ with $\tau = \tau_0 \exp(E/kT)$.

If, as in the McWhorther model, we have a statistical weight $g(\tau) \propto 1/\tau$, we find a pure 1/f spectrum. This means that g(E) is constant for exact 1/f. Dutta and Horn now take a rather broad band, centred round 1 eV for g(E). The summation of the Lorentzian spectra does not give 1/f, but f^{-n} with $n \approx 1$. The slope n is related to the temperature dependence of α. There are no good suggestions as to the nature of the process with the set of activation energies. This process must become visible in the conductivity fluctuations. Since the known scattering mechanisms have only weak temperature dependences, the high values of ~ 1 eV point to some reaction mechanism on the part of the lattice, but not to the direct creation of scattering centres. At this point one must realize that it is only in semiconductors that mobility fluctuations have experimentally been demonstrated.

Fleetwood and Giordano published their results on metal films and wires in a series of papers. The noise of tin films[18] could be described by the temperature-fluctuation model. The magnitude of the noise at room temperature depends on the substrate and on the adhesion to the substrate, hence on the thermal coupling. These experimental findings do not agree with earlier results of other investigators.

It could be shown that the 1/f noise of platinum films and wires had the same bulk origin[19]. The noise can be characterized by $\alpha_{latt} \sim 2 \cdot 10^{-3}$, an agreement which the authors call fortuitous in view of further work on films of 8 different metals[20,24]. Here they found a wild scattering in α-values. A factor of 10 was found for each material. The average value is not far from $2 \cdot 10^{-3}$, but there is a clear trend for samples with a high resistivity to have a low α-value, proportional to $1/\rho$.

In several samples they found that the noise decreased by one or two orders of magnitude in the course of several days[21,24]. They subjected the films to mechanical stress; stress-induced noise decreased with time. This suggests a link between 1/f noise and a process closely associated with strain relaxation. Such processes have activation energies in the 1 eV region. Therefore they find their experiments consistent with the speculation of Dutta and Horn[2] as to the role of vacancy-interstitial diffusion.

A qualitative agreement could be found between measurements on carbon resistors and the Dutta-Horn model of thermal activation by a band of activation energies in the 0·2 to 1 eV region[22]. In a recent paper[23], Fleetwood and Giordano showed that there is a direct link between 1/f noise and defects. A wire of diameter 530 A°, made of $Au_{60}Pd_{40}$ with a high defect density was annealed in several steps. At room temperature α was about 10^{-2} before annealing. Annealing diminished α. After the last annealing step, α at room temperature had decreased to $\sim 5 \cdot 10^{-4}$. Here, too. the slope of the spectrum correlated with the temperature dependence of α.

ΔT: TEMPERATURE FLUCTUATIONS

The theory which explains 1/f noise in condunctance by temperature fluctuations was proposed by Voss and Clarke[8].

In equilibrium we find local temperature fluctuations according to the relation

$$< (\Delta T)^2 > = \frac{kT^2}{C_V} \tag{4}$$

The shape of the spectrum follows from the equations for the diffusion of heat. As the conductance depends on T, the spectrum of the conductance fluctuations follows from the temperature spectrum. Part of the spectrum is $f^{-1/2}$ and part is $f^{-3/2}$. The frequency region where this 1/f-like spectrum is found depends on the dimensions of the sample. The heat transport between sample and support must be taken into account with great care[25]. If one compares the theoretical results for metal films with the experimental relation (1) one finds reasonable agreement. This is due to the fact that the energy per atom is 3 kT. The number of atoms comes in via C_V. The number of atoms in metals is close to N, the number of free electrons. Since the number of electrons N in semiconductors is very much less than the number of atoms, the theory fails completely in the case of semiconductors. Although the theory of temperature fluctuations does not explain the "real" semiconductor 1/f noise, the 1/f-like temperature spectra do exist. They are important in metal films. In the extensive work of Dutta, Eberhard and Horn, this noise is called A-noise[2].

In the early discussion on experiments that could support the temperature theory, Bi played an important part since Bi, as a semimetal, could show whether the number of atoms or the number of electrons appear in the relations.

Interpretations by Clarke and Voss of their own experimental results agreed with their theory. Recent work of Bisschop[26,27] has shown that the mobility model applies.

Δn: NUMBER FLUCTUATIONS

The oldest, yet not obsolate, model of 1/f noise is McWhorter's model of electron trapping in surface states. One assumes a homogeneous concentration of traps in the oxide layer of a semiconductor crystal. If the electrons reach the traps by tunneling, then the trapping probability depends exponentially on the distance to the interface. This exponential dependence, together with the homogeneous concentration, leads naturally to the 1/f spectrum. The 1/f spectrum is the sum of the Lorentzian spectra

$$S \propto 1/(1 + \omega^2 \tau_i^2) \tag{5}$$

with a statistical weight for the relaxation times

$$g(\tau_i) \propto 1/\tau_i \tag{6}$$

There are many variations and refinements to this model, which is still popular with investigators of MOS transistors. There are, however, many experimental findings at variance with this model. The magnitude of the noise sometimes requires unrealistically high concentrations of traps. Above all, it has been demonstrated experimentally, at least for semiconductors, that the 1/f noise is a fluctuation in mobility, whereas McWhorter's model is based on number fluctuations. Modern analyses of the noise of complicated devices often reach to the conclusion that the surface generates only a fraction of the noise, whereas the greater part is bulk noise, probably of the mobility-fluctuation type.

The group of Ralls, Skocpol et al.[28] made extremely small MOSFETs, 1 μm long and 0.1 μm wide. These unique devices make it possible to study the fluctuation in the occupancy of a

single trap. They observed switches of resistance as a function of time. The result is a picture of what we would call burst noise. τ_{on} and τ_{off} depend on temperature and on gate voltage. From these dependences they calculated the activation energy and distance from the surface. The paper presents data on 5 samples where individual traps were seen. They conclude: "The deepest trap seems furthest from the surface".
This is still far away from an experimental demonstration of the correctness of McWorther's model.

Similar results were obtained by Restle et al[12] on somewhat bigger devices. In their samples of about 1 μm^2 they observed randomness in the shape of the spectra and non-gaussian effects. This result is a transition between Rall's observation of a single trap and the usual large-number limit with Gaussian behaviour.

Sputtering of HgCdTe samples eliminated part of the 1/f noise[29]. The low-frequency part became flat below 400 Hz. States with long relaxation times were sputtered away, whereas the states with short times remained.

$\Delta\mu$: MOBILITY FLUCTUATIONS

About 10 years ago, experiments were designed to distinguish between fluctuations in number and in mobility[3]. One started with the equation for the current density

$$j = nq\mu\varepsilon + qD\frac{dn}{dx} + \ldots \qquad (7)$$

with possible terms in ∇T or in the magnetic induction B. Two calculations for S_j (or S_V) were made, one starting from the introduction of number fluctuations, the other from mobility fluctuations. Often the two results are numerically so far apart, that comparison with experiments gives an unambiguous answer as to which the correct noise source is. This comparison was successfully done for the thermo-voltage, the Hall voltage and for hot-electron effects.

In another series of experiments α was measured in samples where, in addition to lattice scattering, other scattering mechanisms were present, such as impurity scattering or surface scattering.
The 1/f noise was found only in the lattice scattering. This leads to the expression

$$\alpha = (\mu/\mu_{latt})^2\,\alpha_{latt} \qquad (8)$$

In these samples we found

$$\alpha_{latt} \approx 2\,.\,10^{-3} \qquad (9)$$

Apart from determining which scattering mechanism fluctuates with a 1/f spectrum, these experiments were an additional argument for the point of view that 1/f noise is a mobility fluctuation. These investigations are rather old, but they have never been refuted. In recent literature, more support for the correctness of (8) had been put forward[30,31,32,41]. In transistors, Mihaila[33,34] observed phonon fine structure in curves of α versus base current and in curves of α versus temperature. This demonstrates the close link between phonons and 1/f noise.

Some people maintain that the low frequencies are in themselves an argument agains mobility fluctuations. Mobility is then seen as a property of a single electron, so that the time an electron stays in the sample determines the lowest frequency of the mobility fluctuations. The mobility must, however, be seen as a property of the lattice. Consider a narrow subband in the conduction band of a small volume of the lattice. An electron in the subband will have a certain average collision time. The lattice properties fluctuate slowly, and this also applies to the average collision time. In the volume and subband, many electrons are scattered at random moments and independently of each other. This group is scattered in the same average collision time. When, at a later instant, other electrons are scattered in that volume and subband, they do so in a somewhat different average collision time. Thus, it is of no importance how long a specific electron stays in the crystal. The lattice stays there, always with electrons in that subband. Therefore it is the mobility of the group of changing electrons in the subband and not that of a specific electron which can fluctuate at very low frequencies.

A final word on this mobility model. It is based on comparing α of the conduction with the magnitude of 1/f noise in other properties. It does not say anything about the value α should have. 2×10^{-3} has often been found. That value was considered as characteristic of lattice scattering. Introduction of other scattering mechanisms gave lower values than $2 . 10^{-3}$. Here also it is only the relative decrease that is of importance in our arguments. Values of α_{latt} lower than $2 . 10^{-3}$ are, on their own, no argument against the idea of fluctuations in mobility.

HANDEL'S THEORY

Handel's theory of 1/f noise is a mobility theory[35]. He considers a beam of electrons that are scattered by some mechanism, the details of which we do not need to know. In the scattering process, radiation is emitted, which slightly diminishes the frequency of the electron wave. Interference of scattered and unscattered electron waves causes low-frequency beats. Since the energy spectrum of the emitted radiation is white, the number of emitted quanta with energy hf will be proportional to 1/f. Calculations show that the spectrum of the beats in the electron waves will also be 1/f. The quantum-mechanical treatment of this model leads to a fluctuation in the current of the electron beam

$$S_I = \frac{4}{3\pi} \frac{e^2}{\hbar c} \left(\frac{\Delta v}{c}\right)^2 \frac{1}{f} I^2$$

where the fine-structure constant $e^2/\hbar c$ has the value 1/137.

The only quantity characteristic of the scattering process is $\Delta\vec{v}$, the difference between the velocity vectors before and after scattering. It is in this way that the angular dependence of the scattering will have a strong numerical influence on the noise parameter α of scattering mechanisms, such as lattice scattering, impurity scattering, etc.

Since there are no further quantities characteristic of this theory, it is very difficult to design experiments that can confirm the model. We can now discuss three such experiments that were put forward in defence of Handel's theory.

A. 1/f noise in the fluctuation of radioactive decay rates[36]. The decay rate of α-particle-emitting ^{291}Am was measured over long intervals of time. The technique chosen gives the results as Allen variances. This can be stated in more familiar terms, such as power spectra. This permits the application of Handel's formula. When the measuring time T is below 100 minutes, we find Poisson noise $\propto$ 1/T. If the measuring time exceeds 100 minutes, up to 3000 minutes or 2 days we find the onset of a constant relative Allen variance, called flicker floor, at 10^{-7}. This floow corresponds to 1/f noise. Handel's formula gives the same order of magnitude when reasonable numerical estimates are used. Because of the singular importance of this fundamental experiment, one would like more independent experimental confirmation, preferably with another, stronger radioactive source. In the experiments reported the flicker floor did not show up free over many decades of measuring time. The very long measuring times required can raise suspicious questions about drift of measuring equipment, etc.

B. Thin gold films[37]. Handel's theory predicts α-values of the order of 10^{-7}. The values measured here agree with that prediction. At low temperatures the Umklapp processes freeze out. They make the highest contribution to the scattering of about 100 K. The observed temperature dependence of α below 200 K agrees with the theoretical expectation. In calculating the α-values from the experimental results one needs the number of free electrons. Kilmer et al.[37] have rightly pointed out that, in metals, only the electrons at the Fermi level count. This gives a substantial correction in the α-values. There is an error of a factor of 2 in their calculations[38], which is not significant here. These experiments of gold films, together with other measurement of low α-values support Handel's theory, which predicts $\alpha \sim 10^{-7}$. The difficulty is now for Handel to explain higher α-values, such as the often quoted value 2×10^{-3}.

C. Pentodes[39,40]. In a classical vacuum pentode we come close to a realization of what Handel chose as his model: a beam of electrons in vacuum. In a pentode, the cathode current is divided into the anode current and the screen-grid current. It is possible to make the cathode current 1/f-noise-free by properly choosing a resistor in the feedback circuit. A fraction of the cathode current goes to the anode and another fraction goes to the grid. This causes partition noise, even if the cathode current is noise-free. Handel's theory is not worked out completely for partition noise. By intelligent guessing how grid and anode currents, and how electron velocity at the grid and at the anode should be introduced into Handel's formulae, agreement can be obtained between experiment and theory. This adaptation leads in ref.[39] to an equation (11), about which the authors say in their own words: "Equation (11) in itself neither proves nor disproves Handel's quantum theory on 1/f noise"[39] p. 303. If the grid were to receive electrons from other parts of the cathode than the anode does, then it cannot be excluded that, although the total cathode current may be 1/f-noise-free, the other two currents contain 1/f noise that stems from the cathode.

In brief, there is experimental support for Handel's theory, especially where low α-values are found. As yet, the experimental evidence is not conclusive. There are no experiments that disprove this theory.

SUMMARY

After having compared theories and experiments, I shall now try to sum up the present chaotic situation in the following scheme. This is the best I can make of it.

Theory	
ΔX	Band of activation energies. Verified in metals. ΔX could be $\Delta\mu$.
ΔT	1/f-like spectra. Part of noise in metals.
Δn	McWhorther. At most a fraction of the 1/f noise in semiconductors.
$\Delta\mu$	Verified in semiconductors. Mechanism ? α ?

REFERENCES

1. A. van der Ziel, Adv. Electronics and El. Phys. 49, 225 (79)
2. P. Dutta, P.M. Horn, Rev. Mod. Phys. 53, 479 (81)
3. F.N. Hooge, T.G.M. Kleinpenning, L.K.J. Vandamme, Rep. Prog. Phys. 44, 479 (81)
4. G.N. Bochkov, Y.E. Kuroviev, Sov. Phys. usp. 26, 829 (83)
5. M.J. Buckingham, Noise in Electron Devices and Systems. Ellis Horwood Ltd. (83)
6. Proc. 7th Int. Conf. "Noise in Physical Systems" and 3rd Int. Conf. "1/f Noise" Montpellier 1983. Edit: M. Savelli, G. Lecoy, J. Nougier, North-Holland 1983
7. R.F. Voss, J. Clarke, Phys. Rev. L 36, 42 (76)
8. R.F. Voss, J. Clarke, Phys. Rev. B 13, 556, (76)
9. H.G.E. Beck, W.P. Spruit, J. Appl. Phys. 49, 3384 (78)

10. R.F. Voss, Phys. Rev. L 40, 913 (78)

11. P.J. Restle, M.B. Weissmann, R.D. Black, J. Appl. Phys. 54, 5844 (83)

12. P.J. Restle, R.J. Hamilton, M.B. Weissmann, M.S. Love, Phys. Rev. B 31, 2254 (85)

13. M.B. Weissmann, J. Appl. Phys. 51, 5872 (80)

14. R.D. Black, P.J. Restle, M.B. Weissman, Phys. Rev. B 28, 1935 (83)

15. R.D. Black, M.B. Weissmann, P.J. Restle, J. Appl. Phys. 53, 6280 (82)

16. R.D. Black, W.M. Snow, M.B. Weissmann, Phys. Rev. B 25, 2955 (82)

17. R.D. Black, P.J. Restle, M. Weissman, Phys. Rev. L 51, 1476 (83)

18. D.M. Fleetwood, N. Giordano, Phys. Rev. B25, 1427 (82)

19. D.M. Fleetwood, J.T. Masden, N. Giordano, Phys. Rev. L. 50, 450 (83)

20. D.M. Fleetwood, N. Giordano, Phys. Rev. B 27, 667 (83)

21. D.M. Fleetwood, N. Giordano, Phys. Rev. B 28, 3625 (83)

22. D.M. Fleetwood, T. Postel, N. Giordano, J. Appl. Phys. 56, 3256 (84)

23. D.M. Fleetwood, N. Giordano, Phys. Rev. B 31, 1157 (1985)

24. D.M. Fleetwood, N. Giordano. In ref. 6 p. 201

25. T.G.M. Kleinpenning, Physica 84B, 353 (76)

26. J. Bisschop, Thesis Eindhoven University of Technology (83)

27. J. Bisschop, A.H. de Kuijper, J. Appl. Phys. 56, 1535 (84)

28. K.S. Ralls, W.J. Skocpol et al., Phys. Rev. L 52, 228 (84)

29. K. Zheng, K.H. Duh. A. v.d. Ziel, Physica 119 B, 249 (83)

30. E. Loh, J. Appl. Phys. 56, 3022 (84)

31. H.C. de Graaff, M.T.M. Huybers, J. Appl.Phys. 54, 2504 (83)

32. J. Kilmer, A. v.d. Ziel, G. Bosman, Solid State Electron. 26, 71 (83)

33. M. Mihaila, Phys. Lett. 104A, 157 (84)

34. M. Mihaila, Phys. Lett. 107A, 465 (85)

35. P.H. Handel, Phys. Rev. A 225, 745 (80)

36. J. Gong, W.H. Ellis, C.M. v. Vliet, Trans. Am. Nuclear Soc. 45, 221 (83)

37. J. Kilmer, C.M. v. Vliet, G. Bosman et al., Phys. St. Sol. b 121, 429 (84)

38. T.G.M. Kleinpenning, J. Bisschop, Physica 128 B, 84 (85)

39. A. v.d. Ziel, C.J. Hsieh et al., Physica 124 B, 299 (84)

40. A. v.d. Ziel, P.H. Handel, Physica 125 B, 286 (84)

41. V. Palenskis, Z. Shobliskas, Solid St. Comm. 43, 761 (82)

NOISE IN PHYSICAL SYSTEMS AND 1/f NOISE - 1985
A. D'Amico and P. Mazzetti (editors)

MEMBRANE NOISE AND EXCITABILITY

Louis J. DE FELICE, William GOOLSBY and Duanxu HUANG

Department of Anatomy and Cell Biology, Emory University School of Medicine
Atlanta, GA 30322

MACROCHAN is a program that simulates the macroscopic behavior of biological membranes from the microscopic properties of ion channels. We illustrate the use of this program by constructing nerve and heart action potentials from simple, stochastic models of K, Na, and Ca channels. MACROCHAN is available on request.

1. INTRODUCTION

Membrane noise, the inherent electrical fluctuations in biological membranes, is ultimately the source of membrane excitability. More specifically, membrane noise signifies the fluctuations in current or voltage caused by the random opening and closing of ion channels.[1-3] The excitability of a membrane refers to its power to generate action potentials, the transient, regenerative changes in the membrane potential. Limiting the discussion to the opening and closing of channels ignores at least two other sources of membrane noise: fluctuations in open-channel conductance and fluctuations in open-channel current.[4] However, a basic explanation of membrane excitability does not necessitate including open-channel noise. Moreover, by considering only three types of channels: K, Na and Ca, and by making the theoretical models that describe these three types of channel less complex than the actual experimental models,[5,6] the relationship between noise and excitability simplifies further. These reduced models contain the essential features of real channels, which are ion selectivity, voltage-gated activation and inactivation, and stochastic two-state conductance. The stochastic nature of the single-channel conductance accounts for the essential feature of excitable membranes - namely, a time-variant conductance.

Beginning with three theoretical models for K, Na, and Ca selective channels, we describe a program that simulates the behavior of these channels under various conditions. The program, MACROCHAN, requires the channel models, the number of each type of channel, and the membrane capacitance. MACROCHAN's output is either the current response to a prescribed change in voltage, the voltage response to a prescribed change in current, or the voltage and current during free-run, with no prescribed stimulus. Using MACROCHAN, we construct a nerve-like action potential composed of K and Na channels and a heart-like action potential composed of K, Na, and Ca channels.

2. CHANNEL MODELS

For MACROCHAN, channels reduce to ion-selective random switches. Ion selection sets a bias for the flow of current through the channel. No current flows when the voltage equals the bias, called Nernst potentials. The Nernst potentials we use are:

$$E_K = -100 \text{ mV} \quad E_{Na} = 50 \text{ mV} \quad E_{Ca} = 150 \text{ mV}$$

The sign is negative for K, and the signs are positive for Na and Ca, because more K exists inside a cell than outside, while the reverse occurs for Na and Ca. The values of the potentials are approximately correct for nerve and heart cells, but actual values depend on specific conditions.

The currents through individual, open K, Na, or Ca channels are:

$$i_K = \gamma_K(V-E_K) \qquad i_{Na} = \gamma_{Na}(V-E_{Na})$$
$$i_{Ca} = \gamma_{Ca}(V-E_{Ca})$$

$$C \underset{\beta_n}{\overset{\alpha_n}{\rightleftharpoons}} O \qquad \text{K}$$

$$\begin{array}{ccc} C & \underset{\beta_m}{\overset{\alpha_m}{\rightleftharpoons}} & O \\ \alpha_h \downarrow\uparrow \beta_h & & \beta_h \downarrow\uparrow \alpha_h \\ I & \underset{\alpha_m}{\overset{\beta_m}{\rightleftharpoons}} & I \end{array} \qquad \text{Na}$$

$$\begin{array}{ccc} C & \underset{\beta_d}{\overset{\alpha_d}{\rightleftharpoons}} & O \\ \alpha_f \downarrow\uparrow \beta_f & & \beta_f \downarrow\uparrow \alpha_f \\ I & \underset{\alpha_d}{\overset{\beta_d}{\rightleftharpoons}} & I \end{array} \qquad \text{Ca}$$

Fig. 1 Three models of ion channels. O represents the open state of the channel, C the closed state, and I the inactive state. Subscripts distinguish the type of transition and the type of channel.

V is the voltage across the channel (the membrane voltage), and γ describes the conductance of the open channel. A closed channel conducts no current, and individual channels open and close randomly. If p stands for the probability that a channel is open, then the macroscopic currents composed of N channels become:

$$I_K = N_K \, i_K \, p_K \qquad I_{Na} = N_{Na} \, i_{Na} \, p_{Na}$$
$$I_{Ca} = N_{Ca} \, i_{Ca} \, p_{Ca}$$

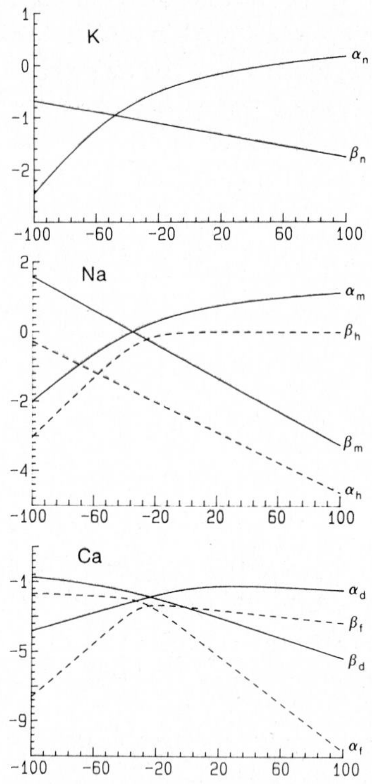

Fig. 2 Voltage dependence[5,6] of the αs and the βs. Solid lines for activation, dashed lines for inactivation. Vertical scales in log units of inverse milliseconds (e.g., -1 stands for 10^{-4} sec^{-1}); horizontal scales in millivolts across the channels.

The theoretical K channel has two states: the closed state C and the open state O (Fig 1). The rate constant for the transition from C to O is α, and the rate for the transition from O to C is β. α and β are functions of voltage only, and they represent the channel's ability to sense electrical field:

$$\alpha = \alpha\ (V) \qquad \beta = \beta\ (V)$$

The meaning of the rate constants is as

follows: If a channel is open, the probability that it will remain open for a time t is $e^{-\beta t}$. If a channel is closed, the probability that it will remain closed for a time t is $e^{-\alpha t}$. For a fixed voltage the average open time is $1/\beta$, and the average closed time $1/\alpha$. Finally, the fraction of time the channel spends in the open state equals $\alpha/(\alpha + \beta)$.

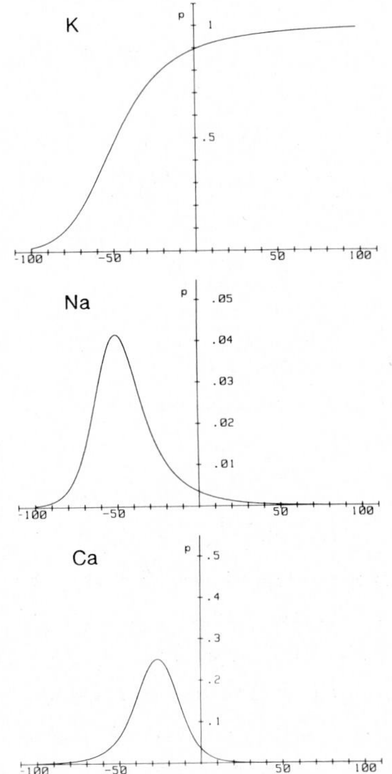

Fig. 3 Steady-state open channel probability.[5,6] p in relative units between 0 and 1; horizontal scales in millivolts.

From Fig. 2, α_n increases with voltage, and β_n decreases with voltage. (The subscript n signifies K.) Therefore, as the membrane voltage goes from negative to more positive values (depolarization), the channel tends to open more often. Figure 3 plots the open-state probability as a function of membrane voltage, and it shows that depolarization activates K channel and increases the K current. Above zero millivolts, the K channel is open virtually all the time.

Na channels are more complex than K channels, not only because depolarization activates Na channels, but also inactivates them. Our theoretical Na channel has four states, an open and closed state (similar to the open and closed state for the K channel) and two inactive states (Fig. 1). All αs and βs are functions of voltage only, and the meaning of the rate constants for the Na channel remains the same as for the K channel. If a Na channel is open, however, it has the option not only to close, but also to inactivate. The distinction is subtle, for the inactive state is also a closed state. The presence of the inactive state, and the dependence of α_h and β_h on voltage, cause the Na channel to open and then close with depolarization. Fig. 2 shows the voltage dependence of the Na channel rate constants, and Fig. 3 shows the open channel probability of the Na channel. In the steady-state, the Na channel remains closed virtually all of the time. We shall demonstrate below that the function of the Na channel is to conduct Na ions transiently to facilitate abrupt changes in voltage.

These remarks about the Na channel apply equally well to the Ca channel with two exceptions: The kinetics of the Ca channel are considerably slower than the kinetics of the Na channel. Thus, while the Ca channel turns on and then off with depolarization, it does so 10 to 100 hundred times more slowly than the Na channel. These slower kinetics are expressed as smaller values of the Ca αs and βs, compared to the Na αs and βs (Fig. 2). The second difference between the Na and Ca channels is that the open-state probability for Ca in the steady-state can be large relative to Na (Fig. 3).

3. PROGRAM

MACROCHAN simulates the macroscopic properties of nerve or heart cell membranes from the microscopic properties of channels. In the voltage-clamp mode, MACROCHAN simulates the current for fixed or changing potential. In the current-clamp mode, MACROCHAN simulates the voltage for fixed or changing current. In the unclamped mode (free-run) MACROCHAN simulates voltage noise, current noise, or action potentials. In the hybrid mode, we may clamp the potential and then release the membrane to free-run.

Many programs use macroscopic currents to simulate membrane activity. These programs use the Hodgkin and Huxley equations[5] for nerve (or similar equations[6] for heart) to describe the time and voltage-dependent gating of the macroscopic currents. For example, the variable n, which controls the K current, ranges between 0 and 1 and obeys a first order differential equation:

$$dn/dt = \alpha\ (1-n) - \beta\ n$$

The macroscopic K current relates to the gating variable n through the equation:

$$I_K = g_K n^4 (V-E_K)$$

g_K is the maximum K conductance and E_K is the Nernst potential.

The macroscopic approach models every aspect of membrane behavior except one - the inherent membrane noise due to the stochastic nature of the ion channels underlying the ionic currents. MACROCHAN, on the other hand, calculates the macroscopic currents and voltages directly from the stochastic behavior of single channels. Unlike previous programs in which the basic model element is a macroscopic ionic current, the basic model element in MACROCHAN is a single channel. The single channel has a number of conductance states, only one of which is occupied at any time. Some of the conductance states (usually one) conduct current, and the others do not. At a given membrane voltage, MACROCHAN calculates the probability of occupancy of each channel state. Thus, in a population of channels, the number of conducting channels varies with time, even at constant voltage. This variation of conductance in time introduces noise into the ionic currents. Basing macroscopic calculations on fluctuating channels constitutes the essential difference between MACROCHAN and previous programs, and is an outcome of previous work.[7]

3.1 System Hardware and Limitations

MACROCHAN runs on IBM PC, XT, or AT microcomputers under the PC DOS operating system, version 2.1 or versions later than 2.1. We use an AT with 80287 math coprocessor, 640K RAM, and 20 megabyte hard disk. The program language is Microsoft PASCAL version 3.3. The Halo graphics library produces high-resolution graphics on an IBM Enhanced Graphics Adapter.

The hardware and software limits the size of the single channel model and the number of channels in any particular simulation. At present, we may have up to 5 channel types and up to 500 channels of each type. The parameters of individual channel models are very flexible; the main restriction is that channels must have 10 or fewer states.

3.2 User Input

MACROCHAN includes an extensive data editor for assembling input parameters. These parameters are put into two primary data structures: Channel Definition Files and Input Data Files. We store these records on disk and retrieve them at the beginning of each simulation.

Channel Definition Files (CDF) include all information necessary to define a channel model: 1). number of states for each type of channel; 2). Nernst potentials (E_K, E_{Na}, etc.); 3). conductance of each state; 4). voltage dependence of the rate constants. Rate constant equations follow a standard form.[6] Each type of channel requires a separate CDF file.

Input Data Files (IDF) describe items other than channel parameters: 1). names of CDF files; 2). number of channels of each type; 3). membrane capacitance per unit area; 4). iteration step size; 5). output step size; 6). total run time; 7). clamp mode (a) voltage clamp; (b) current clamp; (c) unclamped; (d) hybrid (voltage clamp for a given time then unclamp); 8). voltage or current clamp parameters; 9). output data for storage. As part of the program initialization, MACROCHAN reads from disk the specified IDF and CDF files. This allows the program to run in batch mode by specifying an IDF file on the command line. In batch mode, a DOS batch file may contain the specifications for a number of simulations which will run unattended over long periods of time.

3.3 Probability Table

The central data structure in MACROCHAN is the probability table. This is a four-dimensional array of membrane voltage, channel type, present state, and next state. Given these 4 indices, the table returns the probability that a transition will occur from state i to state j within the time Δt.

Table entries are single-precision, 4 byte real. MACROCHAN allows membrane voltages from -150 to +150mV, 5 channel types, and 10 states per channel. Calculating the entire table at once would occupy 600K bytes. 600K bytes exceed memory limitations, so MACROCHAN calculates only portions of the table as the program runs. Of the 4 indices, membrane voltage varies slowest during most simulations; therefore, a table with a small voltage range has the least impact on program performance. At any given time MACROCHAN uses a table with a range of ± 20mV around the present membrane voltage.

Over time the membrane voltage changes, either as a result of present current or voltage stimuli, or as a result of ionic currents charging or discharging the membrane capacitance. Whenever membrane voltage changes, MACROCHAN checks that the probability table is up to date. The program keeps a record of previous membrane voltages to see whether membrane voltage has risen or fallen in the recent past. Presumably this trend continues into the near future, and on this basis we calculate the probability table in either the rising or falling direction, but with a -3mV offset from the present voltage. For example, if voltage has been rising and is presently at 20mV, we calculate the new table over the range 17 to 37mV; these voltages are predominantly higher than the present voltage; we add the -3mV offset to insure the table contains a noise margin below the present voltage. If the voltage had been falling, the new range would have been 23 to 3mV. The noise margin prevents thrashing, a state in which the program recalculates the table with each iteration, because the membrane voltage does not show a clear trend. This simple algorithm for predicting future voltage takes little time to calculate and is effective in reducing the number of probability table recalculations.

The probability table entries are:

$$p(v, c, i, j) = \Delta t \times \text{rate constant}$$

where Δt is the iteration step size, and the rate constant is the one appropriate for membrane voltage v, channel type c, and state transition i to j. Probability p is in the range 0.0 to 1.0. For the transition i to i, that is, a channel remaining in its present state, the probability is

$$p_{ii} = 1 - \Sigma\, p_{ij} \qquad i \neq j$$

3.4 State Transition Algorithm

Channels contain an arbitrary number of states up to ten. For each state transition there is a rate constant vector R_k consisting of seven constants from the standard equation:[6]

$$R_k = \frac{r1 \times \exp\,[r2\,(V + r3)] + r4(V + r5)}{\exp[r6(V + r3)] + r7}$$

In a multi-state model a particle has more than one alternative for state transitions. As an example, consider the model for Na, or for Ca, in Fig. 1.

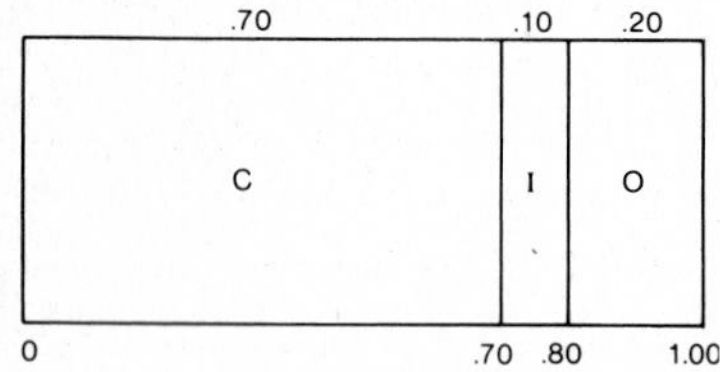

Fig. 4 Line diagram to illustrate method of probability calculation.

To determine the next state of the channels, MACROCHAN needs the present state and the transition probabilities. Given the present state, the channel type, and the membrane voltage, MACROCHAN extracts a value from the probability table for all legal transitions away from this state. Inaccessible states have probabilities of zero. The sum of the probability of all transitions, including remaining in the present state, must equal 1. Fig. 4 visualizes transition probabilities as a number line and shows a possible distribution for the Na model of Fig. 1. The probability of moving to each of the 4 states is a fraction between 0 and 1. Numbers above the block indicate the width of each section and its relative probability; numbers below the block represent the cumulative width of all preceding sections. The channel occupies the closed state. MACROCHAN generates a random number from 0 to 1 and compares it to the cumulative numbers below the block. In this example there is a 0.70 probability of remaining in the closed state (C), 0.10 of moving to the accessible inactivated state (I), .20 to the open state (0), and 0.00 to the inaccessible inactivated state (see Fig. 1). If the random number were 0.76, the new state would be the accessible inactive state I. The new state of each channel is thus determined by a random number falling within a certain number range, where the width of each state's range is proportional to its relative probability.

3.5 Random Number Generator

The congruence method generates pseudo-random numbers. The particular generator MACROCHAN uses is:

$$X[n + 1] = (23 \times [n]) \bmod (10^8 + 1)$$

with a seed value X [0] = 47,594,118. This generator[8] has a period of approximately 5.8×10^6. The power spectrum for 4096 numbers is flat. To illustrate the need for a long period, consider a simulation with 500 Na channels and 50 K channels with a Δt of 2.5 us. For each iteration, we need one random number to determine the state of each channel. For a run of 25 ms there are 10,000 iterations, each needing 550 random numbers, for a total of 5.5×10^6 numbers. Since MACROCHAN uses such a large quantity of random numbers, the generator must be very fast. The congruence method is fast, because multiply and modulo operations use integer arithmetic.

3.6 Determination of Initial State

Before starting a simulation, MACROCHAN must determine the initial state of all channels. Membrane voltage initiates this determination. In voltage clamp, MACROCHAN uses the initial clamp voltage; otherwise it sets the voltage to zero. In the steady state, the voltage determines the probability for each transition from one state to another; this yields a set of simultaneous linear equations, the solution of which gives the occupancy density of each state.

As an example, consider a simple 2-state model with states 1 and 2 and rate constants α and β. Let X1 and X2 be the fraction of all channels which are in states 1 and 2 respectively; X1 and X2 range from 0 to 1. In the steady state, three facts are apparent: (1) the number of channels entering state 1 must equal the number leaving state 1; (2) the number of channels entering state 2 must equal the number leaving state 2; and (3) the fraction of channels in state 1 plus those in state 2 must equal 1.

$$\alpha X2 = \beta X1$$
$$\beta X1 = \alpha X2$$
$$X1 + X2 = 1$$

The first two equations are redundant but are necessary for completeness in a matrix solution. Any channel model may be put in the matrix form:

$$[AA]\ [X] = [BB]$$

where matrix [AA] contains the rate constants; the last row of matrix [AA] consists entirely of ones. [BB] contains zeroes except in the last entry, for which the value is also one. Solving for X gives the proportion of channels in each state.

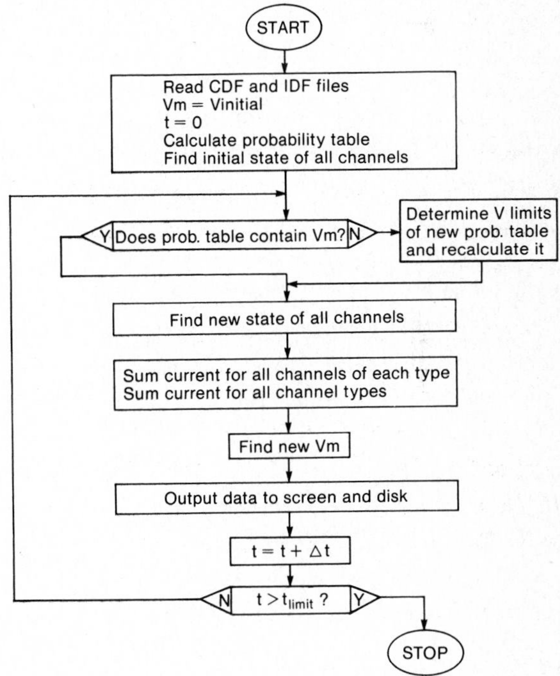

Fig. 5 MACROCHAN Flow Diagram.

3.7 Program Flow

Program initialization includes reading the input CDF and IDF files and setting the intial values of variables.

Membrane voltage is set to the clamp voltage for the voltage clamp mode; otherwise the voltage is set to zero. The program finds the initial state of all channels and calculates the probability table around the initial voltage. Fig. 5 shows the main program loop.

MACROCHAN first checks to insure that the probability table includes present membrane voltage. If not, it recalculates the table. Next, MACROCHAN determines the new state of all channels by using the present state, the probability table, and the random number generator. Once we know the state of a channel, we know its conductance. The microscopic current through each channel is:

$$i = \gamma(V - E)$$

The total current for all channels of a given type is the sum of the microscopic currents. Note that gamma may vary from one channel to the next, as it may even within a given type, because the channels have random properties and may be in different conductance states. Usually, however, only one state conducts ionic current. Total ionic current for the entire membrane is the sum of the current for all channel types.

MACROCHAN stores the conductance and the ion current of each channel type, the conductance and ion current of the entire membrane, and the conductance history of five channels of each type to serve as examples. MACROCHAN then calculates the new membrane voltage if it is running in current clamp or the unclamped mode. In the voltage clamp mode, the program determines the membrane voltage from one of several voltage clamp equations or from an input file of up to 8192 time-voltage pairs.

The preset voltage-clamp options are: 1). constant; 2). step; 3). pulse; 4). sine wave; 5). white noise; 6). ramp. In the unclamped or current clamp mode, MACROCHAN updates voltage with each iteration. If we specify the voltage from an input file, then the file provides up to 8192 time/voltage pairs; each voltage remains in effect for the specified time, after which the voltage jumps to the next value. Upon reaching the end of the file, the file pointer is reset to the beginning, and the program repeats the sequence. We use this option to clamp with complex waveforms such as action potentials. In unclamped or free-run mode MACROCHAN determines the new voltage from:

$$V_{new} = V_{old} + (I_s - I)\ \Delta t/Cm$$

where I_s is the stimulus current; I, total ionic current; Δt, the computation step size; and Cm, the membrane capacitance. In free-run we usually clamp to a given voltage for a small amount of time, then release the clamp (the hybrid mode). Upon release membrane voltage will begin to move up or down based upon the net current (I_s - I). In free-run the system determines its own V_m and is capable of producing action potentials, given the proper types and ratios of channels. Should membrane voltage ever exceed the -150 to +150mV limits, MACROCHAN resets to either the maximum or minimum value and prints a diagnostic message. Output need not occur every Δt. To reduce the volume of output data, we maintain a "modulo N" counter, where N may be any value greater than or equal to one.

MACROCHAN writes one line an output file after passing through the main loop N times and continuously plots the total current and the membrane voltage for each channel type. The user may select any of the following variables to store in ASCII disk file: 1). time; 2). current for all channels of a type; 3). stimulus current; 4). total current; 5). total conductance; 6). membrane voltage; 7). sample current for five channels of a type. We may use the ASCII file in other programs for graphing or further processing. We assign each output variable one column in the output file, and each output iteration writes one line to the file.

The above calculation completes the processing necessary for one Δt. Finally, MACROCHAN increments time by one Δt and repeats the loop.

4. RESULTS

Figs. 1-3 describe the properties of our theoretical channels, and Figs. 6-10 develop the consequences of these models for single channel behavior and for the generation of action potentials.

Figure 6 shows how single K channels respond to a step change in voltage. At -100mV, the K channel is virtually closed, and at -40mV it remains open about half the time (see Fig. 3). However, the channels require time to go from the -100mV steady-state to the -40mV steady-state. We observe this period of adjustment in Fig. 6 for each of the four voltages to which we step. Notice that the single channel current increases as we depolarize the membrane. This increase in current occurs because the channel is farther away from E_K at 20mV than at -40mV.

Figure 7 shows the same simulation as Fig. 6 for 500 K channels rather than one. Although the voltage changes instantaneously to its new value at t = 0, the current requires several milliseconds to reach its new steady-state value. The time dependence of the macroscopic current occurs because of the period of adjustment of the microscopic current (Fig. 6). When the current reaches its new steady-state value it is noisy, i.e., the steady-state is a mean value, about which the current fluctuates. The current noise and the time-variant conductance are both manifestations of the random opening and closing of the channels.

In Figs. 6 & 7 we repeat the calculations for Na and Ca channels. The essential difference between the Na (or the Ca) current, and the K current, is that Na and Ca inactivate. When we depolarize a Na channel, it first has a tendency to open, and it tends to close to the inactive states (Fig. 3). Single Na channels are only likely to conduct during the first few milliseconds following a depolarization (Fig. 6), which accounts for the peak and then the decline of the macroscopic current (Fig. 7). The same is true for the Ca current, except that the kinetics are much slower (Figs. 6 and 7).

Fig. 8 depicts a nerve-like action potential and the underlying ion currents from 500 Na Channels and 50 K channels, and Fig. 9 shows a train of such action potentials in a free-running membrane. Real action potentials from nerve axons require an external stimulus to fire and are not spontaneously active as in this simulation. However, some nerve cells, known as pacemaker neurons, do fire spontaneously.

Fig. 10 illustrates a heart-like action potential and the underlying currents from 500 Na channels, 500 Ca channels, and 300 slow K channels. Slow K channels have the kinetic scheme of Fig. 1, but α_n and β_n are ten times less than in Fig. 2 constructing a heart-like action potential from various combinations of Na, Ca, and K channels, without slowing the K kinetics, proved impossible.

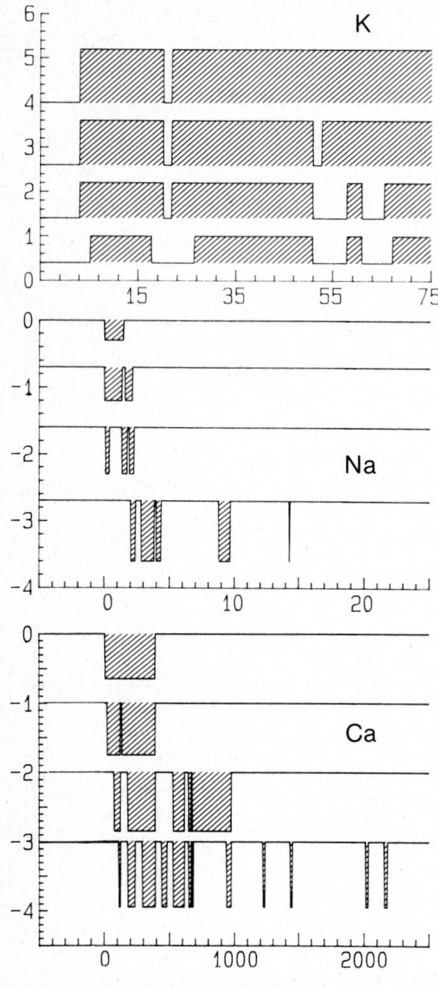

Fig. 6 Behavior of single K, Na, and Ca channels after step changes in voltage. Vertical scales in picoamps; horizontal scales in millisecs. Shaded portions indicate an open channel conducting ions. Positive current indicates ions flowing out of the cell. Each channel, initially at -100 mV, changes to a new voltage at t = 0: -40 (bottom); -20; 0; and 20mV (top).

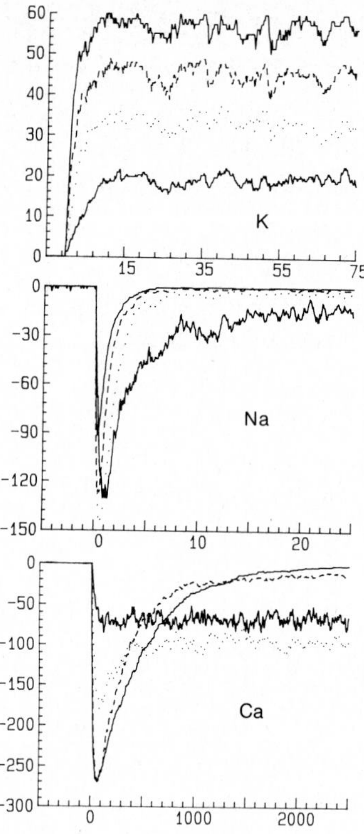

Fig. 7 Behavior of 50K, 500 Na, and 500 Ca channels after a step change in voltage. Voltage steps as in Fig. 6. Repeating random numbers causes correlation between traces. Vertical scales in picoamps; horizontal scales in millisecs.

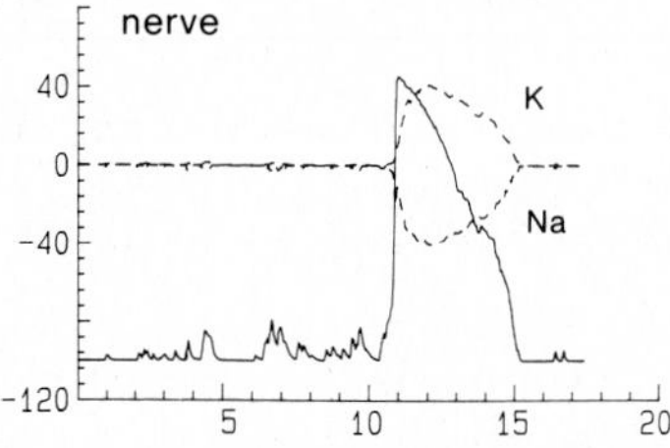

Fig. 8 Nerve-like action potential (solid line) from 50 K channels and 500 Na channels in a 1 μm^2 patch of membrane (10^{-14} farads). The potential is initially -100 mV, but from t = 0 the membrane is in free-run. Inherent membrane noise fires the action potential. Underlying K and Na

currents are dashed lines. Vertical scale in millivolts or picoamps; horizontal scale in millisecs.

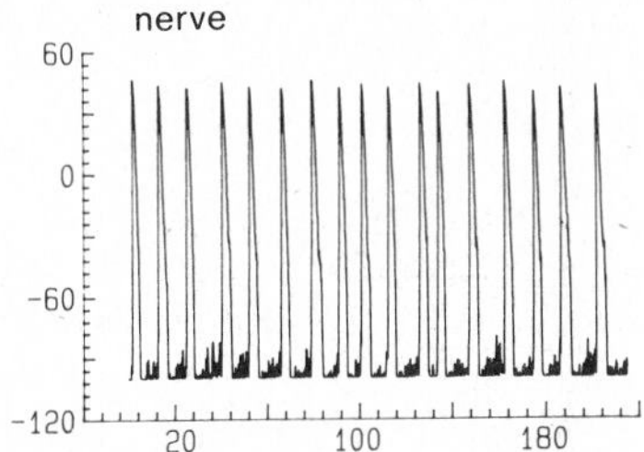

Fig. 9 Nerve-like action potentials, (parameters of Fig. 8) showing variability in baseline noise, interval between action potentials, and action potential shape and height. Verticle scale in millivolts; horizontal scale in millisecs.

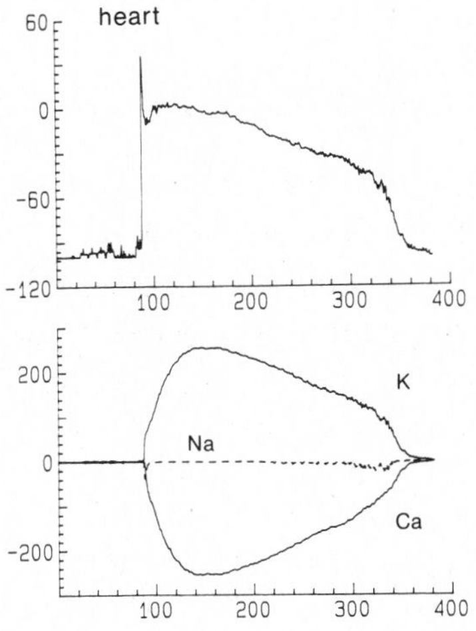

Fig. 10 Heart-like action potential and underlying currents, from 300 slow K channels, 500 Na channels, and 500 Ca channels in a 1 μm^2 patch of membrane (10^{-14} farad). The potential, initially at -100 mV, is in free-run after t = 0. Vertical scale in millivolts (top) or picoamps (bottom); horizontal scales in millisecs.

5. SUMMARY

The ability of membranes to generate action potentials resides in the stochastic properties of ion selective channels located in the membranes of nerve, heart, and other excitable tissues. The nature of ion selective channels is to open and close randomly, and the dependence of the probability of these openings and closings on voltage accounts for the essential feature of excitable membranes: time-variant conductance. Other sources of membrane noise, such as open-channel current noise and Johnson noise, appear to play no direct role in excitability.

ACKNOWLEDGEMENTS

We are grateful for Dr. Renzo Levi's help in developing MACROCHAN, and we wish to thank Dr. Sally Wolff for editing the manuscript. NIH Grant #HL27385 supports this work.

REFERENCES

1. L.J. DeFelice, Introduction to Membrane Noise (Plenum, N.Y., 1981).
2. B. Sakmann and E. Neher, Single-Channel Recording (Plenum, N.Y., 1983).
3. B. Hille, Ionic Channels of Biological Membranes (Sinauer, Sunderland, MA, 1984).
4. F.J. Sigworth, Biophys. J. 47 (1983) 709.
5. A.L. Hodgkin and A.F. Huxley, J. Physiol. 117 (1952) 500.
6. G.W. Beeler and H. Reuter, J. Physiol. 263 (1977) 177.
7. J.R. Clay and L.J. DeFelice, Biophys. J. 42 (1983) 151.
8. M. Abramowitz and I.A. Stegun (Eds.). Handbook of Mathematical Functions (Dover, N.Y., 1972) 949.

NOISE IN PHYSICAL SYSTEMS AND 1/f NOISE - 1985
A. D'Amico and P. Mazzetti (editors)

Squeezed Noise in Precision Force Measurements

Mark F. BOCKO *, Franco BORDONI **, Franco FULIGNI **, Warren W. JOHNSON ***

The effort to build gravitational radiation antennae with sensitivity sufficient to detect bursts of radiation from supernovae in the Virgo cluster of galaxies has caused a consideration of the fundamental limits for the detection of weak forces. The existing Weber bar detectors will be eventually limited, by the phase insensitive transducers now used, to noise temperatures no better than that of the first amplifier which follows the transducer. Even for a quantum limited amplifier this may not give the sensitivity required to definitively detect gravitational radiation. In a "back action evasion" measurement a specific phase sensitive transducer would be used. It is believed that by the technique of measuring one of the two antenna phases it is possible to reach an effective noise temperature for the measured phase which is far below the amplifier noise temperature. This is at the expense of an infinite noise temperature in the unmeasured antenna phase and is thus described as squeezing the noise. We outline the theoretical model for the behavior of such systems and present data from several experiments which demonstrate the main features of a back action evasion measurement. We also briefly describe related work to generate squeezed states of electromagnetic radiation.

1. Introduction

Quantum NonDemolition measurement techniques (QND) open the possibility for significant improvement in the performance of a variety of sensors and receivers. In this paper we will discuss a particular quantum nondemolition technique, Back Action Evasion (BAE), in its application to electromechanical motion sensors. In a back action evasion measurement one has the ability to partially isolate the system being measured from the noise propogating back from the measuring apparatus. In this way it is possible to improve the overall figure of merit for a system beyond that achievable by conventional measurement techniques.

We present a theoretical model in which the connection between back action evading systems and familiar parametric upconvertors is emphasized. We will show that back action evasion is characteristic of the parametric process which is the coherent combination of a phase inverting and non-inverting upconvertor. The predictions of the model are summarized and interpreted without the full theoretical development. We then review the experimental progress on back action evading measurements. Finally we comment on the connection between the effort to generate squeezed states of the electromagnetic radiation field and the role of squeezed states in a QND measurement.

We begin this review with a brief summary of the formal theory of nondemolition measurements and an account of the early proposals for QND measurements.

2. Gravitational radiation antennae and QND

The idea for the quantum nondemolition measurement technique arose in the worldwide effort to detect gravitational radiation from the cosmos.

The strongest likely sources of gravitational radiation bathing the earth are believed to be the birth of a neutron star or black hole accompanied by a supernova . Such events in our galaxy may

* Department of Electrical Engineering, University of Rochester, Rochester, New York, 14627, USA.
** I. F. S. I., C. N. R., PO Box 27, Frascati, Italia.
*** Department of Physics and Astronomy, University of Rochester, Rochester, New York, 14627, USA.

yield millisecond bursts of radiation with an amplitude of 10^{-18} or less. Unfortunately such events seem likely to occur only once every ten to thirty years. To reach a sensitivity which would yield an event rate of one per month we must be able to detect weaker bursts, with an amplitude of 10^{-21} or less, which originate in nearby clusters of galaxies. Thorne (1980) [1] gives a review of the many possible sources of gravitational waves.

Several groups are preparing massive (a few hundred to 5000 Kg) cylinders of aluminum [2], or another low acoustic loss material such as niobium or sapphire, cooled to liquid helium temperatures. See Douglass and Braginsky (1979)[3] for a thorough review of detector research. The detectors are instrumented with sensitive motion sensors and carefully isolated from environmental disturbances. A passing gravitational wave exerts a tidal force and thus may excite the free modes of oscillation of the cylinder. Normally the lowest longitudinal mode of the cylinder will be most strongly excited. In this mode the ends of the cylinder move in opposition and an accelerometer attatched to an end face converts the motion into a measurable electronic signal. In the presence of a wave with an amplitude of 10^{-18} the end faces of a one meter long gravitational wave antenna would be displaced by approximately 10^{-18} meters. Thus to reach the goal of one event per month requires the measurement of displacements of 10^{-21} meters or less!

One may treat the antenna mode of interest as a simple harmonic oscillator coupled to a signal force (the gravity wave) and random forces which model the Langevin force responsible for the antenna's Brownian motion and the effect of the electromechanical transducer upon the motion of the antenna.[4]

Braginsky [5,6] was the first to point out that gravitational radiation detectors would, in the near future, approach a level of sensitivity at which one must attempt to detect changes in the energy of the equivalent simple harmonic oscillator comparable to the quantum of energy. At the kilohertz frequencies that gravitational antennae operate this corresponds to displacements of the antenna end faces of about 10^{-20} meters, an order of magnitude larger than our goal of 10^{-21} meters. In Braginsky's analysis the antenna mode of interest was treated as a quantum harmonic oscillator and uncertainty principle arguments were used to derive the result. This sparked a great deal of theoretical analysis which refined and expanded upon Braginsky's original idea.[7-26]

In present detectors one measures the amplitude and phase of the interesting antenna normal mode rather than the energy. An excellent review and thorough analysis is given in the paper by Caves <u>et al</u> (1980) [27]. In the following we very briefly outline their analysis.

It is convenient to represent the antenna mode by the real and imaginary parts of its complex amplitude.

$$\mathbf{x} + i\mathbf{p}/m\omega = (\mathbf{X}_1 + i\mathbf{X}_2)\, e^{-i\omega t} \qquad (1)$$

Our convention is to use **bold face** type for quantum operators and plain face for expectation values and the corresponding classical observables. The oscillator position and momentum operators are represented by **x** and **p**, m is the mass of the oscillator and ω is the angular frequency. The two phases of the complex amplitude, $\mathbf{X}_1$ and $\mathbf{X}_2$, are the displacement operators in the coordinate system that co-rotates with the oscillator in phase space. The convenience of introducing these variables is that in the absence of forces acting on the oscillator and for vanishing dissipation they are conserved quantities.

The operators $\mathbf{X}_1$ and $\mathbf{X}_2$ have a non-vanishing commutator,

$$[\mathbf{X}_1, \mathbf{X}_2] = i\hbar / m\omega \qquad (2)$$

so the expectation values obey the uncertainty relation

$$\Delta X_1\, \Delta X_2 \geq \hbar / 2m\omega \quad . \qquad (3)$$

This implies that a measurement in which one obtains information about both phases of the complex amplitude, as in a conventional position measurement, is subject to a limit imposed by the quantum nature of the observables. If for example we try to measure X_1 and X_2 with equal precision the measurement is subject to the so called standard quantum limit ,

$$\Delta X_1 = \Delta X_2 \geq (\hbar / 2m\omega)^{1/2} . \qquad (4)$$

The conventional transducers currently under development for gravity wave detectors fall into two classes. First there are capacitive transducers which operate by allowing the *position* of the antenna end face to modulate a capacitance in an electrical circuit. The other class of transducers operates by coupling an inductor to the antenna motion thus making the transducer sensitive to the *velocity* of the antenna end face. In either scheme one obtains information about both phases of the complex amplitude and are thus constrained by the standard quantum limit. If we substitute typical values for the parameters; $m = 10^3$ Kg and $\omega / 2\pi$ = 1KHz we obtain $\Delta X_1 \geq 3 \times 10^{-21}$ meters.

Viewing the problem in this manner it becomes clear that the solution is to measure a single component of the oscillator complex amplitude. The laws of quantum mechanics do not preclude the possibility of an arbitrarily precise measurement of X_1 or X_2 provided the non-commuting observable is correspondingly uncertain. Since X_1 and X_2 are conserved we can monitor one of them continuosly with high precision. In contrast, during a precise measurement of the position, **x**, a large "kick" is given to the momentum **p**. Through the free evolution of the oscillator this disturbance feeds back and makes future values of the position highly uncertain.

Caves et al (1980) [27] suggested practical back action evasion schemes in which one could measure either X_1 or X_2 of a harmonic oscillator without disturbing it. The further developments which we will discuss in this paper have been largely based upon these early suggestions.

3. Parametric Transducer and Equivalent Circuit

A few groups undertook the experimental task of realizing a BAE measurement based upon the Caves et al proposal. [28 - 31] It was soon realized that the proposal was actually a new type of parametric convertor. In Figure 1 we show the electromechanical schematic of the parametric convertor. The motion of a diaphragm is allowed to simultaneously increase the gap of one side of a balanced three plate capacitor while decreasing the other. The three plate capacitor makes up one half of a balanced bridge circuit. The slightly unusual feature of this bridge is the addition of a resonant readout arm.

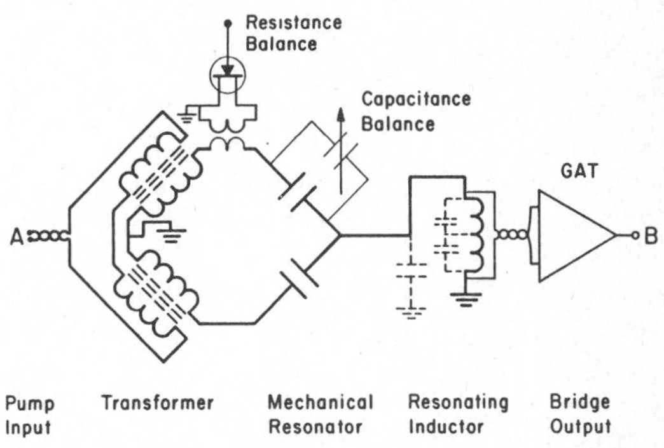

Figure 1 The electromechanical schematic of the BAE transducer. The motion of the diaphragm unbalances the bridge circuit which causes a current in the readout arm. The current is sensed by measuring the voltage across the readout arm inductor with a GaAs transistor (GAT) amplifier.

In terms of the displacement of the accelerometer test mass, x (t), and the charge, q(t), on the center plate of the three plate capacitor, the equations of motion can be written

$$m\ddot{x}(t) + g\dot{x}(t) + k\,x(t) = -E_p(t)\,q(t) \qquad (5a)$$

$$L\ddot{q}(t) + R\dot{q}(t) + 1/C\,q(t) = -E_p(t)\,x(t) \qquad (5b)$$

In the above equations m, g and k are respectively the mass, damping factor and spring constant of the mechanical oscillator; L, R and C are the bridge readout arm inductance, readout arm series resistance and the parallel capacitance of the three plate capacitor. The electric field impressed on the capacitor to couple the mechanical displacement to the electrical mode is denoted by $E_p(t)$.

These equations also describe the dynamics of the parametric device shown in Figure 2. This is the familiar model of a parametric convertor in which two RLC oscillators are coupled by a time varying capacitor. We emphasize that in the electromechanical device the parametric coupling arises from the time varying electric field in the three plate capacitor and not the variation of a real capacitance as is usually the case in electrical parametric convertors.

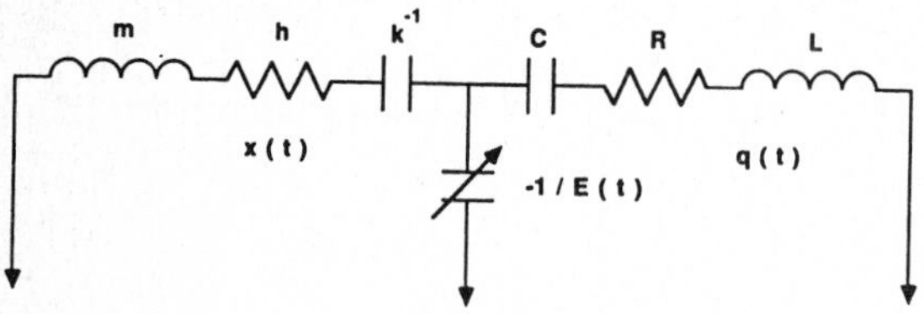

Figure 2 The electrical analog of the BAE transducer is a traditional parametric convertor.

The remaining feature is to specify the nature of the time varying coupling $E_p(t)$. The two familiar ways of parametrically pumping such a device are at either the sum or difference frequencies of the two modes. We choose to write the pump as the superposition of these two possibilities with the parameter f, $(-1 \leq f \leq 1)$, which determines the fraction of the pump amplitude at each frequency.

$$E_p(t) = E_0/2\{(1-f)\cos(\omega_2+\omega_1)t + (1+f)\cos(\omega_2-\omega_1)t\} \quad (6)$$

We use ω_1 and ω_2 for the respective angular frequencies of the mechanical and electrical resonators.

If we substitute this form of the pump into the equations of motion and change variables to complex amplitudes we find Equations 7a - 7d. [31]

$$\left(\frac{d}{dt} + \frac{1}{2\tau_1}\right)X_1 = \frac{E_0}{4}\frac{1}{m\omega_1} f Q_2 - \frac{F_s + f_1}{m\omega_1}\sin\omega_1 t \quad (7a)$$

$$\left(\frac{d}{dt} + \frac{1}{2\tau_1}\right)X_2 = -\frac{E_0}{4}\frac{1}{m\omega_1} Q_1 + \frac{F_s + f_1}{m\omega_1}\cos\omega_1 t \quad (7b)$$

$$\left(\frac{d}{dt} + \frac{1}{2\tau_2}\right)Q_1 = \frac{E_0}{4}\frac{1}{L\omega_2} f X_2 - \frac{v_n}{L\omega_2}\sin\omega_2 t \quad (7c)$$

$$\left(\frac{d}{dt} + \frac{1}{2\tau_2}\right)Q_2 = -\frac{E_0}{4}\frac{1}{L\omega_2} X_1 + \frac{v_n}{L\omega_2}\cos\omega_2 t \quad (7d)$$

We have introduced the relaxation times of the mechanical and electrical oscillators τ_1 and τ_2, a voltage source v_n which represents the random noise driving the electrical oscillator, the signal force F_s and Langevin force f_l which is responsible for the mechanical oscillator Brownian motion.

The form of the above equations make it possible to understand the behavior of the transducer. In all cases we regard Q_1 and Q_2 as the output quantities. In practice if one were to measure the voltage across the bridge readout arm with a lock-in amplifier tuned to the frequency ω_2 the output of the two channels would be proportional to Q_1 and Q_2. In the conventional cases; f = 1 for the non-inverting parametric upconvertor and f = -1 for the inverting upconvertor we see that both phases of the two oscillators are coupled to one another; X_1 is coupled to Q_2 and X_2 is coupled to Q_1. The difference between the two cases is only in a sign change in the coupling coefficients. This difference leads to the familiar behaviors of frequency pulling, resonance peak broadening and parametric instability for the phase inverting upconvertor at sufficiently large coupling.

The interesting case is when f = 0, i. e. equal amplitudes of the lower and upper pump frequencies. We see from Eqn. 7a that X_1 is decoupled from the back action of the electrical mode, but that it is still coupled to the signal force F_s. The forward coupling of X_1 to Q_2 in Eq. 7d is unaffected. Examining the behavior of the other two complex amplitudes described by Eqns. 7b and 7c we see that X_2 feels the full back action from Q_1 but the forward coupling of X_2 to Q_1 vanishes. Thus the signal appears in the Q_2 phase of the output with no noise contamination from the back action while at the same time the other output phase, Q_1, contains only noise.

The price payed for back action evasion is that now the response to a signal force is phase sensitive. In the case of interest to gravity wave detection, impulsive signals, this means that the detector is "on" only half the time. Of course this can be remedied by the construction of a separate detector working in quadrature.

At this point the improvement in performance becomes a quantitative question. Since there are other sources of noise present, Brownian motion and electrical noise at the input and output of the amplifier, we wish to define and evaluate some figure of merit to see under what conditions we may benefit from back action evasion. Although the particular noise sources peculiar to the case at hand may not be of general interest the model may be generalized to represent many different physical systems. We will see that back action evasion is only useful in the cases of strong coupling and when the noise intrinsic to the first mode, here a mechanical oscillator, is sufficiently small.

The figure of merit we define is somewhat different from the usual definitions encountered. In most devices such as linear amplifiers one customarily intends to look for signals which have a narrow bandwidth compared to the operating bandwidth of the amplifier. In such a case one assigns a noise temperature which is a spectral density. In the present case we are interested in broadband burst signals. If we define a noise temperature spectrum in analogy with the usual case we see that it varies greatly over the frequency range of the signal. Thus it is more convenient to define an integrated noise temperature for our case. We assume the use of the optimum filter which will give the maximum signal to noise ratio. In the context of gravity wave detectors this quantity is called the burst temperature.

We define the burst temperature, T_b, by the energy which would be deposited in an initially unexcited antenna from an impulse which gives a signal to noise ratio of 1 in the filtered output. This method is outlined in Ref. 30. The result can be summarized by the simple formula

$$T_b = T_E\,(\omega_1/\omega_2)\,(1/r)\,. \qquad (8)$$

The Manley Rowe frequency ratio factor normalizes the noise temperatures and the factor r is the back action evasion reduction factor. In the conventional cases r is always less than 1 but in the BAE case r may exceed 1. The result of a detailed calculation gives the following result. We may express r in terms of three dimensionless factors :

$$\alpha = (T/T_E)(\omega_2/\omega_1)(1/\omega_1\tau_1) \qquad (9a)$$

$$\beta = E_0^2(\omega_2/\omega_1)(C/m\,\omega_1^2) \qquad (9b)$$

$$\gamma = (1/4)(\omega_2/\omega_1)(Z_c/Z_n) \qquad (9c)$$

where α is a measure of the Brownian noise as compared to the electronic noise, β is a coupling strength parameter and γ is a measure of how closely impedance matched the transducer is to the amplifier. The physical temperature of the transducer is T and Z_c and Z_n are respectively the characteristic impedance of the transducer, $(\omega_2 C)^{-1}$, and the noise impedance of the amplifier. The result is;

$$(1/r) = (f^2 + 8\,\alpha\gamma/\beta)^{1/2} * \{1 + (\beta/4\gamma^2)[(f^2 + 8\,\alpha\gamma/\beta)^{1/2} - f]\}^{1/2} \qquad (10a)$$

when $(\alpha\beta/2\gamma^3)(1 + f\gamma/\alpha) > 1$ and

$$(1/r) = (2)^{-1/2}(f^2 + 8\,\alpha\gamma/\beta)^{1/2} * \{\{1 - f(\beta/4\gamma^2) + [1 - (\beta/2\gamma^2)(f + \gamma/\alpha)]^{1/2}\}^{1/2} + \{1 - f(\beta/4\gamma^2) - [1 - (\beta/2\gamma^2)(f + \gamma/\alpha)]^{1/2}\}^{1/2}\} \qquad (10b)$$

when $(\alpha\beta/2\gamma^3)(1 + f\gamma/\alpha) < 1$. In Figure 3 we plot contours of constant $(1/r)$ in the α, β, γ parameter space. The performance in the conventional modes, $f = \pm 1$, approaches the electronic noise temperature limit, whereas for the BAE case, f = 0, we may exceed the electronic noise temperature.

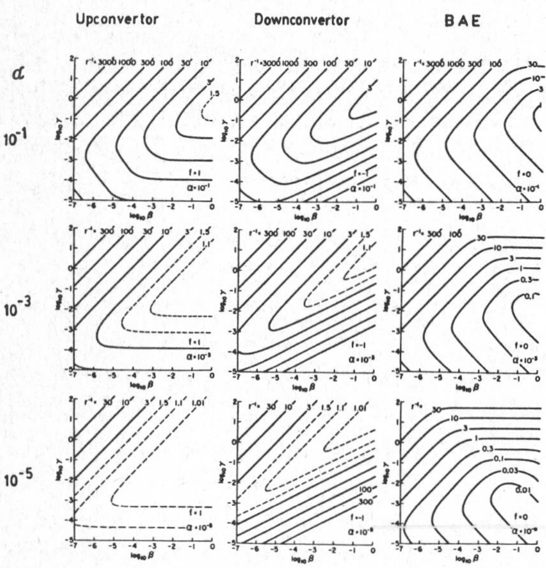

Figure 3 Contours of constant back action evasion reduction factor are plotted in the β ,γ plane for different values of α and the different pumps. See text for definitions.

In Equations 7 we have omitted the terms which vary at the frequency $2\omega_1$. The full equations, including these terms, are included in Ref. 32 in which it is shown that their inclusion gives rise to a small amount of unavoidable back action. This effect may be included by replacing α with $\alpha + 3\beta\gamma/64$.

It is interesting to consider the limiting case when f = 0 in which we assume the Brownian noise is vanishingly small and the electronics are quantum limited, $kT_E / \hbar\,\omega_2 = 1$. In Ref 32 we found

$$kT_b / \hbar\omega_1 \geq (3/32)\,(1/\omega_1 \tau_2). \qquad (11)$$

The factor by which we may exceed the quantum limit is proportional to the ratio of the width of the electrical resonance peak to the separation of the $2\omega_1$ terms from the resonance peak.

This has the following physical interpretation. The electrical noise will have a spectrum with a shape characteristic of the response function of the electrical resonator, a Lorentzian centered at the frequency ω_2. The electrical noise which lies in a band of frequencies extending from $\omega_2 - \omega_1$ to $\omega_2 + \omega_1$ is converted simultaneously by the two pumps to a force noise acting on the mechanical oscillator. Since one pump inverts the phase of the electronic noise and the other preserves it, the coherent superposition cancels in one phase and adds in the other. Thus the magnitude of the noise which acts back on the mechanical resonator will be greatly reduced in one phase and doubled in the other. On the other hand the electrical noise which lies outside this band of frequencies is converted with the same phase by both pumps and thus is not squeezed. Therefore the noise reduction afforded by the back action evasion technique depends strongly upon the electrical Q of the idler circuit.

In a final observation we note that the burst noise temperatures defined independently for the two antenna phases obey a sort of "uncertainty" relation. The calculation of the burst temperature of the " X_2 channel" is identical to the preceeding calculation with the addition of a factor of f in the response function, see Eq. 7c. Thus the burst noise temperature for this phase becomes

$$T_{b2} = T_E\ (\omega_1/\omega_2)\,(1/f^2 r). \qquad (12)$$

When f approaches zero, T_{b2} approaches infinity because the " X_2 channel" contains all noise and no signal. We find the following inequality is obeyed;

$$(T_{b1}\,T_{b2})^{1/2} \geq T_E\ (\omega_1/\omega_2), \qquad (13)$$

where T_{b1} is the burst noise temperature in the "X_1 channel"; Eqn. 8. The equality holds when the Brownian noise vanishes and the electrical relaxation time goes to infinity, i.e. perfect back action evasion.

4. Experimental Results

None of the experiments which have been

performed so far have reached the level of sensitivity at which quantum noise can be observed. In fact the experiments have yet to fully test the classical back action evasion predictions in a quantitative manner. However there are several confirmations of the main features of a back action evasion measurement. Spetz et al [33] and the authors' group in Frascati [34] have seen the reverse coupling characteristics of a BAE measurement in electromechanical transducers. The authors' group in Rochester has measured the forward coupling characteristics and have taken a considerable step toward reaching quantum limited sensitivity in a cryogenic transducer[35]. We begin the discussion of the experiments with the results from a very simple electrical analog of a BAE transducer.

In this experiment we were able to simultaneously demonstrate the forward and reverse coupling which are expected in a BAE measurement. The experiment was performed with two electrical resonators, one at 1 MHz and the other at 5 Mhz, coupled together by a double balanced mixer. We will refer to these respectively as Mode 1 and Mode 2 in the following discussion. There were couplings to both resonators which allowed us to either inject noise to simulate the action of a noisy amplifier or to monitor the responses of the resonators. The two oscillators were attached to the IF and RF ports of the mixer and the LO port was driven with a voltage of the form shown in Eq. 8. In Figure 4 we show the schematic.

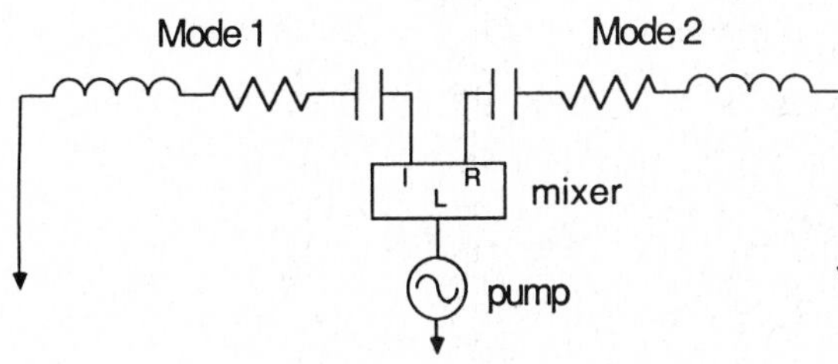

Figure 4 The schematic of two electrical oscillators coupled by a double balanced mixer. By choosing the pump oscillator appropriately, this system displays similar behavior to the BAE transducer.

In the first series of experiments we demonstrated the forward coupling. The 1 MHz resonator was driven on resonance in a series of tests (A - E) in which the phase of the excitation was varied relative to the LO excitation. The response of the 5 MHz resonator was monitored in each case; the results are shown in Figure 5a. When the LO was set at the difference frequency there was a phase preserving upconversion of the 1 MHz signal, i. e. the series of points (A - E) in the Mode 1 phase space are mapped onto the Mode 2 phase plane with the same sense of progression. For LO pumping at the sum of the two frequencies this sense was reversed. When the pumping was the coherent superposition of the sum and difference frequencies the mapping onto the phase plane of mode 2 is the vector sum of the first two mappings. Thus the excitations of mode 1 with the phase of points A and C produce an output signal in mode 2 while the excitations in quadrature are not seen.

At the same time we were able to measure the reverse coupling. Mode 2 was driven with wideband random noise and for the same three choices of the pump we monitored the response of Mode 1. We saw that for the conventional single frequency pumps the noise acts back on both phases of Mode 1 while for the combined frequency pumping the noise acts back on the phase of Mode 1 orthogonal to that which is measured.

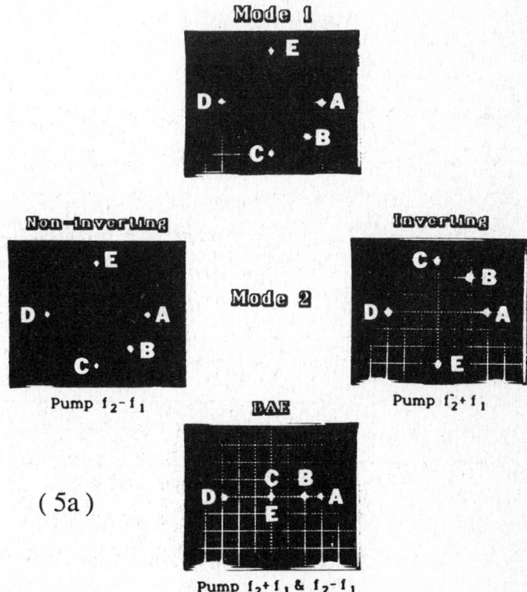

(5a)

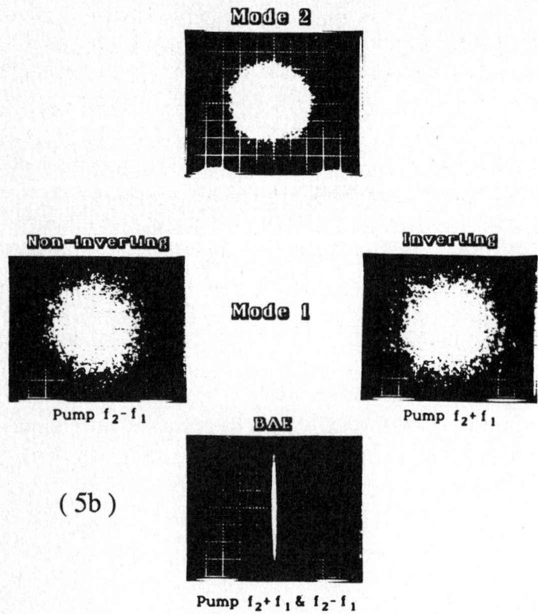

Figure 5 a) The forward coupling behavior of the BAE convertor is displayed by driving Mode 1 with different phases and measuring the response of Mode 2 to each excitation as the pump is varied.

b) The reverse coupling behavior is demonstrated by driving Mode 2 with phase insensitive noise and measuring the respone of Mode 1. Note that the noise is squeezd into the phase of mode 1 which is not measured. There is a reduction of the back acting noise in the measured phase.

The results of Spetz et al [33] and our Frascati group [34] both show the squeezing of the back acting noise in an electromechanical transducer. Spetz et al used a superconducting reentrant microwave cavity transducer under development for their gravitational wave detector and the Frascati device was a KHz frequency bridge coupled to a low frequency mechanical accelerometer. In both cases noise was added to the pump excitation to give a large random component to the electric field coupling the mechanical and electrical modes. The mechanical resonator was monitored with a weakly coupled phase insensitive transducer to monitor its behavior in both phases. The results showed that when BAE pumping was used, the noise acted back on a single component of the mechanical oscillator complex amplitude. See Fig. 6 for the result from the Frascati experiment.

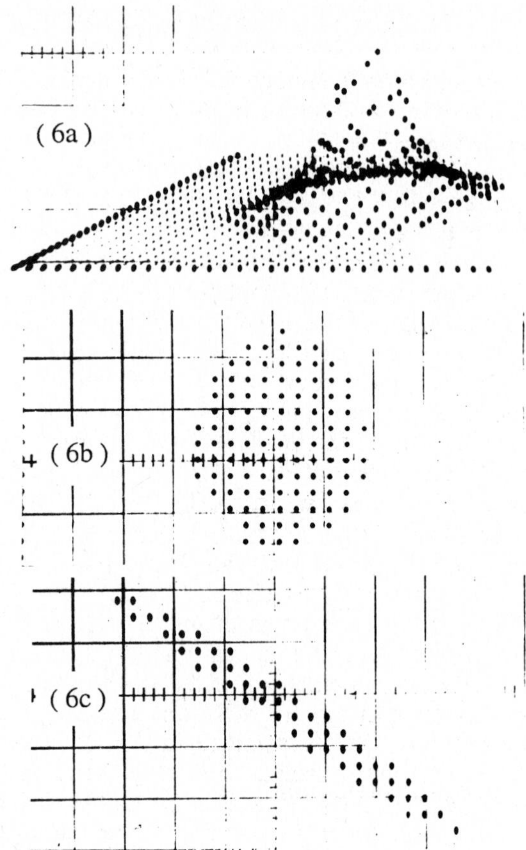

Figure 6 In experiments on a prototype transducer the mechanical oscillator was monitored by a second weakly coupled phase insensitive transducer. The distribution of noise amplitudes was recorded for the two oscillator phases. Shown above are a) The amplitude distribution of the displacements induced in both phases of the mechanical oscillator by the back action in a conventional one pump measurement

b) A cut through the distribution shown in (a) at the half maximum amplitude level

c) A similar cut to (b) when BAE pumping is used.

In an experiment at Rochester, a cryogenic MHz bridge circuit coupled to a low mass, (1 gram) sapphire resonator was tested. The forward coupling characteristics were measured and were found to correspond completely with the purely electrical analog described above. Also we were able to obtain quite high electrical and mechanical quality factors which enabled us to come within a factor of approximately 100 from the quantum noise level in the transducer position sensitivity.[35]

In related efforts the possibility of generating squeezed states of an electromagnetic radiation field is being investigated. An early proposal by Yurke and Denker [36] involved the use of degenerate parametric amplifiers to squeeze the noise in a sort of microwave voltmeter. Various techniques are being investigated in which a nonlinear multiwave mixing process may generate optical radiation which exhibits squeezing of the fluctuations below the coherent state level in one of its phases. We refer the reader to the review article by Walls [37] and several other papers [38-49] for a more complete description of the proposals and experiments.

We wish to emphasize the similarities and differences between this effort and the BAE research. The back action evasion technique will help when the fluctuations of the signal field, in our case the gravitational radiation, are negligible compared to the fluctuations in the detector. There are many "quanta" in the gravitational radiation field and thus the fluctuations are relatively small, but the coupling of gravitational radiation to matter is very weak so few quanta appear in the detector. Conversely, in the work on squeezed radiation states the fluctuations of the signal field itself are the limitation which one is attempting to overcome. In this case the detector, a photodiode for example, has a relatively high quantum efficiency and the observed fluctuations are intrinsic to the radiation field. It is possible that the use of squeezed states of the radiation field may prove of use in communications or metrology. [50]

However, there is a close relationship between back action evasion and squeezed state generation. If one views the back action evasion process as squeezing the field which radiates back from the transducer to the mechanical oscillator then there is a complete equivalence.

To date quantum noise has not been observed in the BAE experiments but efforts to reach that goal are continuing. The use of the BAE technique in other systems is also being investigated. For example, it is possible that this technique may improve the performance of an RF SQUID by avoiding the back action of the following amplifier.[51] In the meantime the BAE technique should prove useful to reduce the classical noise of transducers being developed for the current generation of gravity wave detectors.[52]

References

1) K. S. Thorne, Rev. Mod. Phys., 52, 285 (1980).

2) J. Weber, Phys. Rev. 117, 306 (1961).

3) D. H. Douglass, V. B. Braginsky, in *General Relativity, An Einstein Centenary Survey*, edited by S. W. Hawking and W. Israel, (Cambridge Univ. Press, Cambridge, 1979) p. 90.

4) R. P. Giffard, Phys. Rev. D14, 2478 (1976).

5) V. B. Braginsky, Y. I. Vorontsov, Sov. Phys. Usp. 17, 644 (1975).

6) V. B. Braginsky, A. B. Manukin, *Measurement of Weak Forces in Physics Experiments,* edited by D. H. Douglass (Univ. of Chicago Press, Chicago 1977).

7) W. G. Unruh, Phys. Rev. D17, 1180 (1978).

8) W. G. Unruh, Phys. Rev. D18, 1764 (1978).

9) W. G. Unruh, Phys. Rev. D19, 2888 (1978).

10) O. Von Roos, Phys. Rev. D18, 4796 (1978).

11) Z. Vager, Phys. Lett., 84A, 163 (1981).

12) R. Lynch, Phys. Lett., 87A, 277 (1982).

13) R. Lynch, Phys. Lett., 92A, 9 (1982).

14) L. P. Grishchuk, M. V. Sazhin, Sov. Phys. JETP, 80, 1249 (1981).

15) M. Hillery, M. O. Scully, Phys. Rev D25, 3137 (1982).

16) J. D. Macomber, Phys. Lett., 95A, 19 (1983).

17) R. W. Hayward, in *Proceedings of the Second Marcel Grossman Meeting on General Relativity* , edited by R. Ruffini (North - Holland, Amsterdam, 1982) p. 977.

18) G. J. Milburn, A. S. Lane, D. F. Walls, Phys. Rev. A 27, 2804 (1983).

19) G. J. Milburn, D. F. Walls, Phys. Rev. A 28, 2065 (1983).

20) G. J. Milburn, D. F. Walls, Phys. Rev. A 28, 2646 (1983).

21) H. P. Yuen, Phys. Rev. Lett., 51, 719 (1983).

22) R. Lynch, Phys. Rev. Lett., 51, 1405 (1983).

23) V. B. Braginsky, Y. I. Vorontsov, K. S. Thorne, Science, 209, 547 (1980).

24) K. S. Thorne, R. W. P. Drever, C. M. Caves, M. Zimmermann, V. D. Sandberg, Phys. Rev. Lett., 40, 667 (1978).

25) K. S. Thorne, C. M. Caves, V. D. Sandberg, M. Zimmermann, R. W. P. Drever, in *Sources of Gravitational Radiation,* edited by Larry Smarr (Cambridge Univ. Press, London, 1979) p. 49.

26) J. N. Hollenhorst, Phys. Rev. D19, 1669 (1979).

27) C. M. Caves, K. S. Thorne, R. W. P. Drever, V. D. Sandberg, M. Zimmermann, Rev. Mod. Phys., 52, 341 (1980).

28) F. Fuglini, V. Iafolla, in *Proceedings of the Third Marcel Grossman Meeting on General Relativity,* edited by Hu Ning, (North - Holland, Amsterdam, 1983).

29) W. C. Oelfke, in *Proceedings of the Third Marcel Grossman Meeting on General Relativity,* edited by Hu Ning, (North - Holland, Amsterdam, 1983) p. 1469.

30) D. Blair, Phys. Lett., 91A, 197 (1982).

31) W. W. Johnson, M. F. Bocko, Phys. Rev. Lett., 47, 1184 (1981).

32) M. F. Bocko, W. W. Johnson, Phys. Rev. Lett., 48, 1371 (1982).

33) G. W. Spetz, A. G. Mann, W. O. Hamilton, W. C. Oelfke, Phys. Lett., 104A, 335 (1984).

34) F. Bordoni, S. De Panfilis, F. Fuglini, V. Iafolla, S. Nozzoli, in *Proceedings of the Fourth Marcel Grossman Meeting on General Relativity,* in publication.

35) M. F. Bocko, W. W. Johnson, Phys. Rev. A 30, 2135 (1984).

36) B. Yurke, J. S. Denker, Phys. Rev. A 29, 1419 (1984).

37) D. F. Walls, Nature 306, 141, (10 Nov. 1983).

38) B. Yurke, L. R. Corruccini, Phys. Rev. A 30, 895 (1984).

39) Z. Ficek, R. Tanas, S. Kielich, J. Opt. Soc. Am. B 1, 882 (1984).

40) R. S. Bondurant, J. H. Shapiro, Phys. Rev. D 30, 2548 (1984).

41) P. Kumar, J. H. Shapiro, Phys. Rev. A 30, 1568 (1984).

42) M. J. Collett, C. W. Gardiner, Phys. Rev A 30, 1386 (1984).

43) R. S. Bondurant, P. Kumar, J. H. Shapiro, M. Maeda, Phys. Rev. A 30, 343 (1984).

44) G. J. Milburn, D. F. Walls, M. D. Levenson, J. Opt. Soc. Am. B 1, 390 (1984).

45) S. Friberg, L. Mandel, Opt. Commun. 48, 439 (1984).

46) B. Yurke, Phys. Rev. A 29, 408 (1984).

47) M. S. Zubairy, M. S. K. Razmi, S. Iqbal, M. Idress, Phys. Lett., 98A, 168 (1983).

48) W. Becker, M. O. Scully, M. S. Zubairy, Phys. Rev. Lett., 48, 475 (1982).

49) G. Milburn, D. F. Walls, Opt. Commun., 39, 401 (1981).

50) C. M. Caves, Phys. Rev. D 23, 1693 (1981).

51) F. Bordoni, P. Carelli, V. Foglietti, F. Fuligni, IEEE Trans. Mag., MAG 21, 421 (1985).

52) M. F. Bocko, L. Narici, D. H. Douglass, W. W. Johnson, Phys. Lett., 97A, 259 (1983).

NOISE IN PHYSICAL SYSTEMS AND 1/f NOISE - 1985
A. D'Amico and P. Mazzetti (editors)

NOISE IN SQUIDs

P. CARELLI

Istituto di Elettronica dello Stato Solido-CNR via Cineto Romano 42, 00156 Roma Italy.

The Superconducting Quantum Interference Devices (SQUID) are very sensitive magnetic flux to voltage converters. Large efforts have been devoted in the last years to fabricate very low noise SQUIDs. The developing of high reliability lithographic techniques has allowed the development of dc-SQUIDs of very good performances. In the literature several dc-SQUIDs are reported with a noise temperature value close to the quantum limit for an amplifier.

From a phenomenological point of view it seems clear what are the needed physical parameters in order to obtain the best devices. Many theoretical models that usually cannot take into account all the real conditions were developed. The noise property in the white region appears well explained.

On the contrary the 1/f noise behavior is not yet well understood even if some theories have been proposed. Josephson Junctions and SQUIDs can have problems related to chaos.

In practical devices the noise is still so many orders of magnitude larger than the quantum limit that a large effort must be made to really obtain low noise in practical, usable devices.

1. INTRODUCTION

SQUID is an acronym for Superconducting QUantum Interference Device. The device to a first approximation is made by a superconducting ring interrupted by one or more Josephson junctions (or in general weak links). It is not possible to give here all the general considerations about these devices. More details can be found in good review papers[1,2]. Here I only want to show the general ideas, well known to people working in superconductivity, --ideas that are at the base of the working mechanism of these devices:

Flux quantization: in any superconducting loop the magnetic flux is quantized as anticipated by London[3] and found experimentally later[4]: the value of the flux quantum is approximately ϕ_o=2.07x10^{-15} Wb. It seems that the value of the flux quantum is a more general quantity independent of superconductivity[5]. The quantization of the magnetic flux has as a consequence that coupled fluxes differing by an integer number of flux quanta produce in the loop equal circulating currents. To measure such quantization there is a limit on the loop inductance its value must be less[6] than $(\phi_o/2\pi)^2/K_BT$ otherwise the thermal noise fluctuation will mask the effect.

Josephson junction (JJ): two superconductors weakly connected by a small barrier that permits tunneling of Cooper pairs. JJ were extensively studied in hundreds of papers and of course all their properties are well beyond the purpose of this paper. In the first section I will try to explain phenomenologically the main properties necessary to understand the noise mechanism of a SQUID. A general overview of superconductivity and its recent developments can be found in ref. 7.

In a SQUID an input magnetic flux will produce a variation of some measurable macroscopic quantity that is usually a voltage. So practically the device is a magnetic flux to voltage converter. Many kinds of SQUIDs with exotic geometry and weak links were proposed: of course the only interesting devices are those utilized in some practical experiments.

In fact good SQUIDs have unrivalled properties as magnetometers in the low frequency range (0-10 KHz) and in the same range of frequency the SQUID can have a noise temperature lower than any other existing amplifier.

For many years the most widely used SQUIDs were the rf types. Such a device consists of a superconducting loop interrupted by a weak link (many groups used a sharp point contact on a flat superconducting surface) weakly inductively coupled to a rf biased tank circuit. The input signal modulates the rf signal. The linearized theory of those devices showed[8] that those devices are up-converters as confirmed experimentally[9]. In practice the signal to be detected modulates the amplitude of the pump signal. The best performances are so obtained with higher pumping frequecies. For many years the most used frequency was about 20 MHz for which it is easy to build the device and develop the relative detecting electronics. The best results were obtained at 9 GHz; in such a case the best energy sensitivity was about 1000h[10]. This device was used in a gravitational wave detection experiment as low noise amplifier. The lowest noise temperature so obtained was 20mK[11] (well above that of the device uncoupled), and is still the lowest noise temperature obtained around 1KHz in these experiments. The same group now is using a commercial dc-SQUID. The use of a rf-SQUID in a quantum non demolition scheme[12] has been proposed.

The first developed SQUID was really the dc-SQUID[13] but, due to difficulties in the practical realization of two almost equal weak links, the rf-SQUID has been used more widely. With the diffusion of high quality lithographic techniques the dc-SQUID, that from a theoretical point of view can reach the limit imposed to an amplifier by quantum mechanics, gained new interest and I intend to focus my attention on these devices.

2. JOSEPHSON JUNCTION

It seems clear that to obtain low noise dc-SQUIDs, the best weak links are to be reproducible, with small capacitance, and high quality Josephson Junctions (JJ). The JJ must be shunted by an appropriate shunt resistance of low stray inductance and capacitance. The Langevin equation describing the working mechanism of a JJ current biased is:

$$C\,dV/dt+V/R+I_o\sin\psi=I+I_n(t) \qquad 1)$$

where C is the JJ capacitance, V the voltage across, R the shunt resistance, I_o the JJ critical current, ψ the phase difference, I the applied biasing current, In(t) the noise current.

The II Josephson equation is:

$$V=\phi_o/2\pi\, d\psi/dt \qquad 2)$$

Often the equations are written in reduced form:

$$\beta_c\, dv/d\tau+v+\sin\psi=i+i_n \qquad 3)$$

$$v=d\psi/d\tau \qquad 4)$$

where $i=I/I_o$, $i_n=I_n(t)/I_o$, $v=V/I_oR$, $\tau=2\,I_oRt/\phi_o$. , $\beta_c=2\pi I_oR^2C/\phi_o$ and $\Gamma=\pi K_BT/I_o\phi_o$ are the only parameters that determine the equation. The thermal (Johnson noise) low frequency spectral density of i is 4Γ.

Herewith I synthesize the main features that are derived out from this equation:

2.1. i-$\bar{v}$ characteristics as a function of β_c without noise (Γ=0): It always exists a solution v=0 for i<1. If β_c =0 an analytic solution exists :

$$\bar{v}=(i^2-1)^{1/2} \qquad 5)$$

As β_c increases the solution can be found only numerically. As β_c increases, the characteristics in the high current region approximate the curve v=i faster than 5). For $\beta_c > 1$ there is a region around i=1 where the function is not single-valued, the hysteretic region increasing with β_c. SQUIDs to have low noise properties, must work with $\beta_c < 1$.

2.2. i-$\bar{v}$ characteristics as a function of with $\beta_c = 0$: There is a solution of the Fokker-Plank equation formed by the Langevin equation 3. The solution is an expression with definite integral, with a numerically easy solution[14]. With increasing Γ the i-v characteristic 5) is more and more rounded off in the vicinity of i=1. For $\Gamma=1$ the shape is well described by v=i. Γ is usually very small for low noise SQUIDs operating with large critical currents (e.g. I_0=100uA, T=4.2K, Γ=0.001), so the rounding effect is negligible.

2.3. i-$\bar{v}$ in the general case: the effect is well approximated by a simple combination of a) and b) as shown with numerical simulation[15]. The value of β_c, for which the i-v curve is hysteretic, increases with (the effect is negligible for $\Gamma \ll 1$).

2.4. 3) and 4) are the equation of a non linear oscillator. Therefore togeter with $\bar{v}$, there is a high frequency voltage oscillation with typical pulsation v (in these reduced units). Therefore JJs are voltage controlled oscillators and mixers (due to the non linearity of i-$\bar{v}$ characteristics).

2.5. The total low frequency voltage noise at the end of the JJ is a combination of the low frequency voltage noise (caused by $i_n(0)$) and the high frequency effect of $i_n(v)$ down converted by the mixer mechanism on the JJ. For $\beta_c = 0$ the low frequency spectral density[16] is:

$$S_v(0)=i^2 S_i(0)/\bar{v}^2+S_i(v)/\bar{v}^2 \qquad 6)$$

where the first term is the low frequency component and the second one the down converted. In ref. 15 the expression S_v at various β_c numerically computed is given. Spurious noise, due to the measuring system at high frequency, can be down converted, with the same mechanism, to a low frequency noise. For this reason low frequency measurements on JJs are performed with high quality chock filters.

2.6. The quasiparticle current of a JJ, has shot noise, but the dominant normal current is that one across the shunting resistor, therefore this noise is negligible in the JJs used for dc-SQUIDs. The supercurrent is coherent and not dissipative, so it is not a source of noise. Only the interaction between Cooper pairs and barriers excitations can produce a supercurrent shot noise [17]; this noise is absolutely negligible in small JJs[18], like those used for SQUIDs.

2.7. The equations 3) and 4) are a II order, non-linear system and they can present chaotic solutions in presence of a rf-biasing superimposed with the dc bias (a good review given in ref.19). Even if the rf biasing and the condition $\beta_c > 1$ are not usual for a SQUID (necessary conditions for cahotic solutions), JJs must operate well away from these regions in which the noise voltage can encrease dramatically.

2.8 The complete expression for the spectral current density due to thermal noise[20] is:

$$S_i(\nu)=(2h\nu/R)\coth(h\nu/2kT) \qquad 7)$$

usually this expression is used in the classical approximation $h\nu \ll kT$ for which

$$S_i(\nu)=2kT/R \qquad 8)$$

As JJs are down converters, a term due to the zero point fluctuation can be present in the expression of the total noise at low frequency . This effect has been found experimentally in strongly damped JJ[20]. The effect has been not found in non damped JJ[18]. Probably because in a quantum mechanical situation not a classical description like eq. 3 and 4, but a quantum mechanical approach like that used in ref.21 must be used. All the quantum mechanical corrections are very small terms, not easily detectable at 4.2K and of negligible importance in real operating actual SQUIDs.

2.9. The 1/f noise in JJs is a dramatic problem for SQUIDs. In this field there are more theories than sistematic measurements. Due to the frequency operating range of a SQUID, this noise limits severely the performances of the whole device. It is clear that barriers with a lots of imperfections, like those made of some semiconductor materials or poor oxide layers show a high 1/f noise.

3. DC-SQUID

A superconducting ring interruped by two JJs is called a dc-SQUID. For a particular biasing current and appropriate parameters, the voltage across the JJs is a periodic function of the flux.

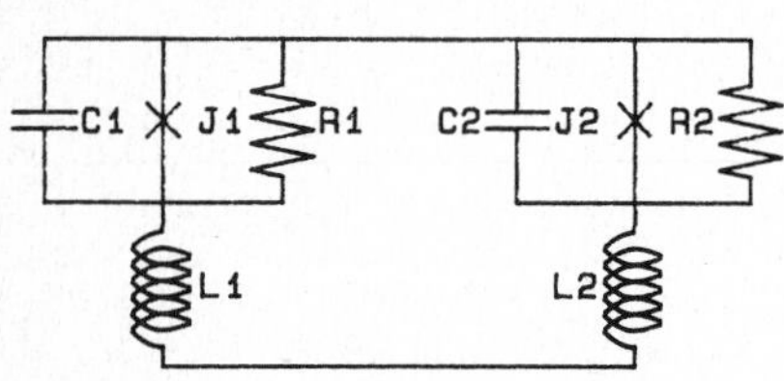

FIGURE 1
Circuit diagram of a dc-SQUID. The total inductance of the device is L_1+L_2.

The device in the simplest form is described in fig.1. The following system of coupled equations describe the dynamic behavior of this device:

$$\frac{2\pi C}{\Phi_0}\frac{d^2\psi_1}{dt^2}+\frac{2\pi}{\Phi_0}\frac{d\psi_1}{dt}/R = I_1 - I_{01}\sin\psi_1 + I_{n1} \qquad 9)$$

$$\frac{2\pi C}{\Phi_0}\frac{d^2\psi_2}{dt^2}+\frac{2\pi}{\Phi_0}\frac{d\psi_2}{dt}/R = I_2 - I_{02}\sin\psi_2 + I_{n2} \qquad 10)$$

$$(\psi_1-\psi_2)\,\Phi_0/2\pi = \Phi_{ext}+L_1I_1-L_2I_2+\text{Cext} \qquad 11)$$

The sum of the current I_1 and I_2 is the impressed current, the half of their difference is the circulating current. Φ_{exc} is the external flux. Cext is the influence of the input circuit on the device, it is a simple inductive coupling in good devices. An adimensional quantity is introduced for the inductance of the device: $\beta_L=\pi(L_1+L_2)(I_{01}+I_{02})/\Phi_0$. If this quantity is negligible it means that the circulating current effect is negligible with respect to the flux quantum.

The influence of the SQUID on the input circuit is caused by the current circulating in the SQUID loop. The circulating current I_R is a function of the dynamic resistance and inductance of the ring.

Also in this ideal case, in which any other stray capacitance,inductance or coupling has been considered negligible, the system of equations is not analitically solvable without some approximations. But of course it is possible to solve numerically any particular case. I think that the noise mechanism and the ultimate performances are well understood at least from a qualitative point of view. A complete theory[22] of the device in the limit of small and large β_L if the β_c terms are negligible exists.

First I will try to explain the main features in a limiting case.

$\beta_c=0$, $\beta_L\ll 1$, Cext negligible: there is an analytic solution[22] if the noise generator is thermal. It can be shown, that in this case the system behavior can be well described by a system of equations formally equal to 3) and 4) :

$2\pi\psi/R\phi_0 + I_c \sin\psi = I + I_n$ 12)

$V = \phi_0/2\pi \frac{d\psi}{dt}$ 13)

where ψ is an effective phase(the phase difference across a junction plus a time indipendent term), I_c is the effective critical current:

$I_c = I_{c1}^2 + I_{c2}^2 + 2I_{c1}I_{c2}\cos 2\pi\phi/\phi_0$ 14)

From 14) I_c is zero at $(n+1/2)\phi_0$. In this hypothetical case the coupling between the junctions is so strong that the system behavior is equivalent to that of a single junction with flux dependent barrier . The phase is an effective one, but its derivative is the voltage across the device (the usual factor apart).

The responsivity can be easily written:

$V_\phi = I_{c1}I_{c2}(R_1+R_2)\sin(2\pi\phi ext/\phi_0)$ 15)

The noise output voltage spectral density is, as in the case of a single junction, the combination of the low noise voltage (linear superimposition of current noise of each junction) and the high frequency down converted noise:

$S_v = (R_1+R_2)\ (S_{i1}+S_{i2})/2V^2 + I_0(S_{i1}+S_{i2})/4V^2$ 16)

Also if both terms are divergent for $V \to 0$, the ratio $S_v/V_\phi^2 2\ L$ is limited; this quantity is the magnetic energy measurable by the SQUID with a signal to noise ratio=1 . The minimum S_ε is found when the critical currents are equal (I_0) and for external flux equal to $(n+1/2)\phi_0$ in this case:

$S_\varepsilon = 18KT\phi_0/\beta_L I_0 R$ 17)

This quantity, that can be straightforwardly measured, is usually given to compair the quality of SQUID devices. Sometimes more properly the magnetic energy is referred to the input so 17) is divided by the coupling constant squared (between input coil and SQUID inductance). S_ε decreases linearly in $I_c R$, so if C is very small (to have β_c negligible) this figure of noise can be even lower than the quantum limit. Of course this is possible because in this approximation, the influence of the SQUID on the measuring system has not been taken into account. An asymmetry in the inductance or resistance does not change the main features. On the contrary an asymmetry in the critical current produces a reduction in V_ϕ an a noise increasing consequently also S_ε increases.

$\beta_c=0$, $\beta_L \ll 1$, Cext negligible: In this case the two JJs are largely uncorrelated. The responsivity is proportional to the inverse of β_L, but the noise output voltage is about half as in 16) because is caused only by a junction each time:

S_ε as inferred from the almost analytical derivation[22] is:

$S_\varepsilon = 3\beta_L\phi_0 K_B T/4\pi^2 I_0 R$ 19)

The ideal β_L inferred from the two formulas is about 3 for which the minimum Energy sensitivity is $S_\varepsilon \simeq \phi_0 KT/4 I_c R$. An asymmetry in the inductances also in this case produces negligible effects. A critical current asymmetry reduces the responsivity and increases S_ε. Also if there are no numerical simulations in the literature, I suspect that an asymmetry in the resistances may increase the noise.

$\beta_c=0$, $\beta_L=1$, Cext negligible: all the main features have been numerically calculated[23] in the case $\Gamma=0.05$, R1=R2=1Ω. The results agree very well with the general model[22] extrapolated to the condition

$\beta c \neq 0$ and Cext=0: If I_0 and R are kept fixed any increase in the capacitance corrisponds to a reduction of the responsivity stronger in the region of high I/I_0. It is found experimentally[24] that with a high β_c the output voltage noise increases. In a dc-SQUIDs the input signal modulates the voltage so the device can be described as a frequency modulator and demodulator, due to the corrispondance between average voltage and frequency.

For high β_L, β_c and $\Gamma=0$ the system presents almost chaotic solutions[25]. In fact in this condition the two non linear oscillators (the JJs) are so uncoupled that they can show an uncorrelated oscillating behavior and only after a few cycles of indipendent oscillations there is again a coupling; consequently together with the Josephson frequency many subharmonics appear.

The back-action of the SQUID on the input circuit is produced by the circulating current in the SQUID loop. The induced noise is given by the fluctuating term of such a current induced in the input circuit. The value of the fluctuating circulating current has been calculated as in the limit $\beta_L<<1$ and $\beta_L>>1$ ($\beta_c=0$) almost analytically[22], as in the case of $\beta_L=1$, $\beta_c=0$, R=1 with a numerical method[26]. The induced voltage in the input circuit is of the order of a few fV for good devices (only to have an idea of the order of magnitude) so it is a very difficult quantity to measure; it has been measured only in a very low-quality SQUID[27] using as amplifier a good SQUID: the experimantal results are in good agreement with the theory. The circulating current noise has a minimum for $n\phi_0$ and it is much more dependent on the high frequency components down converted than the output noise voltage. This current noise is partially correlated with the output noise.

The loop dynamic parameters of the SQUID are also important when the device must be used as a low noise amplifier. The measured dynamic parameters[28] (effective inductance and resistance dependent on applied flux) are in good agreement with the theory[22]. Those quantities are important in the computation of the dc -SQUID optimum performances as low noise amplifier. In fact, for instance, in the gravitational wave detection experiments, devices with a noise temperature close to the quantum limit are requested. In those devices the reached noise temperature is far away from the theoretical one and it has been shown that with present technology it is possible to build quantum limited devices[29]. The correct analysis[30] taking correcly into account the influence on the SQUID loop equations of the Cext term, when the coupling is only inductive, gives similar results.

The 1/f noise found for these devices in the output voltage is usually due to the 1/f noise of the JJs. With high quality junctions[31,32] the best noise performances have been obtained. It seems that in some junctions the 1/f noise is not due to the fluctuating current[33]. There are no measurements on the 1/f noise in the input.

ACKNOWLEDGEMENTS

I am indebted with V. Foglietti for continuous discussions during the preparation of the manuscript. I am grateful with R. Leoni who made all the computer simulations clarifying the various effect . I acknowledge also S. Barbanera for his critical revision of the manuscript. I want also to thank I. Modena and A. Paoletti for continuous encouragements.

REFERENCES

1. R.P. Giffard, in Future Trends in Superconductive Electronics, B. Deaver, C. Falco, J. Harris, S. Wolf, eds. (American Institute of Physics, New York, 1978).
2. J. Clarke, W. M. Goubau and M. B. Ketchen, J. Low Temp. Phys., **25**, 99 (1976).
3. F. London, Superfluids, Vol.1, Dover Public, New York (1960)
4. M. Doll. and M. Nabauer, Phys. Rev. Lett. **7**, 51 (1961).

B. S. Deaver, W. M. Fairbank, Phys. Rev. Lett. 7, 49 (1961).

5. R. A. Webb, S. Washburn, C. P. Umbach, and R. B. Laibowitz, in IC-SQUID '85, Proceedings of the Second International Conference on Superconducting Quantum Devices, Berlin, 1985.
6. V. V. Danilov, K. K. Likharev, O. V. Sniguiriev and E. S. Soldatov, IEEE Trans. on Magn., **MAG-13**, 240 (1977).
7. A. Barone and G. Paterno', "Physics and Application of the Josephson Effect",John Wiley & Sons, New York (1982).
8. G. Ehnholm, J. Low Temp. Phys., **29**, 1 (1977).
9. R. P. Giffard and J. N. Hollenhorst, Appl. Phys. Lett. **32**, 767 (1978).
10. J. N. Hollenhorst and R. P. Giffard, IEEE Trans. Mag. **MAG-15**, 474, (1979).
11. S. P. Boughn, W. M. Fairbank, R. P. Giffard, J. N. Hollenrhost, E. R. Mapoles, M. S. McAshan, P. S. Michelson, H. J. Paik and R. C. Taber, Astrophys. J. **261**, L19(1982).
12. F. Bordoni, P. Carelli, V. Foglietti and F. Fuligni, IEEE Trans. Magn., **MAG-21**, 424 (1985).
13. R. C. Jaklevic, J. Lambe, H. H. Silver and J. E. Mercereau, Phys. Rev. Lett. **12**, 503 (1964).
14. V. Ambegaokar and B. I. Halperin, Phys. Rev. Lett., **22**, 1364 (1969).
15. R. F. Voss, J. Low Temp. Phys., **42**, 151 (1981).
16. L. K. Likharev and V. K. Semenov, JETP Lett. **15**, 442 (1972).
17. M. J. Stephen, Phys. Rev. Lett., **21**, 1629 (1968).
18. R. F. Voss and R. A. Webb, Phys. Rev. B, **24**, 7447 (1981).
19. R. L. Kautz, J. Appl. Phys., **58**, 424 (1985).
20. R. H. Koch, D. J. Van Harlingen and J. Clarke, Phys. Rev. B, **26**, 74 (1982).
21. A. O. Caldeira and A. J. Legget, Phys. Rev. Lett., 46, 211 (1981).
22. V. V. Danilov and K. K. Likharev, Radio Engng . Electron. Phys. **26**, 122 (1981).
23. C. D. Tesche and J. Clarke, J. Low Temp. Phys., **29**,301 (1977).
24. P. Carelli and V. Foglietti, J. Appl. Phys.,**53** ,7592 (1982).
25. J. A. Ketoja, J. Kurkijarvi and R. K. Ritala, Phys. Rev. B, **30**, 3757 (1984).
26. C. D. Tesche and J. Clarke, J. Low. Temp. Phys., **37**, 397 (1979).
27. J. M. Martinis and J. Clarke, IEEE Trans. on Magn., **MAG-19**, 446 (1982).
28. C. Hilbert and J. Clarke, Appl. Phys. Lett., **45** ,799 (1984).
29. R. H. Koch, D. J. Van Harlingen and J. Clarke, Appl. Phys. Lett., **38**, 380 (1981).
30. C. D. Tesche, Appl. Phys. Lett., **41**, 490 (1982).
31. P. Carelli and V. Foglietti, IEEE Trans. on Magn., **MAG-21** ,424 (1985).
32. C. D. Tesche, K. H. Brown, A. C. Callegari, M. M. Chen, J. H. Greiner, H. C. Jones, M. B. Ketchen, K. K. Kim, A. W. Kleinsasser, H. A. Notarys, G. Proto, R. H. Wang, T. Yogi, IEEE Trans. on Magn.,**MAG-21** ,1032 (1985).
33. R. H. Koch, J. Clarke, W. M. Goubau, J. M. Martinis, C. M. Pegrum and D. J. Van Harlingen, J. Low Temp. Phys., **51**, 207 (1983).

NOISE IN PHYSICAL SYSTEMS AND 1/f NOISE - 1985
A. D'Amico and P. Mazzetti (editors)

NOISE MEASUREMENTS ON THICK FILM RESISTORS

András AMBRÓZY

Technical University, Budapest, H-1521 Budapest, Hungary

The development of thick film resistors was mostly empirical. In the possession of the accumulated experiences very good reliability, stability can be achieved even at extreme circumstances. The exact nature of conduction in thick film resistors, however, is not known as yet.
One possibility to enlighten this picture is the measurement of excess noise. Several groups worked out different measuring systems and methodologies to get usable informations. The author of the present paper gives an account on his measuring system and his experimental results.

1. INTRODUCTION

In the last decades we have got used to that the new active devices and passive components of the electronics can be described with comparatively simple models. It is owed mainly to the singlecrystal nature of the new structures.

The structure of the thick film resistors (TFRs) is much more complex. Their development was mostly empirical. In the possession of the accumulated experiences very good reliability, stability and environmental endurance could be achieved. Thick films work reliably in such extreme circumstances as, e.g. long-lasting space voyage or implanting into a human body. It seems timely to perform such experiments which clear the conductivity mechanism of thick film structures.

2. NOISE MEASUREMENTS

Perhaps the earliest attempt to describe the noisiness of TFRs was that of Kuo and Blank [1]. They found that the noise is inversely proportional to the volume of the resistor,which corresponds also to Hooge's statement [2]. However, the coefficient showed a very high discrepancy. It was removed by Vandamme who suggested that the elementary noise sources were located in small volumes where the grains touch each other [3]. Similar reasoning was used by Ambrózy who examined another grainy structure, the semiconducting barium-titanate [4, 5].

To understand better the conduction and noise properties of TFRs, at least three groups of researchers performed exhaustive noise measurements. Prudenziati & al [6] as well as Pellegrini & al [7] - after a long search for possible conduction mechanisms - recently published noise measurements in a rather wide temperature range. Another group in Tampa, Florida emphasized the presence of burst noise [8, 9, 10], over and above the 1/f noise [11]. The third group in Ghent, Belgium devoted its effort mostly to the development of the measuring method [12, 13, 14, 15]. In the next section a fourth approach will be described.

3. HARDWARE AND SOFTWARE OF A NOISE MEASURING SETUP

The standardized measurement [16] is matched to the general 1/f-character of low frequency noise: within constant relative bandwidth the noise power is also constant. Let the nominal frequency of the measurement is 1 kHz and the bandwidth also 1 kHz, then f_{low}=618 Hz,f_{high}= =1618 Hz and

$$f_0 = \sqrt{f_{low} f_{high}} = 1\text{kHz}$$

Higher quality, more up to date components and devices do not show considerable excess noise above the thermal and shot noise at 1kHz, since the 1/f - white corner frequency lies in the range of several ten or several hundred hertz. Therefore the qualification of such components and devices needs another methods. The spectral analysis of noise gives not only whole and perfect data about the magnitude vs. frequency but shows the occasional imperfections in the shape of spectrum. On the other hand, spectrum analyzers supply a plethora of data which cannot be used without subsequent processing.

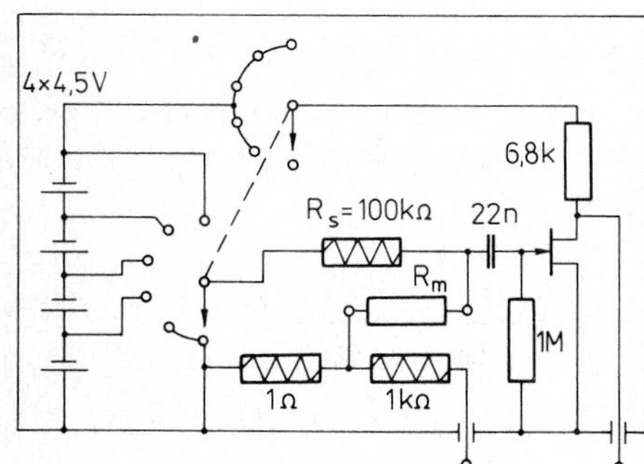

FIGURE 1
Measuring circuit

The noise measuring setup is shown in Fig.1. In series with the resistor R_m to be investigated there is an excess-noise-free (wirewound) resistor of 100 kiloohm. The DC current flowing through R_m is provided by four standard batteries 4,5 V each. The total squared noise voltage across R_m is

$$u^2=4kT(R_s \parallel R_m)\Delta f+C^2I^2R_m^2 \frac{R_s^2}{(R_m+R_s)^2} \frac{\Delta f}{f} \qquad (1)$$

where R_s is the series resistor, C a proportionality factor, I the DC current flowing through R_m. Since the latter is $I=U_{DC}/(R_s+R_m)$

$$u^2=4kT(R_s \parallel R_m)\Delta f+C^2U_{DC}^2 \frac{R_m^2R_s^2}{(R_m+R_s)^4} \frac{\Delta f}{f} \qquad (2)$$

where the first term is the thermal noise and the second the excess one. It has a maximum when $R_m=R_s$; moreover the JFET preamplifier produces the least noise if $R_m \parallel R_s \approx 50$ kΩ. So this setup provides the best results with R_m=30...200 kiloohms.

For the order of 3...30 kiloohms a bipolar transistor preamplifier is the proper selection.

The output of the preamplifier is connected to a main amplifier with a gain of 1000. Both units are battery operated and enclosed in steel boxes.

The central part of our measurement station is a Bruel and Kjaer Type 2131 digital frequency analyzer. It consists of an input amplifier and filter section, an RMS detection and averaging section and the output control and display section. Somewhat more detailed block diagram can be seen in Fig.2.

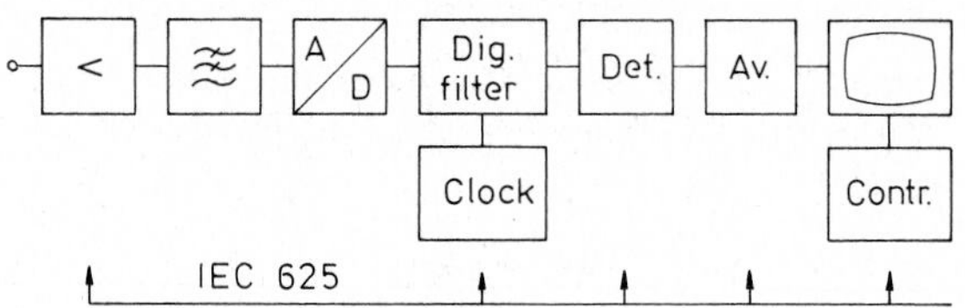

FIGURE 2
Block diagram of Spectrum Analyzer 2131

Because of the properties of digital filters the analyzer works with constant relative bandwidth bands.

Thanks to the digital working principle all setting functions of the instrument are accessible by a standard IEC 625 bus.

The control of the analyzer comes from an

HP 9815 desktop calculator. This same instrument processes all incoming data and controls the graphic plotter which is an HP 9862 A. The total measurement station is shown in Fig. 3.

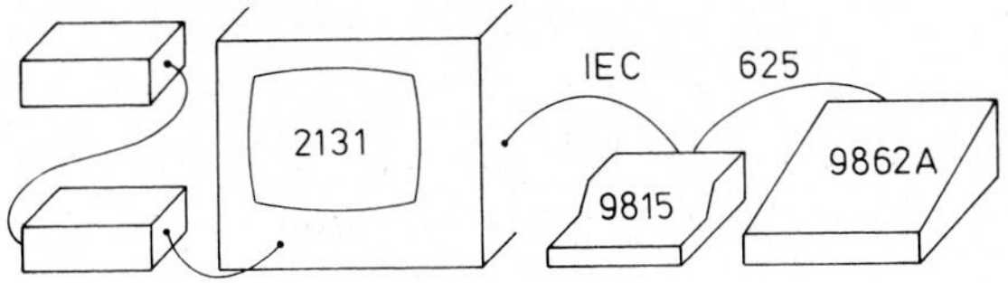

FIGURE 3.
Measurement station

Software was developed for performing the following tasks:

- Standardizing the results of different resistors according to Eq.(2).
- Converting the constant relative bandwidth data to constant absolute bandwidth.
- Removing the artifacts caused by the mains frequency and its harmonics.
- To smooth the resulting curves expecially at low frequencies since the finite averaging time causes fluctuations.
- To command the plotter to plot the NF vs. log f curve in the usual way.

The last but one task is performed by the algorithm

$$y(n) = \frac{y(n-1) + 2y(n) + y(n+1)}{4} \qquad (3)$$

where n is the channel serial number.

Many hundred curves were obtained by using the above described measuring system, in a quick and convenient way.

4. RESULTS

To compare different ink types a series of samples was prepared. The sheet resistances were between 1 kΩ/□ ... 100kΩ/□. The width of the resistor bodies was always constant, the length to width ratio varied between 0,5 and 4, with the middle values of 1 and 2. If Hooge's theory [2] is valid the squared noise voltage accross a resistor R is

$$U_n^2 = \frac{\alpha}{N} \frac{\Delta f}{f} I_{DC}^2 R^2 \qquad (4)$$

where α is the Hooge-constant (in the order of 10^{-3}), N the total number of carriers in the volume of the resistor, Δf the bandwidth and f the center frequency of the measurement, I_{DC} the DC biasing current and R the resistance.

Eq. (4) can be rewritten as

$$U_n^2 = A^2 \frac{\Delta f}{f} U_{DC}^2 \qquad (5)$$

where $A^2 = \alpha/N$. N can be expressed as

$$N = n\ell wt \qquad (6)$$

where n is the carrier number in unit volume, ℓ the length, w the width and t the thickness.

Now

$$A^2 = \frac{\alpha}{n\ell wt} = \frac{\alpha\varrho}{nw^2t^2R} \qquad (7)$$

where ϱ is the resistivity. Knowing that $\varrho/t = R_\square$ we get finally

$$A^2 = \frac{\alpha}{nw^2t} \frac{R_\square}{R} \qquad (8)$$

The parameters w^2 and t depend on the printing (and partly on firing) circumstances which were kept constant during the whole series of experiments. For a given ink α, $R_\square$ and n are also constant. It follows from eq. (8) that A^2 should be inversely proportional to R. Fig. 4. shows that the $A^2(R)$ curves exhibit indeed a negative exponent power law since both axes are logarithmic. The exponent p is $-1<p<-0,5$. Full line curves are taken with higher, broken line ones with lower I_{DC} bias current. The dash and dot line represents the thermal noise limit.

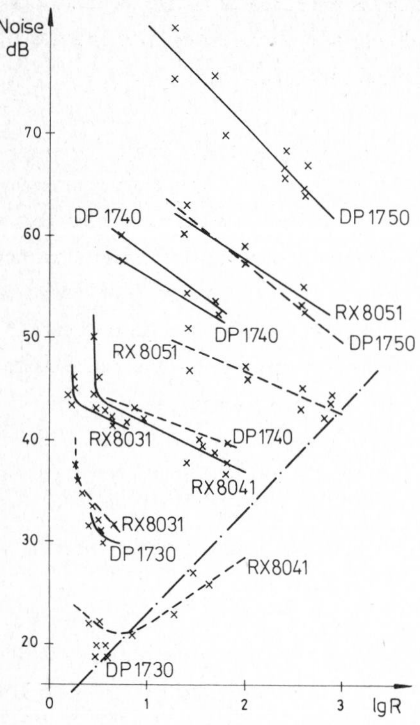

FIGURE 4
Noise figure vs resistance

The DP (DuPont) resistors are generally noisier than the RX(Remex) ones. The curve for RX 8041 is certainly incorrect since it crosses the thermal noise limit.

Some curves bend upward at low resistances. It can be attributed to contact noise: low ℓ/w resistors are more influenced by interfacial layers between the resistor body and contacting conductors.

In our noise measuring setups four different DC bias currents can be set; they are in a ratio of 1:2:3:4. In many cases we observed the characteristical curves in Fig. 5, if the lowest current was set at first. The slope of the lowest curve is markedly different from the others, i.e. proportional to f^{-2}. This is characteristic to the burst noise which can be described as

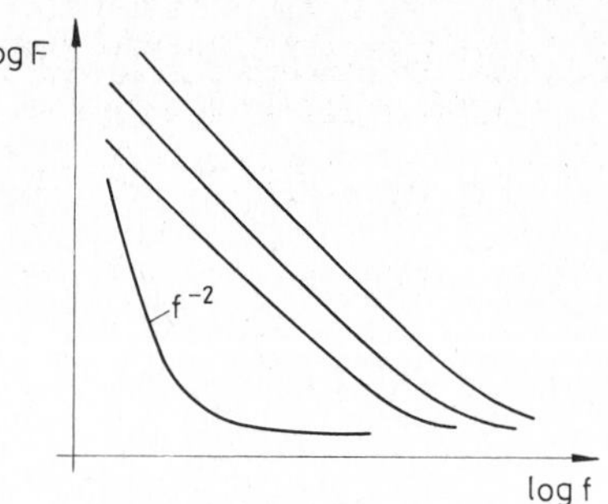

FIGURE 5
Shape of noise curves

$$S(f) = \frac{B}{D^2 + \omega^2} \qquad (9)$$

where B/D^2 is the low frequency power density of this stochastic process. One possible model for 1/f and burst noise can be seen in Fig. 6.

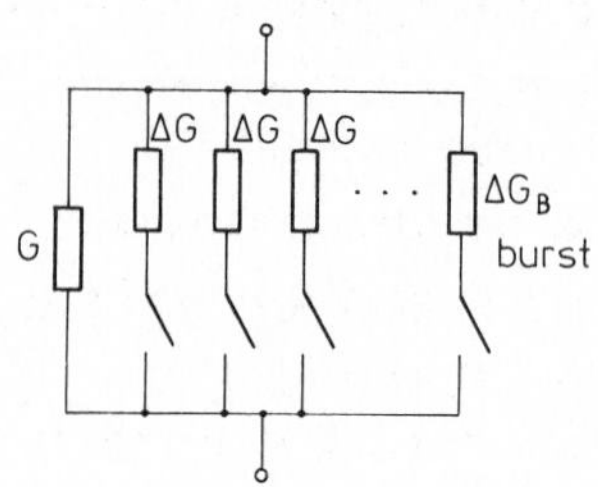

FIGURE 6
Model for 1/f and burst noise

It is probable, that at low bias current not all the grains of the resistor carry the current: there are some sites where burst noise occur. With higher bias current these sites "burn out", leaving pure (or almost pure) 1/f noise. Returning to the lowest bias current a regular 1/f curve can be obtained, the earlier "virgin" curve disappears.

For other experimental results see Ref. [17].

REFERENCES

1. C.Y. Kuo and H.G. Blank, ISHM(1968) 153.

2. F.N. Hooge, Phys.Lett. 29/A, (1969) 139.

3. L.K.J. Vandamme, Electrocomp. Sci. and Technol. 4, (1977) 171.

4. A.Ambrózy, IEEE Trans. ED-26 (1979) 1368.

5. A.Ambrózy, IEEE Trans. ED-28 (1981) 344.

6. M.Prudenziati, B.Morten and A.Massero, Proc. Sixth Intl. Conf.Noise in Physical Systems (1981) 202.

7. B.Pellegrini & al., Physical Review B-27 (1983) 1233.

8. J.G. Cottle and C.T.M. Chen, IEEE Trans. CHMT-6 (1983) 163.

9. T.M. Chen and J.G.Cottle, UGIM Symposium, Texas (1983) 139,

10. J.G.Cottle, T.P.Djeu and T.M.Chen, Proc. ISHM (1984) 293.

11. T.M.Chen, S.F.Su and D.Smith, Solid State Electronics 23 (1982) 821

12. S.Demolder, M.Vandendriessche and A. Van Calster, J.Phys.E. 13 (1980) 1323.

13. S.Demolder, A.Van Calster and M.Vandendriessche, Proc. Third European Hybrid Microel.Conference, Avignon (1981) 19.

14. A.Van Calster & al., Proc.Noise in Physical Systems and 1/f Noise, Montpellier (1983) 193.

15. A.Van Calster & al., Fourth European Hybrid Microel. Conference, Copenhagen (1983) 13.

16. G.T.Conrad, N.Newman and A.P.Stansbury, IRE Trans. CP-7, (1960) 1.

17. A.Ambrózy and G.Wollitzer, Electrocomp.Sci. and Technol. 11, (1984) 203.

NOISE IN PHYSICAL SYSTEMS AND 1/f NOISE - 1985
A. D'Amico and P. Mazzetti (editors)

REVIEW OF NOISE IN PHOTODIODES AND PROSPECTIVE ASPECTS

Michel SAVELLI, Robert ALABEDRA

C.E.M. (UA 391) Centre d'Electronique de Montpellier, Université des Sciences et Techniques du Languedoc, 34060 Montpellier, France

In the first part of this paper the main characteristics of the avalanche photodiodes (A.P.D.) are presented. In the second part, an extension of the Mc Intyre's theory[1] is proposed in taking the variance of the internal gain into account for to obtain the noise spectral density relation. In the last part the prospective aspects for to design the future A.P.D. are reviewed.

INTRODUCTION

The avalanche photodiodes are semiconductor devices that can detect optical signals through electronic processes. The extension of coherent and incoherent light sources into the far-infrared region has increased the need for high-speed sensitive A.P.D. A general behaviour of A.P.D. has basically three processes : carrier generation by incident light, carrier transport with multiplication by impact ionization and interaction of current with the external circuit to provide the output signal. The A.P.D. are very important in optical-fiber communication systems operated in the near-infrared region (0.8 to 2.5 μm). They convert the optical variations into electrical signals. For such applications the A.P.D. must satisfy requirements such as high sensitivity at operating wavelengths, high response speed, and minimum noise[2].

1. GENERAL CONSIDERATION

1.1. The basic structure

The PIN avalanche photodiode is one of the most common photodetectors, because the depletion-region thickness (the intrinsic layer) can be adapted to optimize the quantum efficiency and frequency response. The PIN^+ or NIP^+ for the pure injection of holes or electrons are shown in figures 1-a-b. These devices are in bias reverse voltage and the high-field region thickness w is called avalanche zone.

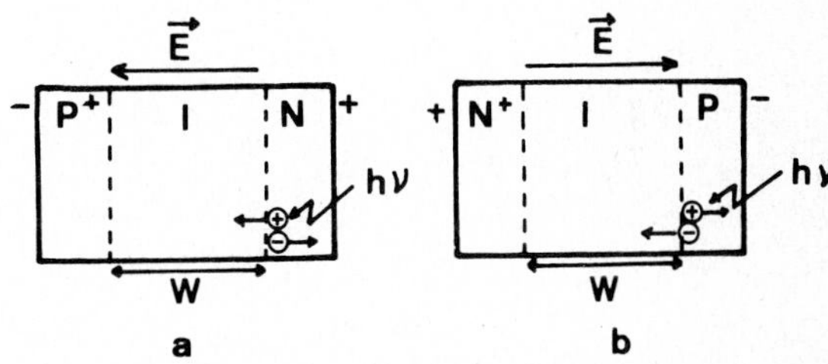

FIGURE 1
Schematic structures of A.P.D. in both cases of holes (1-a) or electrons (1-b) injection

1.2. The I-V characteristics in reverse bias

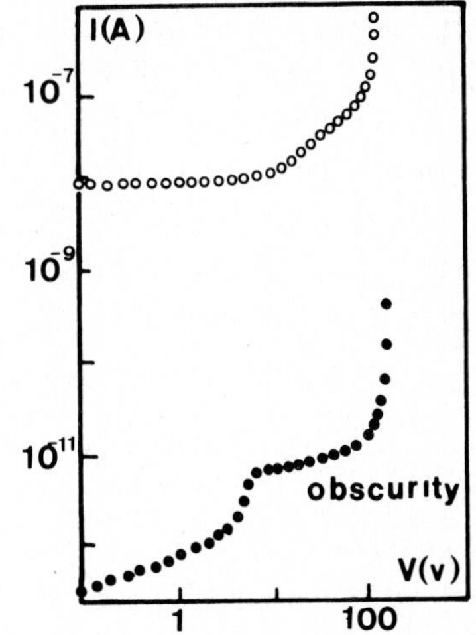

FIGURE 2
Reverse I-V characteristics under darkness and illumination conditions

Both the current-voltage characteristics under obscurity and illumination conditions are shown in figure 2 for one $n^+ \pi p \pi p^+$ Si structure. The multiplication zones in these characteristics are just before the breakdown voltage.

1.3. The responsivity σ

A related figure of merit is the responsivity σ in A/W, which is the ratio of the photocurrent I_p to the optical power ϕ. One of the key factors that determine the responsivity is the absorption coefficient α_a. Since α_a is a strong function of the wavelength, for a given semiconductor, the wavelength range in which appreciable photocurrent can be generated is limited. The long-wavelength cut off λ_c is established by the energy gap of the semiconductor. The short-wavelength cut off of the responsivity comes about because the values of α_a for short wavelengths are very large ($\alpha_a > 10^5\ cm^{-1}$) and the radiation is absorbed very near the surface where the recombination time is short. The photocarriers, thus, can recombine before they are collected in the p-n junction. The definition of the responsivity σ is available as well as in D.C. and A.C. regime by utilizing either the average values or the Fourier Transform (or complex amplitude) of the wide band centred signals (or alternatif signals). In the case of the A.P.D., it is usual to write $I_p = \overline{M}\ I_{pp}$ where I_p is the photocurrent, I_{pp} the primary photocurrent and $\overline{M}$ the average gain. So the responsivity σ is $\overline{M}\ \sigma_0$ where σ_0 is the responsivity at $\overline{M} = 1$. The A.P.D. manufactured in the semiconductor with indirect gap have σ_0 about 0.5 A/W. (For the Si A.P.D. at $\lambda = 0.8\ \mu m$). On the other hand in the semiconductor with direct gap σ_0 is about 0.8 A/W. ($Hg_{1-x}\ Cd_x\ Te$, or In Ga As photodiodes at $\lambda = 1.3\ \mu m$).

1.4. The noise equivalent power (N.E.P.)

In the unit bandwidth centred on f, the noise equivalent power is defined as the ratio of the noise current $[S_{in}(f)]^{1/2}$ over the photocurrent per watt of the incident light. Therefore the N.E.P. is given by :

$$\text{N.E.P.} = \frac{[S_{in}(f)]^{1/2}}{\sigma_f} \qquad (1)$$

The spectral density of noise current $S_{in}(f)$ is the sum of the spectral density $S_{idn}(f)$ of the avalanche photodiode noise current and the thermal noise current of the load resistance R_L at the equivalent temperature T_{en} to take the noise of the amplifier into account.

$$S_{in}(f) = S_{idn}(f) + \frac{4kT_{en}}{R_L} \qquad (2)$$

1.5. The A.P.D. noise conventional expression

The A.P.D. noise conventional expression is to refered to the photomultiplier tube (PMT) noise which presents in the usual behaviour a spectral density of noise current given by :

$$S_i(f) = 2\ q\ I_{pp}\ (\overline{M})^2 \qquad (3)$$

where I_{pp} is the primary photocurrent and $\overline{M}$ the average gain current ($I = \overline{M}\ I_{pp}$). Therefore in the A.P.D. where the multiplication process concerns the both types of carrier, the noise, neglecting the obscurity noise, is given by :

$$S_{idn}(f) = 2\ q\ I_{inj}\ (\overline{M})^2\ F(M) \qquad (4)$$

where $I = \overline{M}\ I_{inj}$ and F(M) is called the excess noise factor.

1.6. Signal-to-noise ratio

In Δf bandwidth at central frequency f, the signal-to-noise ratio S/N is connected to the N.E.P. by :

$$\frac{S}{N} = \frac{1}{2\Delta f}\ \left(\frac{\phi}{NEP}\right)^2 \qquad (5)$$

From equation (5), we see that the signal-to-noise ratio decreases by reducing the N.E.P. We can see that to minimize the signal-to-noise ratio for the small M, a great load charge will be more advisable with a primary photocurrent as small as possible. But for the large M, we must have an avalanche photodiode that presents an excess noise factor F(M) as near about of unit as possible[3]. Equation can be solved for the minimum optical power ϕ_{min} required to produce a given S/N with avalanche gain.

$$\phi_{min} \simeq \sqrt{2\Delta f}\ (N.E.P.)_{min} \qquad (6)$$

where N.E.P. is about 10^{-13} W/$\sqrt{Hz}$ in the unit bandwidth.

2. AVALANCHE MULTIPLICATION NOISE

The avalanche process is stochastic one because every electron-hole pair generated a given distance in the depletion region does not give the same multiplication[4]. In the Mc Intyre's theory[1] this random state of the gain is neglected.

2.1. Mc Intyre's theory digest

In the litterature several authors have proposed[5,6,7,8] various theories relative to the statistical of the avalanche process and the noise in A.P.D. The Mc Intyre[1] theory is interesting because Mc Intyre takes the place of the injection into consideration (laterally or internal injection). This theory permits, statistically speaking, all the possibilities of the impact ionizations compatible with the thickness w, the α and β ionization coefficients of electrons and holes, without any limit of the random impact ionization number, only given by the probability laws. Consider the rate at which pairs are generated in an element dx. So the pair generation-rate results in an increase in the hole current at x + dx which can be expressed in the permannent regime[1] :

$$\frac{dI_p(x)}{dx} = \alpha\, I_n(x) + \beta\, I_p(x) + G(x) = -\frac{dI_n(x)}{dx} \qquad (7)$$

where $I_p(x)$ and $I_n(x)$ are the holes and electrons currents respectively and G(x) the injection sources which in the general case includs the injection currents at x = 0 and x = w and the thermally and/or optically pairs internal generated. G(x) can be written as :

$$G(x) = g(x) + I_p(0)\ \delta(x) + I_n(w)\ \delta(x-w) \qquad (8)$$

If $\overline{M}(x)$ is the average gain obtained when $G(x) = \delta(x)$ we obtain the Mc Intyre's integral equation[1]

$$\overline{M}(x) = 1 + \int_0^x \alpha\, \overline{M}(x')\, dx' + \int_x^w \beta\, \overline{M}(x')dx' \qquad (9)$$

which gives the form of $\overline{M}(x)$. With this result, it is possible to obtain the whole current in the case where the source is defined by (8), thus :

$$I = I_p(0)\overline{M}(0) + I_n(w)\overline{M}(w) + \int_0^w g(x)\overline{M}(x)dx \qquad (10)$$

In the A.P.D. noise modelling proposed by Mc Intyre, the laterally injection sources and the internal current increase give shot noise multiplicated by the square of the average gain :

$$S_i(f) = 2qI_p(0)[\overline{M}(0)]^2 + 2qI_n(w)[\overline{M}(w)]^2 + 2q\int_0^w \frac{dI_p}{dx}\,[\overline{M}(x)]^2\, dx \qquad (11)$$

The relation (11) is the main result of the Mc Intyre's theory[1].

2.2. New formulation of the starting noise relation

The avalanche process is certainly statistical, so M(t,x) is the random function of average value $\overline{M}(x)$.

This point of view has been shown by several authors such as Van Vliet[8], Tager[5], Personick[6] and so on. Thus the injected noise at x is multiplicated with random gain, and the total noise must be equal to the injected shot noise multiplicated by the mean-square $\overline{M^2(x)}$. The spectral density is given by :

$$S_i(f) = 2qI_p(0)\ \overline{M^2(0)} + 2qI_n(w)\ \overline{M^2(w)} + 2q\int_0^w \frac{dI_p}{dx}\ \overline{M^2(x)}\ dx \qquad (12)$$

We must calculate the variance $\sigma^2_{M(x)}$ in order to obtain the supplementary terms due to the fluctuations of the multiplication.

2.3. Random model of M(t,x), random function

Let be a layer Δx, that we can do to tend towards zero, thus enough small to have at maximum one or zero impact ionization with the probability $\gamma\Delta x$, let be an ionization indicator X_γ equal to one or zero in the opposite case. For the electrons we have $\gamma = \alpha$ and for holes $\gamma = \beta$. So the random function M(t,x) is given by : (13)

$$M(t,x) = \lim_{\Delta x \to 0} 1 + \sum_{k=1}^{x/\Delta x} \sum_{j=1}^{M_k} X_{\alpha,j} + \sum_{l=x/\Delta x}^{w/\Delta x} \sum_{m=1}^{M_l} X_{\beta,m}$$

The second generatrice function $\Theta_y(u)$ derived from the characteristic function $\varphi_y(u)$ by the following relation developped in function of increasing power of u :

$$\Theta_y(u) = -i \operatorname{Log} \varphi_y(u) \simeq u\{\bar{y}\} + \frac{iu^2}{2} \sigma_y^2 + \dots \qquad (14)$$

It is deduced the relation :

$$\Theta_{M_x}(u) = u + \lim_{\Delta x \to 0} \sum_{k=1}^{x/\Delta x} \Theta_{M_k}[-i\alpha\Delta x(e^{iu}-1)] + \sum_{l=\frac{x}{\Delta x}}^{w/\Delta x} \Theta_{M_l}[-i\beta\Delta x(e^{iu}-1)] \qquad (15)$$

2.4. $\bar{M}(x)$ and $\sigma^2_{M(x)}$ expressions

From (14) and (15), it is easy to obtain $\bar{M}(x)$ and $\sigma^2_{M(x)}$ by the first and second derivatives of Θ_{M_x} versus u, for u = 0. We find the Mc Intyre's integral equation given by (9) and the following relation given by : (16)

$$\sigma^2_{M(x)} = \bar{M}(x) - 1 + \int_0^x \alpha \sigma^2_{M(x')} dx' + \int_x^w \beta \sigma^2_{M(x')} dx'$$

It is possible to integrate this relation in order to obtain an symetrical form such as :

$$\sigma^2_{M(x)} = \frac{M(x)}{2} \{ M(0)[1-\operatorname{Log}M(0)] + M(w)[1-\operatorname{Log}M(w)] - 2 + \operatorname{Log}M^2(x) + \int_0^w (\alpha+\beta) M(x) \operatorname{Log}M(x) dx \} \qquad (17)$$

This relation in the case $\alpha=\beta$, that is where M(x) is no longer a function of x, becomes :

$$\sigma^2_M = M\,[M-1] \qquad (18)$$

In the case $k = \frac{\beta}{\alpha}$ where k is a constant, we have :

$$\sigma^2_{M(x)} = M(x)\,[\operatorname{Log}M(x) - 1 + D(k)] \qquad (19)$$

where D(k) is given by :

$$D(k) = \frac{kM(w)}{1-k}[\operatorname{Log}M(w)-1] + \frac{M(0)}{k-1}[\operatorname{Log}M(0)-1] \qquad (20)$$

2.5. Noise spectral density with $\alpha = \beta$

From the relations (12) and (18), we obtain the following result :

$$S_i(f) = 2\,q\,I_{inj}(\bar{M})^2\,[2\bar{M} - 1] \qquad (21)$$

The excess noise factor, here, is equal to $2\bar{M} - 1$ instead of $\bar{M}$ in the Mc Intyre's theory. For the great values of $\bar{M}$ the noise is twice higher than the noise calculated by Mc Intyre. We obtain in the case the relation ;

$$S_i(f) \simeq 4\,q\,I_{inj}(\bar{M})^3 \qquad (22)$$

On the figure 3 it is shown the $S_i(f)$ versus current of a Ge A.P.D. where the coefficients α and β are near. Also we have reported on this figure the computed curves with the aid of the Mc Intyre's relation and the relation (21). The agreement is excellent with the relation (21).

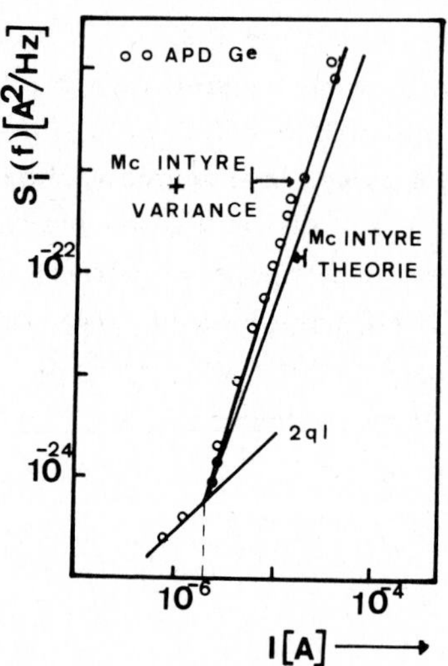

FIGURE 3
Noise current s.d. versus currents in Ge A.P.D.

2.6. Noise spectral density with $k = \frac{\beta}{\alpha}$

The noise spectral density in this case is obtained by calculating the integrals over 0 to w of the functions

$$M(x) \cdot \frac{dI_p}{dx} \text{ and } M(x)\operatorname{Log}M(x)\frac{dI_p(x)}{dx}.$$

In order to simplify the presentation of the calculation we suppose $g(x) = 0$. Thus we consider only the injection of carriers into 0 or w. If the only carriers injected into the avalanche layer are holes [that is $I_p(0) = I_{inj}$ and $\overline{M}(0) = M$ and $I_n(w) = 0$] the noise spectral density is given versus M and k by :

$$S_i(f) = 2qI_{inj}\,M^3\,[1+\frac{1-k}{k}(\frac{M-1}{M})^2] + 2qI_{inj}M^3(\frac{1}{k-1}+\frac{1}{M})[(\frac{1}{k-1}+\frac{1}{2M})\text{Log}A+\frac{1}{M}]\text{Log}A \qquad (23)$$

$$\text{where Log } A = \text{Log}\,\frac{k}{1+\frac{k-1}{M}} \qquad (24)$$

The first term of the second member of this equality is the Mc Intyre's one and the supplementary term is introduced by the calculation of the variance $\sigma^2_{M(x)}$ due to the fluctuations of the gain. We notice when k tends towards 1, we find again the relation (21).

Now if the only carriers injected into the avalanche layer are electrons [that is $I_n(w) = I_{inj}$, $M(w) = M$ and $I_p(0) = 0$] the noise spectral density is of the same form that the relation (23) but with $k = \frac{\alpha}{\beta}$.

The figure 4 shows the computed variations of the $S_i(f)/2qI_{inj}(M)^3$ versus M given the relation (23). We can see that the noise calculated in this work is upper than the noise found by Mc Intyre as it is possible to see it in the figure 2 of the Mc Intyre paper[1].

We notice also for k>>M>>1 that the excess noise factor F(M) becomes $2 + (\text{Log } k)^2/2M$, thus the ideal A.P.D. structures will be always noisier than the P.M.T. because this F(M) limit term which can be several units.

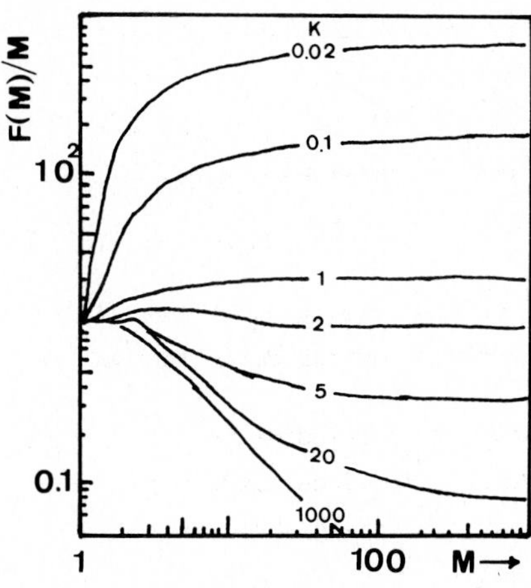

FIGURE 4

3. DESIGN OF LOW NOISE A.P.D.

3.1. The basic principles

The basic principles are the requirement of low reverse current at darkness to allow high responsivity, the requirement of great difference between the ionization coefficients, α for electrons and β for holes, to allow low noise, therefore low N.E.P. The ratio $k = \frac{\beta}{\alpha}$ must be as large as possible in the case of holes injection and as low as possible in the case of electrons injection. There are two ways, the search of high k (or 1/k) materials, and the design of special structures to enhance the ratio ionization avalanche.

3.2. The search of high k (or 1/k) materials

The main means to obtain high k (or 1/k) materials are to use the split-off valence band in III-V[9,10] or II-VI alloys[11]. Indeed when the compound is such that the spin orbit splitting energy Δ is roughly equal to the band gap energy, the impact ionization threshold energy of holes becomes very small (figure 5).

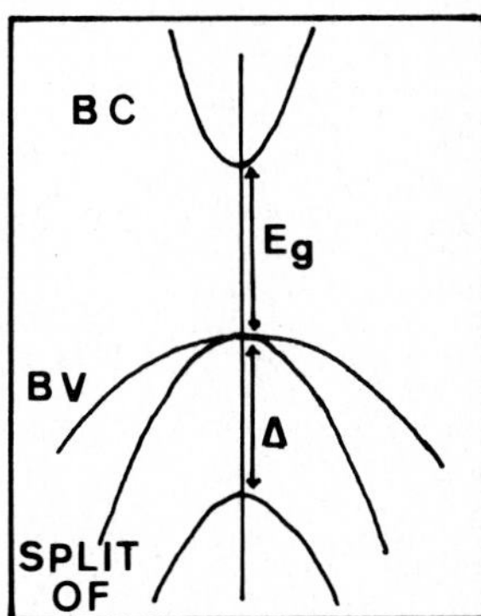

FIGURE 5
The band structure of $Ga_{1-x}Al_xSb$ with x = 6.5 %

Values of k greater than 20 arrive[12,13,14] in Ga_{1-x} Al_x Sb with $x \simeq 0.065$ and in Cd_x Hg_{1-x} Te at $x \simeq 0.65$.

3.3. The graded A.P.D.

The graded A.P.D. is a p^+-i-n^+ structure, with large gap p^+ region, small gap n^+ region and graded gap i region. For example $Ga_{0.55}$ $Al_{0.45}$ As (2eV), Ga_{1-x} Al_x As, Ga As (1.4eV)[15]. The effective injection threshold for electrons is smaller than that for holes because electrons work is lower bandgap region. This structure is interesting only for relatively modest gains.

3.4. The superlattice avalanche detectors

The superlattice avalanche detector is an equivalent structure of graded gap A.P.D., but the i zone is made with a multilayer of successive high gap and low gap thin films (≃ 500Å thick) which presents at the high-low gap material interfaces a layer jump ΔE_c of conduction band well greater than the ΔE_v jump of valence band. By example $\Delta E_c \simeq 0.48$eV and $\Delta E_v \simeq 0.08$eV for the $Al_{0.45}$ $Ga_{0.55}$ As - Ga As lattices. Thus the electrons abrupty gain an energy equal to ΔE_c, and therefore its ionization energy threshold is reduced by ΔE_c. So the ratio $\frac{\alpha}{\beta}$ in As Ga becomes for a S.A.P.D.[16,17] equivalent to 8. Nevertheless the electrons lose one part of their energy in order to come out of the well, to go into the next one where one or several ionizations occur again.

3.5. The stair case A.P.D.

The stair case A.P.D.[18], the most sophistical structure, is the same S.A.P.D. structure, but replacing the i region by a periodical distribution of graded gap wells to eliminate electron trapping effect in the wells. We have a ballistic only by electrons ionization process repeated at each stage.

The noise approach is the same like for the photomultiplier tube (PMT) but the variance of the random gain at each stage is $\delta(\delta-1)$ if δ is the probability that one electron does not impact-ionize in the stage. Thus the excess noise factor is :

$$F(N,\delta) = 1 + \frac{\delta\,[1 - (2 - \delta)^{-N}]}{2 - \delta} \qquad (25)$$

where N is the number of stays, $F(N,\delta)$ versus average gain per stage $M = 2 - \delta$ is shown in figure 6.

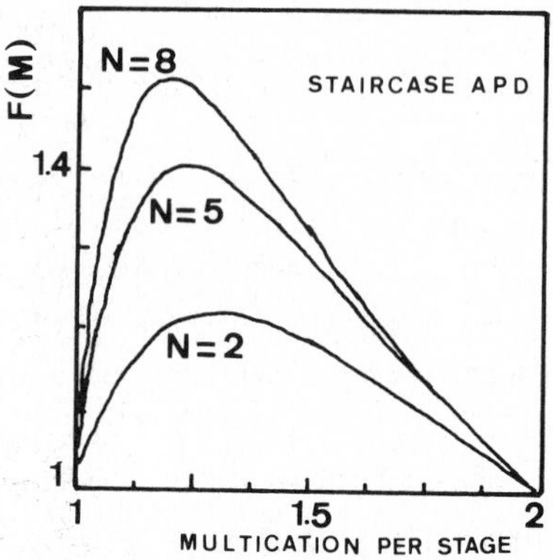

FIGURE 6
Excess noise factor versus average gain per stage for N=2 ; N=5 and N=8 stages (after Capasso[18])

We can see that the excess noise factor is nearly unity if there is one impact-ionization per each stage ($\delta \to 0$; $M \to 2$; $F \to 1$). This main aspect shows that such structures can have comparable performances with the P.M.T. In the material system Hg_x Cd_{1-x} Te the bandgap can be varied from 1.6eV (by changing the composition) and is always direct. Heterojunctions of this material system have essentially all the bandgap jump in the conduction band. By making optimum use of the large conduction band difference at the step, a multiplication per stage greater than two can be achieved with the Hg_x Cd_{1-x} Te material system. Gains as high as 10^5 are possible, similar to those in many P.M.T. with furthermore low bias voltages. Finally other structures such as channeling photodiodes begin to be developped[19,20] where the both types of carriers are injected in the separated regions of different bandgap[21].

1. R.J. Mc Intyre, I.E.E.E. Trans. Electron. Devices, Vol. ED-13, pp 164-168, 1966.

2. S.M. Sze, Physics of Semiconductor Devices 2nd Edition, a Wiley interscience publication John Wiley, 1981.

3. R. Alabedra, C. Maille, T. Saîd, G. Lecoy, Ann. Telecomm. Vol. 35, pp 193-198, 1980.

4. D.J. Robbins, Phys. Stat. Sol. Part I (b) 97-9, 1980.
   Part II, Phys. Stat. Sol.(b) 97 n°2, 1980.
   Part III, Phys. Stat. Sol.(b) 98 n°1, 1980.

5. A.S. Tager, Sov. Phys. Sol. State, vol. 8, pp 1919-1925, february 1965.

6. S.D. Personick, Bell system technical journal, vol. 50, number 10, pp 3075-3095, december 1971.
   S.D. Personick, Bell system technical journal, pp 167-189, january 1971.

7. R.J. Mc Intyre, I.E.E.E. Trans. Electron. Devices, vol. ED-19, n°6, pp 703-713, june 1972.
   R.J. Mc Intyre, I.E.E.E. Trans. Electron. Devices, vol. ED-20, n°7, pp 637-641, july 1973.

8. K.M. Van Vliet, L.M. Rucker, I.E.E.E. Trans. Electron. Devices, part I, vol. ED-26, n°5, pp 746-751, may 1979.
   K.M. Van Vliet, A. Friedmann, L.M. Rucker, I.E.E.E. Trans. Electron. Devices, part II, vol. ED-26, n°5, may 1979.

9. K. Nakajima, A. Yamaguchi, K. Akita, T. Katani, J. Appl. Phys. vol. 49, pp 5944-5950, december 1978.

10. M.P. Mikhaîlova, N.N. Smirnova, S.V. Slobodchikov, Sov. Phys. Semicond. vol. 10, n°5, pp 509-513, may 1976.
    M.P. Mikhaîlova, S.V. Slobodchikov, N.N. Smirnova, G.M. Filaretova, Sov. Phys. Semicond. vol. 10, n°5, pp 578-579, may 1976.
    M.P. Mikhaîlova, A.A. Rogachev, N.I. Yassievich, Sov. Phys. Semicond. Vol. 10, n°8, pp 866-871, august 1976.

11. A. Moritani, K. Taniguchi, C. Hamaguchi, J. Nakaî, J. of Phys. Soc. Japan, vol. 34, n°1, pp 79-87, january 1973.

12. O. Hildebrand, W. Kuebart, K.W. Benz, M.H. Pilkuhn, I.E.E.E. J. Quantum Elec. vol. QE17, pp 284-288, february 1981.

13. D. Kasemset, J. Quantum Elec. Vol. QE17, n°9, september 1981.

14. R. Alabedra, B. Orsal, G. Lecoy, G. Pichard, J. Meslage, P. Fragnon, I.E.E.E. Trans. Electron. Devices, vol. ED 32, n°7, pp 1302-1306, july 1985.

15. F. Capasso, Laser Focus/Electro. Optics, pp 84-101.

16. J.P. Gordon, R.E. Nahory, M.A. Pollack, J.M. Worlok, Electronics letters, vol. 15, n°17, pp 518-519, august 1979.

17. F. Capasso, W.T. Tsang, A.L. Hutchinson, F.G. Williams, Appl. Phys. Lett. 40(1), pp 38-40, january 1982.

18. F. Capasso, W.T. Tsang, G. Williams, I.E.E.E. Trans.Electron. Devices, Vol. ED-30, n°4, pp 381-390, april 1983.

19. F. Capasso, R.A. Logan, W.T. Tsang, J.R. Hayes, Appl. Phys. Lett. 41(10), pp 944-946, november 1982.

20. T. Tanaaîe, H. Sakoki, Appl. Phys. Lett. 41 (1), pp 67-69, july 1982.

21. F. Capasso, B. Kasper, K. Alaui, A.Y. Cho, J.M. Parsey, Appl. Phys. Lett. 44(11) pp 1027-1029, june 1984.

NOISE IN PHYSICAL SYSTEMS AND 1/f NOISE - 1985
A. D'Amico and P. Mazzetti (editors)

ON THE METROLOGICAL ASPECTS OF JOHNSON NOISE, NOISE THERMOMETRY AND PRECISION DETERMINATION OF BOLTZMANN'S CONSTANT

L. STORM

Institut für Angewandte Physik der Universität Münster, Corrensstr. 2/4, 4400 Münster, FRG

By measurement of Johnson noise both the thermodynamic temperature T and the Boltzmann constant k can be obtained. The metrological background of a precision determination of k is outlined. A (not yet quite completed) measuring equipment is described and a thorough theoretical analysis of the measuring procedure is given. It should be possible to measure k with an error of less than 10 ppm.

1. METROLOGICAL BACKGROUND

From statistical thermodynamics the following relations between the entropy S, internal energy U and thermodynamic temperature T are known:

$$S := k \ln W \quad ; \; S,\ k/\text{Joule Kelvin}^{-1} \tag{1}$$

$$\frac{1}{T} := \frac{\partial S}{\partial U} \quad ; \quad U/\text{Joule} \tag{2}$$

$$273.16\ \text{K} := T_{tp}; \quad T/\text{Kelvin} \tag{3}$$

Equ. (1) defines S, equ. (2) defines T. Equ.(3) defines the basic unit Kelvin and furthermore determines the value of the Boltzmann constant k. T_{tp} means the water triple point temperature.

Theorists prefer another set of equations:

$$k := 1 \tag{4}$$

$$S' := \ln W; \; S' \text{ dimensionless} \tag{5}$$

$$\frac{1}{T'} := \frac{\partial S'}{\partial U}; \; T'/\text{Joule} \tag{6}$$

They prefer to set fundamental constants equal to unity. Now the temperature T' is given in Joule. Obviously

$$k = T'/T = T'_{tp}/273.16\ \text{K} \tag{7}$$

The Boltzmann constant is nothing else than the conversion factor between two temperature scales. As we have already disposed of k in (4), the value T'_{tp} at the water triple point cannot be fixed. It must be measured, this is equivalent to a determination of k.

There is a fundamental difference between T and T', from a metrological point of view. The measurement of T and T' are quite different problems. By primary thermometers (gas thermometer, noise thermometer) the ratio T/T_{tp} is measured and nothing else. The only standard one needs is the Kelvin standard, the water triple point cell. The determination of T' means the measurement of an energy, now standards for mass, length and time, or alternatively for voltage, resistance and frequency are required, but not a Kelvin standard. The measurement of T' is considerably more difficult and, up to now, it is less accurate. That is the main reason, besides the historical reason, why experimentalists still use (1), (2), (3).

If we have learned to measure T' sufficiently accurate, (1), (2) and (3) will become superfluous. We can remove the unit Kelvin and the Boltzmann constant from physics. If we like to retain the unit Kelvin for practical reasons we can redefine it by assigning a certain value to the Boltzmann constant. The unit Kelvin will no longer be defined by a material standard. This corresponds to the modern trend in metrology. (Example: New definition of the meter). Up to now the accuracy of temperature measurements is ultimately limited to about 1 ppm by the reproducibility of the water triple point. The above redefinition is conclusive, regardless of the accuracy needed in future. This motivates metrologists to deal with precision determinations of T' and k.

2. MEASUREMENT OF T' AND k

2.1. Measuring equipment

The figure shows an equipment suitable for precision measurements of T' and k. The input ohmic voltage divider D with its output impedance Z (input short circuited, connection 0 in the figure) acts as the thermal noise source.

The noise voltage is connected to two almost identical channels, each consisting of four amplifiers A_{ik}, three ohmic voltage dividers D_{ik}, a sample and hold amplifier S_i and an ADC C_i. Storing the output numbers of the ADCs a time delay Θ or $-\Theta$ can be generated between the channels which can be a multiple of the sampling period $\tau = 1/f_s$. The digital numbers are multiplied by Π and then stored in the summing registers Σ for averaging. Products for 256 different Θ-values can be obtained simultaneously. Of course we must use the cross correlation technique in order to eliminate the major part of the amplifier noise.

During the noise measurements the voltage dividers D_{ik} are switched off and we use the full gain of about 40^4 in each channel. During the calibration with the sine voltage generator U_o, f_o the voltage dividers with attenuation factors of about 1/40 are switched in and the gain is reduced to approximately unity within the pass band, which ranges from 3 kHz to 25 kHz. It is absolutely necessary to realize the required high gain by connecting several ampli-

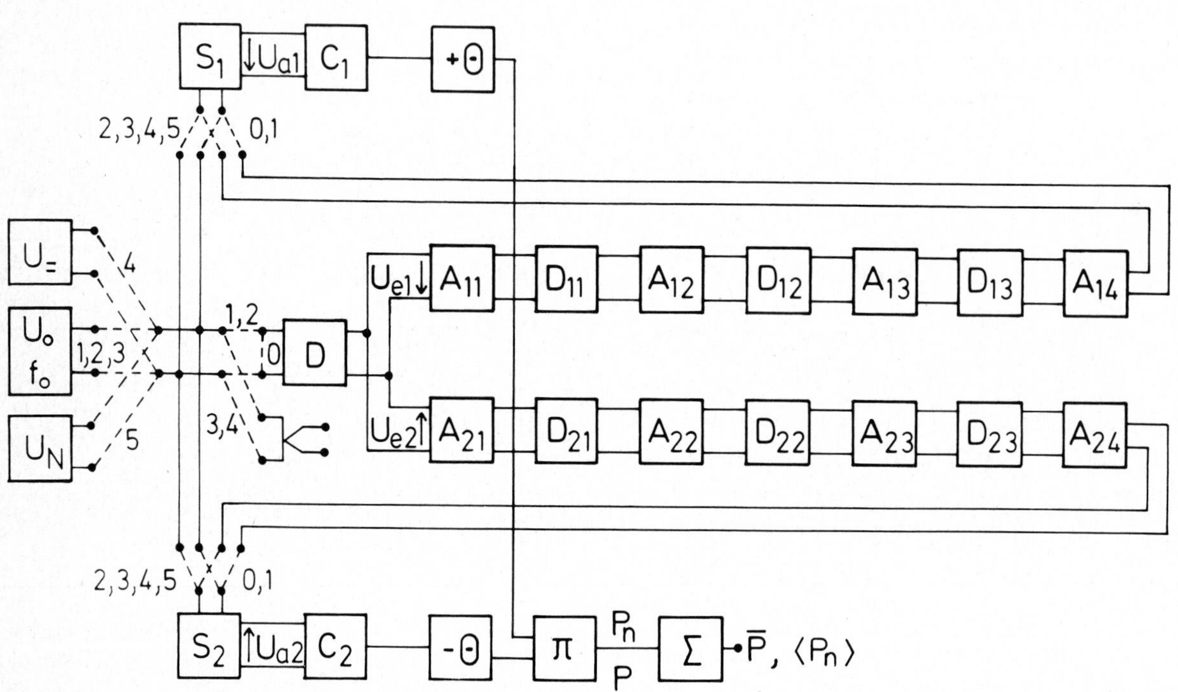

Equipment for the measurement of T' and k. Noise measurement: Connections 0, measurement of $\bar{P}$. Calibration n: Connections n, measurement of $\langle P_n \rangle$, $n = 1 \div 5$.

fiers with moderate gain in series. During the calibration the noise must be small compared with the sine voltage at each point of the circuit.

2.2 Theoretical analysis of the measuring procedure

Let A_{ik}, D_{ik} and S_i denote the complex transfer functions of the corresponding component and C_i the conversion factors of the ADCs. We use the abbreviations

$$A_i = \prod_k A_{ik};\ D_i = \prod_k D_{ik} \qquad V_i = S_i \prod_k A_{ik} D_{ik} \tag{8}$$

W_{e1e2} denotes the complex cross power spectrum of the input voltages U_{e1} and U_{e2}. Then the expectation value $\langle P\rangle$ of the product P is given by

$$\langle P(\Theta)\rangle = C_1C_2 \int_{-\infty}^{+\infty} V_1^*(f)V_2(f)W_{e1e2}(f)\exp(i2\pi f\Theta)df \tag{9}$$

f means frequency, Θ is the applied time delay. The sum $\langle P(\Theta) + P(-\Theta)\rangle$ is given by

$$\langle P(\Theta) + P(-\Theta)\rangle = 2C_1C_2 \int_{-\infty}^{+\infty} \mathrm{Re}(V_1^*V_2W_{e1e2})\cos 2\pi f\Theta df \tag{10}$$

The integrand is an even function of f. Equ. (10) is the basis for the following considerations.

Noise measurement, connections 0, $D_{ik} = 1;\ V_1 := S_1A_1;\ V_2 := S_2A_2$.
A careful theoretical analysis of the input circuit[1] yields the result, that within 1 ppm W_{e1e2} is given by

$$W_{e1e2} = 2kTZ_0(\mathrm{Re}Z/Z_0+Z_0A(f))\Phi_1^*\Phi_2 \tag{11}$$

Z_0 is the output impedance of D at zero frequency. The disturbing contribution $Z_0A(f)$ is mainly due to the gate current noise of the input FETs, which cannot be eliminated by the correlation technique. It can be kept below 10^{-5}, if the FETs are carefully selected for low noise. A is a real and even function of frequency. By means of the functions Φ_1 and Φ_2 we take into account all the influence of the leads between D and A_{11}, A_{21}, for example the loading of Z by the leads capacity.

It is extremely important, that the noise contribution of the amplifiers is proportional to Z_0^2, but not proportional to Z_0. In noise thermometry a contribution $\propto Z_0$ results in a constant temperature error. Low temperature measurements[1] agreed within 0.2 mk with gas thermometric results[2]. So besides the theoretical estimation we have an experimental proof that for $T = T_{tp}$ equ. (11) is correct within 1 ppm, provided that the input circuit is designed according to[1].

From (10) and (11) we get

$$\langle P(\Theta)+P(-\Theta)\rangle = 4kTZ_0C_1C_2 \int_{-\infty}^{+\infty} (\mathrm{Re}Z/Z_0+Z_0A(f))\cdot \cdot\mathrm{Re}(\Phi_1^*\Phi_2A_1^*A_2S_1^*S_2)\cos 2\pi f\Theta df \tag{12}$$

In order to obtain $C_1C_2\mathrm{Re}(\Phi_1^*\ldots S_2)$ the equipment must be calibrated. This will be done in five steps.

Calibration 1, connections 1
$D \approx D_{ik} \approx |A_{ik}|^{-1} \approx 1/40.\ V_1 = A_1S_1D_1;\ V_2 = A_2S_2D_2$. We apply a sine voltage with amplitude $U_0 \approx 1$ V and frequency f_0. Now W_{e1e2} is

$$W_{e1e2} = \frac{U_0^2}{2}\,\frac{\delta(f_0+f)+\delta(f_0-f)}{2}\,|D|^2\Phi_1^*\Phi_2 \tag{13}$$

This together with (10) yields for the case $\Theta = 0$:

$$\langle P_1(\Theta{=}0,f_0)\rangle = (U_0^2/2)C_1C_2|D(f_0)|^2\cdot \cdot\mathrm{Re}(\Phi_1^*\Phi_2A_1^*A_2S_1^*S_2D_1^*D_2)_{f_0} \tag{14}$$

The index f_0 means "at frequency f_0".

When sampling sine functions periodically with the sampling frequency f_s, the condition

$$\frac{f_0}{f_s} = \frac{m}{4n};\qquad m,n = \text{integers without common divisor} \tag{15}$$

must be met, in order to avoid a systematic error. This requires that f_o and f_s are both deduced from the same quartz oszillator by variable frequency dividers. The sequence of samples is periodic with the period $4n/f_s = m/f_o$. $\langle P_1 \rangle$ means the average over a multiple of 4n samples in order to reduce the effect of the residual noise to less than 1 ppm.

When the ac voltage source U_o, f_o is connected, the transfer functions Φ_i must remain unchanged. Because Φ_i depends on Z, which is defined for a short circuited input of D, we need a source with a very low output impedance.

In (12) we need $Re(\Phi_1^* \ldots S_2)$. Because of

$$\begin{aligned} Re(\Phi_1^* \ldots S_2 D_1^* D_2) &= Re(\Phi_1^* \ldots S_2) Re(D_1^* D_2) \\ &- Im(\Phi_1^* \ldots S_2) Im(D_1^* D_2) \end{aligned} \tag{16}$$

we propose

$$\begin{aligned} &|Im(\Phi_1^* \ldots S_2) Im(D_1^* D_2)| \leq \\ &10^{-6} |Re(\Phi_1^* \ldots S_2) Re(D_1^* D_2)| \end{aligned} \tag{17}$$

In order to realize this the phase angles of the amplifiers A_{1k} and A_{2k}, the phase angles of S_1 and S_2 and the phase angles of the dividers D_{1k} and D_{2k} must be as equal as possible. (17) can easily be proven by interchanging the dividers D_{1k} with D_{2k} and repeating the measurement. Now $D_1 D_2^*$ appears in (16) instead of $D_1^* D_2$ and the difference of the two measurements yields $Im(\Phi_1^* \ldots S_2) Im(D_1^* D_2)$. In the following we assume (16) to be fulfilled.

Calibration 2, connections 2

The ac-voltage is connected directly to the sample and hold amplifiers. As we use different leads we have slightly changed transfer functions S_i'. We measure

$$\langle P_2(0,f_o) \rangle = (U_o^2/2) C_1 C_2 Re(S_1'^* S_2') \tag{18}$$

From (14), (17) and (18) we obtain

$$Re(\Phi_1^* \ldots S_2)_f = \frac{\langle P_1(0,f) \rangle Re(S_1'^* S_2')_f}{\langle P_2(0,f) \rangle |D(f)|^2 Re(D_1^* D_2)_f} \tag{19}$$

f_o has been replaced by f. During the consecutive measurement of $\langle P_1 \rangle$ and $\langle P_2 \rangle$ the frequency and amplitude of the ac voltage source must be constant within 1 ppm. A very low distortion factor is also required.

Calibration 3, connections 3

The voltage divider D is replaced by a high precision multi-junction thermal converter with an ac-dc-transfer-error of less than 1 ppm. Such converters are now constructed at the PTB. The transfer functions S_i' must remain unchanged. Therefore the input resistances of the thermal converter and D must be equal, because the transfer function of a transmission line depends on its terminating resistance. We now repeat calibration 2:

$$\langle P_3(0,f) \rangle = (U_o^2/2) C_1 C_2 Re(S_1'^* S_2') \tag{20}$$

The thermal EMF of the converter is compensated by a circuit not shown in the figure.

Calibration 4, connections 4

We connect a dc-source $U_=$ to the converter. $U_=$ is adjusted so that the EMF is again compensated. Then, within 1 ppm, we have

$$U_o^2/2 = U_=^2 \tag{21}$$

We measure $P_4(0,0) = P_4(0,f=0)$:

$$P_4(0,0) = U_=^2 C_1 C_2 S_1'(0) S_2'(0) \tag{22}$$

(20), (21) and (22) yield

$$Re(S_1'^* S_2')_f = \frac{\langle P_3(0,f) \rangle}{P_4(0,0)} S_1'(0) S_2'(0) \tag{23}$$

We have performed an ac-dc-transfer. The S_i must be very broad-band in order to get a very small transient time. So $Re(S_1'^* S_2')$ is only slightly frequency dependent within the pass band of the A_{ik}. All measurements with dc-voltages must be repeated with reversed voltages to eliminate the offset of the S_i and C_i.

Calibration 5, connections 5

We connect a voltage standard U_N:

$$P_5(0,0) = U_N^2 C_1 C_2 S_1'(0) S_2'(0) \tag{24}$$

From (19), (23) and (24) we obtain

$$C_1C_2\mathrm{Re}(\Phi_1^*\Phi_2A_1^*A_2S_1^*S_2)_f = \frac{\langle P_1(0,f)\rangle\langle P_3(0,f)\rangle P_5(0,0)}{\langle P_2(0,f)\rangle P_4(0,0)U_N^2|D(f)|^2\mathrm{Re}(D_1^*D_2)_f} \quad (25)$$

We write

$$|D|^2\mathrm{Re}(D_1^*D_2) = D_0^2D_{10}D_{20}\Psi(f) \quad (26)$$

D_0, D_{10} and D_{20} are the values of the corresponding transfer functions at zero frequency, $\Psi(f) \approx 1$ describes a possible frequency dependence. We introduce the abbreviations

$$M := \frac{8T_{tp}P_5(0,0)}{D_0^2D_{10}D_{20}U_N^2} \quad (27)$$

$$K(f) := \frac{\langle P_1(0,f)\rangle\langle P_3(0,f)\rangle}{\langle P_2(0,f)\rangle P_4(0,0)} \quad (28)$$

From (10), (25) to (28) we obtain

$$\langle P(\Theta)+P(-\Theta)\rangle = kMZ_0 \cdot \int_0^\infty \left(\frac{\mathrm{Re}Z}{Z_0} + Z_0A(f)\right)\frac{K(f)}{\Psi(f)} \cdot \cos 2\pi f\Theta df \quad (29)$$

The integrand is an even function of frequency.

In (29) $\mathrm{Re}Z/Z_0$, $A(f)$ and $\Psi(f)$ are unknown even functions of frequency. The voltage dividers can be constructed in such a manner that the deviations of $\mathrm{Re}Z/Z_0$ and $\Psi(f)$ from unity are less than 10 ppm. $A(f)$ is considerably more frequency dependent, but $Z_0A(f)$ is less than 10^{-5}. So we may assume

$$\left(\frac{\mathrm{Re}Z}{Z_0} + Z_0A(f)\right)\frac{1}{\Psi(f)} \approx 1+A_0Z_0+B(Z_0)f^2 \quad (30)$$

neglecting higher order terms. A_0 is independent of Z_0, the contribution A_0Z_0 is caused by the white part of the gate current noise. We introduce

$$Y(\Theta) := \frac{\langle P(\Theta) + P(-\Theta)\rangle}{\int_0^\infty K(f)\cos 2\pi f\Theta df} \quad (31)$$

$$X(\Theta) := \frac{\int_0^\infty K(f)f^2\cos 2\pi f\Theta df}{\int_0^\infty K(f)\cos 2\pi f\Theta df} \quad (32)$$

and obtain from (29) to (32):

$$Y = kMZ_0(1+A_0Z_0+BX) \quad (33)$$

If (30) is correct, Y is a linear function of X. If we have measured Y and X for different values of Θ, we can obtain

$$Y_0 := kMZ_0(1+A_0Z_0) \quad (34)$$

by a least squares adjustment and we can prove (30) by a χ^2-test. In order to elimimate the systematic error due to A_0Z_0, a second measurement of Y_0' with a much larger impedance Z_0' must be performed. We obtain k from

$$k = \frac{1}{M}\,\frac{Y_0Z_0'/Z_0-Y_0'Z_0/Z_0'}{Z_0'-Z_0} \quad (35)$$

We have eliminated all systematic errors.

2.3 Accuracy requirements

As (10) is only valid for linear, time invariant systems, a linearity and long term stability of better than 1 ppm for the whole equipment are required. The integral in (31) must be computed with an error of less than 1 ppm. The desired function K(f) is

$$K(f) = \frac{K_0}{\left[1+(f_u/f)^8\right]\left[1+(f/f_0)^{12}\right]^2} \quad (36)$$

with f_u = 3 kHz and f_0 = 25 kHz. (36) has been realized within 0.1 % by active Butterworth filters. K(f) must be measured for a sufficiently large number of frequencies, then it can be approximated by more general high pass and low pass functions and the integral can be computed. $X(\Theta)$ may be calculated by the desired function (36) because dY/dX is very small.

$\langle P\rangle$ cannot be measured but only $\bar{P}$, the average over N samples. $\bar{P}$ is a random variable, consequently a statistical error occurs. In computing this error we must take into account that all products for different values of $\Theta=\mu\tau$ are obtained from the same temporal segment of the noise voltage, therefore the errors of the products are correlated. We have to compute the covariance matrix

$C_{\mu\nu} := <(\bar{Y}(\mu\tau)-<Y(\mu\tau)>)(\bar{Y}(\nu\tau)-<Y(\nu\tau)>)>$

Assuming the input noise to be a Gaussian process one obtains

$$C_{\mu\nu} \approx \frac{(kMZ_0)^2}{N\tau}\left[1+\frac{R_e}{Z_0}\left(1+\frac{R_e}{2Z_0}\right)\right] \cdot \tag{37}$$

$$\cdot \frac{\int_0^\infty K^2(f)\cos 2\pi f\mu\tau \cos 2\pi f\nu\tau\, df}{\int_0^\infty K(f)\cos 2\pi f\mu\tau\, df \int_0^\infty K(f)\cos 2\pi f\nu\tau\, df}$$

R_e means the equivalent noise resistance of A_{11} and A_{21}. In order to save measuring time $R_e<Z_0$ is required. The last fraction in (37) has the meaning of a reciprocal equivalent bandwidth. (37) holds true if $f_s>2f_0$, otherwise $C_{\mu\nu}$ becomes larger (loss of information).

The $\bar{Y}(\mu\tau)$ are almost normally distributed with the covariance matrix $C_{\mu\nu}$, so when performing the least squares adjustment, the best we can do is to minimize the expression

$$\sum_{\mu,\nu} (\bar{Y}(\mu\tau)-Y_0-cX(\mu\tau))(\bar{Y}(\nu\tau-Y_0-cX(\nu\tau))C^{-1}_{\mu\nu}$$

where $C^{-1}_{\mu\nu}$ are the elements of the inverse matrix. The adjustment yields the best values of Y_0 and $c = kMZ_0B$ and their covariance matrix, too. In order to reduce the relative standard deviation of Y_0 to less than 3 ppm a measuring time $N\tau$ of several months is needed.

The analog part of the equipment including the ADCs is located inside a shielded room and is connected to the digital part by light guides. The temperature of the analog part must be kept constant within 0.02 K because of the temperature dependence of the capacitors which determine the bandwidth. The input divider D must be kept at the water triple point within 1 ppm. The voltage and resistance standards will be calibrated at the PTB within 1 ppm. The divider factors D and D_{ik} are to be measured with an error of less than 0.5 ppm, because the individual errors must be added linearly to obtain the total error of $D^2D_{10}D_{20}$. It should be possible to measure k with an error of less than 10 ppm.

Although the theoretical analysis is rather simple it is very difficult to construct the electronic devices with the desired somewhat unusual features. G. Klempt has developed the amplifiers, voltage dividers and the sine wave generator. All required features are met. F. Dankwart has developed a 22 bit ADC which is linear and long term stable within 1 ppm, and which needs a conversion time of only 10 µs. The large bit number is useful for linearity and control measurements. When the multiplier is used, the bit number is reduced to 16. For the noise measurement the quantization error is negligible. When measuring dc-voltage the 1 ppm resolution is easily obtained by adding small uncorrelated broad band noise in the S_i.

Now Dankwart is employed on the sample and hold. In the sample mode it is stable and linear within 1 ppm, but when switching from sample to hold a certain kind of nonlinear distortion can occur, which is not yet known. Fortunately the effect of this distortion can be calculated rather exactly. The result is, that a linearity of 1 ppm is not necessary, 100 ppm are already sufficient. The reason is that we really measure mean powers instead of voltages. Each sample must not be perfect but only the mean of the products.

After completing the equipment a total measuring time of about one year will be needed. All the measurements run fully automatic, controlled by a computer, except the calibration 3 and 4, which must not be repeated frequently.

3. NOISE THERMOMETRY

Noise thermometry is much more simple. D is replaced by an almost temperature independent sensing resistor R, which is brought to the unknown temperature T_x and then to T_{tp}. The corresponding products $\bar{P}_x(0)$, $\bar{P}_{tp}(0)$ and resistances R_x, R_{tp} are measured. The following relation holds true[1]:

$$T_{tp} \frac{\bar{P}_x(0)R_{tp}}{\bar{P}_{tp}(0)R_x} (1-\gamma) \approx T_x(1+F(T_x,T_{tp})\cdot R_x)$$

The factor F depends on the temperatures, but not on R_x. γ means a small correction $\leqq 10^{-5}$ due to the residual temperature dependence of R (small change of the transfer functions Φ_i), it can be estimated sufficiently accurate. One obtains T_x by measuring the left hand side for at least two different R_x and extrapolation to $R_x=0$. The equipment must not be calibrated by sine voltages, for the measurement of the ratio R_{tp}/R_x a constant uncalibrated reference resistor is sufficient. The only standard we need is the Kelvin standard.

ACKNOWLEDGEMENT

The financial support by the Minister für Wissenschaft und Forschung des Landes Nordrhein-Westfalen is gratefully acknowledged.

REFERENCES

1. H.H. Klein, G. Klempt, L. Storm, Metrologia 15 (1979) 143-154.

2. K.H. Berry, Document CCT/76-27

NOISE IN PHYSICAL SYSTEMS AND 1/f NOISE - 1985
A. D'Amico and P. Mazzetti (editors)

CHAOS IN NONLINEAR ELECTRONIC CIRCUITS[†]

Leon O. CHUA

Department of Electrical Engineering and Computer Sciences and the Electronics Research Laboratory, University of California, Berkeley, CA 94720

This paper summarizes the main experimental and analytical results from the following recent papers on various chaotic phenomena observed from extremely simple electronic circuits:

1. Y. S. Tang, A. I. Mees, and L. O. Chua, "Synchronization and chaos," IEEE Trans. on Circuits and Systems, Vol. CAS-30, pp. 620-626, September 1983.
2. T. Matsumoto, L. O. Chua, and S. Tanaka, "Simplest chaotic nonautonomous circuit," Physical Review A, Vol. 30, No. 2, pp. 1155-1157, August 1984.
3. T. Matsumoto, "A chaotic attractor from Chua's circuit," IEEE Trans. on Circuits and Systems, Vol. CAS-31, pp. 1055-1058, December 1984.
4. G-Q. Zhong and F. Ayrom, "Experimental confirmation of chaos from Chua's circuit," Int. J. Circuit Theory Appl., Vol. 13, pp. 93-98, January 1985.
5. G-Q. Zhong and F. Ayrom, "Periodicity and chaos in Chua's circuit," IEEE Trans. on Circuits and Systems, Vol. CAS-32, pp. 501-503, May 1985.
6. T. Matsumoto, L. O. Chua, and M. Komuro, "The double scroll," IEEE Trans. on Circuits and Systems, Vol. CAS-32, pp. 797-818, August 1985.

[†]This research is supported in part by the Semiconductor Research Corporation under Contract SRC-82-11-008, and by the Office of Naval Research under Contract N00014-76-0572.

THEORY (GENERAL) - 1

NOISE IN PHYSICAL SYSTEMS AND 1/f NOISE - 1985
A. D'Amico and P. Mazzetti (editors)

ON THE SECOND MOMENTS OF THE PROBABILITY DENSITIES OF LEVEL-CROSSING TIME-INTERVALS

T. MUNAKATA, T. MIMAKI[+], and D. WOLF

Institut für Angewandte Physik der Universität, D-6000 Frankfurt a.M., FRG

The 6-states-model of level-crossing analysis as presented previously[10] has been applied to the estimation of second moments of level-crossing intervals and of the correlation coefficient between adjacent level-crossing intervals. The theory has been evaluated for Gaussian bandlimited processes with different bandwidths and levels. The theoretical results are found in good agreement with measured data.

1. INTRODUCTION

The 2nd moments of the probability density $P_0(\tau,I)$ of the level-crossing time-intervals of a random process $\xi(t)$ and, more generally, the correlation coefficients $\kappa_{i\pm}$ between two level-crossing intervals $\tau_{0\pm}$ and $\tau_{i\pm}$ have been studied particularly by McFadden[1,2], Sato[3], and Mimaki[4,5,6,7].

For the correlation coefficients

$$\kappa_{i\pm}=[E\{\tau_{0\pm}\cdot\tau_{i\mp}\}-E\{\tau_{0\pm}\}\cdot E\{\tau_{i\mp}\}]/\sigma_+\sigma_- \quad \text{(i odd)} \quad (1)$$

$$\kappa_{i\pm}=[E\{\tau_{0\pm}\cdot\tau_{i\pm}\}-E\{\tau_{0\pm}\}^2]/\sigma_\pm^2 \quad \text{(i even)} \quad (2)$$

Mimaki[7] derived the relations

$$A/\beta=(1-\alpha)^2\sigma_+^2(1+2\sum_{i=1}^{\infty}\kappa_{2i+})+\alpha^2\sigma_-^2(1+2\sum_{i=1}^{\infty}\kappa_{2i-})$$
$$-4\alpha(1-\alpha)\sigma_+\sigma_-\sum_{i=0}^{\infty}\kappa_{2i+1,+} \quad (3)$$

$$4(1+2B)/\beta^2 = \sigma_+^2(1+2\sum_{i=1}^{\infty}\kappa_{2i+})+\sigma_-^2(1+2\sum_{i=1}^{\infty}\kappa_{2i-})$$
$$+4\sigma_+\sigma_-\sum_{i=0}^{\infty}\kappa_{2i+1,+}\,. \quad (4)$$

$\sigma_\pm^2 = E\{\tau_\pm^2\}-E\{\tau_\pm\}^2$ denotes the variance of $\tau_\pm$, β is the crossing rate, and α is the probability for $\xi(t)\geq I$.

$$A = \int_0^\infty[r(\tau,I)-(1-2\alpha)^2]d\tau$$

and

$$B = \int_0^\infty[W_+(\tau,I)-\beta/2]d\tau$$

can be determined analytically by means of the Rice function $W_+(\tau,I)$ and the autocorrelation function $r(\tau,I)$ of the process

$$\zeta(t) = \begin{cases} 1 & \text{for } \xi(t)\leqq I \\ -1 & \text{for } \xi(t)<I \end{cases} .$$

In general Eqs.(3) and (4) cannot be solved because of the number of unknown quantities.

In special cases - e.g. with a Gaussian process and I=0 - solutions were derived by introducing specific assumptions on the correlation coefficients[1,2,3,4]. Only in few of the cases the results obtained have been found in agreement with measured data[4,5,7]. Thus, other concepts are required in order to achieve more general representations being not subject to non realistic conditions in a physical sense.

This paper deals with a new approach based on the 6-states-model of level-crossing analysis proposed by the authors in 1983[10]. It is

[+]The University of Electro-Communications, Tokyo, Japan

shown that this approach, without employing any further assumptions, leads to explicit expressions for the variance $\sigma_{\pm}^2$ of level-crossing intervals and the correlation coefficient κ_1 between two adjacent level-crossing intervals.

2. THEORY

As has been reported previously[10] approximate solutions

$$P_{0\pm}^{(6)}(\tau,I;T_1)=\hat{g}_{\pm}(\tau,I;T_1)\exp\{-\hat{G}_{\pm}(\tau,I;T_1)\} \qquad (5)$$

- as well as

$$P_{1\pm}^{(6)}(\tau,I;T_1,T_2) \qquad (6)$$
$$= \hat{h}_{\pm}(\tau,I;T_1,T_2)\exp\{-\hat{H}_{\pm}(\tau,I;T_1,T_2)\}\cdot$$
$$\cdot\int_0^\tau P_{0\pm}^{(6)}(t,I;T_1)\exp\{\hat{H}_{\pm}(t,I;T_1,T_2)\}dt$$

for the probability density of the sum of two adjacent level-crossing intervals - can be derived from a 6-states-model. The functions $\hat{g}_{\pm}(\tau,I;T_1)$, $\hat{h}_{\pm}(\tau,I;T_1,T_2)$, $\hat{G}_{\pm}(\tau,I;T_1)$, and $\hat{H}_{\pm}(\tau,I;T_1,T_2)$ are given by the Rice functions Q and W, and by the two positive relaxation times T_1 and T_2. T_1 and T_2 are determined such that the first moments $m_{1P0\pm}$ of $P_{0\pm}^{(6)}$ and $m_{1,P1\pm}$ of $P_{1\pm}^{(6)}$ are equal to the expectation values $E\{\tau_{\pm}\}$ and $E\{\tau_{\pm\pm}\}$ of the time intervals $\tau_{\pm}$ and $\tau_{\pm\pm}$ between an up- or down-crossing and the subsequent down- or up-crossing, resp.. $E\{\tau_{\pm}\}$ and $E\{\tau_{\pm\pm}\}$ are exactly given by the theory. If T_1 and T_2 are assumed to be equal to zero the 6-states-model reduces to the 4-states-model[8,9].*

Using the relations (5) and (6) one obtains the second moments

$$\tilde{m}_{2,P0\pm}(I) = \int_0^\infty \tau^2 P_{0\pm}^{(6)}(\tau,I)d\tau$$
$$= 2\int_0^\infty t\,\exp\{-\hat{G}_{\pm}(t,I)\}dt\ , \qquad (7)$$

$$\tilde{m}_{2,P1\pm}(I) = \int_0^\infty \tau^2 P_{1\pm}^{(6)}(\tau,I)d\tau \qquad (8)$$
$$= \tilde{m}_{2,P0\pm}(I)+2\int_0^\infty dt[t\cdot\exp\{-\hat{H}_{\pm}(t,I)\}\cdot$$
$$\int_0^t du\cdot P_{0\pm}^{(6)}(u,I)\exp\{\hat{H}_{\pm}(u,I)\}]$$

which may be considered to be estimates for the unknown moments $m_{2,P0\pm}$ or $m_{2,P1\pm}$, resp.. From the moments the variances

$$\sigma_{\pm}^2 = m_{2,P0\pm}(I) - m_{1,P0\pm}^2(I) \qquad (9)$$

$$\sigma_{\pm\pm}^2 = m_{2,P1\pm}(I) - m_{1,P1\pm}^2(I) \qquad (10)$$

and the correlation coefficient

$$\kappa_1 = \{\sigma_{++}^2 - \sigma_+^2 - \sigma_-^2\}/2\sigma_+\sigma_- \qquad (11)$$
$$= \{\sigma_{--}^2 - \sigma_+^2 - \sigma_-^2\}/2\sigma_+\sigma_-$$

can be calculated.

The estimate $\tilde{\sigma}_{++}^2$ based on $P_{1+}^{(6)}$ may differ from the estimate $\tilde{\sigma}_{--}^2$ based on $P_{1-}^{(6)}$. Thus the two expressions in Eq. (11) may lead to different estimates for κ_1.

In the case of a Gaussian process the identities $\sigma_-^2(I)=\sigma_+^2(-I)$ and $\sigma_{++}^2(I)=\sigma_{--}^2(I)=\sigma_{++}^2(-I)=\sigma_{--}^2(-I)$ are valid due to the symmetry of the process with respect to zero level.

3. RESULTS AND DISCUSSION

The variances $\tilde{\sigma}_+^2$ and $\tilde{\sigma}_{++}^2$ and the correlation coefficient $\tilde{\kappa}_1$ have been calculated for a Gaussian process with zero mean and unit variance having a 7-th order Butterworth bandpass spectrum

$$S(\nu) = \frac{1}{1-k^{14}}\left[\frac{1}{1+\nu^{14}} - \frac{1}{1+(\nu/k)^{14}}\right].$$

$\nu=\omega/\omega_H$ denotes the normalized frequency corre-

---

* Attention should be called to the fact that sometimes the solutions $P_0^{(6)}$ and $P_1^{(6)}$ do not exist; in some cases the moment $m_{1,P0}$ is equal to $E\{\tau_+\}$ or $m_{1,P1}$ is equal to $E\{\tau_{++}\}$ but with negative values of T_1 or T_2, resp.. Then the 4-states-model has to be considered in order to preserve causality.

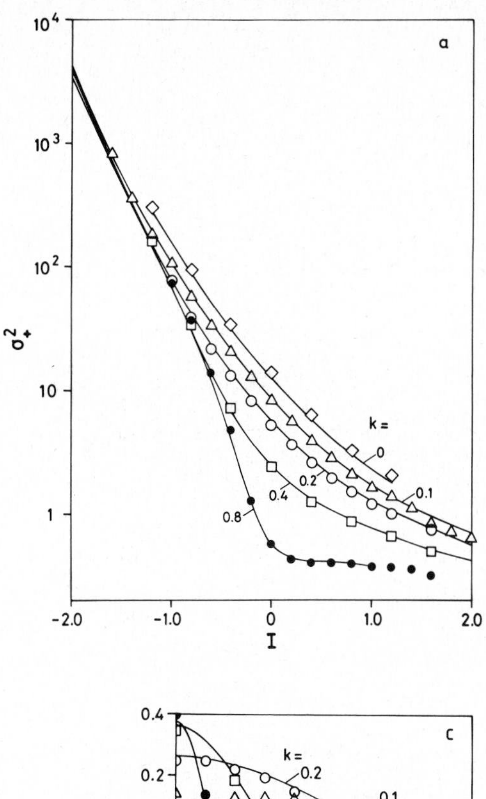

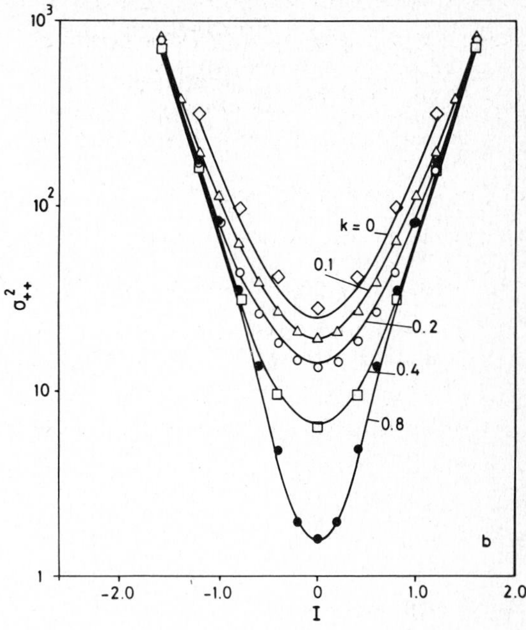

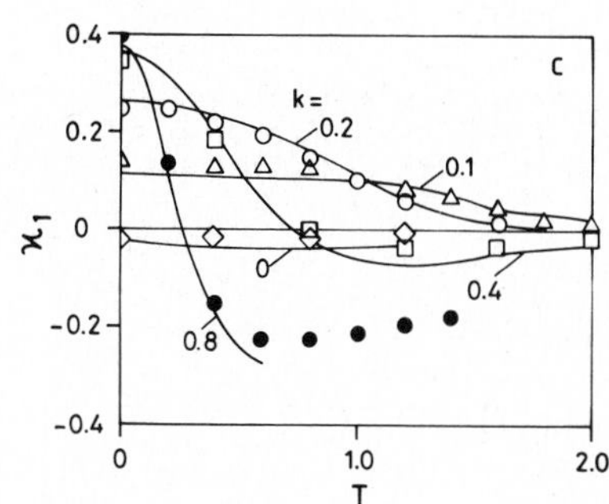

FIGURE 1 a,b,c
Theoretical (lines) and experimental (dots) results for σ^2_+, σ^2_{++}, and κ_1 of the level-crossing intervals of a Gaussian process vs. level I.

sponding to the normalized time $\tau = \omega_H t$.
$k = \omega_L/\omega_H$ is the ratio between the low and the high cutoff frequency.

Figs.1a and b show the theoretical results for $\tilde{\sigma}^2_+$ and $\tilde{\sigma}^2_{++}$ as functions of I in comparison with experimentally determined values [5,7].

For all values of bandwidth k, including the case k=0.8, which is typical for strong correlation between adjacent intervals, a close agreement between theoretical and experimental values can be seen. For k=0.8 we get a correct theoretical representation of the rapid change of $\tilde{\sigma}^2_+$ and $\tilde{\sigma}^2_{++}$ when the level value passes zero.

The behaviour of the correlation coefficient κ_1 in dependence of I is depicted in Fig. 1c for the same Gaussian processes. Sincẽ the calculation of κ_1 according to both forms of Eq. (11) yields slightly different values the mean values are shown in Fig. 1c. The calculated values of κ_1 fit well the measured data in the wideband and the medium band case but fails in the narrowband case (k=0.8) for I>0.6. This is due to the fact that the orders of magnitude of the variances σ_+^2, σ_-^2, σ_{++}^2 and σ_{--}^2 are quite different such that small errors effect large ones in κ_1.

4. CONCLUSION

The results suggest that on the basis of the 6-states-model the variances of level-crossing time-intervals of Gaussian random processes can be estimated quite accurately even in the case of narrowband processes. The model also leads to an appropriate representation of the correlation coefficient between adjacent level-crossing intervals as function of the level value in a wide range of bandwidth. Finally the results confirm the validity of the 6-states-model which does not employ any assumptions on dependencies between level-crossing intervals.

REFERENCES

1. J.A. McFadden, J. Royal Stat. Soc. B 24 (1962) 364.

2. J.A. McFadden, IRE Trans. Inform. Theory IT-4 (1958) 14.

3. H. Sato, M. Yamamoto, and T. Mimaki, J. Phys. Soc. Japan 29 (1970) 253.

4. T. Mimaki, J. Appl. Phys. 44 (1973) 477

5. T. Mimaki and T. Munakata, IEEE Trans. Inform. Theory IT-24 (1978) 515.

6. T. Mimaki and D. Wolf, Fluctuation of Level-Crossing Time Intervals of Noise, in: Noise in Physical Systems, ed. D. Wolf (Springer, Berlin, 1978) pp. 323-326.

7. T. Mimaki, T. Munakata, and D. Wolf, Trans. IECE of Japan E62 (1979) 583.

8. T. Munakata and D. Wolf, Neue Theoretische Lösungen für das Problem der Pegel-Überschreitungen bei Gaußschen Prozessen, in: Theorie und Anwendung bei der Signalverarbeitung, 4. Aachener Koll. (RWTH Aachen, 1981) pp. 151-154.

9. T. Munakata and D. Wolf, A Novel Approach to the Level-Crossing Problem of Random Processes, in: Proc. IEEE Int. Sympos. on Inform. Theory Les Arcs (IEEE-Cat. 82 CH 1767-3 IT, 1982) pp. 149-158.

10. T. Munakata and D. Wolf, On the Distribution of the Level-Crossing Time-Intervals of Random Processes, in: Noise in Physical Systems and 1/f Noise, ed. M. Savelli (North-Holland, Amsterdam, 1983) pp. 49-52.

NOISE IN PHYSICAL SYSTEMS AND 1/f NOISE - 1985
A. D'Amico and P. Mazzetti (editors)

PARAMETRIC EXCITATION IN PHYSICAL SYSTEMS: THE CONNECTION MATRICES METHOD

R. B. PEREZ

Nuclear Engineering Department. University of Tennessee (U.S.A.)

J. L. MUÑOZ-COBO

Catedra of Nuclear Physics, Universidad Politecnica de Valencia (Spain).

V. HERNANDEZ

Mathematics Department. Universidad Politecnica de Valencia (Spain).

Many areas in Physics and Engineering[1,2,3,4] have to deal with problems involving linear stochastic integro-differential equations which arise in fields such as plasma physics, wave propagation and heat transport in random media, among others. The purpose of this paper is to extend previous work in the area of stochastic parametric excitation [1,2,3] to stochastic vector fields, by the introduction of the "Connection Matrices Method." Let $\Psi(x,t,\rho) = [\Psi(x,t,\rho)]_{n \times 1}$, be a random state vector field defined by appropriate boundary and initial conditions and the set of integro-differential equations

$$M(x,t,\rho)\Psi(x,t,\rho) = S(x,t,\rho) \tag{1}$$

where $M(x,t,\rho) = [M_{ij}(x,t,\rho)]_{n \times n}$ is an stochastic integro-differential matrix operator and $S(x,t,\rho) = [S_i(x,t,\rho)]_{n \times i}$ is a random driving source. The variables set (x,t,ρ) in (1) is formed by $x \varepsilon E_m$ (phase space), $t \varepsilon E_1$ (time) and ρ_k which is a set of random variables $(\rho_1 \rho_2 \cdots \rho_k)$ over a measured space Ω. The probability density defined over Ω is designated as $P(\rho)$.

We assume that the matrix operator M, and the source vector S can be split into sure and random parts:

$$M(x,t,\rho) = M_o(x,t)+\varepsilon M_1(x,t,\rho) \tag{2}$$

$$S(x,t,\rho) = S_o(x,t)+\varepsilon S_1(x,t,\rho) \tag{3}$$

Where $\varepsilon (o \leq \varepsilon \leq 1)$ is a parameter different from zero only when the system is stochastically perturbed. Associated with the system (1) we define the following vector of observables:

$$R(x,t,\rho)=\int dt' \int dx' W(x,t|x',t')\Psi(x',t',\rho) \tag{4}$$

where Wnxn is a device weight matrix describing the instrumentation of the experimental set up. From a physical point of view, the quantities of interest to us are the expectation values (ensemble averages)

$$\langle\Psi(x,t)\rangle=\int_\Omega \Psi(x,t,\rho)P(\rho)\, d\rho \tag{5}$$

$$\langle R(x,t)\rangle=\int dt' \int dx'\; W(x,t|x',t')\langle\Psi(x',t')\rangle \tag{6}$$

Where we have assumed that the device weight matrix is deterministic. Let us expand the state and the observables vector in Taylor series in the parameter ε:

$$\Psi(x,t,\rho)=\sum_{n=0}^{\infty} \frac{\varepsilon^n}{n!} \left(\frac{\partial^n \Psi}{\partial \varepsilon^n}\right)_{\varepsilon=0} \tag{7}$$

$$R(x,t,\rho)=\sum_{n=0}^{\infty} \frac{\varepsilon^n}{n!} \left(\frac{\partial^n R}{\partial \varepsilon^n}\right)_{\varepsilon=0} \tag{8}$$

which from equations (1) and (3) and on account of the deterministic nature of M_0 and S_0, leads to the splitting of the state vector into the sure and random parts Ψ_0 and $\delta\Psi$, respectively, eg:

$$\Psi(x,t,\rho) = \Psi_0(x,t)+\delta\Psi(x,t,\rho) \quad (9)$$

The sure part, Ψ_0, is the solution of the deterministic integro-differential system

$$M_0(x,t)\Psi_0(x,t)=S_0(x,t) \quad (10)$$

whereas the stochastically perturbed part is given by:

$$\delta\Psi(x,t,\rho)=\sum_{n=o}^{\infty} \frac{\varepsilon^n}{n!} Y_n(x,t,\rho) \quad (11)$$

where we have introduced the notation:

$$Y_n(x,t,\rho)= \left(\frac{\partial^n\Psi}{\partial\varepsilon^n}\right)_{\varepsilon=o} \quad (n=1.2.3...) \quad (12)$$

Similarly one obtains:

$$R(x,t,\rho)= R_0(x,t)+\delta R(x,t,\rho) \quad (13)$$

where

$$R_0(x,t)=\int dt'\int dx' \; W(x,t|x't')\Psi_0(x,t) \quad (14)$$

and

$$\delta R(x,t,\rho)= \sum_{n=1}^{\infty} \frac{\varepsilon^n}{n!} \cdot \left(\frac{\partial^n R}{\partial\varepsilon^n}\right)_{\varepsilon=o} \quad (15)$$

The next step is to obtain a recursion formula for the coefficients of the Taylor Series (11). To this end, we apply the operator

$$\left(\frac{\partial^n}{\partial e^n}\right)_{\varepsilon=o},$$

to equation (1), arriving at the result

$$M_o(x,t)Y_n(x,t,\rho) = \delta_{1n}\cdot S_1(x,t,\rho) -nM_1(x,t,\rho)\cdot \cdot Y_{n-1}(x,t,\rho) \quad (16)$$

Where $Y_0(x,t,\rho)=\Psi_o(x,t)$, and δ_{1n} is the kronecker delta.

To derive an explicit expression for Y_n, we introduce the advanced propagator $G^+(x,t|x',t')$ defined by

$$M_o^+(x,t)G^+(x,t|x',t')=I\delta(x-x')\delta(t-t') \quad (17)$$

Where, I, is the unit matrix, $\delta(x-x'),\delta(t-t')$ are Dirac deltas and $M_o^+(x,t)$ is the matrix adjoint operator of $M_o(x,t)$. Also to simplify the notation, the inner product of the matrix functions $A(x,t,\rho)_{mxn}$ and $B(x,t,\rho)_{mxn}$ will be expressed as:

$$A*B=\int dt\int dx A(x,t,\rho)B(x,t,\rho) \quad (18)$$

Next, equation (16) is premultiplied by the transposed advanced propagator G^{+T} and the transposed of equation (17) postmultiplied by Y_n, to obtain the relations:

$$G^{+T}*M_oY_n=\delta_{1n}(G^{+T}*S_1)-n(G^{+T}*M_1Y_{n-1}) \quad (19)$$

$$(M_o{}^+G^{+T})^T*Y_n=Y_n \quad (20)$$

Use of the lemma below:

$$G^{+T}*M_oY_n=(M_oG^+)^T*Y_n \quad (21)$$

results, on account of equations (19) and (20), into the recursion relation

$$Y_n=\delta_{1n}(G^{+T}*S_1)-n(G^{+T}*M_1Y_{n-1}) \tag{22}$$

which yields, by induction, the following closed form expression:

$$Y_n=(-1)^{n-1}(n!).(G^{+T}*M_1(G^{+T}*.....M_1(G^{+T}*Q_1))) \tag{23}$$

with

$$Q_1=S_1-M_1\Psi_o \tag{24}$$

Next taking the ensemble average of equation (23) one obtains

$$\begin{aligned}\langle Y_n(0)\rangle=(-1)^{n-1}\ .\ (n!).G^{+T}(n|0)*\langle M_1(n)\cdot\\ \cdot G^{+T}(n-1|n)*M_1(n-1)....*M_1(2)G^{+T}(1|2)*\\ *Q_1(1)\rangle\end{aligned} \tag{25}$$

where we used the notation[1]

$\left((x,t)=0;...(x_n,t_n)=1\right)$

Our goal of finding the expectation value for the vector, R, of observables, can now be achieved by taking the ensemble average of equation (11), with the result

$$\langle\delta\Psi(0)\rangle=\sum_{n=1}^{\infty}\frac{\varepsilon^n}{n!}\langle Y_n(0)\rangle \tag{26}$$

where the expectation value of the Taylor series coefficients is given by equation(25). The use of the present technique will be illustrated within the framework of the following assumptions:

i) The source fluctuation vector S_1 is uncorrelated with the stochastic operator matrix M_1

ii) All the random functions that appear in S_1 and M_1, are Gaussian centered random functions

iii) The operator M_1 does not contain integrodifferential operators in the phase space variables. One obtains in this fashion from equations (25) and (26) the following expression for the expectation value of the component of the stochastic state vector.

$$\begin{aligned}\langle\delta\Psi(0)\rangle_i=\sum_{p=1}^{\infty}\varepsilon^{2p}[G^{+T}_{ij}(2p|0)\langle M_{1jk}(2p)M_{1nm}(2p-1)\cdot\\ ...M_{1qr}(2)M_{1st}(1)\rangle\cdot\\ \cdot G^{+T}_{kn}(2p-1|2p)*...*G^{+T}_{rs}(1|2)*\Psi_{ot}(1)]\end{aligned} \tag{27}$$

where the symbol [ ], indicates a time-ordered product and summation over repeated subindexes is to be understood. Use of the customary cumulant expansion[5], to recast the moments appearing in equation (27), in terms of n-point correlation functions, yields up to fourth order in ε, the result

$$\begin{aligned}\langle\delta\Psi(o)\rangle_i=\varepsilon^2[G^{+T}_{ij}(2|0)*X_{jm}(2,1)*\Psi_{om}(1)]\\ +\varepsilon^4\ [G^{+T}_{ij}\ (4|0)*(X_{jm}(4,3)*G^{+T}_{mn}(2|3)*X_{nr}(2,1)\\ +X_{jr}(4,3,2,1)\\ +X_{jr}(4,3,2,1)+X_{jr}(4,3,2,1))*\Psi_{or}(1)]\end{aligned} \tag{28}$$

where we defined the following two-point and four-point connection matrices

$$X_{jm}(2,1)=C_{jklm}(2,1)G^{+T}_{kl}(1|2) \tag{29}$$

$$X_{jr}(\widehat{4,3,2,1})=C_{jknp}(4,2)C_{lmqr}(3,1)G^{+T}_{kl}(3|4)*$$

$$*G^{+T}_{mn}(2|3)*G^{+T}_{pq}(1|2) \qquad (30)$$

$$X_{jr}(\widehat{4,\widehat{3,2},1})=C_{jkqr}(4,1)C_{lmnp}(3,2)G^{+T}_{kl}(3|4)*$$

$$*G^{+T}_{mn}(2|3)*G^{+T}_{pq}(1|2) \qquad (31)$$

$$X_{jr}(\overline{4,3,2,1})=C_{jklmnpqr}(4,3,2,1)G^{+T}_{kl}(3|4)*$$

$$*G^{+T}_{mn}(2|3)*G^{+T}_{pq}(1|2) \qquad (32)$$

in terms of two-point and four-point correlation functions. Note though that the connection matrices are only function of the two side points, as the intermediate points in phase space were integrated out.

After some manipulations equation (28) can be rewritten in the form of a generalized matrix Dyson Equation[2]

$$\langle\delta\Psi(o)\rangle = G^{+T}(n|o)*A(n|1)*\Psi_0(1)$$

$$+ G^{+T}(n|o)*A(n|1)*\langle\delta\Psi(1)\rangle \qquad (33)$$

where we introduced the "Average" operator

$$A(n,1)=\varepsilon^2X(n,1)+\varepsilon^4\{X(\widehat{n,3,2,1})+X(\widehat{n,\widehat{3,2},1})$$

$$+X(\overline{n,3,2,1})\}+O(\varepsilon^6) \qquad (34)$$

which plays the role of Dyson's mass operator in quantum field theory. Application of the connection matrice method to the covariance between the $i^{\underline{th}}$ and $j^{\underline{th}}$ components of the state vector, leads to the matrix form of the Bethe-Salpeter Equation[2]. We have applied this equation to derive an expression for the cross correlation function between two neutron detectors, located in an infinite non-multiplicative system, driven by a random neutron source. Comparison of our result with the Langevin approach used by Williams[6] for the same problem, shows that the Langevin approximation is correct up to and including second order terms in the series expansion of the state vector. The connection matrices introduced in this work are in fact mathematical representations of perturbation diagrams[2], and as such they facilitate the diagrammatic expansion of random vector fields.

References:

1. Bourret R. C., Nuov. Cim. 26. 1 (1962).

2. Bharucha-Reid A. T., "Probabilistic Methods in Applied Mathematics", Academic Press New York (1972).

3. N. G. Van Kampen, Phys. Reports Vol. 24c 3(1976).

4. J. R. Klauder, Application of Stochastic differential equation in physics MOI-PAE/PTh51/81 (1981).

5. R. F. Fox Phys. Reports Vol. 48.3, (1978) Also R. F. Fox J. Math Phys. 20(12) (1979).

6. M. M. R Williams, "Random Processes in Nuclear Reactors" (Oxford Pergamon Press, 1974).

NOISE IN PHYSICAL SYSTEMS AND 1/f NOISE - 1985
A. D'Amico and P. Mazzetti (editors)

NEW CONTRIBUTIONS TO NEUTRON STOCHASTIC TRANSPORT THEORY IN THE TIME AND IN THE FREQUENCY DOMAIN

G. VERDU, J. L. MUÑOZ-COBO

Cátedra de Física Nuclear, E.T.S.I.I. Universidad Politécnica de Valencia,
Campus camino de Vera, Valencia, SPAIN

J. T. MIHALCZO

Nuclear Engineering Department, University of Tennessee and Oak Ridge National Lab.U.S.

INTRODUCTION

The first theoretical study on neutron fluctuation was performed by Feynman, de Hoffmann and Serber[1], but was Bruno Rossi the first one in suggesting the use of neutron fluctuation for practical applications (The Rossi-αMethod)[2]. The starting models were lumped parameters ones i.e. the system was considered like a point and spatial effects were not included. However the experimentalists in the field were aware of spatial effects in their measurements.[3]

These facts compelled to Pal[4] and Bell[5] to develop a stochastic transport theory including spatial effects, however this methodology was not developed to their point of inmediate application, because in their original form this formalism could not be applied for practical purposes. This situation incited to Muñoz-Cobo and Pérez[6,7] to develop a methodology based on Pal approach that could solve all these problems, obtaining a method free of closure problems, i.e. higher moments of the distributions (neutrons, counts,...) were obtained in terms of lower ones , exactly the opposite occurs when Langevin approach is used. The next step was to compute the C.P.S.D. (cross power spectral density) between two detectors using the Muñoz-Cobo methodology, this stage was accomplished by Verdú et allied[8;9]

In this paper we generalize the stochastic transport theory of Pal and Muñoz-Cobo and Perez methodology, to include the delay neutron effects Then we apply this theory to interpret severals experiments performed by J.T.Mihalczo, who measures the C.P.S.D. $G_{12}(w)$, $G_{13}(w)$, $G_{23}(w)$ of three detectors 1,2 and 3, located in and out of a tank containing a UO_2F_2 solution in water.

We use a standard notation; the set of phase space position coordinates is represented by a number, i.e. $(\bar{r},v,\bar{\Omega})=(1)$, and the inner product between the functions A,B, defined as their product integrated over phase space and time will be represented by<A/B>.The counting number n'th moment of a neutron detector during the time interval $(t_f-\Delta t_c,t_f)$ is obtained by operating with $(\partial^n/\partial Z^n)Z=1$ on the generating function $G_S^D(Z,\Delta t_c,t_f)$, defined in terms of the probability $P_{Nc}(\Delta t_c, t_f)$ of observing Nc counts at a final time ,t_f, upon insertion of a neutron source in the system by :

$$G_S{}^D(Z,\Delta t_c,t_f) = \sum_{Nc=0}^{\infty} Z^{Nc} . P_{Nc}(\Delta t_c,t_f) \quad (1)$$

another ingredient of the theory is the kernel, $K_{nc}^{D}(1,t/\Delta t_c,t_f)$, defined as the probability of observing n_c counts during the counting interval upon injection of one neutron at the phase space point and time (1,t); associated with this last probability we have the kernel generating function $G_K^D(Z,1,t/\Delta t_c,t_f)$.

Because in this theory the delayed neutron are also incorporated, we define the delayed neutron precursors kernel $K_{nc}^{iD}(r,t/\Delta t_c,t_f)$ of the i-kind of precursors, as the probability of having n_c counts, upon injection of one neutron precursor atom of kind i at point $\bar{r}$ and at time t. Associated with this kernel we have the generating function $G_K^{iD}(Z,\bar{r},t/\Delta t_c,t_f)$.

The easiest relationship to deduce is the existing one between $K_{nc}^{iD}(1,t/\Delta t_c,t_f)$ and $K_{nc}^{i}(\bar{r},t/\Delta t_c,t_f)$, that is obtained by means of the probability balance :

$$K_{nc}^{iD}(\bar{r},t/\Delta t_c,t_f) = \delta_{nc0}\,.\exp(-\lambda_i(t_f-t)) + \int_t^{t_f} dt' \int dv'\, d\bar{\Omega}'.\exp(-\lambda_i.(t'-t)).\lambda_i.X^i(v').K_{nc}^{i}(\bar{r},v,\bar{\Omega},t'/\Delta t_c,t_f)\,/4\pi \quad (2)$$

where λ_i and $X^i(v)$ are the decay constant and the normalized neutron spectrum respectively of the i-kind of precursors. The first term takes on account the probability of no precursor disintegration in the (t,t_f) time interval; if this happens we do not inject the neutron into the system and therefore will be zero the number of counts. The second term is formed by : i) $\exp(-\lambda_i(t'-t))\lambda_i dt'$: probability of the i-precursor to disintegrate in the time interval (t',t'+dt'). It is assumed that each precursor disintegration gives one neutron, this is the standard convention followed by many authors[11]. ii) $X^i(v')dv'.d\bar{\Omega}'/4\pi$: probability of the neutron to be isotropically emitted with velocity and direction within the intervals (v',v'+dv') and $d\bar{\Omega}'$ around $\bar{\Omega}'$. iii) $K_{nc}^{i}(\bar{r},v',\bar{\Omega}',t'/\Delta t_c,t_f)$ that has been defined previously. Now multipliying (2) by Z^n and adding up from n=0 to n=∞ it is obtained :

$$G_K^{iD}(Z,\bar{r},t/\Delta t_c,t_f) = \exp(-\lambda_i(t_f-t)) + \int_t^{t_f} dt'\, dv'\, d\bar{\Omega}'\, \exp(-\lambda_i(t'-t)).\lambda_i.X^i(v').G_K^{i}(Z,\bar{r},v',\bar{\Omega}',t'/\Delta t_c,t_f)/4\pi \quad (3)$$

The next steps are :

i) To express $G_S^D(Z,\Delta t_c,t_f)$ in terms of $G_K^D(Z,1,t/\Delta t_c,t_f)$

ii) To find the master equation satisfied by $G_K^D(Z,1,t/\Delta t_c,t_f)$

- Then from i) and ii) it is possible to obtain by a general technique developped by Muñoz-Cobo and Pérez[6,7,9] the average number of counts, the variance and so on ... We generalize the previous calculations including the delayed neutron effect. The connection between the generating function $G_S^D(Z,\Delta t_c,t_f)$ and the kernel generating function $G_K^D(Z,1,t/\Delta t_c,t_f)$ is derived on the basis of a probability balance for $P_{Nc}(\Delta t_c,t_f)$ in terms of $K_{nc}(1,t/\Delta t_c,t_f)$, after some calculations one arrives to :

$$G_S^D(Z,\Delta t_c,t_f) = \exp\left\{\sum_{j=0}^{I_s}\varepsilon_j^s\int_{t_0}^{t_f} dt\, dr\, S_0(\bar{r},t).\left\{\left[\int d\bar{\Omega}\, dv/4\pi\, X_s(v).G_K^D(Z,1,t/\Delta t_c,t_f)\right]^j - 1\right\}\right\} \quad (4)$$

where ε_j^s is the probability of emission of j neutron per source disintegration. $S_0(\bar{r},t)$ is the number of source disintegrations per second and unit volume, $X_s(v)$ is the normalized source spectrum, and I_s is the maximun number of neutron emitted per source disintegration.

Now we derive the equation satisfied by $G_K^D(1,t/\Delta t_c, t_f)$, to this end we set a probability balance for the kernel $K_{nc}^D(1,t/\Delta t_c,t_f)$ and multipliying by Z^{nc}, and adding up from $n_c= 0$ to $n_c = \infty$, it is obtained after some manipulations :

$$\begin{aligned} G_K^D(Z,1,t/\Delta t_c,t_f) = \int^{S_v} ds\, P(1,t,s).\Big\{ & C_{0c}^{(0)}(\bar{r}+s\bar{\Omega},v,t+s/v)+Z.C_{0c}^{(1)}(\bar{r}+s\bar{\Omega},v,t+s/v) + \\ & \int dv'd\bar{\Omega}'(C_{1s}^{(0)}(\bar{r}+s\bar{\Omega},v,\bar{\Omega},t+s/v\,/\,v',\bar{\Omega}') + Z.\ C_{1s}^{(1)}(\bar{r}+s\bar{\Omega},v,\bar{\Omega},t+s/v\,/\,v',\bar{\Omega}')). \\ & G_K(Z,\bar{r}+s\bar{\Omega},v',\bar{\Omega}',t+s/v\,/\Delta t_c,t_f) + \\ & \sum_{j=0}^{I}\sum_{1_1=0}^{L_1}\cdots\sum_{1_6=0}^{L_6}(C_{j1_11_2\ldots1_6}^{(0)}(\bar{r}+s\bar{\Omega},v,t+s/v) + Z\,.\,C_{j1_11_2\ldots1_6}^{(1)}(\bar{r}+s\bar{\Omega},v,t+s/v))\,. \\ & \left[\int dv'\, d\bar{\Omega}'.X(v')/4\pi.G_K^D(Z,\bar{r}+s\bar{\Omega},v',\bar{\Omega}',t+s/v)\right]^j \\ & \prod_{i=1}^{6}\left[G_K^{iD}(Z,\bar{r}+s\bar{\Omega},t+s/v\,/\Delta t_c,t_f)\right]^{1_i}\Big\} + A_d \end{aligned} \quad (5)$$

where $C_{0c}^{(i)}$ is the probability of a neutron capture event leading to i counts in the detector; $C_{j1_11_2\ldots1_6}^{(i)}$ is the probability of a neutron fission event leading to j prompt neutrons ,1_1, 1_2, ..1_6 delayed neutron precursors and to i counts in the detector; $C_{1s}^{(i)}$ is the probability for one scattering event at the phase space point and time given by the first argument, leading to one neutron with velocity and direction

$(v', \bar{\Omega}')$ and to i counts in the detector. $X(v')$ is the normalized fuel fission spectrum. $P(1,t,s).ds$ = probability that a neutron born at the phase space point 1, at time t, interacts with the system in the interval $(\bar{r}+s\bar{\Omega}, \bar{r}+(s+ds)\bar{\Omega})$. A_d gives the probability of no interaction within the system. Finally $S_V = \min\{S_B, (t_f-t).v\}$ is the minimun of the two arguments, where S_B is the distance between the neutron birth-place and the convex boundary of the system along the direction . I and $L_1 \ldots L_6$ are the maximun number of neutron and delayed precursor in a fission event. The physical interpretation of the various terms appearing in equation (5) is quite obvious. The two first terms within the curly brackets come from the probabilities that a interaction in the path interval ,ds, should lead to a neutron capture without neutron emission and leading to zero and one count in the detector respectively. The second term arises from the probability that after an scattering event one neutron emerges with speed and direction $(v', \bar{\Omega}')$ leading to zero or one count at the detector respectively and so on ...

Now appliying Pal's methodology[4,5] and taking on account the relationships between the various C-coefficcients and the cross sections of the system[6,7,8,10] it is obtained the following new result :

$$(1/v.\partial/\partial t + \bar{\Omega}.\bar{\nabla} - \Sigma(\bar{r},v,t))G_K^D(Z,1,t/\Delta t_c,t_f) = -\Sigma(\bar{r},v,t).\Big\{ C_{0c}^{(0)}(\bar{r},v,t) + Z.C_{0c}^{(1)}(\bar{r},v,t) + \int dv'\, d\bar{\Omega}'\,(C_{1s}^{(0)}(\bar{r},v,\bar{\Omega},t/v',\bar{\Omega}') + Z.C_{1s}^{(1)}(\bar{r},v,\bar{\Omega},t/v',\bar{\Omega}')).\ G_K^D(Z,\bar{r},v',\bar{\Omega},t/\Delta t_c,t_f)$$
$$+\sum_{j=0}^{I}\sum_{l_1=0}^{L_1}\cdots\sum_{l_6=0}^{L_6}\big(C_{j,l_1\ldots l_6}^{(0)}(\bar{r},v,t) + Z.C_{j,l_1,\ldots l_6}^{(1)}(\bar{r},v,t)\big)\Big(\int dv'\, d\bar{\Omega}'.X(v')/4\pi.G_K^D(Z,\bar{r},v',\bar{\Omega}',t/\Delta t_c,t_f)\Big)^j \prod_{i=1}^{6} G_K^{iD}(\bar{r},t/\Delta t_c,t_f)^{l_i}\Big\} \quad (6)$$

where $\Sigma(\bar{r},v,t)$ is the macroscopic total cross-section at $(\bar{r},v,t)$ and the rest of the symbols are known.

Now we have all the ingredients that we need to calculate the average number of counts, the variance and so on, to this end we follow the Muñoz-Cobo and Pérez Methodology[6,7,9] appliying the operator $\partial/\partial Z|Z=1$ to equations (3), (4), and (6), it is obtained that we can compute the average number of counts $\bar{N}_c(\Delta t_c,t_f)$ by :

$$\bar{N}_c(\Delta t_c,t_f) = \langle \bar{n}_c(1,t/\Delta t_c,t_f) / S(1,t)\rangle \quad (7)$$

where $S(1,t) = S_o(r,t).X_s(v)/4\pi$ is the strength of the source per phase space and time unit and $\bar{n}_c(1,t/\Delta t_c,t_f)$ can be obtained solving the system : $-\partial/\partial t\, \emptyset^+ + K^+\emptyset^+ = S_D$ (8)

where

$$\emptyset^+ = \begin{bmatrix} \bar{n}_c(1,t/\Delta t_c,t_f) \\ \bar{C}_c^{1+}(1,t/\Delta t_c,t_f) \\ \cdot \\ \bar{C}_c^{6+}(1,t/\Delta t_c,t_f) \end{bmatrix} K^+ = \begin{bmatrix} H^+ & -M_1^+ & \ldots & -M_6^+ \\ -\lambda_1/v & \lambda_1/v & \ldots & 0 \\ \cdot & \cdot & \cdot\ \cdot & \cdot\ \cdot \\ -\lambda_6/v & \cdot & \cdot & -\lambda_6/v \end{bmatrix} S_D = \begin{bmatrix} \Sigma_D \\ 0 \\ \\ 0 \end{bmatrix} \quad (9)$$

$\bar{n}_c(1,t/\Delta t_c,t_f)$ is the average number of counts during the counting interval upon injection of one single neutron at $(1,t)$;$-H^+$ is the adjoint of the standard time independent Boltzmann transport operator; Σ_D is the macroscopic detector cross section; $M_i^+(i=1,2,..6)$ are the adjoints of the precursor production operators[1], and

$$\bar{C}_c^{i+}(1,t/\Delta t_c,t_f) = \int_t^{t_f} \exp(-\lambda_i(t'-t))\lambda_i.dt'.\ \bar{n}_c(1',t'/\Delta t_c,t_f) \quad (10)$$

Now, operating with $\partial/\partial Z|Z=1$ on equations (3), (4) and (6), one derives an expression for the variance of the number of counts. The important result is that using some tricks developped by Muñoz-Cobo and Pérez and which are generalized here, the variance can be computed in terms of $\bar{N}_c$ and $\bar{n}_c(1,t/\Delta t_c,t_f)$, this means that this method is free of the closure problems that arises in the Langevin technique.

In noise analysis one of the most common descriptors is the C.P.S.D.(Cross-Power Spectral Density) between two detectors. It has been shown by Verdú et allied that is possible to obtain two master equations which allow us to calculate the covariance function[8,9] ; this methodology has been generalized in this paper to include the delayed neutron precursors. From the covariance function it is possible to obtain the C.P.S.D., we have obtained in this way that the C.P.S.D. $G_{23}(w)$ between the detectors (2) and (3) in a multiplicative system with two different neutron sources S_1 and S_2 is given by the following new result :

$$G_{23}(w) = \int_{V_s} d^3\bar{r}\, dv\, \overline{\nu_p(\nu_p - 1)}.F(\bar{r},v).\, I^{(2)}(\bar{r},-w).I^{(3)}(\bar{r},w) \; + \sum_{i=1}^{2} \int_{V_s} d^3\bar{r}\, dv\, \overline{\nu_j(\nu_j - 1)}.S_j(\bar{r}).\, I^{(2)}_{sj}(\bar{r},-w).I^{(3)}_{sj}(\bar{r},w) \; + \sum_{i=1}^{6} \int_{V_s} d^3\bar{r}\, dv\, \overline{l_i\,\nu_p}\, F(\bar{r},v).\, (I_i^{(2)}(\bar{r},-w).I_i^{(3)}(\bar{r},w) + I_i^{(2)}(\bar{r},w).I_i^{(3)}(\bar{r},-w)) \tag{11}$$

The meaning of the terms appearing in equation 11 is quite obvious, the first one gives the contribution of the prompt neutrons of the system to the C.P.S.D., where $F(\bar{r},v)$ is the fission rate at point $\bar{r}$ per unit volume and per unit of velocity (or energy); $I^{(k)}(\bar{r},w)$ has the meaning of an spectral weigthed importance for the neutron produced at point $\bar{r}$, when the source of importance is the detector (k), and is given by :

$$I^{(k)}(\bar{r},w) = \int dv\, d\Omega\, X(v).\bar{n}_c(1,w)/4\pi \tag{12}$$

$\bar{n}_c(1,w)$ is the Fourier transform of the count rate at detector k, and $\overline{\nu_p(\nu_p - 1)} = \overline{\nu_p^2} - \bar{\nu}_p$ where $\bar{\nu}_p$ and $\overline{\nu_p^2}$ are the first and second moment of the number of prompt neutron released per fission. The second term appearing in (11) gives the contribution of the two kind of neutron sources to the C.P.S.D. where $S_j(\bar{r})$ is the source strength in source disintegration per unit volume for source j, $\bar{\nu}_j$ and $\overline{\nu_j^2}$ are the first and second moment of the number of neutrons released per source disintegration and the $I_{sj}^{(k)}$ are the spectral weigthed importances. The last term arises from the correlations between prompt and delayed neutrons in group i, l_i, i.e. $\overline{l_i\,\nu_p}$. Due to the time delay between prompt and delayed neutrons, a factor equal to $1/(\lambda_i - iw)$ appears inside $I_i^{(k)}(\bar{r},w)$.

2. RESULTS

Then we have applied this theory to interpret several experiments performed by J.T. Mihalczo, who measures the C.P.S.D. $G_{12}(w)$, $G_{13}(w)$ and $G_{23}(w)$, between three detectors (detector 1, was a fission chamber with a neutron source inside) located in and out of a tank containing a UO_2F_2 solution in water. In figure 1, the theoretical plot of the real part of $G_{23}(w)$ is shown (doted line), the points are the experimental values. We note that we have more dispersion in the experimental values at low frequencies. We have obtained that the theoretical predictions of the real and imaginary parts of the C.P.S.D. agree pretty with the experimental ones. This confirms the validity of the methodology that we have used in our calculations.

3. CONCLUSIONS

Several points are of special interest in this paper :

1. We present here a generalization of the stochastic transport theory including delayed neutrons, which make possible to compute the first moment (average), second moment and so on, of the number of counts of a neutron detector or of the number of neutrons per phase space unit in terms of lower moments.

The solution is free of closure problems, this fact do not occurs in Langevin method where closure problems always arise.

2. We have applied the previous methodology to compute in the frequency domain, the cross power spectral densities, $G_{12}(w)$, $G_{13}(w)$ and $G_{23}(w)$ of a physical system formed by a tank containing a UO_2F_2 solution in water and a Californium source, obtaining very good agreement between theoretical predictions and experimental data. In this way it is possible to get mo-

re information about the system that the obtained from the average values only.

Some of the ideas contained in this paper can be applied to another fields of physics[10] like energy and plasma physics.

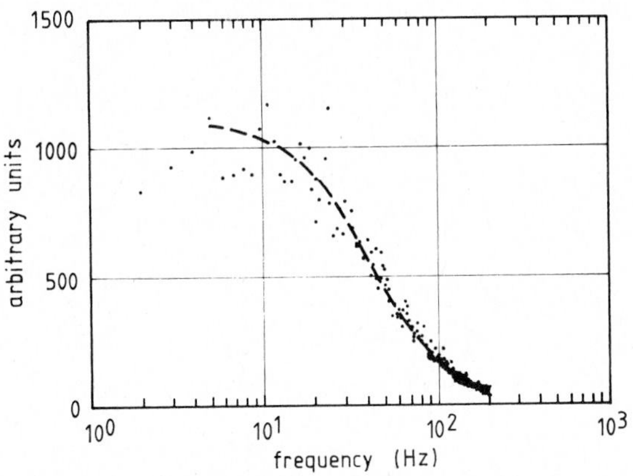

Plot of the part of $G_{23}(w)$ versus frequency

4. REFERENCES

1. R.P. Feynman, de Hoffman and R. Serber, Nuclear Energy 3, 64 (1956).

2. J.D. Orndoff, Nucl. Sci. Eng. 2, 450 (1957).

3. J.T. Mihalczo, R.C. Kryter, W.T. King Trans. Am. Nucl. Soc. 38, 359 (1981), also J.T. Mihalczo Nucl. Sci. Eng. 53, 393 (1974).

4. L. Pal, Il Nuovo Cimento, Supplemento VII, 25 (1958).

5. G.I. Bell, Nucl. Sci. Eng. 2, 450 (1957).

6. J.L. Muñoz-Cobo, R.B. Pērez, Trans. Am. Nucl. Soc. 39, 950 (1981).

7. J.L. Muñoz-Cobo, R.B. Pērez, G. Verdū, "Stochastic Neutron Transport Theory : Neutron Counting Statistics in Nuclear Assemblies". To be published.

8. G. Verdū, J.L. Muñoz-Cobo, R.B. Pērez, " Cross-Correlation between two detectors". To be published.

9. G. Verdū, J.L. Muñoz-Cobo, Energīa Nuclear, Nūmero 143, pāgina 211 (1983).

10. J.L. Muñoz-Cobo, G. Verdū, J. Bolta, " The multiplicity generating function of hadron-nucleus interactions at high energies ". Il Nuovo Cimento, V82A, N2 (1984).

11. Ziya Akcasu, Gerald S. Lellouche, and Louis M. Shotkin, Mathematical methods in nuclear reactor dynamics. Academic Press (1971).

NOISE IN PHYSICAL SYSTEMS AND 1/f NOISE - 1985
A. D'Amico and P. Mazzetti (editors)

NUMERICAL MODELING OF THE NOISE OF ONE DIMENSIONAL DEVICES

J.P. NOUGIER, A. MOATADID, J.C. VAISSIERE, D. GASQUET

Centre d'Electronique de Montpellier*, Université des Sciences et Techniques du Languedoc, 34060 Montpellier Cédex, France+.

Two numerical methods for modeling the noise of one dimensional devices are presented. We briefly recall the technique very recently developped, to be used when the electrical variable of interest is the electric field, and we extend it to the case where the electrical variable of interest is the local potential. Numerical results are in quite good agreement with analytical ones for sclc diodes.

1. INTRODUCTION :

Among the methods used for modeling the noise of devices[1], the impedance field method is one of the most powerfull. It needs the knowledge of the local noise sources, and of the impedance field, which describes the effect of the local noise sources at the electrodes. Unfortunately, the analytical determination of the impedance field $\nabla Z(x')$, at point x', is not easy, and can be achieved in only a few number of simple cases, corresponding to simplified hypothesis on a few number of very simple devices[1,2]. An alternative way is then to perform a numerical simulation of the impedance field method, for example by injecting a small a.c. current at a given point inside the device, and looking at the a.c. voltage induced at the electrodes[3]. The purpose of this paper is to give the mathematical formulation of two such numerical methods allowing the computation of $\nabla Z(x')$, thus allowing modeling the noise of almost every one dimensional device.

2. IMPEDANCE FIELD USING THE LOCAL ELECTRIC FIELD :

This method, very recently developped[4], is a numerical version of the transfer impedance method[2]. We shall briefly recall it here below. One writes the transport equations (conduction, continuity, Poisson, etc...), and eliminates the auxiliary variables, which gives an equation between the local field $E(x)$ and the local current $I(x)$ (which is constant in diodes, but not in transistors), namely :

$$f(E(x)) = I(x) \tag{1}$$

the solution of which gives the d.c. values $E_0(x)$, $V_0(x)$, $I_0(x)$. Now, by setting : $I(x)=I_0(x)+\delta I(x)e^{j\omega t}$, and $E(x)=E_0(x)+\delta E(x)e^{j\omega t}$, this equation, linearized, gives :

$$\hat{L}\,\delta E(x) = \delta I(x) \tag{2}$$

---

* Laboratoire associé au C.N.R.S., UA 391

+ This work was partially supported by the G.C.I.S., France.

where $\hat{L}$ is a linear operator. The method recently developped[4] consists in discretizing the length L of the device, leading to points x_j, and solving numerically eq. (2), which gives :

$$\sum_j a_{ij}\,\delta E_j = \delta I_i \tag{3}$$

where $\delta E_j = \delta E(x_j)$, $\delta I_j = \delta I(x_j)$. The coefficients a_{ij} depend on the operator $\hat{L}$, on the discretization procedure, and on the d.c. bias electrical parameters $E_0(x)$, etc... For a given d.c. bias, they are constant. A numerical solution of eq. (3) gives :

$$\delta E_k = \sum_j T_{kj}\,\delta I_j \tag{4}$$

The impedance field $\nabla Z_j = \nabla Z(x_j)$, at point x_j, is then[4], setting $\Delta x_j = x_j - x_{j-1}$:

$$\nabla Z_j = -\sum_k T_{kj}\,\Delta x_k \,/\, \Delta x_j \tag{5}$$

This method was checked on a p^+pp^+ single space charge limited current diode, with simplified hypothesis, where almost everything can be determined analytically[2]. Figure 1 shows that the analytical and the numerical values of $\nabla Z(x)$ coincide.

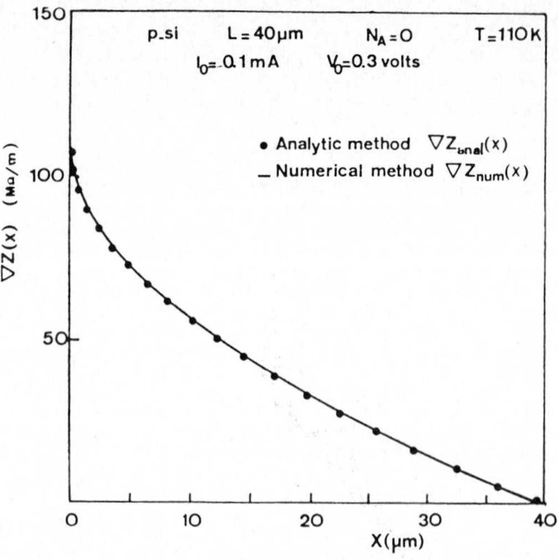

FIGURE 1
Comparison between the analytical (dots) and the numerical (full curve) values of the impedance field Z(x) along the device.

3. IMPEDANCE FIELD USING THE LOCAL VOLTAGE :

In some cases[5,6], the electrical variable of interest is the local voltage V(x) instead of the local electric field E(x). The transport equations, after elimination of the auxiliary variables, write then :

$$\varphi(V(x)) = I(x) \tag{6}$$

The d.c. bias, for a d.c. current $I_0(x)$, gives a local voltage $V_0(x)$. Now, by setting :

$V(x) = V_0(x) + \delta V(x)\,e^{j\omega t}$ and $I(x) = I_0(x) + \delta I(x)\,e^{j\omega t}$,

eq. (6), linearized, gives, for the first order terms :

$$\hat{M}\,\delta V(x) = \delta I(x) \tag{7}$$

where $\hat{M}$ is a linear operator, the Green function of which is $G(x, x')$, solution of :

$$\hat{M}\,G(x, x') = \delta(x-x') \tag{8}$$

when eq. (8) has been solved, the solution of eq. (7) is :

$$\delta V(x) = \int_0^L G(x, x')\,\delta I(x')\,dx' \tag{9}$$

that is, for x=L :

$$\delta V(L) = \int_0^L G(L, x')\,\delta I(x')\,dx' \tag{10}$$

The impedance field is then given by[5] :

$$\nabla Z(x') = G(L, x') \tag{11}$$

Now, dividing the length L of the device into n slices $\Delta x_j = x_j - x_{j-1}$ $(1 \leqslant j \leqslant n)$, eq. (7) writes, setting $\delta V_j = \delta V(x_j)$, etc... :

$$\sum_j m_{ij}\,\delta V_j = \delta I_i \tag{12}$$

that is :

$$[M]\,\delta\vec{V} = \delta\vec{I} \quad (13)$$

the solution of which is :

$$\delta V = [Z]\,\delta I \quad , \quad [Z] = [M]^{-1} \quad (14)$$

which can be written :

$$\delta V_k = \sum_j Z_{kj}\,\delta I_j \quad (15)$$

let $k = n$, then $x_n = L$ and $\delta V_n = \delta V(L)$. Eq. (15) gives :

$$\delta V_n = \delta V(L) = \sum_j Z_{nj}\,\delta I_j \quad (16)$$

The same discretization gives, for eq. (10) :

$$\delta V(L) = \sum_j G(L, x_j)\,\delta I_j\,\Delta x_j \quad (17)$$

Identifying eqs. (16) and (17) shows that :

$$Z_{nj} = G(L, x_j)\,\Delta x_j \quad (18)$$

This value of $G(L, x_j)$, carried into eq. (11), gives :

$$\nabla Z_j = G(L, x_j) = Z_{nj} / \Delta x_j \quad (19)$$

Tridiagonal matrixes :

When the discretization scheme leads to a tridiagonal matrix $[M]$, rather than computing $[Z] = [M]^{-1}$, it is easier to set, in eq. (12) : $\delta I_j = \delta_{j1}$, where δ_{j1} is the Kronecker symbol. Eq. (16) gives then :

$$\delta V_n = \sum_j Z_{nj}\,\delta_{j1} = Z_{n1} \quad (20)$$

which shows that the value Z_{n1}, needed in eq. (19), is the solution of eq. (16) when $I_j = 1$ for $j=1$, and $I_j=0$ for $j\neq 1$. The solution of eq. (12), which is usually obtained by a double sweep method (Choleski method), requires here a single sweep, since δV_k, $k<n$, is not needed.

a.c. impedance and noise :

Once the impedance field $\nabla Z(x')$ has been obtained, one gets the a.c. impedance Z, and the noise voltage S_V, as[7] :

$$Z = \int_0^L \nabla Z(x')\,dx'$$

$$S_V = \int_0^L A(x')\,K(x')\,|\nabla Z(x')|^2\,dx'$$

where $K(x')$ is the local noise source, $A(x')$ is the local cross section of the device. These eqs. can be written, using the discretized points x_j :

$$Z = \sum_j b_j\,\nabla Z_j \quad (21)$$

$$S_V = \sum_j b_j\,A_j\,K_j\,|\nabla Z_j|^2 \quad (22)$$

where $A_j = A(x_j)$, $K_j = K(x_j)$, and ∇Z_j is given by eq. (19).

4. CONCLUSION :

We have presented, in this paper, two efficient numerical methods for modeling the noise of one dimensional devices. These methods are, in fact, suited for determining the Green function of linear operators with boundary conditions. We intend, in the next future, to apply these methods for modeling the noise of devices where no analytical solutions are available.

REFERENCES :

1. J.P. Nougier, Proc. 6th Int. Conf. Noise in Physical Systems, Washington, 1981, NBS 614 397-405, Ed. Meijer, Mountain, Soulen, 1981.
2. K.M. Van Vliet, A. Friedmann, R.J. Zijlstra, A. Gisolf, A. Van der Ziel, J. Appl. Phys.,

46, 1804-1813 and 1814-1823 (1975).

3. B. Carnez, A. Cappy, R. Fauquembergue, E. Constant, G. Salmer, IEEE Trans. El. Dev., ED 28, 784 (1981).

4. J.P. Nougier, A. Moatadid, J.C. Vaissière, D. Gasquet, Proc. 4th Int. Conf. on Hot Electrons in Semiconductors, Innsbruck, 1985, to be published.

5. J.P. Nougier, J.C. Vaissière, D. Gasquet, Proc. 6th Int. Conf. Noise in Phys. Systems, Washington, 1981, NBS 614, p. 42-46, Ed. Meijer, Mountain, Soulen, 1981.

6. P. Hesto, J.C. Vaissière, D. Gasquet, R. Castagné, J.P. Nougier, J. Phys. (Paris), C7, 235-241 (1981).

7. W. Shockley, J.A. Copeland, R.P. James, in "Quantum Theory of Atoms, Molecules and the Solid State", P. Lowdin ed., Acad. Press, New York (1966).

NOISE IN PHYSICAL SYSTEMS AND 1/f NOISE - 1985
A. D'Amico and P. Mazzetti (editors)

EIGENFUNCTIONS AND ENERGY FOR TIME-RESCALED GAUSSIAN PROCESSES

Claudio MACCONE

Dipartimento di Matematica - Politecnico di Torino - Corso Duca degli Abruzzi, 24 - 10129 Torino - Italy

ABSTRACT. The Karhunen-Loève eigenfunction expansion is explicitely given for a class of Gaussian processes whose time is accelerated or decelerated according to an arbitrary law. The eigenfunctions are Bessel functions, and the eigenvalues are the zeros of them. The energy distribution of such time-rescaled processes can be studied analitically, and, in particular, this is done for the two cases where the time-rescaling law is power-like and exponential-like.

1. INTRODUCTION AND MAIN RESULT

Let $B(t)$ denote the ordinary Brownian motion (Wiener process) with mean zero, variance t and initial condition $B(o)=o$. The process

(1.1) $$X(t) = \int_0^t f(s)\, dB(s) \qquad (o \le t \le T)$$

where $f(t)$ is any continuous and non-negative time function, has been studied by this author in ref. [1]. There the Karhunen-Loève expansion

(1.2) $$X(t) = \sum_{n=1}^{\infty} Z_n\, \varphi_n(t) =$$

$$= \sqrt{f(t)\int_0^t f(s)\,ds}\; \sum_{n=1}^{\infty} Z_n\, N_n\, J_{\nu(t)}\left(\gamma_n \frac{\int_0^t f(s)\,ds}{\int_0^T f(s)\,ds}\right)$$

has been proven to hold good. In particular :

1) The order $\nu(t)$ of the Bessel functions $J_\nu(\ldots)$ is given by

(1.3) $$\nu(t) = \sqrt{-\frac{\chi^3(t)}{f^2(t)}\cdot\frac{d}{dt}\left[\frac{\chi'(t)}{f^2(t)}\right]}$$

where we have set

(1.4) $$\chi(t) = \sqrt{f(t)\int_0^t f(s)\,ds} \quad ;$$

2) The constants γ_n are the positive zeros of

(1.5) $$\chi'(T)\; J_{\nu(T)}(\gamma_n) + \chi(T)\cdot \left[\frac{f(T)\;\gamma_n}{\int_0^T f(s)\,ds}\, J'_{\nu(T)}(\gamma_n) + \frac{\partial J_{\nu(T)}(\gamma_n)}{\partial \nu}\,\nu'(T)\right] = o$$

that must be solved numerically to get γ_n ;

3) The normalization constants N_n follow from

(1.6) $$1 = N_n^2 \left[\int_0^T f(s)\,ds\right]^2 \cdot \int_0^1 x\, J^2_{\nu(x)}(\gamma_n x)\,dx \quad ;$$

4) The eigenvalues λ_n are, in terms of the γ_n,

(1.7) $$\lambda_n = \left[\int_0^T f(s)\,ds\right]^2 \frac{1}{(\gamma_n)^2} \quad ;$$

5) The Z_n are orthogonal Gaussian random variables with mean zero and variance λ_n.

The energy of $X(t)$ is the random variable

(1.8) $$I = \int_0^T X^2(t)\,dt = \sum_{n=1}^{\infty} Z_n^2$$

with characteristic function (Fourier transform)

(1.9) $$\Phi_I(\zeta) = \prod_{n=1}^{\infty} \Phi_{Z_n^2}(\zeta) = \prod_{n=1}^{\infty} \left(1 - 2i\lambda_n \zeta\right)^{-\frac{1}{2}} .$$

All the energy cumulants (whence the moments) :

$$\text{(1.10)}\quad K_n = 2^{n-1}(n-1)!\left[\int_0^T f(s)\,ds\right]^{2n}\sum_{m=1}^{\infty}\frac{1}{(\gamma_m)^{2n}}.$$

2. THE POWER-LIKE RESCALED PROCESS

In ref. [2] the author has studied the case

$$\text{(2.1)}\qquad f(t) = \frac{t^{H-\frac{1}{2}}}{\Gamma(H+\frac{1}{2})}\qquad (H>0)\ .$$

The resulting process $X(t) = B_{PH}(t)$ is

$$\text{(2.2)}\qquad B_{PH}(t) = \frac{1}{\sqrt{2H}\ \Gamma(H+\frac{1}{2})}\,B\left(t^{2H}\right)$$

and it is self-similar with exponent H, i.e .

$$\text{(2.3)}\qquad B_{PH}(ct) = c^H\,B_{PH}(t)\qquad (c = constant > 0)\,.$$

The Karhunen-Loève expansion (1.2) becomes

$$\text{(2.4)}\quad B_{PH}(t) = \frac{t^H\sqrt{H+\frac{1}{2}}}{\Gamma(H+\frac{3}{2})}\sum_{n=1}^{\infty} Z_n N_n J_\nu\left(\gamma_n\frac{t^{H+\frac{1}{2}}}{T^{H+\frac{1}{2}}}\right)$$

where :

1) The order ν of $J_\nu(\ldots)$ is the constant

$$\text{(2.5)}\qquad \nu = \frac{2H}{2H+1}\ ;$$

2) The γ_n are the positive zeros of :

$$\text{(2.6)}\qquad J_{\nu-1}(\gamma_n) = 0\ ;$$

3) The normalization constants N_n are given by

$$\text{(2.7)}\quad N_n = \frac{\sqrt{2}\,\gamma_n\,\Gamma(H+\frac{3}{2})}{T^{H+\frac{1}{2}}\sqrt{\gamma_n^2\left[J'_\nu(\gamma_n)\right]^2 + \left(\gamma_n^2-\nu^2\right)\left[J_\nu(\gamma_n)\right]^2}}\ ;$$

4) The eigenvalues λ_n depend on the γ_n as

$$\text{(2.8)}\qquad \lambda_n = \frac{T^{2H+1}}{\Gamma(H+\frac{3}{2})}\cdot\frac{1}{(\gamma_n)^2}\ ;$$

5) The Z_n are Gaussian random variables with mean zero and variance λ_n .

The energy of $B_{PH}(t)$ may be studied well :

$$\text{(2.9)}\qquad I_H = \int_0^T B_{PH}^2(t)\,dt = \sum_{n=1}^{\infty} Z_n^2\,.$$

Its characteristic function (Fourier transform) is derived from (1.9), and put in finite form :

$$\text{(2.10)}\quad \Phi_{I_H}(\zeta) = \left\{\Gamma(\nu)\left[\frac{T^{H+\frac{1}{2}}\sqrt{2\zeta i}}{2\,\Gamma(H+\frac{3}{2})}\right]^{1-\nu}\right\}^{-\frac{1}{2}}\cdot\left\{J_{\nu-1}\left(\frac{T^{H+\frac{1}{2}}\sqrt{2\zeta i}}{\Gamma(H+\frac{3}{2})}\right)\right\}^{-\frac{1}{2}}.$$

The energy cumulant to any order $n = 1,2,\ldots$

$$\text{(2.11)}\quad K_n = \frac{2^{n-1}\,T^{n(2H+1)}}{\left[\Gamma(H+\frac{3}{2})\right]^{2n}}\,(n-1)!\ \sigma_{\nu-1}^{(n)}$$

follows from the $\sigma_{\nu-1}^{(n)}$, as on page 502 of ref. [3]. The energy mean and variance are then :

$$\text{(2.12)}\quad K_1 = E\{I_H\} = \frac{1}{2H\,(2H+1)\left[\Gamma(H+\frac{1}{2})\right]^2}\,,$$

$$\text{(2.13)}\quad K_2 = \sigma_{I_H}^2 = \frac{1}{2H^2\,(2H+1)\,(4H+1)\left[\Gamma(H+\frac{1}{2})\right]^4}\,.$$

The energy skewness and kurtosis are :

$$\text{(2.14)}\quad \text{Skewness} = 2^{\frac{3}{2}}\,\frac{\sqrt{2H+1}\,\sqrt{4H+1}}{3H+1}\,,$$

$$\text{(2.15)}\quad \text{Kurtosis} = \frac{12\,(2H+1)\,(11H+3)}{(3H+1)\,(8H+3)}\,,$$

yielding a narrow energy peak for any $H > o$.

The ordinary Brownian motion case $H = \frac{1}{2}$ has energy distribution (B_n =Bernoulli numbers):

(2.16) $$K_n = T^{2n} \frac{(n-1)!}{(2n)!} 2^{3n-2}\left(2^{2n}-1\right)\left(-1\right)^{n+1} B_{2n},$$

(2.17) Mean energy $= \frac{T^2}{2}$,

(2.18) Energy variance $= \frac{T^4}{3}$,

(2.19) Energy skewness $= \frac{8}{5}\sqrt{3}$,

(2.20) Energy kurtosis $= \frac{1224}{105}$.

The diffusion partial differential equation

(2.21) $$\frac{\partial f(x,t)}{\partial t} = \frac{t^{2H-1}}{2\left[\Gamma(H+\frac{1}{2})\right]^2} \frac{\partial^2 f(x,t)}{\partial x^2}$$

with the initial condition $f(x,o) = \delta(x)$, is fulfilled by the probability density of $B_{PH}(t)$.

Finally, the first-passage time probability density and further results are also given in ref. [2].

3. THE EXPONENTIALLY RESCALED PROCESS

In ref. [4] the author has studied the case

(3.1) $$f(t) = e^{-\varrho t} \qquad (\varrho \geq o)$$

(3.2) $$X(t) = \frac{1}{\sqrt{2\varrho}} B\left(1-e^{-2\varrho t}\right).$$

It is related to the Ornstein-Uhlenbeck process via the diffusion equation (with $f(x,o) = \delta(x)$)

(3.3) $$e^{2\varrho t} \frac{\partial f(x,t)}{\partial t} = \frac{1}{2} \frac{\partial^2 f(x,t)}{\partial x^2}$$

as proved by Feller on page 336, Vol. 2, of ref. [5]. The Karhunen-Loève expansion of $X(t)$ reads

(3.4)

$$X(t) = \frac{e^{-\varrho t}\sqrt{e^{\varrho t}-1}}{\sqrt{\varrho}} \sum_{n=1}^{\infty} Z_n N_n J_{\nu(t)}\left(\gamma_n \frac{1-e^{-\varrho t}}{1-e^{-\varrho T}}\right)$$

where :

1) The order $\nu(t)$ of the Bessel functions is

(3.5) $$\nu(t) = \sqrt{\frac{3}{4} e^{2\varrho t} - \frac{3}{2} e^{\varrho t} + 1} \quad ;$$

2) The constants γ_n are the positive zeros of

(3.6) $$\frac{\varrho}{2} \cdot \frac{2e^{-\varrho T}-1}{1-e^{-\varrho T}} J_{\nu(T)}(\gamma_n) + \left[\frac{e^{-\varrho T}\gamma_n}{\frac{1}{\varrho}\left(1-e^{-\varrho T}\right)} J'_{\nu(T)}(\gamma_n) + \right.$$

$$\left. + \frac{\partial J_{\nu(T)}(\gamma_n)}{\partial \nu} \cdot \frac{3\varrho\, e^{\varrho T}\left(e^{\varrho T}-1\right)}{\sqrt{\frac{3}{4}e^{2\varrho T} - \frac{3}{2}e^{\varrho T} + 1}}\right] = o \quad ;$$

3) The normalization constants N_n read

(3.7) $$1 = N_n^2 \frac{1}{\varrho^2}\left(1-e^{-\varrho T}\right)^2 \int_0^1 x\left[J_{\nu((x))}(\gamma_n x)\right]^2 dx$$

where

(3.8) $$\nu((x)) = \frac{\sqrt{\frac{1}{4}+\frac{1}{2}\left(e^{-\varrho T}-1\right)x+\left(e^{-\varrho T}-1\right)^2 x^2}}{\left(e^{-\varrho T}-1\right)x+1} \quad ;$$

4) The eigenvalues λ_n are related to γ_n by

(3.9) $$\lambda_n = \frac{1}{\varrho^2}\left(1-e^{-\varrho T}\right)^2 \frac{1}{(\gamma_n)^2} \quad ;$$

5) The Z_n are orthogonal Gaussian random variables with mean zero and variance λ_n.

The energy of the process $X(t)$ is given by

(3.10) $$I = \int_0^T X^2(t)\, dt = \sum_{n=1}^{\infty} Z_n^2$$

and its characteristic function $\Phi_I(\zeta)$ reads

$$(3.11)\quad \Phi_I(\zeta) = \prod_{n=1}^{\infty}\left[1-\frac{2i}{\varrho^2}\left(1-e^{-\varrho T}\right)^2\frac{\zeta}{(\gamma_n)^2}\right]^{-1/2} .$$

All the energy cumulants K_n $(n=1,2,\ldots)$ are

$$(3.12)\quad K_n = 2^{n-1}(n-1)!\,\frac{\left(1-e^{-\varrho T}\right)^{2n}}{\varrho^{2n}}\sum_{m=1}^{\infty}\frac{1}{(\gamma_m)^{2n}} .$$

The process mean energy is :

$$(3.13)\quad E=\left\{I\right\} = \frac{2\varrho T+e^{-2\varrho T}-1}{4\varrho^2} ,$$

and similar results can be found.

Formulas for the first-passage time are also found in [4]. Numerical computations of all results are presently being performed.

ACKNOWLEDGMENTS are due to Prof. P. Mazzetti for his help.

REFERENCES

[1] C. Maccone, "Eigenfunctions and Energy for Time-Rescaled Gaussian Processes", Bollettino Unione Matematica Italiana, Series 6, Vol. 3, Section A, (1984), pages 213-219.

[2] C. Maccone, "The Time-Rescaled Brownian Motion B(t^{2H})", Bollettino Unione Matematica Italiana, series 6, Vol. 4, Section C, (1985), pages 363-378.

[3] G. N. Watson, Theory of Bessel Functions, Cambridge University Press, 1966.

[4] C. Maccone, "Sur un mouvement Brownien dont le temps est retardé exponentiellement", Publications de l'Institut de Statistique de l'Université de Paris, Vol. 30 (1985), pages 61-72.

[5] W. Feller, An Introduction to Probability Theory and Its Applications, Wiley, 1971.

NOISE IN PHYSICAL SYSTEMS AND 1/f NOISE - 1985
A. D'Amico and P. Mazzetti (editors)

SOME EXPERIMENTAL RESULTS ON FIRST PASSAGE TIME PROBLEM FOR GAUSSIAN PROCESS

Hector PEREZ Tsutomu KAWABATA T. MIMAKI

University of Electro-Communications, Chofu, Tokyo, 182, Japan

The first passage time probability density function of a random process is experimentally studied for a Gaussian random process having seventh order Butterworth type power spectral density.

1. INTRODUCTION

First passage time problem for stationary random process is one of the most important unsolved problems in noise theory, where the first passage time is a time interval between a point at which the random process starts from a fixed level and the subsequent time point at which it firstly reaches some given boundary. The applications of the problem are found in various fields, such as solid state physics, communication theory and biophysics. In this paper we treat only the case when the boundary is constant. The case when the boundary changes linearly in time is treated in the a accompaying paper in this conference. In this problem the probability density function (pdf) of the first passage time intervals is the essential quantity.

For Gaussian random process Rainal[1], Ricciardi and Sato[2] provided approximation formulas for the first passage time probability density function.

In the present paper we have experimentally determined the first passage time pdf in the case of a constant boundary for Gaussian process having power spectra.

$$W(f)=1/[1+(f/f_c)^{14}] \tag{1}$$

and compared it with theoretical approximations, where f_c denotes the cut-off frequency.

2. FUNDAMENTAL IDENTITIES

As previously described the first passage time pdf is not determined theoretically. This section reviews the fundamental identities obtained by Rainal[1] for the first passage time pdf of a Gaussian process with zero mean and unit variance.

For fixed levels I_1 and I_2 ($I_2<I_1$), we denote $Q(\tau,I_1,I_2)$ as the probability density function that a crossing of level I_2 occurs between $t+\tau$ and $t+\tau+d\tau$ given a crossing of level I_1 at time t. This pdf is a good approximation to the first passage time pdf $P(\tau,I_1,I_2)$ for small values of τ.

Because the amplitud distribution of the random process is symmetrical about its mean value of zero, the case $I_1<I_2$ can always be converted to the case under discussion by considering a random process $-x(t)$.

In order to compute $Q(\tau,I_1,I_2)$, we define the auxiliary probability density functions $Q^{+-}(\tau,I_1,I_2)$ and $Q^{--}(\tau,I_1,I_2)$ as follows.

(i) $Q^{+-}(\tau,I_1,I_2)$: The conditional pdf that a downward crossing of the level I_2

occurs between $t+\tau$ and $t+\tau+d\tau$ given an upward crossing of the level I_1 at time t.

(ii) $Q^{--}(\tau,I_1,I_2)$: The conditional pdf that a downward crossing of the level I_2 occurs between $t+\tau$ and $t+\tau+d\tau$ given a downward crossing of level I_1 at time t.

Rice[3] derived the expression for $Q^{+-}(\tau,I_1,I_2)$ for the case when $I_1=I_2$ and Rainal[1] generalized to the case when $I_1>I_2$ as follows,

$$Q^{+-}(\tau,I_1,I_2) = M_{22}\beta^{1/2}(1-m^2)^{-3/2}$$
$$\times \exp[(I_1^2+I_2^2-2mI_1I_2)/(2(1-m^2))]$$
$$\times J(r,h_3,k_3), \qquad (2)$$

where

$$h_3=m'(mI_1-I_2)/[M_{22}(1-m^2)]^{1/2},$$

$$k_3=m'(I_1-mI_2)/[M_{22}(1-m^2)]^{1/2},$$

$$J(r,h_3,k_3) = [1/2\pi(1-m^2]^{1/2}$$
$$\int_{h_3}^{\infty}dx\int_{k_3}^{\infty}dy(x-h_3)(y-k_3)\exp(z), \qquad (3)$$

$$z=(x^2+y^2-2xy)/2(1-r^2)^{1/2},$$

$$r=(m''(1-m^2)+mm'^2)/(1-m^2)-m'^2),$$

and

$$M_{22}=m(0)(1-m^2)-m'^2.$$

The primes denote differentiations and m(t) is the autocorrelation function of the original Gaussian process.

$Q^{--}(\tau,I_1,I_2)$ is obtained by slightly modifying $Q^{+-}(\tau,I_1,I_2)$. Finally $Q(\tau,I_1,I_2)$ is given by:

$$Q(\tau,I_1,I_2) = Q^{+-}(\tau,I_1,I_2) + Q^{--}(\tau,I_1,I_2). \qquad (4)$$

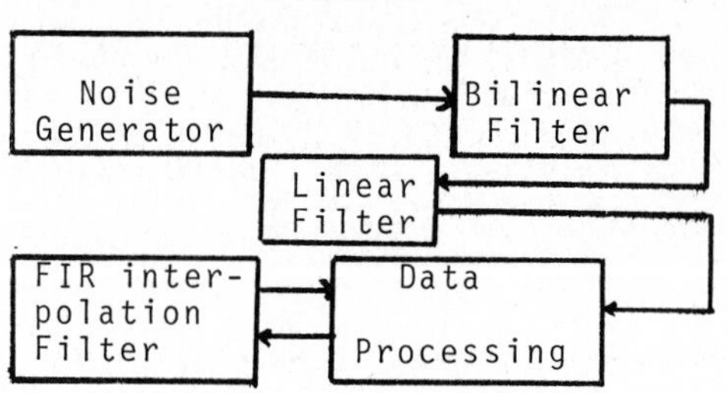

FIGURE 1
Experimental system.

As is noted above, the pdf $Q(\tau,I_1,I_2)$ is an approximation to the first passage time pdf $P(\tau,I_1,I_2)$ for small values of τ.

Moreover we denote $P^{+-}(\tau,I_1,I_2)$ and $P^{--}(\tau,I_1,I_2)$ as first passage time pdfs defined for the same conditions used for pdfs $Q^{+-}(\tau,I_1,I_2)$ and $Q^{--}(\tau,I_1,I_2)$ which are approximations to the pdfs $P^{+-}(\tau,I_1,I_2)$ and $P^{--}(\tau,I_1,I_2)$ for small values of τ. Of course the following equality holds

$$P(\tau,I_1,I_2) = P^{+-}(\tau,I_1,I_2) + P^{--}(\tau,I_1,I_2). \qquad (5)$$

3. EXPERIMENTAL SETUP

The experimental system shown in Fig. 1 is constructed by software running in a super minicomputer, and consist of two parts, that is, Gaussian process generator and measuring system.

(a) Gaussian Process Generator.

The random process is generated by means of a binary random sequence called M-sequence of order 521 using the algorithm by Fushimi and Tezuka[9] which provide a sufficiently independent and uniformly distributed binary sequence.

To generate a Gaussian process from above process the Polar Method[10] is employed.

The power spectrum given by eq. (1) was realized cascading two lowpass filters. The first was anti-aliasing filter designed using the bilinear transform method, the cutoff frequency of which was chosen to be $2\pi/2.4$.

The another filter was designed using the impulse invariance method with a cut-off frequency of $2\pi/8.0$.

(b) Measuring System.

In order to minutely determine the crossing time points, an interpolation filter is used, which was designed using the Lagrange interpolation polynomial of sixth degree. To reduce the computing time this filter is used only when the crossing of levels I_1 or I_2 is detected. The measuring algorithm consist of two parts. One of them computes experimentally $P^{+-}(\tau,I_1,I_2)$ and the other $P^{--}(\tau,I_1,I_2)$.

Each of them was determined using 5×10^5 I_2-level crossings. The first passage time pdf is obtained by summation of $P^{+-}(\tau,I_1,I_2)$ and $P^{--}(\tau,I_1,I_2)$.

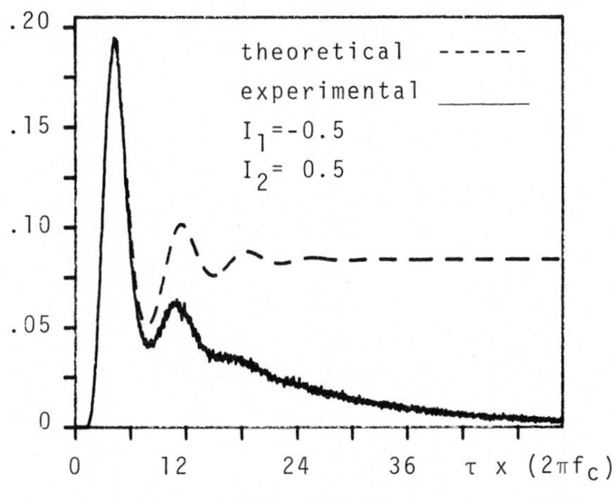

FIGURE 2
Experimental first passage time pdf $P^{+-}(\tau,I_1,I_2)$

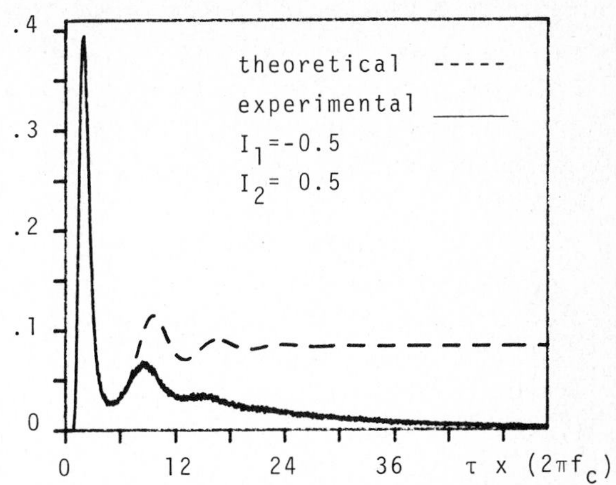

FIGURE 3
Experimental first passage time pdf $P^{--}(\tau,I_1,I_2)$

4. EXPERIMENTAL RESULTS

Fig. 2 shows the measured first passage time pdf $P^{+-}(\tau,I_1,I_2)$ for a Gaussian process having a power spectrum given by eq. (1) and I_1=-0.5 and I_2=0.5, where the theoretical approximation given by eq. (2) is also shown.

Figs. 3 and 4 show, respectively, $P^{--}(\tau,I_1,I_2)$ and $P(\tau,I_1,I_2)$ together with theoretical approximations.

These figures show that eq. (4) gives a very good approximation of the first passage time pdf within a range $0<\tau<10$.

5. CONCLUSIONS

As expected, the coincidence between the experimental curves and $Q^{+-}(\tau,I_1,I_2)$, $Q^{--}(\ ,I_1,I_2)$ and $Q(\ ,I_1,I_2)$ is fairly well for $0<\tau<10$. The remarkable feature of the experimental curves is the multi-peak property found firstly in the level crossing problem[6] which is the special case of the first passage time problem, that is, the case when the starting level is equal to the destination boundary.

The interpretation of this property is given in reference[7] using a random excursion model. The main point of this interpretation depends on the fact that the band limited noise has some kind of regularity.

6. ACKNOWLEDGEMENTS

We are grateful to Prof. S. Sato of Osaka University for helpful discussions and to Prof. M. Tanabe who provided us with his computer program to compute Rainal's expression.

REFERENCES

1. A.J. Rainal, BSTJ, Vol. 47 (1968) 2239.
2. L.M. Ricciardi and S. Sato, IEEE Trans. Inform. Theory, Vol. IT-29 (1983) 454.
3. S.O. Rice, BSTJ, Vol. 37 (1958) 581.
4. D. Slepian, Math. Stat., Vol. 32 (1961) 610.
5. T. Mimaki, J. of Appl. Physics, Vol. 44 (1963) 477.
6. T. Mimaki, M. Tanabe and D. Wolf, IEEE Trans. Inform. Theory, Vol. IT-27 (1981) 525.
7. T. Mimaki, M. Tanabe and H. Sato, Signal Processing, Vol. 7 (1984) 251.
8. R. Crochiere and L. Rabiner, Proc. of IEEE, Vol.69 (1981) 300.
9. M. Fushimi and S. Tezuka, CACM, Vol. 26 (1983) 516.
10. D.E. Knuth, 'The Art of Computer Programming' Vol. 2 (1969) 104.

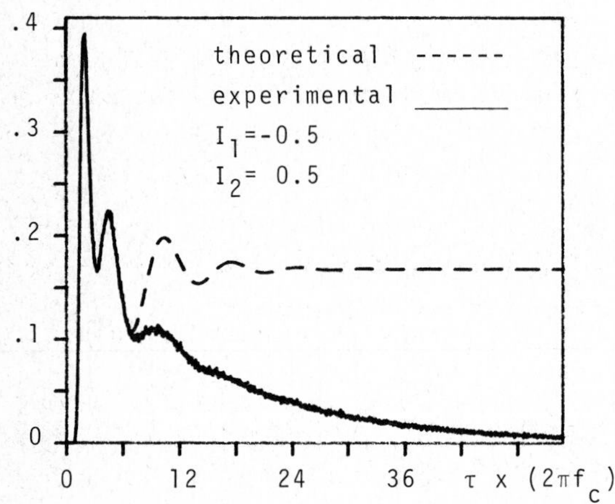

FIGURE 4
Experimental first passage time pdf $P(\tau, I_1, I_2)$

NOISE IN PHYSICAL SYSTEMS AND 1/f NOISE - 1985
A. D'Amico and P. Mazzetti (editors)

COMPUTER SIMULATION OF BARKHAUSEN JUMP FIELD DISTRIBUTION

T. MIMAKI

University of Electro-Communications, Chofu, Tokyo, 182 Japan

The distribution of the starting field of individual reproducible Barkhausen jumps measured on thin iron whiskers during repeated hysteresis cycles is reported to fit to Gaussian profile. Also the dependence of the mean jump field on the sweep rate is said to be approximately proportional to the logarithm of sweep rate. The experimental situation is considered to correspond to a modified first passage time problem. Since this problem is not solved analytically, the computer simulation is performed. The result shows that the distribution is not necessarily Gaussian and that the mean jump field may be proportional to the logarithm of sweep rate.

1. INTRODUCTION

Heiden et al[1] investigated the behavior of the hysteresis loop of iron whiskers, in which most of the magnetization reversal takes place in a few large Barkhausen jumps. These jumps occur to a considerable extent in a reproducible manner during repeated hysteresis cycles. The starting field H_{si} for a given jump i however is observed to fluctuate around a mean value $\langle H_{si} \rangle$. Their main results are that the distribution of the starting field H_{si} fits to Gaussian profile, and that the dependence of the mean jump field $\langle H_{si} \rangle$ on the sweep rate dH/dt is approximately equal to A+B*log(dH/dt), where A and B are constants.

On the other hand, the first passage time problem is one of the most important unsolved problem in the field of noise theory. The problem is to investigate the distribution of the time intervals between time point at which the random process starts from some fixed level and the subsequent time point at which it firstly reaches some given boundary.

The forementioned Heiden's experimental situation corresponds to the modified first passage time problem where the boundary changes linearly in time and the random process starts from arbitrary level, not from fixed level. Since this modified problem as well as original one is not solved analytically, we have done computer simulation and compared the results with experiment.

2. SIMULATION

In the simulation, a Gaussian random noise having 7-th order Butterworth low-pass power spectrum is used, which is generated by smoothing a shift-register sequence. This spectrum is selected as

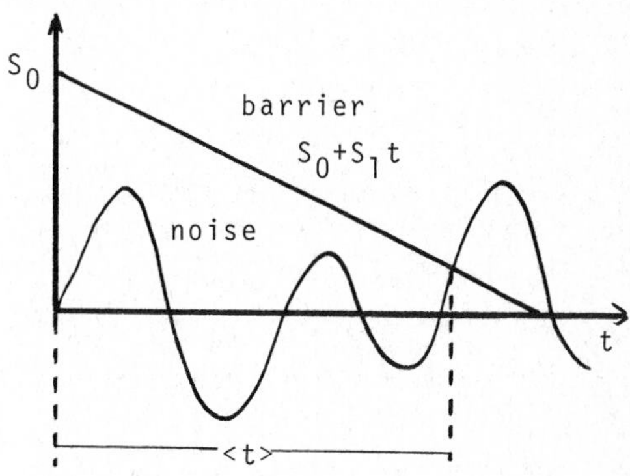

FIGURE 1
An example of the sample.

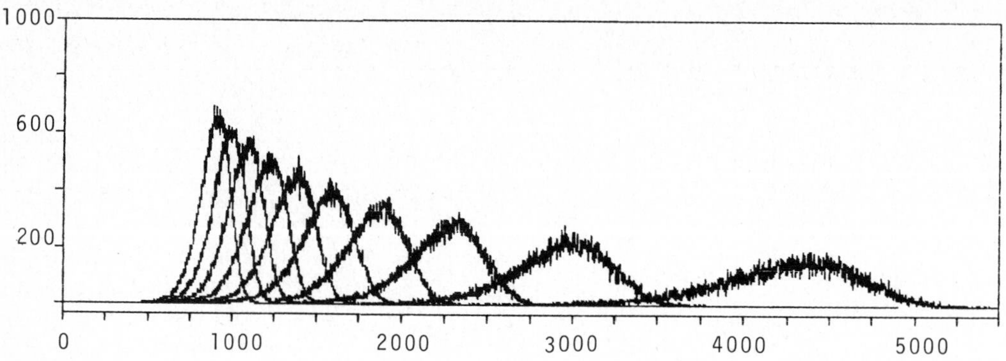

FIGURE 2
The distribution of the first passage times. The slopes of barriers becomes steeper from the right to left.

approximation of the sharp-cut-off band-limited white noise.

The procedure of the simulation is as follows: Initially, the boundary is set to be very hight level and then it is linearly decreased in time. The distribution of the time points at which the noise firstly crosses the boundary is recorded. Fig. 1 shows an example of the sample.

3. RESULTS

Fig. 2 shows a set of the distribution densities where the slope of the boundary is used as parameter. Each curve is determined by about seventy thousand crossings. When the slope is not so steep, the situation is similar to the level-crossing problem of the Gaussian process. Therefore, the profile of the distribution is somewhat predictable by referring to the results of the level crossing problem[2]. When the slope is relatively steep (leftside curves in Fig. 2), the profile of the distribution looks like Gaussian, which is the suggested form by Cobb[3] for the level-crossing problem of noise. On the contrary, it is obviously asymmetry distribution for gentle slope (rightside curves in Fig. 2). Since in this case distribution becomes broad, the exact profile of it is still vague, so that it seems to be necessary to increase the number of data at least to the order of millon.

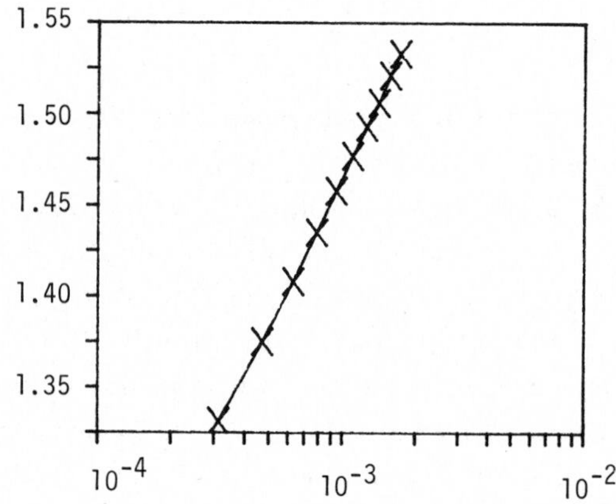

FIGURE 3
Mean values of the barrier height versus logarithm of slope of the barrier.

Fig. 3 shows average values of the distribution of Fig. 2 versus the logarithm of slope of boundary. The average

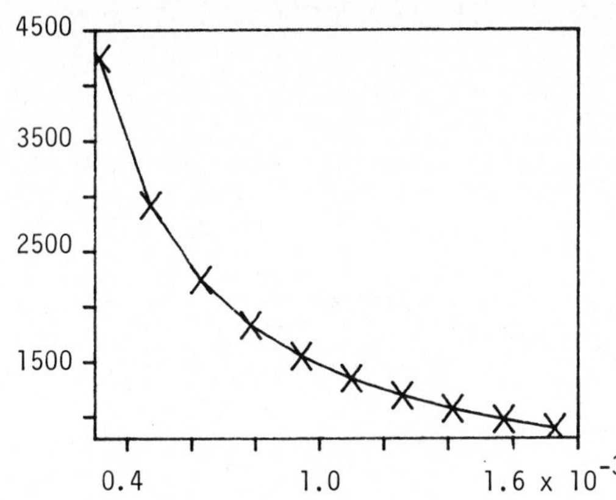

FIGURE 4
Mean values of the first passage times versus slope of the barrier.

value of the distribution and the slope correspond to a mean field value $<H_{si}>$ and the sweep rate dH/dt, respectively. From this figure, it seems that the observed relation A+B*log(dH/dt) may hold. However, since the range of slope used for simulation is not so wide as Heiden's experiment, rigid conclusion is difficult to derived from Fig. 3.

Fig. 4 shows the mean values of the first passage times versus the slope of barrier, from which the mean values are considered to be proportional to the reciprocal of the slope.

Although the simulation is also performed for the noise having bandpass power spectrum, we can not find out any difference from the lowpass case.

4. DISCUSSIONS

Ricciardi and Sato[4,5] derived the first passage time probability density function $g[S(t),t/x_0]$ for one dimensional stationary Gaussian process X(t) conditional on $X(0)=x_0$ through a time dependent boundary S(t) in the form of a series expansion. The first term $W_1(t/x_0)$ of the expansion is explicitly given and can be used as an approximation of $g[S(t),t/x_0)]$. In our case, $S(t)=S_0+S_1t$ with $S_1<0$ and the condition $X(0)=x_0$ is omited. If the second derivative of the covariance of the noise evalued at t=0 is much smaller than one, they derived an approximation for large t,

$$W_1(t/x_0) \sim -(S_1/(2\pi)^{1/2}) \times \exp[-(S_0+S_1t)^2/2].$$

The expression suggest that for large t $W_1(t/x_0)$ becomes independent of the starting point x_0 and that the distribution approach to the Gaussian profile. Furthermore, the maximum of $W_1(t/x_0)$ occurs at $t=S_0/-S_1$, which is close to the mean time of the distribution. If this situation can be applied to our case, the distribution of starting field of Barkhausen jumps may be approximated by Gaussian distribution. The mean values of the first passage time is inversely proportional to slope. In fact, the above relation holds as is shown in Fig. 4.

Figs. 5(a) to (c) show the plots of the first term $W_1(t/x_0)$ of $g[S(t),t/x_0]$ for $S(t)=S_0+S_1t$ with various values of S_1. For relatively rapidly changing barrier, that is, large S_1 the curve is simple and almost symmetrical about its peak, while for slowly changing barrier it becomes asymmetrical and shows the complex shape for small values of τ.

Fig. 5(c) contains the extended portion for small values of τ.
Neither Heiden's experiment nor our simulation does not reveal this strange structure for small τ. In order to experimentally obtain the structure sug-

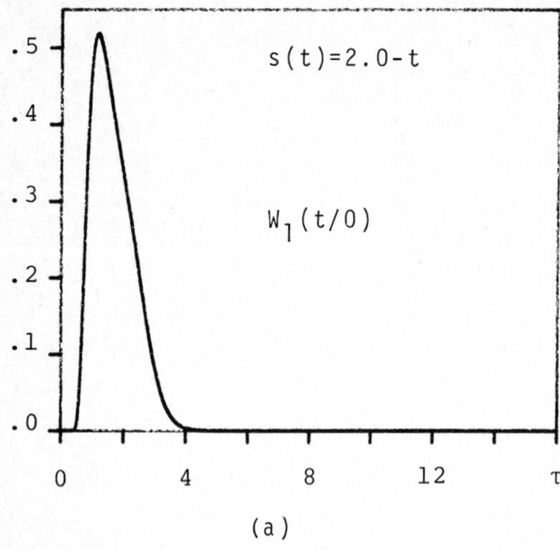

(a)

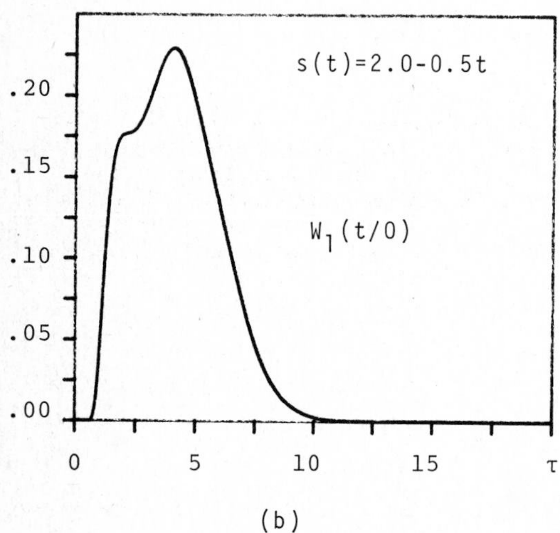

(b)

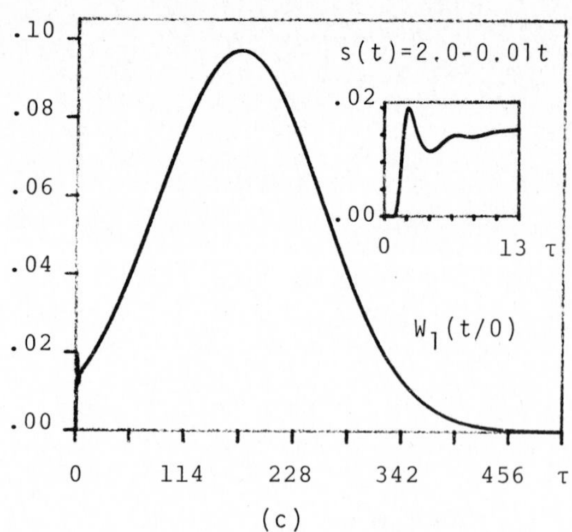

(c)

FIGURE 5
The first term $W_1(t/x_0)$ of $g[S(t),t/x_0]$ for $S(t)=S_0+S_1t$. Seventh order Butterworth spectrum.

gested in Figs. 5, it seems to be necessary to increase the number of data up to several millions.

In summary, the distribution of the first passage time for linearly decreasing barrier is obtained by computer simulation for Gaussian random process having 7th order Butterworth spectrum. The profile of the distribution seems to be Gaussian for relatively rapidly varying barrier, although it is different for the slowly varying barrier. The expression given by Ricciardi and Sato suggest that the distribution may have some structure for small τ. Up to now, however, it seems to be difficult to obtain this structure by simulation because of the very long computing time.

5. ACKNOWLEDGEMENTS

The autor is grateful to Prof. S. Sato of Osaka University for helpful discussions.

REFERENCES

1. C. Heiden and H. Rogalla, J. of Magnetism and Magnetic Materials, Vol. 26 (1982) 275.
2. T. Mimaki and T. Munakata, IEEE Trans. Inform. Theory, Vol. IT-24 (1978) 515.
3. S. Cobb, IEEE Trans. Inform. Theory, Vol. IT-11 (1965) 220.
4. L.M. Ricciardi and S. Sato, IEEE Trans. Inform. Theory, Vol. IT-29 (1983) 454.
5. L.M. Ricciardi and S. Sato, to appear in Signal Processing.

NOISE IN PHYSICAL SYSTEMS AND 1/f NOISE - 1985
A. D'Amico and P. Mazzetti (editors)

BISTABILITY DRIVEN BY COLORED NOISE: THEORY AND EXPERIMENT

Peter HANGGI

Polytechnic Institute of New York, Brooklyn, NY 11201, USA

Frank MOSS

University of Missouri at St. Louis, Saint Louis, MO 63121, USA

P. V. E. McCLINTOCK

University of Lancaster, Lancaster, LA1 4YB, UK

We consider a nonequilibrium, bistable system exposed to external, additive, Gaussian, colored noise. A new approximate theory, which results in a nonlinear Fokker-Planck type equation capable of accurately modeling the long time dynamics is constructed, and the mean sojourn time (MST) (in a potential well) is evaluated in the limits of weak noise and small correlation time. An electronic circuit which accurately mimics the bistable system has been constructed. Experimental measurements on this circuit are in good quantitative agreement with the predictions of the theory, which indicate an exponential increase of the MST with increasing noise correlation time. This observed exponential dependence is the clear signature that an Arrhenius law governs the process of escape from the well.

Recent years have witnessed an increasing interest in the stochastic behaviour of non-equilibrium systems coupled to a fluctuating external environment.[1] The influence of these fluctuations on nonlinear systems is not trivial.[2] For example, it has recently been realized that the usual Fokker-Planck descriptions are generally beset with difficulties for colored noise problems (or systems with state space dimension $d > 2$).[3]

We have studied the bistable system defined by

$$\dot{x} = ax - bx^3 + \xi(t) \qquad (1)$$

with the time correlated noise

$$\langle\xi(t)\ \xi(s)\rangle = \frac{D}{\tau} \exp - |t-s|/\tau \qquad (2)$$

both experimentally, as described below, and theoretically, using a new approximate procedure suggested by previous work.[4] The rate of change of the single event probability $p_t(x)$ of the bistable, colored dynamics obeys the exact equation[5]

$$\dot{p}_t(x) = -\frac{\partial}{\partial x}(ax - bx^3)p_t + \frac{D}{\tau}\frac{\partial^2}{\partial x^2}\int_0^t ds \exp - (t-s)/\tau \cdot \langle\delta(x(t) - x)\frac{\delta x(t)}{\delta\xi(s)}\rangle \qquad (3)$$

where the functional derivative $(\delta x(t)/\delta\xi(s))$ obeys the integral equation

$$\frac{\delta x(t)}{\delta\xi(s)} = \Theta(t-s)\{1 + \int_0^t d\tau\ (a - 3bx^2(\tau))\ \frac{\delta x(\tau)}{\delta\xi(s)}\} . \tag{4}$$

Upon repeated use of (4) we can rewrite the average in (3) in the form

$$\langle\delta(x(t) - x)[1 + \int_0^t d\tau_1(a - 3bx^2(\tau_1))\ [1 + \int_s^{\tau_1} d\tau_2(a - 3bx^2(\tau_2))[1 + \ldots]]\ldots]\rangle. \tag{5}$$

If we now neglect transients $(t\to\infty)$, and make use of a decoupling ansatz, thereby neglecting induced correlations among the successive functionals in (5), we end up with an effective, "nonlinear" Fokker-Planck dynamics

$$\dot{p}_t(x) = -\frac{\partial}{\partial x}(ax - bx^3)p_t(x) + \frac{D}{[1 + \tau(3b\langle x^2\rangle - a)]}\frac{\partial^2}{\partial x^2}p_t(x), \tag{6}$$

where $\langle x^2\rangle$ depends itself on the (unknown) solution $p_t(x)$. This causes no trouble, because the stationary 2-nd moment varies only weakly with D in this bistable. (Indeed, $\langle x^2\rangle \simeq a/b$, which insures that the effective diffusion constant in (6) is always > 0). This approximation will clearly work best for small τ and converges for $\tau \to o$ to the proper white noise limit (the Smoluchowski equation).

We have used this theory to calculate the mean sojourn time (MST), a quantity which is sensitive to the wings of the stationary distribution.[5] An example (measured) time trajectory is shown in Fig. 1(b), where a clear time separation between the time spent in the well (sojourn time) and the time in transit between wells is evident. Thus transitions between wells can be considered as independent random events. The sojourn times are thus exponentially distributed

$$p(T) = (1/\langle T\rangle)\exp-T/\langle T\rangle \tag{7}$$

The MST now depends only on the ratio of the stationary density evaluated at either of its maxima at $x_{1/2} = \pm\ (a/b)^{1/2}$ to the unstable point at x = 0,

$$\langle T\rangle = p_{st}(X_{1/2})/p_{st}(x = 0) \tag{8}$$

Because (6) accurately models the long-time dynamics of x(t), one obtains for the leading behavior of the MST[4] (by the method of steepest descent)

$$\langle T\rangle = \frac{\pi}{a\sqrt{2}}\exp\frac{\Delta\phi(\tau)}{D} \tag{9}$$

where

$$\Delta\phi = (a^2/4b)[1 + \tau(3b\langle s^2\rangle - a)]. \tag{10}$$

thus the approximation scheme in (6) predicts an exponential increase of $\langle T\rangle$ with noise correlation time.

We found (9) and (10) to be in quite good agreement with measurements of ⟨T⟩ using an analog electronic circuit similar to one used previously to study noise induced phase transitions[7] and other bistable systems.[8] The circuit is shown in Fig. 1(a). It was designed to obey (to an accuracy of a few percent) the integral of equation (1) with a = b = 1, and with an integrating time constant which scales the time and noise intensity in the experiment as explained in Ref. 7. A measured time trajectory x(t) is shown in Fig. 1(b). Typically 5000 such trajectories were digitized by the computer from which the mean time between zero crossings (the MST) and the distribution of these times were assembled.

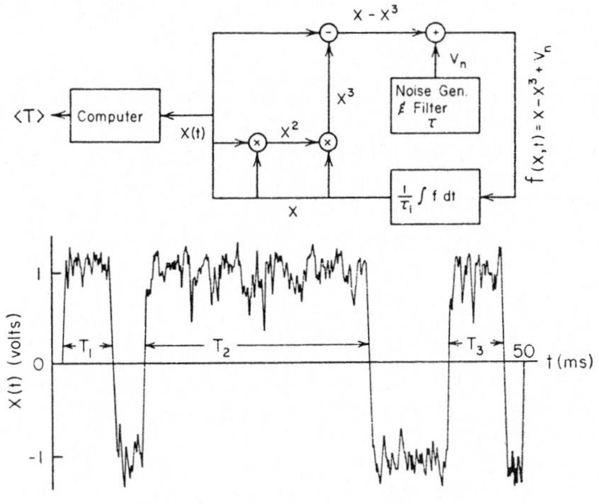

Figure 1. (a) the circuit, and (b) an example trajectory

Measurements of ⟨T⟩ were obtained for four different noise intensities, held constant as the noise correlation time was varied.

Equation (9) was then used to obtain the Arrhenius factor ($\Delta\phi$), and the results are shown in Fig. 2. The theory, as given by (9) and (10) is represented by the solid lines for the two values of $\langle x^2 \rangle$ shown. Arrhenius behavior (straight lines on this plot) is clearly indicated by the agreement evident over a surprisingly wide range of τ.

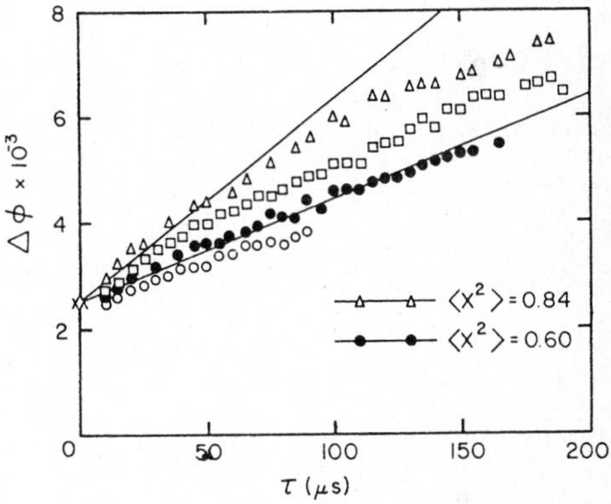

Figure 2. Data derived from MST measurements via Eq. (9). Triangles, squares, solid and open circles are dimensionless noise intensities: 0.21, 0.15, 0.11 and 0.083 respectively.

1. For a recent review, see Fluctuations and Sensitivity in Nonequilibrium Systems, W. Horsthemke and D. K. Kondepudi, eds. (Springer, Berlin, 1984).

2. P. Hanggi and H. Thomas, Phys. Rep. 88, 207 (1982).

3. R. Graham and T. Tel, J. Stat. Phys. 35, 729 (1984).

4. P. Hanggi, F. Marchesoni and P. Grigolini, Z. Physik B56, 333 (1984).

5. P. Hanggi, in Stochastic Processes Applied to Physics, L. Pesquera and M. Rodriquez, (World Scientific, 1985).

6. P. Hanggi, T. Mroczkowski, F. Moss and P. V. E. McClintock, Phys. Rev. A32 695 (1985).

7. J. Smythe, F. Moss and P. V. E. McClintock, Phys. Rev. Lett. 51, 1062 (1983).

8. J. D. Robinson, F. Moss and P. V. E. McClintock, J. Phys. A18, L89 (1985).

NOISE IN PHYSICAL SYSTEMS AND 1/f NOISE - 1985
A. D'Amico and P. Mazzetti (editors)

BRANCH SELECTIVITY AT A PITCHFORK BIFURCATION IN THE PRESENCE OF NOISE

Frank MOSS

University of Missouri at St. Louis, Saint Louis, MO 63121, USA

D. K. KONDEPUDI

Center for Studies in Statistical Mechanics, University of Texas at Austin, Austin TX 78712, USA

P. V. E. McCLINTOCK

University of Lancaster, Lancaster, LA1 4YB, UK

The problem of branch (or state) selection, initiated by a slowly changing external parameter in a bifurcating system has wide applications to far-from-equilibrium processes in Physics, Chemistry and Biology. In macroscopic systems such as laser and fluid dynamic transitions, bifurcating chemical reactions and possibly some selection processes in certain biological applications, for example, state selection always takes place in the presence of noise and a nonzero (though possibly quite small) symmetry breaking field. We consider a quite general representation of a transition from a monostable to a bistable state, driven by a linearly time varying control parameter in the presence of additive, Gaussian, white noise with a nonzero (though small compared to the noise variance) mean. We have constructed an electronic analog of this system, and developed methods for measuring the time evolution of its statistical densities (the first such measurements on a physical system). The probabilities of branch selection are extracted from these measurements and are compared to the predictions of a recent theory with applications to the problem of prebiotic selection of molecular chirality.

During the past decade a large interest has developed in studies on nonequilibrium, macroscopic systems coupled to fluctuating environments.[1] Certainly, one reason for this interest is that many such systems exhibit a remarkable sensitivity at some point in their development to one or a few fluctuations which cause the selection of a single, final evolutionary path out of many very different possibilities.[2] The effects of this "external" noise on nonlinear systems are diverse, ranging from "noise induced" phase transitions[3,4] to the quite practical problems of the operation of switches in a noisy environment.[5] In this paper we present measurements of the time evolution of the statistical density in a model bistable system,[6] along with the outline of a simple, physically motivated theory of state selection at a noisy bifurcation.[7,8] Other bistable systems have been studied using similar techniques.[9] Quite recently, the first measurements of two dimensional statistical densities for a bistable system exposed to colored noise have been reported.[10]

We consider a bistable representation which previous arguments[8] have demonstrated to be generic to the process of branch selection in the presence of noise:

$$dx/dt = -x^3 + \lambda(t)x + g + V_n(t) \quad (1)$$

where λ is the time varying control parameter, and $V_n(t)$ is a Gaussian, white noise with mean equal to g. In the steady state ($dx/dt = 0$), and for zero noise, the system has only one solution: $x_o = 0$ for all $\lambda < 0$, but bifurcates to $x_{ss} = \pm(\sqrt{\lambda})$ for all $\lambda > 0$. In the presence of noise, but with zero mean ($g = 0$), any dynamical system which obeys (1) will choose the upper (+) or lower (-) branch with equal probability as λ sweeps past the origin.

It is worth emphasizing that no actual, macroscopic, physical system can possibly have $g = 0$. If $g > 0$ (or < 0), then the selection probability will be biased in favor of the positive (or -) branch. This process is shown in Fig. 1(b) for $g < 0$ (negative branch more likely), where the measured trajectories are shown superimposed on the steady states of (1). Note that the effect of nonzero g on the steady state is to open a gap at a critical value λ_c. Even though, in this example, negative branch selection is the more probable (as shown by the figure on the left) a trajectory will sometimes choose the improbable branch (depicted on the right) depending on the magnitude and sign of a few chance fluctuations which occur near λ_c. Our results demonstrate that, depending on the speed with which λ is swept past its critical value, g can have a very large influence on the branch selection probabilities. If λ is swept slowly enough, the system behaves like a "noise averaging switch" with strong branch selection in accord with the sign g even for g << the r.m.s. value of V_n.

The electronic circuit, which integrates (1) with a time constant τ_i, is shown in Fig. 1. The circuit can be operated in either of two modes: 1. as a noise averaging switch with λ swept from some negative, initial value to a positive final value; and 2. as a "conventional" switch with λ fixed at a positive value, and with $x(t = 0)$ set equal to zero (on the unstable branch) then released at $t = 0$. In mode 2, there is also a statistical bias in favor of the branch indicated by the sign of g, but as we show below, the randomizing influence of the noise is much greater than in mode 1, so that the system switches to the "wrong" branch more often. The sweep generator shown in Fig. 1, is operated for mode 1 while the switch on the integrator remains open. Mode 2 operation is achieved by fixing λ, momentarily closing the switch, then opening it at $t = 0$ when the computer is triggered. In order to compare the statistical performance of the two modes of operation, we define the "selectivity" S, which is the ratio of the probability that the "correct" branch will be chosen to the probability that an "error" will occur. An error is the selection of the branch with sign opposite to that of g.

The computer, shown in Fig. 1, digitizes the output voltage x(t) from the electronic circuit into a 1000 point time series, then assembles the probability densities at a sequence of times after t = 0. Typically 5000 such trajectories

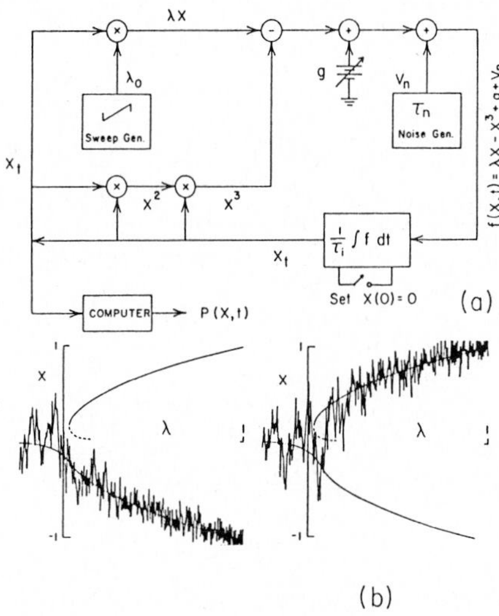

Fig. 1(a) Schematic diagram of the circuit, composed entirely of analog components except for the computer which functions only as a measuring instrument of P(x,t). (b) Example trajectories measured for g = −0.016 and for V_n = 0.5 V r.m.s. The left hand trajectory is the more probable.

were obtained in order to produce a suitably averaged, time evolving density. Examples for the two modes of operation are shown in Fig. 2. Note that the sign of g has been reversed in order to demonstrate alternate branch selection, but its magnitude was the same for both the measured P(x,t) shown.

Experimental values for S were obtained as shown from the final densities by fitting Gaussians to the two peaks, then obtaining their areas. Fig. 2 demonstrates the remarkable improvements in S which can be obtained by using the swept control parameter method of operation (mode 1) where for the example shown in (b)

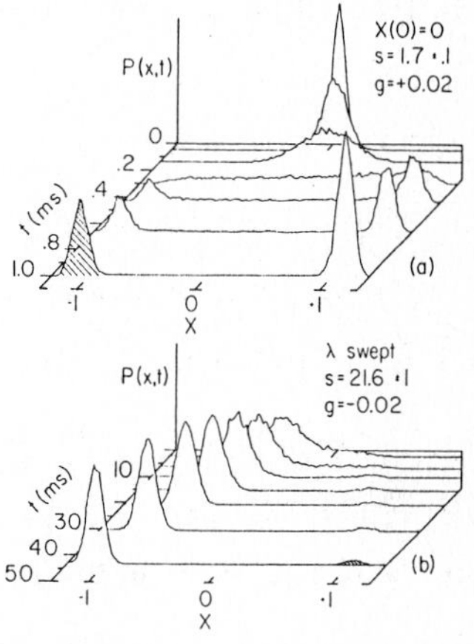

Fig. 2. Measured P(x,t) for (a) fixed control parameter with x (t = 0) = 0 (mode 2); and (b) swept control parameter (mode 1). S is the ratio of the unshaded to shaded area in the final density. In both cases V_n = 0.25 V r.m.s. so that the "noise/signal" ratio was 12.5.

selection "error" occurs less often than once in 20 operations even when the noise is some 12 times greater than the "signal" represented by the magnitude of g.

Only the briefest outline of the theory can be given here. For details, Ref 8 may be consulted. For all $\lambda < \lambda_c$, the density is assumed to be a Gaussian which is drifting and diffusing at rates given by the white noise Fokker-Planck equation. When $\lambda = \lambda_c$, the width and location of the maximum can be estimated. One then slices this Gaussian vertically at the origin and assumes that all area in the slice to the left (or right) of the origin will eventually evolve to the negative (or positive) branch. The ratio of these two areas is S. The results of the theory are compared with this experiment in Ref. 6.

REFERENCES

1. I. Prigogine, Science 201, 777 (1978); I. Prigogine and I Stengers, Order Out of Chaos (Bantam, New York, 1984).

2. Fluctuations and Sensitivity in Nonequilibrium Systems, W. Horsthemke and D. K. Kondepudi, eds. (Springer, Berlin 1984).

3. W. Horsthemke and R. Lefever, Noise Induced Transitions (Springer, Berlin, 1984).

4. J. Smythe, F. Moss, and P. V. E. McClintock, Phys. Rev. Lett. 51, 1062 (1983).

5. F. Moss in Chaos in Nonlinear Dynamical Systems, J. Chandra, ed. (SIAM, Philadelphia, 1984).

6. D. K. Kondepudi, F. Moss and P. V. E. McClintock, Physica D, forthcoming.

7. D. K. Kondepudi and I Prigogine, Physica 107A, 1 (1981).

8. D. K. Kondepudi and G. W. Nelson, Nature 314, 438 (1985) and Phys. Rev. Lett. 50, 1023 (1983).

9. S. D. Robinson, F. Moss and P. V. E. McClintock, J. Phys. A 18, L89 (1985).

10. F. Moss and P. V. E. McClintock, Z. Phys. B, forthcoming.

NOISE IN PHYSICAL SYSTEMS AND 1/f NOISE - 1985
A. D'Amico and P. Mazzetti (editors)

CHAOTIC OSCILLATIONS IN JOSEPHSON JUNCTIONS

Zdravko IVANOV, Zhivko GENCHEV, and Bogdan TODOROV

Institute of Electronics of the Bulgarian Academy of Sciences, 1784 Sofia, Bulgaria

This paper is concerned with an analytical and experimental tool for measuring the presence of chaos in the dynamics of the generalized resistively shunted junction (GRSJ) model of a Josephson superconducting junction. For such a model the phase $\Phi(t)$ obey the dimensionless differential equation

$$\ddot{\Phi} + \delta(1 + \gamma\cos\Phi)\dot{\Phi} + \sin\Phi = i_0 + i_1\sin\Omega t, \tag{A}$$

where $\dot{\Phi} = \frac{d\Phi}{dt}$, $\delta > 0$, $|\gamma| \leq 1$, $i_0 \geq 0$, $i_1 \geq 0$, $\Omega > 0$.

We study the qualitative behavior of this non-linear oscillator by using both the analytical method of Melnikov and the technique of electronic analog modeling.

1. INTRODUCTION

A basic role in the electrodynamics of the superconductor junctions belongs to the oscillations of the phase function which are conveniently described in terms of the GRSJ model [1]. The majority of cases in which chaos has been reported for the equation (A) are based on the simplifying assumption $\gamma = 0$ (see, for example, [2]), i.e. the usual non-generalized RSJ model /which neglects the interaction between the supercurrent and the quasiparticle current/.

Our aim in this paper is to include the influence of the new parameter γ on the criterion describing the existence of chaos in the dynamics of (A).

Consider (A) as a system

$$\dot{x} = f(x) + \varepsilon g(x,t), \tag{1}$$

where f from R^2 to R^2; g from R^3 to R^2 is $\frac{2\pi}{\Omega}$ periodic in t.

In the case when

$$\delta \sim i_0 \sim i_1 << 1,$$

we may replace δ by $\varepsilon\bar{\delta}$, i_0 by $\varepsilon\bar{i}_0$, i_1 by $\varepsilon\bar{i}_1$ with ε small and from (A),(1) we have

$$f_1(x_1,x_2) \equiv x_2,$$

$$g_1(x_1,x_2,\Omega t) \equiv 0,$$

$$f_2(x_1,x_2) \equiv -\sin x_1,$$

$$g_2(x_1,x_2,\Omega t) \equiv \bar{i}_0 + \bar{i}_1\sin\Omega t - \bar{\delta}(1 + \gamma\cos x_1)x_2. \tag{2}$$

Obviously, the homoclinic orbit of the "unperturbed system (ε =0)" is given by [2]

$$\bar{\Phi} = \bar{x}_{01}(t) = \pm 2\operatorname{arctg}(\sinh(t)),$$

$$\dot{\bar{\Phi}} = \bar{x}_{02}(t) = \pm 2/\cosh(t). \tag{3}$$

The Melnikov integral $\Delta(t_0)$ (see p.705 in [3]) is introduced

$$\Delta(t_0) = \int_{-\infty}^{\infty} f(\bar{x}_0(t-t_0) \wedge g(\bar{x}_0(t-t_0),t)dt, \tag{4}$$

where the wedge product $\wedge$ is defined by

$$\begin{bmatrix} f_1 \\ f_2 \end{bmatrix} \wedge \begin{bmatrix} g_1 \\ g_2 \end{bmatrix} = f_1 g_2 - f_2 g_1. \tag{4a}$$

Due to $\pm$ signs in (3) we have two different branches of the Melnikov function $\Delta(t_0)$ .After evaluating (4)explicity we get

$$\varepsilon\Delta(t_0) \equiv m^{\pm}(t_0) = \pm 2\pi\{ i_0 + (i_1 \sin\Omega t_0)/\cosh(\pi\Omega/2)\} - 8\delta(1 + \gamma/3). \tag{5}$$

If the functions $m^{\pm}(t_0)$ have transversal zeros, i.e.if $\exists\ t_0^1$ such that

$$m^{\pm}(t_0^1) = 0,$$

$$\frac{dm^{\pm}}{dt_0}(t_0^1) \neq 0,$$

than horseshoe chaos exists [2,3].

It is easy to show that there exists a minimal (critical) value of the radio-frequency external amplitude ($i_1 = i_1^{cr}$) which assures transversal zero of the $m^{+}(t_0^{'})$. This value of i_1 defines the so called threshold curve

$$i_1^{cr} = \{4(1 + \gamma/3)/\pi\sqrt{\beta_c} - i_0\} \times \cosh(\pi\Omega/2). \tag{6}$$

The curve (6) divides the parameter space (i_0, i_1, $\beta_c \equiv \delta^{-2}$, γ, Ω) into two regions (if $i_1 \geqslant i_1^{cr}$ than Smale horseshoe chaos exists; but in the domain $i_1 < i_1^{cr}$ the oscillations are regular).

2. EXPERIMENTAL AND THEORETICAL RESULTS

Fig.1 shows the influence of the McCumber number $\beta_c \equiv \delta^{-2}$ on the threshold curve $i_1^{cr} = f(\Omega)$ under the assumptions: $i_0 = 0$, $\gamma = 0$ (β_c = 3,5; 7; 9) (.... - theory, ——— - analog modeling).

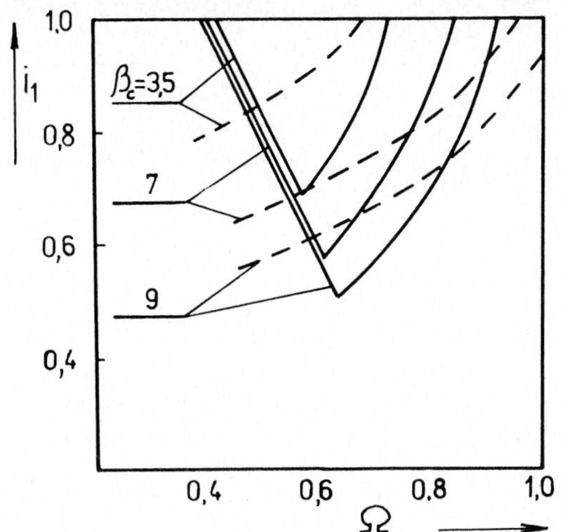

FIGURE 1

Threshold curve $i_1^{cr} = f(\Omega)$ which determines the beginning of chaotic behaviour ($i_1 > i_1^{cr}$).

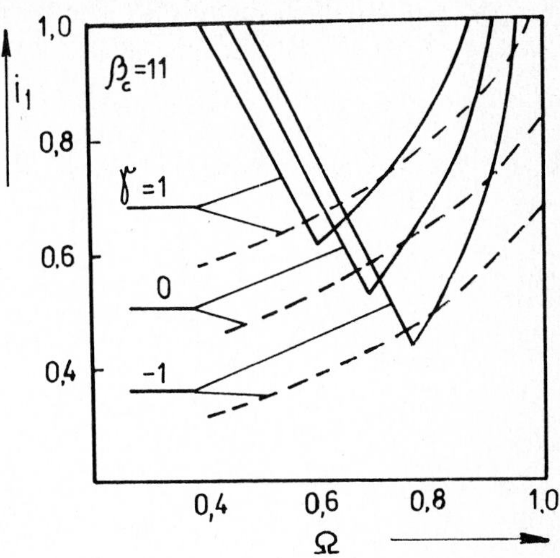

FIGURE 2

The same as on the Figure 1 but now γ is varied.

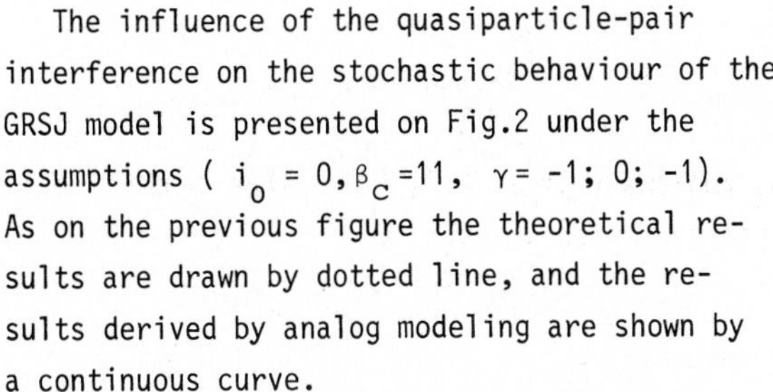

The influence of the quasiparticle-pair interference on the stochastic behaviour of the GRSJ model is presented on Fig.2 under the assumptions ($i_0 = 0, \beta_c = 11, \gamma = -1; 0; -1$). As on the previous figure the theoretical results are drawn by dotted line, and the results derived by analog modeling are shown by a continuous curve.

A conclusion may be drawn that the analytical criterion is in agreement with the electronic modeling especially in the high-frequency region ($\Omega \gtrsim 0,6$).

ACKNOWLEDGMENTS

We want to thank especially Prof.A.Spassov for continuous encouragements during all the work and for many useful discussions.

REFERENCES

1. S.Rudner and T.Claeson, J.Appl.Phys. (1979) 50, 7070.

2. Z.D.Genchev, Z.G.Ivanov, and B.N.Todorov, IEEE Trans.Circuits and Systems, v.CAS-30 (1983) 633.

3. F.M.A.Salam, J.E.Marsden, and P.Varaiya (1983) Ibid., p.697.

NOISE IN PHYSICAL SYSTEMS AND 1/f NOISE - 1985
A. D'Amico and P. Mazzetti (editors)

RESPONSE OF A CUBIC BISTABLE SYSTEM TO PARAMETRIC MODULATION IN THE EXTREME COLOUR LIMITS OF EXTERNAL NOISE

R.C.M. DOW*, C.J. LAMBERT*, R. MANNELLA*, P.V.E. McCLINTOCK* and Frank MOSS+

*Department of Physics, University of Lancaster, Lancaster LA1 4YB, U.K.
+Department of Physics, University of Missouri-St. Louis, MO 63121, U.S.A.

We examine the possibility of fluctuation induced transitions in a cubic bistable electronic circuit. Soft transitions are found to occur in the presence of a dichotomous modulation, but appear to be absent when the system is subject to a harmonic modulation. Despite these differences, the behaviours of the supports of the probability densities arising in the two cases are rather similar. In particular they each exhibit an anomaly over a limited range of the amplitude of fluctuation V_0, over which the supports shrink as V_0 increases.

1. INTRODUCTION

The effect of a randomly fluctuating environment on non-linear systems has been the focus of much attention recently[1]. Examples of systems studied are an Ising model with a randomly time varying coupling constant[2], a laser with a noisy cooperativity parameter[3], a nematic liquid crystal with a randomly changing externally applied magnetic field[4] and a cubic bistable system with multiplicative noise[5]. In each of these examples, some macroscopic property x of the system obeys a first order evolution equation, in which the state of the environment is modelled by a parameter λ. In the presence of a noisy environment, λ exhibits random temporal fluctuations and the equilibrium state is characterized by a stationary probability density $\rho(x)$. Such systems are particularly convenient to study theoretically, because if λ is a white noise or dichotomous Markov process, an expression for $\rho(x)$ is readily obtained. Such studies have shown that a noisy environment can produce marked changes in the the equilibrium phases; under some circumstances, for example, bistability can be induced in an otherwise monostable system[5].

The fact that random fluctuations can dramatically alter the equilibrium state of a non-linear system has naturally led to investigations of the effect of a periodically varying external environment[6,7]. In this paper we examine the response of a cubic bistable electronic circuit to two types of parametric modulation. First we obtain results for a square wave modulation and compare with the theoretical predictions obtained by extending the analysis of reference 6. The response to sinusoidal modulation is also obtained. In this case no exact theoretical predictions are available and so a numerical integration of the evolution equation is carried out.

2. THE NON-LINEAR CIRCUIT

The electronic system is designed to model

$$\frac{dx}{dt} = -x^3 + \lambda x^2 - Qx + R = F(x,\lambda) \quad (1)$$

with Q = 3 and R = 0.7. This system is described in detail by Robinson et al[7] who studied the steady state behaviour of x under the influence of a randomly varying control parameter λ. In present paper the variations in λ of interest are as follows:

For harmonic modulation

$$\lambda = \lambda_0 + V_0 \sin 2\pi t/\tau \quad (2)$$

and for square wave modulation

$$\lambda = \lambda_0 + V_0 \quad 0 \leqslant t \ (\text{modulo } \tau) < \tau/2 \quad (3)$$

$$\lambda = \lambda_0 - V_0 \quad \tau/2 \leqslant t \text{ (modulo } \tau) < \tau \qquad (4)$$

In both cases, following any adjustment to V_0 or the transition frequency $\nu = \tau^{-1}$, the system was allowed to settle. The stationary density $\rho(x)$ was then determined by means of a Nicolet 1080 minicomputer system, whose role, it should be emphasised, was simply that of a measuring instrument. The overall scale-factor of the circuit was unity, so that x = 1 was equivalent to 1 volt, but the design was such that maximum permissable slew-rates were not exceeded even when relatively large amplitude modulation was being applied. In what follows, the quoted frequencies and transition times have all been scaled to the frequency corresponding to unity time constant.

3. RESULTS

Some typical experimental densities are shown for the case of dichotomous modulation in figure 1. They have been scaled vertically in order to demonstrate that they form portions of a common curve but, otherwise, are exactly

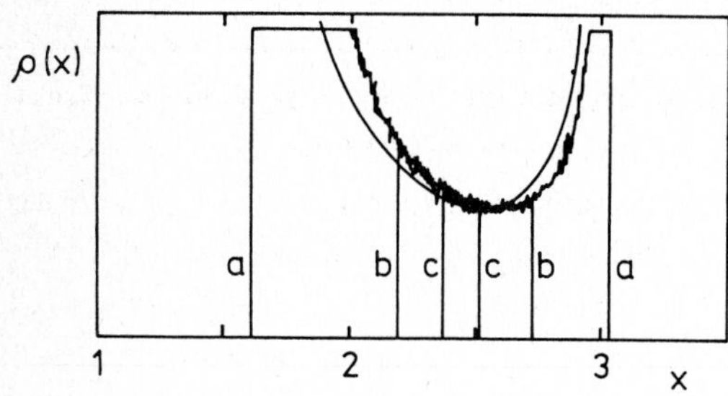

FIGURE 1
Stationary density functions for (1) under cyclical dichotomous modulation (3) with $\lambda = 3.5$, $V_0 = 0.41$ and periods (a) 4.8, (b) 0.43 and (c) 0.11 s. The horizontal truncation of the maxima of (a) arises from clipping by the graph plotting system. The smooth full curve represents the theoretical prediction (4) of the envelope.

as measured. The noise that is evident was introduced by the digitiser. The continuous full curve represents the theoretical result[6]

$$\rho(x) = \nu\left\{\left(\left|f(x,\lambda_0+V_0)\right|^{-1} + \left(\left|f(x,\lambda_0-V_0)\right|\right)^{-1}\right\} \qquad (5)$$

which is trivially obtained from the observation that the average time spent by the solution of equation (1) in some small interval $[x, x + \Delta x]$ is proportional to the inverse velocity at x. The support of $\rho(x)$ is given by the upper and lower cut-off points x_+ and x_- of the density. These are obtained from the solution of the simultaneous equations

$$\frac{\tau}{2} = \int_{x_-}^{x_+} dx\, f^{-1}(x, \lambda_0 + V_0) \qquad (6)$$

$$\frac{\tau}{2} = \int_{x_+}^{x_-} dx\, f^{-1}(x, \lambda_0 - V_0) \qquad (7)$$

The experimentally measured values for $x_\pm$ are indicated by vertical lines in figure 1. A comparison is made with the solution of equations

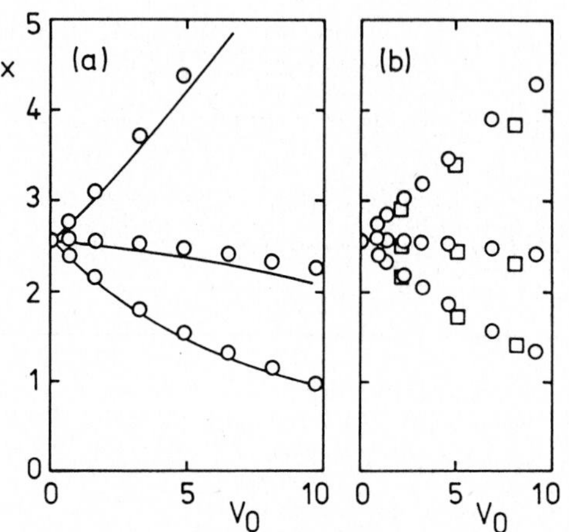

FIGURE 2
Loci of the support and first moment $\bar{x}$ for (1) under (a) dichotomous and (b) harmonic cyclical modulation plotted as functions of amplitude V_0. The circles represent experimental measurements. The full curves in (a) are theoretical predictions derived from solutions of (5) and the squares in (b) indicate the results of the digital simulation.

(6) and (7) in figure 2(a), where measured and calculated values of the first moment $\bar{x}$ are also plotted. Figure 1 shows results for a variety of frequencies with $\lambda_0 = 3.5$, $V_0 = 0.41$ and figure 2(a) for a range of V_0, with $\lambda_0 = 3.6$, $\nu = 5.8$. The small disagreements apparent in the figures 2(a) can be accounted for in terms of the known non-idealities of the analogue components of the circuit.

In the case of harmonic modulation no detailed theory exists with which the experimental results can be compared. Some typical experimental densities are shown in figure 3. For small values of V_0 or τ, the density is similar to that of a sine wave, with a support framed by a pair of singularities. As a result of the discrete nature of the digitiser these are resolved as sharp maxima of approximately equal height. When either V_0 or τ is increased, there is a corresponding increase in the separation of the maxima, but in each case, the behaviour of

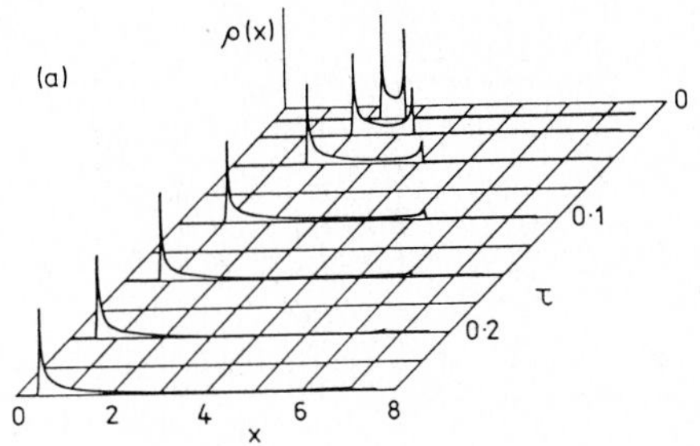

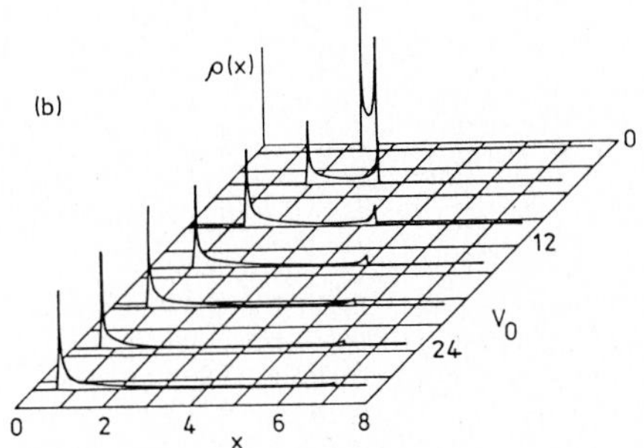

FIGURE 3
Stationary density functions measured for (1) under harmonic modulation : (a) for $V_0 = 16.3$ and various periods τ; (b) for $\tau = 0.17$ and various amplitudes V_0.

the lower maximum is markedly different from that of the upper one. The upper maximum shrinks, broadens and, at the same time, moves rapidly towards larger x; whereas the lower peak remains relatively insensitive to changes in V_0 and τ. Clearly this behaviour is a reflection of the very different nature of the positive roots of $f(x, \lambda_0 + V)$ and $f(x, \lambda_0 - V_0)$ as $V_0 \to \infty$. The latter vanishes in this limit and for large V_0 is therefore insensitive to changes in this parameter. In contrast the former approaches V_0 in this limit and consequently remains sensitive to further changes.

To confirm these reuslts in the absence of a theoretical description, we have carried out a digital simulation of equation (1) under harmonic modulation. The densities obtained compare favourably with the experimental results of figure 3. A detailed comparison of the upper and lower bounds of the support is given in figure 2(b), where the first moment $\bar{x}$ is also shown. Despite the differences in the densities shown in figures 1 and 3 for dichotomous and harmonic modulations respectively, the variation in the supports with V_0 are remarkably similar. This similarity is emphasised by figure 4 which shows the variation of $x_{\pm}$ and $\bar{x}$ with V_0 arising from dichotomous and harmonic modulations. The first of these is obtained from the solution of equations (6) and (7) and the second from a digital simulation. The range of V_0 shown in the figure extends far beyond the tolerances of the electronic circuit so that experimental results are not available. Apart from the similarity in the shapes of these curves they reveal the remarkable feature that a range of V_0 exists in which an increase of V_0 causes the support to shrink.

3. DISCUSSION

The aim of this paper has been to examine the response of a cubic bistable system to both dichotomous and sinusoidal modulations and to determine the existence or otherwise of fluctuation induced transitions. Following Doering and Horsthemke, we distinguish between two possible types of transition. A "hard" transition is said to occur at values of τ and V_0, where $\rho(x)$ ceases to diverge at one or more of the boundaries $x_\pm$. We have found that for finite τ and V_0, hard transitions do not occur for either of the two modulations examined. This can be understood by considering a continuity argument. A sinusoidal modulation renders dx/dt continuous. This implies that for any finite V_0 and τ, dx/dt must vanish at $x_\pm$, yielding a divergence in $\rho(x)$ at the boundaries of the support. In contrast a square wave modulation yields a discontinuity in dx/dt at $x_\pm$. As a result for any finite V_0 and τ, $\rho(x)$ does not diverge at the end points.

A "soft" transition is said to occur

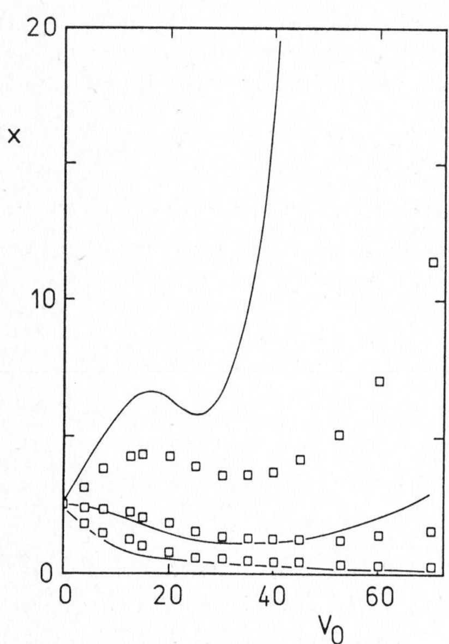

FIGURE 4
Loci of the support and $\bar{x}$ for $\nu = 6.0$, plotted as functions of V_0 for a dichotomous modulation (full curves) and a harmonic modulation (squares). Note the anomalous region for intermediate V_0 in which the support shrinks as V_0 increases.

when there is a change in the number of extrema of $\rho(x)$ within the support. Figure 1 shows that these transitions do indeed occur in the presence of dichotomous modulation. However as shown in figure 3, no such transitions are observed with harmonic modulation.

Despite the differences in the densities, the behaviours of the supports for each type of modulation are rather similar. The most striking feature is that there exists a range of V_0 for which $d(x_+-x_-)/dV_0$ is negative. This unexpected and counter-intuitive behaviour is currently under investigation and will be discussed more fully in a future publication.

ACKNOWLEDGEMENTS

One of us (R.C.M.D.) would like to thank the University of Lancaster Science and Engineering Research Committee for the provision of a research Fellowship. We are very grateful to W. Horsthemke and D.K. Kondepudi for commenting on an earlier version of this paper. The experiments were supported in part by the Science and Engineering Research Council.

REFERENCES

1. For a recent review see W.Horsthemke and R. Lefever, Noise Induced Transitions : Theory and Applications in Physics, Chemistry and Biology (Springer-Verlag, Berlin 1984).
2. E. Gudowska-Nowak, to be published in Phys. Lett. A.
3. Moore S.M., Nuovo-Cim. 79B (1984) 125.
4. W. Horsthemke, C.R. Doering, R. Lefever and A.S. Chi, to be published.
5. G.V. Welland and F. Moss, Phys.Lett. 89A (1982) 273 and 90A (1982) 222.
6. C.R. Doering and W. Horsthemke, to be published in J.Stat.Phys.
7. C.R.Doering in Fluctuations & Sensitivity in Non-equilibrium Systems, ed. W.Horsthemke D.K.Kondepudi (Springer-Verlag, Berlin, 1984) pp 461-492.
8. S.D. Robinson, F. Moss and P.V.E. McClintock, J. Phys. A18 (1985) L89.

NOISE IN PHYSICAL SYSTEMS AND 1/f NOISE - 1985
A. D'Amico and P. Mazzetti (editors)

STOCHASTIC POSTPONEMENT OF CRITICAL ONSETS IN A BISTABLE SYSTEM

P. V. E. McCLINTOCK* and Frank MOSS†

* Department of Physics, University of Lancaster, Lancaster, LA1 4YB, U.K.
† Department of Physics, University of Missouri-St. Louis, St. Louis, MO63121, U.S.A.

We report the results of an experimental investigation of the effect of external, multiplicative, bandwidth-limited, white noise upon a cubic nonlinear electronic circuit modelling $\dot{x} = f(x, t)$ which exhibits bistability within a certain range of a control parameter λ. Measurements of the statistical density of x are found to be in excellent agreement with theoretical predictions based on the use of the Stratonovic stochastic calculus for integration of the stochastic differential equation. In addition, we have observed pronounced stochastic shifts of the region of bistability towards larger values of λ. These shifts are found to be in gratifyingly good agreement with the theoretical prediction. Thus, although an unambiguous demonstration of stochastic postponement in a natural system has yet to be achieved, the reality of this remarkable phenomenon in relation to physical systems is no longer open to doubt.

1. INTRODUCTION

The application of multiplicative white noise to systems described by certain one dimensional nonlinear differential equations is expected[1] to give rise to diverse phenomena that are often, at first sight, quite unexpected and which are frequently also counterintuitive in character. A case in point is the remarkable prediction[2] that the critical onset of the transitions between the two states of the bistable

$$\frac{dx}{dt} = -x^3 + \lambda_t x^2 - Qx + R \tag{1}$$

would be shifted to larger mean values λ of the noisy control parameter

$$\lambda_t = \lambda + \sigma\xi_t \tag{2}$$

as the variance σ^2 of the external white noise was increased. On the basis of a white noise Fokker-Planck approach[1] it has been predicted[2] that the statistical density of x should be given by

$$\rho(x) = \frac{N}{x^{2(\nu + 1/\sigma^2)}} \exp\left[\frac{2}{\sigma^2}\left(-\frac{\lambda}{x} + \frac{Q}{2x^2} - \frac{R}{3x^3}\right)\right] \tag{3}$$

with maxima at x_m given by the solutions of

$$(1 + \nu\sigma^2)x_m^3 - \lambda x_m^2 + Qx_m - R = 0 \tag{4}$$

where N is a normalising constant and $\nu = 1$ or 2 for the Stratonovic or Ito versions of the white noise Fokker-Plank equation respectively. The interest and relevance of the topic springs mainly from the existence of a wide range of natural systems (see, for example, references given in Robinson et al[3]) that are believed to be described, at least approximately, by (1).

2. EXPERIMENTS

The use of a natural system for a quantitative experimental test of the above predictions is fraught with difficulties, however, in particular because we are usually ignorant of the accuracy with which (1) may be expected to model the system in question. To avoid this complication we have chosen, in the first instance, to study the effect of white noise upon an electronic circuit designed specifically to correspond to (1) as closely as possible. It is of crucial importance to note that, notwithstanding the fact of its having been deliberately contrived to follow (1), the electronic circuit can still quite properly be regarded as a real physical system in its own right. A block diagram of the circuit (in somewhat simplified form) is shown in Fig. 1. In the absence of noise ($\sigma = 0$) it models (1) for Q = 3 and R = 0.7 with an accuracy typically of a few percent, as indicated by the circled points and the corresponding theoretical curve of Fig. 2. The external white noise is bandwidth-limited such that its correlation time τ_N = 12μs is much smaller than the time constant τ_N = 1.2 ms of the integrator in the circuit: the noise is therefore perceived by the circuit as though white;

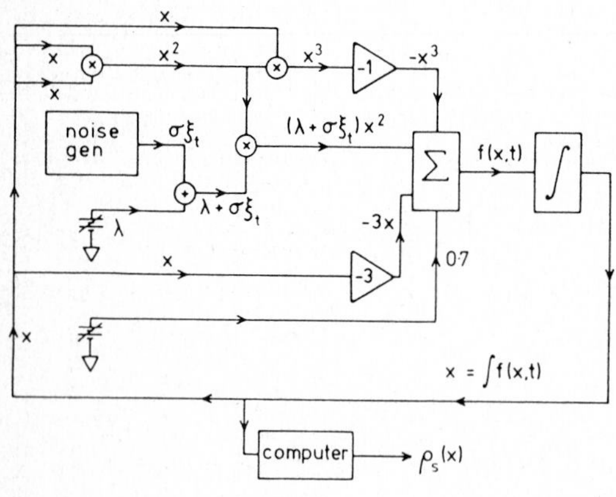

FIGURE 1
Block diagram of the electronic circuit corresponding to (1).

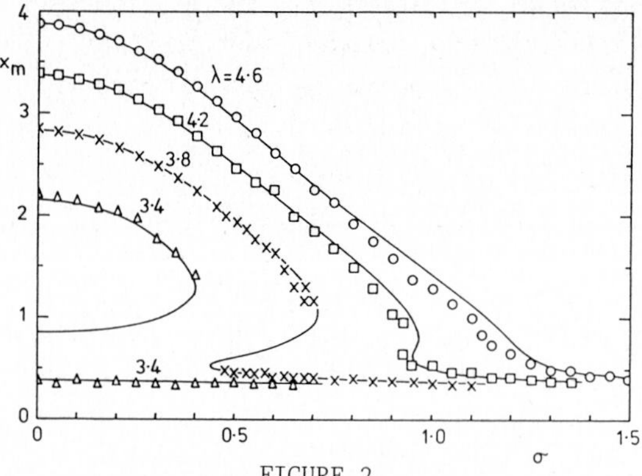

FIGURE 2
Loci of the maxima x_m of the densities as functions of the control parameter λ for various noise intensities σ. The full curves represent solutions of (4) with ν = 1.

subject, however, to a rescaling[4] of the noise intensity such that we must then use an effective value

$$\sigma = (2\tau_N/\tau_I)^{\frac{1}{2}} V_N \tag{5}$$

where V_N is the rms value of the external noise in the white noise equations (1-4).

When noise is being applied to the circuit, *x* is, of course, no longer constant for any given value of λ, but it can still be described by its statistical density. We have used a Nicolet 1080 computer to measure stationary density functions for *x* under various conditions. Some typical examples are shown in Fig. 3 for λ = 3.6 and various values of σ. For small σ, the distribution approximates to a delta-function (not shown) centred on the deterministic stationary value of *x*. As σ is increased, however, it can be seen that the density gradually broadens, shifts, becomes bimodal and finally, for very large σ, converges towards a delta function close to lower root of the deterministic (σ = 0) version of (1). The dashed lines in Fig. 3, representing plots of the Stratonovic (ν = 1) version of (3) with N chosen arbitrarily to give agreement at *x* = 1.5, have been raised slightly above the experimental densities in the interests of clarity. The agreement between experiment and theory, though not perfect, is remarkably close. We have located the maxima of densities such as those of Fig. 3 by the local fitting of parabalae through the method of least squares. The loci of x_m as functions of λ and of σ are shown respectively in Figs. 2 and 4 where the full lines in each case represent solutions of the Stratonovic

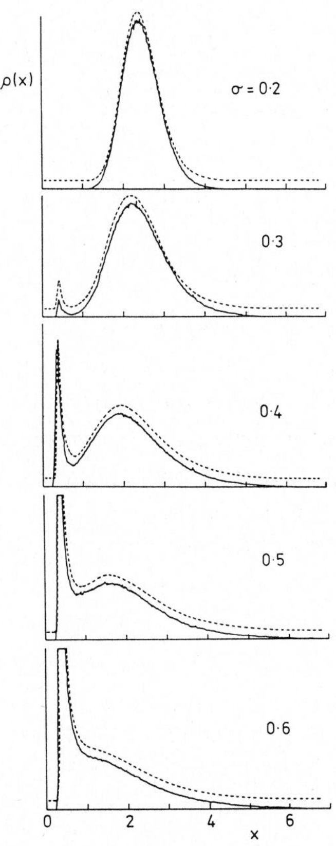

FIGURE 3
Experimental density functions $\rho(x)$ measured for the electronic circuit of Fig. 1. with different intensities of the external noise. The dashed curves represent (3) with ν = 1: in the interests of clarity, they have been raised slightly above the (noisy) experimental curves.

version of (4). Again, encouragingly close agreement is obtained between experiment and theory; the observed discrepancies can readily be accounted for in terms of the known small non-idealities of the components in the circuit.

3. DISCUSSION

We wish to make two main comments about the significance of these results. First, the gratifying agreement between experiment and theory embodied in Figs. 2 - 4 is dependent on the use of the Stratonovic calculus for integration of the stochastic differential equation (1). If, instead, the Ito interpretation is followed, the theoretical curves that are then obtained exhibit a clear and definite disagreement[5] with the experimental data. One can, of course, comment dismissively that, in the light of the theoretical discussion given by van Kampen[6] and others[7,8], such a result was only to be expected, that it is unsurprising and therefore, by implication, unimportant; but we believe that such a reaction would

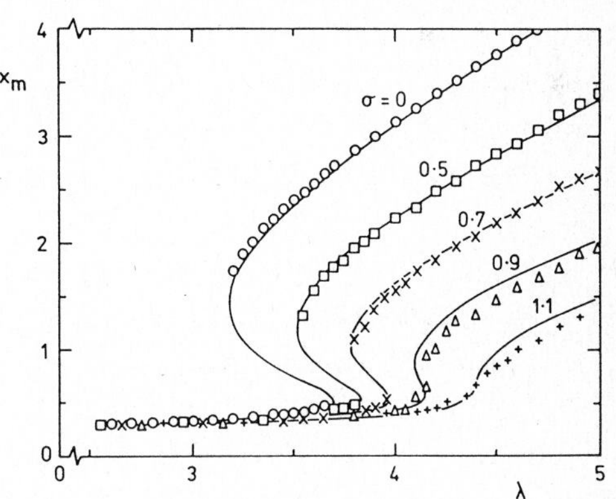

FIGURE 4
Loci of the maxima x_m of the densities as functions of the noise intensity σ for various values of the control parameter λ. The full curves represent solutions of (4) with ν = 1.

be entirely misguided. We would argue very strongly, on historical grounds, that theory unsupported by experiment is never to be relied upon: rather, it is the fruitful interplay of theory and experiment that has been of seminal importance in the development of scientific understanding. The successful completion of a series of *post hoc* experimental tests for every new theory should be regarded as an essential prerequisite for general acceptance of its validity. To our knowledge, the results described above represent only the second such test to have been performed that can distinguish unambiguously, on experimental grounds, whether it is the Ito or the Stratonovic stochastic calculus that should be used to describe a real physical system. (The first such test[9] related to a noise-induced phase transition is the so-called genetic model equation: a system very different in character to the cubic bistable studied here.) It is gratifying, therefore, that the results of Figs. 2 - 4 suggest the conclusion consistent with the theoretical predictions and with the results of the single earlier experiment, that, although the Ito and Stratonovic interpretations are both formally correct in a purely mathematical sense, it is the Stratonovic stochastic calculus that accurately describes what actually happens in nature.

Secondly, it is immediately evident from the data of Fig. 3 that the remarkable stochastic postponement effect referred to in the opening paragraph, does indeed occur in reality: as σ increases, the range of bistability shrinks and simultaneously moves towards larger values of λ. An important *caveat* must, however, be entered in relation to the application of these ideas to certain natural systems. It must be emphasized that most of our discusion above (and, in particular, the theoretical and experimental results of Figs. 3 and 4) refers to density maxima, that is, to the "most probable" values of x. There are many circumstances in practice where, instead, it is only the first or second moments of x that can be measured. In such cases, the details of the exotic phenomena that we have been considering will very largely and unavoidably be concealed from the view of the experimenter.

The work has been supported in part by the Science and Engineering Research Council (U.K.).

REFERENCES

1. W. Horsthemke and R. Lefever, Noise Induced Transitions: Theory and Applications in Physics, Chemistry and Biology, (Springer, Berlin, 1984).

2. G. V. Welland and F. Moss, Phys. Lett. 89A (1982) 273 and 90A (1982) 222.

3. S. D. Robinson, F. Moss and P. V. E. McClintock J. Phys A: Math. Gen. Phys. 18 (1985) L89.

4. D. K. Kondepudi, F. Moss and P. V. E. McClintock, to be published in Physica D.

5. P. V. E. McClintock and F. Moss, Phys. Lett. 107A (1985) 367.

6. N. G. van Kampen, J. Stat. Phys. 24 (1981) 175 and 25, (1981) 431 and Stochastic Processes in Physics and Chemistry, North-Holland, Amsterdam, 1981.

7. T. Morita, Phys. Lett. 82A, (1981) 215.

8. P. Hanggi and H. Thomas, Phys. Rep. 88C, (1982) 207.

9. J. Smythe, F. Moss, and P.V.E. McClintock, Phys. Rev. Lett. 51 (1983) 1062; J.Smythe F. Moss, P.V.E. McClintock and D. Clarkson, Phys. Lett. 97A (1983) 95.

NOISE IN PHYSICAL SYSTEMS AND 1/f NOISE - 1985
A. D'Amico and P. Mazzetti (editors)

CORRELATION TIMES IN THE CUBIC BISTABLE SYSTEM

P. V. E. McCLINTOCK*, R. MANNELLA*, J. M. SANCHO≠ and Frank MOSS§

* Department of Physics, University of Lancaster, Lancaster, LA1 4YB, U.K.
≠ Departament de Física Teòrica, Universitat de Barcelona, Barcelona, 08028, Spain.
§ Department of Physics, University of Missouri-St. Louis, MO 63121, U.S.A.

We have carried out an experimental and theoretical study of the effect of external multiplicative white noise on the relaxation time of a general representation of a bistable system. The experiments were based on a bistable electronic analogue circuit; the calculations employed a new technique recently introduced by Jung and Risken. Satisfactory agreement is obtained between the results of these two very different approaches. We demonstrate that, by acting as an indicator of critical slowing down, the relaxation time can be used to provide a correct characterisation of a noise-induced transition.

1. INTRODUCTION

The dynamical behaviour or nonlinear systems driven by parametric noise[1-3] can be described in terms of a characteristic relaxation time[4], T. One may anticipate that this type of description will be especially valuable in the case of multistable systems because the behaviour of T can be expected to yield a rather direct indication of the critical slowing down that must always occur in the vicinity of an instability. In this paper, we report the results of an experimental and theoretical investigation of T for the cubic bistable system

$$x = -x^3 + \lambda_t x^2 - Qx + R \qquad (1)$$

where the noisy multiplicative parameter

$$\lambda_t = \quad + \sigma\xi_t \qquad (2)$$

has a mean value of λ and $\sigma\xi_t$ is Gaussian white noise of zero mean and variance σ^2. The experimental measurements were carried out on an electronic circuit designed to model (1) that has been described elsewhere[5]; the calculations were based on the novel approach recently developed by Jung and Risken[6].

It should be noted that (1) can be regarded as representative of numerous bistable systems occuring in diverse areas of physics and other branches of science[1,7-10] that are believed to be described, if not by (1) itself, then by other stochastic differential equations that are closely similar in form. Potentially, therefore, the discussion that follows is of very wide relevance and applicability.

2. EXPERIMENTS

In the electronic circuit[5] used for the experiments, Q and R were set equal to 3 and 0.7 respectively. The circuit models the deterministic ($\sigma = 0$) form of (1) with an accuracy typically of a few percent. In practice, of course, the externally applied noise cannot be white but is exponentially correlated with a correlation time, τ_N. Provided that the integrator time constant $\tau_I >> \tau_N$, then the noise is still perceived as white by the circuit and the effective value of σ for noise of rms amplitude V_N is given[11] by

$$\sigma = (2\tau_N/\tau_I)^{\frac{1}{2}} V_N \quad (3)$$

The response of the circuit to external noise has been investigated through measurements of the statistical density $P(x)$ which has been found[5] to be in excellent agreement with predictions derived from a Fokker-Planck equation on the assumption[12] that the Stratonovic stochastic calculus must be used for the integration of (1).

The relaxation time, T, of the system is determined for different values of λ and σ by first digitising a block of 1024 discrete points on $x(t)$ and then computing the autocorrelation function $c(s)$ of $\{x(t) - \langle x(t)\rangle\}$ by means of a standard FFT technique[13]; the process is repeated typically 200 times, the results being added and averaged so as to enhance their statistical reliability. For small σ, $c(s)$ is found to be exponential within experimental error as shown in Fig. 1 (a) and (b). For larger σ, however, $c(s)$ broadens and becomes decidedly non-exponential as shown in Fig. 1 (c) and (d), because of the switching that then occurs between the upper and lower states. In order to be able to characterise correlation functions of this kind, we have calculated the relaxation time

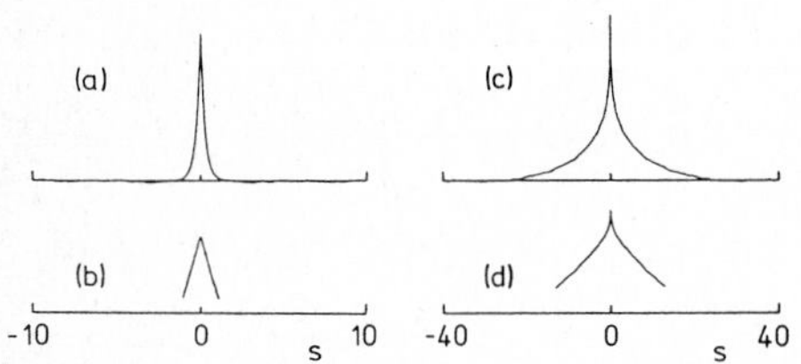

FIGURE 1
Autocorrelation functions c(s) of $\{x(t) - \langle x(t)\rangle\}$ measured with $\lambda = 3.6$: (a) for $\sigma = 0.1$ and (b) its logarithm; (c) for $\sigma = 0.6$ and (d) its logarithm.

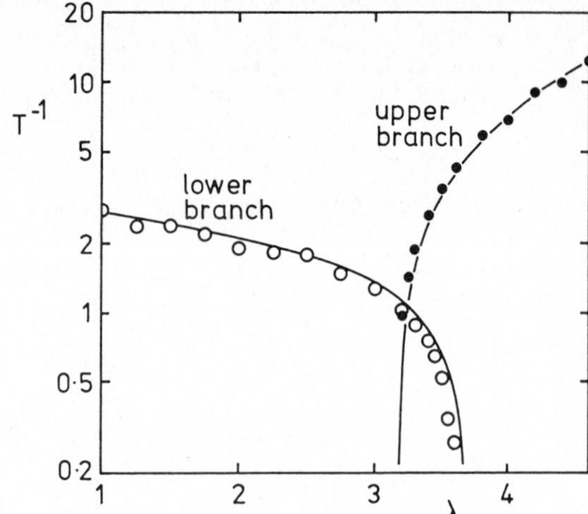

FIGURE 2
Experimental inverse relaxation times T^{-1} (points) as functions of λ compared with theorical curves representing (6) in the limit $\sigma \to 0$.

$$T = \int_0^\infty \{c(s)/\, c(o)\}\, ds \quad (4)$$

by integration: in practice we truncate the integral at a value of s where $c(s)$ has fallen to zero. To allow for the non-unity time constant of the integrator, which introduces a scaling of real time as well as of σ via (3), we have divided all our measured T values by τ_N.

Experimental values of T^{-1} are plotted for selected values of σ and λ in Figs. 2 - 4. In Fig. 2 T^{-1} determined for very small σ is compared with the predictions (curves) of a deterministic approach that is expected to be valid in this limit (see below). For these measurements no switching occurred during the determination of T^{-1}. For larger σ, a distinct minimum in $T^{-1}(\lambda)$ is seen, becoming shallower with increasing σ, as indicated by the data of Fig. 3. A minimum also occurs in $T^{-1}(\sigma)$ at fixed λ, as shown in Fig. 4. The curves in Figs. 3 and 4 represent calculations (see below) based on the theory of Jung and Risken [6].

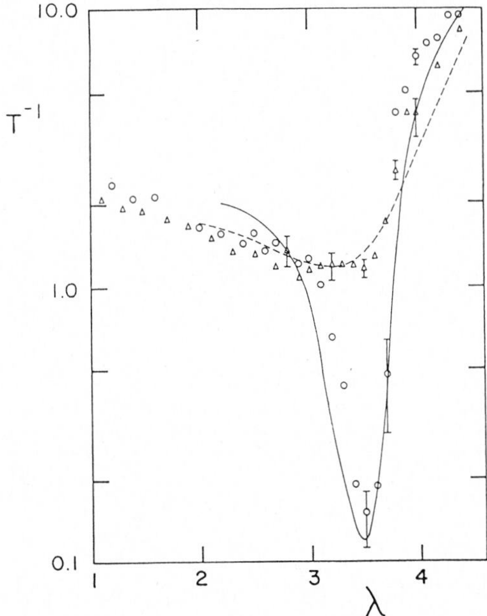

FIGURE 3
Experimental inverse relaxation times T^{-1} as functions of λ for $\sigma = 0.5$ (open circles) and $\sigma = 1.0$ (filled circles) compared with the prediction of the Jung-Risken theory (curves).

3. CALCULATIONS

When σ is sufficiently small, the system makes only minor fluctuations

$$\Delta x = x - x_i \tag{5}$$

about the deterministic solution, x_i. In this case, it is straightforward to substitute (5) into (1), linearise the result and show that the relaxation time is just

$$T^{-1} = 3x_i^{\,2} - 2\lambda x_i + 3 \tag{6}$$

leading to the two curves of Fig. 2. Strictly, (6) cannot be used at all within the bistable regime because x_i is then not unique; in practice it gives good agreement with the experimental data provided σ is so small that switching does not occur (and so x_i remains unique) during the period of time necessary to complete a measurement of T, as can be seen in Fig. 2.

For larger σ, the assumptions underlying (6) break down and a more sophisticated approach is needed. We have used two techniques for calculating T under these conditions: the Stratonovic decoupling approach[14]; and the recently introduced Jung-Risken method[6]. Of these, the latter has turned out to be much the more satisfactory. Its application to (1) yields the exact expression

$$T = \{c(o)\}^{-1} \int_o^\infty I^2(x)dx/\{x^4\ P(x)\} \tag{7}$$

where

$$I\ (x) = -\int_o^x \Delta(x)\ P(x)dx \tag{8}$$

$$\Delta x = x - \langle x\rangle \tag{9}$$

and $P(x)$ is the stationary statistical density[5,15] of x. The numerical algorithm that we have used in the evaluation of (7) will be described in detail elsewhere [16].

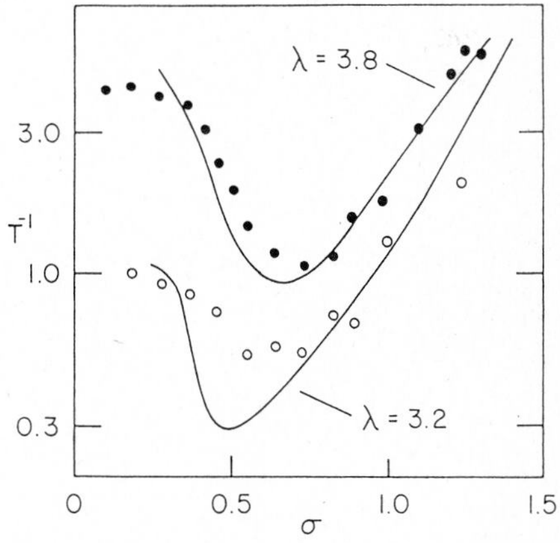

FIGURE 4
Experimental inverse relaxation times T^{-1} as functions of σ for the indicated values of λ, compared with the prediction of the Jung-Risken theory (curves).

It leads to the full curves shown in Figs. 3 and 4 which, gratifyingly, exhibit very much the same general behaviour as the experimental data with which they are compared.

4. DISCUSSION

We may regard the quality of agreement between the theoretical curves and the experimental data as gratifying. The observed discrepancies are attributable to known nonidealities of the analogue components that comprise the electronic circuit, and especially to "clipping" of the largest excursions of x, an effect that is nonnegligible when $\sigma > 0.7$.

In conclusion, we would make three comments. First, the dominant feature predicted by the theory and observed in the experiments, the minimum in T^{-1}, provides a good dynamical characterisation of a noise-induced transition since it clearly acts as an indicator of the critical slowing down that can be induced by variation of either σ or λ. Secondly, it would appear that the exact approach of Jung and Risken[6] offers the best method for the calculation of relaxation times because explicit results can be obtained, albeit numerically, with a high degree of accuracy and only minor computational difficulty. Moreover, unlike the more usual perturbative methods, their approach is not limited to small values of σ. Finally, it is of particular interest to note that, whereas the depth of the minimum in T^{-1} increases as σ decreases, and (in principle) $T^{-1} \to 0$ as $\sigma \to 0$, no corresponding singularity occurs in the moments [17-19].

The work was supported in part by the Science and Engineering Research Council (U.K.).

REFERENCES

1. W. Horsthemke and R. Lefever, Noise Induced Transitions (Springer-Verlag, Berlin, 1984).
2. P. Grigolini, J. Stat. Phys. 27 (1982) 283.
3. P. Hanggi and H. Thomas, Phys. Rep. 88 (1982) 207.
4. A. Hernandez-Machado, M. San Miguel and J.M. Sancho, Phys. Rev. A 29 (1984) 3388.
5. S.D. Robinson, F. Moss and P.V.E. McClintock, J. Phys.A 18 (1985) L89; also, P.V.E. McClintock and F. Moss, in these Proceedings.
6. P. Jung and H. Risken, Z. Phys. B to be published.
7. F. Moss and G. Welland, Phys. Rev. A 25 (1982) 3389.
8. S.N. Dixit and P.S. Sahni, Phys. Rev. Lett. 50 (1983) 1273.
9. P. Lett, R. Short and L. Mandel, Phys. Rev. Lett. 52 (1984) 341.
10. P. Hanggi and F. Moss in Proc. Int. Conf on Instabilities in Lasers and Optical Systems, Rochester, N.Y., 1985; to be published by Cambridge University Press.
11. D.K. Kondepudi, F. Moss and P.V.E. McClintock, Physica D. to be published.
12. P.V.E. McClintock and F. Moss, Phys. Lett. 107A (1985) 367.
13. K.G. Beauchamp and C.K. Yuen, Digital Methods for Signal Analysis (Allen and Unwin, London 1979).
14. R.L. Stratonovic, Topics in the Theory of Random Noise, vol. I (Gordon and Breach, New York, 1963).
15. G.V. Welland and F. Moss, Phys. Lett. 89A (1982) 273 and 90A (1982) 222.
16. J.M. Sancho, P.V.E. McClintock, R. Mannella and F. Moss, to be published.
17. J. Smythe, F. Moss and P.V.E. McClintock, Phys. Rev. Lett. 51 (1983) 1064.
18. H.R. Brand, Phys. Rev. Lett. 54 (1985) 605.
19. F. Moss, P.V.E. McClintock and W. Horsthemke, Phys. Rev. Lett. 54 (1985) 606.

NOISE IN PHYSICAL SYSTEMS AND 1/f NOISE - 1985
A. D'Amico and P. Mazzetti (editors)

A MODEL FOR THE ELECTRICAL CONDUCTION AND 1/F-NOISE IN CERMET THICK FILM RESISTOR-SYSTEMS

S. DEMOLDER and A. VAN CALSTER

Laboratory of Electronics, Ghent State University, Sint Pietersnieuwstraat 41, B-9000 Gent, Belgium

In this paper we report conduction and noise measurements on a cermet and a metallo-organic thick film resistor system. On the basis of a possible conduction mechanism through the resistors, and the justified application of Hooge's formula, we can explain the trends in all parameter dependencies of resistance and noise.

I. INTRODUCTION

In a previous paper[1] we reported 1/f-noise measurements on metanet and cermet thick film Resistor systems. In this paper a more complete analysis of the cermet and metanet systems is presented. The dependence of 1/f-noise on the temperature, the volume and the sheet resistivity is investigated.

Some insight in the noise data is obtained by linking the theoretical flicker noise considerations of Pellegrini with a possible conduction mechanism for thick film resistors.

II. MEASUREMENTS

In this paper we report measurements on two thick film resistor systems: the first system is a cermet one, namely the DP 1400 series. Here we tested all sheet resistivities between 10 and 10^5 Ω/□. The second system concerns a metallo-organic paste (a product of Sprague Electromag Ronse-Belgium) where we tested a 10, 150 Ω/□ and a 7 kΩ/□ paste.

The measurements of the conduction properties and the 1/f-noise of the pastes were carried out on a test-vehicle with current and voltage tabs as desribed in reference [1]. Such a configuration enables us to eliminate the influence of contact resistance and noise at the current terminals as well as other anomalies in the contact regions caused by the Pd-Ag contacts (thickness,diffusion of Ag...).

In order to obtain a complete documentation on the different resistor pastes, we will review the most important measurements (from the literature and own experiments) concerning conduction and 1/f-noise in both systems.

* The different sheet resistivities are obtained from different blends of insulating and conducting material. Data for the DP 1400 series are well known[2] (see also figure 1). From this figure it is seen that the resistivity strongly depends on the volume-fraction (V_f) of the conductor (we plotted ρ versus ($V_f^{-1/3}$ -1)=the mean distance between cubic grains r_{pp}, normalised by the mean grain size Φ). Although no information is available on the metanets, we know that the volume fractions after firing are nearly 100 % and that the layers are very thin (about 100 nm) and encapsulated in glass.

* From the data-sheets of DP1400, it follows that higher firing temperatures lead to nearly identical or smaller values for the resistivity. This holds in exactly the same way for the influence of the firing period.

* TCR-curves on the DP1400 series have frequently been published[2]. There is always a minimum

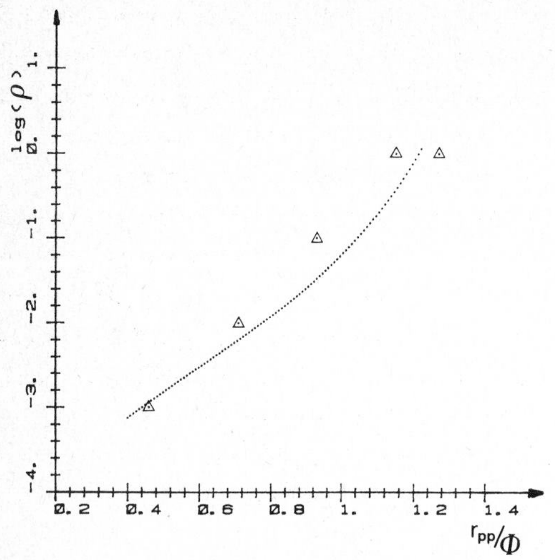

Figure 1 : Resistivity as function of the volume-fraction $[(V^{-1/3}-1)=r_{pp}/\Phi]$

△ : DP1400 (ρ=R□*10^{-5})

... : computer simulation

around 300 K in the curves $R_{\square}$ versus T. The variation of the $R_{\square}$ with T from 100K to 400K is less than 5%. The TCR-curves for three metanet pastes are presented in figure 2. For $R_{\square}$= 7KΩ/□ there is a minimum as has been reported in [1]. For R□=150Ω/□ we see that the slope is decreasing around 100K, which probably results in a minimum around 60K. Data for the lowest sheet resistivity are completely different and suggest that this resistor consists of a very thin

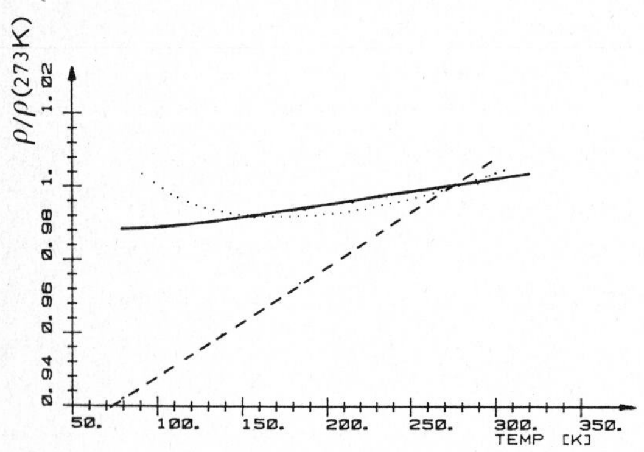

Figure 2 : ρ versus T for metanet resistors

... : R□= 7 kΩ/□

—— : R□= 150 Ω/□

--- : R□= 10 Ω/□

metallic film, with a resistivity increasing linearly with the temperature.

* Mobility measurements on the DP1400 series indicate a mobility of charge carriers smaller than 0.03 cm^2/Vs [2]. For the high sheet resistivities of metanets, we found, due to noise and drift, in the worst case upper limits of 0.4 cm^2/Vs (R□=150 Ω/□) and 1.8cm^2/Vs (R□ = 7KΩ/□), while for R□=10Ω/□ we obtained a value of 1.9 ± 0.2 cm^2/Vs. This results correspond well with the TCR-curves, where we suggested a metallic behaviour for the R□=10Ω/□.

* For all pastes we verified that the spectral density S_v of the 1/f-noise voltage increases with the square of the DC-voltage V_{DC} across the sample [1].

* The level of S_v was always inversely proportional to the resistor area as has been shown in reference [3].

* We also investigated the influence of the temperature on the 1/f-noise level. As has been reported in [3], there are only small variations of the order of magnitude of 2 in the whole temperature range from 80K to 350K.

To compare the different measurements of 1/f-noise we normalised the results as follows, in agreement with the empiric formula of Hooge [4] (see figure 3).

$$C^* = \frac{S_v}{V_{DC}^2} * l * w * t * f \qquad (1)$$

where l*w*t = volume of the sample [m^3]. C^* is the noise power spectral density (thus in a frequency band of 1Hz) at 1Hz for a unity volume of 1m^3 and 1V across the sample. According to Hooge, C^* equals α/n where n is a carrier concentration and α some dimensionless constant. In figure 3, we see that for the DP1400 series the value for C^* for R□≤1 kΩ/□ is about the same. With higher R□, the value of C^* increases nearly linearly with R□. For the metanets, a nearly invariant value for C^* is found for R□=150Ω/□ and 7KΩ/□. Again, as was remarked for the mobi-

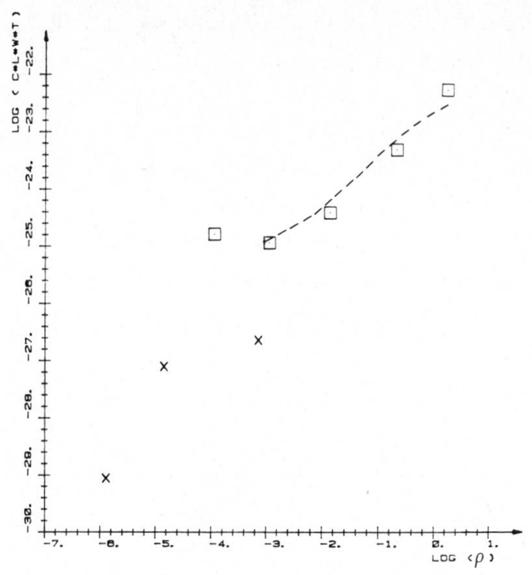

Figure 3 : Normalised noise spectral densities as function of the resistivity

□ : DP1400 $\rho = R\square * 10^{-5}$

x : Metanets $\rho = R\square * 10^{-7}$

lity and TCR, the behaviour of the paste with R□= 10Ω/□ is totally different from the other sheet resistivities. C^* has a very low value according to the formula of Hooge[4], what can be understood by the large concentration of free charge carriers.

A general remark is that all data for the metanets are at least 1.5 decades smaller then those for the cermets. This also corresponds to Hooge´s formula as the metal fractions for metanets are much larger and hence the carrier concentrations.

III. DISCUSSION.

The problem with thick film resistors is that the conduction model is still a point of discussion. It is established that a thick film resistor consists of conducting grains embedded in glass[5]. During firing of the resistor it is assumed that material from the conducting grains diffuses into the glass, creating an deep impurity band where conduction is dominated by hopping[5,6]. As the conducting grains are believed to be far more conducting than the glass, we may model the TFR as a random network, where the nodes stand for the conducting grains and the the resistors for the glass separating the grains. In the case of complete firing of the resistor, the glass is uniformly doped by the diffusion from the grains. In this way the resistors of the random network have all nearly the same resistance. In the case of uncomplete firing,the diffusion is nonuniform, and there is a large spread in the resistances, making the current flow become dominated by percolation paths.

To link the 1/f-noise with the conduction in TFR´s, we refer to the work of Pellegrini et al.[7,8], where the noise may be generated by localised states (in the glass). The noise spectrum S_V is given by the relation:

$$S_V = \frac{\langle \Delta V^2 \rangle}{f} D \qquad (2)$$

where $\langle \Delta V^2 \rangle$ the variance of the voltage fluctuations and D a distribution function of the logarithm of the conductance G of the localized states. In the case of a macroscopically homogeneous material, Pellegrini shows that $\langle \Delta V^2 \rangle$ becomes equal to

$$\langle \Delta V^2 \rangle = \frac{F\, V_{DC}^2}{N} \qquad (3)$$

where F = the volume fraction of the carriers generating 1/f-noise, V_{DC} the applied DC voltage and N the total number of charge carriers. Expression (2) and (3) give then the Hooge formula:

$$S_V = \frac{\beta\, V_{DC}^2}{N\, f} \qquad (4)$$

It seems acceptable to apply the Hooge formula (4) in the case of a complete firing, from which it follows that the noise is inversely proportional to the volume of the resistor as was experimentally observed. Moreover from (4) it is seen that S_v is also inversely proportional to the carrier density. In our case of a hopping dominated conduction, the conductivity of the glass becomes :

$$\sigma = q\ n\ \mu_{hop} \qquad (5)$$

where the carrier density n (≈ density of the localised states) is proportional to the volume fraction of the conducting grains. The mobility μ_{hop} is exponentially dependent on the separation of the localised states, and hence of the separation of the grains for a homogeneous doping from a finite source[6] (= $(V_f^{-1/3} -1)*\Phi$ where Φ is the mean grain dimension). Epression (5) explains why the sheet resistance becomes exponentially related to $V_f^{-1/3}$, but from (5) and (4) it is expected that S_v is only weakly dependant on R□, due to the weak variation of of the volume fraction with R□. This means that from figure 3, the behaviour of the cermets for R□< 1kΩ/□ and for the metanets for R□> 150 Ω/□ can be explained. In the case of cermets with R□> 10KΩ/□ the paste is uncompletely fired, resulting in a more nonuniform, more percolation type of conduction, and thus it is expected to generate more noise then normally calculated with (4). In the case of metanets with R□=10Ω/□ we believe that a metal non-metal transtion has occured, which is confirmed by the TCR measurements of figure 2 and the mobility data. It seems logically that in this case the density of carriers has increased, resulting in a low noise level.

IV. CONCLUSIONS

In this paper we reported 1/f-noise measurements on cermet and metanet resistors. On the basis of the conduction through the glass by a mechanism of nearest neighbour hopping, we can explain the trends in the variation of 1/f-noise with the resistivity. The justification of the application of the empirical formula of Hooge is given by a nearly homogeneously doped glass and the theory of Pellegrini et al.. In the case of uncomplete firing, the inhomogenous doping generates supplementary 1/f-noise. The extremely low value of the noise of the metanet with R□=10 Ω/□ is attributed to the formation of a metallic layer, which is justified by the results of the TCR and mobility measurements.

ACKNOWLEDGEMENT

The authors are grateful to the companies BTMC-Ghent and Sprague Electromag Ronse for their technical assistence and discussions resulting in a better understanding of the thick film technology.

REFERENCES.

1. A. Van Calster, L. Van Den Eede, S. Demolder and A. De Keyser : Proceedings of the 7th International Conference on 'Noise in Physical Systems' and the 3th International Conference on '1/fnoise' - Montpellier, may 1720 1983 - p 193

2. G.E. Pike and C.H. Seager : J. Appl. Phys. Vol 48, pp 5152-5169, dec 1977

3. S. Demolder, A. Van Calster, A. Lesquesne and E. Delen: ISHM - EUROPE - Stresa 1985 p 265.

4. F.N. Hooge : Physica 60 (1972) 130,144.

5. T.V. Nordstrom and C.R. Hills : Proc. 1979 Int. Microelectronics Symposium, Los Angeles 1979, pp 40,45

6. S. Demolder and A. Van Calster : Conduction in thick-film resistors: to be publised.

7. B. Pellegrini : Phys. Rev. B : Vol 24 number 12 - dec 1981 - p 7071.

8. B. Pellegrini, R. Saletti, P. Terrini and M. Prudenziati - Phys. Rev. B , Vol 27 number 2 , 15 jan 1983.

DEVICES - 1

NOISE IN PHYSICAL SYSTEMS AND 1/f NOISE - 1985
A. D'Amico and P. Mazzetti (editors)
Elsevier Science Publishers B.V., 1986

GATE DEPENDENT RANDOM TELEGRAPH NOISE IN SILICON MOSFETs

Michael J. UREN, Michael J. KIRTON and Derek J. DAY*

Royal Signals and Radar Establishment, Great Malvern, Worcestershire, U.K.

We demonstrate that the 1/f spectra found in large-area MOSFETs operating at room temperature are decomposed into individual Lorentzian components as the device areas are reduced. In the time domain, devices of area $0.4\mu m^2$ produce Random Telegraph Signals due to localised and discrete modulations of the channel resistance caused by individual carrier trapping events. We show that the trapping kinetics are thermally activated and that the traps reside in the oxide.

Recently, it has become evident that new insights into the 1/f problem can be obtained through the study of the noise properties of small area devices[1-3]. In this paper, we present noise spectra taken from conventional n-channel silicon MOSFETs operating at room temperature with active gate areas spanning the range $0.4\mu m^2$ to $350\mu m^2$. Large area devices gave the ubiquitous 1/f spectrum, while small area devices exhibited Random Telegraph Signals (RTSs) due to individual carrier trapping events. We shall demonstrate quite unequivocally that the deviations from large-scale noise behaviour found in the sub-micron devices are the result of the decomposition of the 1/f spectrum into its constituent Lorentzian components. In addition, we demonstrate that an analysis of the kinetics of carrier capture and emission points to a strongly thermally activated process.

Conventional CMOS devices with gate oxide thicknesses of 40nm were investigated. Both n and p channel devices showed similar effects but we will only present results for n-channel devices, all from the same wafer. Gate dimensions ranging from 20x20μm to 2x2μm were available, although the active area was less than the drawn area. Drain current noise and RTSs were measured using an HP3561A spectrum analyser/transient recorder for drain voltages between 4mV and 5V and drain currents between 100pA and 100μA.

In figures 1a,b,c we show the effect on the noise power spectral density of reducing device area, A. All these measurements were made in the linear regime for a fixed gate voltage in strong inversion. For comparison between differently sized devices, the scaled function $A.S_{Id}(f)/I_d^2$ was plotted against frequency. For devices from the same wafer, this should give 1/f spectra which superimpose one on top of the other. For the largest devices shown in figure 1a this behaviour was observed with a variation of less than 20% across the wafer. However, smaller devices with active areas below about $30\mu m^2$ departed significantly from the expected behaviour. Figure 1b depicts the scaled noise spectra from three devices with active areas of $15\mu m^2$. Two points are worth noting: first, the spectra no longer exhibit smooth 1/f behaviour, but significant and consistent irregularities are present so that they no longer superimpose; second, the spectrum is fixed in time for each

*Present address: Bell Northern Research, Ottawa, Canada.

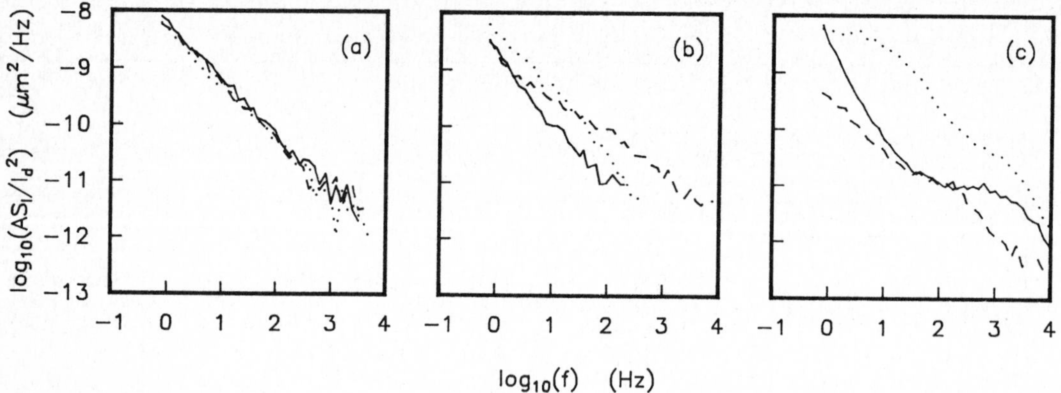

FIGURE 1
Noise power spectral density of n-channel MOSFETs for V_g =2V, T=293K. White noise has been subtracted. (a) three 20x20μm devices, active area 350μm^2, V_d=100mV, I_d~3.5μA; (b) three 20x2μm devices, active area 15μm^2, V_d=20mV, I_d~15μA; (c) three 2x2μm devices, active area 0.4μm^2, V_d=20mV, I_d~500nA.

device. However, it is clear that by averaging over many such devices the resultant spectrum would be 1/f, smooth and would exactly match those of figure 1a. This is shown even more dramatically in figure 1c where the spectra for three 0.4μm^2 devices are presented. Assuming a carrier trapping model, in these small-area devices one would expect only a handful of traps to be lying within a few kT of the surface Fermi level, and thus to be contributing to the noise. The spectrum of each device should then consist of the sum of a few individual Lorentzians. Inspection of figure 1c shows this hypothesis to be true. Again, averaging these scaled spectra over many such devices would lead to the spectra shown in figure 1a. The features depicted in figures 1a,b,c are, therefore, an elegant demonstration of the full decomposition of the 1/f spectrum into its constituent Lorentzian components and the sample-to-sample variation in the power spectra expected from very small systems[3].

In order to fully investigate the carrier trapping model, the variation of drain current versus time was measured. As expected, RTSs were observed[4]. Figure 2 shows representative signals from a 0.4μm^2 area device in moderate inversion and for five different gate voltages. These devices were sufficiently small that normally only one trap signal was visible at any one time; though by varying the gate voltage from weak to strong inversion it was possible to scan through several traps with different occupancy level positions. The discrete nature of the change in channel resistance observed is consistent with the effect being brought about by the trapping of individual electrons. The low current state of figure 2 can thus be associated with the trap containing one more electron than in the high current state, thereby giving rise to a localised modulation of the electrostatic potential at the Si:SiO_2 interface. This in turn produces a local reduction in electron density or, if preferred, a local increase in channel resistance over a small region of the channel. (A more complete discussion of the relationship between the fluctuation in trap occupancy and the magnitude of the fluctuation of channel resistance will be

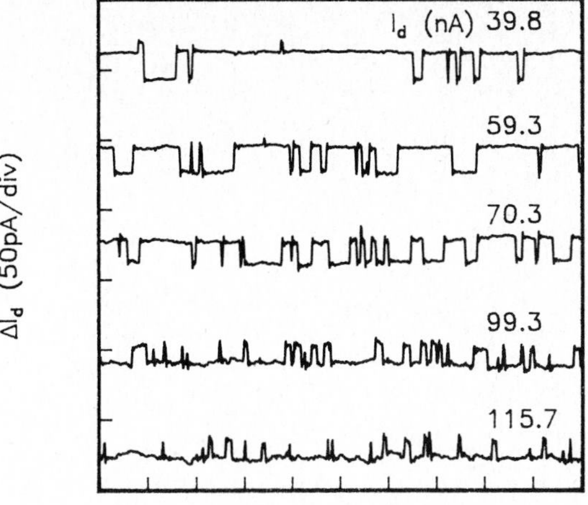

FIGURE 2
Random Telegraph Signals. Change in current vs. time measured at the indicated currents. Active area is $0.4\mu m^2$, V_d=4mV, T=293K.

given elsewhere[5].) For a given RTS, this analysis then allows us to define a mean capture time, $\bar{\tau}_c$, from the distribution of high current times and, similarly, a mean emission time, $\bar{\tau}_e$, from the low current times.

The RTSs shown in figure 2 are a manifestation of the equilibrium occupancy fluctuations of a trap at E_T on the source-drain current. The mean fractional occupancy, f, of the trap is given by

$$f = \frac{1}{1 + \exp(E_T - E_F)/kT} \quad . \tag{1}$$

It is evident from figure 2 that the time $\bar{\tau}_c$ was a sensitive function of gate voltage and decreased as the gate voltage increased - i.e. as the Fermi level moved up to and crossed the trap occupancy level. On the other hand, the emission time was essentially constant.

A measure of the position of the trap can be obtained by first re-arranging equation 1 and differentiating with respect to the gate voltage, V_g, giving[1]

$$\frac{dE_T}{dV_g} = \frac{kT}{q}\frac{d}{dV_g}\left[\ln\frac{\bar{\tau}_c}{\bar{\tau}_e}\right](\text{eV/Volt}) \quad . \tag{2}$$

Thus by evaluating the variation of $\bar{\tau}_c$ and $\bar{\tau}_e$ with respect to the gate voltage we can determine the movement of the trap occupancy level. For a given δV_g, the change in surface potential, $\delta\phi_s$, can be estimated using standard MOSFET analysis[6]; accurate values were difficult to obtain, however, due to poorly characterised short-channel effects. The depth into the oxide, d, is then simply given by

$$d = t_{ox}(\delta\phi_T - \delta\phi_s)/(\delta V_g - \delta\phi_s) \quad , \tag{3}$$

where t_{ox} is the oxide thickness. Preliminary values recovered from our data place the traps in the oxide, typically up to 5nm from the $Si:SiO_2$ interface.

The temperature dependences of three RTSs were preliminarly investigated. From Arrhenius plots of the variation of emission time with inverse temperature, we deduced activation energies in the range 0.6eV to 0.7eV: see, for example, figure 3. Since in these devices the Fermi level at the interface was between 0.2eV and 0.15eV from the conduction band edge of the silicon, this requires that capture and emission was via an intermediate state whose energy was ~0.6-0.7eV greater than the occupancy level E_T. Presumably, after electron capture into the intermediate state the total energy of the system was reduced by lattice relaxation processes bringing the energy of the trapped electron to the occupancy level, E_T. Carrier capture and emission is thus a two-stage process involving both tunnelling and thermally activated features. In addition, it is worth

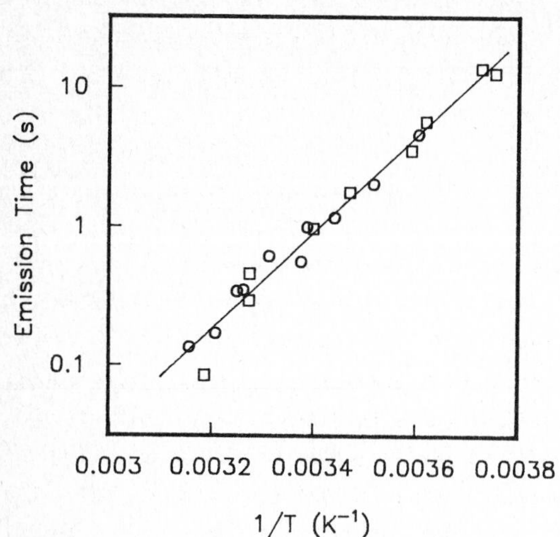

FIGURE 3
Arrhenius plot of mean emission time for a device of active area 0.4μm^2 measured at fixed drain current. ○ I_d=2nA; □ I_d=6.3nA; V_d=10mV. Activation energy was 0.69eV (solid line).

noting that the observation of thermally activated kinetics is consistent with recent findings in the 1/f spectra of silicon-on-sapphire resistors[7].

In conclusion, we have demonstrated that 1/f noise in our MOSFETs is the result of a superposition of Lorentzians due to individual carrier trapping events. Furthermore, we were able to follow the variation of capture and emission times of traps both as a function of gate voltage and temperature. We have shown that the trapping centres reside in the oxide and that the trapping kinetics are thermally activated.

ACKNOWLEDGEMENTS

Thanks are due to C.E.J. Kilgour for computational assistance, to R. Lambert of the GEC Hirst Research Centre for the devices and to J.P. Gillham for helpful discussions.

REFERENCES

1. K.S. Ralls, W.J. Skocpol, L.D. Jackel, R.E. Howard, L.A. Fetter, R.W. Epworth and D.M. Tennant, Phys. Rev. Lett. 52 (1984) 228.

2. C.T. Rogers and R.A. Buhrman, Phys. Rev. Lett. 53 (1984) 1272.

3. P.J. Restle, R.J. Hamilton, M.B. Weissman and M.S. Love, Phys. Rev. B31 (1985) 2254.

4. S. Machlup, J. Appl. Phys. 25 (1954) 341.

5. M.J. Uren, D.J. Day and M.J. Kirton Appl. Phys. Lett. in print.

6. H.C. Pao and C.T. Sah, Solid State Electron. 9 (1966) 927.

7. R.D. Black, P.J. Restle and M.B. Weisman, Phys. Rev. B28 (1983) 1935.

NOISE IN PHYSICAL SYSTEMS AND 1/f NOISE - 1985
A. D'Amico and P. Mazzetti (editors)

THERMAL NOISE IN A HIGH INDUCTANCE SQUID

A. MIOTELLO[+], M. MAZZER[++], M. CERDONIO[&] and S. VITALE[&]

+ Centro Studi CNR and Istituto per la Ricerca Scientifica e Tecnologica, 38050 Povo (TN) Italy
++ Dipartimento di Fisica, Università di Padova
& Dipartimento di Fisica, Università di Trento

A general description of the thermal noise in a SQUID, without limits in the choice of the physical parameters is presented. The central point of the numerical integration of Fokker-Planck equation in the theoretical analysis of the thermal noise is performed in this work. Indicative numerical solutions are presented.

1. INTRODUCTION

A quantitative description of the SQUID thermal noise is a relevant problem both from a theoretical and experimental point of view. Indeed experimental people working in this field are generally interested at the control of the magnetic flux fluctuations while theoretical people apply some interest to establish for fixed values of the typical SQUID's parameters, the regimes where quantum tunneling is dominant with respect to normal fluctuations due to thermal noise. A theoretical description of the thermal noise has been performed by Kurkijärvi[1] only for the physical situations where fluctuations may be described, in the framework of the rate theory, as a jumping process of a particle having a kinetic energy in a sinusoidal potential. This approach holds when the inductance L associated to the SQUID is low. Here we try to extend the model calculation to the case of high inductance value to provide a general picture of the intrinsic fluctuations in a SQUID due to the Johnson noise and to show more precisely the limits in which the model of Kurkijärvi can be used. We describe the fluctuations process by the Fokker-Plank equation, for the magnetic flux, including the thermal noise by Langevin equations and appropriate boundary conditions. Some results connected to the numerical integration of Fokker-Plank equation are presented and discussed.

2. STATEMENT OF THE PROBLEM

The time variation of the admitted flux Φ for a ring with an ideal Josephson junction at T=0 (T is the temperature) is governed in the RSJ scheme by eq. 12.2.5 of ref. 2. According to ref. 1, this equation may be written as:

$$\ddot{\Phi} + \frac{1}{RC}\dot{\Phi} + \frac{1}{C}\frac{\partial U}{\partial \Phi} = 0 \qquad (1)$$

where $U = \frac{1}{2}\frac{(\Phi-\Phi_x)^2}{L} - i_c\frac{\Phi_0}{2\pi}\cos\left(2\pi\frac{\Phi}{\Phi_0}\right)$

i_c is the critical current, C the capacitance, R the resistence L the inductance, Φ_x the applied external flux and Φ_0 the flux quantum $(\frac{h}{2e})$. Eq. (1) is homologous with the equation for a particle moving in a sinusoidally modulated parabolic potential and having a damping coefficient $\frac{1}{RC}$. Having in mind this classical analog, when $\beta_e \equiv \frac{2\pi L\, i_c}{\Phi_0} > 1$

the flux "jumps" only for fixed value of $\Phi_X \equiv \Phi_{XC}$ (see e.g. fig. 1 of ref. 1). Assuming that the jump occurs reasonably close to the critical flux Φ_{XC}, Kurkijärvi calculates the potential barrier to the lowest order in $\Delta\Phi_X = \Phi_{XC} - \Phi_X$ and evaluates the height of the "potential barrier" which must be overcome in order to have a flux transition as

$$\Delta U = \frac{4}{3L}\left(\frac{\Phi_o}{2\pi}\right)^{\frac{5}{2}}\left(\frac{2}{L\, i_c\{1-(1/\beta_e)^2\}^{1/2}}\right)^{\frac{1}{2}}\left[\frac{2\pi(\Phi_X-\Phi_{XC})}{\Phi_o}\right]^{\frac{3}{2}} \quad (2)$$

Finally by utilizing the Chandrasekhar's[3] results for the jumping process of a particle having a mean thermal energy of $k_B T$ trying to get over a barrier ΔU, Kurkijärvi describes the dynamical process of the flux fluctuations. However Chandrasekhar's results hold only for $\Delta U \gg k_B T$ and so they cannot be applied for high values of L parameters in eq. (2). In order to give a more general treatment we note that, in the RSJ scheme when $T > 0$, a fluctuating current $\tilde{I}(t)$ flows through the resistive element. $\tilde{I}(t)$ is such that

$$<\tilde{I}(t)> = 0, \quad <\tilde{I}(t+\tau)\tilde{I}(t)> = \frac{2k_BT}{R}\,\delta(\tau)$$

Taken into account this fluctuating current for which we know only the statistical properties, the dynamics of flux Φ is completely described by the Fokker-Plank equation, for the magnetic flux

$$\frac{\partial W}{\partial t} = -V\frac{\partial W}{\partial \Phi} + \frac{\partial}{\partial V}\left[\left(\frac{1}{C}\frac{\partial U}{\partial \Phi} + \frac{1}{RC}V\right)W + Rk_BT\,\frac{\partial W}{\partial V}\right] \quad (3)$$

coupled with Langevin's equations

$$\begin{cases} V = -\dfrac{\partial \Phi}{\partial t} \\[2ex] \dfrac{\partial V}{\partial t} = -\dfrac{V}{RC} + \dfrac{1}{C}\dfrac{\partial U}{\partial \Phi} + \dfrac{1}{C}\tilde{I}(t) \end{cases} \quad (4)$$

By integration of equations (3,4) we may give both the probability density W (Φ, v, t) and the probability for the transition P $(\Phi, v, t, \Phi', v'$ o). General analytical solutions of eqs. (3), (4) are not available and so numerical integration must be performed while the values for $\tilde{I}(t)$ are given by a computer routine generating random numbers which have a gaussian distribution around their mean value $(<\tilde{I}(t)>=0)$ and appropriate physical values for the variance.

The numerical integration of the Fokker-Planck equation is "per se" a non trivial mathematical problem. In the following we concentrate on this computer numerical problem which is essential for a quantitative description of the thermal noise in a SQUID in which no limits for its physical parameters are imposed.

3. NUMERICAL INTEGRATION OF FOKKER-PLANCK EQUATION

In order to solve the Fokker-Planck equation we made use of a numerical method mainly based on a finite differences scheme[4]. The probability density function W (Φ, v, t) depends on three variable quantities and so it would be necessary to consider a three-dimensional lattice in which Δt, $\Delta\Phi$ and ΔV would be the distance between a given point and his nearest neighbours along the three directions. In order to satisfy the stability conditions for the numerical solution, $\Delta\Phi$ ΔV, and Δt cannot be independent on each others and on SQUID's parameters but they are subject to some constraints. As we need to obtain significant results in a reasonable computation time we follow a method which enables us to work with a lattice spanned by larger $\Delta\Phi$ and ΔV.

We can achieve this purpose evaluating the derivatives

$$\frac{\partial W}{\partial \Phi}, \quad \frac{\partial W}{\partial V}, \quad \frac{\partial^2 W}{\partial V^2}$$

analitically after fitting the function values corresponding to the points in the lattice by means of the cubic spline method [5]. This is a useful method for interpolating points by means of cubic polinomials which incorporates mathematical constraint that would prevent the appearence of undesirable inflection points. Eq. (3) must be solved with appropriate boundary conditions in order to have a meaningful numerical solution. First of all we must fix the values W (t=0, Φ , V) of the probability density function. Furthermore the behaviour at any instant t of W (t, Φ , V) must satisfy the conditions:

$$\begin{cases} \lim\limits_{\Phi \to \pm\infty} W(t, \Phi, V) = 0 \\ \lim\limits_{V \to \pm\infty} W(t, \Phi, V) = 0 \\ \lim\limits_{V \to \pm\infty} \frac{\partial W}{\partial V}(t, \Phi, V) = 0 \end{cases} \tag{5}$$

which are enough to fix the value of the derivatives

$$\frac{\partial W}{\partial \Phi}, \quad \frac{\partial W}{\partial V}, \quad \frac{\partial^2 W}{\partial V^2}$$

at the boundary of the lattice's plane for a constant value of the time parameter. As far as the step Δt is concerned we can say that less is the inductance L (for fixed values of R and C) the more we must lower Δt so as to satisfy the stability conditions.

4. NUMERICAL RESULTS

In this section we give an idea of the kind of the results which have been obtained for the W (t, Φ , V) function in a simple indicative situation. As we have just noted, the behaviour of the analyzed physical system is like that of a classical particle (having a "position" Φ a "velocity" $\frac{\partial \Phi}{\partial t} = -V$ and mean "kinetic energy" k_B T) moving in a sinusoidally modulated parabolic potential and with a damping coefficient $\frac{1}{RC}$. The more the external flux approaches the critical value Φ_{xc} the lower is the potential barrier to be overcome by the particle in order to make a transition. Starting at t=0 with an external flux Φ_x such that the particle is unlikely to make a transition, we suddenly move it towards the critical value Φ_{xc} so as to lower the potential barrier. The shape of the surface describing the probability density at $t=0=t_0$ is gaussian with a maximum at V=0 and $\Phi = \Phi_{min}$, Φ_{min} being the flux corresponding to a local minimum of the potential U. Beginning from the instant $t_0 + \Delta t$ if we observe any section of the surface at V=const., we note the presence of a new peak corresponding to another value of the flux Φ .

Moving away from V=0, i.e. increasing the "kinetic energy" of the "particle" we can see that the amplitude of this new peak becomes soon relevant. As t increases the amplitude of this new peak becomes larger than the first one's untill the transition becomes complete and the system reaches the final "equilibrium" situation.

Fig. 1 a, b, c shows the time evolution of three sections of the W (t, Φ , v) for three values of the V parameter.

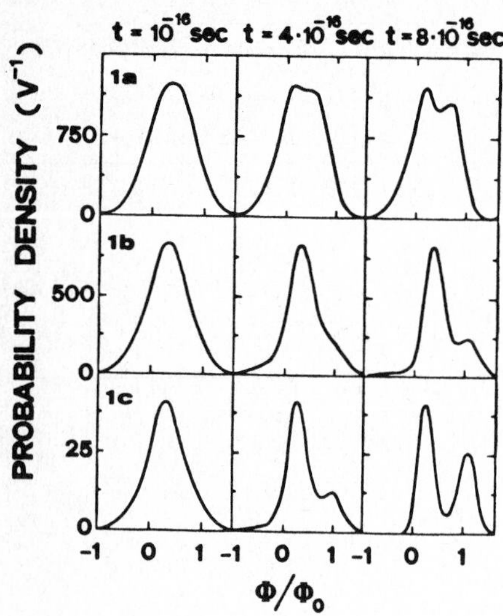

FIGURE 1a,b,c
Time evolution of three sections of the $W(t,\Phi,V)$ function: 1a) $V = 1.875.10^{-4}V$; 1b) $V=-2.5.10^{-4}V$; 1c) $V = -5.10^{-4}V$.
$R = 50\,\Omega$, $L = 10^{-12}$ H, $C = 10^{-14}$F.

In order to chek the validity of the results we evaluated for appropriate values of the time parameter the integral $\iint W(\Phi,V,t)d\Phi dV$: its value remains very close to initial unity value.

Fig.1a,b,c gives a clear idea of the trend of the solutions of Fokker-Planck equation for a simple experimental situation. However by the same numerical procedure, more complicated (but realistic) physical situation may be described. In particular working with a sinusoidally variable external flux with an RF - SQUID characteristic frequency, we can analyze the behaviour of the system within half a period of a flux oscillation and so we can extend the analysis of the thermal noise in an RF - SQUID at high value of L's parameter where the Kurkijärvi approximation fails to work.

5. CONCLUSIONS

Integration of Fokker-Planck equation coupled with Langevin equation give the general solution of the thermal noise problem in a SQUID, without limits in the choice of the physical parameters (in particular the L values). This is not the case of the Kurkijärvi model which holds only for low values of the inductance parameter. The central point of the numerical integration of Fokker-Planck equation, in the theoretical analysis of thermal noise, has been performed in this work. Physical situations closer to realistic experimental conditions (as well as the evaluation of the probability density correlation functions) will be presented elsewhere.

REFERENCES

1. J. Kurkijärvi, Phys. Rev. B, 6 (1972) 832.

2. A. Barone and G. Paternò, Physics and Applications of the Josephson Effect (J. Wiley, New York, 1982) p. 365.

3. S. Chandrasekhar, Rev. Mod. Phys. 15 (1943)1.

4. William F. Ames, Numerical Methods for Partial Differential Equations II edition (Academic Press, New York, San Francisco, 1977).

5. Yehnda Taitel and Abraham Tamir, Aiche Journal 29 (1983) 153.

NOISE IN PHYSICAL SYSTEMS AND 1/f NOISE - 1985
A. D'Amico and P. Mazzetti (editors)

MODE FLUCTUATIONS IN SEMICONDUCTOR LASER

Tatsuyuki KAWAKUBO, Hideo AKABANE and Tohru TSUCHIYA

Department of Applied Physics, Tokyo Institute of Technology, Oh-okayama, Meguro, Tokyo, Japan

The mean value and the noise variance of output power of each mode of a double heterostructure laser diode have been measured as a function of injection current. With increasing current, the lasing mode is found to surpass the adjacent nonlasing modes near threshold and to grow up alone, while the nonlasing modes once decrease above threshold and again increase at a low rate. As for the noise variance, no conspicuous critical enhancement is observed for the lasing mode, but the nonlasing modes show an enhancement near threshold. A model simulation based on the nonlinear mode coupling is compared with the experimental result.

1. INTRODUCTION

The noise of output intensity from laser diodes is a crucial problem for optical application systems. Intensity and frequency noise measurements on laser diodes have been made from various aspects during the last two decades. The statistical behaviors of the intensity fluctuation in the emission from various modes of a cw GaAs laser was first examined by Armstrong and Smith[1]. Paoli[2] measured the frequency spectrum of the fluctuation for double heterostructure junction lasers. The fluctuation in each longitudinal mode was measured by Ito et al[3] who found that the low frequency noise for each mode is remarkably large compared with the noise for total modes above the threshold current. The mode hopping noise, in particular, is a serious problem in laser application because of its macroscopic scale and it seems to be strongly concerned with the nonlinear coupling between modes.

In this report, the mean value and the fluctuation of the output intensity of the lasing and nonlasing modes of a laser diode have been measured and a comparison with a model simulation based on the nonlinear mode coupling has been made.

2. EXPERIMENTAL

The output of AlGaAs double heterostructure laser diode was divided into each mode by a plane reflection grating. The whole spectrum was observed with a charge coupled device (CCD) and the fluctuation was measured with avalanche photodiodes. Figure 1 shows spectra obtained from CCD at 25°C for several injection currents.

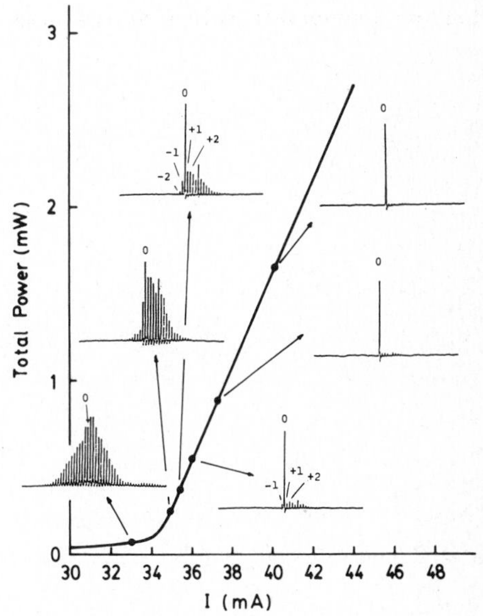

FIGURE 1
Spectra of output intensity

Here the lasing mode is labeled 0 and the nonlasing modes with longer wavelengths are numbered successively 1, 2, 3,....; those with shorter wavelengths are numbered -1, -2, -3,..... The wavelength difference between two adjacent modes is 0.31 nm. The solid line represents the total output power vs injection current curve. The intensity of each spectrum in Fig.1 is normalized by its maximum value, but the absolute value of spectral intensity rapidly increases with injection current.

The injection current dependence of the intensity for several modes is shown in Fig.2. With increasing current, the lasing mode(0) is found to surpass the adjacent nonlasing modes and to grow up alone at 35 mA; while some nonlasing modes rather once decrease above 35 mA and again increase at a low rate.

The intensity of each mode fluctuates around its mean value. Figure 3 shows the root mean square of intensity noise for several modes which was measured with a 15 kHz band pass filter of 2 kHz half-width. An arbitrary unit is used for the ordinate axis but it is common to both of Figs 2 and 3. No conspicuous critical fluctuation is observed for the lasing mode, however, the nonlasing modes exhibit a slight enhancement near threshold and increase again at higher current. In comparison with the result of Fig.2, the noise intensity of each mode is seen to be roughly proportional to its mean value.

The cross correlation of fluctuations between different modes was also measured at several temperatures[4]. The cross correlation coefficient between the lasing mode and its adjacent nonlasing modes shows a small positive value, while it changes to large negative values above threshold. The negative value of the cross correlation coefficient indicates that the energy transfer occurs between the two modes, so that the lasing mode seems to have a direct

FIGURE 2
Current dependence of the intensity for lasing (0) and nonlasing(+1, -1, -2) modes

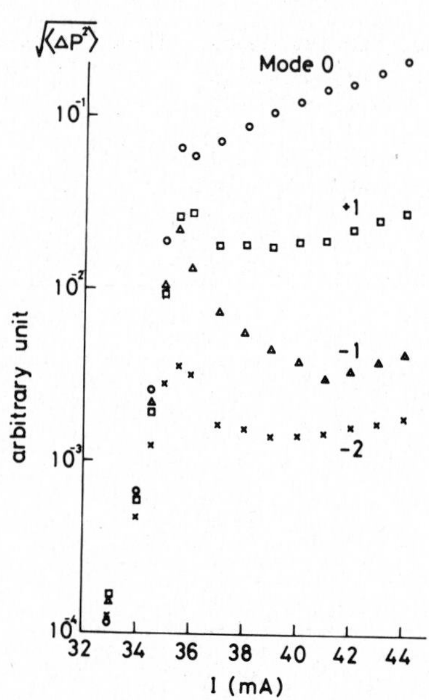

FIGURE 3
Current dependence of the root mean square of intensity noise for lasing and nonlasing modes

coupling with its adjacent nonlasing modes.

3. MODEL SIMULATION

Every mode has a tendency to enhance its intensity near threshold with increasing current, but only one mode surpasses all the adjacent modes above threshold. The noise intensity of each mode varies with injection current roughly in parallel with the mean value. To clarify these behaviors of the mode competition, we have made a model simulation based on the nonlinear mode-mode coupling.

We start a van der Pol type Langevin equation for electric field E,

$$\frac{dE}{dt} = \alpha E - \gamma E^3 + f(t) \quad (1)$$

where α is the pumping parameter determined from the gain and the feedback facter, γ is the nonlinear coefficient and f(t) is the random force which may be due to the spontaneous emission. The gain value is taken to be different for each mode. If we make a Fourier transformation of E,

$$E = \sum_n(\tilde{E}_n e^{in\Delta\omega t} + \tilde{E}_n^* e^{-in\Delta\omega t}) \quad (2)$$

where $\Delta\omega$ is the angular frequency difference between two neighboring modes. Then the rate equation for $\tilde{E}_n$ is given by

$$\frac{d\tilde{E}_n}{dt} = \alpha_n\tilde{E}_n - 6\gamma\sum_{m\neq n}|\tilde{E}_m|^2\tilde{E}_n - 3\gamma|\tilde{E}_n|^2\tilde{E}_n + \tilde{f}_n \quad (3)$$

where $\tilde{f}_n$ is the Fourier component of the random force.

In computer simulation, we took only five modes; $n = n_0-2$, n_0-1, n_0, n_0+1 and n_0+2, centering around the mode n_0 which has the highest gain, and hereafter we abbreviate them to -2, -1, 0, +1 and +2, respectively. The pumping parameters for submodes -2, -1, +1 and +2 were chosen slightly less than the parameter for central mode α_0; i.e. $\alpha_{\pm1}=\alpha_0-0.01$ and $\alpha_{\pm2}=\alpha_0-0.02$. The nonlinear coefficient γ was chosen 0.1. As the random force, we used a series of normal random numbers generated from uniform random numbers.

Starting from adequate initial values of $\tilde{E}_n$'s, numerical calculations of (3) were carried out by the Runge-Kutta method and each mode power was obtained from the steady state value of $\tilde{E}_n$ by $P_n=|\tilde{E}_n|^2$. If we neglect the random force $\tilde{f}_n$ in (3), only mode 0 which has the largest pumping parameter grows up in the active region ($\alpha_0>0$) and all the submodes tend to zero in the steady state. In the presence of the random force, however, a small power presents itself for submodes. The α_0 dependence of the power for each mode is shown in Fig.4.

The mode 0 is seen to surpass the adjacent modes ±1 and ±2 above threshold, though the difference in pumping parameter between them is very small. The time average of the Fourier component of electric field for submodes is found to vanish, i.e., $\langle\tilde{E}_{\pm1}\rangle=\langle\tilde{E}_{\pm2}\rangle=0$; but the average power which is expressed by $\langle P_{\pm1}\rangle=\langle|\tilde{E}_{\pm1}|^2\rangle$ or $\langle P_{\pm2}\rangle=\langle|E_{\pm2}|^2\rangle$ is not zero. Hence, the emission from submodes is not a coherent ray but may be referred to a thermal ray. A marked difference between the model simulation (Fig.4) and the experimental result (Fig.2) is that above threshold, the submode power once remarkably decreases

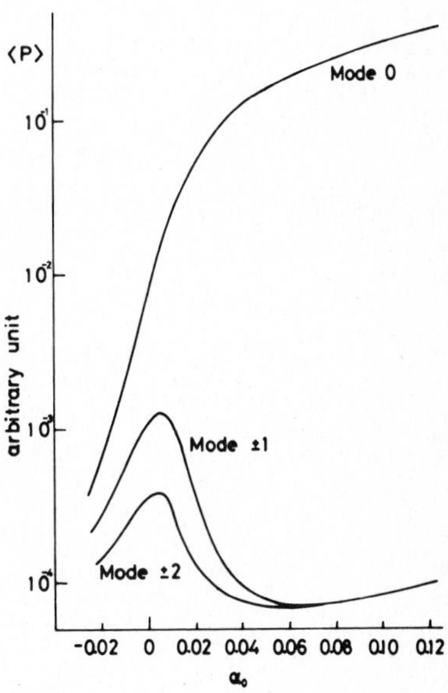

FIGURE 4
α_0 dependence of each mode intensity by model simulation
$\alpha_{\pm1}=\alpha_0-0.01$ and $\alpha_{\pm2}=\alpha_0-0.02$

with increasing α_0 in the model simulation, while the decrease is not so remarkable in the experimental result. In the model simulation, we changed the values of pumping parameters keeping the superiority of α_0 over the others, but in actual laser diodes the gain profile is known to shift its maximum to longer wavelength, so that the adjacent modes with longer wavelength become unstable with increasing injection current and a hopping to a neighboring mode occurs.

Figure 5 shows the α_0 dependence of the root mean square of the power fluctuation. It should be noticed here that the noise experimentally measured (Fig.3) is a frequency component of 15 kHz, while that of model simulation (Fig.5) is

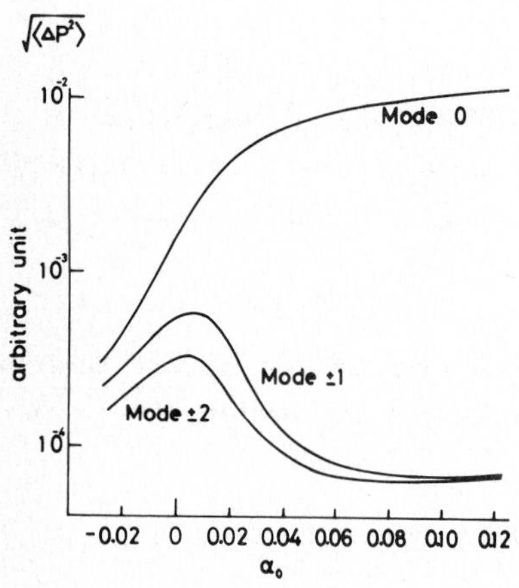

FIGURE 5
α_0 dependence of the root mean square of the intensity fluctuation by model simulation

the total fluctuation. In comparison between Figs 4 and 5, we can see that the root mean square of the power fluctuation varies in parallel to the mean value; the same aspect is observed in experimental results (Figs 2 and 3).

4. DISCUSSION

Even if the pumping parameters of several modes are very close, only one mode which has the largest pumping parameter grows up above threshold. An establishment of this kind of "autocracy" seems to be achieved by nonlinear coupling. The submode intensity is suppressed by the lasing mode and will vanish if there is no random force. According to the present model, the mean power of each mode is the second moment of the fluctuating electric field; $\langle P_m \rangle = \langle |E_m|^2 \rangle$, and the variance of the power noise is expressed in terms of the fourth and the second moments of the electric field; $\langle \Delta P_m^2 \rangle = \langle |E_m|^4 \rangle - \langle |E_m|^2 \rangle^2$. Thus the enhancement of the submode intensity near threshold is a critical fluctuation due to an instability.

REFERENCES

1. J.A. Armstrong and A.W. Smith: Phys. Rev. 140 (1965) A155.

2. T.L. Paoli: Appl. Phys. Lett. 24 (1974) 187.

3. T. Ito, S. Machida, K. Nawata and T. Ikegami: IEEE J. Quantum Electron. QE-13 (1977) 574.

4. H. Akabane, T. Tsuchiya, T. Kawakubo and H. Abe: Jpn. J. Appl. Phys. 24 (1985) L501.

NOISE IN PHYSICAL SYSTEMS AND 1/f NOISE - 1985
A. D'Amico and P. Mazzetti (editors)

MONTE CARLO CALCULATIONS OF THE DIFFUSION COEFFICIENT AND THE DRIFT VELOCITY OF ELECTRONS IN AN AlGaAs-GaAs QUANTUM WELL DEVICE

A.D. van RHEENEN and G. BOSMAN

Dept. of Electrical Engineering U. of Florida, Gainesville, FL 32611, USA

The charge transport properties of a two-dimensional, GaAs square quantum well have been studied as a function of well width and electric field strength. The results are interpreted in terms of the field dependent mobility, diffusion coefficient, and valley occupancy.

1. INTRODUCTION

Recent progress in submicron lithography and the M.B.E. growth technique has led to the development of very fast, ultra low-noise ALGaAs-GaAs modulation-doped FETs (MODFETs). Charge transport in these FET structures takes place predominantly in the high-mobility GaAs channel and under the influence of electric field strengths well into the hot electron regime and parallel to the interface. One of the approaches to a better understanding of the transport mechanism is the simulation of an electron traveling through the channel under the influence of an electric field parallel to the interface employing the Monte Carlo method.

Due to the difference in electron affinities of the GaAs and the AlGaAs and the phenomenon of real space-charge transfer, the conduction band profile of a MODFET structure can be represented by a finite height, quasi-triangular quantum well. Hence the description of charge transport should involve solving simultaneously Poisson's equation and the Schroedinger equation in addition to the Monte Carlo simulation. To avoid the complications associated with this simultaneous solution, we adopt the simple model of an infinitely high, square quantum well for which the wave functions and energy eigenvalues are well known. Note that this model adequately describes the conduction band of AlGaAs-GaAs heterostructures. In addition, this simplification allows us to study the effect of quantization on the hot electron properties by varying the width of the quantum well.

2. THE MONTE CARLO METHOD

The small width of the GaAs layer results in the quantization of the wave vector perpendicular to the interface (x-direction). The electron transport occurs along the interface (z-direction).

The wave function for such a quantum well with infinitely high barriers is expressed as follows[1]:

$$\psi(\bar{r}) = \sqrt{\frac{2}{V}}\, u_{\bar{k}}(\bar{r})\exp(i[k_y \cdot y + k_z \cdot z])\sin\left(\frac{n\pi x}{L}\right) \quad (1)$$

and the dispersion relation is

$$E_{\bar{k}} = E_{\bar{k}} + E_n = \frac{\hbar^2}{2m^*}(k_y^2+k_z^2) + n^2 E_0 \quad (2)$$

where

$$E_0 = \frac{\hbar^2}{2m^*}\left(\frac{\pi}{L}\right)^2. \quad (3)$$

The width of the well is L, h is Planck's constant and m* is the effective mass of the electron depending on the valley (central or satellite) in which the electron is.

Also,

$$\bar{k} = (k_x, k_y, k_z) = \left(\frac{n\pi}{L}, k_y, k_z\right). \quad (4)$$

In Fig. 1 we show the dispersion relation eq. (2) (not on scale). The arrows indicate probable phonon scattering processes. The density of states per unit energy range is

$$N(E_{\bar{K}}) = \frac{nm^*}{2\pi\hbar^2 L} \times 2 \qquad (5)$$

where the last factor 2 accounts for spin degeneracy. The scattering rates are then given by

$$W_{\bar{K}} = \frac{2\pi}{\hbar} \int |\langle\bar{K}'|H_{ep}|\bar{K}\rangle|^2 \delta(E_{\bar{K}'} - E_{\bar{K}}) dN_{\bar{K}'} \qquad (6)$$

where $\langle\bar{K}'|$ and $|\bar{K}\rangle$ denote the final and initial state respectively, $\delta(E_{\bar{K}'} - E_{\bar{K}})$ provides for energy conservation, and H_{ep} is the electron-phonon interaction.

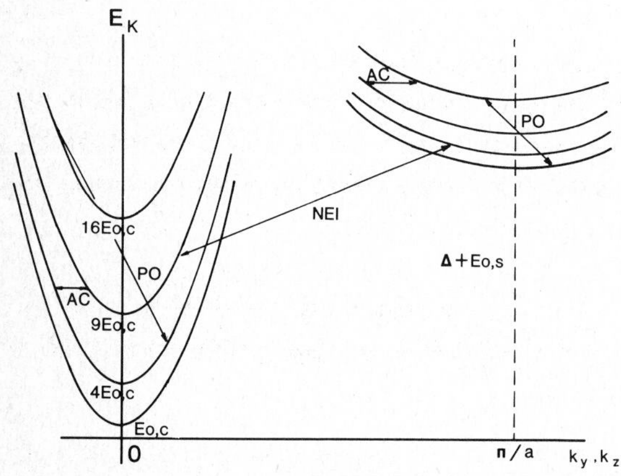

FIGURE 1
The band structure we considered in our model. The quantization is in the x-direction. The arrows denote the transitions due to electron phonon scattering. PO = polar optical, AC = acoustic, NEI = nonequivalent intervalley scattering.

These scattering rates are to be calculated and inserted in the Monte Carlo simulation program. As a basis for our computer program, we used the program as it was developed by Boardman[2], which accounts for acoustic and polar-optical intravalley scattering and for equivalent and nonequivalent intervalley scattering. We do not consider impurity scattering here because in practical heterostructures the electrons in the GaAs are spatially separated from their donors in the AlGaAs by a spacer layer.

As stated, one of the main modifications that had to be made was the insertion of the quasi-two-dimensional scattering rates. Another modification that had to be made was the selection procedure of the final wave vector after scattering. In wave vector space only planes with distinct values for the x-component ($k_x = \frac{n\pi}{L}$) are available. This significantly reduces the number of available wave vectors after scattering compared to bulk GaAs. Therefore, we expect the mobility to be higher in a 2-D system than in a 3-D system since less scattering events will occur.

In our calculations we made the following two assumptions. Firstly, the phonon gas is treated three dimensionally. This seems to be justified by the close lattice matching of the GaAs and the AlGaAs. Secondly, we assume that the momentum conservation approximation (MCA)[1] holds. This approximation is made in order to get analytic expressions for the scattering rates. For our quantum well the MCA results in an error of 60% for L = 50Å and in an error of less than 10% for $L \geq 200$Å.

We calculated the drift velocity, the diffusion coefficient and the occupation numbers of the subbands in both the central and satellite valleys as a function of the electric field strength for values ranging from the low-field ohmic regime up into the hot electron regime.

3. DISCUSSION

In Fig. 2 we present the drift velocity-field curves for different well widths (L = 200Å, 600Å) as calculated, using the model described above. For comparison we also included the bulk drift velocity-field curve. As expected, due to the smaller number of final states available after scattering we see that as the well width is decreased, the low-field mobility increases significantly, resulting in a steeper velocity-field curve.

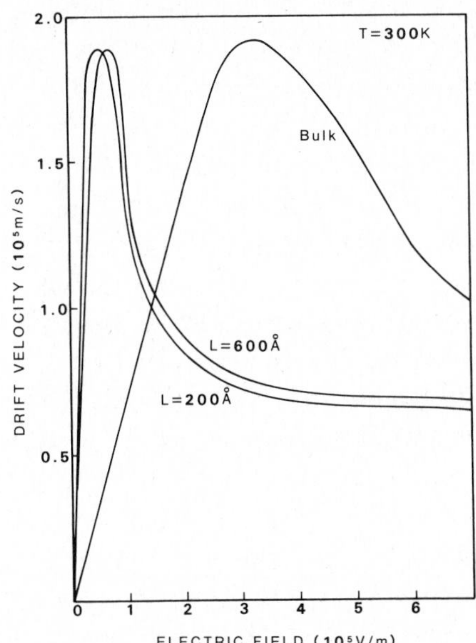

FIGURE 2
The drift velocity vs. electricfield strength for different quantum wellwidths. For comparison we included the curve for bulk GaAs.

All of the curves show to have the same peak velocity. As we look at Fig. 3, we see that as the well width is decreased, the occupation of the satellite valleys occurs at lower fields, suggesting that due to the higher mobility the electrons are heated up more quickly, resulting in a transfer from the central to the satellite valleys. The electrons lose kinetic energy in that process, resulting in a lowering of the electron temperature. From this we conclude that the intervalley scattering process is limiting the drift velocity.

As the well width is decreased, the saturation velocity is reduced. This can again be explained by the occupation ratio of the satellite and central valleys. As mentioned before, for a fixed electric field the occupation of the satellite valleys increases as the well width decreases (see Fig. 3). The occupation of the

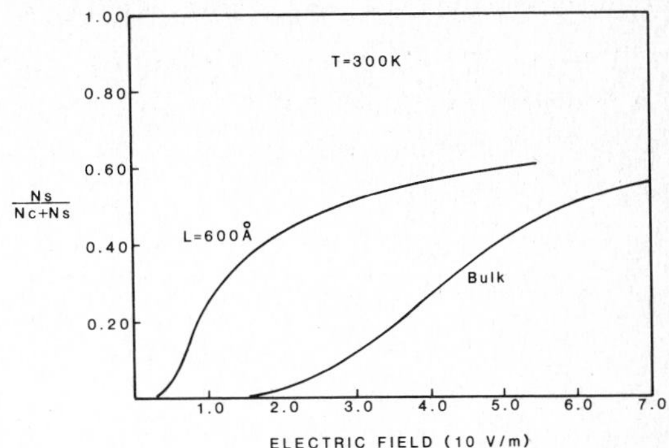

FIGURE 3
The relative occupancy of the satellite valleys vs. electric field strength for different quantum well widths. We included the curve for bulk GaAs.

higher effective mass valley is increased, resulting in a lower mobility and a lower drift velocity. Results for the diffusion coefficient are shown in Fig. 4. The low-field values are consistent with the Einstein relation for a two-dimensional gas:

$$\frac{D}{\mu} = \frac{k_BT}{q} \frac{\sum_{n=1}^{\infty} n \,\ell n\{1 + \exp(-\frac{n^2E_0-E_F}{k_BT})\}}{\sum_{n=1}^{\infty} [n/\{\exp(\frac{n^2E_0-E_F}{k_BT}) + 1\}]} \tag{7}$$

where E_F denotes the Fermi level. For a non-degenerate gas, $E_F \ll E_0$ or $E_0 - E_F \gg k_BT$ eq. (7) complies with the well-known result:

$$\frac{D}{\mu} = \frac{k_BT}{q} . \tag{8}$$

Although the shape of quantum wells in practical AlGaAs devices differs from the one we have studied, the general trend of obtaining a higher low-field mobility and diffusion coefficient and a lower saturation velocity for decreasing dimensionality is probably generally valid. This phenomenon can be used to enhance the device performance of, for example, a high-frequency FET. A quantum well will lower the value of the parasitic source and drain distance in which only small electric fields are preserved. In addition, a Gunn diode employing a 2-D channel will operate efficiently at lower bias voltages than a 3-D device, since the 2-D negative differential mobility regime extends to lower field strengths.

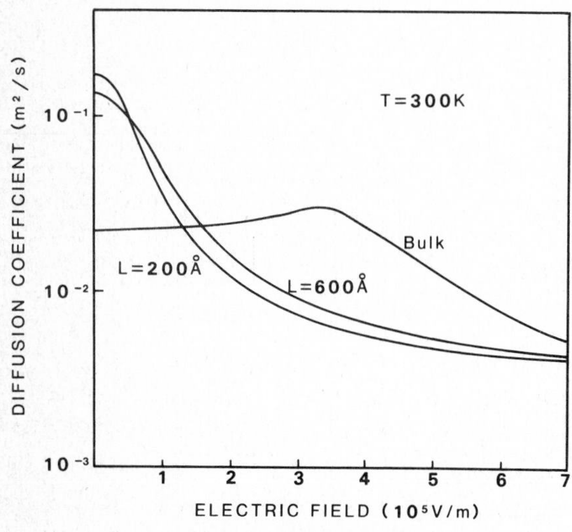

FIGURE 4
The diffusion coefficient vs. electric field strength for different quantum well widths. We also show the curve for bulk GaAs.

In minority carrier devices, the high value of the 2-D diffusion coefficient at low fields can lead to ultra short transit times in the base of bipolar transistors and to short switching times. A severe disadvantage of quantum wells, however, is their low saturation velocity. Specifically, short channel FETs used in an amplifier configuration operate in a regime where the carrier velocity is saturated. The lower 2-D saturation velocity degrades the transconductance and consequently the gain bandwidth product and the noise figure.

In conclusion, a selective use of quantum wells in device fabrication can significantly enhance device performance.

ACKNOWLEDGEMENT

We thank the National Science Foundation for supporting this research under grant #ECS-83-00222.

REFERENCES

1. B.K. Ridley, J. Phys. C: Solid State Phys. 15 (1982), 5899.

2. A.D. Boardman, Physics Programs (Chichester, John Wiley & Sons, 1980).

NOISE IN PHYSICAL SYSTEMS AND 1/f NOISE - 1985
A. D'Amico and P. Mazzetti (editors)

HIGH-FREQUENCY CHANNEL NOISE OF MODULATION-DOPED AlGaAs-GaAs FETs

C.F. WHITESIDE, G. BOSMAN, A. van der ZIEL,

Department of Electrical Engineering, U. of Florida, Gainesville, FL 32611, USA

W. KOPP and H. MORKOC

E.E. Dept. and Coordinated Science Laboratory, U. of Illinois, Urbana, IL 61801, USA

The channel noise of a normally-on MODFET has been measured at T = 300K as a function of bias and frequency. The results are in good agreement with theoretical predictions for velocity fluctuation noise if the correct dependence of the two-dimensional carrier concentration on channel voltage is used.

1. INTRODUCTION

In recent years much attention has been paid to GaAs-AlGaAs Modulation Doped Field Effect Transistors (MODFETs) for potential use in high-speed logic circuits. The very high transconductance g_m and high cut-off frequencies f_T also make them of interest for low-noise microwave amplification. Excellent articles by Solomon et al.[1] and Linh[2] have been written reviewing the characteristics of these new transistors.

Since the first report of the noise figure of these devices[3] in the microwave frequency range, an interest in the noise behavior has been developed. The noise figures of various MODFETs have been reported recently and show improvements over conventional GaAs MESFETs of comparable gate lengths. As an example of the low-noise capability, a minimum noise figure of 1.3 dB with an associated gain of 8.2 dB at 18 GHz was recently reported[4].

Up to now only noise figure measurements have been reported in the microwave frequency range. In this paper we will not focus on the noise figure, but instead report on the noise characteristics of the FET channel. A good understanding of the noise behavior of the channel is important, since at intermediate frequencies ($0.5 \leq f \leq 10$ GHz) the channel noise is the major contributor to the overall device noise.

2. DEVICE DESCRIPTION AND THEORY

The device configuration is given in Fig. 1. Note that the gate has been recessed so that the space charge regions of the Schottky barrier and the AlGaAs-GaAs interface overlap.

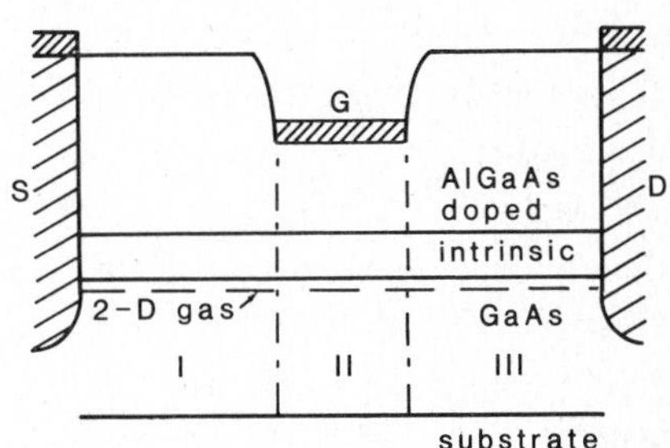

FIGURE 1
Device configuration. The length of regions I, II, and III is 1.5×10^{-6}m, 1.0×10^{-6}m, and 1.5×10^{-6}m respectively. The AlGaAs doping concentration is $2.5 \times 10^{24}/\text{m}^3$, and its Al mol. fraction is 0.28. The thicknesses of the doped AlGaAs, intrinsic AlGaAs and undoped GaAs layers are about 600Å, 30Å and 1 μm respectively. The width of the gate is 145 μm.

For the purpose of modeling, we subdivide the channel into three regions. Regions I and III represent the access resistances R_{ss} and R_{dd}, whereas Region II is the active gate region.

To calculate the noise, we follow the impedance field method as outlined by Nougier[5]. Our starting equation is the current equation. Neglecting diffusion and displacement current, it reads for a MODFET

$$I = qwn_s[V(x)]v[E(x)] \tag{1}$$

where w = gatewidth, $q = 1.6 \times 10^{-19}$, E = electric field parallel to the interface, V = potential in the channel, n_s = voltage dependent 2-D. electron charge in the channel, and v = electric field dependent drift velocity. The sign convention is as follows. The source is chosen at x = 0, the drain at x = L > 0, q > 0, V(x) > 0, E(x) < 0, v[E(x)] > 0, and I > 0.

Upon linearization eq. (1) gives for the a.c. terms

$$\delta I = -qwn_s(V_0)\left(\frac{dv}{dE}\right)_0 \frac{d}{dx}\delta V + qw\left(\frac{dn_s}{dV}\right)_0 v(E_0)\delta V, \tag{2}$$

where we used $\delta E = -\frac{d}{dx}\delta V$.

Some terms in eq. (2) deserve extra attention. To calculate the velocity and its derivative for a specific field value, we use an analytical expression which describes accurately the velocity field dependence of a 2-D. and 3-D.[8] electron gas in GaAs at T = 300K for fields up to the value where the peak velocity is reached (E_{peak}):

$$v(E) = -\mu_0 E/(1 - E/E_c). \tag{3}$$

E_c is a critical field equal to $+11.4 \times 10^5$V/m, and μ_0 is the low-field mobility, equal to $0.8m^2$/Vsec. At present there is no satisfactory model available to calculate the 2-D. carrier concentration and its derivative for a specific value of channel potential. The models[3] based on quasi-triangular potential wells at the interface explain correctly the value of the saturation current, but fail to predict correct values for the small signal parameters in the region of interest. However, by using an iterative procedure, which we outline later on, the correct $n_s(V)$ dependence can be obtained.

Equation (2) can be solved using Green's functions, and the expressions for the impedance field $\nabla Z(x')$, the impedance Z_{L_1}, and the voltage noise $S_{\Delta V,L_1}(f)$ follow:

$$\nabla Z(x') = -\frac{E_0(L_1)}{I_0}\left[1 - \frac{E_0(x')}{E_c}\right], \tag{4}$$

$$Z_{L_1} = -\frac{E_0(L_1)\cdot L_1}{I_0}\left(1 + \frac{V(L_1)}{E_c \cdot L_1}\right), \tag{5}$$

$$S_{\Delta V}(f,L_1) = \int_0^{L_1} 4q^2wD[E(x')]n_s[V(x')] \, |\nabla Z(x',f)|^2 dx' \tag{6}$$

where D[E(x')] = diffusion coefficient, L_1 is the spacing between the source and the position in the channel where the impedance and noise are calculated. Our modeling is restricted to field strengths below E_{peak}. Therefore the electric field strength inside the device is monitored during our calculations and is not allowed to exceed this value (3.5×10^5V/m). The drain side of the gate is chosen for this purpose since at this point the field inside the device will reach a maximum for any value of applied bias.

To calculate $S_{\Delta V}$, $n_s[V(x)]$ has to be determined. In regions I and III n_s is a constant, not controlled by the gate potential, and can be calculated using the Anderson model for a heterostructure[6]. To determine n_s in region II the following numerical interation procedure has been developed. For our model it holds that $Z_{L_1} = \frac{dV}{dI} - R_{dd}$. Hence, using the measured

I-V characteristic and its derivative $E_0(L_1)$ follows from eq. (5). Then, with the help of eq. (1), $n_s(V_{L_1})$ can be determined. By iterating through the I-V characteristic up to values where E_{L_1} equals E_{peak}, the $n_s(V_L)$ dependence follows.

In the gradual channel approximation n_s only depends on the voltage difference between gate and channel. Consequently, $n_s(V_L) = n_s(V(x))$ and eq. (6) can be evaluated.

3. EXPERIMENT AND RESULTS

The current-voltage characteristic and the thermal noise were measured under pulsed bias conditions from the linear regime up into saturation. The gate was a.c. and d.c. shorted to cancel any induced gate noise. Spectral noise analyses were done at frequencies between 0.5 and 1.0 GHz, and they showed that the spectral noise density was independent of frequency.

The current-voltage characteristic is shown in Fig. 2, where the dashed line represents the results, using a triangular quantum well[3]. The solid line represents the results, using the model outlined in this paper. The model satisfactorily explains the I-V characteristic up to 26 mA, where E_{L_1} becomes equal to E_{peak}. For higher currents a negative differential mobility regime will form at the drain, and subsequently velocity saturation will occur. These phenomena are not included in our model.

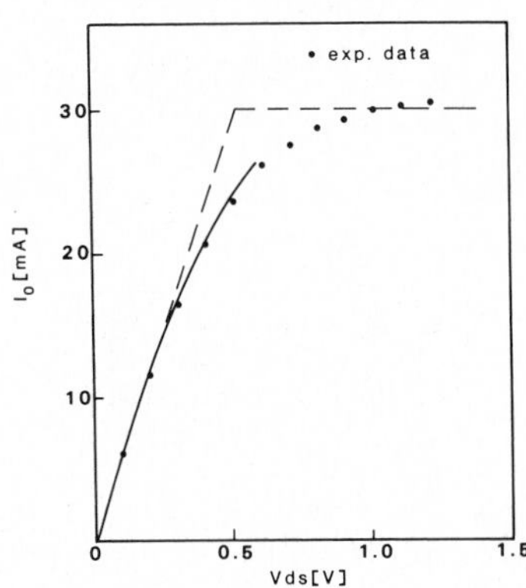

FIGURE 2
I-V characteristic. The dots indicate the measured data. The solid line results from calculations based on the calculated $n_s(V)$ dependence, whereas the dashed line represents the quasi-triangular model.

Figure 3 shows $n_s(V)$ as calculated from the I-V characteristic and its derivative. The solid and dashed line indicates the results employing the triangular quantum well model, using gate-electron gas spacings of 290Å and 460Å respectively. Note the nonlinearities in $n_s(V)$, which were also observed by Vinter[7].

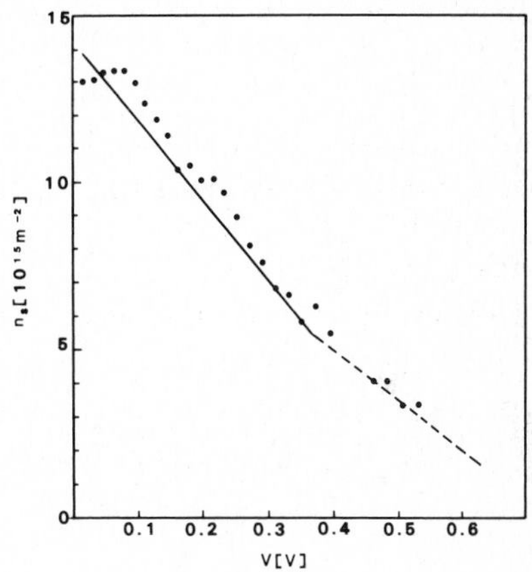

FIGURE 3
The 2-D carrier concentration as a function of channel potential. The dots indicate the results of the iteration procedure. The solid and dashed lines correspond to gate-interface spacings of 290Å and 460Å respectively.

In Fig. 4 we present the calculated and measured voltage and current spectral densities. We used $D(E) = D_0$ to obtain the excellent agreement between the calculated and measured $S_{\Delta V}$. For device operation considerations we also plotted $S_{\Delta I}$. The discrepancy between calculated and measured values of this quantity can be attributed to an error of 5% in the calculated value of Z_{L_1}.

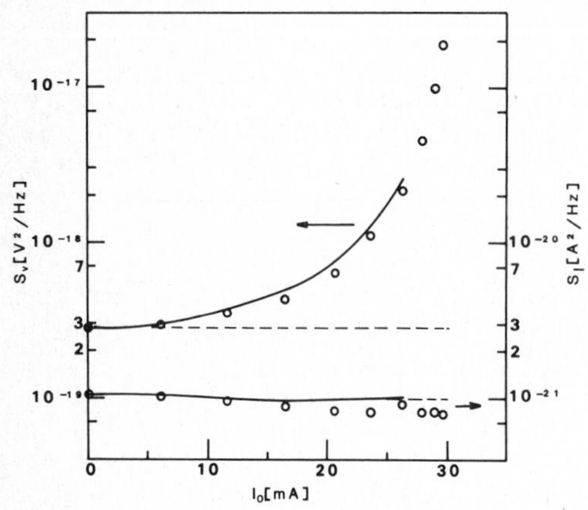

FIGURE 4
Current and voltage noise spectral density as a function of bias current. The solid line results from calculations based on the $n_s(V)$ dependence, whereas the dashed line represents the quasi-triangular model.

ACKNOWLEDGEMENTS

We thank the National Science Foundation for supporting this research under grant #ECS-83-00222.

REFERENCES

1. P.M. Solomon and H. Morkoc, IEEE Trans. ED31 (1984), 1015.

2. N.T. Linh, Semiconductors and Semimetals (Academic Press, to be published).

3. D. Delagebeaudeuf and N.T. Linh, IEEE Trans. ED-29 (1982), 955.

4. O.K. Mishra, S.C. Palmateer, P.C. Chao, P.M. Smith and J.C.M. Hwang, IEEE EDL-6 (3) (1985), 142.

5. J.P. Nougier, J.C. Vaissiere and D. Gasquet, Proceedings of 5th Inter. Conf. of Noise in Physical Systems (Washington, 1981), p. 42.

6. T.C. Hsieh, K. Hess, J.J. Coleman and D.P. Daphus, Solid-State Electr. 26 (1983), 1173.

7. B. Vinter, Appl. Phys. Lett. 44 (1984), 307.

8. A. Cappy, A. Vanoverschelde, M. Schortgen, C. Versnaeyen and G. Salmer, Physica 129B (1985), 380.

NOISE IN PHYSICAL SYSTEMS AND 1/f NOISE - 1985
A. D'Amico and P. Mazzetti (editors)

HOT ELECTRON DIFFUSION NOISE ASSOCIATED WITH INTERVALLEY SCATTERING IN SHORT GaAs DEVICES

J. ANDRIAN, G. BOSMAN, A. van der ZIEL and C.M. Van VLIET

Department of Electrical Engineering, U. of Florida, Gainesville, FL 32611, USA

The velocity fluctuation noise of a low impedance, 1.1 μm $N^+N^-N^+$ GaAs device is measured as a function of applied bias voltage employing a tuned step-up transformer. The experimental results are expressed in terms of the electron mobility and diffusion coefficient, and agree with Monte Carlo simulation data.

1. INTRODUCTION

The progress in semiconductor device technology has made it possible to fabricate devices of ever-decreasing size. As a consequence, relatively low bias voltages can easily produce high electric fields inside these small devices. Therefore, it is important to understand the high-field properties of the transport mechanisms in these devices. In this paper we focus on the voltage dependence of electrical noise observed at the terminals of the devices at 11 and 17 MHz.

At low bias voltages the devices operate in the ohmic regime, whereas at higher bias voltages, space-charge accumulation and electron heating[1] occur. As Gunn discovered, the latter phenomenon results in a transfer of electrons from the central high-mobility valley to a low-mobility satellite valley.

The electrical noise observed at 11 and 17 MHz is velocity fluctuation noise and is caused by inter valley and intravalley electron scattering. The method used to measure the noise, and the results will be described in Section 2. The last section is devoted to discussions and comparisons between our experimental data and the theoretical calculations.

2. DEVICE DESCRIPTION AND EXPERIMENTAL SETUP

The measurements have been performed on $N^+N^-N^+$ GaAs mesa structures at T = 300K. The active region (N^- layer) has a thickness of 1.1 μm, a cross-sectional area of $7.8 \times 10^{-9} m^2$ and a Si doping of $10^{21} m^{-3}$. The N^+ layers are needed to provide good ohmic contacts at both ends of the device. A pulsed bias method was used to avoid excessive Joule heating and possible destruction of the device at high electric fields. The pulse generator is synchronized with a switch at the input of the power detector so that one only measures the noise when the bias is applied.

Bareikis et al.[2] have used this method before; however, a detailed description of a tuned step-up transformer at the input of the amplifier is given in this paper. This network is very important because the device has a very small equivalent noise resistance (≈ 2Ω) compared to the 60Ω equivalent noise resistance of the amplifier. As a consequence, the device noise would drown in the amplifier noise without this network.

2.1. Step-up transformer

The noise equivalent circuit of the transformer is shown in Figure 1, where r is the series resistance of the inductor, r_d is the device small signal resistance, and $\sqrt{4kT_0 r\Delta f}$ and $\sqrt{4kT_d r_d \Delta f}$ (k is the Boltzmann constant) are the noise sources associated with r and r_d respectively. The temperature T_0 and T_d are the equivalent noise temperature of the series resistance and the device under test respectively.

During our experiments the inductor was kept at room temperature.

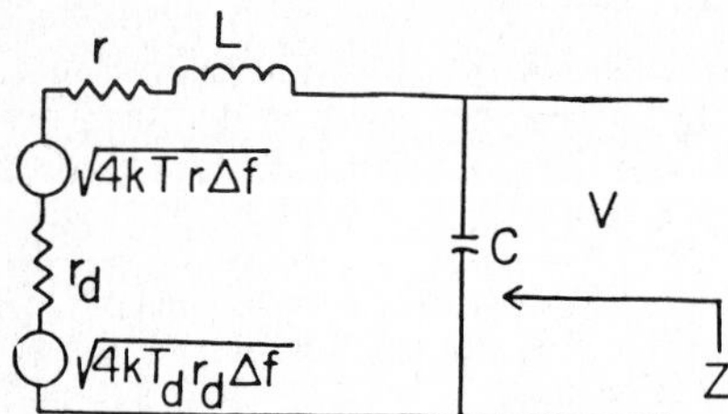

FIGURE 1
Noise equivalent circuit of the step-up transformer.

The impedance at the terminals is given by

$$z = \frac{j\omega L+r+r_d}{1-\omega^2 CL+j\omega C(r+r_d)} . \tag{1}$$

If we operate at the tuning frequency $\omega_0 = 1/\sqrt{LC}$, and if we choose ω_0 so that $r+r_d << |j\omega_0 L|$, then

$$z \approx \frac{L}{C} \cdot \frac{1}{(r+r_d)} . \tag{2}$$

The spectral density S of the AC open-circuited voltage fluctuations at the output of the network is

$$S = 4k(T_0 r+T_d r_d) \left| \frac{\frac{1}{j\omega_0 C}}{\frac{1}{j\omega_0 C} +r+r_d+j\omega_0 L} \right|^2 . \tag{3}$$

Carrying out the algebra and again using the fact that $\omega_0^2 CL = 1$, equation (3) becomes

$$S = 4k(T_0 r+T_d r_d) \frac{L}{C} \frac{1}{(r+r_d)^2} . \tag{4}$$

The condition for "up" transformation of the noise equivalent resistance of the device is easily derived from equation (4) if one neglects r. Then S becomes

$$S = 4kT_d \frac{L}{Cr_d} . \tag{5}$$

"Up" transformation takes place if

$$\frac{L}{Cr_d} > r_d \text{ or } \frac{L}{C} > r_d^2 . \tag{6}$$

For the series resonance step-up transformer, it holds that

$$\frac{L}{C} = r_d^2 Q^2 \tag{7}$$

where Q is the quality factor of the transformer. Hence, condition (6) is fulfilled as long as Q > 1. However, a high Q circuit has a narrow passband; therefore, transformers for each frequency of interest (11 and 17 MHz) were built.

2.2. Measurement procedure

The measurement of the device noise involves the following three steps.

i) The device is replaced by a short circuit, and one obtains

$$S_1 = 4kT_0 \frac{L}{Cr} . \tag{8}$$

Next, a known resistance r_0 is used. Equation (4) gives

$$S_2 = 4kT_0 \frac{L}{C(r+r_0)} . \tag{9}$$

Equations (8) and (9) yield

$$\frac{S_1}{S_2} = 1+ \frac{r_0}{r} . \tag{10}$$

From equation (10) one obtains a value of the series resistance r of the inductor at the tuning frequency, which accounts for possible high-frequency skin effects in the inductor wires.

ii) The value of the small signal resistance r_d as a function of the bias voltage was obtained by numerically differentiating the measured current-voltage characteristic of the device. This method is justified since the small contact spacing of the device under test allows us to neglect the effects of the transit time on the impedance of the device.

iii) Finally, a noise measurement S_3 with the device under test is done. One obtains from equation (4) the noise resistance R_n:

$$R_n = \frac{S_3}{S_1}\frac{(r+r_d)^2}{r} - r. \tag{11}$$

The components of the bias network were chosen in such a way that their contribution to the overall noise could be neglected.

In Figure 2 R_n is plotted as a function of the applied voltage for an ambient temperature of 300K and $f_0 = \frac{\omega_0}{2\pi}$ = 11.0 MHz and f_0 = 17.0 MHz. The value of R_n is independent of frequency, and for bias voltages up to 0.2V agrees within the experimental error with the small signal resistance of the device. For bias voltages above 0.2V, R_n increases sharply with the voltage applied.

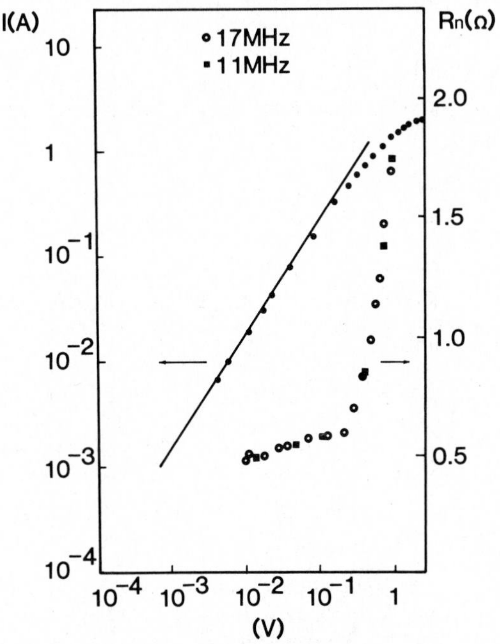

FIGURE 2
Current-voltage characteristic and R_n as a function of the applied voltage.

Also shown in Figure 2 is the measured I-V characteristic. In this short device, space-charge injection from the n^+ contacts and electron heating compete. The first phenomenon causes the current to increase above its ohmic value, whereas the second phenomenon causes the current to decrease. The measured I-V curve in Figure 2 illustrates that above 0.2V the electron heating starts to dominate the charge transport mechanism.

3. DISCUSSION

Using Poisson's equation and the current equation, neglecting the displacement current and the diffusion current, one can express the length of the device L and the applied voltage at the anode V_0 as a function of the applied field, in an integral form,

$$V_0 = \frac{\varepsilon A}{I_0}\int_0^{E_L} \frac{E^2\mu(E)dE}{\left(1-q\frac{AN_D E\mu(E)}{I_0}\right)} \tag{12}$$

$$L = \frac{\varepsilon A}{I_0}\int_0^{E_L} \frac{E\mu(E)dE}{\left(1-q\frac{AN_D E\mu(E)}{I_0}\right)} \tag{13}$$

where V_0 = bias voltage, I_0 = DC current, A = cross-sectional area, L = contact spacing, E_L = electric field strength at the anode, N_D = concentration of fully ionized donors, and ε = permittivity. In these two equations the field dependence of the electron mobility is taken into account. Moreover, Gisolf and Zijlstra[3] derived an integral form of the noise spectral density, taking space-charge injection and carrier heating into account:

$$S_{\Delta V}(f) = \frac{4\varepsilon qA}{I_0^2}\int_0^{E_L} \frac{D(E)(E_L-E)^2dE}{\left(1-q\frac{AN_D\mu(E)E}{I_0}\right)^3} \tag{14}$$

where D(E) is the electron diffusion coefficient. Equation (14), together with equations (12) and (13), form a system of integral equations with three unknowns, namely E_L, $\mu(E)$, and D(E). This system is difficult to solve analytically. Numerical techniques are required to circumvent such difficulty. This was discussed by G. Bosman[4].

The diffusion coefficient and the electron drift velocity obtained from this decomposition

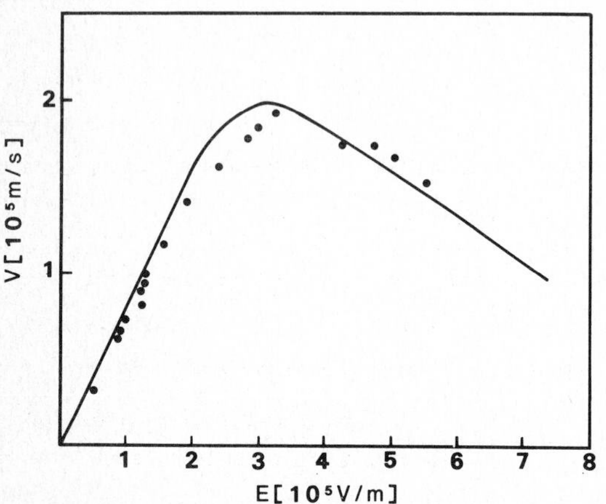

FIGURE 3
Drift velocity as a function of electric field obtained by Monte Carlo simulation and numerical deconvolution.

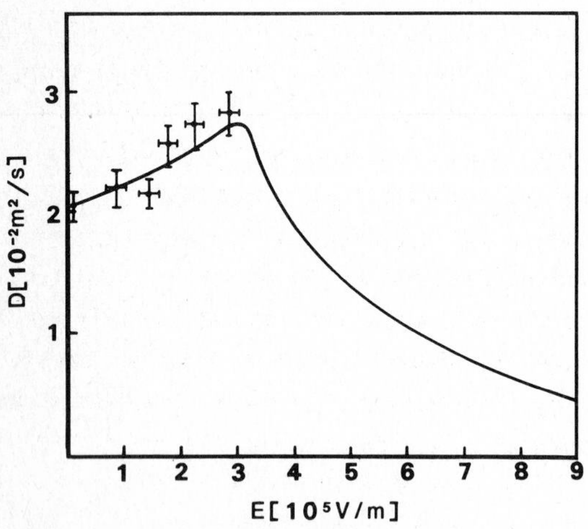

FIGURE 4
Diffusion coefficient as a function of electric field obtained by Monte Carlo simulation and numerical deconvolution.

are plotted along with the values obtained by Monte Carlo simulation in Figures 3 and 4.

The Monte Carlo program we use is a slightly modified version of the program developed by Boardman[5]. It considers a central and six satellite valleys located at the main axes of the Brillouin zone and intravalley as well as intervalley scattering with acoustic and polar-optical phonons. Note that the agreement between measured and calculated mobility is quite satisfactory, indicating that a hot-electron, collision-dominated model adequately describes charge transport in this short 1 μmGaAs device. The calculated and measured diffusion coefficients agree within the experimental error of 15% in the range of electric field strength where a comparison can be made. For electric fields larger than 3.10^5 V/m, the deconvolution program indicates that the magnitude of the terminal voltage noise is no longer critically determined by the value of the diffusion coefficient D. Consequently, no high field values of D can be obtained.

ACKNOWLEDGEMENTS

We thank the National Science Foundation for supporting this research under grant #ECS-83-00222.

REFERENCES

1. M.P. Shaw, H.L. Grubin and P. Solomon, The Gunn-Hilsum Effect (Academic Press, New York, 1979).

2. V. Bareikis, A. Galdikas, R. Miliusyte and V. Viktoravicius, Program and papers of 5th Int. Conf. on Noise in Physical Systems (Bad Nauheim, W. Germany, March 1978), p. 212.

3. A. Gisolf and R.J.J. Zijlstra, J. Appl. Phys. 47 (1976), 2727.

4. G. Bosman, Charge Transport Noise in Monocrystalline Silicon, Ph.D. dissertation, U. of Utrecht, The Netherlands (1981).

5. A.D. Boardman, Physics Program vol. 4 (Wiley Interscience Publication, New York, 1980).

NOISE IN PHYSICAL SYSTEMS AND 1/f NOISE - 1985
A. D'Amico and P. Mazzetti (editors)

CURRENT NOISE IN N-CHANNEL Si-MOSFETS AT 4.2 K.

E.A. HENDRIKS, R.J.J. ZIJLSTRA

Laboratory for Experimental Physics, University of Utrecht, Princetonplein 5, 3584 CC Utrecht, the Netherlands.

J. WOLTER
Philips Research Laboratories
Eindhoven, the Netherlands.

Abstract
Current noise was measured of (100) n-channel Si-MOSFET's at liquid Helium temperatures as a function of drain- and gate-voltage, in the frequency range of 10 Hz to 200 kHz. Three contributions can be distinguished. (i) 1/f-noise, (ii) g-r-noise, (iii) intervalley and diffusion noise. The noise as well as the drain-current versus voltage relationship were affected by hot carrier effects while the intervalley noise turned out to be due to electric field induced transitions.

1. INTRODUCTION

Although the MOSFET device has been subject to many investigations under cryogenic conditions [1] little is known about the noise properties at 4.2 K. As far as we know, the only data, reported in literature are of Sesnic and Craig [2] who measured the equivalent voltage noise spectrum of a MOSFET in saturation. They observed excess thermal noise which they ascribed to Joule-heating of the device. In this paper we report current noise measurements on n-channel Si-MOSFETs at 4.2. K.

2. DEVICES

The n-channel MOSFETs were grown in the (100) direction of a p-type Si substrate with a Chemical Vapour Disposition technique at Philips Research Laboratories. The MOSFETs have a channel width w of 4 μm, a channel length L of 40 μm and an oxide thickness d of 50 nm. The p-type substrate was doped with Boron and has a specific resistance of 17-23 Ω/cm. The n-type contact-areas, called source and drain are heavily doped with Phosfor ($>5.10^{18}$ cm^{-3}).

3. CURRENT-VOLTAGE-CHARACTERISTICS

First, the I_d-V_d characteristics were measured for several values of gate voltage, where I_d and V_d are the drain current and source-drain voltage respectively. From a plot of the conductance at low V_d against gate voltage, the threshold voltage (V_T) was determined to be 0.8V. We ascertained that our results were not affected by Joule heating. For low V_d the I_d-V_d characteristics obey the theory for MOSFETs in the gradual channel approximation if modified for hot-carrier effects. The latter were formally described by an electric-field dependent mobility [3].

$$\mu(E) = \mu_o/(1 + E/E_c) \quad (1)$$

where E and E_c are the normal source-drain field and the critical source-drain field for hot carrier effects resp. Note that E varies with position in the channel. With this assumption Van der Ziel et al [4] derived the following expression for the drain current.

$$I_d = \mu_o wC_{ox}(V_g-V_T-V_d/2)V_d/(L+V_d/E_c) \quad (2)$$

where V_d is the source-drain voltage, V_g is the gate-source voltage, V_T is the threshold voltage, C_{ox} is the oxide-capacitance per unit area, w the channel width and L the channel-length. In this derivation the conduction electrons are assumed to have the same mobility associated with the lowest 2-fold degenerated valley [5]. For higher V_d the second 4-fold degenerated valley will become populated. Taking this into account we have

$$I_d = q(\mu_1 N_1 + \mu_2 N_2)E \text{ and } dN_1/dt = N_2/\tau_2 - N_1/\tau_1 \quad (3)$$

where N_1 and N_2 are the number of electrons per unit length in the lowest and highest valley resp. and $\frac{1}{\tau_1}$, $\frac{1}{\tau_2}$ the transition rates.

$$\text{If we assume } 1/\tau_1 = \alpha E;\ \tau_2 \text{ independent of E} \quad (4)$$

$$\text{and } \mu_i = \mu_{i0}/(1+E/E_{ci}) \ (i = 1,2) \quad (5)$$

$$\text{with } \mu_{10}/\mu_{20} = m_{c2}/m_{c1} = 1.5 \quad (6)$$

where m_{ci} the conductivity mass in valley i [6], then

$$I_d = qN \left\{ \frac{\mu_{10}}{(1+E/E_{c1})} + \frac{\mu_{20}E}{(1+E/E_{c2})E_{c3}} \right\} \frac{E}{(1+E/E_{c3})} \quad (7)$$

$$N(x) = N_1(x) + N_2(x) = wC_{ox}(V_g - V_T - V(x))/q \quad (8)$$

$$E_{c3} = 1/(\alpha\tau_2) \quad (9)$$

Using equations (3), (4), (6), (7) and (8) it was possible to obtain a computer fit to our data with the critical fields as fitting parameters, up to saturation.

4. EXPERIMENTAL SET-UP FOR NOISE MEASUREMENTS.

The devices were used with ac shorted gates and were essentially one-ports with the source and drain contacts as terminals. Therefore the noise could be described by a noise current generator circuited in parallel with spectral noise intensity $S_{Id}(f)$. The MOSFET was mounted in a liquid Helium bath. All resistors used were metal-film resistors. The drain signal was fed into a Brookdeal 5004 preamplifier which was connected to a spectrum analyser covering the range of 1Hz-200 kHz. Bias voltages were supplied by dry batteries. In order to correct for background noise a measurement was done with the MOSFET replaced by a dummy resistor.

5. EXPERIMENTAL RESULTS.

In most of the measured current noise spectra three contributions could be distinguished; a $1/f^{\beta}$ -spectrum with $0.8 \leqslant \beta \leqslant 1.2$, a Lorentz spectrum and a white spectrum. We therefore write for the spectral current noise intensity

$$S_{Id} = S_{Id}^{I} + S_{Id}^{II} + S_{Id}^{III} \quad \text{where} \quad (10)$$

$$S_{Id}^{I} = C^{I}/f^{\beta},\ S_{Id}^{II} = C^{II}/(1+2\pi f\tau)^2,\ S_{Id}^{III} = C^{III} \quad (11)$$

The parameters C^{I}, C^{II}, C^{III}, β,τ could be determined from a computerfit of equation (10) to the experimental data. A typical result is shown in figure 1. If we plot the 1/f com-

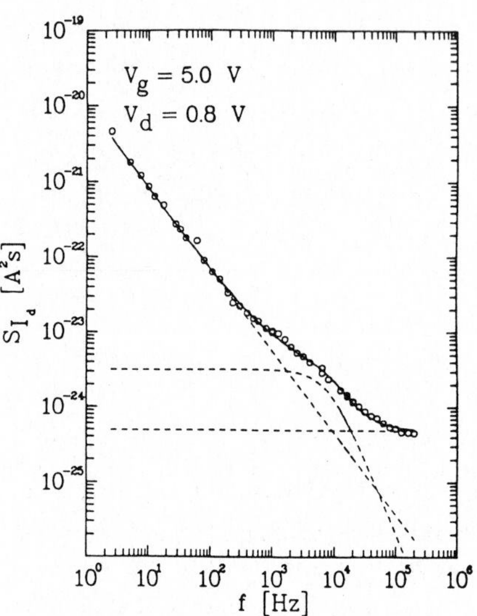

Fig 1. A typical result of the spectral current-noise intensity versus frequency. The solid line represents the fit to the measured values.

ponent, at 100 Hz, versus V_d it initially varies with V_d^2 whereas for higher values it was found to vary nearly linearly with V_d (fig.2). As a function of V_g the 1/f-component doesn't vary very much. Differences of more than an order of magnitude were found in the Lorentz component by varying V_g. $1/\tau$ initially varies linearly with I_d whereas $S(o)/\tau$ was found to be proportional to I_dV_d. For V_g just above threshold the white current noise level versus I_d ini-tially varies linearly with the DC conductance whereas for higher values of I_d different behaviour was found (fig. 3). For high values of V_g compared with V_T the white current noise increases linearly with increa-sing drain current then saturates and subsequently increases again (fig. 4).

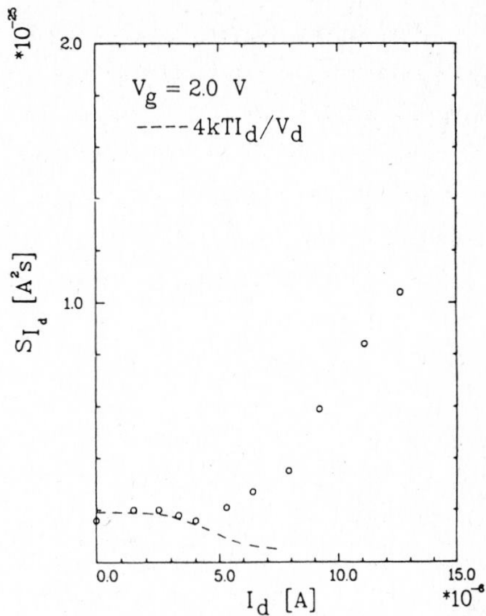

Fig 3. The white current-noise density versus drain-current for V_g = 2V. The dashed line represents the thermal noise.

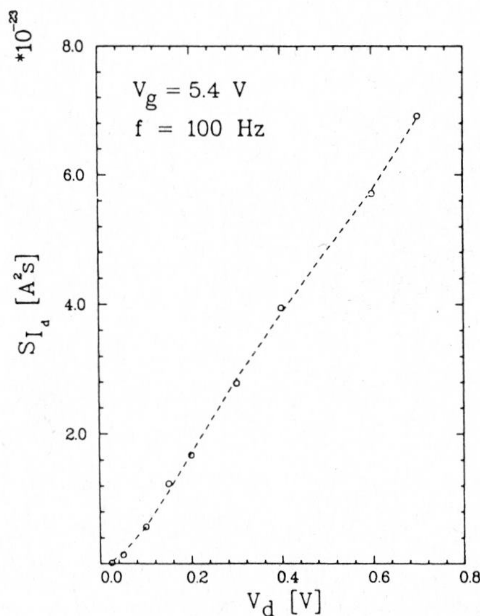

Fig 2. The spectral 1/f-current-noise intensity versus drain-voltage. The dashed line represents the 2-valley-fit.

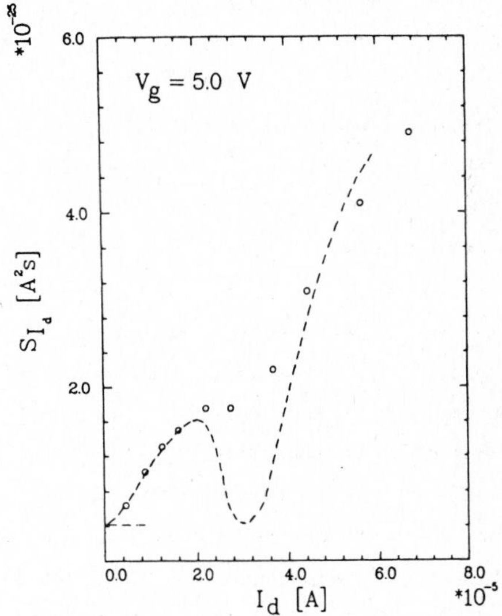

Fig 4. The white current-noise density versus drain-current for V_g = 5V. The dashed line represents the 2-valley-fit.

6. INTERPRETATION

The 1/f noise could be due to mobility fluctuations as well as to number fluctuations. Our measurements, however, cannot discriminate between the number fluctuation model [7] and the mobility fluctuation model [8]. We have analysed our data in terms of the mobility fluctuation model. Park et al [9] derived an integral expression for the drain current 1/f-noise in MOSFETs for the one-valley model.

If we modify this for the two-valley model we find

$$S_{Id}(f) = \frac{qI_d}{fL^2} \int_o^{V_d} \frac{\alpha_1\mu_1^2 + \alpha_2\mu_2^2 E/E_{c3}}{\mu_1+\mu_2 E/E_{c3}} dV \qquad (12)$$

where α_i is the Hooge parameter associated with the i-th valley.

The low field mobility μ_{10} was obtained from the I_d-V_d characteristics and with the help of equation (6) μ_{20} was determined. The parameters $Ec_1, Ec_2, Ec_3, \alpha_1, \alpha_2$ were obtained by a computer-fit of equation (12) to the experimental data. In our analyses α_1 and α_2 turned out to be independent of E and was found to be of the order of 10^{-5}. This value can be accounted for by using $\alpha_i(T) = (\frac{\mu(T)}{\mu_\ell(T)})^2 .2.10^{-3}$ [10] if we extrapolate the lattice mobility $\mu_\ell(T)$ to 4.2 K. Van der Ziel and Handel have derived an expression for α on basis of incoherent quantum 1/f noise [11]. If we apply this to our devices we obtain a value of 1.3 10^{-11} which is several orders of magnitude lower than our results.

The Lorentzspectrum can be ascribed to generation- recombination noise with field induced transitions. Suppose that the g-r-process is a simple 2-level process then we can write for the current noise density

$$S_{Id}(f) = \frac{4q^2}{L^2} \int_o^L \frac{\mu^2(x)E^2(x)N(x)}{(\omega^2+1/\tau_t^2(x))\tau_t(x)} dx \qquad (13)$$

where $1/\tau_t$ is the transition probability. If we assume that $1/\tau_t = \alpha E$ and no hot carrier effects occur (low source-drain voltage) then equation (13) reduces to

$$S_{Id} \simeq S(o)/(1+\omega^2\tau^2) \qquad (14)$$

where

$$1/\tau = \alpha I_d/(C_{ox} w\mu(V_g - V_t))$$
$$S(o) = 4q\mu V_d I_d \tau/L^2 \qquad (15)$$

It follows that $1/\tau$ is proportional to I_d whereas $S(o)/\tau$ is proportional to $I_d V_d$ in agreement with our experimental results. Deviations can be ascribed to the occurence of hot carrier effects; for higher values of V_d μ decreases so $1/\tau$ increases and $S(o)/\tau$ decreases. For low values of V_g and of V_d the observed white noise can be ascribed to thermal noise. In that case it can be derived that

$$S_{Id,\ thermal} = 4kT\ I_d/V_d \qquad (16)$$

where k is the Boltzmann constant and T the temperature of the device. The increase of the white current noise level can be explained by electric field induced intervalley-scattering. The following expression was derived.

$$S_{Id}(f) = \frac{4q^2}{L^2} \int_o \frac{L(\mu_1-\mu_2)^2\ E^2\tau^3}{(1+\omega^2\tau^2)\tau_1\tau_2} Ndx \qquad (17)$$

where

$$N_1 = N\tau/\tau_2\ ;\ N_2 = N\ \tau/\tau_1\ ;\ 1/\tau = 1/\tau_1 + 1/\tau_2 \qquad (18)$$

As is shown in figure 4 the fit of our data to equations (4) (5) and (17) shows a local minimum as a function of I_d instead of a plateau as found experimentally. This is due to the fact that $E_{c_1} < E_{c_2}$. Therefore, for low values of V_d $\mu_1 > \mu_2$ but for higher values $\mu_1 < \mu_2$. Apparently a third level is involved; e.g. the first excited subband of the lower valley, which is almost degenerated with the ground state of the upon valley. However, this makes the situation much more complicated and at this moment we have not obtained a consistent set of parameters, but this is subject to further investigation.

LITERATURE

1. T. Ando, A.B. Fowler and F. Stern, Rev. Mod. Phys., 54, 437 (1982).
2. S.S. Sesnic and G.R. Graig, IEEE Trans. on Elec. Dev., 19, 933 (1972).
3. G. Bosman, R.J.J. Zijlstra and A. van Rhenen, Phys. Lett. 78 A, 385 (1980).
4. A. van der Ziel, R.J.J. Zijlstra and H.S . Park, J. Appl. Phys. 52, 4300 (1981).
5. F. Stern and W. Howard, Phys. Rev. 163, 3 (1967).
6. F. Stern, Phys. Rev. B. 5, 12 (1972).
7. A.L. McWorther, Semiconductor Surface Physics, Ed. R.H. Kingston, University of Pennsylvania, 207 (1956).
8. F.N. Hooge, Phys. Lett. 29 A, 139 (1969).
9. H.S. Park, A. van der Ziel, R.J.J. Zijlstra and S.T. Liu, J. Appl. Phys. 52, 296 (1981).
10. F.N. Hooge and L.K.J. Van Damme, Phys. Lett. 66 A, 315 (1978).
11. A. van der Ziel and P.H. Handel, Physica 129 B, 578 (1985).

NOISE IN PHYSICAL SYSTEMS AND 1/f NOISE - 1985
A. D'Amico and P. Mazzetti (editors)

NOISE AND IMPACT IONIZATION RATES FOR ELECTRONS AND HOLES IN $Hg_{0.3}Cd_{0.7}Te$ AVALANCHE PHOTODIODES

R. ALABEDRA, B. ORSAL, M. VALENZA, G. LECOY and M. SAVELLI

C.E.M. (C.N.R.S. UA 391), Université des Sciences et Techniques du Languedoc, 34060 Montpellier Cedex, France

Preliminary experimental results on the $Hg_{0.3}$ $Cd_{0.7}$ Te avalanche photodiodes are shown. In the first part of the paper, both multiplication coefficients M_n and M_p are determined. The ionization coefficients α and β are calculated so that the ratio k = β/α is deduced. In the second part, the noise characterization is presented. The experimental results are in good agreement with the value of k deduced from the first order measurements.

1. INTRODUCTION

The Hg Cd Te system has been widely investigated in the mercury rich composition for its well-known properties of far infrared intrinsic detection. The Cd rich composition range of these alloys has regained considerable importance because of its potential for opto-electronic components in near infrared optical communications (mainly at λ = 1.3 µm and λ = 1.55 µm). This alloy has been chosen because there is a very large spin-orbit coupling exhibited by these materials[1,2]. Indeed the ratio k = β/α is optimum[3,4] when the spin-orbit splitting is roughly equal to the band gap energy.

2. DESCRIPTION OF THE DEVICE

The growth of Hg Cd Te is performed by the travelling heater method. The main characteristics of the obtained ingots are the following : holes concentration of 10^{16} cm^{-3} and the dispersion within composition (expressed in $\Delta\lambda$) is $\pm$ 0.01 µm on a ϕ 30 mm diameter. The Hg Cd Te photodiodes, manufactured by S.A.T. (France), are made by a conventional planar photolitographic process[5].

The main characteristics of $Hg_{0.3}$ $Cd_{0.7}$ Te APD's (see figure 1) are :

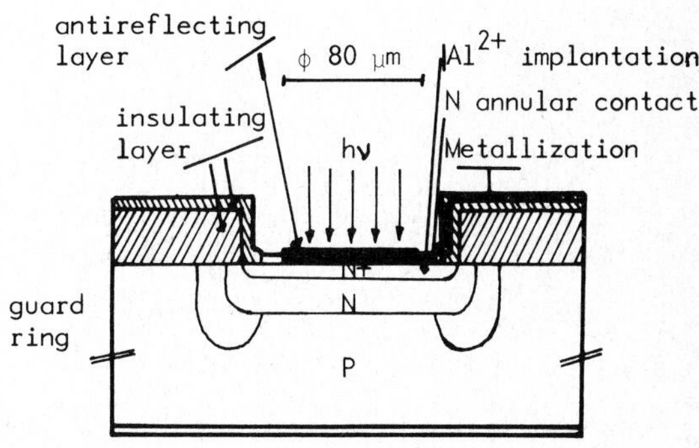

FIGURE 1

Diameter of sensitive area = 80 µm
Dark current at -10 Volts = 1 nA
Sensitivity at λ = 1.3 µm = 0.8 A/W
Spectral sensitivity range = 0.9 to 1.35 µm
Capacitance at -10 Volts < 1 pF
Bandwith at M = 5 : 850 MHz
Quantum efficiency ≃ 70 %

3. FIRST ORDER ELECTRICAL MEASUREMENTS

The electrical measurements carried out on those devices are the capacitance-voltage (C-V), the current-voltage (I-V) characteristics in forward and reverse bias voltages, under obscurity and illumination conditions.

3.1. C-V characteristics

The C-V characteristics show a $V^{-1/3}$ variation in the low bias range up to 2 Volts and a $V^{-1/2}$ between 2 and 10 Volts. This agrees with previous results[5]. The concentration profile is shown in figure 2 :

The n^+ type layer is formed by aluminium ion implantation (60 keV, 10^{14} cm^{-2}) with $R_p \simeq 0.3$ µm and $\Delta R_p \simeq 0.3$ µm.

The n type layer is due to the propagation of defects, following the implantation ; the extension of this layer[6] is about 2 µm. At the n-p junction, the transition between the n and p sides is gradual with a gradient of about 5.1×10^{19} cm^{-4}. The effective doping remains relatively constant $N_D = 9 \times 10^{15}$ cm^{-3}.

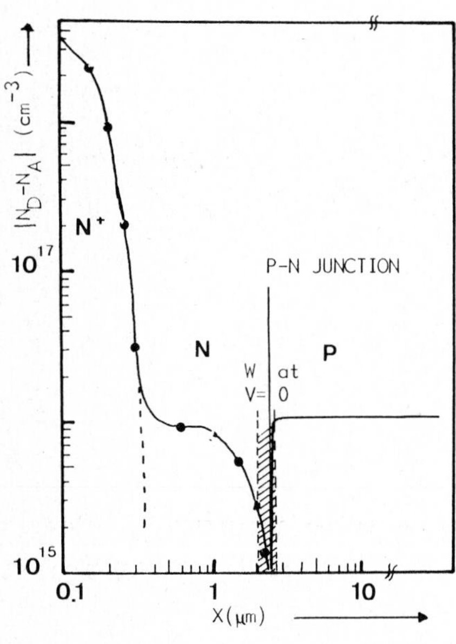

FIGURE 2

The p-type side corresponds to the substrate with a doping density of 10^{16} cm^{-3}.

3.2. I-V characteristics

Both the I-V characteristics under obscurity and illumination conditions are shown in figure 3.

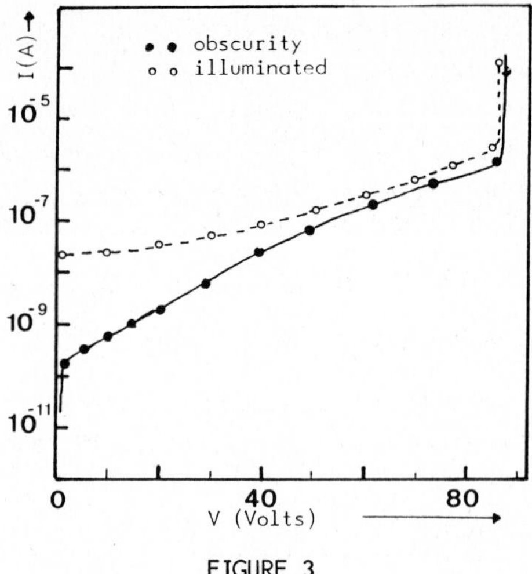

FIGURE 3

The increase of the dark current up to 10 volts is due to the increase of the depletion layer W versus bias voltage. Experimental and theoretical values of the dark current are in good agreement.

Above 30 volts the dark current increase is due to the impact ionization. The breakdown voltage appears at 86 V. It is due to the avalanche phenomenon.

3.3. Spectral response and multiplication coefficients

Figure 4 shows the external efficiency η_{ex} versus λ. The maximum of η_{ex} is obtained at λ = 1.3 µm.

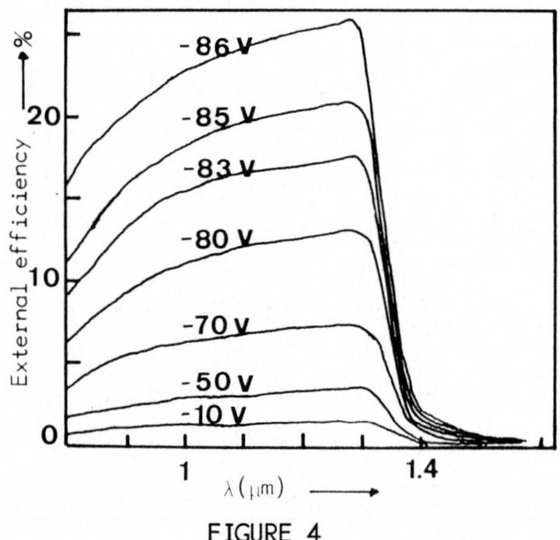

FIGURE 4

The wavelength cut off $\lambda_c \sim 1.38$ μm corresponds to the semiconductor band gap energy (0.9 eV) of the alloy.

At $\lambda = 1.3$ μm, the absorption coefficient is 2×10^4 cm^{-1} so that a hole injection occurs in the multiplication region.

At $\lambda = 1.38$ μm, due to the value of the absorption coefficient ($\sim 10^3$ cm^{-1}) the electrons of the p type side are injected in the multiplication region.

M_p and M_n can be respectively deduced at λ 1.3 μm and $\lambda \leqslant 1.38$ μm versus electric field (figure 5).

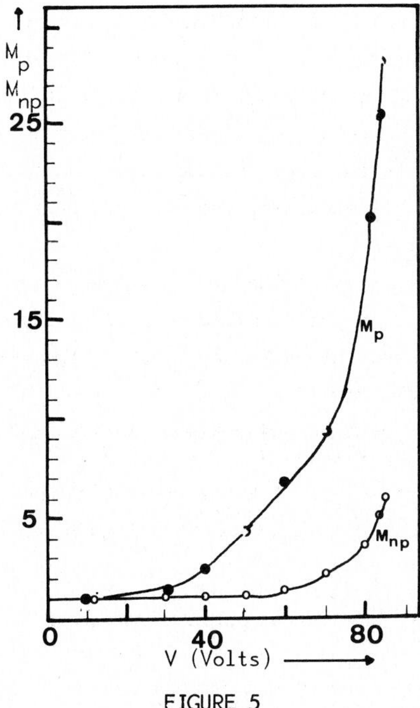

FIGURE 5

By using relationships given by Stillman[7] for p-i.n APD's :

$$\alpha(E) = \frac{1}{w} \left\{ \frac{M_p(V) - 1}{M_p(V) - M_n(V)} \right\} \mathrm{Ln} \frac{M_p(V)}{M_n(V)} \quad (1)$$

$$\beta(E) = \frac{1}{w} \left\{ \frac{M_n(V) - 1}{M_n(V) - M_p(V)} \right\} \mathrm{Ln} \frac{M_n(V)}{M_p(V)} \quad (2)$$

Where w is the width of the space charge region determined from the C-V characteristics, it is possible to deduce the values of α and β. For $E \sim 2.5 \times 10^5$ V/cm, $k = \frac{\alpha(E)}{\beta(E)}$ is about 20. This result has been given in literature[8].

4. NOISE MEASUREMENTS

Figure 6 shows the current noise spectral density versus I under obscurity and illumination conditions[9]. In the both cases we observe a pure shot noise followed by a KM^x law.

For the dark current this law can be written as :

$$S_i(f) = 2\, q\, I_{obsinj}\, M^x \quad (3)$$

and under illumination :

$$S_i(f) = 2\, q\, (\, I_{obsinj} + I_{phinj}\,)\, M^y \quad (4)$$

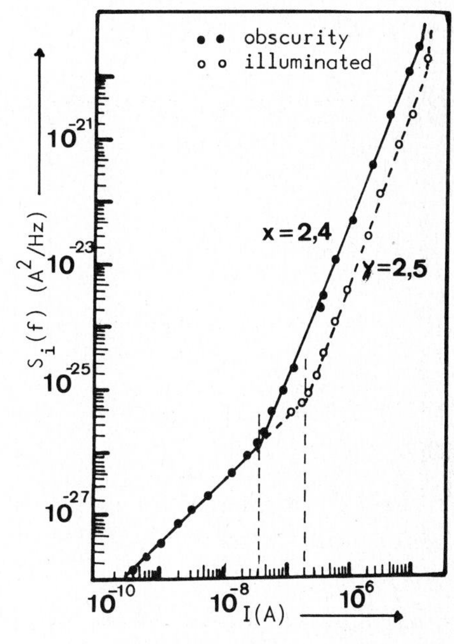

FIGURE 6

From these curves it is possible to deduce the experimental curve $S_i(f)$ of the net photocurrent (see figure 7).

The noise is given by :

$$S_{iph}(f) = 2qI_{phinj}\, M^{xB} = 2qI_{phinj}\, M^2\, F(M) \quad (5)$$

where F(M) is the excess noise factor.

As the multiplication is initiated mainly by the holes generated in the n^+ region, F(M) can be written as[10]:

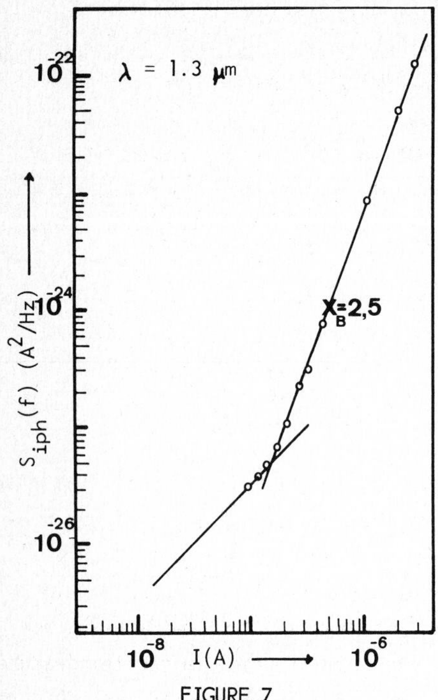

FIGURE 7

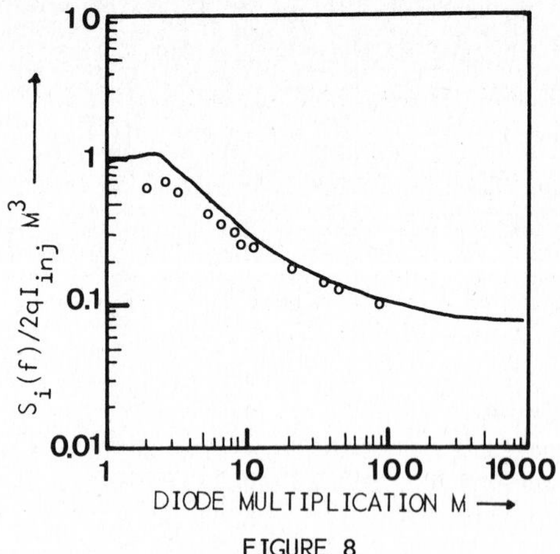

FIGURE 8

$$F(M_p) = M_p \left[1 + \frac{1-k}{k}\left(\frac{M_p - 1}{M_p}\right)\right] +$$

$$M_p \left[\frac{1}{k-1} + \frac{1}{M_p}\right] \times \left\{\frac{1}{k-1} + \frac{1}{2M_p}\left(\log\frac{\frac{1}{k}}{1+\frac{1-k}{k}\frac{1}{M_p}} + \frac{1}{M_p}\right)\right\} \times$$

$$\log\frac{\frac{1}{k}}{1+\frac{1-k}{k}\frac{1}{M_p}} \qquad (6)$$

Experimentally F(M) is given by the ratio :

$$F_{exp}(M) = \frac{S_{iph}(f)}{2\,q\,I_{phinj}\,M^2} \qquad (7)$$

$$\text{where } M_{exp} = \frac{I_{ph}(V)}{I_{phinj}} \qquad (8)$$

The figure 8 represents the both theoretical and experimental variations of F(M)/M versus M for $k \simeq 20$.

5. CONCLUSION

This paper has shown that the ratio $k = \beta/\alpha$ deduced from the spectral responses is in good agreement with the value of k deduced from the noise measurements. Furthermore it has been shown that the Hg Cd Te alloys are promising candidates for detectors of 1.3 μm optical communications.

REFERENCES

1. A. Moritani, K. Taniguchi, C. Hamaguchi and J. Nakaï, Phys. Soc. of Japan, 34(1973) p.73.
2. C. Vérié, F. Raymond, J. Besson and T. Nguyen Duy, J. Cryst. Growth 59 (1982) p. 342.
3. M.P. Mikhaïlova, N.N. Smirnova, S.V. Slobodchikov, Soviet Phys. Semicond. 10 (1976) 509. M.P. Mikhaïlova, S.V. Slobodchikov, N.N. Smirnova and J.M. Filaredova, Soviet Phys. Semicond. 10 (1976) 578.
4. O. Hildebrand, W. Kuebart, K.W. Benz, M.H. Pilkuhn, IEEE J. Quantum Electron. QE 17 (1981) 284.
5. M. Royer, T. Brossat, P. Fragnon, J. Meslage, G. Pichard, T. Nguyen Duy, Ann. Télécom. 98 (1983) 62.
6. C. Van Opdorp, Solid State Electron. 11 (1968) 397.
7. G.E. Stillman and C.M. Wolfe, Vol. 12, K.Willardson and A.C. Beer, Academic Press (1977).
8. R. Alabedra, B. Orsal, G. Lecoy, G. Pichard, J. Meslage, P. Fragnon, IEEE Trans. Electron. Devices, Vol. ED 32, n°7, july 1985.
9. M. Valenza, Thesis Ph D, C.E.M. U.S.T.L. Montpellier (1985).
10. M. Savelli, R. Alabedra, 8th international conference on noise in physical systems, Roma, september 9-13th 1985.

NOISE IN PHYSICAL SYSTEMS AND 1/f NOISE - 1985
A. D'Amico and P. Mazzetti (editors)

PREPARATION AND NOISE CHARACTERIZATION OF Ga Al (As) Sb DIFFUSED P^+N DIODES

B. ORSAL*, R. ALABEDRA*; T. BELATOUI**, A. JOULLIE** ; A. SCAVENNEC***

*Centre d'Electronique de Montpellier (CEM) associé au CNRS (UA 391), U.S.T.L., 34060 Montpellier Cedex ; **Equipe de Microoptoélectronique de Montpellier (EM²) associée au CNRS (UA 392, U.S.T.L., 34060 Montpellier Cedex ; ***Centre National d'Etude des Télécommunications (C.N.E.T.), 196, rue de Paris, 92220 Bagneux.

GaAlAsSB / GaSb is a suitable system for fiber optical communications in the long wavelength detection (1.3 - 1.6 µm). GaAlSb and GaAlAsSb MESA p^+n photodiodes were prepared by LPE and Zinc diffusion. The noise currents of both devices, measured in the different polarization range, are shot noise, 1/f noise, G.R. noise and tunneling component multiplication noise.

1. DIODE FABRICATION AND CHARACTERISTICS

Epitaxial layers of GaAl(As)Sb were grown by liquid phase epitaxy on Te doped (111)B GaSb substrates. The carrier concencentration and the etch pits density of the substrate were about 3×10^{18} cm^{-3} and 10^{6} cm^{-2}. Undoped layers were p type, with a residual carrier concentration $N_A = 6\text{-}8\times10^{16}$ cm^{-3} ; n type was achieved by Tellurium doping. MESA p^+n photodiodes were obtained by Zinc diffusion and Argon ion etching[1,9]. The carrier concentration of active layer was $n = 1\text{-}2\times10^{16}$ cm^{-3}. Its composition was $Ga_{0.85}Al_{0.15}Sb$ for ternary alloys and lattice matched $Ga_{0.85}Al_{0.15}As_{0.013}Sb_{0.987}$ for quaternary alloys.

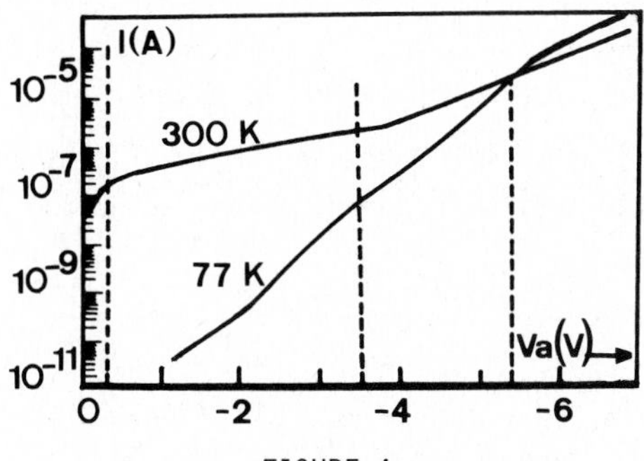

FIGURE 1

Reverse bias dark current of a GaAlAsSb/GaSb photodiode at T = 77°K and 300°K

Figure 1 shows reverse current-voltage characteristics of the quaternary photodiode at 77°K and 300°K. The dark current is 0.5 µA at - 1 Volt. Soft breakdown is observed under reverse bias voltage. In this region a crossing of the reverse characteristics is noted. Before the crossing, the breakdown voltage was observed to decrease with the increase in the temperature. After the crossing, it is the reverse. This result indicates that the reverse current in the vicinity of the avalanche region is governed first by tunnel effect, next by the impact ionization effect.

Analysis of C-V characteristics showed abrupt junctions with diode capacitance values about 10 pF at zero bias voltage.

The short circuit current was measured under illumination by calibrated light. The spectral photoresponse, at zero voltage bias, has a maximum value of 0.6 A/W at a wavelength of 1.3 µm.

2. NOISE CHARACTERISTICS

2.1. Noise at $V_a = 0$

We plotted in fig.2 the shot noise density S_{id} of the photodiode at 0 bias in terms of the photocurrent I_{ph}, as well as the theoretical curve whose equation is $2q\,I_{ph}$. From Van Der Ziel law [2] we can write :

$$S_{id} = 4\,kT < g_d > + 2q\,I_{ph} \qquad (1)$$

where I_{ph} is the photocurrent generated at 0 bias and $\langle g_d \rangle$ is the photodetector conductance.

From (1) we found $\langle g_d \rangle = 4.4 \times 10^{-7}\ \Omega^{-1}$.

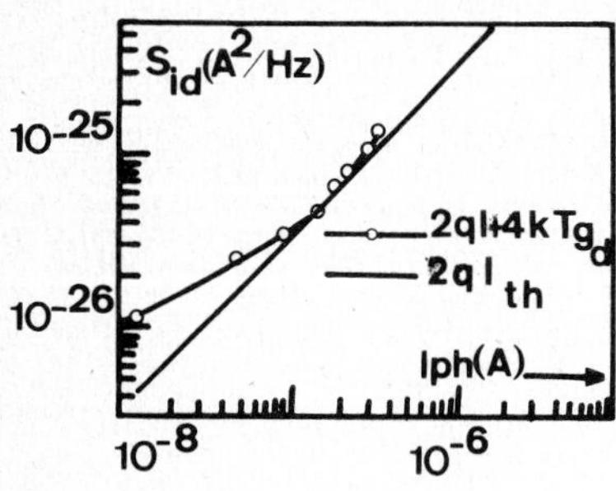

FIGURE 2

Spectral density versus photocurrent at $V_a = 0$ of GaAlAsSb photodiode

This result perfectly coincides with the value derived in the previous paragraph ($4.63 \times 10^{-7}\ \Omega^{-1}$). This means that the total photocurrent effectively flows through the junction and is collected as its terminals.

For photocurrents higher than 3×10^{-7}A the experimental curve does not fit along the theoretical shot noise straight line. This can be due to thermal generation of carriers caused by hight concentration (focalisation effect) and temperature effect.

2.2. Noise analysis at $V_a \neq 0$

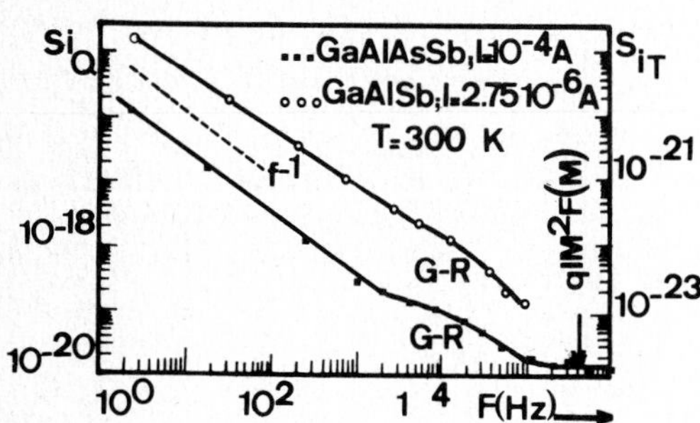

FIGURE 3

$S_i(f)$ versus frequency of ternary and quaternary diode at obscurity

The dark current noise density spectra $S_i(f)$ of two ternary GaAlSb and quaternary GaAlAsSb diodes are represented on fig. 3 from 1 Hz up to 1 MHz. They are dominated by :

- the 1/f noise around $F = 10^3$ Hz
- the G-R noise, with one trap level in the case of the ternary alloy ($F \sim 50$ kHz) and two trap levels in the case of the quaternary ($F \sim 50$ kHz and 300 Hz). It also shows in fig. 3 a flat spectral density between 100 kHz and 1 MHz proportional to $2q\,I\,M^2\,F(M)$.

2.2.1. 1/f noise

1/f noise variations in terms of reverse bias dark current at 1 kHz are presented in fig. 4. The quaternary diodes show noise variations in agreement with those described by Kleinpenning[3] as long as we make the assumptions listed below:

- noise is due to mobility fluctuations
- two different carriers mobilities are not correlated
- diffusion mecanism are random
- interactions between carriers are instantaneous and random
- the tunnel current components flowing across the junction do not contribute to 1/f noise since they cannot be considered as mobility fluctuations. In the case of a tunnel current larger than 10^{-6}A we will substract its component.

Therefore, the noise current spectral density can be described using the model proposed by Kleinpenning in the case of P^+n diodes when dominated by generation recombinaison currents[3]. If one considers that carrier concentration inside the space charge region is the intrinsic n_i concentration at T = 300°K one can write :

$$S_i(f) = \frac{2}{3}\,\frac{\alpha\, I^2_{GR}}{AW\, n_i F} \qquad (2)$$

Where α is the Hooge constant (3×10^{-5}), A the diode area (1.5×10^{-4} cm^2) w the space charge region thickness (0.58 µm).

For $I_{GR} = 10^{-6}$ A, we obtain $S_i(F) = 7.4 \times 10^{-23}$ A^2/Hz which is in good agreement with experimental data.

However we must note a very large difference

from this law at currents larger than 5×10^{-6}A, current level which corresponds to impact ionization built-up. The shape of the curve requires some more explanation and needs to be interpretated in terms of the G.R. current multiplication since assume that the tunnel current is a priori not related to mobility fluctuations.

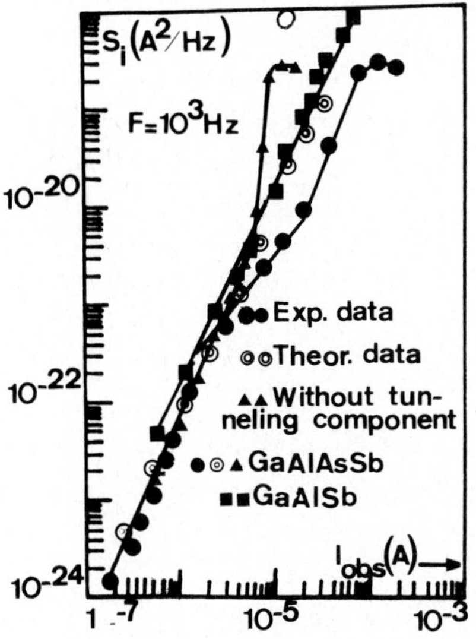

FIGURE 4

The 1/f $S_i(f)$ noise versus the dark current at $F = 10^3$ Hz of the GaAl(As)Sb diode

Furthermore, we note a 1/f noise saturation for currents as high as 10^{-4}A which might correspond to the raise of high multiplication regime ($M \sim 4$) close to the breakdown regime[8].

2.2.2. Generation recombinaison noise

In the frequency range (between 40 KHz and 55 KHz) of G-R noise density maxima, the reverse dark current noise spectral density was found proportional to I^2 and predicted by the law[4] :

$$S_i(f) = \frac{4I^2}{N_0^2}\ \Delta N^2 \left[\frac{\tau}{1+\omega^2\tau^2}\right] \qquad (3)$$

where N_0 is the equilibrium carrier number and $\Delta N = (N - N_0)$ stands for the carrier numer fluctuations. The time constant τ responsible for G-R phenomenon is a function of temperature and the activation energy of the trap level under consideration. A temperature analysis turns out to be necessary in order to derive the value of the activation energy E_A and the relaxation time τ of these traps in the two alloys, ternary and quaternary.

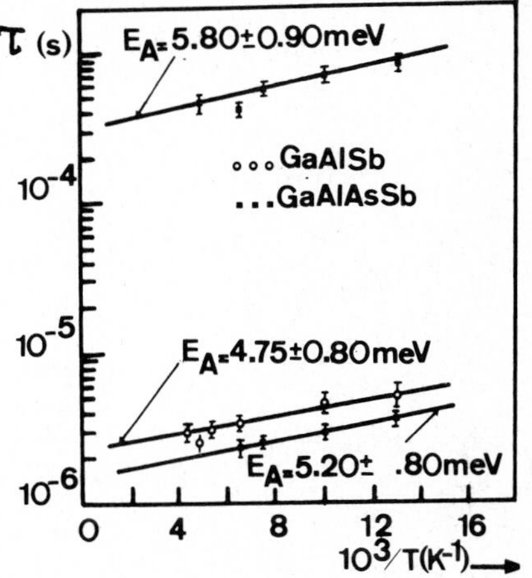

FIGURE 5

Variation of the time constant wersus temperature

The time constant τ is bound to the activation energy E_A as expressed by :[4,5]

$$\tau = \alpha \ . \ \exp \frac{E_A}{kT} \qquad (4)$$

and from the plot of τ versus $\frac{1}{T}$ (fig. 5) we have obtained one level ($E_A = 4.75 \pm 0.90$ meV, $\tau = 3.8 \pm 0.5$ µs) in the case of GaAlSb and two levels ($E_{A1} = 5.8 \pm 0.9$ meV, $\tau_1 = 0.61 \pm 0.02$µs; $E_{A2} = 5.2 \pm 0.8$ meV, $\tau_2 = 3.0 \pm 0.3$ µs) in the case of GaAlAsSb.

These levels are not very deep and close to the conduction band. We have to remind that the material is initially Tellurium doped and we relate these trap levels to Tellurium atoms[6].

3. NOISE MULTIPLICATION

At f = 200 kHz (white part of the spectra)

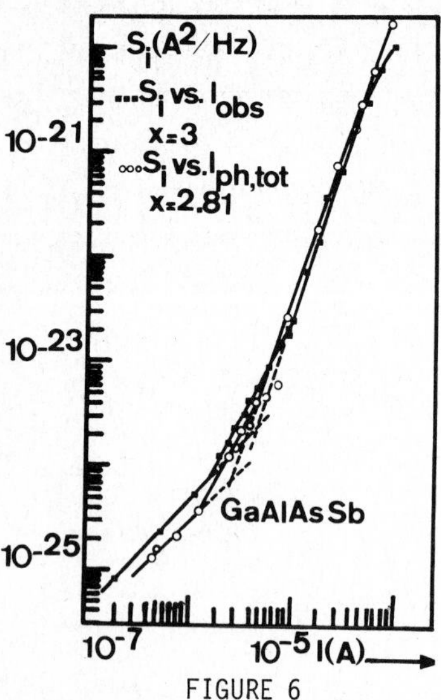

FIGURE 6

White multiplication noise $S_i(f)$ versus the current of GaAlAsSb photodiode

it is shown on the figure 6 the $S_i(f)$ versus dark current. The shot noise appears up to the values of the bias current about $2x10^{-6}$A. From this value we observe a strong increase of the noise due to the impact ionization. There is a transient zone from $2x10^{-6}$A to 10^{-5}A. Beyond 10^{-5}A the impact ionization is well etablished. At obscurity it is the tunneling component which is multiplied and the noise is given by the following relation :

$$S_i(f) = 2\ q\ I_{tuninj}\ M^2\ F(M) \quad (5)$$

where, I_{tuninj} is the component tunneling injected in the multiplication zone, $F(M) \simeq M^x$, where x excess noise coefficient is about 1 at obscurity.

The cross point obtained by extrapolating the shot noise curve and the noise multiplication curve is about $5x10^{-6}$A near of the crossing point deduced of the figure 1.

The noise curves $S_i(f)$ are of the same form in the dark and under illumination (fig. 6). But the noise, when the device is illuminated, is the sum of two components : the noise of the multiplied tunneling component and the noise of the photocurrent also multiplied. The noise curve over the multiplication zone is given by the law[7].

$$S_i(f) = 2q\left[I_{tuninj} + I_{phinj}\right] M^2\ F_{tot}(M) \quad (6)$$

In this case we have : $I_{tuninj} \simeq 5x10^{-6}$A ; $I_{phinj} = 4x10^{-7}$A ; $F_{tot}(M) \simeq M^x$ with x = 0.81 under illumination.

The presence of the tunneling component is prejudical for the photodetection performances. Indeed we can evaluated the noise equivalent power (N.E.P.) and the optimum multiplication coefficient M_{opt} from :

$$N.E.P. = \frac{1}{\sigma_o}\left[\frac{4kTe}{R_L M^2} + 2q(I_{tuninj} + I_{phinj})M^x\right]^{1/2} \quad (7)$$

$$M_{opt} = \left[\frac{4kTe}{R_L q\ (I_{tuninj} + I_{phinj})x}\right]^{1/x} \quad (8)$$

At λ = 1.3 µm it is found M_{opt} = 4 to 5, N.E.P. = $5x10^{-12}$ W/Hz$^{1/2}$. If one assumes the tunneling effect is absent, M_{opt} and N.E.P. become $M_{opt} \simeq 8$ and N.E.P. $\simeq 3x10^{-12}$ W/Hz$^{1/2}$.

CONCLUSION

In conclusion the energy levels are determined by the noise measurements versus temperature. They are found about 5 ± 1 meV. Also the noise have allowed to identify the impact ionization process of the tunneling component when the bias voltage increases in good agreement with the I = f(V) characterization at the different temperatures. In order to reduce this tunneling component one would have to obtain the absorption zone with net donor concentration lower than 10^{16} cm^{-3}.

REFERENCES

1. A. Joullié, F. De Anda, P. Salsac and M. Mebarki, Revue Phys. Appl. 19 (1984) 223.

2. A. Van Der Ziel, Sources, characterization, measurement, Prentice Hall (1970).

3. T.G.M. Kleinpenning, J. Vac. Sci. Technol.A3 (1), (Jan/Feb 1985) 176.

4. A. Van Der Ziel, Noise in Measurements, John Wiley & Sons, Inc. New York (1976).

5. G. Bosman, R.J.J. Zijlstra, Solid State Electron.,25, n° 4 (1982) 273-280.

6. L. Konczewicz, E. Litwin-Staszewska, S. Porowski, A. Iller, R.L. Aulombard, J.L. Robert, A. Joullie, Physica 117B et 118B, North-Holland Publishing Company (1985) 92-95.

7. R. Alabedra, C. Maille, S. Tariq, G. Lecoy, Ann. Télécomm. 35 (1980) 193-198.

8. J.A. Ringo, P.O. Lauritzen, Solid State Electronics, 16 (1973) 327-328.

9. T. Belatoui, Thèse de 3ème Cycle, U.S.T.L., Montpellier, France (Juillet 1985).

NOISE IN PHYSICAL SYSTEMS AND 1/f NOISE - 1985
A. D'Amico and P. Mazzetti (editors)

HIGH FREQUENCY NOISE IN SCHOTTKY BARRIER DIODES

M. TRIPPE, G. BOSMAN and A. van der ZIEL

Department of Electrical Engineering, U. of Florida, Gainesville, FL 32611, USA

A program of systematic noise measurements on Schottky barrier diodes was carried out at microwave frequencies, in order to determine the noise temperature as a function of both frequency and bias. The diode model was extended to include the effects of the transit time on the noise. This takes into account that some electrons have insufficient kinetic energy to cross the barrier of the junction. Excellent agreement was found between the measured results and the theoretical analysis.

1. INTRODUCTION

Schottky barrier diodes have found widespread use at very high frequencies for the purpose of mixing, as well as the direct detection of signals[1]. Their operation extends through the mm-wave range and into the infrared portion of the spectrum. This paper discusses some limitations of Schottky barriers in terms of their noise performance, especially those due to transit time effects.

The theory of noise in Schottky barrier diodes has been developed for some time. Van der Ziel[2] applied the results from a vacuum tube analysis to the problem of transit time effects in Schottky barriers, and here we present experimental results which agree well with his analysis.

We assume that the DC junction current in the diode can be described by the simple equation

$$I_d = I_s \exp(-V_{dif}/mV_T) \exp(V_d/mV_T) \quad (1)$$

where $V_T = kT/q$ is the thermal voltage, m is the diode ideality factor, V_{dif} is the diffusion potential and I_s is the saturation current. This is valid for the case of bias voltages V_d greater than approximately five times mV_T.

Assuming that this current gives full shot noise and that the device conductance can be found from the derivative of I_d with respect to V_d, evaluated at the operating point, it follows directly that the noise temperature of the device is

$$T_n = mT/2. \quad (2)$$

This is the well-known, low-frequency result and has been widely used; note that it does not depend on the bias under the stated assumptions. This result does not include the effects of the series resistance or reverse saturation current. By including the reverse saturation current, the noise temperature of the junction would be just the ambient temperature at zero bias, as it should be.

A more serious omission is to neglect the transit time effects on the noise performance when the diode is operated at high frequencies. It will be shown that the measured results are well explained by extending the model to include the effect of the transit time.

2. THEORY

As previously mentioned, van der Ziel has already provided the theoretical background for this type of device. The case to be considered here is for an electric field varying linearly with position in the junction region. In his work the initial velocities of the electrons are divided into three groups for the purpose of calculating the small-signal conductance at high frequencies. The spectral intensity of the noise as a function of frequency can be calculated by dividing the electron velocities into two cases; that which gives sufficient kinetic energy to cross the junction and that which gives insufficient kinetic energy to carry the electron across the junction. The spectral intensity of the current is calculated by taking the Fourier transform of each current pulse due to a particular initial velocity. Finally, a summation is made over all possible initial velocities. This is Carson's theorem.

The conductance due to those electrons which have insufficient kinetic energy to cross the barrier is given in ref. 2, eq. 32a, as

$$g_{11} = \frac{I_s - I_d}{V_{dif} - V_d} \frac{1}{2} \frac{c}{1-c^2} \left(\frac{1-\cos \pi(1-c)}{1-c} - \frac{1-\cos \pi(1+c)}{1+c} \right) \quad (3)$$

where the parameter c is the normalized angular frequency defined as ω/a. The constant "a" is the plasma frequency, which equals $\sqrt{eE_{max}/m^*d}$ and ω is the frequency of interest. E_{max} is the maximum electric field in the junction region and m* is the effective mass of the electrons. Note that I_s is very much greater than I_d, for most cases. This conductance dominates at high frequencies for low values of DC bias current.

For low frequencies the device conductance is dominated by those electrons which only have sufficient energy to just reach the metal contact. With an applied signal these will sometimes be collected at the metal contact, and sometimes they will return to the neutral semiconductor, hence dI/dV is large for this group of electrons. Their contribution to the device conductance is

$$g_{33} = g_0 \frac{(1-2c \sin \frac{1}{2} c\pi - c^2 \cos c\pi)}{(1-c^2)^2} \quad (4)$$

where g_0 is the low frequency small-signal conductance.

Finally, the third group of electrons must be considered, and these are the ones which have sufficient energy to always reach the opposite contact. Since this group of electrons is always collected, they carry a constant DC current and are not much affected by an applied small-signal voltage (dI/dV is approximately zero for this group of electrons). Nonetheless, their contribution can be calculated as

$$g_{22} = \frac{I_d}{V_{dif} - V_d} \int_0^\infty \exp - \left(\frac{eV_0'}{kT}\right) d\left(\frac{eV_0'}{kT}\right) \times \frac{1}{2} \frac{c}{1-c^2} \left(\sin\phi \sin c\phi + \frac{(1-\cos(1-c)\phi)}{1-c} - \frac{(1-\cos(1+c)\phi)}{1+c}\right) \quad (5)$$

where $\phi = a\tau$, eV_0' is the energy of the arriving electrons and τ is the DC transit time.

The spectral intensity, S_{11}, of those electrons which do not reach the metal contact and then return to the neutral semiconductor region can be found from the equation of motion of the electrons.

$$S_{11} = \int_0^{v_{00}} 2\lambda C_1^2 8\omega^2 \left\{\frac{(1+\cos \omega\tau_r)}{(a-\omega)^2 (a+\omega)^2} dv_{x0}\right. . \quad (6)$$

and v_{00} is the minimum initial velocity required to just reach the opposite contact. The number of electrons emitted with a given initial velocity is λ.

The characteristic time required for the electrons to return is

$$\tau_r = \pi/a, \quad (7)$$

which is the same for all returning electrons regardless of their initial velocity. The constant $C_1 = q\, v_{x0}/2d$, where v_{x0} is the initial x-directed velocity and d is the space-charge width.

The spectral intensity of those electrons which cross is

$$S_{22} = \int_{v_{00}}^{\infty} 2\lambda C_1^2 \frac{2}{(a+\omega)(a-\omega)} \{(1-\cos(a-\omega)\tau_F)[\frac{a+\omega}{a-\omega} - 1] + (1-\cos(a+\omega)\tau_F)[\frac{a-\omega}{a+\omega} - 1] + (1-\cos 2a\tau_F)\}\, dv_{x0} . \quad (8)$$

The transit time is found from the equation of motion to be

$$\tau_F = \frac{1}{a} \text{Arcsin} \left(\frac{ad}{v_{x0}}\right). \tag{9}$$

The spectral intensities are shown in Fig. 1. The noise due to those electrons which return does not change significantly with bias, but the noise of those electrons crossing the junction depends linearly on the device current, or exponentially on the device voltage. Thus, for the ideal junction operated at a fixed frequency (e.g. 12 GHz), with a very low bias, the noise and the device conductance will be dominated by the returning electrons. The noise temperature which will be measured is that of the returning electrons for this case.

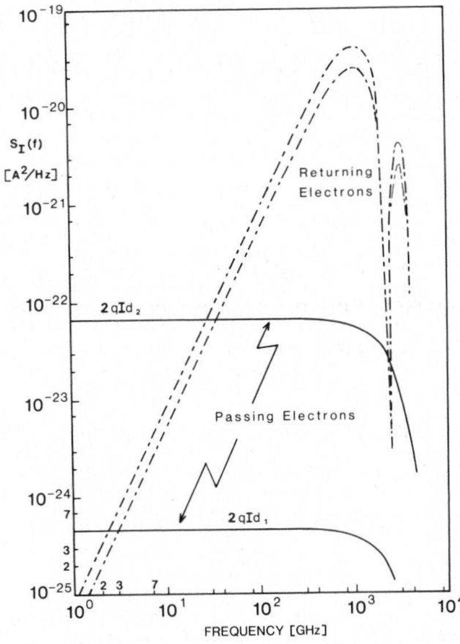

FIGURE 1
Spectral intensities vs. frequency. Lower solid and lower dot-dash curves correspond to $I_{d1} = 1.4 \times 10^{-6}$ A, $V_{d1} = 5\ V_T$. Upper solid and upper dot-dash curves correspond to $I_{d2} = 2.1 \times 10^{-4}$ A, $V_{d2} = 10\ V_T$.

Once the diode current has been increased to the point that the noise of the passing electrons dominates, the noise temperature will drop to mT/2. Thus, there are two distinct regions of operation for the diode junction.

Combining the different components of the device conductance and the spectral intensities of the current, the device noise temperature can be calculated as a function of both bias and operating frequency.

The results presented so far only describe the behavior of the ideal exponential junction. The final results should include the reverse saturation current, which we assume to be constant, and the series resistance, which we assume to have a noise temperature equal to the ambient temperature.

The complete results of the theoretical calculations are presented in Fig. 2. An ideality factor of one was assumed to simplify the calculations since this is a reasonable assumption for the diode which we measured (m = 1.09). The Schottky barrier height of 0.585 eV and series resistance of 10 Ohms were also found from the DC I-V measurements. The diode was assumed to have a diameter of 2 micrometers and a doping of 10^{23} m^{-3}.

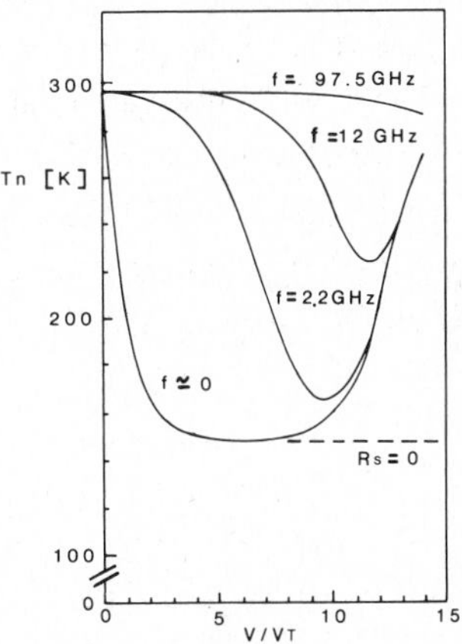

FIGURE 2
Theoretical noise temperature vs. bias voltage.

At low frequencies, with no series resistance, the noise temperature can be seen to start at 295 K and then drop to 147.5 K, as it should. By including the series resistance, it can be seen that for high biases the series resistance dominates over the junction resistance and the noise temperature rises again to the temperature of the series resistance. This is the well-known result for low frequencies.

However, when operated at higher frequencies the noise temperature does not begin to drop until a higher bias is reached, since the noise and conductance of the returning electrons is dominant at low bias. At 97.5 GHz there is almost no decrease in the noise temperature before the series resistance begins to become important. Thus, at high frequencies the noise temperature of the diode remains nearly at the device temperature, due to the transit time effect[3].

3. MEASUREMENTS

The noise measurements on a Hughes diode detector (model 47316H-1111) were performed using the circulator method described by Gasquet et al.[3] This method yields both the noise temperature and reflection coefficient of the device. One major advantage is that it does not require any tuning elements.

The results of the measurements are presented in Fig. 3. It is seen that they agree quite well with the expected theoretical

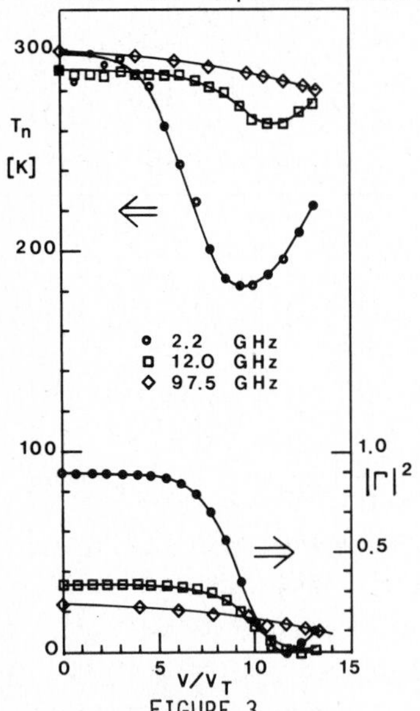

FIGURE 3
Measured noise temperature vs. bias voltage.

results. Also shown is the reflection coefficient. If the junction had been behaving as in the low-frequency approach, then the reflection coefficient would be close to unity at low bias. It is seen that the reflection coefficient at low biases goes down as the frequency is increased. This is tantamount to an increase in the device admittance.

By measuring the noise from a diode in this fashion it is possible to determine exactly the value of DC bias which should be applied in order to operate the diode for lowest noise.

4. CONCLUSION

Noise measurements on Schottky barrier diodes have been presented and the results have been compared with a theoretical analysis which includes the main effect of the transit time. Also included in the analysis was the reverse saturation current and the series resistance. By combining these effects, an accurate model of the diode can be formed which explains the noise behavior with both frequency and bias. We expect that this type of modeling should be especially important to those doing high frequency mixer and detector work.

ACKNOWLEDGEMENT

We thank the National Science Foundation for supporting this research under grant #ECS-8007623.

REFERENCES

1. M.T. Faber, J.W. Archer and R.J. Mattauch, "A very low-noise, fixed tuned mixer for 240-270 GHz," IEEE MTT-S International Microwave Symposium Digest (1985), 311.
2. A. van der Ziel, "Infrared detection and mixing in heavily doped Schottky-barrier diodes," J. App. Phys. 47 (5) (1976), 2059.
3. D. Gasquet, J.C. Vaissiere and J.P. Nougier, "New method for wide band measurement of noise temperature of one-port networks at high pulsed bias," Sixth Inter. Conf. on Noise in Physical Systems, NBS Special Publication 614 (1981), 305.

NOISE IN PHYSICAL SYSTEMS AND 1/f NOISE - 1985
A. D'Amico and P. Mazzetti (editors)

CURRENT NOISE IN GaAlAs/GaAs HETEROJUNCTION BIPOLAR TRANSISTORS

G. BLASQUEZ, C. LIMSAKUL, J.P. BAILBE, A. MARTY and C. ANTARASANA
C.N.R.S. - L.A.A.S., 7, avenue du Colonel Roche, 31077 TOULOUSE CEDEX, France

Experiments carried out within a wide frequency range indicate that the intrinsic current noise in GaAlAs/GaAs heterojunction bipolar transistors is shot noise. In addition, actual devices, may display a small 1/f noise at low frequencies and/or an unusual level of recombination noise at intermediate and high frequencies.

1. INTRODUCTION

The studies carried out on GaAlAs/GaAs heterojonction bipolar transistors highlight an excess noise reaching high frequencies. Consequently, the legitimacy of the applicability of the classical theories to this particular type of transistor is, in actual fact, put into question. In the following, an answer is provided to that fundamental question as regards the future of this transistor as a phototransistor and low level amplifier for high frequencies and microwaves.

2. INTRINSIC NOISE IN BIPOLAR TRANSISTORS

The elementary theory of noise in bipolar transistors shows that the intrinsic noise is the shot noise. If a high source impedance is connected to the transistor input, the internal sources of shot noise are equivalent to a current noise generator i_N connected between the base and the emitter and defined by :

$$S_{i_N}(f) = 2\,q\,I_B + \frac{2\,qI_C}{\left|h_{fe}\right|^2} \qquad (1)$$

where $S_{i_N}(f)$ is the spectral density, h_{fe} refers to the small signal a.c. current gain in the common emitter configuration and q, I_B and I_C have their usual meanings.

3. EXPERIMENTAL RESULTS

The technological process, the doping profile and the electrical characteristics of the transistors were given elsewhere [1]. The specifications of the measurement system have been described in [2].

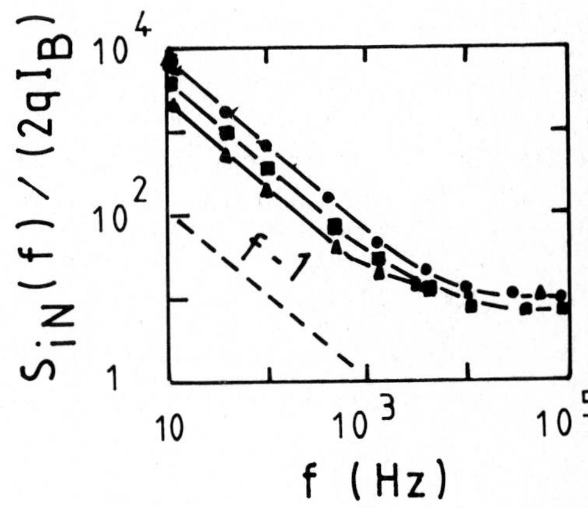

FIGURE 1

Input current noise at low and intermediate frequencies. Experimental conditions : T = 300°K, V_{CE} = 5 V, I_C = 30 µA (▲), I_C = 300 µA (■), I_C = 3 mA (●).

Figure 1 shows a typical aspect of current noise in low noise transistors. At low frequencies, the spectrum is proportional to f^{-1}. At

This work was partially supported by D.A.I.I. grant 82.35.239.00.790.75.00.

intermediate and high frequencies, a white noise occurs.

Figure 2 shows the result of the comparison of measurements to the analytical expression (1)

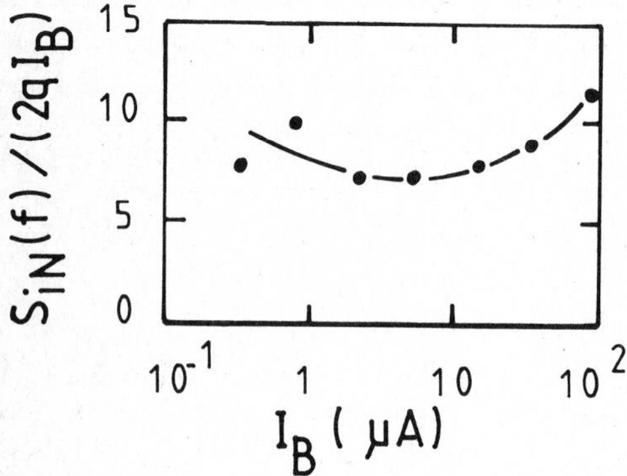

FIGURE 2

Excess white noise at intermediate frequencies. Experimental conditions :
T = 300°K, f = 100 KHz, V_{CE} = 5 V.

in the case where noise is white and $2qI_C/|h_{fe}|^2$ is negligible. It appears that measured noise is much greater than $2qI_B$. Such a type of behavior has never been observed on silicon bipolar transistors. It belongs specifically to the GaAlAs/GaAs heterojunction transistors. In practice, if such an excess noise exists, it greatly reduces the sensitivity of the transistors driven by a current source as in the case of phototransistors.

Figure 3 depicts the high frequency behavior of this excess noise. To facilitate the interpretation of the results the contribution of $2qI_C/|h_{fe}|^2$, has been plotted. Its increase as a function of the frequency f results from the decrease of h_{fe} . The good agreement between theory and experiments at the highest frequencies indicates that the noise produced by the collector base junction is shot noise. Such a feature has been confirmed by additional measurements of the input voltage noise at low level biasing conditions. The excess "white" noise shown in Figure 2 and occuring at the intermediate frequencies therefore results from the heterojunction. Figure 3 also shows that the excess noise has a Lorentzian spectrum. Thus, it may be assumed that is is the results of a recombination process occuring either in the vicinity of or within the heterojunction. Observations carried out with a C R T indicated that this noise is not burst noise.

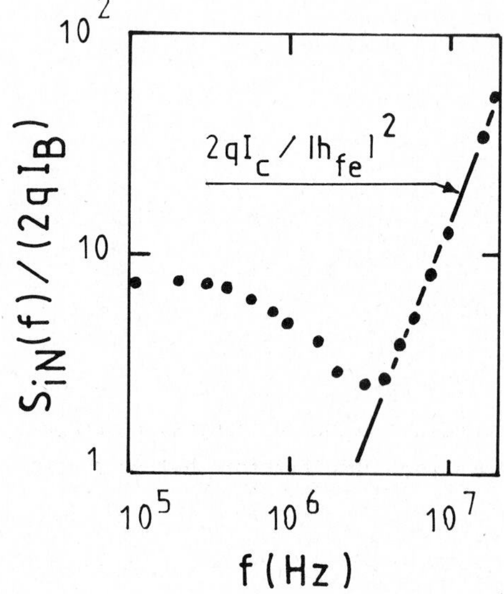

FIGURE 3

Lorentzian behaviour of excess "white" noise in the high frequency range
Conditions : T = 300°K, V_{CE} = 5 V, I_C = 100 μA.

Figure 4 illustrates the case where the excess "white" noise is low. For 3.5 μA < I_B < 9 μA, it can be seen that $S_{i_N}(f) \approx 2qI_B$. In these conditions, the theory enables us to state that the collector base junction and the emitter base heterojunction produce shot noise. Given the state of the art, this type of observation is rarely made. Nevertheless, it demonstrates that the shot noise theory is applicable to the GaAlAs/GaAs bipolar transistor.

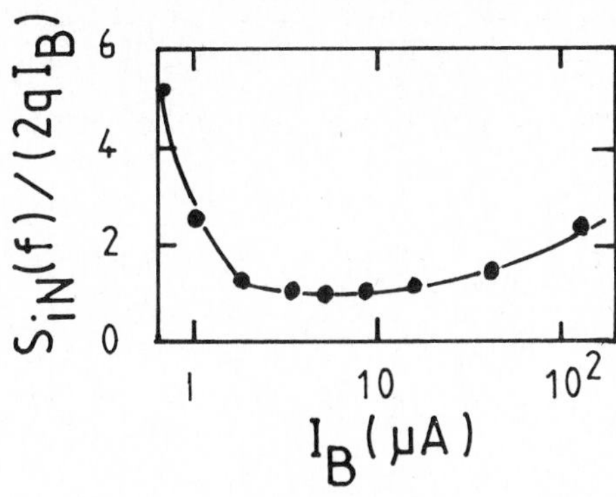

FIGURE 4

Intrinsic shot noise in the range 3 $\mu A > I_B$ < 9 μA.

4. POTENTIAL PRACTICAL APPLICATIONS

It is well-known that the injection efficiency of heterojunctions is noticeably greater than that of homojunctions. Consequently, for fixed biasing conditions (I_C Constant), I_B can be much lower in heterojunction transistors. For low level applications in the low and intermediate frequency ranges, heterojunction transistors are therefore better current amplifiers than homojunction transistors.

In addition, owing to the high mobility of electrons in GaAs, the transition frequency f_T of N P N heterojunction transistors can be three to four times greater than that of silicon bipolar transistors. By expliciting h_{fe} in (1) it can be easily shown that, the input current noise increases in the high frequency range from a frequency f_N approximately equal to $f_N \approx f_T/\sqrt{h_{FE}}$, where h_{FE} is the d.c. current gain. Consequently, NPN GaAlAs/GaAs heterojunction bipolar transistors can have a frequency f_N three to four times greater than that silicon transistors. For low level wideband applications, this means that the frequency range over which the noise remains low can be three to four times wider.

REFERENCES

1. A. Marty, G. Rey and J.P. Bailbe, Solid State Electronics, 22, 549, 1979.

2. G. Blasquez, A high performance digital sys- for noise characterization in electronic devices, 9th IMEKO congress, Berlin, 1982.

NOISE IN PHYSICAL SYSTEMS AND 1/f NOISE - 1985
A. D'Amico and P. Mazzetti (editors)

NOISE PHENOMENA IN SUBMICRON CHANNEL LENGTH SILICON NMOS TRANSISTORS

R.P. JINDAL

AT&T Bell Laboratories, Murray Hill, New Jersey 07974 USA

A variety of equilibrium and transport noise phenomena in submicron channel length NMOS FETs have been studied. These include the distributed gate and substrate resistance noise, substrate current supershot noise, excess channel thermal noise and 1/f noise. Of most interest in this paper is the excess channel thermal noise. This is believed to be a result of carrier heating due to the high electric field existing in the submicron FET channel.

1. INTRODUCTION

Silicon MOS devices provide an excellent tool for studying noise in semiconductors. The submicron device structure and other technology constraints result in the manifestation of a variety of physical phenomena responsible for generating noise. Noise measurements were carried out to study these effects in the frequency range from 1 KHz to 1.5 GHz. Some of these results which have already been discussed elsewhere will be briefly touched upon. Others will be described in detail.

2. DISTRIBUTED GATE RESISTANCE NOISE[1,2]

In a typical Integrated Circuit layout consisting of large devices, the gate of an FET is in the form of a matrix of stripes of submicron widths with the input potential defined at one or several end points. Due to the small cross section area, long lengths and resistive nature of the gate material these stripes add a random noise voltage to the input signal. This noise gets coupled to the FET channel producing excess fluctuations in the drain current. For uniform gate structures we have an analytical expression for the equivalent input noise resistance amenable to optimization.

3. DISTRIBUTED SUBSTRATE RESISTANCE NOISE[3]

Analogous to gate resistance, distributed substrate resistance gives rise to another noise source in the FETs. This resistance arises due to the finite conductivity of the substrate material and acts between the substrate contact and the FET channel. The noise contribution to the drain current is given by

$$S_I(f) = 4K_BTR_{sub}g_{mb}^2 \quad (1)$$

where R_{sub} is the distributed substrate resistance and g_{mb} is the substrate transconductance.

4. SUBSTRATE CURRENT NOISE[4]

Impact ionization in the channel gives rise to substrate current flowing between the substrate and the drain of the FET. It is found that the noise associated with this current corresponds to full shot noise at low drain voltages and steadily increases above the shot noise level as the drain voltage is increased. This super shot noise behavior can be explained by a multistep ionization process involving both holes and electrons and is dominated by a single ionization event per carrier traversal of the ionizing region, emphasizing the need for discrete avalanche noise theories.[5]

These substrate current fluctuations couple to the FET channel through the substrate transconductance producing drain current fluctutations. For large values of substrate current where the ohmic drop across the substrate resistance approaches 0.7 volts, due to feedback built in the device structure, these fluctuations are suppressed to below shot noise levels.

5. CHANNEL HOT ELECTRON NOISE

Having characterized the above sources and virtually eliminating some of them it is now possible, because of the magnitudes involved, to study the remaining excess FET noise. This excess noise is believed to be generated by carrier heating due to high electric fields in the submicron FET channel.

5.1. Design of Experiment

A special test chip with varying channel length devices was laid out. The coded (W/L) ratio of all devices was kept the same (800/1) in order to facilitate performance comparison. The coded channel length of the devices varied from 1.5μ to 0.5μ. For measurement purposes a <100> epi wafer was selected. TEM study and electrical measurements both yielded a net compensation of 0.25μ resulting in devices with electrical channel lengths varying from 1.25μ to 0.25μ.

5.2. Theoretical and Experimental Considerations

For long channel FETs the channel thermal noise is given by[6]

$$S_I(f) = 4K_B T\gamma g_{do} \qquad (2)$$

where γ is a bias dependent parameter and g_{do} is the drain conductance at zero drain bias. Ideally, the value of γ is unity at zero drain bias and approaches 2/3 under device saturation. Now the noise current spectrum and g_{do} can be calculated experimentally and hence γ can be estimated. Larger values of γ would indicate the presence of excess noise.

The bias point was chosen such that all devices had approximately equal g_{do}. Noise measurements[7] were then done at various drain to source voltages in the frequency range from 10 MHz to 1.5 GHz and the noise in the flat portion of the spectrum was used for further analysis. At each bias point, all the small signal device parameters were also measured in order to extract a value for the noise parameter γ accurately.

5.3. Experimental Results and Discussion

All the four devices have the same coded (W/L). However, irrespective of their coded channel lengths they all suffer the same compensation $\Delta L = -0.25\mu$. Hence the $(W/L)_{eff}$ for the four devices is not the same any more. It therefore turns out that the devices have the same g_{do} at different $V_{GS}-V_t$. Thus the shorter channel length devices are biased closer to threshold.

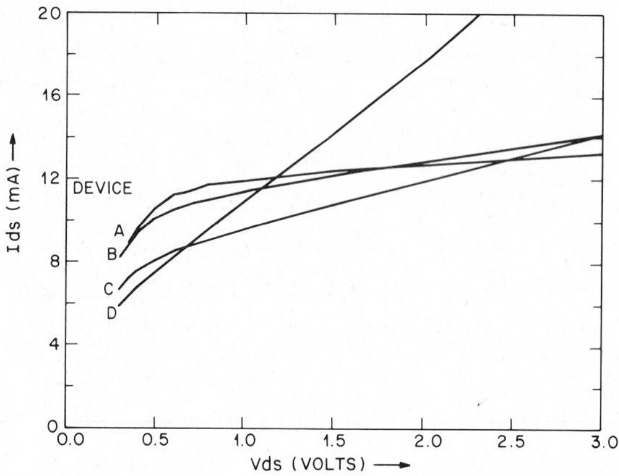

Fig. 1 Drain current versus drain to source voltage at different gate to source voltages for devices A,B,C and D.

(a) Small Signal Parameter Behavior

In Figure 1 we plot the drain current for the four devices as a function of the drain to source voltage V_{ds} at fixed gate voltages. We observe, according to expectation, that the slope (output conductance g_{ds}) gets progressively larger as we go to shorter channel length devices. Further, we see that for a given device the slope gets progressively smaller as we increase V_{ds}. The only exception to this is device D for which we observe inflection around V_{ds} = 1V.

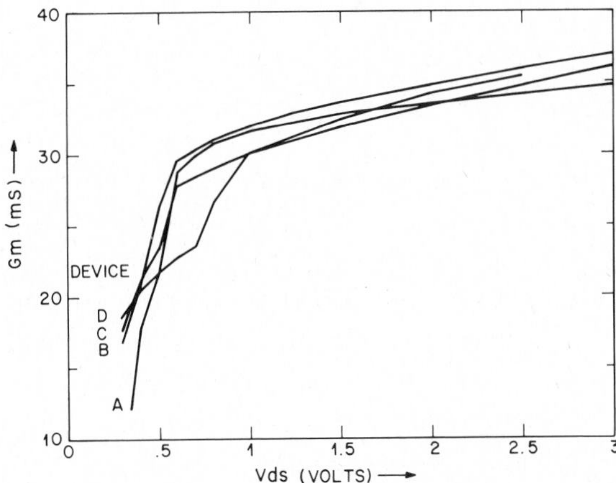

Fig. 2 Gate transconductance versus drain to source voltage at a different gate to source voltage for devices A,B,C and D.

In Figure (2) we plot the device transconductance g_m as a function of V_{ds} at fixed gate voltages. All the four devices have a comparable g_m throughout the measurement range with the exception of device D. This device exhibits a kink at around V_{ds} = 0.7 volts.

The peculiar behavior of g_{ds} and g_m for device D around V_{ds} = 1V is typical for these devices and is ascribed to the extremely small channel length of 0.25μ where the influence of the high drain electric field on the unpinched portion of the channel becomes important.

(b) Noise Behavior

In Figure (3), we plot noise factor γ (defined by Eq. 2) as a function of V_{ds} for the four devices. In general, we observe that the curves start around unity and increase as a function of V_{ds}. This increase in γ is an

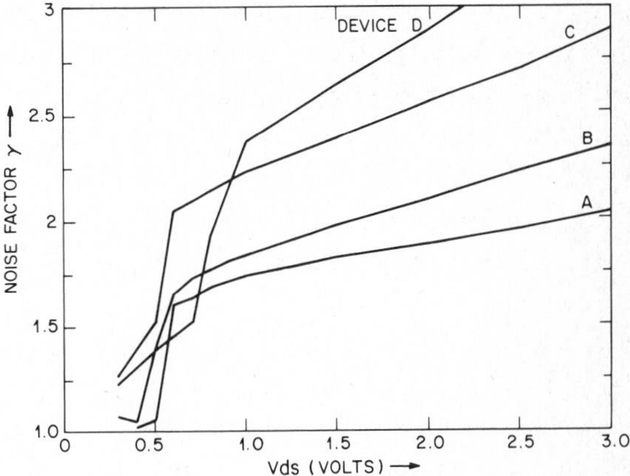

Fig. 3 Noise factor γ versus drain to source voltage at different gate to source voltages for devices A,B,C and D.

indication of the presence of an excess noise source in the FETs. We can divide the curves into three regions which we shall call the high field region, the transition region and the low field region.

The high field region is defined for V_{ds} > 1 volt. In this region the devices are well in saturation exhibiting a small g_{ds} thus making them useful for normal circuit applications. The noise factor γ is a monotonically increasing function of V_{ds} for a given device and monotonically decreasing function of the channel length at a fixed V_{ds}. Thus, the higher the drain to source voltage and the smaller the channel length, the higher is the noise factor. For example, we find that device B (Leff = 0.75μ) exhibits noise power a factor of 2.8 above the ideal (γ = 2/3) at V_{ds} = 1V and a factor of 3.5 above the ideal at V_{ds} = 3V. Also, at a fixed V_{ds} = 1.5V, device A (Leff = 1.25μ) exhibits 2.7 times the ideal noise power while device D (Leff = 0.25μ) exhibits 4.0 times the ideal noise power.

The transition region is characterized by a sharp increase in γ as a function of V_{ds}. This region coincides with the transition of the device from triode into the saturation region. Hence, the closer the device is biased to V_t, the earlier the jump. (This has also been verified by noise measurements on a device at different gate voltages.) The only exception to this is device D which is biased only 0.4V above low field V_t. This implies that as the devices enter saturation resulting in an increase in the electric field in the channel, the noise goes up.

The low field region is bounded by V_{ds} = 0 on one side and the transition region on the other side. We observe here that devices A and B exhibit $\gamma \simeq 1$ close to the ideal behavior. Devices C and D show a monotonic increase in γ. This difference in behavior is partly due to the shorter channel lengths and hence higher electric fields existing in the channel even in the triode region for devices C and D. In order to further explore and understand this behavior of the noise factor γ we are doing a Monte Carlo simulation of the high field effects on electron transport and noise.

6. 1/F NOISE

At low drain voltages where the FET channel is uniform, one can show[8] that for 1/f noise generated in the channel, the equivalent noise current is given by

$$I_{eq} = \frac{\alpha_H}{2fL^2} \mu I_{ds} V_{ds} \qquad (3)$$

where α_H is the Hooge parameter, L is the channel length, μ is the carrier mobility and the rest of the symbols have their usual meaning. Measurements on enhancement mode n - channel device, indicate $\alpha_H = 4 \times 10^{-5}$ which is below the bulk value of 2×10^{-3} but well above the fundamental limit of 2.3×10^{-8} predicted by the quantum 1/f noise theory.[9]

7. CONCLUSIONS

Several noise phenomena in fine line silicon NMOS devices have been studied. They consist of the distributed gate and substrate resistance noise, substrate current supershot noise, 1/f noise and excess channel thermal noise. Any attempt to increase the channel electric field either by increasing the drain to source voltage or by reducing the channel length results in an increase in the excess channel noise. This suggests that this excess noise is a manifestation of hot electron effects in the FET channel. In order to fully understand this noise we are currently studying this phenomena using Monte Carlo techniques.

ACKNOWLEDGEMENT

I would like to thank Marty Lepselter, George Smith and Harry Boll for support in initiating and carrying out this study, Larry Nagel for encouragement during the later phase of the study and Don Fraser for helpful interactions.

REFERENCES

1. K. K. Thonber, IEEE-JSSC, SC-16 (4) (1981) 414.
2. R. P. Jindal, IEEE-TED, ED-31 (1984) 1505.
3. R. P. Jindal, IEEE-TED, ED-32 (1985), to be published.
4. R. P. Jindal, IEEE-TED, ED-32, (1985) 1047.
5. K. M. van Vliet, A. Friedman and L. M. Rucker, IEEE-TED-26 (1979) 752.
6. See for example A. Van der Ziel, Noise: Sources Characterization, Measurement, Chapter 5 (Prentice Hall, New Jersey, 1970).
7. A brief description of the measurement system is given in Ref. 3.
8. A. van der Ziel, Private Communication.
9. A. van der Ziel, P. H. Handel, X. C. Zhu and K. H. Duh, IEEE-TED-32 (1985) 667.

NOISE IN PHYSICAL SYSTEMS AND 1/f NOISE - 1985
A. D'Amico and P. Mazzetti (editors)

NOISE IN SHORT n^+-n-n^+ GaAs DIODES

V.BAREIKIS, K.KIBICKAS, J.LIBERIS, R.MILIUŠYTĖ, P.SAKALAS

Semiconductor Physics Institute of the Lithuanian SSR Acad. Sci., Vilnius, K.Pozhelos 52, 232600 USSR

Experimental microwave noise temperature and drift velocity dependences on electric field strength for the different lengths of GaAs n^+-n-n^+ diode active region at room temperature are presented. It was shown that the noise temperature, current-voltage characteristics and diffusion coefficient depend on diode active region length (in range $L < 10\ \mu m$) due to differences of the electron heating and intervalley scattering conditions in short devices compared with the long ones.

I. INTRODUCTION

In the last few years due to the appreciable progress in semiconductor technology considerable attention has been focused on the transport phenomena in short devices. In these devices electrical properties as well as electrical fluctuations may be considerably different due to the near-ballistic behaviour from those in the long ones[1,2,3]. Some progress has been achieved in the understanding of the noise in submicron devices. It was shown that noise is considerably effected there by the space correlation[2] as well as contact properties[4,5]. It was found out experimentally that near-ballistic n^+-n-n^+ GaAs devices exibit extremely low 1/f noise. At low frequencies thermal noise of these devices is slightly higher than the Nyquist one[6,7]. However charge transport characteristics as well as electrical fluctuation properties in short devices depend on electron energy and the efficiency of electron scattering. Thus in GaAs structures in the range of electric field where electron energy exeeds the energy gap between valleys the intervalley scattering becomes effective and strongly infuences drift velocity and noise[3,8]. However the highest voltages used in the experiments[6,7] were too low to give rise to intervalley scattering. Besides the measurements of the samples 0,4 μm long were carried out. The investigations of the device length influence on the noise are not carried out as yet although they are of interest. Namely the length of active region in micron devices is much greater than the mean electron free path, but the electron transit time through the device may be compared to the characteristic time of intervalley scattering. In this paper we present the experimental results of electron intervalley transition influence on the noise and drift velocity in micron n^+-n-n^+ GaAs devices.

2. EXPERIMENTAL

For electric noise measurements in microwave range the samples were mounted in the middle of a rectangular waveguide section. Noise temperature T_n is determined by somparing the sample-radiated noise power P with that of the standard noise generation. T_n is known to be associated with P by relationship $P=kT_n(1-\Gamma^2)$. Γ is the reflection coefficient determined by usual technique. T_n was found determining P from the readings of radiometer taking into account its modulator as well as waveguide losses and thermal radiation. In order to avoid the thermal heating of the sample all experiments were performed using voltage pulses as short as 3 μs with low repetition rate. Drift velocity of electrons was determined by the means of microwave integral technique[9].

The active region of the devices had the following parameters: length L = = 7,5; 1,5; 1,0 μm; electron concentration 1÷3·10^{15} cm^{-3}; mobility 0,76 $m^2V^{-1}s^{-1}$. Voltage U across the active region was determined including potential drop on the contact resistance. The latter was evaluated by the angle-dependent geometrical magnetoresistance method[10].

3. RESULTS AND DISCUSION

Noise temperature and drift velocity dependences on the electric field strength for the devices of different length are presented in Fig. 1 and Fig. 2 respectively. From T_n and v_d measurement data the diffusion coefficient may be determined[11]. According to the experimental results the shorter is the diode the higher is the value of drift velocity as well as the threshold electric field for the intervalley noise and negative differential conductivity, and the lower noise temperature and diffusion coefficient at the fixed electric field.

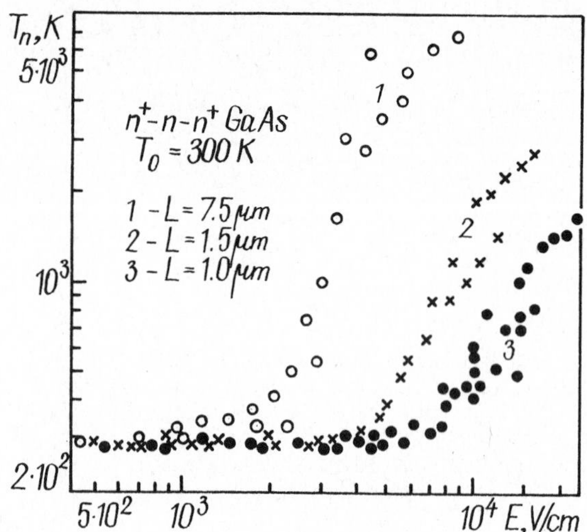

FIGURE 1
Noise temperature vs effective electric field ($E=U/L$)

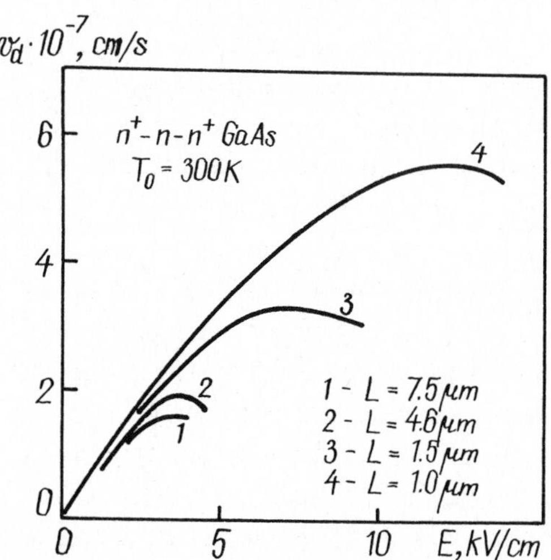

FIGURE 2
Drift velocity vs microwave electric field

The numerical calculations by Monte Carlo technique for electrons in GaAs were carried out to explain T_n and v_d dependences on the length of diode active region. It is known that the transitions of electrons into higher valleys is possible only when electron energy is equal or higher than the energy gap between valleys. The mean transit length necessary for the electron in the homogeneous electric field to get the energy equal to the energy gap between Γ and L valleys is calculated. The parameters for the one valley GaAs model including acoustic and polar optical scattering are taken from[12]. Initial energy of electrons was taken to be equal to 0,001 eV. The data of calculation presented in Fig. 3 qualitatively explains the ex-

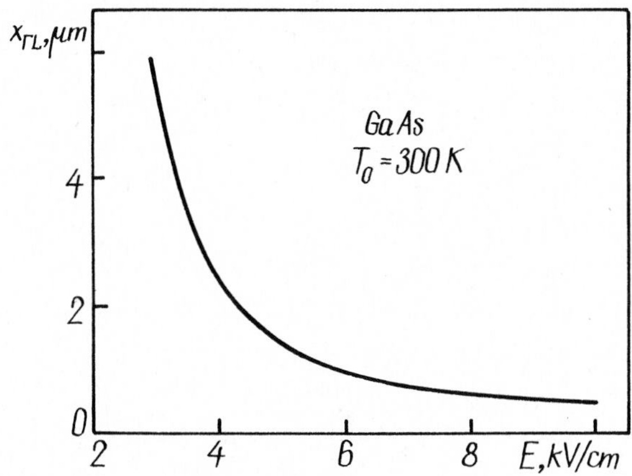

FIGURE 3
Calculated distance $x_{\Gamma L}$ vs electric field

perimentally determined pecularities of intervalley noise manifestation in micron n^+-n-n^+ GaAs devices: the smaller is the electric field the longer is the distance $x_{\Gamma L}$ necessary to get energy $\varepsilon_{\Gamma L}$. Consequently for the intervalley noise to appear in the devices with short active region the strength of electric field must be higher than that of the long devices. The same stands for the threshold electric field of the negative differential conductivity.

4. CONCLUSION

At room temperature the noise temperature, current-voltage characteristics and diffusion coefficient in GaAs n^+-n-n^+ diodes depend on the diode active region length (in range $L < 10\ \mu m$) due to the differences in the electron heating and intervalley scattering conditions in short devices compared with the long ones.

5. REFERENCES

1. A.Van der Ziel, Thermal, shot, diffusion and 1/f noise in GaAs, in: Proc. 7-th Int. Conf. Noise in Physical Systems and 1/f Noise, Montpellier, eds. M.Savelli, G.Lecoy, J.-P.Nougier (Nort-Holland, Amsterdam, Oxford, New York, Tokyo, 1983) 147.

2. J.-P.Nougier, Noise of submicron devices, in: Proc. 7-th Int. Conf. Noise in Physical Systems and 1/f Noise, Montpellier, eds: M.Savelli, G.Lecoy, J.-P.Nougier (Nort-Holland, Amsterdam, Oxford, New York, Tokyo, 1983) 153.

3. V.I.Rygij, N.A.Bannov, V.A.Fedirko. Fizika i technika poluprovodnikov 18, 5 (1984) 769.

4. J.Zimmermann, E.Constant. Sol.St. Electr. 23 (1980) 915.

5. B.Boittiaux, E.Constant, A.Ghis. Simulation of diffusion noise in a device, in: Proc. 7-th Int. Conf. Noise in Physical Systems and 1/f Noise, Montpellier, eds. M.Savelli, G.Lecoy, J.-P.Nougier (Nort-Holland, Amsterdam, Oxford, New York, Tokyo, 1983) 19.

6. R.R.Schmidt, G.Bosman, C.M.Van Vliet, L.F.Eastman, M.Hollis, Solid-St.Electr. 26 (1983) 437.

7. A.Peczalski, A.Van der Ziel, M.A. Hollis, IEEE Trans. Electron. Dev. ED-29 (1982) 1837.

8. V.Bareikis, V.Viktoravičius, A.Galdikas, R.Miliušytė, Fizika i technika poluprovodnikov 14, 7 (1980) 1427.

9. V.Dienys, A.Dargys, Journal de Physique 42 (1981) C7-33.

10. L.Gutai, I.Mojses, Appl. Phys. Lett. 26 (1975) 325.

11. J.P.Price, Fluctuation phenomena in Solids (Acad. Press. New York, 1965) 355.

12. J.Požela, A.Reklaitis, Solid-St. Electron, 23 (1980) 927.

BIOLOGICAL SYSTEMS - 1

NOISE IN PHYSICAL SYSTEMS AND 1/f NOISE - 1985
A. D'Amico and P. Mazzetti (editors)

SINGLE CHANNEL CURRENTS IN RETINAL CELLS

Davide LOVISOLO^, Renzo LEVI^, Dario CANTINO", Elena STROBBIA"

^Department of Animal Biology, Laboratory of General Physiology, and "Institute of Human Anatomy, University of Torino, Italy+

Using the "patch clamp" technique, we have performed recordings of ionic currents flowing through single channels in membranes of neuronal cells obtained from chick embryo retina. This approach, giving a direct measurement of events at the molecular level, allows to obtain information about channel properties that was more difficult to deduce from noise analysis. In these cells, that are differentiating into nerve cells, our aim was to study electrophysiological characteristics at the single channel level, in order to compare them with fully differentiated neurons. In this paper we describe some of the properties of one population of channels, likely to conduce potassium.

1. INTRODUCTION

Application of noise analysis to the voltage or current fluctuations that occur across the membrane of many living cells has allowed, in the last 15 years, to obtain relevant information about the mechanisms that regulate ionic fluxes through these membranes. In a preceding paper[1] we have analyzed the voltage fluctuations that can be measured in the photoreceptor cells of the honeybee drone (*Apis mellifera*) in dark adapted conditions and at different levels of light adaptation. An increase in noise amplitude with intermediate light intensities has been observed, while at greater light intensities noise decreased.

These data, together with a spectral analysis and some limiting assumptions, led to the conclusion that the change in transmembranic potential caused by light ("receptor potential") can be ascribed to the opening of ionic channels of 40 pS average conductance. While average amplitude did not change and opening rate increased linearly increasing the light intensity, at least for low levels of adaptation, mean open time was found to decrease, thus accounting for the decrease in amplitude and duration of the macroscopic response following light adaptation.

However, the limitation of this approach (mainly the difficulty of obtaining complete information from voltage noise analysis) have practically reduced the usefulness of noise analysis to preparations in which it is possible to perform current measurements; and this is not technically feasible in many cases.

The "patch clamp" technique, introduced by Neher, Sakmann, and colleagues[2,3], allows to cope in a straightforward way with the problem of measuring the elementary (molecular) events that give rise to the changes in membrane conductances and, as a consequence, to the transmembranic ionic fluxes, and it has drawn the attention of many researchers in this field.

It consists essentially in placing a micro-

+This research was partially financed by MPI and by CNR, GNCB.

pipette, filled with a conducting saline solution, against a "clean" (i.e. without connective tissue) membrane, and, by gentle suction, to create a tight seal between the pipette and the membrane, with a seal resistance of 10 - 100 GΩ. By this way, almost all the current that flows through the membrane area sealed to the pipette (a few μm^2) can be recorded by the pipette itself, and it is possible to resolve currents al low as .5 pA. Unitary events caused by the opening - or closing - of single ionic channels can be directly observed.

We have applied this technique to the membranes of neuronal cells obtained from chick embryo retina. One of the aims of the present work is to assess their level of differentiation, that is to study if, at this stage of embryonal development, they show the membrane properties characteristic of nerve cells[4].

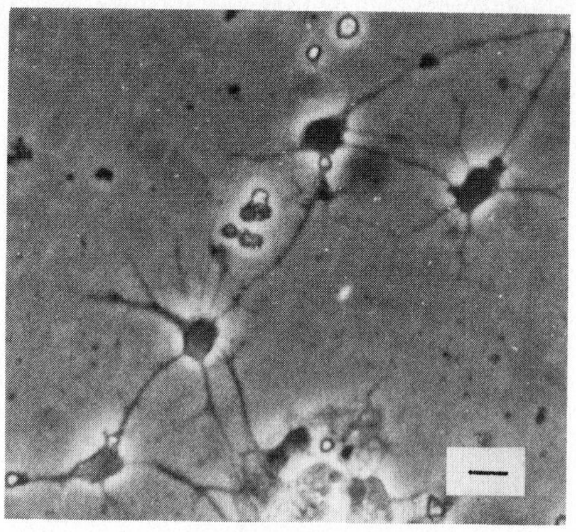

FIGURE 1

A culture of neuroretinal cells from an embryo of 7 days after 5 days of culture. Calibration bar: 10 μm

2. MATERIALS AND METHODS

The cells, obtained by mechanical or enzymatical dissociation from embryos of 7 days, were cultured in standard Dulbecco's medium in plastic Petri dishes, where they adhered to a coating of polylysin. Cells free from glia, isolated or connected in small networks, survive and grow for 5 - 8 days in the dishes. Many of them are bi- or multi-polar while some are monopolar in shape. Cell somata are 5 to 8 μm in diameter (Fig. 1).

Electrodes were pulled on a modified Narishige PG-1 puller from borosilicate glass pipettes (Garner Glass Co. 7052); after moderate fire--polishing a tip of about 2 μm diameter was obtained. In order to reduce noise, the pipettes were coated with Sylgard (Dow Corning 184) as nearest to the tip as possible, working under a x100 dissection microscope.

The electrode was connected to a List PC-5 amplifier and signals were recorded on a HP 3694A magnetic recorder, and analyzed with a Nicolet 4094 digital oscilloscope and a HP 9816 computer.

3. RESULTS

Figure 2 shows a typical single channel current recording at different patch potentials. The pipette was filled with a solution containing 133 mM KCl, 10 mM HEPES, 5 mM EGTA, in order to have nearly simmetrical K^+ concentrations on the two sides of the membrane patch. Downwards deflection represents an inward current, that is positive ions flowing through the channel into the cell.

An I-V relation for this channel was obtained

and it is shown in Figure 3.

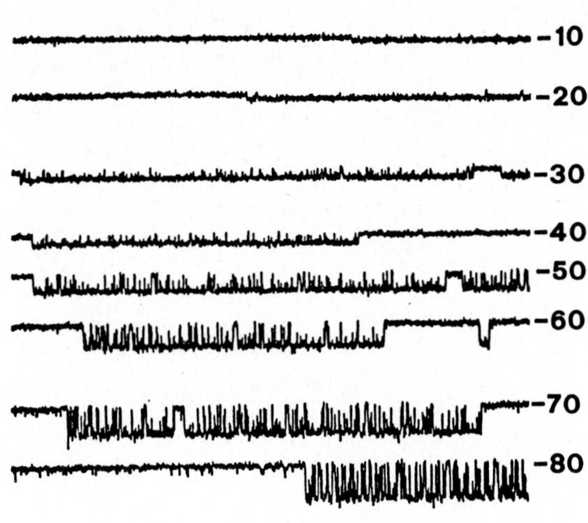

FIGURE 2
A typical single channel recording at different potentials of the membrane patch (numbers on the right). Filter: 1 kHz, 4 pole Bessel

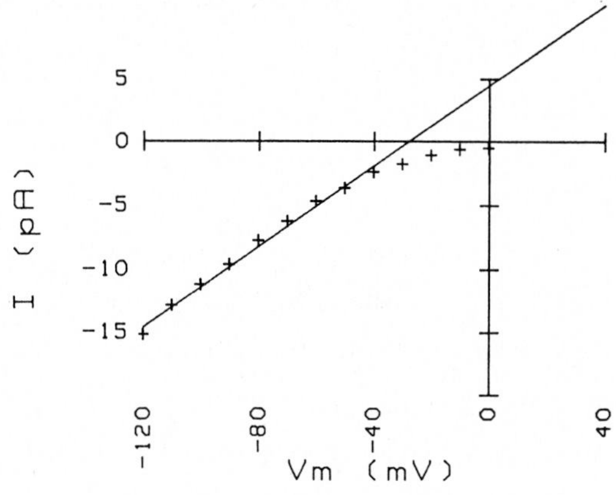

FIGURE 3
I-V relation of the channel shown in Fig.2. Slope con ductance (from straight line fit): 160 pS

It has been drawn assuming that the membrane potential at rest is about -50 mV (assumption based on some measurements in the whole-cell configuration, in which the pipette has direct access to the cell interior): membrane potentials more negative than -50 mV are hyperpolarizations (obtained by giving positive potentials to the pipette) and potentials more positive are depolarizations (pipette potential negative). With simmetrical K^+ concentrations, a reversal of current polarity near 0 mV can be assumed as a first evidence for a K^+ conducting channel, under the assumption that the intracellular concentration of Cl^- is lower than K^+ concentration.

Further analysis will deal with open and close time distributions and with changes in channel kinetics depending on changes of the physico-chemical environment (voltage, ionic concentrations, etc.).

In many instances, however, the experimental situation may become much more complicated and a quantitative analysis more difficult.

Figure 4 shows a recording from another cell in which the same channel shown in Figure 2 can be observed, but a smaller one is also present.

Figure 4
A stretch of recording showing two different channel types (arrows) opening in the same patch. Patch potential: -60 mV

In these cases, the different behaviour of the two channels at different voltages may create problems in discriminating between the two: the only approach is to chemically dissect the channels, that is to use substances that specifically block one population in order to isolate and study the other.

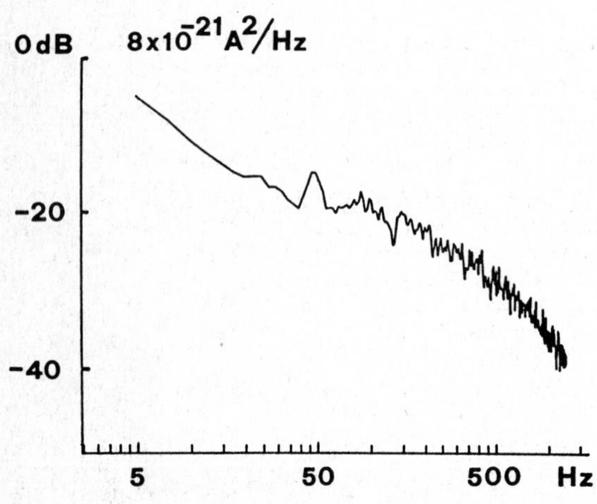

Figure 5
Average power spectrum (15 averages) from the same recording as in Fig. 2. Patch potential: -90 mV. Signal sampled at 200 μsec per point. Filter: 1 kHz, 4 pole Bessel. Background noise was not subtracted

Figure 5 deals with a different approach to channel analysis: here a power spectrum has been obtained from a stretch of recording in which only one channel was present. The use of power spectra, together with the direct knowledge of the number of channels and of the unitary conductance, allows a quick measurement of some kinetic parameters (in the oversimplified case of a two state channel, the measure of the sum of the two rate constants), nevertheless requiring a more detailed statistical study, using histograms from single channel measurements, to build a kinetic model.

REFERENCES

1. M. Ferraro et al., Biophys. Struct. Mech. 10 (1983) 129.
2. E. Neher et al., Pflügers Arch. 375 (1978) 219.
3. O.P. Hamill et al., Pflügers Arch. 391 (1981) 85.
4. R. Adler et al., J. Cell Biol. 99 (1984) 1173.

NOISE IN PHYSICAL SYSTEMS AND 1/f NOISE - 1985
A. D'Amico and P. Mazzetti (editors)

RESISTIVITY FLUCTUATIONS IN IONIC SOLUTIONS

R.J. VAN DEN BERG, A. DE VOS, P. VAN DER BOOG and J. DE GOEDE

Department of Physiology and Physiological Physics, State University of Leiden, The Netherlands

Electrical resistivity fluctuations in aqueous solutions of $CuSO_4$ and $AlCl_3$ at 20.0°C were measured. The sample consisted of a micro-capillary separating two compartments filled with the same concentration of one of these electrolytes. The spectral densities, estimated in the frequency range from 0.5 to 5000 Hz are characterized by a Lorentz-like function. These spectra can be explained by convective diffusion.
Using high concentrations of $CuSO_4$, the electro-osmosis was so small, that a diffusion dominated spectrum was measured. Its corner frequency is near the predicted frequency $D/\pi l^2$ (D is the diffusion constant and l the length of the capillary) while the high frequency slope obeys the "universal $f^{-3/2}$ law".
No 1/f noise is present in our measurements. Estimated upper limits for the relative and normalized 1/f noise parameter α range from 10^{-4} to 10^{-5}.

1. INTRODUCTION

Studies in electrolyte-filled capillaries have revealed the existence of several types of electrical noise. Besides Johnson noise there have been reported noise from a chemical reaction [1,2], noise due to temperature fluctuations [3] and noise due to fluctuations in the total number of ions [1,4-7]. These excess noises are characterized by spectral densities, which are flat at low frequencies and decline proportional to $1/f^x$ with $x \geq 3/2$.
The existence of 1/f noise in ionic solutions is highly controversial. In our search for 1/f noise in strong electrolytes we have found that concentration fluctuations are the sole excess noise source [4-7].
In this paper we report on the spectral densities of the noise voltage across micro-capillaries filled with $CuSO_4$ and $AlCl_3$ solutions. We also estimated upper bounds for possible 1/f noise intensities in these electrolytes.

2. EXPERIMENTAL PROCEDURE

Experimental methods have been described in [5-6]. The length and radius of the micro-capillaries were about 1 µm. After filling, the streaming potential (V_{str}) was measured. Fulfilment of the relation

$$V_{str} = a\ P \tag{1}$$

where a is a proportionality constant depending on the resistivity of the solution and on the zeta-potential (ζ) of the electrolyte-glass interface and P the pressure drop across the capillary, guaranteed that the capillary was completely filled with the electrolyte [6].
Difference spectra (double log, in the range from 0.5 to 5000 Hz) were calculated from spectral densities obtained with and without current injection.
Since the noise voltage is filtered by the impedance of the capillary we multiplied these spectra by $|Z(+0)/Z(f)|^2$, where $|Z(f)|$ is the modulus of the impedance, measured by white noise cur-

rents and $|Z(+0)|$ its low frequency value. These spectra were subjected to a non-linear least squares curve fitting utilizing a simple physical model [7].

3. CONCENTRATION FLUCTUATIONS

3.1. Variance

The variances, obtained from integration of the spectral densities of $CuSO_4$ and $AlCl_3$ solutions were in good agreement with:

$$\langle \Delta V^2 \rangle = I^2 R^2 \lambda^2 / (n\, V_{eff}) \qquad (2)$$

where I is the applied constant current, R the resistance of the capillary, λ the coefficient relating resistance fluctuations to concentration fluctuations, n the total ion density and V_{eff} the effective volume in which the noise is generated [7]. Eq. (2), obtained from thermodynamic fluctuation theory, indicates that the measured excess noise is caused by fluctuations in the ionic concentration only.

3.2. Spectral density

The frequency spectra of the noise depend on the dynamics of the concentration fluctuations, which is governed by two mechanisms: diffusion and volume flow through the capillary. Volume flow can originate from a pressure difference between the two reservoirs (Poiseuille flow) and from an applied electric field (electro-osmotic flow).
An example of the spectral density obtained from a capillary filled with 10 Mol/m^3 is depicted in fig. 1. At low frequencies the spectral density is flat and above a corner frequency f_o (point of intersection of the low and high frequency asymptotes) the spectral

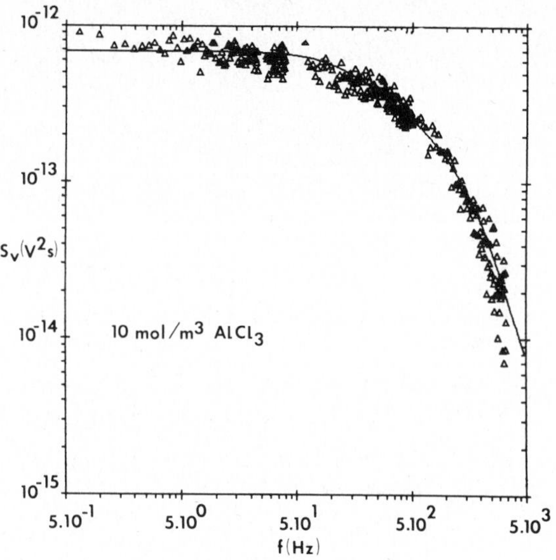

FIGURE 1
Spectral density of the voltage fluctuations across a capillary with a length and a radius of 2 μm as directly measured with a light microscope and filled with 10 Mol/m^3 of $AlCl_3$ at 20.0°C. The pressure drop was approximately 50 N/m^2. The applied voltage was 1.20 V and the resistance 1.65 MΩ. From the measured V_{str} we calculated $\zeta = +$ 11.4 mV, resulting in a v_e in opposite direction of v_p.
The solid line represents the fit of the physical model to the experimental data. This yielded $f_D = 115 \pm 12$ Hz and $f_F = 495 \pm 18$ Hz, which is in fair agreement with values predicted by eq.'s (3) and (4) with $D = 1.3\times10^{-9}$ m^2/s, $\eta = 1.0\times10^{-3}$ Pa.s and $\varepsilon = 7.1\times10^{-10}$ C^2/Nm^2: 123 Hz and 613 Hz respectively. The parameters f_o and x were 537 ± 27 Hz and 1.67 ± 0.05 respectively.

density declines proportional to $1/f^x$, with $x > 3/2$.
From a physical one-dimensional model [7] two characteristic frequencies are obtained, viz. for diffusion

$$f_D = D/\pi l^2 \qquad (3)$$

and for volume flow

$$f_F = v_F/2\pi l \quad (4)$$

where D is the diffusion constant, l the length of the capillary and v_F the volume flow velocity given by

$$v_F = v_P + v_E \quad (5)$$

In eq. (5), v_P and v_E are the velocities of the mean Poiseuille flow and of electro-osmosis respectively. The characteristic frequencies, obtained from the model by curve fitting, agree within a factor of two with the calculated ones, using the measured dimensions of the capillary and the necessary physical constants and variables. Furthermore to a fair approximation

$$f_o = f_D + f_F \quad (6)$$

In all our previous experiments volume flow has dominated diffusion. It proved to be extremely difficult to realize a condition in which the spectral density is determined by diffusion only. To this end the value of the zeta-potential has to be very small i.e. < 1 mV to reduce electro-osmotic flow as much as possible. In fig. 2 the spectral density is shown for a capillary filled with 500 Mol/m^3 $CuSO_4$. This spectrum shows the characteristics of a diffusion dominated spectrum: $f_o = f_D = D/\pi l^2$ and slope x = 3/2. However our physical model now yields unrealistic high values for f_F. This is due to the one-dimensional character of the model where the low frequency asymptote goes as $f^{-1/2}$, while the observed one is flat. To obtain a more realistic description of the diffusion noise a three dimensional model has to be developed.

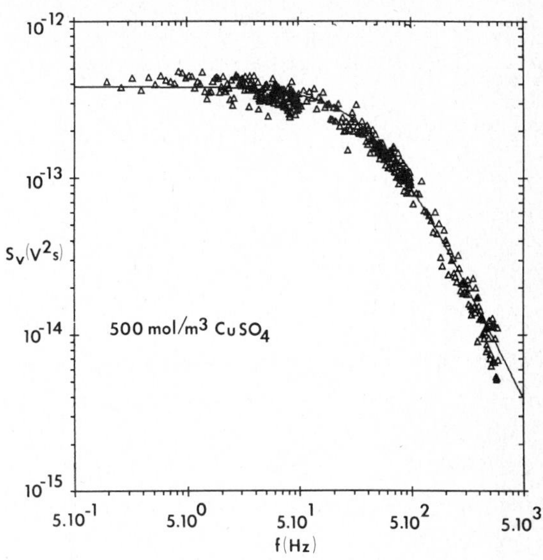

FIGURE 2
Spectral density of the voltage fluctuations of a capillary filled with 500 Mol/m^3 $CuSO_4$ solution at 20.0°C. Its length is 0.80 μm and its radius 0.45 μm, which were directly measured by methods given in [7]. The pressure drop was less than 5 N/m^2. The applied voltage was 0.40 V and the resistance was 0.74 MΩ. The V_{str} < 1 μV with P = 3.10^3 N/m^2, corresponding to ζ < 1 mV. The solid line represents the fit of $S(f) = S(+0)/[1+(f/f_o)^x]$ to the data. This fit yielded f_o = 211 ± 4 Hz and x = 1.44 ± 0.02, which is close to the predicted values for f_D and x_D, 199 Hz and 3/2 respectively (D = 0.4 x 10^{-9} m^2/s).

4. 1/f FLUCTUATIONS

No 1/f noise has been detected under all our experimental conditions. An upper limit for possible 1/f noise intensity can be estimated by setting it equal to twice the standard deviation of the low frequency asymptote of the concentration fluctuations. In analogy to solid state materials the 1/f noise intensity (S_F) is written as [8]:

$$S_F = \alpha\ I^2R^2/(n\ V_{eff}f) \qquad (7)$$

where α is a dimensionless parameter. The estimated upper limits for the α-parameter range from 10^{-4} to 10^{-5}, the same range of the α-values recently reported for electrons in samples of metals and semiconductors [8].
Our results contradict those of other laboratories cited in [4] and those recently reported in [9] and [10]. However, we are confident that our results are correct, because they can be explained by fluctuation theory using known transport mechanisms in ionic solutions.

This work was supported by the Netherlands Organization for the Advancement of Pure Research (ZWO).

REFERENCES

[1] G. Feher and M. Weissman. Proc. Nat. Acad. Sci. USA 70 (1973) 870-875.

[2] K. Ohbayashi, M. Minoda, K. Murakami and T. Yasunaga. J. Chem. Phys. 78 (1983) 2008-2012.

[3] M. Weissman and G. Feher. J. Chem. Phys. 63 (1975) 586-587.

[4] R.J. van den Berg, A. de Vos and J. de Goede. In: Sixth Intern. Conf. on Noise in Physical Systems, eds. P.H.E. Meyer, R.D. Mountain and R.J. Soulen, National Bureau of Standards, S.P. 614 (1981) 217-220.

[5] R.J. van den Berg and A. de Vos. Phys. Lett. 92A (1982) 203-206.

[6] A. de Vos, R.J. van den Berg and J. de Goede. Phys. Lett. 102A (1984) 320-322.

[7] J. de Goede, N. Roos, A. de Vos and R.J. van den Berg. In: Proc. 7th Intern. Conf. on Noise in Physical Systems, eds. M. Savelli, G. Lecoy and J.P. Nougier. Elsevier Science Publishers (1983) 393-396.

[8] L.K.J. VanDamme. In: Proc. 7th Intern. Conf. on Noise in Physical Systems eds. M. Savelli, G. Lecoy and J.P. Nougier. Elsevier Science Publishers (1983) 183-192.

[9] T. Musha and K. Sugita. J. Phys. Soc. Japan 51 (1982) 3820-3825.

[10] H.A. Kolb and D. Woermann. J. Chem. Phys. 80 (1984) 3781-3784.

NOISE IN PHYSICAL SYSTEMS AND 1/f NOISE - 1985
A. D'Amico and P. Mazzetti (editors)

MEASUREMENT AND APPLICATIONS OF NOISE GENERATED BY A METAL-ELECTROLYTE INTERFACE

C. GABRIELLI, F. HUET, and M. KEDDAM

L.P.15 du CNRS, "Physique des Liquides et Electrochimie", associé à l'Université Pierre et Marie Curie, Tour 22, 4 place Jussieu, 75230 PARIS CEDEX 05.

The power spectral density of the electrochemical noise generated by a metal-electrolyte interface is measured with a two measurement channel arrangement. Two noise applications will be presented. The electrochemical noise generated during the discharge of a battery allows a noise based state-of-charge test to be devised. The analysis of the voltage fluctuations generated by a gas bubble evolution allows an identification of the characteristic parameters of the evolution.

1. INTRODUCTION

Many metal-electrolyte interface phenomena, either at a microscopic scale (e.g. electrochemical reactions, diffusion processes), or at a semi-macroscopic scale (e.g. gas bubble nucleation, pitting corrosion) can be considered as stochastic processes. These processes can be investigated by analysing the random fluctuations of the electrolysis current (potential) at constant potential (current). So far a few electrochemical systems have been investigated either for analysing fundamental processes or for obtaining some efficiency tests (batteries, corrosion, ...)[1,2,3].

In this paper we shall describe the measurement technique of the noise power spectral density (p.s.d.) and point out the specific characters of this technique in electrochemistry coming, first from the out-of-equilibrium polarization of the interface, secondly from the poor stationarity of the system.

Two types of application will then be shown : the first concerns the noise generated during the discharge of a battery (a few microvolts amplitude noise) in order to search for a noise based state-of-charge test. The second one deals with the identification of the characteristic parameters of the evolution regime of electrolytically generated hydrogen bubbles (nucleation rate, life time, etc...) through the noise p.s.d. measured during the evolution (a few millivolts amplitude noise).

2. MEASUREMENT TECHNIQUE[1]

The p.s.d. was measured by using a Fourier Analyser (Hewlett Packard 5451C) based on a F.F.T. algorithm, in the 0.001 Hz - 50 kHz frequency range at constant current, by calculating the cross-spectrum of the outputs of two identical but independent amplification channels (Fig.1).

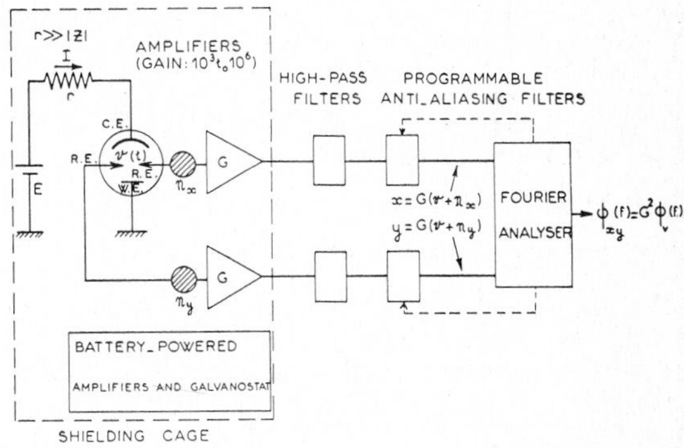

FIGURE 1
Experimental arrangement used for measuring the voltage noise p.s.d. ψ_V (W.E. : working electrode, R.E. : reference electrode, C.E. : counter electrode, n_x, n_y : parasitic noises generated by the amplification channels with gain G).

This technique allows the parasitic noises of the measurement channels (reference electrodes, amplifiers, filters) to be eliminated. For a resistive galvanostat made of a battery E and a resistance r much higher than the working electrode impedance Z ($r > 100\ |Z|$) the regulation noise can be neglected.

The measured spectrum accuracy $\varepsilon(f)$ has been investigated : it essentially depends on the number N of averaged elementary spectra, and on the signal-to-noise ratio $\eta(f) = \psi_v(f)/\psi_n(f)$ where $\psi_n(f)$ and $\psi_v(f)$ are respectively the p.s.d. of the parasitic measurement channel noise, and the electrochemical voltage noise[1] :

$$\varepsilon^2(f) = \frac{1}{N}\left(1 + \frac{1}{\eta} + \frac{1}{2\eta^2}\right)$$

The most important measurement difficulty arises when the signals present a non-stationarity character. Two cases are usually encountered : first a continuous drift superimposed over the voltage fluctuations can be observed (for example, coming from the natural discharge of the batteries) ; secondly, when two electrochemical processes simultaneously take place on the electrode, one in the low frequency range, the other one in the medium frequency range, the evolution of the former perturbs the p.s.d. measurement of the latter : the signal can be supposed stationary over a long time, but over a shorter time, sufficient to measure the medium frequency range spectrum, it appears as non-stationary.

The influence of a linear drift $v_1(t) = at + b$ superimposed over the voltage fluctuations v is given by[1] :

$$\psi_{v,\ \text{measured}}(f) = \psi_v(f) + \frac{Ta^2}{4\pi^2 f^2}$$

(T: acquisition time of an elementary spectrum) so, the measured p.s.d. presents a fallacious f^{-2} tendency. Such a drift limits the p.s.d. measurement at low frequencies : if the drift amplitude aT during one signal acquisition is much lower than the voltage fluctuations variance σ in the studied frequency range ($aT \ll \sigma$) the correctly analysed minimum frequency $\Delta f = 1/T$ will be such as : $\Delta f \gg a/\sigma$ (e.g. for $a = 10\ \mu V/mn$ and $\sigma = 10\ \mu V$ the measurement is altered at frequencies lower than 0.1 Hz). The low frequency non-stationarity, either a deterministic drift or a random fluctuation, may be an important part of the processes under investigation. However, in order to analyze the higher frequency range with a sufficient accuracy (dynamic range problem of the analyzer), it has to be eliminated by high-pass filtering.

3. APPLICATION TO BATTERIES

The impedances of several types of electrochemical storage cells have been measured in order to look for some information about the fundamental aspects of the kinetics of these devices for improving their performances, or to obtain a state-of-charge test[4]. However the measurements were made for cells at rest (no discharging currents), because the authors were mainly interested in self-discharge consequences on the life time of the cells as ready-to-use devices. In that case the noise is generally very low (Fig.2, curve A).

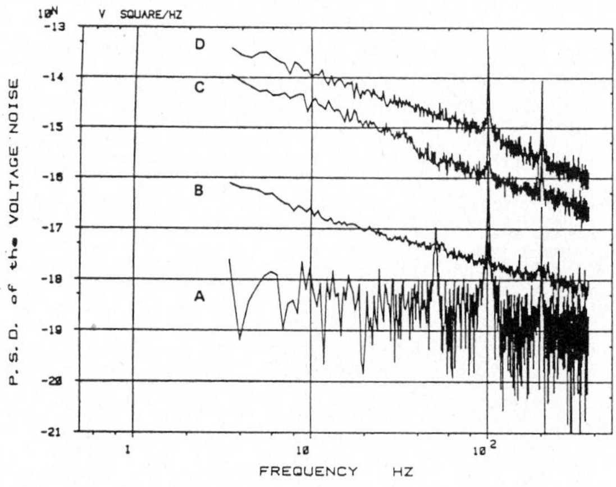

FIGURE 2
P.s.d. of the voltage fluctuations generated by a fully charged alkaline battery (Duracell MN1500) during the discharge at a constant current : A) I = 0 B) I = 14 mA C) I = 125 mA D) I = 260 mA.

The measurements have been made on alkaline zinc-manganese dioxide cells (Duracell, MN1500, LR6, E = 1.5 V, capacity : 1.8 Ah) during the discharge. Fig.2 shows that for a fully charged device and at a given frequency (10 Hz) the noise p.s.d. is proportional to I^{α} (where $\alpha \# 2$), and that, for a given current, the noise p.s.d. is of the $1/f^{\beta}$ type (where $\beta \# 1$).

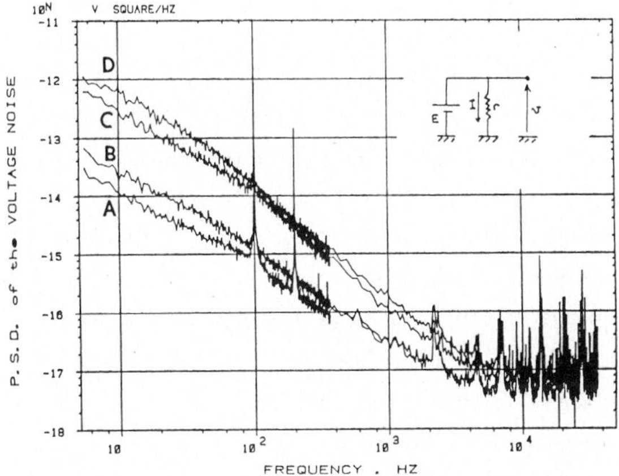

FIGURE 3
P.s.d. of the voltage fluctuations generated during the discharge of the same alkaline battery at different state-of-charge : A) Fully charged B) 66% charged C) 33% charged D) Fully discharged. (Experimental arrangement for p.s.d. measurements in the insert).

The time-evolutions of the voltage, the impedance and the voltage noise p.s.d. have been measured for a 260 mA discharging current. As shown in Fig.3 the p.s.d. of the electrochemical noise generated during the discharge increases as the battery charge decreases (the p.s.d. are of the $1/f^{\gamma}$ type (where $1 \leqslant \gamma \leqslant 1.5$ depending on the state-of-charge). The voltage, the impedance modulus and the noise p.s.d. at 10 Hz are depicted in Fig.4 in terms of the battery state-of-charge. The voltage noise p.s.d. is more sensitive to the battery state-of-charge variations than the impedance or the voltage[5].

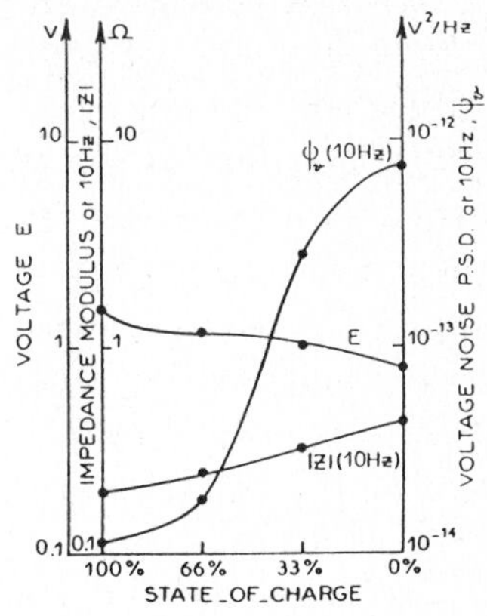

FIGURE 4
Time-evolutions of the voltage, the impedance modulus and the voltage noise p.s.d. at 10 Hz, for the same alkaline battery (discharge current 260 mA).

4. APPLICATION TO GAS EVOLUTION[6]

At low anodic current density, for polarization close to the corrosion potential, the dissolution of an iron electrode in aqueous sulphuric acid medium takes place together with hydrogen formation. Under current control, voltage random fluctuations, due essentially to the hydrogen bubble growths and departures, can be observed.

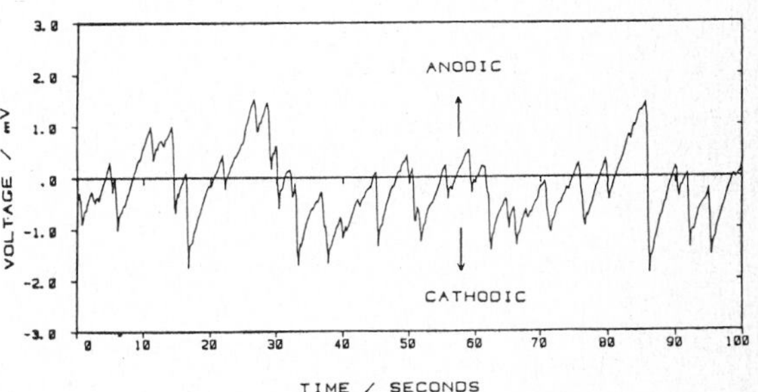

FIGURE 5
Iron dissolution in 1M sulphuric acid medium. Voltage time recording at a 3.7 mA/cm^2 anodic current density (1 mm Johnson-Matthey iron disc electrode facing upwards).

A recording of the voltage noise versus time under galvanostatic control is shown in Fig.5. Optical observation of the electrode surface shows that the steep jumps are related to the detachment of the bubbles and that the slow linear voltage increase is due to their growth.

The observed fluctuations of the voltage can be supposed to be the sum of elementary voltage transients which are linear functions of time (Fig.6). The total voltage fluctuation is :

$$V(t) = \sum_{i=1}^{N(t)} v(t - s_i, u_i)$$

where s_i, u_i are the random nucleation time and the random life time of the ith bubble respectively, and N(t) the total number of bubbles nucleated at time t. When the number of bubbles is not too high, the elementary events can be considered as independent : hence s_i is assumed to follow a homogeneous Poisson law of intensity λ. u_i is supposed to have an exponential probability distribution with parameter α (where $1/\alpha$ is the mean lifetime)[6].

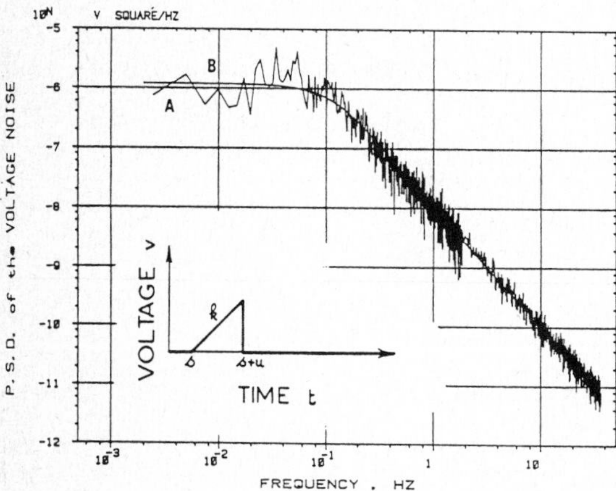

FIGURE 6
Experimental (A) and theoretical (B) p.s.d. of the voltage fluctuations corresponding to the recording of Fig.5 (Shape of the elementary transient used in the model, in the insert).

The mean and the p.s.d. of V(t) can be calculated using the generalized shot noise theory[6] :

$$\langle V(t)\rangle = \frac{\lambda k}{\alpha^2}, \quad \psi_V(f) = \frac{2\lambda k^2}{\alpha^2}\,\frac{3\alpha^2 + 4\pi^2 f^2}{(\alpha^2 + 4\pi^2 f^2)^2}$$

The theoretical and experimental p.s.d. have been plotted in Fig.6. From Fig.5, the slope of the voltage increase k is determined : k = 400 μV/s. From the p.s.d. $\psi_V(f)$, we can calculate : $\alpha = 1.1\ s^{-1}$, $\lambda = 1.5\ s^{-1}$.

The characteristic parameters of the bubble evolution regime are thus determined through the analysis of the electrochemical noise generated during the evolution. These results which cannot be obtained by deterministic techniques (dc-voltage, impedance) because of the random nature of the involved processes present a practical interest in electrochemical engineering where bubble evolution is often a limitative phenomenon (screening effect on the electrode surface, increase of the electrolyte resistance).

5. CONCLUSION

These two applications show the interest of a general noise investigation carried out through a spectral analysis of the processes. On the practical point of view, further noise-based tests, which would have the advantage to be non-destructive and non-perturbative, could be devised from the measurement of the variance or of the power spectral density at an adequate frequency.

REFERENCES

1. F. HUET, Thèse de Doctorat d'Etat, Paris VI, 1984.
2. C. GABRIELLI, F. HUET, M. KEDDAM, in "Noise in physical systems and 1/f noise", ed. by M. Savelli, G. Lecoy, J.P. Nougier (Elsevier Science Pub., Amsterdam 1983) pp. 385 - 388.
3. C. GABRIELLI, F. HUET, M. KEDDAM, R. OLTRA and C. PALLOTTA, in "Passivity of Metals and Semiconductors", ed. by M. Froment (Elsevier Science Publishers, Amsterdam, 1983) pp. 293 - 298.
4. S.A.G.R. KARUNATHILAKA, N.A. HAMPSON, R. LEEK, J.J. SINCLAIR, J. Appl. Electrochem., 11 (1981) 365 - 372, 11 (1981) 715 - 721, 13 (1983) 217 - 220.
5. C. GABRIELLI, F. HUET, M. KEDDAM, T. KOSSEV, to be published.
6. C. GABRIELLI, F. HUET, M. KEDDAM, J. Appl. Electrochem., 15 (1985) 503 - 508.

VARIOUS PHYSICAL SYSTEMS - 1

NOISE IN PHYSICAL SYSTEMS AND 1/f NOISE - 1985
A. D'Amico and P. Mazzetti (editors)

NOISE AT SEMICONDUCTOR GRAIN BOUNDARIES: A QUANTITATIVE MODEL

A.J. MADENACH, J. WERNER* and H. STOLL**

Max-Planck-Institut für Festkörperforschung, D-7000 Stuttgart 80, Heisenbergstr.1
Federal Republic of Germany

We report a systematic analysis of noise measurements at grain boundaries in p-type Czochralski-grown silicon bicrystals. Such bicrystals are characterized by deep interface states that capture and emit holes. The resultant potential barrier controls the current flow of free holes across the boundary. The capture and emission processes at the boundary modulate the barrier height and the voltage drop across the boundary. This stochastically modulated voltage is observed as noise. A description of this mechanism by a Lorentz spectrum with a single time constant fails to explain the frequency dependence of the measured noise spectra. We resolve this discrepancy by taking into account the effect of spatial disorder within the interface plane. The inhomogeneous distribution of trapped charge causes spatial fluctuations of the barrier height that naturally lead to a distribution of time constants. The consideration of these spatial potential fluctuations allows us to evaluate the measured noise spectra quantitatively. The density of interface states, capture cross section and the standard deviation of the potential fluctuations are obtained from our analysis. These results are in excellent agreement with those obtained by independent methods.

1. INTRODUCTION

Artificially grown bicrystals allow an intentional variation of the geometrical parameters that determine the structural properties of grain boundaries. Therefore, such bicrystals are a prerequisite for a systematic investigation of current transport across grain boundaries. Semiconductor grain boundaries are characterized by interface traps[1] which result in a potential barrier for the majority carrier flow. The ac-response of semiconductor grain boundaries clearly shows the effect of disorder within the boundary plane on the electrical transport.[1]

The significance of disorder on the transport properties of semiconductor interfaces was first discussed by Nicollian and Goetzberger for the case of metal-oxide-semiconductor (MOS)-interfaces.[2] Whereas in the case of MOS-interfaces the influence of disorder on the noise properties seems to be masked by additional effects,[3] is the frequency behavior of grain boundary noise here shown to be completely explainable by considering the influence of disorder.[4]

2. POTENTIAL FLUCTUATIONS MODEL

In our experiments the current through the grain boundary is kept constant while the voltage fluctuations are monitored. The measurements have been performed on p-type silicon bicrystals.

We explain the grain boundary noise by the modulation of the voltage drop eU and the barrier height $e\phi$ through the stochastic capture and emission current j_T of holes at localized states as indicated in Fig.1.

* Present address: IBM T.J. Watson Research Center, Yorktown Heights, New York 10598
** Max-Planck-Institut für Metallforschung, D-7000 Stuttgart 80, Heisenbergstr.1

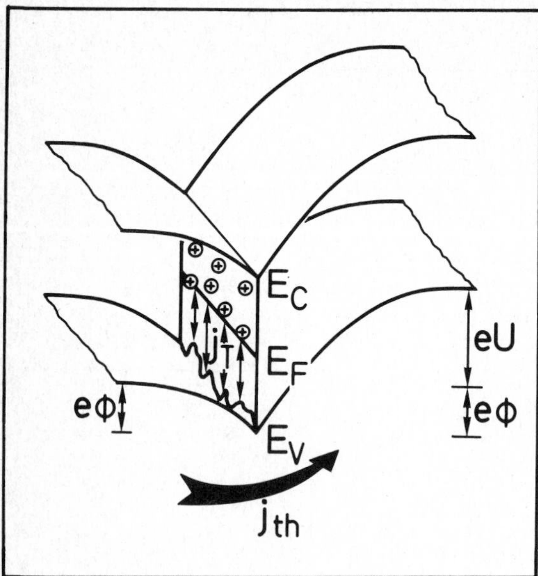

FIGURE 1
Band diagram at the grain boundary of a p-type silicon bicrystal. The capture and emission current j_T between deep interface states and the valence band stochastically modulates the barrier height $e\phi$ and the voltage drop eU. The random spatial distribution of trapped positive charges leads to potential fluctuations along the boundary as indicated by the wiggled line.

The appropriate Langevin equation which describes the fluctuations of the trapped charge eQ by considering the capture and emission processes in terms of Shockley-Read-Hall statistics reads:[5]

$$\frac{d}{dt} Q = v_{th} S_p N_T \cdot \left\{(1-F)p - \frac{1-F_0}{F_0} p_0 F\right\} + \xi(t) \quad (1)$$

The relaxation of excess charge is proportional to the capture cross section for holes S_p and the density of interface states N_T; v_{th} is the thermal velocity of free charge carriers. F denotes the occupation function of traps by holes whose equilibrium value F_0 is given by the Fermi-Dirac distribution; p is the density of free holes at the boundary which has the equilibrium value p_0. Fluctuations are introduced by the stochastic force ξ which has a white spectrum with power spectral density $S_\xi = 4\langle \delta Q^2 \rangle / \tau$, where τ is the relaxation time constant of the process described by Eq.1:

$$1/\tau = \frac{rp_0}{F_0} \cdot \frac{e^2 N_T F_0 (1-F_0)}{kT \{C_\Sigma + C_R\,(\exp(eU/kT)-1)\}} \quad (2)$$

Here C_R denotes the high frequency capacitance of the right hand side reverse-biased space charge region, C_Σ the sum of the capacitances of the right and of the left hand side space charge regions, U is the voltage drop across the boundary, and kT/e is the thermal voltage. In deriving Eq.(2) we used the relation between the two stochastic variables δp and δQ:[5]

$$\delta p = -\frac{e}{kT} \cdot \frac{1}{C_\Sigma + C_R\{\exp(eU/kT)-1\}}\, e\delta Q \quad (3)$$

Fluctuations in the trapped charge eQ are related to the measured voltage fluctuations δU through:[5]

$$\delta U = C_U \cdot e\delta Q \quad (4)$$

with:

$$1/C_U = \frac{\exp(eU/kT)-1}{C_\Sigma + C_R\{\exp(eU/kT)-1\}} \quad (5)$$

We obtain the spectrum of the voltage fluctuations $S_U(f)$ with the help of Eqs.(1-5) and by using standard methods:[6]

$$S_U(f) = \frac{e^2}{C_U^2} \cdot \frac{4}{A} \cdot \frac{\tau^2 rp_0 N_T (1-F_0)}{1 + (2\pi f\tau)^2} \quad (6)$$

where A is the area of the boundary. Voltages $eU \gg kT$ allow a simplification of Eq.(6) by the approximations: $\tau \simeq \tau_p := F_0/rp_0$ and $C_U \simeq C_R$.

The case of a continuum of energy levels with density $N_{SS}(E)$ which is observed at silicon grain boundaries,[1,7] can be discussed in the same limit: for $eU \gg kT$ the spectrum of a multivariate process is given by the integral of Lorentzian spectra as given by Eq.(6) over energies E within the forbidden gap.[5]
The resultant spectral density

$$S_U(f) = \frac{e^2}{C_R^2} \cdot \frac{2kTN_{SS}(E_F)}{A \cdot \tau_p (2\pi f)^2} \cdot \ln\{1+(2\pi f\tau_p)^2\} \quad (7)$$

does not deviate significantly from the spectrum of Eq.(6) since only levels a few times kT around the Fermi level E_F contribute to the noise.[8] Lorentz spectra of the form of Eq.(6) or Eq.(7) clearly fail to explain the measured noise, as demonstrated in Fig.2.

The discrepancy between a Lorentz type spectrum and the measured spectra is resolved by considering the influence of spatial disorder on the electronic transport across the grain boundary. The random spatial distribution of trapped charge eQ results in spatial fluctuations of the band edges as sketched by the wiggled line in Fig.1. In a first approximation, these potential fluctuations have a Gaussian distribution $P(\phi)$ around a mean barrier height $\bar{\phi}$ with standard deviation σ. This random distribution of barrier heights naturally introduces a distribution of time constants which completely describes the frequency behavior of the measured spectra:

We account for the potential fluctuations by integrating the spectrum for a single time constant of Eq.(7) over the distribution of potentials $P(\phi)$:

$$S_U^{PF} = \int_{-\infty}^{+\infty} P(\phi)\ S_U(f)\ d(\phi-\bar{\phi}) \quad (8)$$

Note that $S_U(f)$ depends on the barrier height ϕ through the carrier density at the boundary $p_0 = p_G \exp(-e\phi/kT)$, where p_G is the free charge carrier density in the grains. The integration in Eq.(8) cannot be done analytically. A procedure proposed by Simonne[9] and extended in Ref.1 allows us a straightforward evaluation of the measurements by taking into account only two points of each measured spectrum.

Figure 2 shows the excellent agreement between measured and calculated spectra. Variation of the voltage U or the temperature T allows us to sweep the Fermi level E_F through the forgidden gap. The resultant energy dependence of the density of states N_{SS} agrees excellently with N_{SS} as obtained from admittance

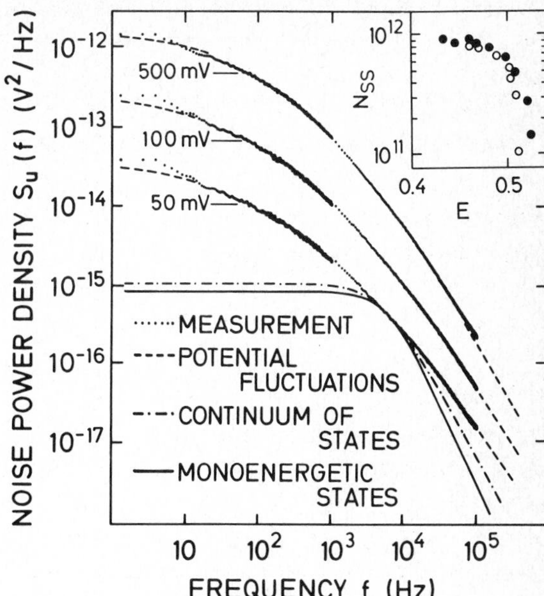

FIGURE 2
Noise power density $S_U(f)$ of a p-type silicon grain boundary at T = 295K for different voltage drops U after subtraction of shot noise. The inset compares the density of deep interface states $N_{SS}(E)$, in $eV^{-1}cm^{-1}$, obtained from noise measurements using the potential fluctuations model (full circles) with results from admittance spectroscopy (open circles). The energy E is measured in eV, the energy zero point is at the valence band edge at the boundary.

spectroscopy as demonstrated in the inset of Fig.2. Similar agreement is also obtained for the capture cross section and the standard deviation of the potential fluctuations: The capture cross section varies from $2\cdot10^{-14}$ cm^2 to $2\cdot10^{-13}$ cm^2 and depends on temperature. The standard deviation σ has a value of 50±5 meV, independent of bias voltage and temperature. This value is to be compared with a typical mean barrier height $\bar{\phi}\simeq$ 300 meV.

CONCLUSIONS

Semiconductor grain boundaries turn out to represent a model system which allows us to investigate the influence of disorder within the interface plane on electrical transport properties. The disorder is caused by the random spatial distribution of trapped interface charge. This random distribution results in potential fluctuations along the interface. Potential fluctuations naturally lead to a spatial distribution of time constants and thus cause a significant broadening of the Lorentz spectra anticipated for a spatially homogeneous system. This broadening of Lorentz spectra in spatial disorderd systems gives an explanation for the transition to 1/f-noise.

The extension of the potential fluctuations model to polycrystalline material has been shown to be straightforward.[10]
The significance of disorder for the noise properties of other interfaces, such as Schottky contacts or heterojunctions, and for disorderd bulk material, such as amorphous or compensated semiconductors will be the objective of future work.

ACKNOWLEDGMENTS

The authors deeply appreciate the continuous encouragement and support of H. J. Queisser. We thank J. Trost for his help with the computer programs and F. J. Stützler for fruitful discussions.

REFERENCES

1. J. Werner in "Polycrystalline Semiconductors", G. Harbeke ed., Springer, Berlin 1985, p. 76

2. E. H. Nicollian and A. Goetzberger, Bell Syst. Techn. J. 46, 1055 (1967)

3. K. L. Ngai and S. T. Liu, Thin Solid Films 93, 321 (1982)

4. A.J. Madenach and J. Werner, Phys. Rev. Lett. 55, 1212 (1985)

5. A. J. Madenach, unpublished

6. See for example: A. van der Ziel, "Noise: Sources, Characterization, Measurement", Prentice-Hall, Englewood Cliffs N.J. 1970, chapt. 2

7. J. Werner, W. Jantsch and H. J. Queisser, Solid State Comm. 42, 415 (1982)

8. A similar result has been obtained for a homogeneous semiconductor by K. Lee, K. Amberiadis and A. van der Ziel, Solid State Electron. 25, 999 (1982)

9. J. J. Simonne, Solid State Electron. 16, 121 (1973)

10. A.J. Madenach, J. Werner and F.J. Stützler, in Proceedings of the 18th IEEE Photovoltaic Specialists Conference, Las Vegas 1985, to be published.

NOISE IN PHYSICAL SYSTEMS AND 1/f NOISE - 1985
A. D'Amico and P. Mazzetti (editors)

DIFFUSION NOISE OF HOT ELECTRONS IN GaAs AT 300 K

D. GASQUET, M. de MURCIA, J.P. NOUGIER, M. FADEL
Centre d'Electronique de Montpellier * Universite des Sciences
et Techniques du Languedoc , 34060 Montpellier Cedex , France

We present noise and conductivity measurements of electrons in GaAs at 300 K in the frequency range 220 MHz - 10 GHz. The noise is white. Partition noise is shown to occur above 2 kV/cm. We give a method to determine the differential conductivity versus the electric field, and we deduce the diffusion coefficient versus the electric field, up to 3 kV/cm.

1. INTRODUCTION

For modeling shorts devices, it is important to know the transport parametres in hot carrier regime, such as the drift velocity and the diffusion coefficient versus electric field. The pupose of this paper is to determine experimentaly the longitudinal diffusion coefficient D(E), below the threshold field, for electrons in GaAs.

D(E) can be obtained using either a time of flight technique [(1)] or noise and conductivity measurements [(2)]. We report here results obtained at 300 K by noise and conductivity measurements.

2. NOISE CURRENT

The spectral density $S_i(E,\omega)$, of the current fluctuation, is defined as:

$$S_i(E,\omega) = 4\,k\,T_n(E,\omega)\,g'(E,\omega) \qquad (1)$$

where k is the Boltzmann constant, T_n is the noise temperature, and g' is the real part of the a.c. conductance around the d.c. bias corresponding to a uniform electric field E.

When the noise is due to the fluctuation of the velocities of the carriers (diffusion noise), the diffusion coefficient D(E) can be defined as:

$$S_i(E,\omega) = 4\,q^2/L^2\;n\;D(E,\omega) \qquad (2)$$

and equation (1) gives:

$$D(E,\omega)/D_o = T_n(E,\omega)/T_o\;g'(E,\omega)/g'(E_o,\omega) \qquad (3)$$

where the subscript 0 stands for ohmic values.

GaAs is a multivalley semiconductor, and intervalley transfers occur. However, up to the threshold field, only two groups of valleys are involved (Γ and L valleys), since the population of the X valleys can be neglected. The noise current density is then given [(3)] by:

$$S_i(E,\omega) = 4q^2n/L^2\;(\beta\, D_\Gamma+(1-\beta)D_L)+$$
$$4q^2nE^2/L^2(\mu_\Gamma-\mu_L)^2\,\beta(1-\beta)\;\tau/(1+\omega^2\tau^2) \qquad (4)$$

where subscripts Γ and L stand respectiveley for the Γ and the L valleys, $\beta=n_\Gamma/n$, $n=n_\Gamma+n_L$.

In the right hand side of eq.(3), the first line represents the diffusion of the carriers within each of the two groups of valleys, whereas the second line represents the contribution of the carrier transfers between these two groups: the latter is called 'partition noise'. τ is the time constant associated with the intervalley scattering. This term may be considered either as diffusion, or as generation-recombination noise.

Eqs. (2) and (3) give the diffusion coefficient, obtained by taking into account the whole mechanism of velocity fluctuations. This is the diffusion coefficient given by Monte-Carlo simulations.

3. EXPERIMENTAL METHOD

The noise temperature measurements were performed at 220,320,650,850 MHz, and at 4 and 10.5 GHz, using pulsed technique [(4)], in order to avoid the thermal heating. The sample used were mesa $n^+\,n\,n^+$ Gunn diodes. The active layer was L= 11µm thick, and 10^{+15} cm^{-3} doped. The ohmic mobility was about 5200 cm²/Vs at 300 K, and the cross section A= 1.2 10^{-4} cm^2. The bias pulses had 20 µs duration, and the repetition rate was 100 Hz. We observed Gunn oscillations

*Laboratoire associe au Centre National de la Recherche Scientifique, UA 391

for a bias voltage of 3.8 V, corresponding to an electric field threshold of 3.5 kV/cm. We verified, using a simulation, that, up to 3.5 kV/cm, the field could be considered as being uniform in the device.

The differential conductance was measured by three different ways.

i) The first one was performed using a G.R. admittance bridge. This method gives directly the differential conductance and susceptance in the frequency range 150-400MHz, but the accuracy is not very good. This experiment allowed us to verify that the susceptance in whole frequency range for all the electric fields studied was null.

ii) In the second one we obtain $g'(E)$ by differentiating the pulsed (I,V) characteristic.

iii) Thirdly we measured the a.c. microwave power reflected by the sample using the apparatus describe on fig. 1.

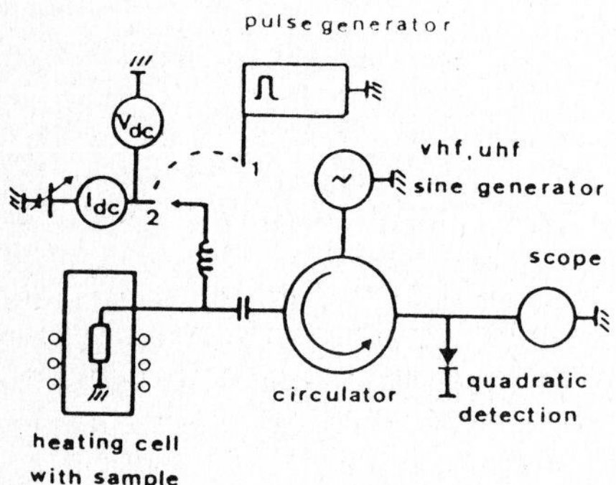

Figure 1

Under pulsed bias we measured, at T_o, the reflected power $P_r(E,T=T_o)$ define by:

$$P_r(E,T_o) = \left| \frac{Z'(E,T_o)-Z_c}{Z'(E,T_o)+Z_c} \right|^2 P_i \qquad (5)$$

Then we increased the temperature of the sample to obtain at zero bias the same reflected power as previously.

$$P_r(E=0,T) = \left| \frac{Z'(E=0,T)-Z_c}{Z'(E=0,T)+Z_c} \right|^2 P_i \qquad (6)$$

by identification we obtain $Z'(E=0,T)=Z'(E,T_o)$.

Assuming that, in the frequency range of the measurements (220 - 850 MHz), the impedance of the sample is real (see (i)), we obtain at very low bias $Z'(E=0,T) = R(E=0,T) = V/I$. This method allowed us to determine the differential conductivity at high field by a d.c. resistance measurement.

4. EXPERIMENTAL RESULTS AND DISCUSSION

Fig. 2 shows the variation of the conductance $g'(E,\omega)$ versus the applied voltage. We found a good agreement between the two methods (white squares for dI/dV, black squares for $g'(E,\omega)$ derived from the reflected power). It is plotted also on fig. 2 data performed with the admittance bridge at 2 V bias (circle). The real part of the differential conductivity can be considered as being frequency independent, at least up to 400 MHz.

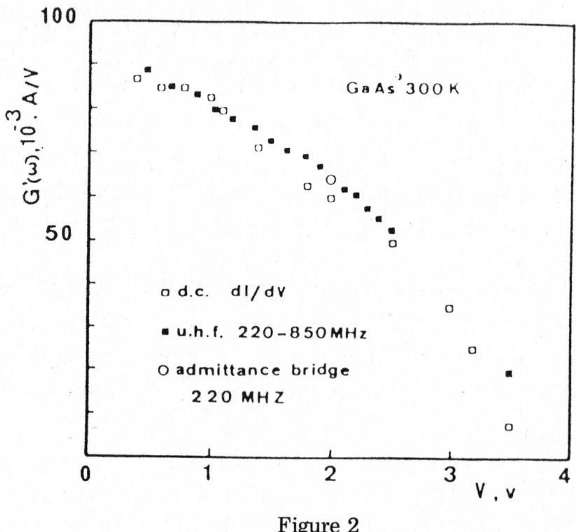

Figure 2

On fig. 3 are plotted the measured excess noise temperatures T_n-T_o versus electric field at various frequencies in the range 220 MHz - 10.5 GHz. The noise is white, up to 10 GHz, and can be considered as being intervalley diffusion noise and/or intervalley partition noise below its cut-off frequency. Fig. 3 shows a sharp increase of T_n-T_o for $E > 2$ kV/cm. This increase is attributed to electron transfers between the Γ and L valleys (partition noise). In order to prove this we measured the excess noise at 77 K fig. 4. The shape of the noise temperature curve versus electric field is similar at 300 K and 77 K, but the behaviour with frequency is quite different. The noise below the kink, is white, and becomes frequency dependent at higher field. This frequency dependence is explained by eq. 4. At low field, the number of transferred electrons in the L valley is very small, $\beta \approx 1$, and the partition noise is negligible. At higher field, intervalley transfers occur, $\beta \neq 1$, the partition noise becomes important, and a frequency dependence appears; τ increases when the temperature decreases and at 77K $1/2\pi\tau$ is in the experimental frequency range. It is thus clear that the sharp increase of T_n with E is associated with the partition noise, this one at 300K is white up to at least 10.5 GHz. Therefore its cut-off frequency f_c, is such that $f_c \geq 50$GHz, and thus $\tau \lesssim 3$ps. This important result shows that the diode is long enough, so that one may consider that the partition noise is constant over the whole volume. Assuming a drift velocity of 10^5 m/s, the distance over which the transfer occurs is less than $3 \times 10^{-12} \times 10^5 = 3 \times 10^{-7}$ m, that is much smaller than 11 µm.

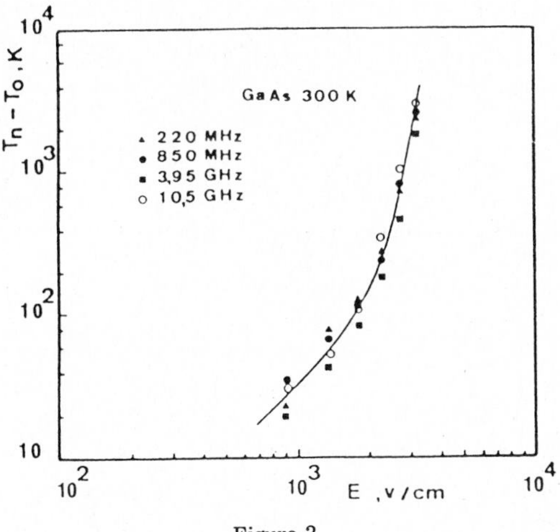

Figure 3

The diffusion coefficient can be derived from eqs. (1) and (2). $D(E)/D_o$ has been plotted on fig. 5, versus the electric field E. Fig.(5) shows that, below 1.8 kV/cm, D(E) slightly decreases with increasing E. Such a behaviour is typical of single valley semiconductor, and is related to the variation of the diffusion coefficient in the Γ valley, the L valley being then almost empty of free carriers. When intervalley transfers occur, D(E) increases, and reaches 2 D_o at 3 kV/cm. The results of Bareikis et al.[(2)] are qualitativley similar (see fig. 5), the quantitative difference can

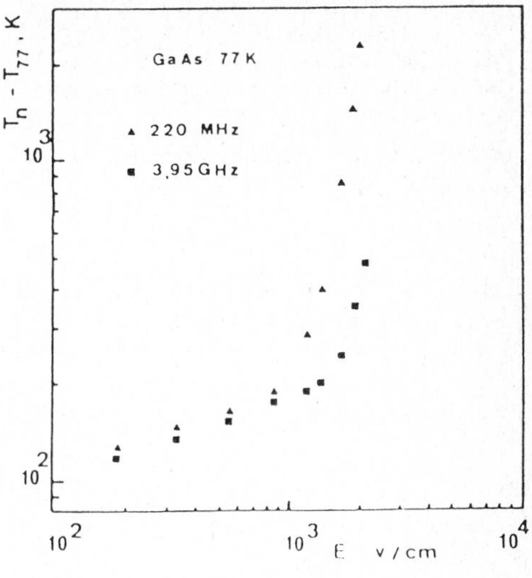

Figure 4

be explained by the method used by this authors for determining $g'(E,\omega)$. In this method the heating of the sample is provided by the d.c. bias [(5)]. The magnitude of the heating current must be located in the I range of the ohmic regime, assuming that $g'=I/V$. We could not respect this condition when we done measurements on our samples above 1.5 kV/cm. So the d.c. conductance measured was greater than the a.c. one. In fact we obtained the same results than Bareikis concerning the noise temperature and the disagreement on D/D_o could be only explained by the $g'(E,\omega)$ measurement. Ruch and Kino results [1]exhibit a sharp increase of D(E), starting at low field (see fig. 5), which is far too strong : this result may be due to the problems encountred by time of flight technique at low field, and also to the compensation between the acceptors and the donnors in the high resistivity samples used.

5. CONCLUSION

In this paper, we showed that the diffusion noise in n type GaAs is white up to 10 GHz. The diffusion coefficient begins to decrease with increasing field, below 2 kV/cm: this behaviour is typical of single valley semiconductors, thus showing that, up to 1kV/cm, the satellites valleys of GaAs are almost empty of free carriers. At higher field, above 2 kV/cm, the

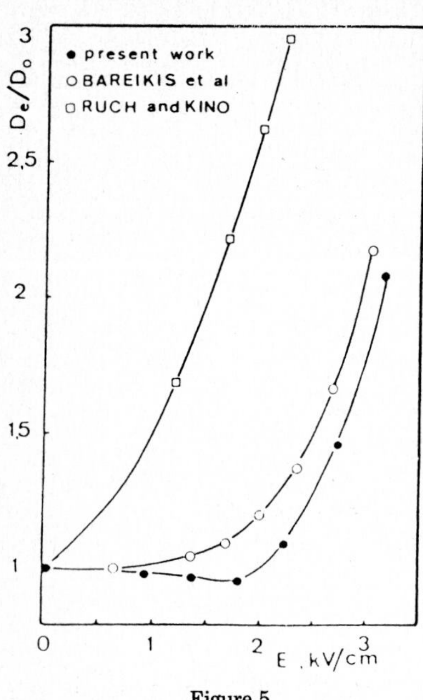

Figure 5

strong increase of the diffusion coefficient is due to the partition noise linked to intervalley transfers, thus showing that the electrons begin to populate the satellite valleys (in practice, only the L valley). From the fact that the noise is white up to 10 GHz, we conclude that the cut-off frequency of the partition noise is greater than 50 GHz, which means that the intervalley relaxation time in GaAs, at 300 K, is lower than 3 ps.

REFERENCES

1. J.G. Ruch, G.S. Kino, Phys. Rev. 174, 921, 1968
2. V. Bareikis, V. Viktoravichyus, A. Gal'dikas R. Milyushite, Sov. Phys. Semicond. 14 (7), 847, 1980
3. P.H. Handel, A. Van Der Ziel, S.S.E. Vol. 25 541, 1981
4. D. Gasquet, J.C. Vaissiere, J.P. Nougier 6th Int. Conf. on noise in physical systems, Washington (USA) 1981 , National Bureau of Standards, special publication No 614, p305
5. J. Vindevoghel, Y. Leroy, C. Bruneel and J. Zimmermann, Rev. Sci. Instrum , Vol. 45 No. 7,pp 920-921 1974

NOISE IN PHYSICAL SYSTEMS AND 1/f NOISE - 1985
A. D'Amico and P. Mazzetti (editors)

GENERATION RECOMBINATION NOISE IN p-Si AT 77 K

D. GASQUET, B. AZAIS, J.C. VAISSIERE, J.P. NOUGIER
Centre d'Electronique de Montpellier* Universite des Sciences et Techniques du Languedoc
34060 Montpellier Cedex France

Noise and conductivity measurements in p-Si at 77K, up to 2.5kV/cm electric field strength, shows that three noise sources are significant: diffusion, G. R., and 1/f noise. The diffusion coefficient versus electric field is given. The 1/f noise is important up to 10 GHz. The G.R. noise cannot be explained assuming two energy levels (impurity level and the valence band).

1. INTRODUCTION

The large excess noise of hot electrons observed in p-type silicon at 77 K and its frequency dependance suggests the existence of G.R noise due to the fact that part of the impurities are not ionized at this temperature.Such a phenomenon was encountred at the same temperatre in n-Si [1]. The frequency dependance of the noise exhibits a cut-off frequency around 1GHz. So using a pulsed technique described in [2] we performed excess noise temperature measurements in the range 220MHz -10.5GHz, on $p^+p\ p^+$ silicon diodes.

2. EXPERIMENTAL RESULTS AND DISCUSSION

On fig. 1 is plotted the excess noise temperature $T_n - T_o$ versus frequency at various electric fields. Obviously, the noise is not white.

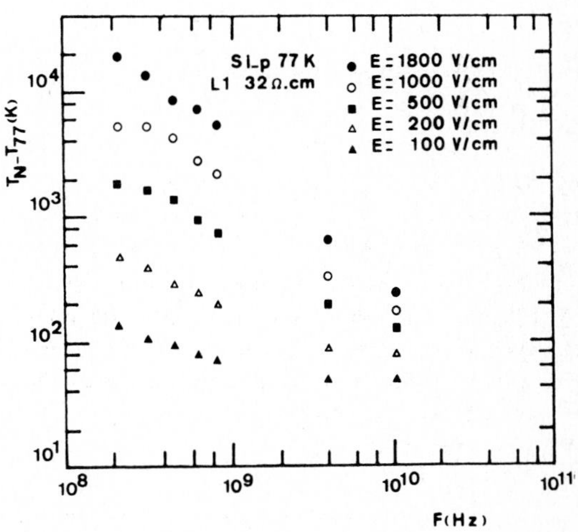

Fig.1

The noise current (spectral density) is given by:

$$S_i(E,f) = 4\ k_B\ g'(E)\ T_n(E) \quad (1)$$

where k_B is the Boltzmann constant and g' the differential conductivity around the bias point $g'(E)$ is obtained by differentiating the (I,V) characteristics. We verified that $g'(E)$ is frequency independant in the range studied. Fig 2 shows $S_i(f)$ for a bias current of 1.2 A. The shape of this curve is typical of the existance of generation recombination noise and diffusion noise. If only two levels are involved, according to Van der Ziel[3] the total spectral current density can be written as:

$$S_i(f) = 4q^2 N_a u\ D \frac{A}{L} + \frac{I^2 \tau (1-u)}{N_a AL[4\pi^2 f^2 u^2 + \tau^2 (2-u)^2]} \quad (2)$$

$$S_i(f) = S_{idiff} + S_{iGR}$$

where: $u = u/N_a$ is the fraction of ionized imp-

* Laboratoire associe au Centre National de la Recherche Scientifique, UA 391

urities, and $\gamma = u/[(2-u)\tau]$ where τ is the life-time of the carriers

Equation (2) involves three parameters u, τ, D which can be obtained, by a least square fitting of equ. (2) to the experimental results. The curve obtained is drawn (full line) on fig. 2. We can see that the fit is not good, mainly at 4 and 10 GHz. This result suggests the existance of a third kind of noise, namely 1/f noise. This yields:

$$S_i(f) = S_{idiff} + S_{iGR} + CI^2/f \qquad (3)$$

where C is the 1/f constant. Now the least mean square fitting has to be performed using the four unknown parameters u, τ, D and C. The curve obtained is drawn (dashed line) on fig. 2. We can see an excellent agreement and this clearly shows that 1/f noise at both low and high frequencies (up to 10 GHz!) is significant.

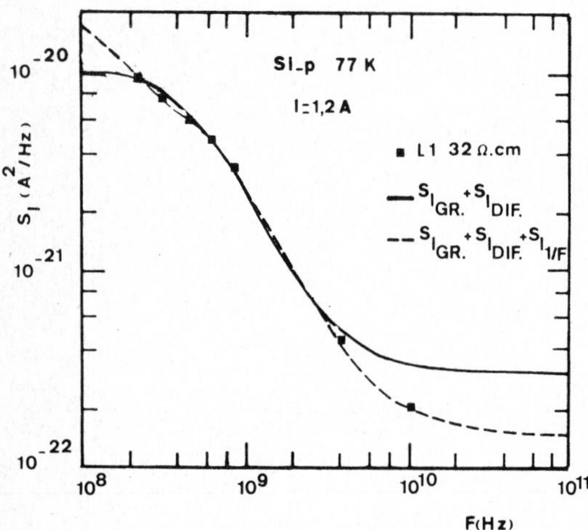

Fig.2

Fig.3 presents the values of D plotted versus the electric field, obtained by the fit (black squares), compared with experimental results obtained by time of flight measurements (white circles) at Modena university [4]. The agreement shows that we are able to get bulk diffusivity.The diffusion coefficient decreases with increasing electric field.

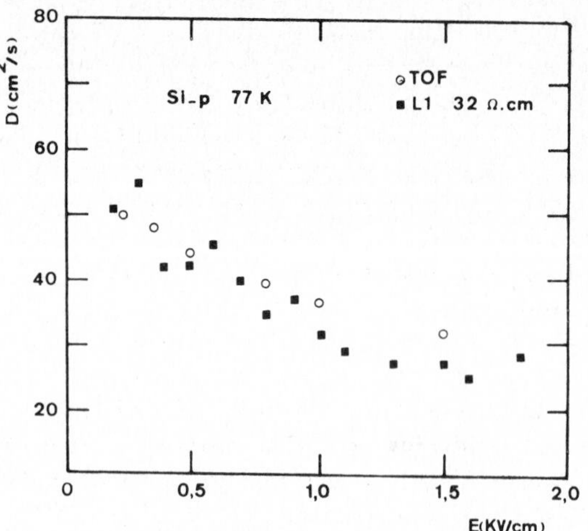

Fig.3

On fig.4 we plotted the fraction of ionized impurities u versus the electric field. These results are not in agreement with those obtained in n-Si [1]. In this material u increased quickly with the electric field and tended toward 1 at about 2.5 kV/cm (this behaviour was explained by impact ionisation). On fig. 4 we can see that u is nearly constant and the mean value is about 0.6 for the sample L1 (32 Ωcm). We have done similar experiments on the sample L2 (4.8 Ωcm) and found for u a value of about .3. These results cannot be explained by impact ionisation or Poole-Frenckel effect and they suggest that the two levels hypothesis is wrong.

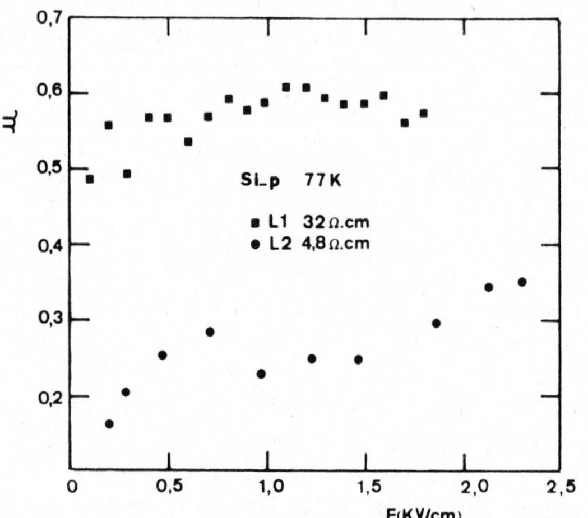

Fig.4

3. CONCLUSION

In this study we show that, at 77 K, p-Si exhibits three noise sources: diffusion, generation recombination, and 1/f. The influence of 1/f noise is important up to 10 GHz. However this 1/f noise source is probably due to the contacts. On the other hand we show that the hypothesis of a two levels G.R. noise cannot explain the value found for the fraction of ionized impurities and a third trapping level likely exist, the population of which does not depend on the electric field. Nevertheless the plateau obtained at high frequencies seems to be due to diffusion noise, since we obtain a value of D(E) in agreement with other measurements.

REFERENCES

1 D. Gasquet, H. Tijani, J.P. Nougier, A. Van der Ziel, Proceeding of 7th conf. on noise in physical systems, Montpellier 1983 edited by M Savelli North Holland, Elsevier Publications, pp 165-171
2 D. Gasquet, J.C. Vaissiere, J.P. Nougier 6th conf. on noise in physical systems, Washington 1981, N.B.S. special publication No 614, p305
3 A. Van der Ziel, R. Jindal, S.K. Kim, J.P. Nougier, S.S.E.,22, p77, 1978
4 L. Reggiani , Private Communication

NOISE IN PHYSICAL SYSTEMS AND 1/f NOISE - 1985
A. D'Amico and P. Mazzetti (editors)

ELECTRON-ELECTRON INTERACTION EFFECT ON THE SPECTRAL DENSITY OF CURRENT FLUCTUATIONS OF HOT ELECTRONS IN SI

Paolo LUGLI and Lino REGGIANI

Dipartimento di Fisica e Centro Interuniversitario di Struttura della Materia dell'Universita' di Modena, Via Campi 213/A, 41100 Modena, Italy

The effect of the electron-electron interaction on the current spectral density due to velocity fluctuations is quantitatively calculated through an ensemble Monte Carlo simulation. Results for the case of electrons in Si are discussed in terms of the autocorrelation functions of the single-particle velocity fluctuations and of the drift-velocity fluctuations.

1. INTRODUCTION

For an homogeneous two terminal device of length L, the autocorrelation function of the current fluctuations $\overline{\delta I\,\delta I(t)}$ can be analyzed in terms of the velocity fluctuations of the single particle, δv, as:

$$\overline{\delta I\,\delta I(t)} \equiv \left(\frac{e\,N}{L}\right)^2 \overline{\delta v_d \delta v_d(t)} =$$

$$= \left(\frac{e}{L}\right)^2 \left[N\overline{\delta v \delta v(t)} + \sum_{i \neq j} \overline{\delta v_i \delta v_j(t)} \right] \qquad (1)$$

where e is the electron charge, $v_d(t) = (1/N) \sum_i v_i(t)$ the drift velocity, the sums are over all carriers N and the bar denotes time average.

The second term in the squared brackets describes the cross correlations and satisfies the following properties.

(i) Under equilibrium conditions it vanishes in accordance with the Nyquist and the Einstein laws, and the following relationship holds:[1]

$$S_I = \frac{4\,e^2}{L^2} N\,D \qquad (2)$$

where the spectral density of the current fluctuations S_I and the diffusion coefficient D are given by:

$$S_I = 4\int_0^\infty \overline{\delta I\,\delta I(t)}\,dt \qquad (3)$$

$$D = \int_0^\infty \overline{\delta v \delta v(t)}\,dt \qquad (4)$$

(ii) Under far from equilibrium conditions (hot-electrons), in accordance with Ref.(1), the cross correlations are responsible for the appearance of an extra-term, $\theta(\vec{E})$, in Eq.(2) which now takes the form:

$$S_I(\vec{E}) = \frac{4\,e^2 N}{L^2} \left[D(\vec{E}) + \theta(\vec{E}) \right] \qquad (5)$$

Owing to the far from equilibrium conditions, S_I, D, and θ are functions of the applied electric field $\vec{E}$.

When electron-electron (e-e) collisions are negligible $\theta(\vec{E}) = 0$, and Price's relationship[2] is recovered:

$$S_I(\vec{E}) = 4 \frac{e^2}{L^2} N D(\vec{E}) \qquad (6)$$

The aim of this paper is to study the importance of the cross-correlation term in Eq.(1) which is responsible for the breaking of the diffusion noise relationship given by Eq. (6). To this end, an original Monte Carlo procedure, first proposed by Jacoboni[3] has been used.

2. THEORY AND RESULTS

The e-e interaction is accounted for by incorporating a molecular – dynamics – type simulation into the standard Ensemble Monte Carlo algorithm. Under stationary conditions, this algorithm enables the autocorrelation functions of both current fluctuations and single – particle velocity fluctuations to be evaluated separately[4]. Thus the difference which arises between these functions is naturally ascribed to the cross correlation contribution. In the simulation each electron interacts simultaneously via a bare Coulomb potential with all the other electrons of the ensemble. The real space trajectories are calculated for the electrons embedded in a uniform positive background. Calculations are modeled for the case of electrons in Si with a field applied along the <111> direction; scattering with phonons are included in the usual Monte Carlo algorithm[4]. The strength of the method comes from the fact that it avoids simplifying and often not justifiable assumptions on the form of the electron distribution function and on the screening process. Furthermore, particle density fluctuations are naturally accounted for.

The results of the Monte Carlo simulation are reported in Figs. 1-3. There, the value of the autocorrelation function of the drift velocity fluctuations multiplied by the number of carriers, $\Phi_{v_d}(t) = N\overline{\delta v_d \delta v_d(t)}$, (dotted curves) and of the velocity fluctuations,

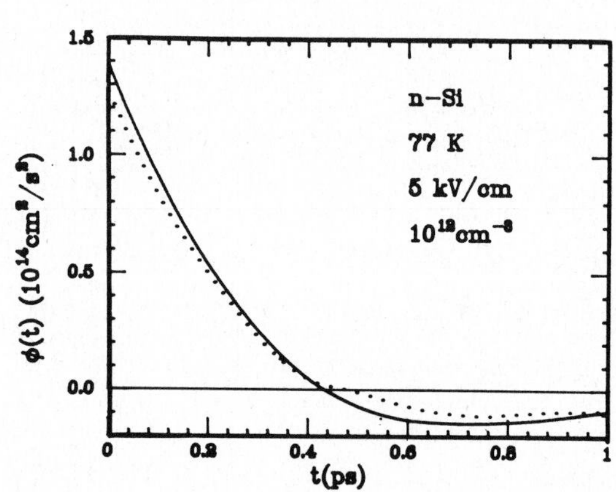

FIGURE 1
Correlation functions of drift – velocity fluctuations (dotted curves) and velocity fluctuations (continuous curves) for the case of electrons in Si at 77 K with E = 5 kV/cm and n = 10^{12} cm^{-3}. The Monte Carlo simulation has been performed with N = 128, L = 8 μm

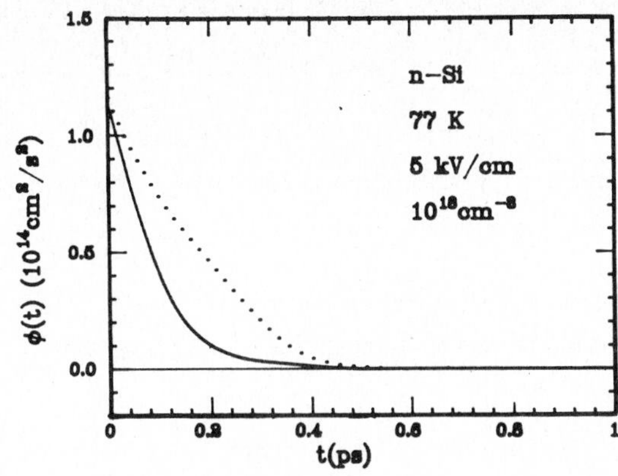

FIGURE 2
The same as in Fig. 1 with $n = 10^{18}$ cm^{-3}

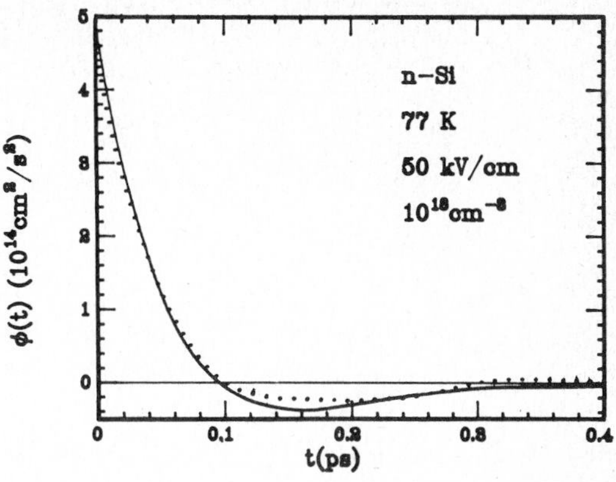

FIGURE 3
The same as in Fig. 2 with E = 50 kV/cm

$\Phi_v(t) = \overline{\delta v \delta v(t)}$, (continuous curves) are reported as a function of time. To verify the reliability of the calculations two tests have been performed. First, for a given concentration the results have been checked to be independent from different choices of the electron number and of the device volume (typical values of the electron number ranges between 16 and 128). Then, within the same set of parameters, several simulations have been performed with different series of random numbers. From these tests we conclude that Φ_{v_d} is affected by a larger uncertainty (about 20%) than Φ_v (about 5%).

Figs. 1-2 analyze the two correlation functions for different electron concentrations n. Calculations show that a significative difference shows up at $n = 10^{18}$ cm^{-3} where, as a result of the e-e interaction, Φ_v is found to decay more rapidly than Φ_{v_d}. It is interesting to notice that, for the highest carrier concentration, the negative region of the autocorrelation functions disappears. This should be correlated with the observed maxwellian shape of the energy distribution function, a consequence of the dominant role played by e-e scattering. Indeed, the negative region originates from some streaming character along the field direction of the electron distribution function which, in absence of e-e scattering, is due to the optical-like intervalley phonon emission.

The calculation of the correspondent $S_I(\vec{E})/(4e^2N/L^2)$ shows that (within the statistical uncertainty) it remains equal to its low-concentration value of 18 cm^2/s, going from the lowest to the highest carrier concentration. On the other hand, the diffusion coefficient is found to change significantly. Its value, which at low concentration is about 18 cm^2/s reduces to about 10 cm^2/s at the highest carrier concentration. This result implies the violation of Price's relationship, as predicted in Refs.(1,5).

Fig. 3 reports the calculations performed with the highest carrier concentration at 50 kV/cm. Within the uncertainty of the calculations, no difference is found between Φ_{vd} and Φ_{v}. This follows from the competitive increase in the efficiency of phonon over e-e scatterings. Indeed, owing to its Coulomb nature, e-e interaction is known to lose importance at increasing carrier energy.

3. CONCLUSIONS

The effect of e-e interaction on the spectral density of current fluctuations and the diffusion coefficient of electrons in Si has been analyzed with a Monte Carlo simulation of molecular-dynamics type under far from equilibrium conditions. At intermediate field strength, the results show that for an electron concentration of 10^{18} cm^{-3} the diffusion coefficient is significantly reduced on accounts of e-e interaction. Accordingly, the present quantitative calculations give a further confirmation to the theoretical predictions of the Leningrad group[1] that no simple relation exists between noise and diffusion when e-e interaction is present.

ACKNOWLEDGEMENTS

We wish to thank Profs. D.K. Ferry and C. Jacoboni for clarifying discussions on the subject. The financial support of the Computer Center of the Modena University and of the European Research Office (ERO) is gratefully acknowledged.

REFERENCES

1. S.V. Gantsevich, V.L. Gurevich and R. Katilius, Sov. Phys. JEPT 30, 276 (1970); Rivista del Nuovo Cimento, 2, 1 (1979).

2. P.J. Price in "Fluctuation Phenomena in Solids", edited by K.E. Burgess (Academic Press, New York, 1965) p. 355

3. C. Jacoboni, in: Proc. of 13th I.C.P.S., Ed. by G. Fumi (Marves,Roma) 1976, p. 1195.

4. P. Lugli, Ph. D. Dissertation, Colorado State University (1985) unpublished.

5. K.M. Van Vliet, J. Math. Phys. 12, 1998 (1971).

NOISE IN PHYSICAL SYSTEMS AND 1/f NOISE - 1985
A. D'Amico and P. Mazzetti (editors)

SPACE-CHARGE-LIMITED VOLTAGE NOISE IN THE PRESENCE OF TRAPS IN SiC

S. TEHRANI, G. BOSMAN, C.M. VAN VLIET and L.L. HENCH

E.E. and Materials Science Depts., U. of Florida, Gainesville, FL 32611, USA

Current-voltage measurements on a 16 μm nitrogen-doped $n^+n^-n^+$ α-SiC sample indicate single-carrier space-charge-limited current flow in the presence of shallow traps. At T = 77K a low bias Ohmic, a low bias quadratic, a fast-rising trap-filling, and a quadratic Mott-Gurney regime are clearly observed. Generation-recombination dominates the noise spectrum in the frequency range from 10 Hz to 10 MHz. The experimental results are in excellent agreement with calculations based on a model which accounts for the n^+n^- interface properties and the effect of diffusion on charge transport and noise.

1. INTRODUCTION

Silicon carbide (SiC) is a natural layered semiconductor due to the phenomenon of polytypism[1]. This implies that the same chemical compound of silicon and carbon atoms crystallizes into different crystallographic modifications known as polytypes. These polytypes are all similar in the plane perpendicular to the symmetry axis (c-axis), but differ from each other in the direction parallel to the c-axis.

Silicon carbide (SiC) is also known to have one of the largest energy bandgaps (~ 3.0 eV) of common semiconductor materials. This property makes it valuable for high-temperature device applications and the fabrication of blue light-emitting diodes[2].

Many observations on several samples cut from single crystals led to the conclusion that the layered structure which makes up the device contains a highly resistive (strongly compensated) polytype, sandwiched between low resistive polytypes[3]. This creates an n^+nn^+-like structure. The low resistive polytypes act as injecting and extracting "contacts" for the high resistive polytype, introducing into the latter single-carrier, space-charge limited flow.

The I-V characteristic measured at 77K temperature shows four different regimes of operation (Fig. 1): an ohmic regime; a low-bias quadratic regime; a fast-rising trap-filling regime (TFL); and finally, a quadratic Mott-Gurney regime.

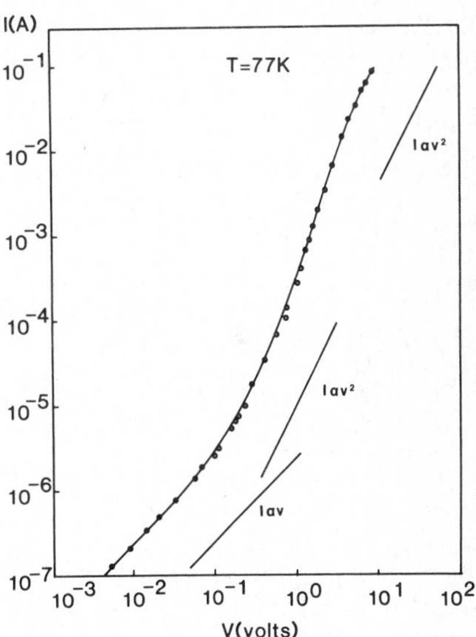

FIGURE 1
I-V characteristics at T = 77K. The dots indicate the measurements. The solid line represents the results of computer calculations.

Expressions for the impedance and the noise of a two-terminal device, operating under conditions of SCL flow, were derived by Van Vliet et al.[4], employing the transfer impedance method. Neglecting the diffusion term in the current equation, closed analytical expressions were obtained for the first two regimes, i.e., the ohmic and the low-voltage quadratic regime.

From our DC computer calculations, however, we conclude that diffusion plays an important role, specifically at low-voltage bias levels. Therefore, we include the diffusion term in our present study and apply the transfer impedance method to obtain the noise in the ohmic regime and three SCL regimes.

2. THEORETICAL MODEL

The pertinent transport equations, accounting for DC, AC, as well as Langevin noise sources, are:

$$J(t) = q\mu nE + qD\frac{dn}{dx} + H(x,t) + \varepsilon\varepsilon_0\frac{dE}{dt} \quad (2.1)$$

$$\frac{dE}{dx} = -(q/\varepsilon\varepsilon_0)(n + \sum_i n_{ti} - n_c) \quad (2.2)$$

$$\frac{\partial n_{ti}(x,t)}{\partial t} = -\beta_i n_{1i} n_{ti} + \beta_i n(N_{ti} - n_{ti}) - \gamma_i(x,t). \quad (2.3)$$

Here n is the free electron density, n_t is the trapped density, n_c is the equilibrium charge, n_1 is the Shockley-Read parameter, β is a capture constant, H is the thermal noise source, γ is the trapping noise source, and subscript i indicates different trap level. The other symbols have their usual meaning. These equations are split into DC and AC parts or in the noise analysis into DC and fluctuating parts, so that $n = n_0 + \Delta n$, $I = I_0 + \Delta I$, etc. Note that the suffix zero now denotes DC or average values, $n_0 = \langle N \rangle$, etc. The equations describing the DC part are related to the quasi-Fermi potential and the electrostatic potential and solved simultaneously to give the DC values of the different physical quantities involved.

To obtain the spectral density of the voltage fluctuations we proceed as follows. Equations (2.2) - (2.3) are used to express Δn_{ti} and Δn in terms of ΔE. Then, after a Fourier transformation the equation for the current ΔI reads

$$-\frac{\Delta I}{A} = qn_0\frac{d(\mu E_0)}{dE_0}\Delta E - \left(q\mu E_0 + qD\frac{d}{dx}\right)\left(\frac{\varepsilon\varepsilon_0}{q}\frac{d(\Delta E)}{dx}\right)$$
$$\cdot\left[1 + \sum_i \frac{1/\tau_{1i}}{1/\tau_{2i} + j\omega}\right]^{-1}\Big) + \left(q\mu E_0 + qD\frac{d}{dx}\right)\left(\sum_i \frac{\gamma_i}{1/\tau_{2i} + j\omega}\right)$$
$$\cdot\left(1 + \sum_i \frac{1/\tau_{1i}}{1/\tau_{2i} + j\omega}\right)^{-1} + j\omega\varepsilon\varepsilon_0\Delta E + H(x,\omega) \quad (2.4)$$

where $\tau_{1i} = 1/\beta_i(N_{ti} - n_{ti})$, $\tau_{2i} = 1/\beta_i(n_{1i} + n_0)$, and $1/\tau_i = 1/\tau_{1i} + 1/\tau_{2i}$.

To obtain the transfer impedance tensor, we set $H = \gamma = 0$ and discretize eq. (2.4). Formally, the current density Δj_α can be related to the field ΔE_β by linear conductivity operator σ_{op}

$$\Delta j_\alpha = \sigma_{op} \cdot \Delta E_\beta. \quad (2.5)$$

The inverse of eq. (2.5) defines the impedance operator Z_{op} and the transfer impedance tensor $Z_{\alpha\beta}$, i.e.,

$$\Delta E_\alpha = Z_{op} \cdot \Delta j_\beta = A\sum_\beta Z_{\alpha\beta} \cdot \Delta j_\beta \Delta x \quad (2.6)$$

where A is the device cross-sectional area. Due to the fact that we include diffusion in our transport equation, the perturbation of the electric field caused by a current fluctuation at α propagates to the cathode as well as to the anode. Omitting diffusion would cause the perturbation to propagate only towards the anode as has been shown by several authors[4].

In general, the noise spectral density of the electric field is defined as[4]

$$S_{\Delta E(x)\Delta E(x')} = \iint dx_n dx_m Z(x, x_n, j\omega)$$
$$\cdot S_\xi(x_n, x_m) Z^*(x_m, x', j\omega) \quad (2.7)$$

where the impedance elements are defined in discrete form by equation (2.6). To obtain the noise source $S_\xi(x_n, x_m)$, we define

$$\xi = A\gamma + BH + C\frac{d\gamma}{dx}, \quad (2.8)$$

where A, B, and C are constants, given by eq.

(2.4), and diffusion is accounted for by the derivative $\frac{d\gamma}{dx}$. The noise spectral density of ξ, S_ξ is then related to the shot noise in the transition rate $S_\gamma(x_n,x_m)$ and the velocity-fluctuation noise in the conduction band $S_H(x,x_m)$. The spectral density of the voltage fluctuations observed at the terminals is given by

$$S_{\delta V}(L) = \iint dxdx' \, S_{\Delta E(x)\Delta E(x')}. \qquad (2.9)$$

3. I-V CHARACTERISTIC

Fig. 1 shows that the computer simulation program explains the measured I-V characteristic at 77K very well. The overflow of carriers from the n^+ regions into the n region plays an important role at low bias voltage levels. This large overflow of carriers is due to the large gradient in the carrier concentration at the n^+n interface. In the linear regime the traps in the n^- region are mainly empty and the trap energy lies above the Fermi level for most of the n^- region (Fig. 2). In this regime the diffusion plays a dominant role in charge transport. As the biasing voltage increases, the potential minimum shifts towards the n^+ cathode contact. The quasi-Fermi level lies below the trap level for most of the device. Consequently the traps are mainly empty. The width of the section of the n^- region dominated by the drift current increases, and the current becomes space-charge-limited and proportional to the square of the voltage. A further increase of bias voltage causes the quasi-Fermi level to pass through the trap level as is indicated in Fig. 3. The traps are being filled and the I-V characteristic shows a fast rise in the current. The drift current dominates current flow in the major part of the n^- region. Finally, the quasi-Fermi level lies completely above the trap level, and almost all of the traps are filled. Drift dominates the sample, and the I-V characteristic shows the quadratic dependence between current and voltage ($I \alpha V^2$).

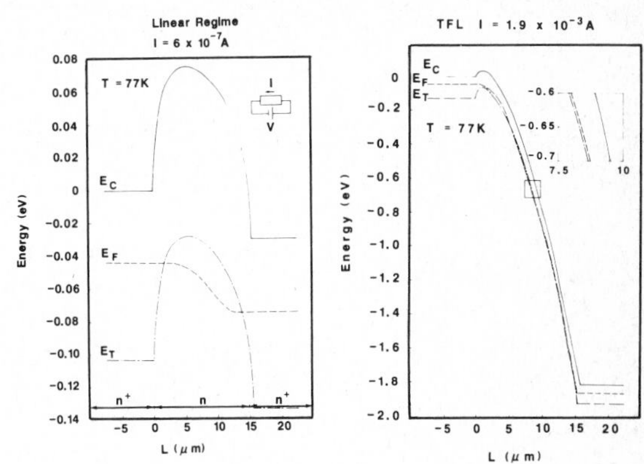

FIGURE 2
Energy band diagram in the linear regime.

FIGURE 3
Energy band diagram in the trap-filling regime.

4. NOISE CHARACTERISTIC

The results of the voltage noise calculation in the Ohmic regime are shown in Fig. 4.

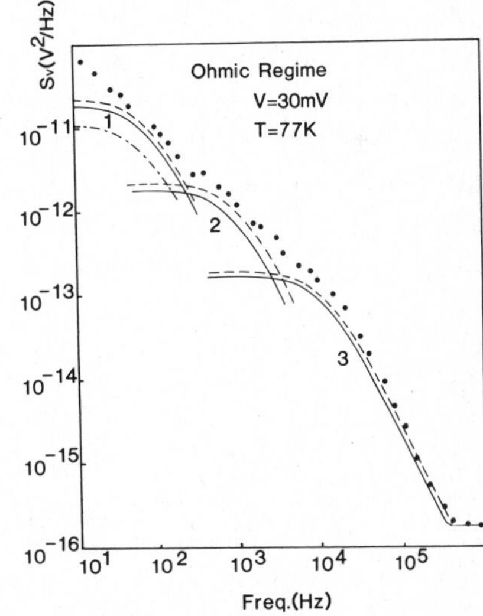

FIGURE 4
Voltage noise magnitude in the ohmic regime. Dots: measured data. Dashed lines: resolution into Lorentzians. Solid lines: computer calculations. Dot-dashed lines: the effect of the diffusion part of the noise source term, $d\gamma/dx$, for trap level one.

The parameters of the trap levels observed in the noise spectral density are given in Table 1. The capture cross sections σ_i are within the normal range for neutral or negatively charged single ion trap centers.

TABLE 1
Trap parameters used for computer calculations

Trap #	E_{act}(meV)	$N_t(m^{-3})$	$\sigma(m^2)$
1	103	5×10^{18}	2.5×10^{-21}
2	85	6×10^{18}	5.4×10^{-21}
3	67	1×10^{20}	5.4×10^{-21}

To obtain a better understanding of the variation of the noise spectral density with bias for the different regimes of the I-V characteristics, a plot of the characteristic time constants τ_i vs. V_0 is shown in Fig. 5 where

$$1/\tau_i = \beta[N_{ti}-n_{ti}+n_{1i}+n_0]. \qquad (3.1)$$

In Fig. 6 a plot of voltage noise spectral (S_V) as a function of bias voltage is shown.

For the Lorentzians involved the statistical factor $\frac{\langle\Delta n^2\rangle_i}{n_i}$ is

$$\frac{\langle\Delta n^2\rangle_i}{n_0} = \frac{\tau_i}{\tau_{1i}} = \frac{N_{ti}-n_{ti}}{N_{ti}-n_{ti}+n_0+n_{1i}}, \qquad (3.2)$$

where n is the number of carriers in a section Δx. In the Ohmic regime the changes in n_t and n_0 with bias are small, and since $n_t \ll N_t$ and $n_0 \ll n_1$, the noise magnitude and the time constant changes with bias will be small.As the biasing increases, the carrier concentration n_0 increases. For the trap closest to the quasi-Fermi level (traps with E_{act} = 103 meV), the trap carrier concentration n_t will increase

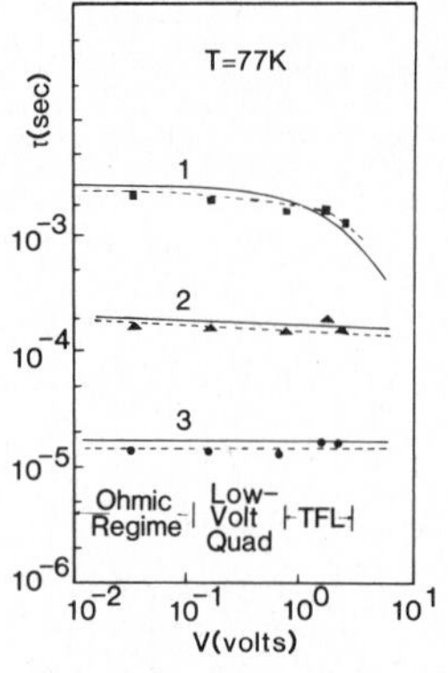

FIGURE 5
Characteristic time constants of the Lorentzian spectra vs. V_0. Solid lines: computer calculations.

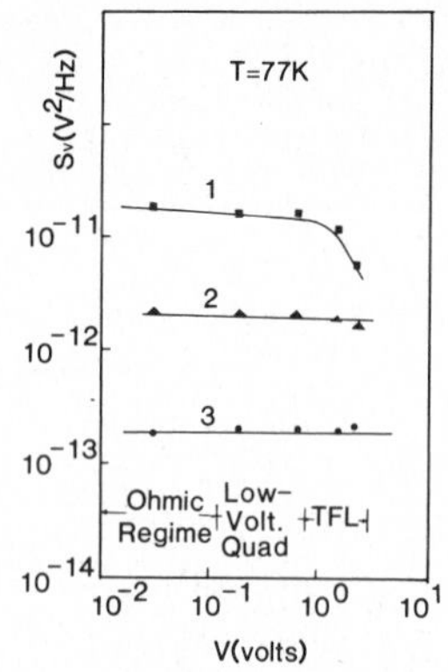

FIGURE 6
Voltage spectral density vs. V_0. Solid lines: computer calculations.

rapidly, and consequently the noise magnitude and time constant will show a rapid decrease.

ACKNOWLEDGEMENTS

We thank the Air Force Office of Scientific for supporting this research under grant #F49620-83-C-0072.

REFERENCES

1. R.C. Marshall, J.W. Faust, Jr and C. E. Ryan, Silicon Carbide (U. of South Carolina Press, Columbia, S.C., 1974).

2. E. Pettenpaul, W. von Munch and G. Ziegler, Silicon carbide devices, Inst. Phys. Conf., Ser. No. 53 (1980), 21-35.

3. S. Tehrani, J. Kim, L.L. Hench, C.M. Van Vliet and G. Bosman, Parts I & II, Appl. Phys. 58 (August 1985), 1562-1577.

4. K.M. van Vliet, A. Friedman, R.J.J. Zijlstra, A. Gisolf and A. van der Ziel, Parts I & II, J. Appl. Phys. 46 (1975), 1804-1823.

NOISE IN PHYSICAL SYSTEMS AND 1/f NOISE - 1985
A. D'Amico and P. Mazzetti (editors)

NOISE IN SODIUM β" ALUMINA SINGLE CRYSTALS

James J. BROPHY and Steven W. SMITH

Physics Department, University of Utah, Salt Lake City, Utah 84112, U.S.A.

Experimental current noise spectra for single crystal sodium β" alumina are essentially identical to those previously reported for polycrystalline ceramic specimens. The measured noise power is proportional to the square of the dc sample current and varies as $f^{-1.5}$, which suggests conductivity fluctuations arising from diffusion of the mobile ions. Only a small fraction of the mobile ions appear to participate in the diffusion noise process and the number is thermally activated with an activation energy of 0.8eV, in agreement with data from ceramic specimens.

1. INTRODUCTION

Conductivity fluctuations in polycrystalline ceramic sodium β" alumina have been observed and interpreted in terms of diffusion noise of the mobile sodium ions[1]. This interpretation implies, however, that the granular nature of the ceramic does not infuence diffusion noise and that only a small fraction of the mobile ions participate. The present work examines electrical noise in single crystal sodium β" alumina under conditions essentially identical to those used to study ceramic specimens. In addition, preliminary data on noise of single crystal silver β" alumina are compared to sodium conductors.

2. EXPERIMENTAL TECHNIQUE

Single crystal sodium β" alumina samples grown by the flux evaporation method are cut to an approximately square shape, 1x1x0.2 cm^3. The corners are sealed into the sides of four plastic test tubes to provide diagonally opposing corner current electrodes and transverse noise terminals. Two mole percent amalgam elecrodes prepared by electrodeposition from an 80% aqueous solution of NaCl into mercury are introduced into all four test tubes. This technique provides ohmic, low noise contacts and is essentially the same configuration previously employed to examine ceramic specimens[1]. Aqueous 5-M silver nitrate electrodes are used in the case of single crystal silver β" alumina samples[2].

Transverse noise voltages are measured with a PAR 113 preamplifier. The preamplifier output is analyzed digitally using an A/D converter and FFT routine developed for the Apple IIe personal computer[3]. The system accurately measures Nyquist noise of resistances from 10^4 to $2x10^8$ ohms over the frequency range 10^{-4} to 10^4 Hz. Current is supplied to the current terminals from a filtered battery source through a 10^5 ohm metallic resistor.

3. EXPERIMENTAL RESULTS

Representative transverse noise voltage spectra are shown in Figure 1. The zero-current spectrum is attributed to chemical reaction noise at the crystal-amalgam interface similar to that seen in ceramic specimens[1,4]. The contact noise magnitudes and spectral shapes are the same for both types of samples and the single crystal contact noise decreases with time as previously reported for ceramics. At high frequencies, the measured noise level agrees with Nyquist noise of the sample resistance, as determined from the slope of the (linear) current-voltage characteristic[4].

Current noise spectra are proportional to the square of the dc sample current and show spectral shapes ranging from $f^{-1.5}$ to $f^{-2.0}$, depending upon the duration of current in the sample. Lower values are seen accompanying the initial current levels and values of f^{-2} are noted after the extended current durations required to obtain full spectra. As in ceramic samples[1], the change is attributed to electrochemical changes accompanying current in the sample. Significantly, the spectral slope reverts to $f^{-1.5}$ upon current reversal, Figure 1.

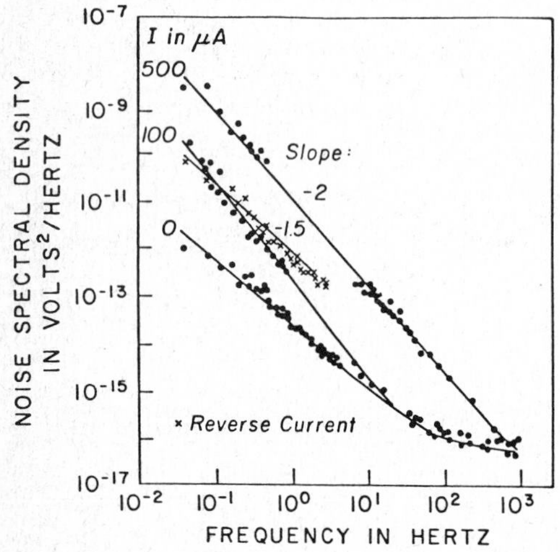

FIGURE 1
Transverse noise spectra of a sodium β" alumina single crystal.

Only one single crystal silver β" alumina sample has been examined. The small size, 0.5x0.3x0.1 cm^3 precludes the use of the four-terminal method, but evidence from ceramic specimens[2] indicates that contact current noise is negligible, so that a two-terminal measurement may be acceptable. The observed noise is about the same as for ceramic samples and the spectral shape, $f^{-1.5}$, is very stable with both time and current.

The temperature dependence of the noise is different in different current and frequency regimes, Figure 2. At zero current and high frequencies the sodium β" alumina Nyquist noise is thermally activated with an activation energy of 0.33eV, in agreement with ceramic sample[1] and conductivity data[5]. Similarly, the silver β" alumina activation energy, 0.16eV, agrees with ceramic[2] and conductivity results[6].

The current noise is also thermally activated, with activation energies of 1.3eV and 0.49eV for sodium and silver β" alumina, respectively. The former value is in good agreement with ceramic sample data[1], as is the corresponding figure for the silver ceramic[2]. Figure 2 also shows an activation energy of -0.76eV for contact noise in the case of silver β" alumina. The negative sign is consistent with a thermally-activated chemical reaction at the interface.

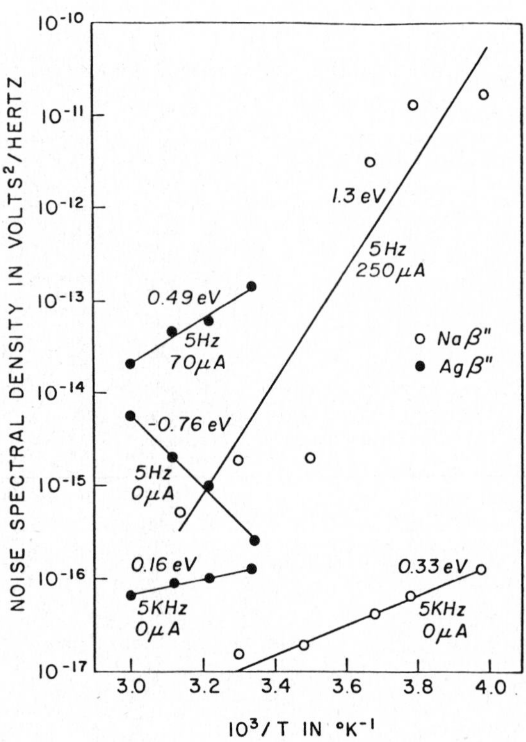

FIGURE 2
Temperature dependence of Nyquist noise (0μA, 5 kHz), current noise (70 or 250μA, 5 Hz), and contact noise (0μA, 5 Hz), for sodium and silver β" alumina single crystals.

4. DISCUSSION

The observed spectral shape, $f^{-1.5}$, is consistent with ceramic sample data, and with diffusion noise[7]. The noise voltage spectral density for conductivity fluctuations arising from diffusion can be written as

$$\frac{S(V,f)}{V^2} = 4\,\frac{\langle \Delta N^2 \rangle}{N}\left(\frac{D}{2L^2}\right)^{1/2} \omega^{-3/2} \qquad (1)$$

Where V is the dc voltage across the sample, $\langle \Delta N^2 \rangle$ and N are the variance and average number of the diffusing ions, D is the diffusion constant, L is a characteristic length, and ω is the angular frequency. This expression is valid for frequencies greater than a characteristic frequency ω_0,

$$\omega_0 = 2D/L^2 \qquad (2)$$

Below ω_0, the spectrum flattens.

Taking $D = 1\times10^{-6}\,cm^2/sec$ at room temperature[6] and L equal to the sample length, 1 cm, the characteristic frequency is calculated to be 3.2×10^{-7} Hz, well below the range of present data. No characteristic frequency is observed for ceramic specimens either[1], which indicates that the polycrystalline granularity is not important in determining the characteristic length, L.

Using experimental values of $S(V,f)/V^2$, and assuming Poisson statistics, $\langle \Delta N^2 \rangle = N$, a value for the mobile ion density of 1.2×10^8 ions/cm^3 is calculated from Equation 1. This is much smaller than the known concentration of mobile sodium ions, about 10^{22} ions/cm^3, but is in agreement with data from ceramic samples.

Following the approach taken in the case of the ceramic specimens[1], the temperature dependence of the noise can be accounted for by assuming

$$\begin{aligned} D(T) &= D_o \exp(-E_d/kT) \\ R(T) &= R_oT \exp(E_d/kT) \\ N(T) &= N_o \exp(-E_n/kT) \end{aligned} \qquad (3)$$

so that Equation (1) becomes

$$S(V,f,T) = \frac{4I^2R_o^2}{N_o}\left(\frac{D_o}{2L^2}\right)^{1/2} T^2\omega^{-3/2} \times$$
$$\times \exp[(3E_d/2 + E_n)/kT]$$

Now, from Figure 2, $(3E_d/2 + E_n) = 1.3$eV and $E_d = 0.33$eV, so that $E_n = 0.8$eV. Inserting this and the calculated room temperature value of the ion density into Equation 3, $n_o = 3.5\times10^{21}$ ions/cm^3. This is close to the expected number of mobile Na ions.

5. CONCLUSIONS

This experimental data indicates that conductivity fluctuations in single crystal sodium β'' alumina and silver β'' alumina are very similar to those observed in polycrystalline ceramic specimens. In particular, it appears that the granularity of ceramic specimens does not influence the diffusion noise, so that future studies may concentrate on ceramic samples, which are more plentiful than single crystals. The observed experimental noise level can be accounted for by assuming that the number of mobile ions participating in the diffusion noise process is thermally activated.

ACKNOWLEDGMENTS

This research is supported in part by the Office of Naval Research. The authors express their appreciation to G. R. Miller and J. M. Viner for many helpful suggestions and advice.

REFERENCES

1. J. J. BROPHY and S. W. SMITH, J. Appl. Phys. (1958), 58, 351.
2. S. W. SMITH and J. J. BROPHY, Eighth International Conference on Noise in Physical Systems (Rome, Sept. 1985).
3. S. W. SMITH, Rev. Sc. Inst., (1985) 56, 159.
4. J. J. BROPHY and S. W. SMITH, J. Appl. Phys. (1984), 56 , 801.

5. R. STEVENS and J. G. P. BINNER, J. Matl. Sc. (1984) 19, 695.

6. J. L. BRIANT and G. C. FARRINGTON, J. Sol. St. Chem., (1980) 33, 385.

7. K. M. VAN VLIET and J. R. FASSETT, "Fluctuation Phenomena in Solids", R. E. Burgess (ed), Academic Press, New York (1965), p. 268.

NOISE IN PHYSICAL SYSTEMS AND 1/f NOISE - 1985
A. D'Amico and P. Mazzetti (editors)

NOISE IN SILVER β'' ALUMINA CERAMICS

Steven W. SMITH and James J. BROPHY

Physics Department, University of Utah, Salt Lake City, Utah 84112, U.S.A.

Voltage fluctuations at ohmic electrodes to silver β'' alumina ceramic samples are observed both at contacts and in the bulk over the frequency interval 10^{-3} to 10^4 Hz. Contact noise power in the absence of current varies as f^{-2} at low frequencies and is dominated by Nyquist noise of the sample at frequencies greater than 100 Hz. Bulk current noise measured at transverse contacts has a $f^{-3/2}$ power spectrum and increases with the square of the current. Low-frequency contact noise, sample Nyquist noise and bulk current noise are all thermally activated with activation energies of -0.96, 0.18 and 0.47 eV, respectively. These experimental results are very similar to those previously reported for sodium β'' alumina ceramics.

1. INTRODUCTION

Electrical noise voltages attributed to diffusion noise of the mobile ions have been observed in superionic sodium β'' alumina ceramics[1]. This interpretation leads to conclusions that only a small fraction of the mobile ions participate in the noise process and that the granularity does not influence the diffusion noise magnitude. It is of interest to examine conductivity fluctuations in silver β'' alumina ceramics to determine the influence of the nature of the mobile ions on the noise processes. This study is facilitated by the ease with which the mobile ions can be exchanged in the β'' alumina structure[2].

2. EXPERIMENTAL PROCEDURE

Commercial sodium β'' alumina ceramic (90.4% Al_2O_3, 8.85% Na_2O, 0.75% Li_2O) specimens[3], 1x1x0.3 cm^3, are converted to silver β'' alumina by ion exchange in molten 50% $AgNO_3/NaNO_3$ at 300° C for 8 hours. Weight change of the samples indicates 98% of the mobile sodium ions are replaced by silver ions. Ohmic contacts are 5-M $AgNO_3$ solution in water or 0.05-M in glycerin. Silver amalgam contacts are ohmic, but have very high resistance[4]. The corners of the sample are sealed into the sides of four plastic test tubes containing the liquid electrode material to provide diagonally opposing corner current terminals and transverse noise contacts. This is essentially the same configuration used to study sodium β'' alumina ceramics[1]. The noise measurement technique is the same used for both ceramic and single crystal specimens[1,5].

3. NOISE RESULTS

Typical noise spectra, Figure 1, show con-

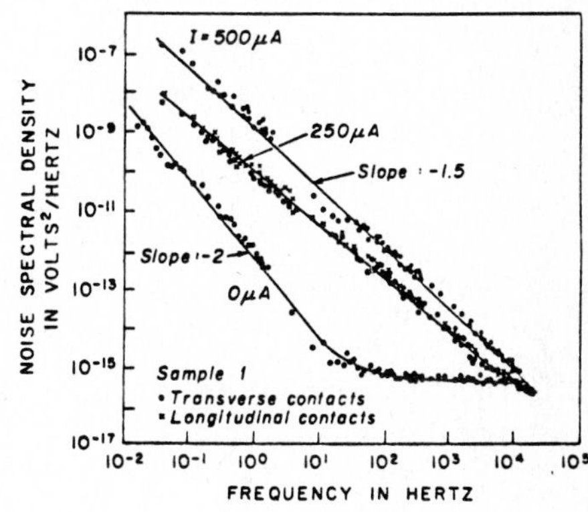

Figure 1.
Noise spectra of silver β'' alumina ceramic.

tact noise[6] in the absence of current and current noise characterized by an $f^{-1.5}$ trend. The slopes of the spectra tend to be smaller, $f^{-1.3}$, at low currents. At high frequencies the observed noise is accounted for by Nyquist noise of the sample resistance calculated from the linear current-voltage characteristic. In general, all three noise properties are very similar to sodium β'' alumina[1]. The range of current noise magnitudes exhibited by the silver samples overlap those of the sodium conductors while the Nyquist noise level is greater, in agreement with the greater resistivity of silver β" alumina[2]. Similarly, contact noise levels tend to be greater, although still sufficiently low to permit reliable current noise measurements.

The contact noise levels of aqueous and glycerin solutions are identical after each ages for several hours. Both contact materials exhibit noise levels in excess of the $2x10^8$ ohm preamplifier input resistance Nyquist noise at frequencies below 10^{-1} Hz and the spectral shapes are near f^{-2}. These characteristics are attributed to non-equilibrium chemical reaction noise at the sample-electrode interface[6]. Since both electrode solutions have the same noise level, chemical contact noise must be associated with the mobile silver ions.

Current noise in silver β" alumina is much more stable with respect to time and current than is the case for sodium β" specimens. Relatively minor changes attributable to electrochemical effects[1] are observed. Also, transverse and longitudinal (two-terminal) noise levels are the same for low-noise contacts. This means that contact current noise levels are small compared to bulk conductivity fluctuations. Both the stability and the absence of contact current noise suggest that silver β" alumina is a better material than sodium β" alumina to use in the study of noise in superionic conductors.

A few samples inadvertently heated to high temperatures (800°C) for several hours experienced an increase in resistivity of four to five times, together with a darkening of the surface. This behavior has been previously reported[7], but no chemical or structural change has been detected to account for the increase in resistivity. Noise properties of such "darkened" samples are not noticeably different from normal specimens, although the temperature dependence changes somewhat.

The various noise processes are thermally activated, Figure 2. The activation energy for

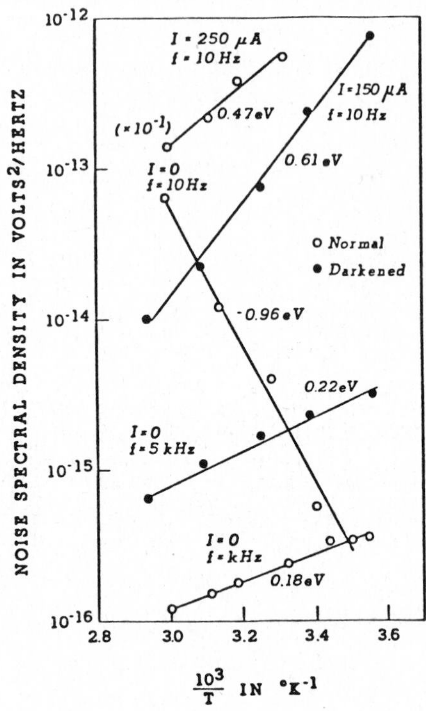

Figure 2.
Temperature dependence of Nyquist (0μA, 5 kHz) current noise (150 μA, 10 Hz), and contact noise (0μA, 10 Hz) of normal and darkened silver β" alumina ceramic.

Nyquist noise of the normal sample is a little lower than literature values[2,7], but the increase upon darkening is consistent with conductivity data[7]. The current noise activation energy is about one-half of that for sodium β" alumina, while the activation energy for contact noise is significantly greater. This is consistent with the greater contact noise level of silver β" alumina and a thermally-activated chemical reaction at the contact.

4. DISCUSSION

As in the case of sodium β'' alumina ceramics, the $f^{-1.5}$ spectral shape suggests diffusion noise. The same analysis applied to sodium β'' ceramics[1] and single crystals[5] can be used here.

Taking $D=10^{-7}$ cm^2/sec at room temperature[2,7], the turnover frequency is calculated to be 3×10^{-8} Hz if the characteristic length L is the sample length and 1.3×10^{-1} Hz if L is equal to the average gain size[8], 5×10^{-4} cm. No departure from the $f^{-1.5}$ trend is noticed down to 10^{-3} Hz, so that it appears that the granular structure does not influence the diffusion noise. This is consistent with data from sodium β'' alumina ceramics[1]. Observed noise levels are very much larger than those predicted, as in the case of sodium β'' alumina[1]. Furthermore, the activation energy seen in Figure 2 is greater than can be accounted for, unless the number of diffusing ions is thermally activated. This approach leads to satisfactory numerical agreement in the case of sodium β'' alumina[1], but does not appear to apply to silver β'' alumina.

5. CONCLUSIONS

These experimental results show that conductivity fluctuations in silver β'' alumina ceramics are very similar to those previously observed in sodium β'' alumina. The spectral shape suggests diffusion noise and the noise does not appear to be influenced by the granularity of the samples. It is not possible to account quantitatively for the observed magnitude of the noise by assuming that the active ion density is thermally activated, as in the case of sodium β'' alumina.

ACKNOWLEDGMENTS

The authors express their appreciation to G. R. Miller and J. M. Viner for many helpful suggestions and advice. This work is supported in part by the Office of Naval Research.

REFERENCES

1. J. J. BROPHY and S. W. SMITH, J. Appl. Phys. (1958), 58, 351.

2. J. L. BRIANT and G. C. FARRINGTON, J. Solid State Chem. (1980), 33, 385.

3. Obtained from Ceramatec, Inc., Salt Lake City, Utah 84115.

4. J. J. BROPHY, "Noise in Physical Systems and 1/f Noise", M. Savelli, G. Lecoy, and J-P, Nougier (eds), Elsevier Science Publishers, B.V., Amsterdam, (1983) p. 351.

5. J. J. BROPHY and S. W. SMITH, Eighth International Conference on Noise in Physical Systems (Rome, Sept. 1985)

6. J. J. BROPHY and S. W. SMITH, J. Appl. Phys. (1984), 56, 801.

7. M. W. BREITER and B. DUNN, Sol. State Ionics, 9&10, (1983) 227.

8. A. D. JATKAR, I. B. CUTLER, A. V. VIRKAR and R. S. GORDON, "Processing of Crystalline Ceramics", H. Palmour, R. F. Davis and T. M. Hare (eds), Plenum Publishing, New York (1978), p. 421.

NOISE IN PHYSICAL SYSTEMS AND 1/f NOISE - 1985
A. D'Amico and P. Mazzetti (editors)

STOCHASTIC BEHAVIOUR OF INDIVIDUAL BARKHAUSEN DISCONTINUITIES IN IRON WHISKERS

K. ZEIBIG, S. STIELER and C. HEIDEN

Institut für Angewandte Physik der Justus-Liebig-Universität Giessen, D-6300 Giessen, FRG

The magnetization reversal of iron whiskers often takes place in several large Barkhausen jumps, if the direction of the applied magnetic field is parallel to the 100- axis of the single crystal. The jump fields of the individual jumps are approximately Gauss-distributed. We measured the standard deviation of and the correlation between these jumps. The results lead to the conclusion, that the distribution of the jump fields is not caused by thermal spin fluctuations but by stochastic magnetic precursor events, which via internal stray field affect the unpinning of domain walls.

1. INTRODUCTION

The magnetization reversal of iron whiskers in an external magnetic field parallel to an easy axis often takes place in one large individual Barkhausen jump (BJ). The field value, at which this discontinuity occurs is not sharp but found to be distributed around a mean value $\langle H_s \rangle$[1]. The distribution function for H_s often to good approximation is Gaussian. Looking for a mechanism leading to such a distribution we tried to explain our results with thermal activation of domain walls anchored at pinning centres inside the whisker: the time dependent thermal fluctuation field [2] with rms value H_f is superimposed on the external field and leads to a distribution of the jump field (JF) around the pinning field. H_f is a monotonously increasing function of temperature T as long as the activation volume does not depend on T. Following Néel for instance, H_f is proportional to $T^{1/2}$. A computer simulation of the thermal activation process of a BJ shows, that the thermal activation therefore should lead to a broadening of the JF-distribution with increasing temperature (c.f. fig. 1 and 2). The standard deviation s of the observed distribution function however was found to decrease with increasing temperature [3]. Furthermore for the present samples we could not detect any aftereffect i.e. a variation of the sweep rate of the external field did not change the average JF. Therefore thermal activation does not seem to generate the observed JF distribution.

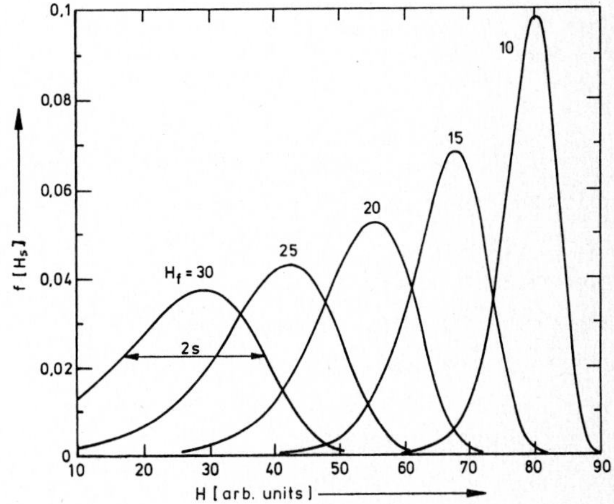

FIGURE 1
Calculated jump field distribution $f(H_s)$ for different temperatures ($H_f \sim T^{1/2}$), thermal activation model. For details see [4].

Another less fundamental mechanism, which may explain the observations can be the generation of a stray field h(r) inside the sample, originating from other domains or nuclei of domains. h(r) is superimposed on the external field. Its stochastic nature is postulated to be

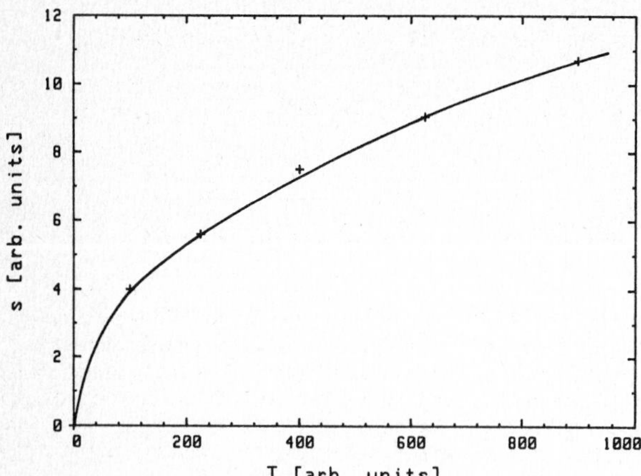

FIGURE 2
Calculated temperature dependence of the standard deviation s of the distribution $f(H_s)$, thermal activation model.

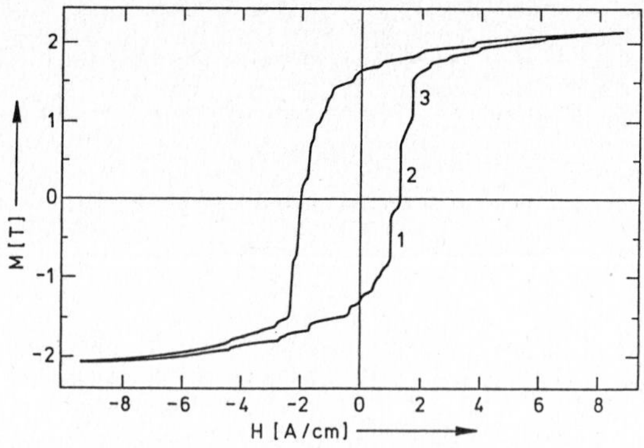

FIGURE 3
Magnetization curve of whisker W39. The three investigated jumps are denoted by numbers.

caused by the varying number and spatial distribution of nuclei for each sweep starting from saturation. This paper describes experiments, carried out to support the hypothesis, that the uncertainty of the JF is caused by stochastically varying stray fields.

2. EXPERIMENTAL PROCEDURE

A high slew-rate SQUID system was used in our experiments [1]. We studied the magnetization reversal of two iron whiskers W39 and W42 which here takes place by several BJ's, separated by continuous magnetization processes (c.f. fig. 3 and 4).

The external field is controlled by a microcomputer. A BJ can be detected by a comparator, which interrupts the computer and causes it to read and store the JF.

At a field excursion of ±9.1A/cm and a temperature of 4.2K we collected data of 3 and 5 BJ's over 400-1000 sweeps along the lower branch of the magnetization curve of W39 and W42, respectively. The mean JF $<H_{si}>$ and the standard deviation s_i were calculated for each individual jump i. The correlation coefficient

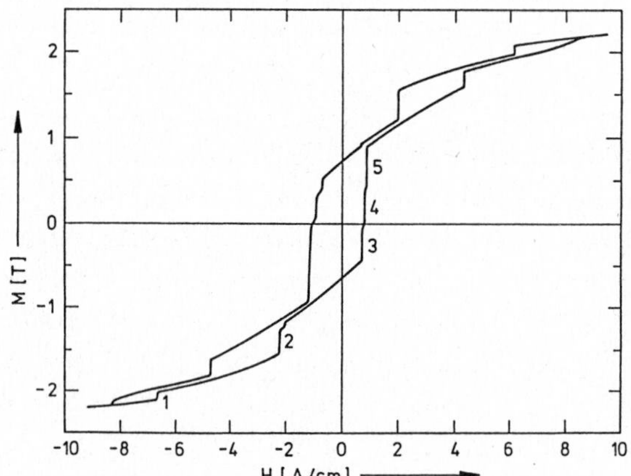

FIGURE 4
Magnetization curve of whisker W42.

$$C_{ij}=<\Delta H_{si}\cdot\Delta H_{sj}>/(<\Delta H^2_{si}>\cdot<\Delta H^2_{sj}>)^{1/2}$$

of the deviation $\Delta H_{si}=H_{si}-<H_{si}>$ also was determined.

3. RESULTS

Table 1 and 2 present the measured values for M_{ij}, which is the difference of the magnetizations at the starting points of BJ's j and i, the correlation coefficient C_{ij}, the mean value $<H_{si}>$ and the standard deviation s_i for the two whiskers W39 and W42, respectively. For

TABLE 1
Data measured on W42 at T=4.2K and a field excursion of ±9,1A/cm.

i,j	M_{ij} (T)	C_{ij}	i	$<H_{si}>$ (A/cm)	s_i (A/cm)
1,2	0.59	0.18	1	-6.56	0.172
1,3	1.71	0.18	2	-2.18	0.136
1,4	2.15	0.04	3	0.74	0.147
1,5	2.68	0.04	4	0.86	0.089
2,3	1.13	0.51	5	0.87	0.052
2,4	1.56	0.14			
2,5	2.09	0.10			
3,4	0.44	0.41			
3,5	0.97	0.25			
4,5	0.53	0.68			

both samples s_i tends to decrease from jump to jump (c.f. fig. 5). The correlation coefficients, which are found to be always positive, are highest for neighbouring jumps and decrease with increasing separation of the jumps. More general, C_{ij} shows a decreasing tendency for increasing M_{ij} as can be seen from fig. 6.

4. DISCUSSION

A model will help to understand the experimental results qualitatively. The magnetization reversal of a whisker with very simple domain structure in an axial external magnetic field shall take place by Bloch wall motion of a single rigid and straight 180° wall dividing the sample into two parts (c.f. fig. 7). The displacement of the wall can be stopped at pinning centers indicated by crosses in fig. 7. At saturation in a high external field H the whole whisker is magnetized in direction of this field (a). When H is lowered, randomly distributed reverse domain nuclei form at imperfections of the crystal and the 180° domain wall is created at one side (b). Further decrease and then reversal of H lets the wall jump from one to the next pinning center (c), (d), (e). At (f) saturation in the opposite direction is reached.

Without any stochastic process i.e. without the

TABLE 2
Data measured on W39 at T=4.2K and a field excursion of ±9,1A/cm.

i,j	M_{ij} (T)	C_{ij}	i	$<H_{si}>$ (A/cm)	s_i (A/cm)
1,2	0.71	0.48	1	1.67	0.229
1,3	1.80	0.35	2	1.83	0.188
2,3	1.09	0.67	3	1.94	0.124

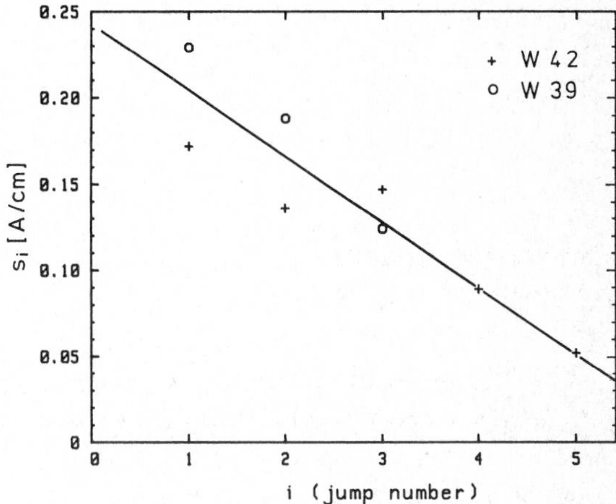

FIGURE 5
Standard deviations s_i of the JF-distributions of the individual BJ's of W39 and W42.

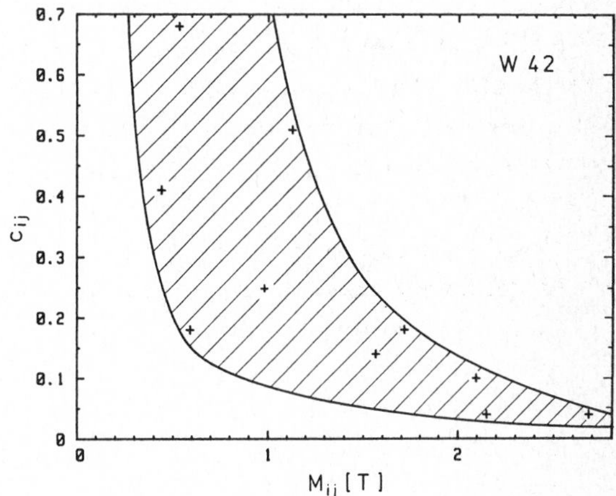

FIGURE 6
Correlation coefficients C_{ij} versus magnetization change M_{ij} for W42.

formation of the randomly distributed nuclei, the JF's of the three BJ's would be the same for every cycle and equal to the individual (sharp) pinning fields of the three pinning centers. With the random formation of the nuclei another situation arises. Their stray fields are superimposed on the external field and therefore modify the JF's stochastically. Because the number of nuclei is reduced when the Bloch wall moves on (c.f. fig. 7 c...e), the random stray field h becomes the smaller, the more of the sample is magnetized in the opposite direction. This explains qualitatively the experimental result, that the standard deviation s_i decreases with the jump index i (c.f. fig. 5).

The correlation coefficient C_{ij} is the larger, the less the stray fields at pinning centers i and j differ from each other, immediately before the jumps occur. Equal stray fields at both wall

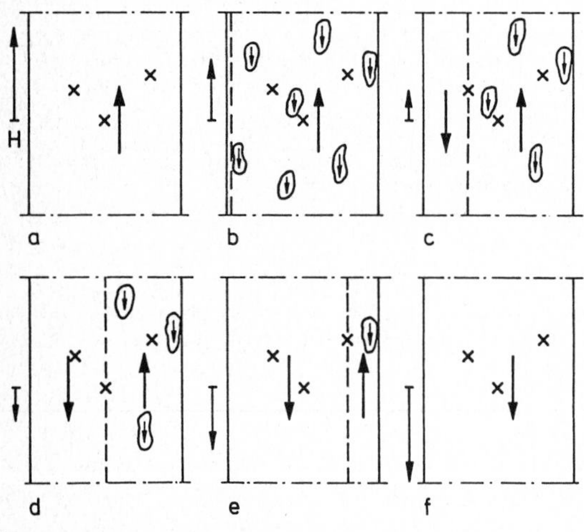

FIGURE 7: See text.

locations would lead to C_{ij}=1, whereas a uniform spatial distribution of sources of h(r) should cause a monotonous decrease of the C_{ij} with the distance between the two pinning centers i and j. Supposed, all pinning centers lie on a straight line, this distance is proportional to the magnetization change M_{ij}, so that C_{ij} should decrease with increasing M_{ij}. The experiment shows, that the plotted (C_{ij};M_{ij}) values cannot be fitted by a single curve, but the general tendency of the C_{ij} is decreasing for increasing M_{ij}.

Because all measured values for C_{ij} are positive, the origin of the deviations of the individual JF's from their mean values must be the same for all jumps. Therefore a stochastic variation of the pinning fields while sweeping the field can be excluded as mechanism. So the experiments give strong support to the assumption, that internal stray fields are responsible for the uncertainty of the JF's. Thermal activation does not play a significant role for activating the jumps, but may affect the random formation of the sources of the stray field h(r).

REFERENCES

1. C. Heiden and H. Rogalla, J. Magn. Magn. Mat. 19 (1980) 240

2. L. Néel, J. Phys. Rad. 11 (1950) 49 and 12 (1951) 339

3. S. Stieler and C. Heiden, J. Magn. Magn. Mat. 41 (1984) 289

4. S. Stieler, Diploma thesis, University of Giessen (1982)

NOISE IN PHYSICAL SYSTEMS AND 1/f NOISE - 1985
A. D'Amico and P. Mazzetti (editors)

BARKHAUSEN NOISE: A LINK BETWEEN THE MACROSCOPIC DYNAMICS OF THE MAGNETIZATION PROCESS AND THE MACROSCOPIC PROPERTIES OF FERROMAGNETIC MATERIALS

Fausto FIORILLO and Giorgio BERTOTTI

Istituto Elettrotecnico Nazionale Galileo Ferraris and GNSM-CISM, 10125 Torino, Italy

The relationship between the statistical parameters of the Barkhausen noise and the macroscopic behavior of ferromagnetic materials is discussed. Results are presented concerning a SiFe single crystal, where a power law dependence of such parameters on the differential irreversible permeability is observed. A detailed study of the stochastic motion of the 180° domain walls puts in evidence the role of the demagnetizing field in the space-time evolution of the magnetization reversals. Such a field, simply related to the permeability, explicitly enters into a stochastic differential equation of motion of the domain wall, whose predictions are in good agreement with experiments.

1. INTRODUCTION

The macroscopic behavior of a ferromagnetic material during the magnetization process is characterized in general by a relatively simple and well reproducible phenomenology, described by a set of few parameters, like permeability, coercive field, remanence, area of the hysteresis cycle. On a microscopic scale, however, this regular behavior appears as an extremely complex evolution of the magnetic domains, which is intrinsically stochastic and unpredictable in character and must be managed through statistical methods. From this point of view the most natural method of investigation of the magnetization process is the study of the Barkhausen (B.) noise, the random e.m.f. induced by the motion of the magnetic domain walls. According to the fact that the reproducibility of the macroscopic features of the material is the result of a proper statistical average of the microscopic magnetization processes, we expect a direct connection between the parameters of the magnetization noise and the macroscopic magnetic behavior. It comes then to light the appealing property of the B. noise of being a unique source of information about the dynamic properties of the domain walls and, at the same time, about the mechanisms by which such properties ultimately give rise to a specific macroscopic magnetic behavior. The final goal which is pursued is the development of a microscopically founded comprehensive theory of magnetic hysteresis[1], which, as yet, has been achieved at a rudimentary level only. In this paper we shall discuss in particular how the long range action of the demagnetizing fields leads to a direct, quantitative connection between the B. noise and the irreversible permeability of the material. It will be shown how in SiFe single crystals a very simple domain structure permits a direct interpretation of the noise features, in terms of the parameters characterizing the dynamic behavior of the individual domain walls.

2. EXPERIMENTAL

In the conventional B. noise experiments, a ferromagnetic sample is cyclically excited by a low frequency triangular field waveform and the random e.m.f.s induced on a uniformly wound coil are suitably analyzed[2,3]. The shortcomings of this method (different induction rates along the magnetization curve, poor spatial resolution) can be overcome by investigating the local B. signals detected by narrow windings and/ /or couples of microelectrodes placed on the sample surface and by traversing the hysteresis cycle at a constant magnetization rate, through the use of suitably shaped field waveform[4,5]. One is then able to accurately follow the dynamics of the domain walls along the magnetization curve, as well as the transverse and longitudinal correlation effects associated with the development of the B. reversals. The case of a (110) [001] 3% SiFe single crystal, characterized by a system of few longitudinal 180° domain walls, is of great interest, for we can trace here the macroscopic features of the material back to the level of the individual wall dynamics. We can sense directly, for instance, the motion of the domain walls by placing the microelectrode couples transversally with respect to the crystal axis (i.e. magnetization direction)[6]. Possible correlation effects between different moving walls can be put in evidence by measuring the coherence function of the noise signals originating in two different portions of the sample cross-section. It is found that this function is generally << 1, indicating that each B. reversal is, to a very large extent, generated by the motion of a single wall at a time. Thus, different 180° walls act as independent noise sources and one can obtain reliable information on the dynamics of the individual walls by analysing the noise signal collected by localized windings, suitably placed along the sample length. Moreover, by exciting the crystal at an extremely low rate, the B. reversals occur neatly separated in time and the spectral density measurement can be supplemented by counting and signal time averaging. In the present experiments we have taken the sample through the hysteresis cycle at a rate as low as $2\cdot10^{-4}$ $T\cdot s^{-1}$ (i.e. magnetizing frequency $6.5\cdot10^{-5}$ Hz). The crystal, in the form of a strip (200x10.2x0.24 mm), had three to four 180° walls in the demagnetized state and was placed in a massive NiFe yoke, having a variable gap at one of the ends. By changing suitably the width of the gap, the relative differential permeability μ_{irr} could be changed from $3\cdot10^3$ to $1.5\cdot10^5$. The noise, measured upon an interval of 0.3 T centered around the magnetization value I=0, was correspondingly analyzed. The power spectra, related to the same cross-section where the field is applied, are shown in Fig. 1, as a function of μ_{irr}. One can notice how the low frequency behavior of the spectra is dramatically influenced be the same magnetostatic counterfields which, at a macroscopic level, determine the actual slope of the magnetization curve. This result is reminiscent of early experiments by Storm et al.[7] on SiFe strips, but, as discussed in the following, a more quantitative interpretation can now be pursued. As previously stressed, the noise consists, at the very low investigated induction rate, of a sequence of isolated B. avalanches, whose space-time propagation features can be determined from the signals simultaneously collected by a series of windings placed along the crystal axis. Figure 2-a shows the average B. avalanche shape, obtained by a triggering and signal enhancement procedure, in correspondence of three different values of μ_{irr}. We can see how the previously remarked strong abatement of the low frequency noise in correspondence of low values of μ_{irr} is associated with an inhibiting effect on the wall motion by the magnetostatic fields which grow up during

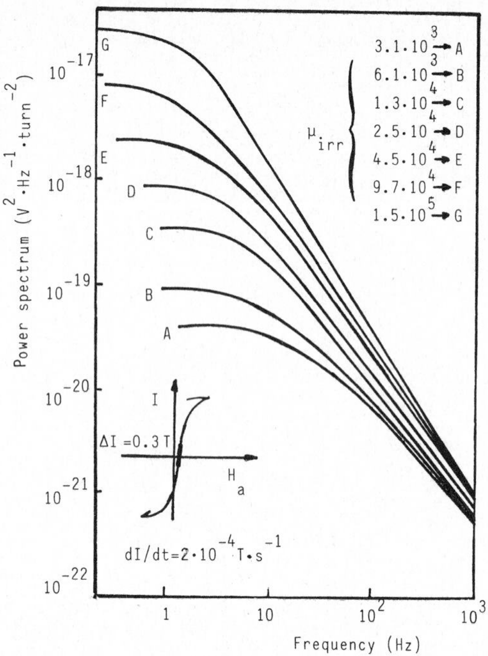

FIGURE 1

B. noise power spectra, detected about I=0 in a 3% SiFe single crystal, vs. differential permeability. Maxima of the spectra and cut-off frequencies are related to the domain wall dynamic parameters.

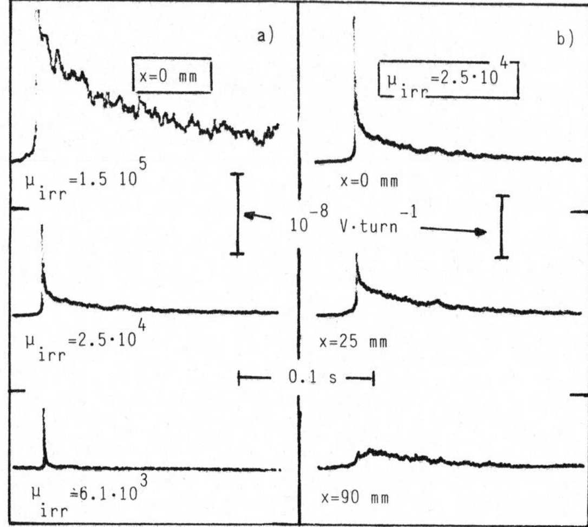

FIGURE 2

Average B. avalanches: a) as a function of differential irreversible permeability; b) as a function of distance from the nucleation site.

the development of a B. reversal. But the forward motion of the wall requires longitudinal propagation of the magnetization, a process strongly hindered by the demagnetizing fields, so that at increasing distances from the nucleation cross-section (where the external field is applied) more and more flux lines tend to leak outside the sample. A clear illustration of this is given by the behavior, reported in Fig. 2-b, of the average B. avalanche vs. distance from the nucleation site.

3. DISCUSSION

The quantitative connection between permeability and microscopic domain wall dynamics can be discussed on the basis of a general expression for the power spectrum of the B. noise, formulated according to the theory of clustering of elementary magnetization jumps[4,8]

$$S(f) = 2\dot{\phi}(1+\sigma^2)\cdot\frac{\hat{\rho}f^2}{(v_g/2\pi)^2+f^2}\cdot\frac{1}{2\alpha f\cdot 2\pi\hat{\rho}\langle\hat{\tau}\rangle}\cdot \left|\mathrm{arctg}\, 2\pi\hat{\rho}\langle\hat{\tau}\rangle(1+\alpha)f - \mathrm{arctg}\, 2\pi\hat{\rho}\langle\hat{\tau}\rangle(1-\alpha)f\right| \quad , \tag{1}$$

where $\hat{\rho}$, v_g and $\langle\hat{\tau}\rangle$ are the average amplitude (i.e. flux variation), number in unit time and duration of the B. avalanches. $\dot{\phi}$ is the macroscopic flux rate and α a parameter ($0\leq\alpha\leq 1$) which accounts for the distribution of the avalanche durations[4]. Finally, the term $(1+\sigma^2)$ is related to the distribution of the B. reversal amplitudes, with $\sigma^2=1$ in the case of Poisson distribution. By fitting the experimental spectra with Eq.(1), one obtains the behavior of the fundamental parameters $(1+\sigma^2)\cdot\hat{\rho}$ and $\langle\hat{\tau}\rangle$ vs. μ_{irr}. As already suggested by previous experiments performed at a much higher induction rate[9], $\langle\hat{\tau}\rangle$ is found to be proportional to μ_{irr}. Moreover, by evaluating the area of the average B. avalanches shown in Fig. 2-a, we get for $\hat{\rho}$ (expressed in Wb)

$$\hat{\rho} \simeq 8\cdot 10^{-8}\,(\mu_o\mu_{irr})^{3/2} \quad . \tag{2}$$

From comparison with the term $(1+\sigma^2)$ obtained from the noise spectra, we conclude that generally $\sigma^2>1$ (e.g. $\sigma^2 = 2.5$ when $\mu_{irr} = 1.5\cdot 10^5$). Obviously, $\hat{\rho}$ and $\langle\hat{\tau}\rangle$ have a direct meaning in terms of domain wall parameters, namely the mean free path $\lambda = \hat{\rho}/2I_s d$, I_s and d being the saturation induction and sample thickness respectively (200 Å $\leq \lambda \leq$ 5 μm), and the approximate wall velocity in a reversal $v_w = \lambda/\langle\hat{\tau}\rangle$. We can also evaluate the average amplitude $\langle\Delta H_i\rangle$ of the fluctuations of the internal stochastic hindering fields that a wall experiences during its motion,

$$\langle\Delta H_i\rangle = N_w\,\hat{\rho}/S\mu_{irr}\mu_o \quad , \tag{3}$$

where S is the cross-sectional area of the sample and N_w the number of walls (three in the present case). $\langle\Delta H_i\rangle$ varies as $\mu_{irr}^{1/2}$ and is found to range between $6\cdot 10^{-3}$ and $2.5\cdot 10^{-2}$ A/m, values which are from two to three orders of magnitude lower than the macroscopic coercive field. Actually, the wild fluctuations, of the order of the coercive field, of the local pinning fields are smoothed out by the magnetostatic interactions between different parts of the distorted wall, so that the wall dynamics upon time intervals $\geq\langle\hat{\tau}\rangle$ is essentially the one of a fairly rigid object, characterized by a smooth, global value of coercivity, to which slight variations, arising from the statistical fluctuations of the number and strength of the pinning centers, are superimposed[10]. Let us therefore consider a small portion of a 180° longitudinal wall which is being pulled forward in the nucleation cross-section by the pressure of a constant applied field H_a. Such a field will be balanced at each instant by a stochastic pinning field $\tilde{H}_{pin}$, a demagnetizing field H_{dem} and a local field H_{loc}, due to the free poles arising at the distorted wall surface

$$H_a = H_{loc}(t) + H_{dem}(z(t)) + \tilde{H}_{pin}(z(t)) \quad , \tag{4}$$

where z represents the distance covered by the wall since inception of the B. reversal at t=0. H_{dem} substantially plays the role of a feedback on the wall motion, the irreversible permeability μ_{irr} being the linking parameter,

$$H_{dem}(z(t)) = k\,z(t)/\mu_{irr} \quad , \tag{5}$$

with k a constant. H_{loc} is in turn supposed to depend explicitly on time according to a rate equation for the local free poles, also incorporating the dynamic effects of eddy currents

$$dH_{loc}(t)/dt = a\,dz(t)/dt - b\,H_{loc}(t) \quad . \tag{6}$$

The first and second term on the right hand side of Eq.(6) respectively describe the rates of local free pole generation and propagation along the wall. Combining Eq.s (5) and (6) we finally obtain the stochastic differential equation for wall motion in a B. reversal

$$(k/\mu_{irr}+a)\, dz(t)/dt+(bk/\mu_{irr})\, z(t)-bH_a + \\ + b\tilde{H}_{pin}(z(t)) + d\tilde{H}_{pin}(z(t))/dt=0 \quad . \tag{7}$$

By averaging Eq.(7) upon the statistical ensemble at a given time t, we get

$$(k/\mu_{irr}+a)\, d\langle z(t)\rangle/dt+(bk/\mu_{irr})\langle z(t)\rangle - \\ - b(H_a-\langle\tilde{H}_{pin}\rangle)=0 \quad , \tag{8}$$

having assumed $\langle d\tilde{H}_{pin}(z(t))/dt\rangle=0$.
Eq.(8) predicts an exponential time behavior for the average B. avalanche

$$\langle\dot{z}(t)\rangle = (\mu_{irr}/k\langle\hat{\tau}\rangle)\,(H_a-\langle\tilde{H}_{pin}\rangle)\cdot \\ \cdot\exp(-t/\langle\hat{\tau}\rangle), \tag{9}$$

with

$$\langle\hat{\tau}\rangle = (1+a\mu_{irr}/k)/b \tag{10}$$

and a mean free path

$$\lambda\equiv\int_0^\infty \dot{z}(t)dt=(Ha-\langle\tilde{H}_{pin}\rangle)\,\mu_{irr}/k = \\ =\langle\Delta H_i\rangle\,\mu_{irr}/k \tag{11}$$

The experimentally determined average B. avalanches of Fig. 2 consistently follow a near exponential behavior and are in good agreement with the prediction of Eq.(9). It must be noticed that, in view of the measured linear dependence of $\langle\hat{\tau}\rangle$ on μ_{irr}, Eq.(10) implies everywhere $a\gg k/\mu_{irr}$, so that $\langle\hat{\tau}\rangle$ will simply be proportional, for a given μ_{irr}, to a/b. Finally, let us note that Eq.(11) and the measured behavior of λ imply $\langle\Delta H_i\rangle\propto\mu_{irr}^{1/2}$. It has been verified by computer simulation that this dependence is not consistent with a wall potential function model[1] with Gaussian distributed inflexion slopes[11]. This suggests that correlation effects between successive slopes might be present[12], or that scaling concepts have to be introduced in the description of the random function $\tilde{H}_{pin}(z)$[13]. This is a point which is pursued at the moment and will be the subject of a future paper.

REFERENCES

1. J.L. Porteseil and R. Vergne, J. de Physique 40 (1979) 871.
2. M. Celasco, F. Fiorillo and P. Mazzetti, Il Nuovo Cimento 23B (1974) 376.
3. M. Komatsubara and J.L. Porteseil, J. de Physique, in press.
4. G. Bertotti, F. Fiorillo and M.P. Sassi, J. Magn.Magn.Mat. 23 (1981) 136.
5. G. Bertotti, F. Fiorillo and M.P. Sassi, Proc. 6th Intern. Conf. Noise in Physical Systems, Washington (1981) 324.
6. G. Bertotti and F. Fiorillo, J. Magn.Magn. Mat. 41 (1984) 303.
7. L. Storm, C. Heiden and W. Grosse Nobis, IEEE Trans.Magn. MAG-2 (1966) 434.
8. P. Mazzetti and G. Montalenti, Proc. Intern. Conf. Magnetism, Nottingham (1964) 701.
9. G. Bertotti, F. Fiorillo and A.M. Rietto, IEEE Trans.Magn. MAG-20 (1984) 1481.
10. J.A. Baldwin, Jr. and G.M. Pickles, J.Appl.Phys. 43 (1972) 1263.
11. L. Néel, Cahiers Phys. 12 (1942) 1.
12. J.L. Porteseil, Phys.Stat.Sol. 51 (1979) 107.
13. J.L. Porteseil and R. Vergne, C.R. Acad.Sc. Paris 288 (1979) 343.

NOISE IN PHYSICAL SYSTEMS AND 1/f NOISE - 1985
A. D'Amico and P. Mazzetti (editors)

SURFACE BARKHAUSEN NOISE MEASUREMENTS BY MEANS OF OPTICAL TECHNIQUES

Marcello CELASCO(*), Aldo MASOERO, Piero MAZZETTI(°), Giorgio MONTALENTI(§) and Aurelia STEPANESCU

Istituto Elettrotenico Nazionale Galileo Ferraris, Corso M.D'Azeglio 42, 10125 Torino, Italy and Gruppo Nazionale Struttura della Materia del C.N.R., U.R. 24, Torino, Italy.

For the first time,by means of an optical set-up making use of the Kerr-effect and of a cross-correlation technique it has been possible to measure the power spectrum of the velocity fluctuations of a single Bloch wall at the surface of a ferromagnetic lamination during the magnetization process.Experiments were done on a single crystal Fe-Si 3% lamination characterized by a structure of antiparallel domains in a frequency range from a few Hz to 1 kHz.On the same specimen the power spectrum of the conventional Barkhausen noise has been detected by means of a coil linked to the whole lamination.Experimental proof that different Bloch walls have uncorrelated speed fluctuations is reported and allows to compare the two spectra. From these results it is possible to evaluate the average amplitude of an elementary Barkhausen jump at the lamination surface (0.1 μm) and to show that a Bloch wall during the jump does not behave as a rigid wall but bows inwardly. These conclusions have also a strong practical interest because they show the importance of the surface on the magnetic loss.

1. INTRODUCTION

An important approach to the study of the magnetization process of a ferromagnet is the statistical one.

The process is in itself highly non linear and it is characterized by stochastic events which must be properly taken into account if the properties of the material are to be correctly understood.

An important property which is strongly related to the stochastic character of the magnetization process is, for instance, the magnetization power loss. The development of models which assume a regular motion of the Bloch walls, not only fails to account for the hysteresis loss, but also cannot explain many important features of the dynamic loss.[1]

Irregularities in the motion of Bloch walls are related to a random distribution of internal fields and are responsible also for the presence of the so called Barkhausen noise. This magnetization noise is generated by the irreversible jumps of the Bloch walls arranging themselves in successive equilibrium positions under the action of the external field, the internal stray fields and the pinning forces created by defects, stresses, impurities.
During the jumps in metallic ferromagnets there is also another type of field acting on the wall, called the eddy current field, and generated by the local transient currents induced by the jumping wall. It is just the power dissipated in the material by these currents that constitutes the magnetization power loss.
The existence of a relation between this last quantity and the Barkhausen noise is at this point quite obvious, even if it is in general not very straightforward. This is particularly true when the Barkhausen noise is studied in a polycrystralline material, because it involves many Bloch walls whose motion is strongly correlated.
Single wall dynamics studies can be performed only under special conditions and require rather sophisticated equipment.

One possibility is to use iron wiskers which are in many cases characterized by a single Bloch wall whose motion can be studied by means of a small winding[2] or by means of squid.[3]

On monocrystalline laminations one possibility concerns the use of point contacts to detect surface emf generated by the wall motion[4] or the use of an optical technique to monitor directly the Bloch wall motion on the lamination surface by means of the Kerr effect upon reflection of a polarized laser beam.

It is about the use of this last technique that the present paper is concerned. Measure-

---

(*) Dipartimento di Fisica, Università di Genova - (°) Dipartimento di Fisica, Politecnico di Torino
(§) Istituto di Fisica Generale, Università di Torino

ments of surface Barkhausen noise, defined as the wall speed fluctuation noise during magnetization, are reported here for the first time and compared with results concerning the bulk Barkhausen noise. Some interesting conclusion concerning the behaviour of a Bloch wall during a Barkhausen jump is also drawn.

2. EXPERIMENTAL SET-UP

The measurement of the power spectrum of the surface Barkhausen noise by optical means in a sufficiently large bandwidth requires the use of a cross-correlation technique to reduce photon noise. Typically, by using He-Ne lasers of moderate power (20 mW), owing to the strong reduction of the light intensity after crossing the second polarizer (which must be rotated very near to the extinction angle) the power density of the Bakhausen noise above 1000 Hz may be well 20 dB below the power density of the photon noise. Cross-correlation is also useful to reduce the effect of the intensity fluctuations of the laser beams if two different lasers are used as light sources.

The actual experimental set-up is shown in Figure 1. Magnetic domains can be seen in the ocular fields of the stereo microscope suitably modified with the insertion of high extinction

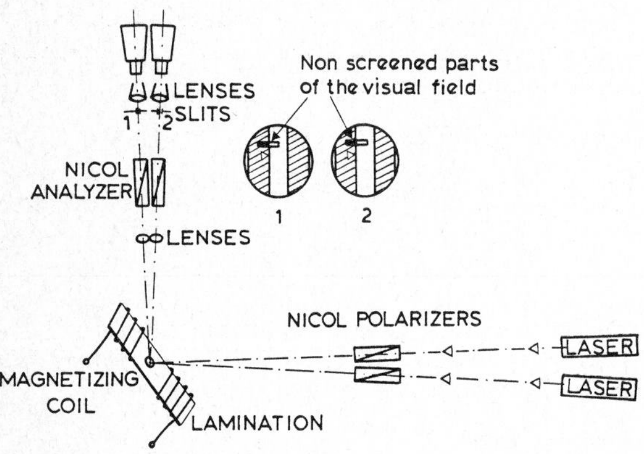

FIGURE 1
Optical arrangement for the measurement of the power spectrum of the surface Barkhausen noise through a cross-correlation technique. The inserts 1 and 2 show the optical fields of the oculars of the stereo microscope on which the same part of a single Bloch wall (or parts of two different Bloch walls, according to the type of experiment) is selected

Nicol prisms. Two removable caps containing pin photodiodes are fitted above the oculars, the photodiodes being in correspondence of the second focal point of the ocular lenses.

By means of adjustable slits it is possible to select on the visual field of the oculars the same portion of a given wall,so that the cross-spectrum of the signals coming from the two photodiodes during wall displacement coincides with the power spectrum depurated from the spurious noises not related with the displacement itself.

The same set-up can be used also to study correlations between the Barkhausen noise associated with different walls or different portions of the two adjustable slits.

Analysis is performed by means of a Nicolet Mod. 660B 2 channels Signal Analyzer.

The specimen is a lamination whose surface is optically polished before annealing. In single wall dynamics studies a monocrystalline lamination is used which is characterized by a regular structure of antiparallel domains. As shown in the following, it can be experimentally proved that in these conditions the motion of different walls are very nearly uncorrelated, so that each wall behaves as if the other walls did not exist. Comparison with the bulk Barkhausen noise, which is detected through a winding linked to the whole specimen, becomes thus possible.

Finally it should be noted that the whole set-up is assembled on a air suspended antivibration table, because vibrations give rise to correlated light intensity fluctuations.

3. EXPERIMENTAL RESULTS

Results concerning the surface Barkhausen noise generated by a single wall on a single cristal Fe-Si 3% lamination are reported here for the first time. As shown in Figure 2, the power spectrum of the noise obtained by taking the cross-spectrum of the same position of a single Bloch wall is about a decade higher than the cross-spectrum relative to two different contiguous walls. This result shows that, for what concerns the Barkhausen noise, in a well oriented monocrystal, Bloch walls behave very nearly independently. It also proves that the measured noise (which is well below photon and laser intensity fluctuation noise, particularly in the high frequency range) is really generated by the moving wall and it is not due to spur

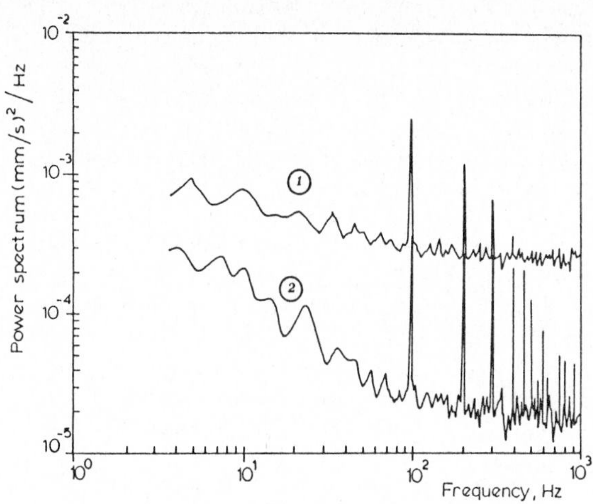

FIGURE 2

Curve 1: power spectrum of the surface Barkhausen noise generated by a single Bloch wall during the magnetization of a Fe-Si 3% single crystal lamination obtained by cross-correlating two optical signals monitoring the movement of the same Bloch wall.
Curve 2: Cross spectrum of two optical signals monitoring the movement of two different walls. Lines are due to correlated light intensity fluctuation from the two lasers.

ious effects related to the experimental set-up.

The nearly independent behaviour of the different Bloch walls allows to make also measurements of the power spectrum of the bulk Barkhausen noise generated by a single wall by means of a coil linked to the whole specimen.

The detected power spectrum is simply divided by the number of Bloch walls, after that the electronically integrated electrical signal is suitably normalized in terms of the optically observed total wall displacement.
As shown in Figure 3, the power spectra of the surface and of the bulk Barkhausen noise show a remarkably different behaviour towards frequency: the power spectrum of the bulk Bakhausen noise falls approximately with a $1/f^2$ law, while the power spectrum of the surface Barkhausen noise tends to remain constant.

As discussed in the following paragraph, these results cannot be accounted for by the screening effect of eddy currents within the lamination bulk, but are consistent with an interpretation which takes into account the

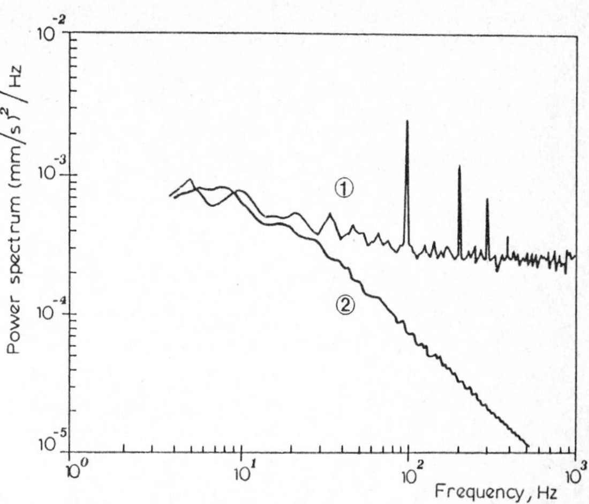

FIGURE 3

Comparison between surface (curve 1) and bulk (curve 2) Barkhausen noise. Bulk Barkhausen noise intensity is normalized to a single wall by assuming that the noises generated by different walls are uncorrelated.

flexibility of a Bloch wall during the Barkhausen jumps.

The picture arising from the development of this model is a new one in the study of the Barkhausen noise and it seems fundamental in order to understand the complicated behaviour of the magnetization dynamics of a ferromagnet at the microscopic level.

4. DISCUSSION OF THE RESULTS AND CONCLUSIONS

The interpretation of the power spectrum of the surface Barkhausen noise is consistent with the model developed in a previous paper[5] of a train of small and rapid jumps clustered in large Barkhausen discontinuities. However, the possibility to compare the surface Barkhausen noise with the bulk noise, together with other information concerning the dynamics of the Bloch walls obtained by means of the optical technique, allows to develop a much clearer model of the physical process underlying the noise generation during the motion of a Bloch wall.

The fact that the power spectrum of the surface Barkhausen noise tends to become flat in the high frequency range allows a calculation of the average amplitude of the elementary jump. Actually according to an analysis made in a previous paper[6], the power spectrum $\Phi(\omega)$ of a train of correlated events can be expressed as the superposition of the power spectrum of the same train of events in absence of any correlation, $\varphi(\omega)$, and of a lower frequency component dependent on the correlation type.

In the case of clustering into avalanches of average duration $\tau_0\rho_0$ as a good approximation[5]

$$\Phi(\omega)=\varphi(\omega)\left[1+2\,\frac{\rho_0(1-\nu\tau_0)}{\omega^2\,\tau_0^2\,\rho_0^2(1-\nu\tau_0)^2+1}\right] \qquad (1)$$

where ρ_0 is the average number of pulse in a cluster and ν the average number of pulses per unit time and τ_0 the average time interval between successive pulses within the avalanche. From eq.(1) it can be seen that the flat, high frequency part of the spectrum of the surface B.noise should be attributed to the component

$$\varphi(\omega) = \nu<|S(\omega)|^2> \approx 2\nu<A^2>, \quad (\omega\ll\omega_c) \qquad (2)$$

In eq.(2) $S(\omega)$ is the Fourier transform and A the area of the elementary pulse, coincident with the wall displacement during a single jump. Taking into account that $\nu<A>$ represents the total wall displacements per unit time, and assuming as an approximation $<A^2> \approx 2<A>^2$, from the data given in the preceding section:

$\nu<A> = 0.4$ mm/s

$2\nu<A^2> = 3\cdot10^{-4}$ $(mm/s)^2/Hz$

one gets:

$<A> \simeq 2\cdot10^{-4}$ mm

$\nu \simeq 2000$ pulses per second.

Another important result concerns the fact that during a Barkhausen jump the wall does not behave as a rigid object, even along a direction perpendicular to the lamination surface. Actually it can be readily shown that in a lamination of Fe-Si 3%, 0.24 mm thick, the much steeper slope of the power spectrum of the bulk Barkhausen noise, as compared with the power spectrum of the surface noise, cannot be accounted for in terms of eddy currents screening effects. The time constant of the decay of eddy currents due to an elementary jump in a lamination of thickness D can be calculated by the approximate expression

$$\tau \leqslant \frac{\sigma\mu_0\,(D/2)^2}{1.55} \qquad (3)$$

where σ is the material conductivity and $\mu_0 = 1.256\ 10^{-6}$. Eq.(3) has been obtained as an approximation of a more complicated but exact expression given in paper 7. In the present case, where $\sigma=2.13\cdot10^{6}\,\Omega^{-1}\,m^{-1}$ and $D=0.24\cdot10^{-3}$ m, a value of $\tau=2.5\cdot10^{-8}$ s is obtained, by far too small to give any effect on the measured spectra.

The difference in slope of the power spectra relative to the surface Barkhausen noise and the bulk Barkhausen noise can be explained on the basis of a model taking into account the flexibility of the Bloch wall and the fact that pinning is stronger at the lamination surface than in the bulk[8].

Exact calculations should be made starting from the differential equation

$$\gamma\frac{\partial^2 f(y,t)}{\partial y^2} - \frac{8J_s^2 D}{\varrho}\sum_n{}'\,\frac{1}{n(n\pi)^2}\sin\frac{n\pi}{2}\,\cdot$$

$$\cdot\cos\left(\frac{n\pi}{2}\cdot\frac{2y}{D}\right)\cdot\frac{\partial f(y,t)}{\partial t} - 2H(t)J_s = 0 \qquad (4)$$

where $f(y,t)$ is a function which describes the shape of the wall, γ is the specific wall energy, ϱ the material resistivity, $H(t)$ the applied field and the sum is over all the odd values of n.

A full analysis of the problem is beyond the scope of the present paper, but it can be anticipated by means of a very rough calculation that the time constant τ_c concerning the readjustement of the wall within the bulk after a surface jump is long enough to explain the roll-off shape of the bulk noise spectra[7]:

$$\tau_c \approx \frac{\sigma\cdot D<A>\,2.6\cdot I_s}{\pi H_c} \approx 0.5\cdot10^{-3}\,s \qquad (5)$$

This time constant corresponds to a cut-off frequency of about 300 Hz.

REFERENCES

1. G.Bertotti, J Mag.Mag.Mat. 41 (1984) 253
2. U.Hartmann and H.H.Mende Proc.Int. S.M.M.7 Conf., Blackpool (1985) in print
3. K.Zeibig, S.Stieler, C.Heiden Proc. 8th Int. Conf.on Noise in Physical Systems Roma, (1985) in print
4. J.E.L.Bishop, J.Mag.Mag.Mat. 42 (1984) 233
5. P.Mazzetti and G.Montalenti Proc. Int. Conf. on Magn.Nottingham (1964) 701
6. P.Mazzetti, Il Nuovo Cimento 31, (1964) 88
7. P.Allia and F.Vinai, J. Appl. Phys. 48, (1977), 4649
8. M.Celasco,A.Masoero,P.Mazzetti,A.Stepanescu Proc.Int. S.M.M.7 Conf., Blackpool (1985) in print

NOISE IN PHYSICAL SYSTEMS AND 1/f NOISE - 1985
A. D'Amico and P. Mazzetti (editors)

NOISE IN THE QUASI-ONEDIMENSIONAL CONDUCTING $(TCNQ)_2$ SALTS: $MEM(TCNQ)_2$, $MEtM(TCNQ)_2$ and $MBM(TCNQ)_2$

H.F.F. JOS, R.J.J. ZIJLSTRA
Fysisch Laboratorium, Rijksuniversiteit Utrecht, Princetonplein 5,
3584 CC Utrecht, The Netherlands

J.L. de BOER, Material Science Centre, Laboratorium voor Anorganische Chemie,
University of Groningen, The Netherlands

We report noise measurements on three types of $(TCNQ)_2$ salts that can be explained by assuming electrical transport through solitons with charge ± ½e. These solitons reflect at the contacts, while changing the sign of their charge, unless trapped solitons are present at the ends of the TCNQ stacks in which case they recombine. This leads to frequency independent current noise levels equal to eI at small current and proportional to $I^{2-\beta}$ beyond a critical current I, where $0 \leq \beta \leq 1$ and β is larger then the conductivity is smaller.

INTRODUCTION

The studied compounds were $MEM(TCNQ)_2$, $MEtM(TCNQ)_2$ and $MBM(TCNQ)_2$. The noise measurements were carried out in the linear, ohmic regime at fields below 100V/m. However accurate measurements of harmonic distortion show that in this regime non-linearities are present. The ac.-impedances of all compounds were frequency independent up to 5 MHz in the pseudo-ohmic regime.

$MEM(TCNQ)_2$ IN THE DIMERIZED PHASE.

The current noise of $MEM(TCNQ)_2$ in the dimerized phase can be decomposed into a 1/f contribution and a Lorentzian $(1/(1+\omega^2\tau^2))$ contribution. The Lorentzian has a low frequency level which equals eI, where e is the electron charge. We have interpreted the Lorentzian as shot noise in which charges of $\pm$ ½e are involved. The relaxation time of the Lorentzian is the transit time of the charge carriers, which we believe are solitons. Although the idea of solitons in quasi-onedimensional organic conductors is not new [1,2], the connection with electrical transport and noise is made here for the first time. Details about our measurements and their interpretation will be published elsewhere [3]. There we have also argued that current-voltage characteristics essentially measure the relation between the detrapping rate of solitons from the contact boundary (S^+ solitons from the anode and S^- solitons from the cathode) and the electric field E. In fig. 1 we give a schematic representation of such solitons. In the pseudo-ohmic regime the detrapping rate $g=\alpha E$ where α determines the conductivity. This also means that the conductivity is independent of the velocity of the solitons.

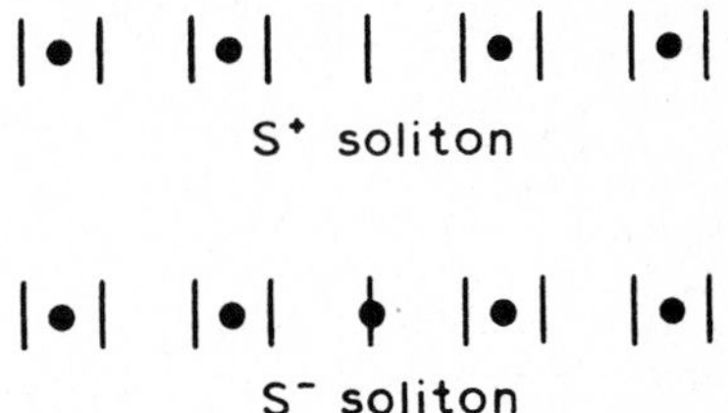

Fig 1: Schematic representation of S^+ and S^- solitons. The TCNQ molecules are represented by the strokes and the electrons by the dots.

MEM $(TCNQ)_2$ IN THE UNIFORM PHASE.

In fig. 2 we have plotted the low frequency level and the relaxation time of the Lorentzian in uniform MEM$(TCNQ)_2$ as a function of the current I. At currents above 10^{-5}A the low frequency noise level is above eI and proportional to $I^{1.5}$ whereas the relaxation time is propertional to $I^{-0.5}$. An explanation of this phenomenon is found in the reflection of solitons at the contact, changing the sign of their fractional charge. An S^- soliton, when reflected at the anode, changes into an S^+ soliton, while the electron goes into the contact. The S^+ soliton then moves in the opposite direction. If the probability that a soliton will be absorbed at the contact is k, the noise is [3]:

$$S_I(\omega) \simeq \frac{2eI\tau/\tau_t}{1+\omega^2\tau^2} \qquad (1)$$

where $\tau = \tau_t(2-k)/2k$
and $\tau_t = L/v$, the transit time

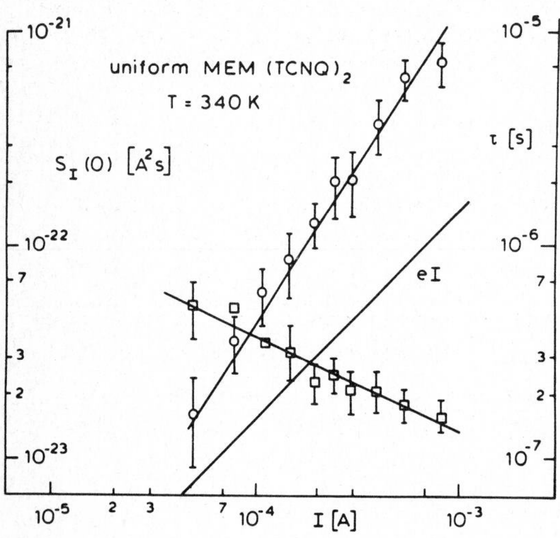

Fig 2: The low frequency noise level (circles) and relaxation time (squares) versus the current for uniform MEM$(TCNQ)_2$. Also the eI line is given.

Note that for k = 1, $S_I(o)$ = eI.
Assuming that τ_t is proportional to E^{-1},and hence to I^{-1}, the results of fig. 3 can be explained by assuming $(2-k)/2k \sim I^{+0.5}$.
A soliton in an uniform TCNQ stack is only present in the electron expectation value and not in the lattice as depicted in fig. 1. Nevertheless the same arguments for both types of solitons hold.

THEORY

In our picture [3] we cannot distinguish between S^+ and S^- solitons if they have the same drift velocities but with opposite sign in an electric field. We assume that the total number of free S^+ solitons equals the total number of free S^- solitons N. In equilibrium the generation rate g of free solitons at the contact equals the absorption rate Nvk/L of incoming solitons at the contact: Nvk/L = g.
Using I = 2N. ½e.v/L we find I = eg/k.
This means that g/k is proportional to E in the pseudo-ohmic regime. We found from fig. 3 that $(2-k)/2k \sim I^{0.5}$. So there must be a mechanism that couples the generation of solitons and the absorption. We propose the following model: from the cathode an electron can easily go into the crystal end where it forms two trapped S^- solitons. In an analoguous way two trapped S^+ solitons can be formed at a stack end near the anode. There cannot be more than two trapped solitons at a stack end due to the Coulomb repulsion. The rate of creation of trapped solitons at a contact is $2g_t$ (M_t-N_t), where M_t is the total number of TCNQ stacks and N_t is the number of stacks ends at that contact that are occupied by trapped solitons. The generation of free solitons from the contact is: $\alpha N_t E$. When an S^+ soliton moving along a TCNQ stack in the direction of the electric field comes to the cathode and one or two trapped S^- solitons are present at the end of the stack, the S^+ and an

S^- soliton recombine. If there are no trapped solitons the S^+ soliton reflects as described before. The chance for an incoming soliton to recombine is $k = N_t/M_t$. We have seen before that $I = eg/k = \alpha E M_t e$. Note that now I is dependent of N_t and g_t!

In equilibrium the rate of creation of trapped solitons at a contact equals the sum of the detrapping rate and the recombination rate:

$2g_t(M_t - N_t) = \alpha N_t E + Nvk/L$.

Also it holds that $\alpha N_t E = Nvk/L$, hence:

$$k = \frac{N_t}{M_t} = \frac{g_t}{g_t + \alpha E} \qquad (2)$$

Equation (1) reduces to:

$$S_I(\omega) = \frac{g_t + 2\alpha E}{g_t} \frac{eI}{1+\omega^2\tau^2} \qquad (3)$$

$$\text{where } \tau = \frac{g_t + 2\alpha E}{2g_t} \tau_t$$

Equation (3) shows that reflection of solitons becomes important above a critical current density $J_{crit} = \frac{1}{2} g_t e m_t$ where m_t is the surface density of the TCNQ stacks at the crystal end. Around the critical current we expect a transition region in which the relation between $S_I(o)$ and I can be described by $I^{2-\beta}$ where $0 \leq \beta \leq 1$. This transition region can be extended when g_t has a slight field dependence, for instance like: $g_t = g_t' + g_t''E$. Substituting this in equation (3) yields:

$$S_I(\omega) = \frac{g_t' + g_t'' I/A\sigma + 2I/Am_t e}{g_t' + g_t'' I/A\sigma} \cdot \frac{eI}{1+\omega^2\tau^2} \qquad (4)$$

where A is the cross section of the crystal. We expect for very poor conducting crystals $S_I(o) = eI$. For better conducting crystals β will be smaller when the conductivity is better.

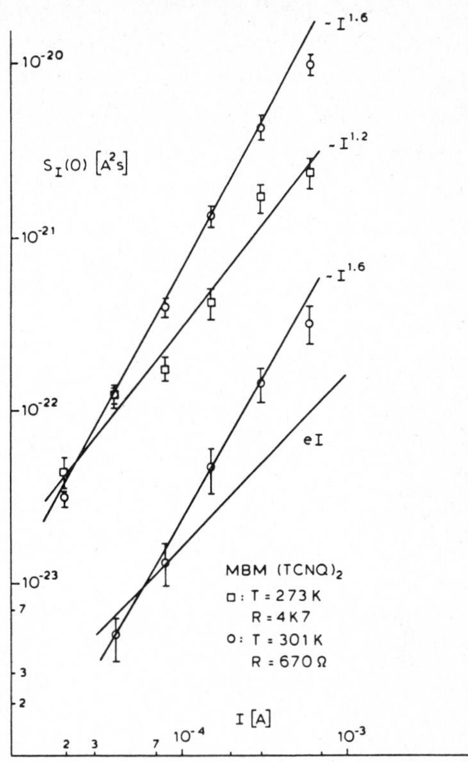

Fig 3: The low frequency levels of $MBM(TCNQ)_2$ at 301K (circles) and the higher low frequency level at 273K (squares) versus the current.

$\underline{MBM(TCNQ)_2}$

$MBM(TCNQ)_2$ has two Lorentzian noise components. In figure 3 we have plotted the low frequency levels at two different temperatures as functions of the current. At 273 K only the higher Lorentzian could be measured. The lower one was submerged in the background noise. We find a higher slope in the current dependence of the low frequency level when the conductivity is larger. In fig. 4 we have plotted the relative low frequency levels $S_I(o)/I$ and relaxation times τ of the two Lorentzians versus 1000/T at $I = 100\mu A$. The level of the lower Lorentzian is independent of T while its relaxation time becomes larger at higher temperatures. The level of the higher Lorentzian as well as its relaxation time show an activation energy of 0.7 eV. $MBM(TCNQ)_2$ has two inequivalent types of TCNQ stacks with a different crystal energy [4]. In our model this means that in the low current limit stack type 1, which has the lower potential energy, has trapped solitons at every stack end and stack type 2 has a probability of having a trapped soliton at its end of $\exp\{-\varepsilon(T^{-1}-T_o^{-1})/k_B\}$ for $T < T_o$, where ε=0.7 eV is the difference in potential energy between the two types of stacks. Preliminary Madelung calculations have established potential

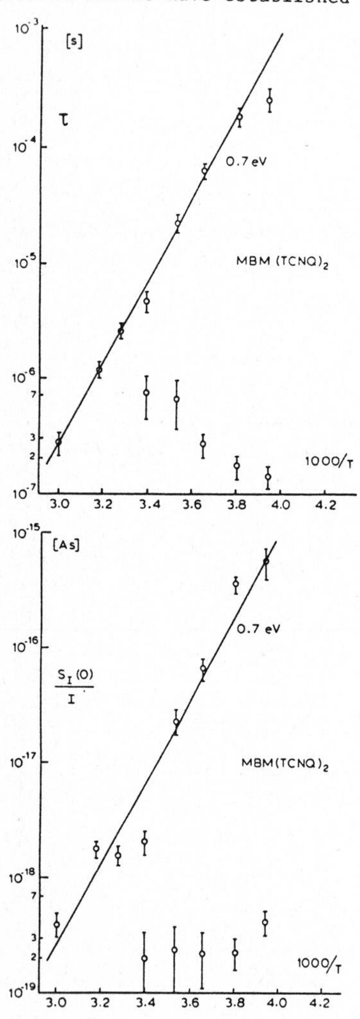

Fig 4: The relative low frequency levels $S_I(0)/I$ and relaxation times τ of $MBM(TCNQ)_2$ versus 1000/T.

energy differences of 0.2 eV[4]. In fact $k_2/k_1 = N_{t2}/N_{t1} = \exp\{-\varepsilon(T^{-1}-T_o^{-1})/k_B\}$ is always true at every current. This explains why both low frequency levels show the same dependence on the current and also explains the activation energy in fig. 4.

Note that the temperature dependences of the relaxation times of both Lorentzians need not be the same because of possible differences in the soliton velocities in both stack types.

In addition note that both stack types contribute to the current in an unknown ratio, which is the cause of finding a low frequency level in stack type 1 at 40μA and 301 K below eI, where I is the total current and a relatively high critical current compared to uniform $MEM(TCNQ)_2$.

$MEtM(TCNQ)_2$

In $MEtM(TCNQ)_2$ we found similar results as for $MBM(TCNQ)_2$. The critical current is 20μA. Here also two Lorentzians are present. The nature of the higher Lorentzian is still being studied.

This work was performed as a part of the research programme of the Stichting voor Fundamenteel Onderzoek der Materie (FOM) and with financial support from the Nederlandse Organisatie voor Zuiver Wetenschappelijk Onderzoek (ZWO).

LITERATURE

[1] V.G. Makhankov, V.K. Fedyanin, Physics Reports 104 (1984) 1.

[2] A.J. Epstein, J.W. Kaufer, H. Rommelmann, J.S. Miller, R. Comès and J.P. Pouget, Journal de Physique C3 (1983) 1479.

[3] H.F.F.Jos, R.J.J. Zijlstra Shot noise due to Soliton transport in Methyl-Ethyl Morpholinium $(TCNQ)_2$, to be published.

[4] R.J.J. Visser, Ph.D. Thesis, Groningen (1984).

NOISE IN PHYSICAL SYSTEMS AND 1/f NOISE - 1985
A. D'Amico and P. Mazzetti (editors)

STOCHASTIC STRESS-STRAIN RELAXATION PROCESSES IN THE VORTEX LATTICE OF A TYPE-II SUPERCONDUCTOR

R. DITTRICH and C. HEIDEN

Institut für Angewandte Physik der Justus-Liebig-Universität Gießen, D-6300 Gießen, FRG

For transport currents I below their critical value I_c nonstationary local rearrangements of the vortex lattice in a thin type-II superconducting sample have been observed. They can be regarded as stress-strain relaxation processes similar to those detected by current noise measurements in metal films.

1. INTRODUCTION

Passing a transport current I, which exceeds the critical value I_c, through a type-II superconductor in the mixed state leads to a flux flow voltage V, which in general contains a noisy component δV. This flux-flow noise was subject to many experimental and theoretical approaches in the last two decades[1]. Using movable contact pairs[2] as voltage probes, strong stationary flux flow noise can be found for $|I|>|I_c|$ for certain settings of the voltage probes. In this case, also instationary noise was observed when sweeping the transport current with amplitudes smaller than I_c. Stress-strain relaxation processes in the vortex lattice are regarded as underlying mechanism for this effect.

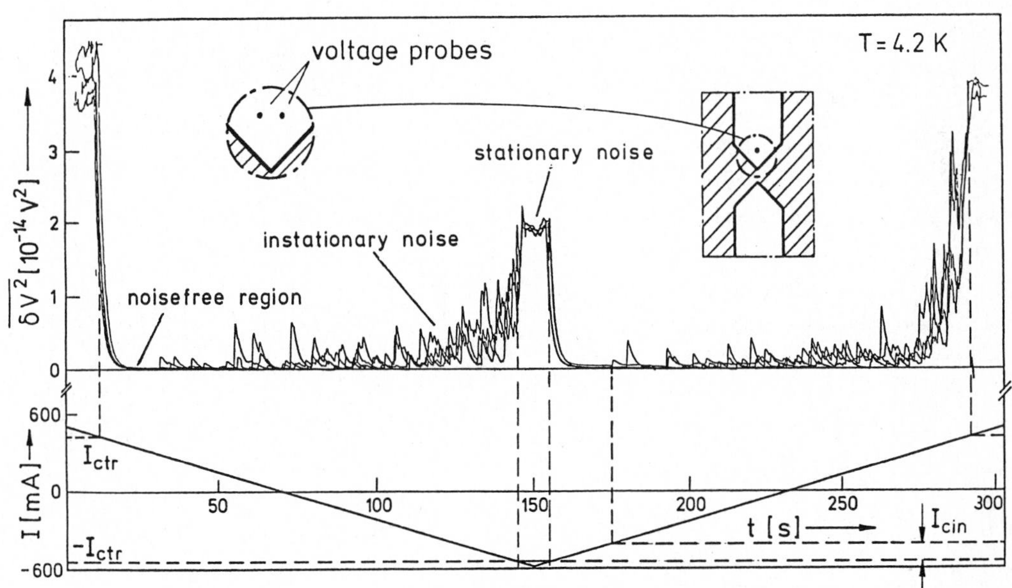

FIGURE 1

Mean square noise voltage $\overline{\delta V^2}$ and transport current I as function of time. The superposition of four current cycles is shown. B = 0.18T.

2. EXPERIMENTAL

The standard soccer goal type of the measuring circuit was used[1]. The samples were made out of single crystalline Nb-foils with low intrinsic pinning (critical current densities of the order of $10^2 A/cm^2$ at B=1kG). In order to control the flow pattern of flux motion, artificial pinning structures were deposited on the foils using cold sputtered NbN[3]. In contrast to earlier experiments using Nb_3Sn[4], no annealing with the danger of increasing the volume pinning force of the Nb-foil is necessary.

3. RESULTS

Fig.1 shows the mean square noise voltage measured, while sweeping the transport current, from a Nb-foil whose pinning structure is sketched in the inset. One can separate stationary from instationary noise and a noisefree region. The sharp transition from the stationary to the noisefree region defines a critical current I_{ctr}. Another critical current (I_{cin}) is defined by the difference between I_{ctr} and the current corresponding to the onset of the instationary noise. In fig.2 the same is plotted for another pinning structure and different field values. By averaging over 4 current cycles, and multiplying the critical currents with the corresponding field values, data of fig.3 is obtained. The values are equal to the critical volume forces per unit length. Additionally in fig.3 the pinning force (F_{cdc}), derived from the onset of dc voltage, is shown. F_{ctr} and F_{cdc} are found to be approximately equal. For low field values F_{cin} is also approximately equal to F_{ctr}.

With $|I| > |I_c|$ stationary noise is observed. The autocorrelation functions are of the exp. or slightly oszillating type (cf. fig.4). The time constant is very sensitive to the variation of the transport current. The dependence can be described to good approximation by:

$$\tau_c \sim \frac{1}{(I - I_c)^k} \quad , \; k \approx 2.$$

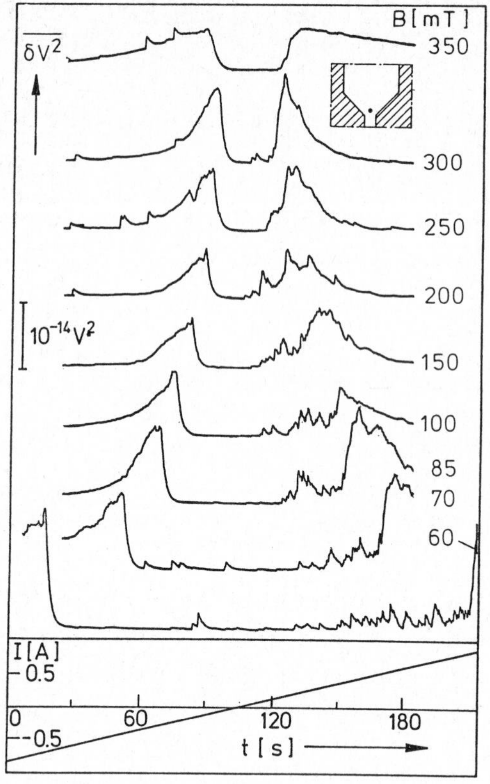

FIGURE 2
Same data as in fig.1 at different fields for another sample.

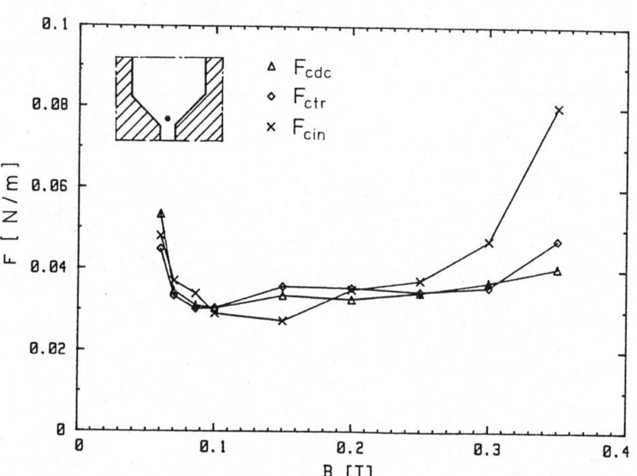

FIGURE 3
Critical forces derived from I_{ctr}, I_{cin} and I_{cdc} as functions of the magnetic field.

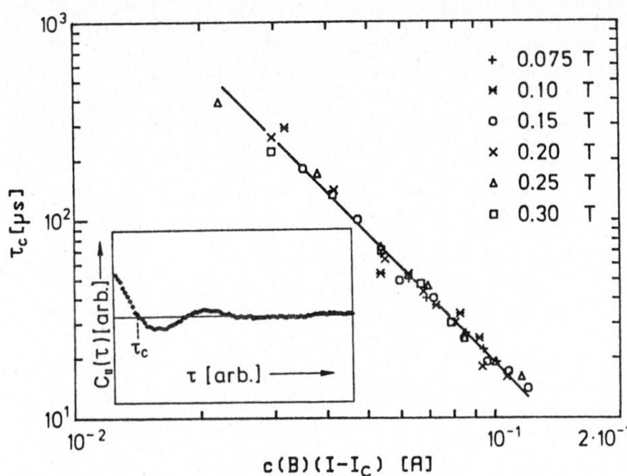

FIGURE 4
Time constant τ_c as function of $(I-I_c)$. c(B) has the following values: c(0,075T)=1, c(0.1T)=1.25, c(0.15T)=1.5), c(0.2T)=1.58, c(0.25T)=1.94, c(0.3T)=1.56.

It should be mentioned, that these measurements were performed for transport currents very close to the cirtical current.

The noise (stationary and instationary) was found to be very sensitive to the adjustment of the contacts. Fig.5 shows the stability of the mean square voltage of the stationary noise after putting down the voltage probes onto the sample. After mechanical vibration (kick with a screwdriver against the cryostat) the noise often assumes new values (cf. right hand side of fig.5).

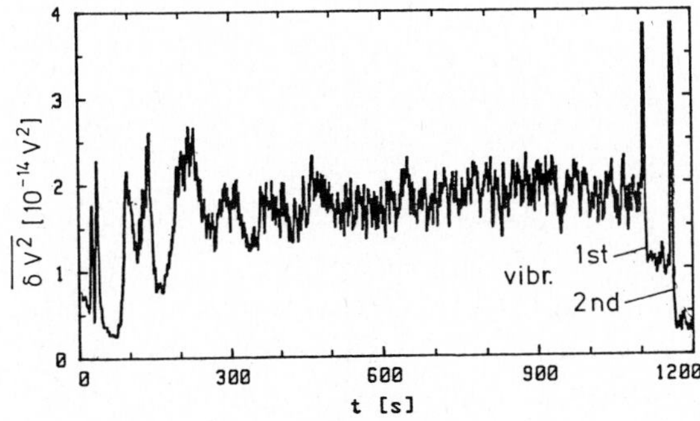

FIGURE 5
Stability of the stationary noise after putting down the voltage probes and sensitivity to mechanical vibration.

This can lead to such a low signal, that it disappears in the noise of the amplifier system. The shape of the autocorrelation function of the flux flow noise however is not affected by this. According to theory of signal generation[1] this observation suggests, that flux motion in the immediate vicinity of the contacts, which after pressing down may have a very small diameter, is detected.

4. DISCUSSION

The observed behavior can be understood qualitatively with the following simple model: For $I>I_c$ there is stationary flux flow whose fluctuating component produces the observed stationary noise. When the transport current is reduced and reaches I_c, flux motion stops. The deformation of the fluxon array associated with the critical pinning force F_{ctr} is frozen in. Assuming, that no flux motion occurs when reducing the current further then will lead to the generation of internal currents (I_{int}) such that the deformation in the flux line lattice is maintained. Locally the current densities may obey $j_{int}= j_c- j$. For this case, I=0 yields $j_{int}= j_c$. For $I<0$, j_{int} at least locally will exceed j_c, which corresponds to F_{cin}. Lattice relaxations in turn are produced which reduce j_{int}. This process, which gives rise to the instationary noise, is repeated until I reaches $-I_c$ and stationary flux flow again occurs. At first sight $F_{ctr} \approx F_{cin}$ seems to be in contrast to the data of fig.1 (obviously $I_{ctr} \neq I_{cin}$). But in this sample an artificial barrier was deposited on the foil. Therefore the critical current is increased. But F_{cin} reflects the local pinning properties of the foil, which means that I_{cin} (0.175A) must be compared with I_c (0.19A) of the undestorted foil.

The above described instationary processes can be regarded also as a sequence of stress-strain relaxation processes similar to those observed via current noise measurements in metal films[5].

REFERENCES

1. J.R.Clem, Phys. Rep. 75(1) (1981) 1.

2. H.Dirks and C.Heiden, in: Noise in Physical Systems, ed. D.Wolf (Springer-Verlag, Berlin, 1978) pp. 229-233.

3. D.D.Bacon, A.T.English, S.Nakahara, F.G. Peters, H.Schreiber, W.R.Sinclair, R.B.van Dover, J. Appl. Phys. 54 (1983) 6509.

4. K.Beckstette, H.Dirks, C.Heiden, Phys. Letters 81A (1981) 351.

5. G.Bertotti, M.Celasco, F.Fiorillo, and P. Mazzetti, J. Appl. Phys. 50 (1979) 6948.

NOISE IN PHYSICAL SYSTEMS AND 1/f NOISE - 1985
A. D'Amico and P. Mazzetti (editors)

AGEING EFFECTS AND EXCESS NOISE IN ION BEAM IRRADIATED POLYMER FILMS

K F KNOTT[+], I BELLO[*] and L HAWORTH[+]

[+]University of Salford, Salford M5 4WT, England.

[*]Department of Microelectronics, Slovak Technical University, 81219 Bratislava, Czechoslovakia.

Thin films of HPR-204 polymer have been bombarded with 1.5 MeV, Ar ions to doses in the range 10^{14} to 10^{17} cm^{-2}. Resistivity and excess noise was found to change dramatically with dose. This paper concentrates on the excess noise results. Spot measurements at 10 Hz and spectral measurements up to 100 KHz were performed as a function of dc bias on four-contacts, rectangular samples. The excess noise of samples in the intermediate dose range 1×10^{15} to 2×10^{16} cm^{-2} was initially found to obey neither an I^2 law nor a 1/f law. However, for some samples, the excess noise was observed to relax over a period of time to an I^2 law, at the same time relaxing to a 1/f spectrum. Calculations of Hooge's parameter, α, based on published values of carrier concentration indicates a figure of about 5×10^2 which is five orders of magnitude higher than that observed in metals and semiconductors!

INTRODUCTION

In a recent paper[1] some preliminary results were published concerning excess noise and resistivity of conducting paths formed in thin polymer films by high energy Ar ion bombardment. The action of the ion beam eventually carbonises the initially insulating film typically reducing its resistivity by 12 orders of magnitude at doses of 10^{16} cm^{-2}. Production of conducting patterns in polymer films in this way may prove attractive[2] in integrated circuit manufacture. In this present paper more results of this work will be discussed. The films investigated were of HPR-204 polymer of thickness approximately 1 μm.

EXCESS NOISE RESULTS

The conductivity modulation theory[3] of excess noise predicts that its spectral intensity obeys the relationship

$$f.\frac{S_x(f)}{x^2} = C_{1/f} \qquad (1)$$

where: f = frequency

$S_x(f)$ = spectral intensity of fluctuations in x, where x can be resistance, voltage or current

$C_{1/f}$ = 1/f noise constant for a particular sample.

If this holds it is only necessary to perform measurements at one frequency and one bias to calculate the value of $C_{1/f}$. However in this case some of the samples do not satisfy equation (1) so perhaps the best way to present the results is to plot excess noise in dB relative to thermal as a function of current. Typical results are shown in Figure 1. At low and high doses the excess noise is very nearly proportional to I^2 but at intermediate doses the law index is less than 1.4. The spectra of such samples also showed departures from a 1/f law. Subsequent measurements revealed that some of the intermediate specimens were exhibiting ageing effects. An example of this result is illustrated by the dashed and dotted lines in Figures 1 and 2.

It is seen from Figure 2 that sample 2 initially shows evidence of a conduction process which involves at least two distinct time constants or lifetimes but these relax

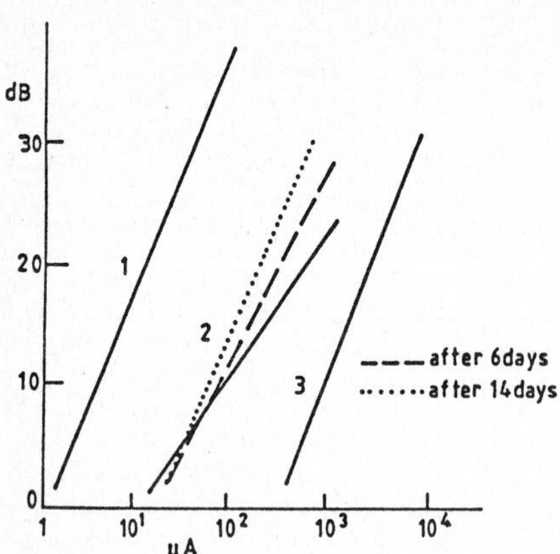

Excess noise at 10Hz as a function of current

1 — 1x 10^{15} cm^{-2}
2 — 2x 10^{16} cm^{-2}
3 — 3x 10^{17} cm^{-2}

FIGURE 1

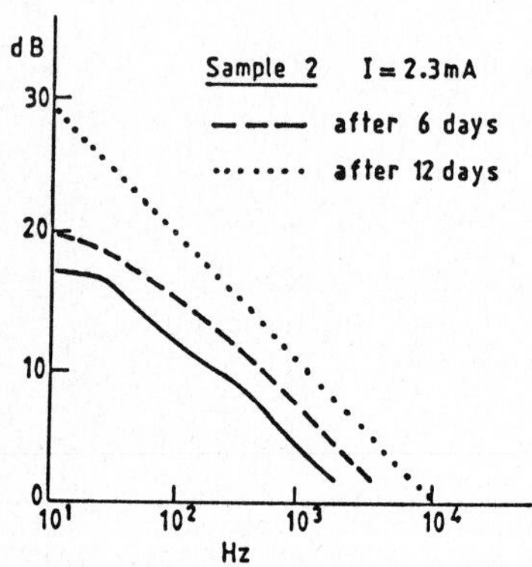

Noise spectra of sample 2

FIGURE 2

out to the more usual straight line spectrum. At the same time the law index increases to 2.

ESTIMATION OF α

Experimental evidence[4] leads to the conclusion that conduction in irradiated polymers involves a very high concentration of very low mobility charge carriers. This makes it difficult to obtain reliable values for carrier concentration using say Hall effect. For this reason the present authors have been unable to measure Hall coefficients with their existing equipment. However, if one takes the carrier concentration of 5×10^{22} cm^{-3} given in the above reference and calculates Hooge's universal parameter, α, defined by

$$\alpha = C_{1/f} \cdot N \qquad (2)$$

where N = total number of free charge carriers in the sample.

One obtains a value of 5×10^{2}. This is in contrast with Hooge's value of 2×10^{-3} and other workers[2] values of less than this for materials where mobility is decreased significantly by scattering processes.

DISCUSSION

It is apparent, once again, that electrical noise is a useful tool in the study of conduction in materials. In the material being studied here i.e. an irradiated polymer it is evident that large differences exist between it and say a homogeneous semiconductor as far as electrical conduction is concerned. Judging by the estimated value of α the conduction process is extremely noisy compared with that in many other materials. Although no detailed amplitude distribution measurements were taken no burst noise was detected during observations of the noise waveform on a storage oscilloscope. Presumably therefore the non 1/f spectra of the intermediate dose samples were indicative of conduction by a hopping process.[5]

To obtain more information about the

relaxation mechanism observed in some samples a much more controlled series of experiments would be required.

REFERENCES

1. K F Knott, I Bello and L Haworth. Excess noise in ion beam irradiated polymer films. Radiation Effects 89 (1985) 157-163.

2. T Venkateson, M Feldmann, B J Wilkens and W E Willenbrock. Ion beam irradiated via connect through an insulating polymer layer. J Appl Phys 55 (1984) 1212.

3. L K J Vamdamme. 1/f noise theories. IEE Colloquium Digest No 1984/26 (Mar 1984).

4. B Wasserman, G Braunstein, M S Dresselhaus and G E Wnek. PImplantation-induced conductivity of polymers. Proc Mat Res Soc Annual Meeting, Boston, MA, USA (Nov 1983).

5. R A Lomas, M J Hampshire, R D Tomlinson and K F Knott. Hall effects and noise measurements in epitaxial, polycrystalline and amorphous Ge. Phys Stat Sol 16 (1973) 385-394.

QUANTUM NOISE - 1

NOISE IN PHYSICAL SYSTEMS AND 1/f NOISE - 1985
A. D'Amico and P. Mazzetti (editors)

PROPERTIES OF LOW TEMPERATURE QUANTUM NOISE

Roland JUNG, Gert-Ludwig INGOLD, Peter SCHRAMM, and Hermann GRABERT*

Institut für Theoretische Physik Universität Stuttgart, D-7000 Stuttgart 80, Germany

The nature of the noise described by the c-number quantum Langevin equation is investigated. Subtleties in the stochastic calculus are shown to originate from the differences between quantum noise and classical noise. For a harmonically bound particle the zero temperature correlation functions are shown to display long time tails. At finite temperatures an algebraic decay at intermediate times is followed by an exponential decay with time constant $\tau=\hbar/2\pi k_B T$. Some general conclusions for nonlinear systems are drawn.

Recent experiments on Josephson devices have given rise to renewed interest in the influence of dissipation on quantum systems at low temperatures[1] where standard weak coupling theories fail. In this context it has been suggested that quantum Brownian motion can be described by a c-number quantum Langevin equation (QLE)[2]. For the coordinate q of a particle of mass M moving in a harmonic potential $V(q)=\frac{1}{2}M\omega_0^2q^2$ while coupled to a heat bath environment consisting itself of harmonic oscillators the QLE is formally exact and takes the form

$$\ddot{q} + \gamma\dot{q} + \omega_0^2 q = \xi(t)/M. \qquad (1)$$

Here $-M\gamma\dot{q}$ is a frictional force caused by Ohmic environmental dissipation. $\xi(t)$ is a Gaussian random noise with zero mean whose coloured spectrum

$$\langle\xi(\omega)\xi(-\omega)\rangle = 2\gamma\hbar M\omega\left[1-\exp(-\beta\hbar\omega)\right]^{-1} \qquad (2)$$

reflects the quantum nature of the process. In the classical limit $1/\beta=k_BT\gg\hbar\omega$ the correlation $\langle\xi(t)\xi\rangle$ reduces to real valued white noise. To render (1) a proper c-number representation of quantum mechanics, operators in equilibrium averages $\langle\ \rangle$ are time ordered such that time arguments increase from right to left. In case of need operators at equal time are symmetrized. In previous work[2] the power spectrum of the random force has been taken to be the symmetrical part of (2). In this case only the real part of the correlation $\langle q(t)q\rangle$ is given by the QLE.

It is well known that in the theory of classical stochastic processes the noise of the Langevin equation is always assumed to be uncorrelated with the macroscopic variables at earlier times. A striking feature of quantum mechanics is the loss of this property in the c-numer QLE. In fact we obtain a correlation[3]

$$\langle\xi(t)q\rangle = -\frac{\hbar}{\pi}\frac{\gamma}{\tau}\sum_{n=1}^{\infty}\frac{(n/\tau)e^{-nt/\tau}}{(n/\tau)^2+(n/\tau)\gamma+\omega_0^2} \qquad (3)$$

which involves a thermal time scale $\tau=\hbar/2\pi k_BT$. This shows that even in the c-number representation quantum noise is substantially different from classical noise. It should be noted that the nontrivial low temperature behavior of dissipative systems, which is analyzed in the following, is intimately connected with this dependence of the noise on the history of the system. In the classical limit the correlation (3) properly vanishes.

By Fourier transform it is easy to show that the QLE (1) with (2) is in accordance with the fluctuation dissipation theorem[4]

$$J(\omega) = 2\hbar\left[1-\exp(-\beta\hbar\omega)\right]^{-1}\chi''(\omega) \qquad (4)$$

which relates the spectral density

$$J(\omega) = \int_{-\infty}^{\infty}dt\ e^{i\omega t}\langle q(t)q\rangle \qquad (5)$$

of displacement fluctuations in thermal equilibrium to the imaginary part, $\chi''(\omega)$, of the dynamical susceptibility

$$\chi(\omega) = M^{-1}[\omega_0^2-\omega^2-i\omega\gamma(\omega)]^{-1} \quad (6)$$
$$= \chi'(\omega) + i\chi''(\omega).$$

For later use we introduced a frequency dependent damping coefficient $\gamma(\omega)$. Using (4) the real time displacement autocorrelation function $J(t)$ of the damped harmonic oscillator with Ohmic dissipation can be evaluated exactly[5]. Because of $\dot{q}=p/M$, where p is the momentum, all other pair correlataions are obtained from $J(t)$ as time derivatives, while higher order correlations can be factorized into pair correlations due to the Gaussian property of the process.

At zero temperature we obtain from (4)

$$J(t) = \frac{\hbar}{\pi}\int_0^\infty d\omega\, e^{-i\omega t}\, \chi''(\omega). \quad (7)$$

Using (6), the asymptotic expansion of $J(t)$ for large t is found as

$$J(t)=-(\hbar/\pi M\omega_0)(2\alpha(\omega_0 t)^{-2}+3\mu(\omega_0 t)^{-4} \quad (8)$$
$$+O[(\omega_0 t)^{-6}])$$

where

$$\alpha = \gamma/2\omega_0 \quad (9)$$

$$\mu = 16\alpha^3 -8(1-\gamma')\alpha -\omega_0\gamma'' \quad (10)$$

in which $\gamma=\gamma(\omega=0)$, $\gamma'=-i(\partial\gamma/\partial\omega)_{\omega=0}$, and $\gamma''=(\partial^2\gamma/\partial\omega^2)_{\omega=0}$. Hence, for a system with an Ohmic dissipative mechanism, i.e. $\gamma\neq 0$, the zero temperature displacement correlation function has a long time tail[5] $\propto t^{-2}$. On the other hand, if the frictional influence of the environment has no Ohmic part, $J(t)$ fades out more rapidly. For instance, a dissipative mechanism characterized in the classically accessible region by a frictional force $\eta\dddot{q}$ leads to the long time behavior $J(t)=-(6\hbar\eta/\pi M)(\omega_0 t)^{-4}$.

At finite temperatures (4) gives

$$J(t)=\frac{\hbar}{\pi}\int_{-\infty}^{\infty}d\omega\, e^{-i\omega t}\chi''(\omega)[1-\exp(-\beta\hbar\omega)]^{-1} \quad (11)$$

In the lower half plane $\chi''(\omega)$ has poles at $\omega=-i\lambda_j$. Now, classically $J(t)$ decays as $\exp(-\Omega t)$ where $\Omega=\min(\mathrm{Re}\lambda_j)$. For instance, for a system with Ohmic dissipation one has $\Omega=\gamma/2$ for $\alpha\leq 1$ and $\Omega=\omega_0[\alpha-(\alpha^2-1)^{1/2}]$ for $\alpha>1$. In the quantum regime, $k_BT\ll\hbar\Omega$, the long time behavior of $J(t)$ arises from the poles of $[1-\exp(-\beta\hbar\omega)]^{-1}$ at $\omega=-in\nu$, where $\nu=2\pi k_BT/\hbar$ is the inverse of the thermal time τ. For large times $t\gg\Omega^{-1}$, we find

$$J(t)=-i(\hbar\nu/\pi)\sum_{n=1}^{N}\chi''(-in\nu)\exp(-n\nu t) \quad (12)$$

where N is of order Ω/ν and where terms of order $\exp(-\Omega t)$ have been disregarded. Expanding $\chi''(-in\nu)$ about the origin, we obtain by virtue of (9) and (10)

$$J(t) =-(\hbar/\pi M\omega_0)(2\alpha\psi_1(t) + 3\mu\psi_2(t) \quad (13)$$
$$+O[\Theta^6 e^{-\nu t}/(1-e^{-\nu t})^6])$$

where $\Theta=2k_BT/\hbar\omega_0$ and where

$$\Psi_1(t)=-\Theta^2(\partial/\partial x)[\exp(x)-1]^{-1} \quad (14)$$

$$\Psi_2(t)=-\tfrac{1}{6}\Theta^4(\partial^3/\partial x^3)[\exp(x)-1]^{-1}$$

in which $x=\nu t$. For T=0 one has $\Psi_1=(\omega_0 t)^{-2}$, $\Psi_2=(\omega_0 t)^{-4}$ and we recover our previous result (8). Contrary, for finite temperatures $\Psi_1(t)$ and $\Psi_2(t)$ decay asymptotically as

$$\Psi_1(t) \simeq \Theta^2\exp(-\nu t) \quad (15)$$

$$\Psi_2(t) \simeq \tfrac{1}{6}\Theta^4\exp(-\nu t)$$

leading to an exponential decay of $J(t)$ for $t\gg\nu^{-1}$. On the other hand, for intermediate times $\Omega^{-1}\ll t\ll\nu^{-1}$ we obtain from (14)

$$\Psi_1(t) \simeq (\omega_0 t)^{-2},\ \Psi_2(t) \simeq (\omega_0 t)^{-4}. \quad (16)$$

Now, for $k_BT\ll\hbar\omega_0$ there is a finite time interval where $\omega_0 t>>1$ but $\nu t<<1$. Hence for sufficiently low temperatures $k_B T<<\hbar\Omega,\hbar\omega_0$, J(t) displays an intermediate algebraic decay before it crosses over near $t=\nu^{-1}$ to the asymptotic exponential decay. This behavior is illustrated in the following figure for a system with Ohmic dissipation:

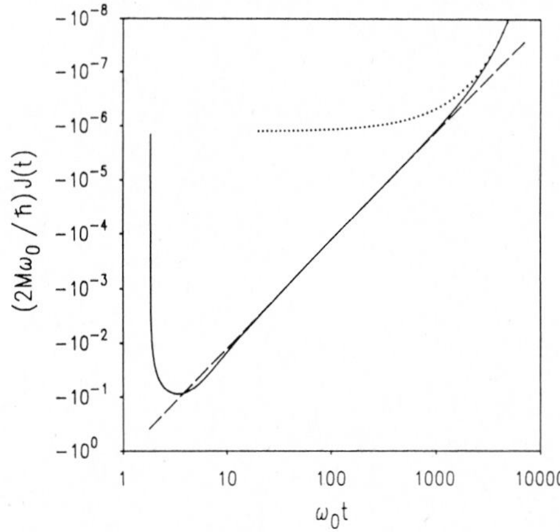

The real part of the dimensionless displacement correlation function $(2M\omega_0/\hbar)J(t)$ is shown as a function of $\omega_0 t$ for a harmonic oscillator with Ohmic damping $\gamma=2\omega_0$ (i.e. $\alpha=1$) at temperature $T=10^{-3}\hbar\omega_0/2\pi k_B$ (i.e. $\Theta=10^{-3}$). The intermediate algebraic decay $(2M\omega_0/\hbar)J(t)\simeq-(4\alpha/\pi)(\omega_0 t)^{-2}$ is shown as a dashed line, and the asymptotic decay $(2M\omega_0/\hbar)J(t)\simeq-(4\alpha/\pi)\Theta^2\exp(-\nu t)$ as a dotted line. Note that the real part of J(t) is negative for large $\omega_0 t$ so that J(t) approaches 0 from below as $t\to\infty$. In the strongly overdamped case at finite but low temperatures one finds a region where J(t) decays logarithmically, followed by a region where the decay is algebraic $\propto t^{-2}$ before the asymptotic exponential decay $\propto\exp(-\nu t)$ sets in[6].

So far we have considered linear systems only. However, the long time tails of zero temperature correlation functions are a rather general feature of dissipative quantum systems for the following reason. As a consequence of the fluctuation dissipation theorem, spectral densities of correlation functions have poles at the Matsubara frequencies $\omega=n\nu$ which come closer together as T is lowered and give rise to a cut contribution at zero temperature. This cut determines the long time behavior of the correlation function and leads to algebraic decay. For instance, the result (8) likewise applies to a particle moving in a stable nonlinear potential provided that $\gamma(\omega)$ and ω_0 are replaced by the corresponding renormalized quantities[7]. The discussion of the behavior at finite but low temperatures can be extended to the nonlinear case accordingly. However, it must be noted that the determination of the low frequency behavior of the response function, which determines the amplitudes of the asymptotic decay laws derived here, may itself be a complicated problem for nonlinear systems near T=0.

ACKNOWLEDGEMENTS

We would like to thank Fritz Haake, Peter Talkner, and Ulrich Weiss for valuable discussions and acknowledge partial support by the Deutsche Forschungsgemeinschaft.

REFERENCES

1. A.O. Caldeira and A.J. Leggett, Ann.Phys.(N.Y.) 149(1983)374, and Physica 121A(1983)587
2. R.H. Koch, D.L. van Harlingen, and J. Clarke, Phys.Rev.Lett 45(1980)2132; A. Schmid, J.Low Temp.Phys. 49(1982)609
3. P. Schramm, R. Jung, and H. Grabert, Phys.Lett. 107(1985)385
4. H.B. Callen and T.A. Welton, Phys.Rev. 83(1951)34
5. H. Grabert, U. Weiss, and P. Talkner, Z.Phys. B55(1984)87
6. R. Jung, G.L. Ingold, and H. Grabert, Phys.Rev. A October (1985), in print
7. K.S.J. Nordholm and R. Zwanzig, J.Stat.Phys. 11(1974)143; H. Grabert, in Springer Tracts in Modern Physics Vol. 95 (Springer-Verlag, Heidelberg, 1982), p.87

*also: Institut für Festkörperforschung der Kernforschungsanlage Jülich

NOISE IN PHYSICAL SYSTEMS AND 1/f NOISE - 1985
A. D'Amico and P. Mazzetti (editors)

MACROSCOPIC QUANTUM COHERENCE

Ulrich WEISS and Hermann GRABERT*

Institut für Theoretische Physik, Universität Stuttgart, D-7000 Stuttgart 80, Germany
*Institut für Festkörperforschung, Kernforschungsanlage Jülich, D-5170 Jülich, Germany

The dynamics of an asymmetric double-well system coupled to a heat bath is examined. Macroscopic quantum oscillations by barrier penetration are shown to be sensitively affected by thermal fluctuations and biasing forces. We report on recent analytical and numerical work on the dynamical behavior of the macroscopic variable and discuss Q-factor and amplitude of the quantum oscillations as a function of temperature, bias and dissipation strength. The results should directly be applicable to the experimental search of quantum oscillations in rf SQUIDs.

Tunneling rates at low temperatures are sensitively affected by dissipation arising from the coupling of the tunneling coordinate to its environment. Such damping phenomena are always present when the tunneling system is part of a macroscopic system. Specifically we consider a system with an environmental coupling leading in the classical limit to an equation of motion which is isomorphic to that of a Brownian particle moving in a nonlinear potential $V(q)$, i.e., $m\ddot{q}+\eta\dot{q}+\partial V/\partial q=0$. This case of Ohmic dissipation presents many special features at T=0 [1] and at finite temperatures[2,3].

Here we consider the phenomenon of coherent tunneling between macroscopically distinct states, which is usually referred to as 'macroscopic quantum coherence' (MQC). The most likely systems for an experimental observation of MQC[4,5] are radio-frequency superconducting quantum interference devices (SQUID). A SQUID consists of a superconducting ring of selfinductance L closed by an ideal Josephson junction shunted by small capacitance C and resistance R and with an external flux Φ_x being applied. The dynamics of the total flux $\Phi=\Phi_i+\Phi_x$, where Φ_i is the flux associated with the internal current, is given by the equation of motion

$$C\ddot{\Phi} + \dot{\Phi}/R = - \partial U/\partial\Phi \qquad (1)$$

where $U(\Phi)=(\Phi - \Phi_x)^2/2L-(I_0\Phi_0/2\pi)\cos(2\pi\Phi/\Phi_0)$ and where $\Phi_0=h/2e$ is the flux quantum. For an external flux of approximately $\Phi_0/2$ and for $\beta_L=2\pi LI_c/\Phi_0>>1$ $U(\Phi)$ essentially consists of two slightly asymmetric wells separated by a distance $\delta\Phi=\Phi_0$ and with a bias $\hbar\sigma=\Phi_0(\Phi_x-\Phi_0/2)/L$. The two different minima correspond to the two different senses of rotation of the supercurrent $I_c\cdot\sin(2\pi\Phi_i/\Phi_0)$.

At extremely low temperatures excitations in each well may be neglected. Then the double well behaves effectively like a biased dissipative two-state system[6] and the dynamics of tunneling transitions is characterized conveniently by $P(t)=P_L(t)-P_R(t)$, where $P_L(t)$ and $P_R(t)$ are the occupation probabilities of the left-hand and of the right-hand well, respectively. The calculation of $P(t)$ is based on real time path integral techniques which are convenient because they allow the inclusion of dissipation as an influence functional in the effective action. For low temperatures and small bias the functional integral is dominated by multi-instanton trajectories.

In a previous work[7] we have summed the instanton contributions to $P(t)$ in the dilute bounce gas approximation (DBGA) and have obtained a Laplace integral representation for $P(t)$ describing the time evolution into equilibrium as a function of temperature T, bias σ and damping parameter $\alpha = \delta\Phi^2/2\pi\hbar R$.

It can be shown that deviations from the DBGA are not relevant for the MQC phenomenon. The Laplace transform is given by

$$\tilde{P}(z)=\frac{1}{z}\,\frac{z-i\,\sin(\pi\alpha)[J_+(-z)-J_-(-z)]}{z-\;\cos(\pi\alpha)[J_+(-z)+J_-(-z)]} \qquad (2)$$

where

$$J_\pm(\lambda)=-\frac{\Delta^{2-2\alpha}}{2}\cdot\left(\frac{\hbar\beta}{2\pi}\right)^{1-2\alpha}\cdot\frac{\Gamma(\alpha-\hbar\beta(\lambda\pm i\sigma)/2\pi)}{\Gamma(1-\alpha-\hbar\beta(\lambda\pm i\sigma)/2\pi)} \qquad (3)$$

and where $\Gamma(z)$ is the Gamma function. Further, $\Delta=\Delta_{eff}$ is the effective tunneling frequency introduced in Ref. 6 and $\beta=1/k_BT$. For finite α , Δ is diminished as compared to the tunneling frequency Δ_o of an undamped system with zero bias. The phenomenon of MQC is restricted to the parameter window $\alpha<1/2$ and both k_BT and $\hbar|\sigma|$ of the order of $\hbar\Delta$ or smaller. In the following we restrict most of our attention to the case where $\alpha\lesssim 0.2$.

Let us first briefly discuss the case $\sigma=0$. At T=0 $\tilde{P}(z)$ has a pair of complex conjugate poles and a branch point at z=0. Hence P(t) is conveniently decomposed according to $P(t)=P_o(t)+P_1(t)$, where $P_o(t)$ is determined by the poles and $P_1(t)$ is determined by the cut.

The poles lead to a damped oscillatory behavior of $P_o(t)$

$$P_o(t) = p\,\exp(-\Gamma t)\cos(\Omega t), \qquad (4)$$

whereas the cut leads to an algebraic decay of $P_1(t)$. The amplitude $p=1+d$ deviates from 1 because of the incoherent contribution which has an initial value $P_1(0)=-d$, $d>0$. For T=0 we have $d=\alpha/(1-\alpha)$ and the frequency Ω and the damping rate Γ are given by

$$\Omega = \cos(\tfrac{\pi}{2}\,\tfrac{\alpha}{1-\alpha}\Delta)$$
$$\Gamma = \sin(\tfrac{\pi}{2}\,\tfrac{\alpha}{1-\alpha}\Delta)\;. \qquad (5)$$

The frequency Ω decreases as damping increases and it vanishes for $\alpha=1/2$. On the other hand, the damping rate Γ increases with growing α. When α exceeds 1/2 the system becomes overdamped and the oscillation disappears.

At finite temperatures the cut resolves into a set of poles on the negative real z-axis, $z_n=-(n+v_n)2\pi k_BT/\hbar$, n=1,2,3,..., and where $-\alpha\leq v_n\leq\alpha$. Each of them gives a contribution to $P_1(t)$ describing exponential relaxation. To leading order in α and for arbitrary temperatures we find for the initial value

$$P_1(0) = \alpha\varepsilon\,\mathrm{Im}\,\Psi'(1+i\varepsilon), \qquad (6)$$

where Ψ' is the Trigamma function and where $\varepsilon=(2\pi k_BT/\hbar\Delta)^{\alpha-1}$. This yields at high temperatures ($\varepsilon<<1$) $P_1(0)=-2\alpha\varepsilon^2\zeta(3)$, where $\zeta(3)$ is a Riemann number, and $P_1(0)=-\alpha$ at T=0. Hence for $\alpha<<1$ $P_1(t)$ has a very small initial value for all temperatures and therefore is negligible in the MQC region.

As the temperature is increased, the quality factor of the oscillation $Q=(\Omega^2+\Gamma^2)^{1/2}/\Gamma$ rapidly decreases after a small initial increase[8]. The oscillation vanishes at a temperature T* which for small α is given by

$$T^* = \hbar\Delta[2(1-\gamma\alpha)/\alpha]^{1/(1-\alpha)}/2\pi k_B\;; \qquad (7)$$

where γ is Euler's constant. This formula becomes exact in the limit $\alpha\to 0$ and has an error of less than 1% for $\alpha\lesssim 0.2$. In the region of very small α we simply have $T^*\simeq\hbar\Delta/\pi k_B\alpha$. Above T* the systems tunnels incoherently. At temperatures $T\gtrsim 3T^*$ the only relevant contribution to $P_o(t)$ comes from the pole moving to the origin whereas the contribution of the other pole is negligible. This precisely is the region of incoherent tunneling discussed in Refs. 7 and 9.

In the temperature range $0\leq T<T^*$ where oscillations occur we find

$$P_o(t) = p\,\exp(-\Gamma t)\cos(\Omega t-\Phi)/\cos\Phi. \qquad (8)$$

In general the initial value p=1+d, the frequency Ω, the damping rate Γ and the phase shift Φ are given by rather complicated formulas, but they reduce to simple expressions for $\alpha\lesssim 0.2$ and $\hbar\Delta/k_B\pi\lesssim T\leq T^*$. This range of parameters is particularly promising regarding an experimental search for the MQC effect. We find

$$\Gamma = \alpha\pi k_B T/\hbar ,$$
$$\Omega = \Gamma[(T^*/T)^{2-2\alpha}-1]^{1/2} ,$$
$$\Phi = \arc\tan(\Gamma/\Omega) ,$$
$$p = 1 + 2\alpha\varepsilon^2\zeta(3). \qquad (9)$$

At T=0 the phase shift Φ vanishes and Γ and are given by (5) and $p=1/(1-\alpha)$.

The situation is more complex for a biased system. Now two additional real poles are relevant for $P_0(t)$, one of which is located at the origin and hence determines the equilibrium distribution, while the other determines the relaxation to equilibrium.

Picking up the contributions of the four poles we obtain

$$P_0(t) = P_\infty + [p(1-A_1)-P_\infty(1+Z_1)]e^{-\Gamma_r t} \qquad (10)$$
$$+[(pA_1+P_\infty Z_1)\cos(\Omega t)+(pA_2+P_\infty Z_2)\sin(\Omega t)]e^{-\Gamma t},$$

where $P_\infty=-\tanh(\hbar\sigma/2k_BT)$ and where p=1+d. The first terms describe relaxation into the lower well at a rate Γ_r and vanish for $\sigma=0$. The last term describes the damped oscillation of macroscopic quantum coherence. We have obtained simple experessions for all quantities occuring in (10)[8].

At T=0 we find to leading order in α

$$\Omega = \Delta_b \; ; \; \Gamma = \alpha\pi\Delta^2/2\Delta_b$$

$$\Gamma_r = \alpha\pi|\sigma|\Delta^2/\Delta_b^2 \qquad (11)$$

where $\Delta^2_b = \Delta^2+\sigma^2$ and the factors defining the amplitudes are given by $p=1+\alpha$, $A_1=\Delta^2/\Delta^2_b$, $A_2=Z_1=Z_2=0$.

In the relevant parameter window for experiments on MQC, $\hbar\Delta_b/k_B\pi \lesssim T \lesssim T^*\Delta^2/\Delta_b^2$ and $\alpha \lesssim 0.2$ and σ of order Δ or smaller, both Γ and Γ_r depend almost linearly on αT. In this region we find

$$\Omega = [\sigma^2 + (2\pi\tau\Delta\,\Lambda)^2 - (\Delta^2 + 4\sigma^2)\Theta^2]^{1/2} ,$$
$$\Gamma = (\Delta^2 + 2\sigma^2)\Theta/\Delta ,$$
$$\Gamma_r = 2\Theta\Delta , \qquad (12)$$

where $\Lambda = \varepsilon(1-\gamma\alpha)$ and where $\Theta=\alpha\pi k_B T\Delta/\hbar\Delta_b^2$. The amplitude factors in (10) are given by

$$A_1 = [\Omega^2 - \sigma^2 + (\Delta^2 - 4\sigma^2)\Theta^2]/N ,$$
$$A_2 = [(\Delta^2+2\sigma^2)\Omega^2 + (\sigma^2+\Delta^2\Theta^2)(\Delta^2-2\sigma^2)]\Theta/N\Omega\Delta ,$$
$$Z_1 = 8\sigma^2\Theta^2/N ,$$
$$Z_2 = -2[\Delta\Omega^2+(\Delta^4-4\sigma^4)\Theta^2/\Delta]\Theta/N\Omega ,$$
$$N = \Omega^2 + (2\sigma^2-\Delta^2)^2\Theta^2/\Delta^2 ,$$
$$p = 1 + 2\alpha\varepsilon^2\zeta(3). \qquad (13)$$

The formulas (10), (12) and (13) provide very accurate predictions of the influence of dissipation, temperature and bias on coherent oscillations of a double well system in the window of parameter space where the search for the MQC effect is most promising.

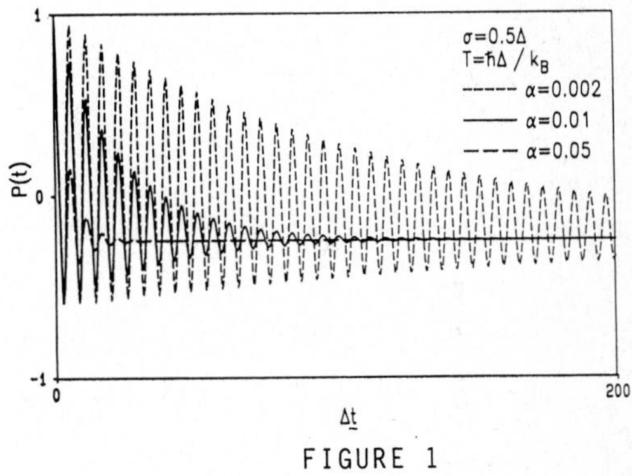

FIGURE 1

Figure 1 shows plots of P(t) for various values of α while σ and T are kept fixed. The strong influence of dissipation on the coherent oscillation is evident.

In Figure 2 the temperature is varied while σ and α are kept fixed. The plots show the strong influence of temperature on the quality factor of the oscillation and also the influence of temperature on the equilibrium state for a biased system.

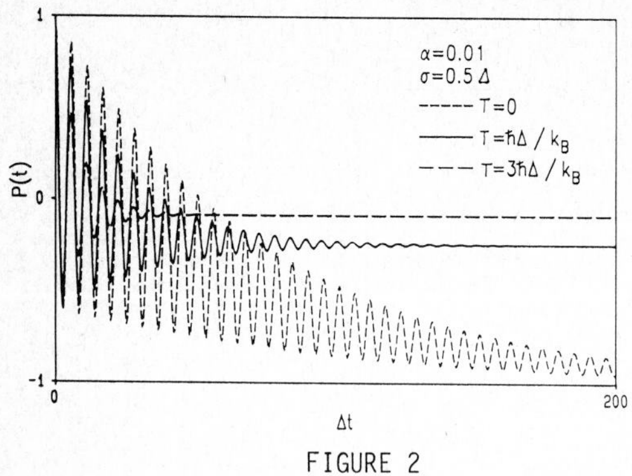

FIGURE 2

Figure 3 shows the influence of a bias on the oscillations. Like in an undamped system the amplitude of the oscillation is reduced as compared with an unbiased system.

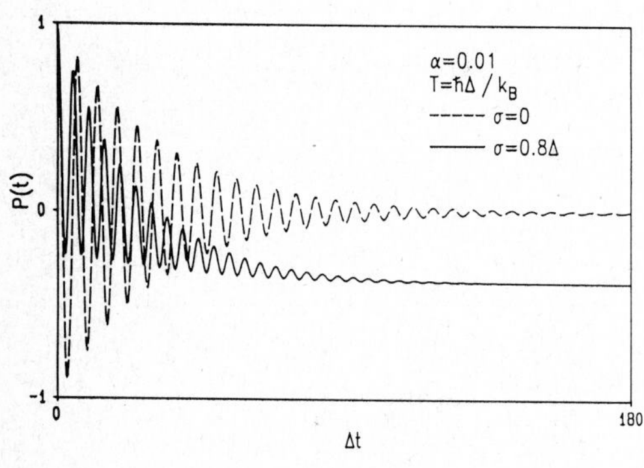

FIGURE 3

These results should be directly applicable to the design and analysis of MQC-experiments with SQUID rings.

ACKNOWLEDGEMENT

We are grateful to thank S. Chakravarty, A.J. Leggett, S. Linkwitz and C. Tesche for useful discussions and acknowledge partial support by the Deutsche Forschungsgemeinschaft.

REFERENCES

1. A.O. Caldeira and A.J. Leggett, Phys.Rev. Lett. 46(1981)211, and Ann.Phys.(N.Y.)149, (1983)374.

2. H. Grabert, U. Weiss, and P. Hanggi, Phys. Rev.Lett. 52(1984)2193.

3. H. Grabert and U. Weiss, Phys.Rev.Lett.53 (1984)1787.

4. R. de Bruyn Ouboter, Physica 126B(1984)423.

5. H. Grabert, Macroscopic Quantum Tunneling and Quantum Coherence, in: SQUID' 85 Superconducting Quantum Interference Devices, eds. H.D. Hahlbohm and H. Lubbig (de Gruyter, Berlin, 1985).

6. S. Chakravarty and A.J. Leggett, Phys.Rev. Lett. 53(1984)5.

7. H. Grabert and U. Weiss, Phys.Rev.Lett. 54 (1985)1605.

8. U. Weiss and H. Grabert, to be published.

9. M.P.A. Fischer and A.T. Dorsey, Phys.Rev.Lett. 54(1985)1609.

NOISE IN PHYSICAL SYSTEMS AND 1/f NOISE - 1985
A. D'Amico and P. Mazzetti (editors)

MACROSCOPIC QUANTUM TUNNELING

Hermann GRABERT

Institut für Festkörperforschung, Kernforschungsanlage Jülich, D 5170 Jülich, Germany

and

Peter OLSCHOWSKI and Ulrich WEISS

Institut für Theoretische Physik, Universität Stuttgart, D 7000 Stuttgart 80, Germany

The temperature dependence of the decay rate of a metastable state of a system coupled to a dissipative environment is determined. Kramers' classical theory of the rate of decay by thermal activation is extended to lower temperatures where quantum tunneling prevails. By means of a functional integral approach a universal treatment of both the classical and the quantum regimes is achieved. Analytical methods for high and intermediate temperatures are combined with an accurate numerical method for low temperatures. Some analytical results are obtained in the latter region, too. The calculations cover the entire range of parameters of interest in the problem of macroscopic quantum decay in Josephson systems.

In the classical regime a macroscopic system in a metastable state undergoes transitions to states of lower energies via thermal activation over the intervening barrier. At very low temperatures, however, the metastable state can only decay by macroscopic quantum tunneling /1/. This transition was recently observed in the decay of the zero voltage state of current biased Josephson junctions /2/ and in fluxoid quantum transitions in SQUID rings /3/. For these systems the part of the potential relevant for the decay problem can well be represented by a potential V(q) of the form as sketched in Fig. 1

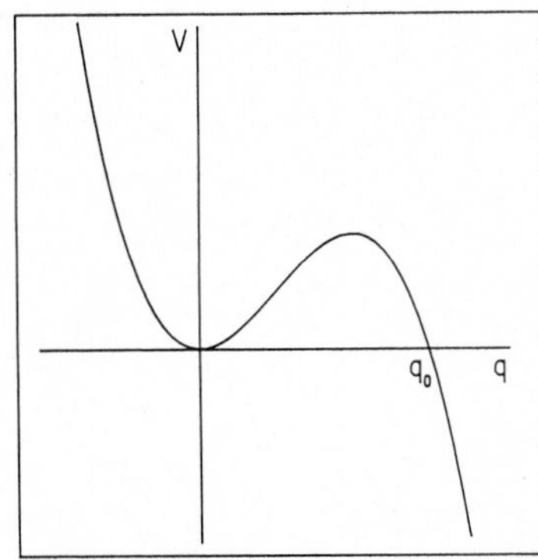

FIGURE 1: A metastable well

The basis of our calculation of the decay rate is a functional integral approach which is convenient because it allows the inclusion of environmental influences as an additional nonlocal term in the action. The decay rate is calculated by means of the 'bounce-technique' developed by Langer /4/. In the following we give a brief summary of our results.

There is a typical <u>crossover temperature</u> /5/

$$T_o = \hbar\omega_R/2\pi k_B \tag{1}$$

where roughly the transition between thermal hopping and quantum tunneling occurs. The frequency ω_R is the largest positive root of the equation

$$\omega_R{}^2 + \omega_R\eta(\omega_R)/m = \omega_b{}^2, \tag{2}$$

where $\eta(\omega)$ is the frequency-dependent damping coefficient and $m\omega_b{}^2 = -V_b''$ is the curvature of the potential at the barrier top. For frequency-independent (Ohmic) damping, i.e., $\eta(\omega)\equiv\eta$, we have

$$\omega_R = \omega_b[(1+\alpha_b^2)^{1/2} - \alpha_b] \tag{3}$$

where $\alpha_b = \eta/2m\omega_b$ is a dimensionless damping parameter. In general, it must be noted that T_0 is influenced by the damping coefficient at the frequency ω_R which is of the order of the barrier frequency ω_b and may differ from the low-frequency damping $\eta(\omega=0)$.

For temperatures above T_0 the decay rate takes the form

$$\Gamma = f\,\exp(-\Delta V/k_BT) \tag{4}$$

where ΔV is the barrier height while f is an attempt frequency. The exponential factor is of the familiar Arrhenius form typical for thermally activated processes. The attempt frequency may conveniently be written as

$$f = f_{c\ell} \cdot q\,, \tag{5}$$

where

$$f_{c\ell} = \frac{\omega_0}{2\pi}\,\frac{\omega_R}{\omega_b} \tag{6}$$

is the classical attempt frequency in which $m\omega_0^2 = V_0''$ is the curvature of the potential at the metastable minimum, while q describes quantum effects. The classical attempt frequency (6) includes the effects of frequency-dependent damping /6/ and it reduces to the familiar result of Kramers /7/ for Ohmic damping where ω_R is given by (3). The quantum correction factor reads /5,8/

$$q = \prod_{n=1}^{\infty} \frac{n^2\nu^2+\omega_0^2+n\nu\eta(n\nu)/m}{n^2\nu^2-\omega_b^2+n\nu\eta(n\nu)/m} \tag{7}$$

where $\nu = 2\pi k_BT/\hbar$. The factor q approaches unity for $T \gg T_0$ where the decay becomes purely classical. On the other hand, q diverges at the crossover temperature T_0 in the vicinity of which an improved result (see below) is needed. It should be noted that q can be quite large even at temperatures of the order of a few T_0. For Josephson systems this complicates the determination of junction parameters from 'classical' activation data. The leading quantum corrections are given by /9/

$$q = \exp\{\tfrac{1}{24}\hbar^2(\omega_0^2+\omega_b^2)/(k_BT)^2\}. \tag{8}$$

For moderately damped systems this approximation describes the enhancement of the rate by quantum effects very accurately down to temperatures where the rate Γ deviates from the classical result already by an order of magnitude.

The crossover between thermal hopping and quantum tunneling occurs in a transition region of size /5,10/

$$|T - T_0| \lesssim (\hbar\omega_0/d\Delta V)^{1/2}\,T_0 \tag{9}$$

where d is a numerical factor which depends on the form of the potential and the dissipation strength. For Josephson systems the relevant part of the potential can well be represented by a cubic potential and d assumes the form

$$d = 12\pi\,\frac{(1+\alpha_b^2)(1+2\Theta^2)\Theta}{1 + 4\,\Theta^2} \tag{10}$$

where $\Theta = (1+\alpha_b^2)^{1/2}-\alpha_b$. In the crossover region it is convenient to multiply the rate by the Arrhenius factor and to consider the quantity

$$y = \Gamma\,\exp(\Delta V/k_BT) \tag{11}$$

as a function of $x = T-T_0$. Then, there is a frequency scale y_0 and a temperature scale x_0 such that

$$y/y_0 = F(x/x_0) \tag{12}$$

where $F(u) = \mathrm{Erfc}(u)\exp(u^2)$ is a universal function which is independent of the form of the metastable potential and also independent of the form of the dissipative mechanism /5/. The scale factors x_0 and y_0 are nonuniversal but depend on the particular system under consideration. For a system with a cubic potential and Ohmic damping one has

$$x_0 = (\hbar\omega_0/d\Delta V)^{1/2}\,T_0$$

$$y_0 = \frac{\omega_0}{2\sqrt{\pi}} \frac{T_0}{x_0} \frac{1}{\Theta+\alpha_b} \prod_{n=1}^{\infty} \frac{n(n+1)\Theta^2+n+2}{n(n+1)\Theta^2+n} \qquad (13)$$

where d and Θ have been introduced above. The rate follows the universal law (12) in the temperature range (9). Fig. 2 shows the rate in the crossover region together with the high tempera-

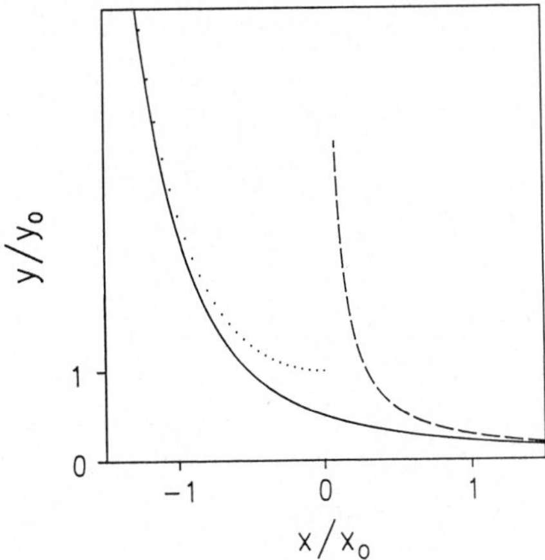

FIGURE 2: The crossover function

ture formula (4) [dashed line] and the low temperature formula (14) [dotted line] discussed below. Observe that the crossover function (12) smoothly matches onto these formulas valid outside the crossover region.

For temperatures below T_0 the decay rate Γ is obtained in the form

$$\Gamma = f_{qm} \exp(-S_B/\hbar) , \qquad (14)$$

where S_B is the action of a saddle point trajectory (the so-called bounce) with period $\hbar/k_BT$ of the path integral representing the partition function of the system /1,11/. The quantum mechanical prefactor f_{qm} is related to the spectrum of small fluctuations about the bounce. An explicit analytical evaluation of formula (14) is generally not possible except for temperatures in the range $(\hbar\omega_0/d\Delta V)^{1/2}\ T_0 \lesssim T_0 - T \ll T_0$ directly below crossover region /5/. A detailed analysis of the zero temperature rate $\Gamma(0)$ is due to Caldeira and Leggett /1/. They have emphasized the strong suppression of the decay by dissipation.

At finite temperatures the rate is enhanced by thermal fluctuations. Grabert et al. /11,12/ have shown that at low temperatures the rates of all systems with finite damping at zero frequency follow a universal T^2-law

$$\ln[\Gamma(T)/\Gamma(0)] = sT^2 \qquad (15)$$

In the case of a cubic potential and Ohmic damping the slope s is approximately given by /11/

$$s = \frac{2}{\pi}(18k_B)^2(\hbar\omega_0)^{-3}\Delta V\alpha_b[(\alpha_b^2+9/\pi^2)^{1/2}-\alpha_b]^{-2}$$

Very recently, the T^2-law has been observed experimentally in Josephson systems /2,3/.

A numerical calculation of the zero temperature decay rate $\Gamma(0)$ as a function of α_b has been performed by Chang and Chakravarty /13/ and for finite temperatures by the present authors /14/. Fig. 3 shows our numerical results for a system with cubic potential with barrier height $\Delta V=5\hbar\omega_0$. The circles show the decay rate of an undamped

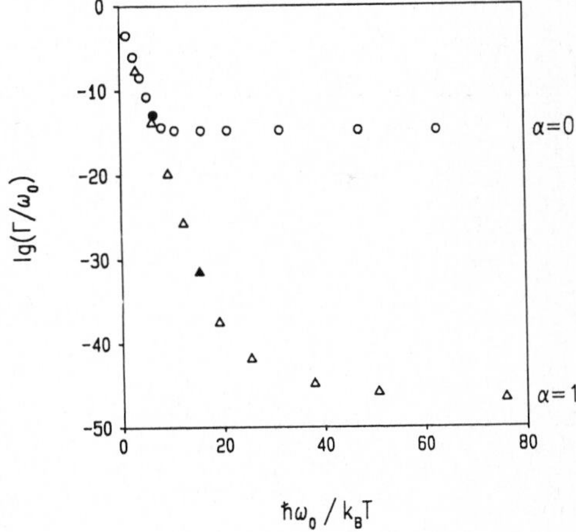

FIGURE: 3

system and the triangles of a system with an Ohmic damping coefficient $\eta=2m\omega_0$. The values at the crossover temperature are indicated by closed symbols.

We see that theoretical predictions of macroscopic quantum tunneling rates are now available for the entire range of temperatures.

ACKNOWLEDGEMENT

The authors wish to thank C.N. Archie, S. Chakravarty, J. Clarke, M.H. Devoret, D. Estere, P. Hanggi, A.J. Leggett, J.E. Lukens, D.B. Schwartz, R.F. Voss, R.A. Webb, for helpful discussions or correspondence.

REFERENCES

1. A.O. Caldeira and A.J. Leggett, Phys.Rev.Lett. 34(1981)211; Ann.Phys.(N.Y.) 149(1983)374; H. Grabert, in: SQUID' 85 Superconducting Quantum Interference Devices, H.-D.Hahlbohm H. Lübbig, eds.) de Gruyter, Berlin, 1985

2. R.F. Voss and R.A. Webb, Phys.Rev.Lett. 47(1981)265; L.D. Jackel et al. Phys.Rev.Lett. 47(1981)697; S.Washburn, R.A. Webb, R.F. Voss and S.M. Faris, Phys.Rev.Lett 54(1985)2696

3. D.B. Schwartz et al., preprint (1985)

4. J.S. Langer, Ann.Phys.(N.Y.) 41(1967)108

5. H. Grabert and U. Weiss, Phys.Rev.Lett 53(1984)1787

6. R.F. Grote and J.T. Hynes, J.Chem.Phys. 73(1980)2715; P.Hanggi and F. Mojtabai, Phys.Rev. A26(1982)1168

7. H.A. Kramers, Physica 7(1940)284

8. P.G. Wolynes, Phys.Rev.Lett 47(1981)968

9. P. Hanggi, H. Grabert, G.L. Ingold and U. Weiss, Phys.Rev.Lett 55(1985)761

10. A.I. Larkin and Yu.N. Ovchinnikov, Zh.Eksp. Teor.Fiz. 86(1984)719

11. H. Grabert and U. Weiss, Z.Phys. B56(1984)171

12. H. Grabert, U. Weiss and P. Hanggi, Phys.Rev.Lett. 52(1984)2193

13. L.D. Chang and S. Chakravarty, Phys.Rev. B29(1984)103, B30(1984)1566(E)

14. H.Grabert, P. Olschowski and U. Weiss, Phys.Rev. B32(1985)3348

NOISE IN PHYSICAL SYSTEMS AND 1/f NOISE - 1985
A. D'Amico and P. Mazzetti (editors)

THERMAL AND QUANTUM NOISE IN OVERDAMPED JOSEPHSON JUNCTIONS

Antonio BARONE, Carlo CAMERLINGO, Roberto CRISTIANO and Paolo SILVESTRINI

Istituto di Cibernetica del Consiglio Nazionale delle Ricerche
Via Toiano, 6 - 80072 Arco Felice (Napoli) ITALY

The present work deals mainly with the effect of fluctuations on the current-voltage (I-V) characteristics of Josephson tunnel structures. In particular the role played by the junction parameters is investigated in connection with both thermal activation and Macroscopic Quantum Tunneling (MQT) processes. Using the analysis of ref. 6 results concerning the effect of fluctuations on the low voltage region of I-V curves of Josephson junctions are reported in the case of the overdamped regime corresponding to values of the β_C parameter ($\beta_C = (2e/\hbar)I_C R^2 C$) smaller than unity. In the quantum limit, in particular, curves of the supercurrent lifetimes vs the bias current are reported, and the role of the junction parameters in connection with the possibility to observing quantum phenomena is briefly discussed. Finally it is also shown that the M.Q.T. approach gives results which confirm those by R.H. Koch, D.J. Van Harlingen and John Clarke obtained in the framework of quantum activation.

Josephson Tunnel Junctions (J.T.J.) are the basic circuital elements of various superconducting devices. Among other advantages, they are competitive in comparison with other solid state components either for their fast switching properties or for their high sensitivity to very small signals as magnetic fields, microwave radiation and so on[1].

The ultimate sensitivity of these devices is often set by the intrinsic noise, either of thermal[2] or quantum[3] nature. Moreover, noise phenomena in J.T.J. are also of particular interest from a fundamental point of view[4]. A simple electronic model of J.T.J. will be used here to understanding the role played by junction parameters[2]. The model separates the total current I in various parallel components as shown in fig. 1.

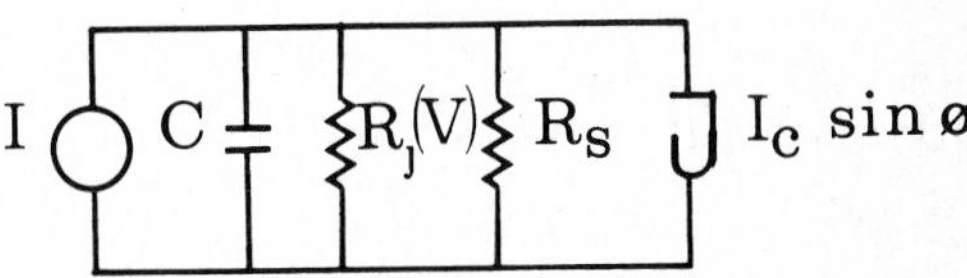

FIGURE 1
A sketch of the circuital model of a current biased J.T.J.

An ideal Josephson element carries the supercurrent $I_J = I_C \sin\phi$, where I_C is the critical Josephson current and ϕ is the phase difference between the two superconductors. The expression of I_J is the first Josephson relation and together with $\dot{\phi} = 2eV/\hbar$ (the second Josephson relation) accounts for the Josephson effect. The dissipative current V/R is a parallel combination of the intrinsic, voltage-dependent, tunneling resistance $R_J(V)$ and of the external linear shunt R_S.

Although most of our results are general we will confine ourselves to the case $R \backsim R_s$, neglecting non-linearity and capacitance renormalization. Finally, the intrinsic junction capacitance is taken into account by the term CdV/dt.

In the noise free case, depending on the values of I_c, R and C, hysteresis effects can be present in the current-voltage (I-V) characteristics as shown in Fig. 2. In normalized units a useful parameter to classify junctions with respect to the hysteresis is $\beta_c = (2e/\hbar) I_c R^2 C$, which contains all the above mentioned junction parameters. If $\beta_c < 1$ hysteresis effects can be practically neglected.

Thermal noise can induce significant modifications of I-V curves from the one shown in Fig. 2. If $\beta_c > 1$ (underdamped regime) the influence of noise can lead either to a random distribution of the I_c and I_r values, or to the hysteresis removal, depending on the bias current frequency and noise conditions. The former case can be better understood if the non-stationary nature of the process is considered together with the fact that increasing values of β_c lead to more stable V=0 and V≠0 branches of the I-V curves, depending on the initial bias conditions.

For $\beta_c<1$ (overdamped regime) such a stability decreases and the noise produces a rounding of the I-V curves. Nevertheless, also in this regime, as it will be shown, increasing β_c values lead to smaller voltage (more stable supercurrent) at very low voltage, whereas the opposite occurs at high voltage, in agreement with ref. 5. In a recent work[6] we have investigated such a low voltage region, obtaining an expression for the mean voltage which is inversely proportional to the life-

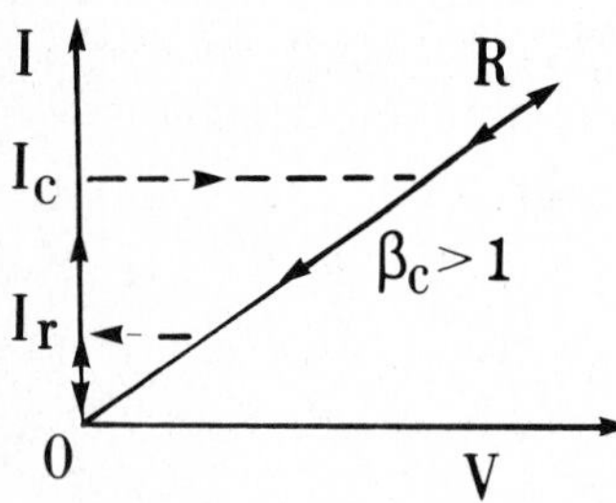

FIGURE 2
Hysteretic I-V characteristic. Arrows indicate the path of the current in the junction, first increasing and then decreasing the bias current.

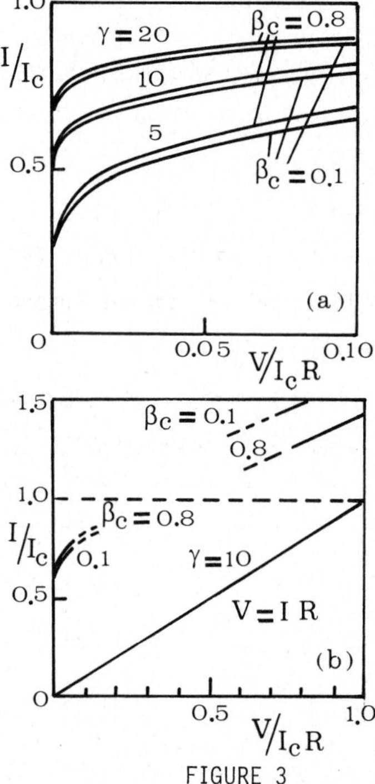

FIGURE 3
I-V characteristics in presence of thermal noise for $\beta_c = 0.8$ and 0.1

a) The low voltage region at various noise levels ($\gamma = 5, 10, 20$);

b) low and high voltage regions at $\gamma = 10$

time of the Josephson critical current τ_{th}: $(2e/\hbar)\langle V\rangle = 2\pi/\tau_{th}$. The results of this analysis are shown in Fig. 3, where the low voltage region of the I-V curves at various noise levels $\gamma = \hbar I_c/ekT$ (fig. 3a) and, for $\gamma = 10$, the low and high voltage regions (fig. 3b) are shown. Unfortunately, it is difficult to calculate I-V curves at intermediate voltages where the crossing of curves with different β_c values should occur. The opposite behavior of the curves as function of β_c, in the two voltage regions, recalls, in some sense, what happens for $\beta_c \gg 1$.

We have extended our investigation at temperatures below $T_o = (eI_cR/2\pi k\beta_c)\left\{\left[1 + 4\beta_c(1-I^2/I_c^2)^{\frac{1}{2}}\right]^{\frac{1}{2}} - 1\right\}$ where quantum noise becomes larger than thermal one[7]. Our analysis considered the so called macroscopic quantum tunneling (MQT) as a quantum noise mechanisms. To obtain the low voltage region of the I-V curves, the lifetime τ_q of the supercurrent has been computed. Fig. 4 shows τ_q^{-1} normalized to I_c versus I/I_c for various junction parameters. In this case the expression of τ_q^{-1} can not be simply handled just in terms of β_c.

It is worth observing that, the parameters of curve a) correspond to the experimental conditions characterized by T = 50mK, $I_c = 1\,\mu A$, R = 50 ohm, C∼ .13pF and $T/T_o(I/I_c=.92) = .7$. In Fig. 5 we report T_o/I_c vs I/I_c for various β_c and R values. These curves together with the expression of β_c provides an estimate of the values of the junction parameters, temperatures and bias conditions where quantum effects are relevant and experimental conditions could be satisfied. For instance, considering the previous example, it is easy to see that bias values less than $I/I_c = .92$ increase T_o, improving experimental possibilities.

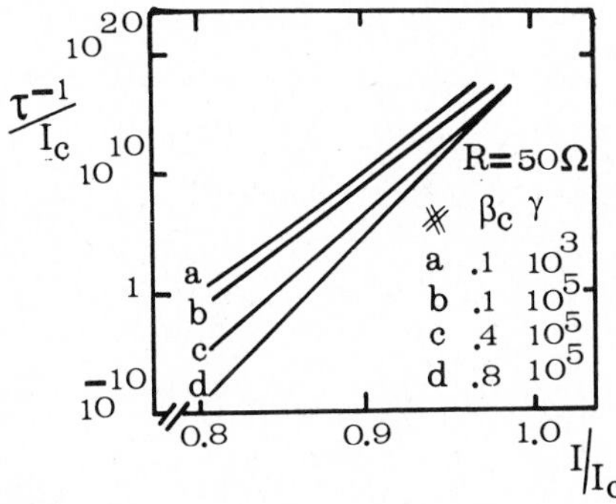

FIGURE 4

Ratio of the tunnel rate τ_q^{-1} on the Josephson critical current vs the normalized current I/I_c at various β_c and γ values.

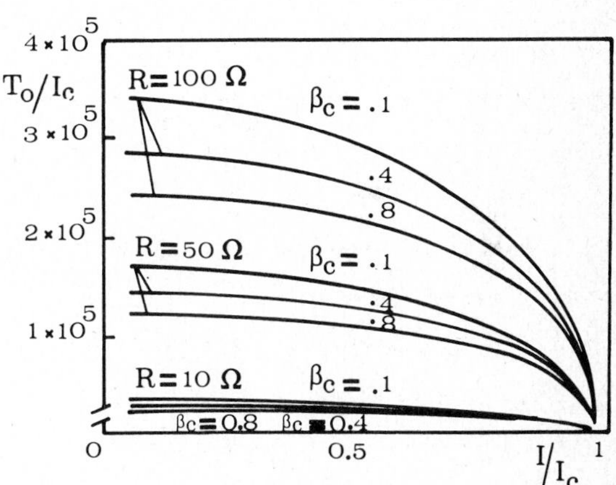

FIGURE 5

Ratio of the temperature T_0 on the Josephson critical current vs the normalized current at various β_c and R values

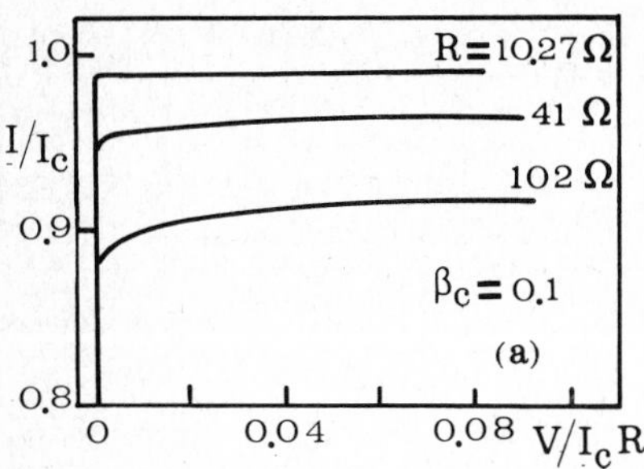

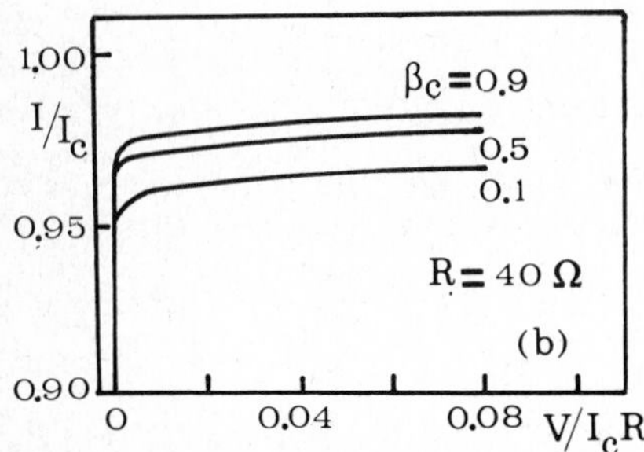

FIGURE 6
a) Low Voltage region on I-V characteristics in presence of M.Q.T. at fixed β_c value, with R = 10.27; 41; 102 Ω. R values correspond respectively to the values 0.005; 0.02; 0.05 of the parameter k Γ of Ref. 8;
b) I-V characteristics in presence of M.Q.T. at a fixed R value (k Γ = 0.0194) and for various β_c.

Finally, Fig. 6 shows the low-voltage I-V curves at various junctions parameters. An interesting feature of these curves is that they are in good agreement with those obtained in ref. 8, based on a different approach, i.e., on the quantum activation as a noise mechanism. In the literature on quantum noise it is still an open question whether such two mechanisms describe the same phenomenon. Our results, valid only in the overdamped regime, seem to give an affirmative answer[9]. Concerning the behavior of I-V curves as function of β_c, I-V curves with higher β_c show more stable supercurrent branches, similarly as in the thermal region.

REFERENCES

1. A. Barone and G. Paternò, "Physics and Application of the Josephson Effect" - Wiley, New York, 1982

2. See chapt. 6 of ref. 1

3. A.I. Larkin, K.K. Likharev and Yu. N. Ovchinnikov - Physica, 126B, 414 (1984)

4. A.J. Leggett- Annals of Physics 149, 374 (1983)

5. P.A. Lee - J. Appl. Phys. 42, 325 (1971) J. Kurkijarvi and V. Ambegaokar Phys. Lett. 31A, 314 (1970)

6. Yu. N. Ovchinnikov, R. Cristiano and A. Barone - J. Appl. Phys. 56, 1473 (1984)

7. A.I. Larkin and Yu. N. Ovchinnikov, Pis'ma Zh. Exsp. Teor. Fiz. 37, 322 (1983) JETP Lett. 37, 382 (1983)

8. R.H. Koch, D.J. Van Harlingen and John Clarke - Phys. Rev. Letts. 45, 2132 (1980)

9. A. Barone, C. Camerlingo and R. Cristiano Phys. Rev. Letts. 54, 157 (1985)

NOISE IN PHYSICAL SYSTEMS AND 1/f NOISE - 1985
A. D'Amico and P. Mazzetti (editors)

WHITE NOISE AND 1/F NOISE IN OPTICAL OSCILLATORS: STATE-OF-THE-ART IN RINGLASERS

H.R. BILGER and M.R. SAYEH

School of Electrical and Computer Engineering, Oklahoma State University, Stillwater, OK 74078, U.S.A

The theoretical power spectral density of frequency fluctuations δf in optical oscillators, $S_{\delta f} = hf_o^3/Q^2P$, has been closely realized in ringlasers, where two (or more) contracirculating modes produce a low-frequency beat, the noise of which is then measured. This represents a constant power spectral density i.e. white noise. Steady progress has been made in the meantime to lower this white spectrum as well as to extend it to lower Fourier frequencies. Power densities $S_{\delta f} < 10^{-4}$ Hz^2/Hz are now being reached by using mirrors with finesses up to F = 50,000 and by using larger cavities (rings) to increase the quality factor. The fractional frequency fluctuations $\delta f/f = y$(at f_o = 474 THz) have a corresponding power spectral density $S_y < 4.5 \times 10^{-34}$ Hz^{-1}, which is the smallest fractional frequency fluctuation yet observed in nature. There is, as yet, no comprehensive theory of the 1/f noise available in laser oscillators, although 1/f noise, or an equivalent flicker floor in the Allan variance, has been consistently observed in the data which we have analyzed. Experimentally, $S_{\delta f} = (\bar{A} f_o^2/Q^4)(1/\nu)$ has been found for the dependence of the 1/f noise with $\bar{A} \approx 4$. The Q^{-4} dependence is readily explained by loss (or gain) fluctuations, using a Van der Pol oscillator.

1. INTRODUCTION

One of the more exciting applications of lasers is the ring laser gyro (RLG). The RLG is simply a ring-type cavity laser in which its radiation field travels in a closed path. In principle, the laser beam oscillates in the clockwise and the counterclockwise directions. The frequency difference (beat frequency) between these oscillations depends on the rotation of the cavity with respect to an inertial frame.

Application of a RLG for a number of precision measurements requires thorough study of the various factors affecting the stability of the beat frequency of the counterpropagating waves. Noise causing fluctuations of the beat frequency is a very crucial factor by which the RLG performance limits are defined.

White noise: The main cause of this type of frequency fluctuation is due to interaction between the quantized radiation field and atoms. This shows a finite probability that atoms in the upper energy state jump to the lower state without presence of photons. This phenomenon gives rise to spontaneous emission. The one-sided power spectral density (PSD) of the frequency fluctuations due to spontaneous emission can be obtained by a noise-driven Van der Pol oscillator. This is reviewed in Sec. 2. White noise was observed experimentally in RLG's.[1]

1/f noise: Recently, measurements of frequency fluctuations for a long period of time showed additional noise (1/f noise or Flicker noise) at low Fourier frequencies.[2,3] In these observations the 1/f noise level is an inverse function of the quality factor. In Sec. 3 we show this can be explained by loss mechanisms using the Van der Pol oscillator as a model for the laser oscillator.

2. WHITE NOISE IN FREQUENCY FLUCTUATIONS, REVIEW

Noise in lasers has been a subject of research for the last two decades. The PSD per mode vs Fourier frequency ν is

$$S_{\delta f}(\nu) = hf_o^3/Q^2P \qquad (1)$$

[h is Planck's constant, $f_o = \omega_o/2\pi$ is the

(average) laser frequency, Q is the quality factor of the passive cavity, and P is the power loss per mode]. Here the frequency fluctuation $\delta f(t)$ due to spontaneous emission noise gives an approximate white noise level which has been verified experimentally (see Ref. 1). Equation (1) has been generalized.[4]

Let us consider a noise-driven Van der Pol oscillator as a model for a laser oscillator:[5]

$$\ddot{x} + [r - (g - \gamma x^2)]\dot{x} + \omega_0^2 x = N(t) \qquad (2)$$

where x is the mode amplitude, r is the energy decay rate, g is the unsaturated gain, γ is the saturation parameter, and N(t) is the noise source due to the spontaneous emission processes. The mode amplitude x is chosen such that x^2 is the mode energy. Under conditions $\omega_0 \gg (r, g, \text{ and } \gamma x^2)$, x(t) can be written as

$$x(t) \simeq [x_0 + \xi(t)] \cos[\omega_0 t + \Theta(t)] \qquad (3)$$

where $x_0^2 = \frac{4}{\gamma}(g - r)$, $\xi(t)$ is a slow varying amplitude (compared to x(t)), and $\theta(t)$ accounts for phase fluctuations. It can be shown that the frequency fluctuations $\delta f(t) = \frac{1}{2\pi}\frac{d\Theta}{dt}$ have a PSD given by

$$S_{\delta f}(\nu) = \frac{\gamma r h f_0}{8\pi^2 (g-r)} \left[\frac{N_2 g(\nu + f_0)}{(N_2 - N_1 g_2/g_1) g(f_0)} + n_{th}\right]. \qquad (4)$$

Some of the parameters in Eq. (4) are difficult to relate to experimentally observed quantities, for example quantity (g - r). We can avoid this difficulty by introducing the passive quality factor Q and the power loss per mode P. Under limiting conditions ($N_2 \gg N_1$, $\nu \ll f_0/Q$, $hf_0/kT \gg 1$), Eq. (4) converges to Eq. (1).

The frequency fluctuations $\delta f(t)$ due to spontaneous emission noise give thus an approximately white noise level.

3. 1/f NOISE IN FREQUENCY FLUCTUATIONS

The 1/f noise is a random process defined in terms of its PSD. The PSD of 1/f noise is, roughly speaking, proportional to the reciprocal Fourier frequency, viz.

$$S(\nu) = \frac{B}{\nu^\beta} \qquad (5)$$

where B is a constant and β is close to unity.

1/f noise has recently been reported in RLG's (see Refs. 2 and 3). Figure 1 shows a measured PSD of a four-mode RLG versus frequency($Q \simeq 5\times10^8$, $P \simeq 80$ μW, and $\lambda = 633$ nm). The white noise dominates down to 6×10^{-4} Hz, 1/f noise is from 6×10^{-4} Hz to 4×10^{-5} Hz, and $1/f^2$ noise predominates at frequencies less than 4×10^{-5} Hz.

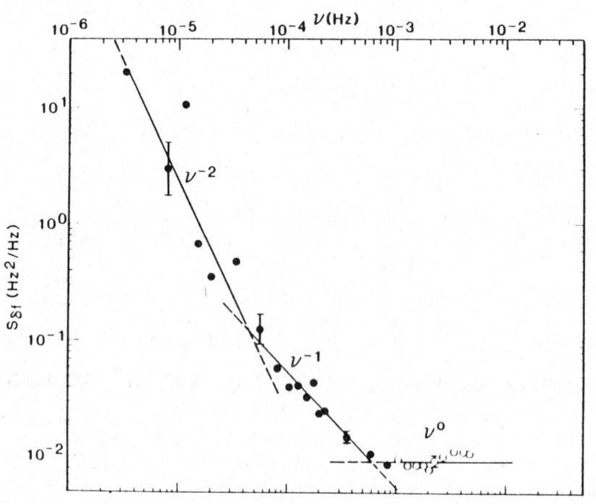

FIGURE 1
Power spectral density of frequency fluctuations of a four-mode RLG ($Q \simeq 5\times10^8$, $P \simeq 80$ μW, and $\lambda \simeq 633$ nm). Open circles depict the white noise level after removing quantization noise.

The PSD of a RLG with a higher quality factor ($Q \simeq 10^{10}$, $P \simeq 100$ μW, and $\lambda = 633$ nm) is shown in Fig. 2. This shows a substantially reduced level for both 1/f noise and white noise. This PSD corresponds to the 54 day run. The PSD components were averaged every 10 points to reduce the power density uncertainty.

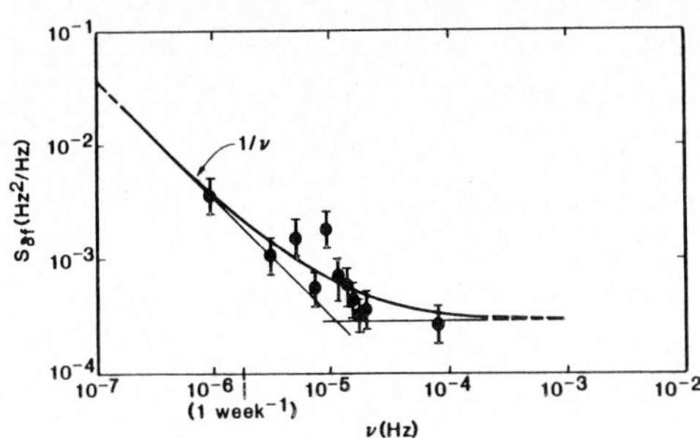

FIGURE 2

Power spectral density of frequency fluctuations of a two-mode RLG ($Q \simeq 10^{10}$, $P \simeq 100$ μW, and $\lambda \simeq 633$ nm).

Figure 3 summarizes the measured values of B, in Eq. (5), versus the passive quality factor Q for different types of RLG's. The least squares fitted line in Fig. 3 shows a Q^{-4} dependence, with $B \simeq 9\times10^{29}/Q^4$ Hz^2.

We now show that the Q^{-4}-law can be realized by loss fluctuations.

Let us assume that there exist fluctuations in the loss r in Eq. (2), independent of the existence of white noise, i.e. for the purpose of this derivation, N(t) in Eq. (2) is set to zero. Now we establish the PSD of the frequency fluctuations due to loss fluctuations.

An approximate solution to Eq. (2) ($N(t) = 0$) for small $(g - r)/\omega_0$ using Poincare's method is[6]:

$$x(t) \simeq x_0 \cos(\mu\omega_0 t) \tag{6}$$

where $\mu = 1 - [(g-r)/4\omega_0]^2$. From this, we find the fluctuation δf with respect to the fluctuation δr as

$$\delta f/f = (1/8Q_a^2)\,\delta r/r \tag{7}$$

where $Q_a = \omega_0/(g - r)$.

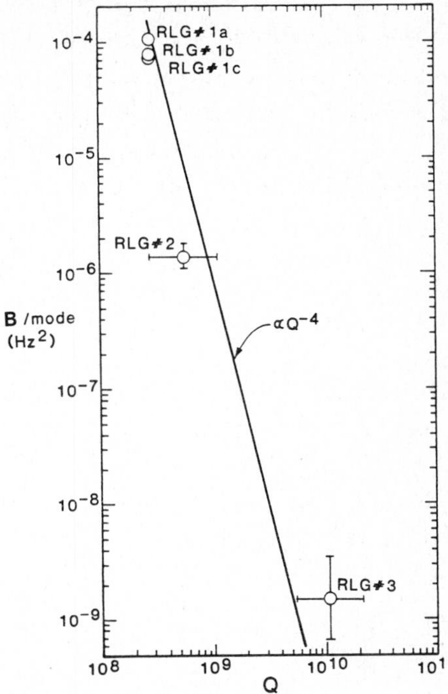

FIGURE 3

1/f noise level versus passive quality factor Q for three ring laser gyros. Q^{-4}-law is fitted through the points using the least-square fit criterian.

The PSD of the fractional frequency fluctuation can be related to the PSD of the fractional loss fluctuations by using Eq. (7) as

$$S_{\delta f/f}(\nu) = (\sqrt{8}\,Q_a)^{-4} S_{\delta r/r}(\nu) . \tag{8}$$

The PSD of δf/f is simply related to the PSD of δr/r by a Q_a^{-4} law. This law has been observed experimentally in quartz oscillators over six decades of the PSD.[7] The proportionality to Q_a^{-4} appears to be independent of the specific assumptions on the types of loss. Loss fluctuations may originate from losses through the mirrors, coupling between modes, or scattering at mirrors or at atoms and so on.

Using Handel's quantum theory of 1/f noise,[8] one would expect loss fluctuations to originate from loss processes inside the cavity whose elementary cross sections of interaction with the electromagnetic field fluctuate with a $1/\nu$ spectrum. In this case we expect

$$S_{\delta r/r}(\nu) = \frac{A}{\nu} \tag{9}$$

where A is a constant depending on the nature of the interaction.

By using Eq. (9), the PSD of the frequency fluctuation per mode can be written as

$$S_{\delta f}(\nu) = Af_0^2/64Q_a^4\nu \tag{10}$$

In Eq. (5) B is then

$$B = Af_0^2/64Q_a^4 \tag{11}$$

for $\beta = 1$.

The experimental result in Fig. 3 gives a value 256 for the constant A in Eq. (11) if we assume $Q_a = Q$. Therefore, the experimental evidence suggests for the 1/f noise

$$S_{\delta f}(\nu) \simeq 4f_0^2/Q^4\nu \tag{12}$$

4. CONCLUSION

The power spectral density of frequency fluctuations of a ring laser gyro can be summarized as

$$S_{\delta f}(\nu) = h_0 + h_{-1}\nu^{-1} + h_{-2}\nu^{-2}$$

where $h_0 = hf_0^3/Q^2P$, $h_{-1} = B$, and h_{-2} are constants with respect to the Fourier frequency ν. The white noise level h_0 due to the spontaneous emission was reviewed using the noise-driven Van der Pol oscillator. This noise level has been reached by using a four-mode RLG.

1/f noise has already been observed in RLG's. It has been demonstrated experimentally that this noise is not due to temperature fluctuation of the RLG (see Ref. 2, in that reference the 1/f noise levels were the same for $\delta T \simeq 100\mu K$ and $\delta T \simeq 1K$). It was shown that the Q^{-4} dependence of 1/f noise level (h_{-1}) is due to loss mechanisms in the laser oscillator as has been observed in quartz oscillators.

The 1/f noise level sets a limit for the performance of RLG's because the averaging of data containing 1/f noise does not reduce the noise level whereas that of white noise would be reduced by averaging.

REFERENCES

1. T.A. Dorschner, H.A. Haus, M. Holz, I.W. Smith, and H. Statz, "Laser gyro at quantum limit," IEEE J. Quantum Electronics QE-16, pp. 1376-1379, 1980.

2. H.R. Bilger and M. Sayeh, "Noise phenomena in ring lasers," in Noise in Physical Systems and 1/f Noise (eds. M. Savelli, G. Lecoy, and J.-P. Nougier), Elsevier Science Publ. B.V., 1983, pp. 325-328.

3. H.R. Bilger, "Low frequency noise in ring laser gyros," Proc. of SPIE, vol. 487, Physics of Optical Ring Gyros, 1984, pp. 42-48.

4. H. Haken, "Laser Theory," Springer-Verlag, New York, 1984.

5. A. Yariv, "Quantum Electronics," John Wiley and Sons, Inc., New York, 1975, Chpt. XIII.

6. A.A. Andronov, A.A. Vitt, and S.E. Khaikin, "Theory of oscillators," Addison-Wesley Publishing Company, Massachusetts, 1966.

7. J.J. Gagnepain, J. Uebersfeld, G. Goujon, and P.H. Handel, "Relation between 1/f noise and Q-factor in quartz resonator at room and low temperature, first theoretical interpretation," Proc. 35th Ann. Freq. Cont. Symp., pp. 476-483, 1981.

8. P.H. Handel, "Quantum approach to 1/f noise," Phys. Rev. A 22, pp. 745-757, 1980.

NOISE IN PHYSICAL SYSTEMS AND 1/f NOISE - 1985
A. D'Amico and P. Mazzetti (editors)

NOISE IN NON-LINEAR AMPLIFIERS : QUANTUM LIMIT IN A SQUEEZED OPERATING STATE

G. D'ARIANO*+, M. RASETTI°+ and M. VADACCHINO°

*Dipartimento di Fisica, Universitá di Pavia, Italy
°Dipartimento di Fisica, Politecnico di Torino, Italy
+G.N.S.M. del C.N.R., Italy

A quantum non-linear amplifier is analysed in which the operating state is coherent, squeezed and multi-boson squeezed. The quantum noise characterization of the amplifier is crucially connected to its phase sensitiveness: different types of amplifiers are available depending on their operating states.

The quantum mechanical description of amplifiers is customarily based on the following ingredients: a set of (bosonic) input modes, a set of (bosonic) output modes, the amplifier's degrees of freedom (which can be either bosonic or fermionic) and an interaction scheme of the latter which transfers input to output[1].

Restricting for simplicity our attention to the single mode case and denoting by a,a^+ and A,A^+ the input and output bosons respectively, the evolution equations of the amplifier - averaged over the internal degrees of freedom - are given by the mapping

$$A=\mathscr{D}(a,a^+) \qquad /1/$$

where the functional $\mathscr{D}$ is constrained to belong to a realization of the automorphism group of the Weyl algebra (namely $\left[a,a^+\right]=1$ should imply $\left[A, A^+\right]=1$).

The other relevant property of an amplifier is its operating state $|\varphi\rangle$ - which is assumed to be independent of the input state $|\varphi_{in}\rangle$ - describing how the amplifier's internal modes are prepared.

The relevant physical quantities describing how the amplifier actually operates are the mean-square fluctuation of the output amplitude

$$(\Delta A)^2= \langle\varphi_{in}|A^2|\varphi_{in}\rangle-\langle\varphi_{in}|A|\varphi_{in}\rangle^2 \qquad /2/$$

where the averages $\langle\bullet\rangle = \langle\varphi|\bullet|\varphi\rangle$ have been explicitly written in terms of input states only, because of the assumed factorization properties of the density matrix and in view of the averaging performed on the evolution equations; and the gain

$$G = |\langle A\rangle / \langle a\rangle|^2 \qquad /3/$$

In present communication we shall consider a variety of evolution functionals $\mathscr{D}$, both polynomial and analytic, as well as a set of different choices of input states selected to be coherent states and study the phase sensitivity of the amplifier, namely how the quantum input noise is processed.

More precisely Caves' definition of phase-insensitive linear amplifier will be extended to non-linear evolution equations, essentially by dropping the requirement of invariance of $\langle A\rangle$ under arbitrary phase transformations. The most general polynomial $\mathscr{D}$ of degree 2 reads[2].

$$\mathscr{D} :P(a+Qa^+)^2+Ma+Na^+ \qquad /4/$$

where the complex parameters P,Q,M,N satisfy the conditions

$$|M|^2 - |N|^2 = 1 = |Q|^2 \qquad /5/$$

$$\frac{P}{\bar{P}} = -\bar{Q}\frac{M-N\bar{Q}}{\bar{M}-\bar{N}Q} \qquad /6/$$

Eqs./5/,/6/ are the constraints whereby /4/ indeed realizes an automorphism of Weyl's algebra.

We consider three different choices for $|\varphi_{in}\rangle$:

i) Coherent states[3]

$$|\alpha\rangle = D(\alpha)|0\rangle$$
$$D(\alpha) = \exp(\alpha a^+ - \bar{\alpha} a) \qquad /7/$$

where $|0\rangle$ is the vacuum state and $\alpha\in\mathscr{C}$.

Such states are interesting in present application because they correspond to phase insensitive input noise, namely $(\Delta a)=0$.

ii) Squeezed states[4]

$$|z,\alpha\rangle = D(\alpha)S(z)|0\rangle$$
$$S(z) = \exp\{z(a^+)^2 - za^2\} \qquad /8/$$

where $z\in\mathscr{C}$ is the squeezing parameter. They can never give a phase insensitive input noise, in that

$$(\Delta a)^2 = (z/2|z|)\sinh(4|z|) \qquad /9/$$

can be zero only for z=0, in which case /8/ transform into /7/.

iii) Multi-boson squeezed states[5]

$$|z,w;\alpha\rangle_k = D(\alpha)S_{(k)}(z,w)|0\rangle \qquad /10/$$

$$S_{(k)}(z,w) = \exp\{zb^+_{(k)} + iwa^+a - zb_{(k)}\}$$

$$w\in\mathscr{R}\ ;\ z\in\mathscr{C}\ ;\ k\in\mathscr{N}$$

where $b_{(k)}$, $b^+_{(k)}$ are multi-boson operators

$$\left[b_{(k)}, b^+_{(k)}\right] = 1\ ;\ \left[a^+a, b_{(k)}\right] = -kb_{(k)} \qquad /11/$$

which can be realized by the normal ordered series[6]

$$b_{(k)} = \sum_{j=0}^{\infty} \mathbf{a}_j^{(k)} (a^+)^j a^{j+k} \qquad /12/$$

where

$$\mathbf{a}_j^{(k)} = \sum_{t=0}^{j} \frac{(-)^{j-t}}{(j-t)!}\left(\frac{1+[\![t/k]\!]}{t!(t+k)!}\right)^{1/2} e^{i\vartheta_t} \qquad /13/$$

$[\![x]\!]$ denoting the maximum integer $\leqslant x$ and ϑ_l, $l=0,\ldots,j$ a set of arbitrary phases. Note that $a^k \neq b_{(k)}$.

For the state /10/ the input noise is characterized by variance

$$(\Delta a)^2 = \sqrt{2}\ \delta_{k,2}\ \varphi_{(k)} zF(|\varphi_{(k)}|^2|z|^2) \qquad /14/$$

where

$$\varphi_{(k)} = \frac{\sin(kw/2)e^{ikw/2}}{kw/2} \qquad /15/$$

and the function F, whose properties where thoroughly discussed in reference 6, is defined by

$$F(x) = e^{-x}\sum_{n=0}^{\infty}\frac{x^n\sqrt{2n+1}}{n!} \qquad /16/$$

(Δa) as given by /16/ can vanish only for z=0 or $w=\frac{2m\pi}{k}$, $m\in\mathscr{Z}-\{0\}$ or $k>2$.

In the former two cases the state /10/ turns into the coherent state[7]. The states $|z,w;\alpha\rangle_k$ are thus non-Gaussian wave packets[7], which, for k>2, are phase insensitive as well as coherent in a non-trivial way.

In case ii) the output noise is characterized by mean-square fluctuation

$$(\Delta A)^2 = 2P^2SC(2T^2+SC) + \sinh(2|z|)(Mz^2+N\bar{z}^2)/2|z| +$$
$$+2PT\{(\sinh|z|\,(MSz+NC\bar{z})/|z| + \cosh|z|(MC+NS)\} +$$
$$+MN\cosh(2|z|)$$
$$T = \alpha + Q\bar{\alpha} \qquad /17/$$

and gain

$$G = |P(T^2+SC)/\alpha \quad +M+N\,\bar{\alpha}/\alpha\,|^2 \qquad /18/$$

where

$$S=\cosh|z|+Q\,\bar{z}\,\sinh|z|/|z| \qquad /19/$$

$$C=Q\,\cosh|z|+z\,\sinh|z|/|z|$$

case i) is recovered from /17/,/18/ simply setting z=0, and the linear case for P=0.

One can check how the non-linearity forbids phase intensitivity. In the linear case, Caves' results $(\Delta A)^2 = MN$, $G = |M+N\;\bar{\alpha}/\alpha\;|^2$ are straightforwardly recovered adopting the vacuum as operating state. It is also interesting to notice how the gain given by /18/ has state dependence hinting (minimum of G vs. α) to different regimes of stability of the amplifier.

In case iii) the results are much more complicated and rich of features.

In view of the special role ascribed by eq./14/ to the choice k = 2, we shall confine here our attention to it.

The mean square fluctuation of the output amplitude is:

$$(\Delta A)^2 = 2P^2\Big[4\sqrt{2}\,R\;\Omega_1F' + \Omega_2H - \{\,\Omega_2 + 4\sqrt{2}\,R(\,\zeta^2+Q^2\,\bar{\zeta}^2\,)+2QR\,\}\,F^2+2\sqrt{2}\,\Big(\Omega_1(R+T^2+Q) + MT(\,\zeta+Q\,\bar{\zeta}\,)/P+\frac{(M^2\,\zeta+N^2\,\bar{\zeta}\,)}{2P^2}\Big)F+4\,\Big(R^2+|\,\zeta\,|^2(5Q^2+2\boldsymbol{T})\Big)+\boldsymbol{T}\Big].$$

$$\Omega_1 = \zeta+Q^2\bar{\zeta} \;\; ; \;\; \Omega_2 = \zeta^2+Q^4\,\bar{\zeta}^2 \qquad /20/$$
$$R=Q|\,\zeta\,|^2 \;\; ; \;\; \xi = |\,\varphi_k|\,|z|$$

$$\boldsymbol{T} = Q(2T^2+Q)+\,T(N+MQ)/P+MN/2P^2$$

where $F = F(\,\xi^2\,)$ is given by /16/ and $H = H(\,\xi^2)$ is defined by

$$H(x)=e^{-x}\sum_{n=0}^{\infty}\frac{x^n}{n!}\sqrt{(2n+2)^2-1} \qquad /21/$$

The complex structure of the dependence of $(\,\Delta\,A)^2$ on P,Q,M,N and z,w , hints to the possibility of changing the characteristics of the amplifier (switching from phase-insensitive to phase-sensitive) by a suitable choice of the state parameters (z,w,k).

Another interesting case is that of the evolution equation given by:

$$A=Lb_{(k)}+Jb^+_{(k)}+Ma+Na^+ \qquad /22/$$

The more outstanding property of /22/ is that unlike /4/ it is possibly to put M=N=0 without infringing the restriction for $\mathcal{D}$ to be a realization of the automorphism group of the Weyl algebra. The calculations are a direct application of the method outlined above and the result are too long to be reported here.

We gratefully acknowledge a valuable discussion with C. Buzano.

REFERENCES

1. C.C. Caves, Phys. Rev. D26, 1817 (1982).
2. R.K. Colegrave and Ponnudurai Bala, J. of Phys. A 18, 779 (1985).
3. R.J. Glauber, Phys. Rev. 131, 2766 (1963).
4. J.N. Hollenhorst, Phys. Rev. D19, 1669 (1979)
5. G. D'Ariano, M. Rasetti and M. Vadacchino, Phys. Rev. D 32, 1034 (1985).
6. R.A. Brandt and O.W. Greenberg, J. Math. Phys. 10, 1168 (1969).
M. Rasetti, Int. J. Theor. Phys. 5, 377 (1972).
7. G. D'Ariano, M. Rasetti and M. Vadacchino, in preparation.

APPLICATIONS AND MEASUREMENT TECHNIQUES - 1

NOISE IN PHYSICAL SYSTEMS AND 1/f NOISE - 1985
A. D'Amico and P. Mazzetti (editors)

MEASUREMENT OF TEMPERATURE FLUCTUATION SPECTRA IN CURRENT-CARRYING MICROBRIDGES

Sumihisa HASHIGUCHI and Masao YAJIMA*

Faculty of Engineering, Yamanashi University, Kofu, JAPAN

*Sankyo Seiki Mfg. Co., Ltd., Suwa, JAPAN

Spectra of temperature fluctuation in metal microbridges were determined for 4 decades of frequency ranging from 0.0001 to 1Hz. This was accomplished by measuring the correlation between the variance fluctuation of thermal noise and the resistance fluctuation in samples driven by a constant current source. The spectra of the temperature fluctuation is of the 1/f type and fluctuation levels were about $50K^2/Hz$ at 0.01Hz and $4.3K^2$ for the frequency band ranging from 0.0001 to 1Hz. The resistance fluctuation is closely correlated with the variance fluctuation of the thermal noise, as well as with the temperature fluctuation. It was concluded that most of the 1/f resistance fluctuation in the metal microbridge samples originates from the temperature fluctuation in the frequency range investigated here.

1. INTRODUCTION

We have reported that the temperature fluctuation in a current-carrying resistor can be estimated from the correlation coefficient between the variance fluctuation of thermal noise and the 1/f resistance fluctuation[1]. It has been shown that there is a close correlation between the fluctuations within 2 decades of frequency ranging from 0.0001 to 0.01Hz.

The frequency range of the measurement of the temperature fluctuation can extended by preparing samples which generate extraordinally high 1/f fluctuation, and also by extending the bandwidth of the detection of the variance fluctuation of thermal noise.

We report here the experimental estimation of the temperature fluctuation in the microbridge type resistor within 4 decades of frequency ranging from 0.0001 to 1Hz.

2. EXPERIMENTS

Samples are prepared by evaporation of indium on glass substrates as shown in Fig.1.

A microbridge is formed between the evaporated indium layers on the substrate and at the bottom of the dent made on the deposited SiO step.

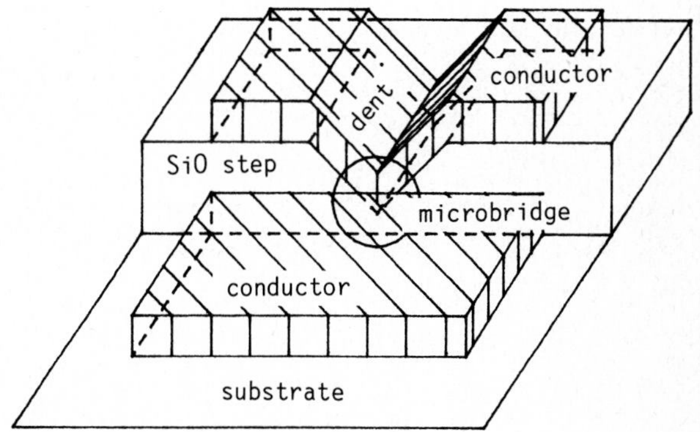

FIGURE 1
A Microbridge sample

The microbridge samples were annealed in a oxygen-rich atmosphere at 543K for several hours and then covered with nail enamel. By controlling the quantity of the evaporated indium, the height of the SiO step on the substrate, and the duration of heat treatment, we can obtain samples with the resistance of 10k-100kΩ and the relative 1/f resistance fluctuation level of more than -70db at 1Hz.

This research was supported by the grant of Nippon Seimitsu Kogyo Co.,Ltd., Kofu, JAPAN.

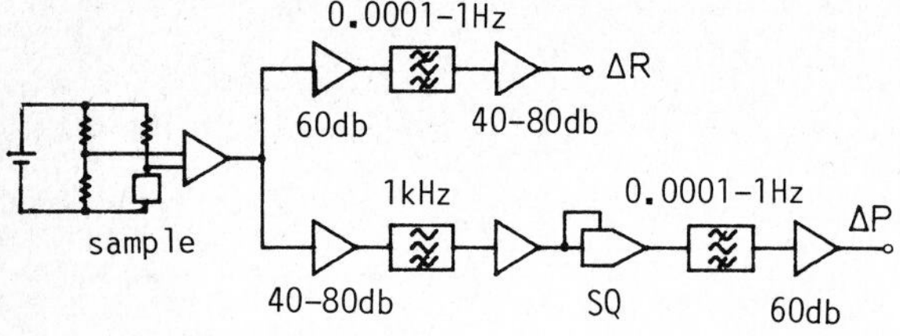

FIGURE 2
Measuring setup

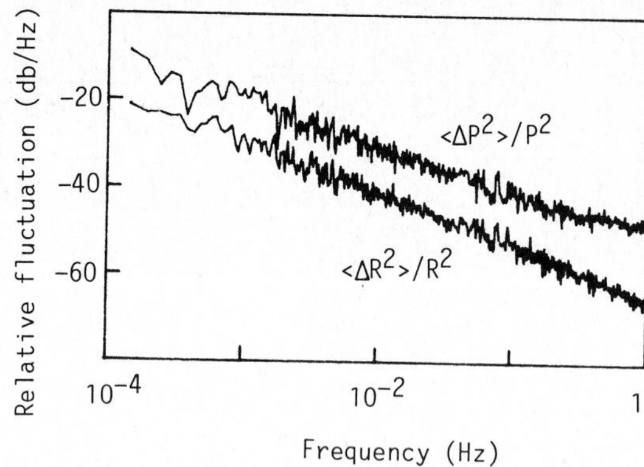

FIGURE 3
Spectra of the variance fluctuation and the resistance fluctuation

The measuring setup for fluctuations is shown in Fig.2. The sample is put in an arm of a bridge consisting of low noise resistors, and is driven by a constant current supplied from batteries. The magnitude of the driving current is determined so as to make the spectral density of the generated 1/f noise spectral density at about 500Hz equal to the thermal noise spectral density.

The variance fluctuation of thermal noise is obtained by squaring the amplified thermal noise voltage across the sample within the band ranging from 1kHz to 1MHz.

Both the variance fluctuation and 1/f fluctuation, in the frequency range from 0.0001 to 1Hz, are digitized and analized with a computer. The mean-square value of each fluctuation and the cross-product between them are obtained from the complex Fourier transform of each fluctuation.

The spectra of the temperature fluctuation, and of the correlation between 1/f fluctuation and temperature fluctuation are computed according to the following equations, respectively.

$$\frac{\langle\Delta T^2\rangle}{T^2}=\frac{\langle\Delta P^2\rangle}{P^2}+\frac{\langle\Delta R^2\rangle}{R^2}-2\frac{\langle\Delta R\Delta P\rangle}{RP}, \quad (1)$$

$$\rho_{RT}=\left(\frac{\langle\Delta R\Delta P\rangle}{RP}-\frac{\langle\Delta R^2\rangle}{R^2}\right)/\frac{\langle\Delta R^2\rangle\langle\Delta T^2\rangle}{RT}, \quad (2)$$

where T, P, and R are the average of the values of the temperature, the variance and the resistance; Δ denotes a fluctuating quantity, and <> denotes the time average.

The thermal time constant of the sample is measured by observing the transient in the current when a voltage step is applied to the sample.

3. RESULTS AND DISCUSSIONS

Figure 3 shows the typical spectra of the relative resistance and variance fluctuations. The relative variance fluctuation is 10db higher than the relative resistance fluctuation. The slopes of both spectra are -1.1. The sample resistance is 10.2kΩ and the relative resistance fluctuation is -67db/Hz at 1Hz. The driving current is 17.6nA.

Figure 4 shows the spectrum of the correlation coefficient between the variance fluctuation and the resistance fluctuation. There exists a close correlation between them.

Figure 5 shows the spectrum of the temperature fluctuation obtained by using eq.(1). The level of the spectrum is -32db/Hz at 0.01Hz and its slope is -1.1. These values correspond to the magnitudes of the mean square temperature fluctuation of $53K^2$/Hz at 0.01Hz, and $4.3K^2$ for the frequency band ranging from 0.0001 to 1Hz.

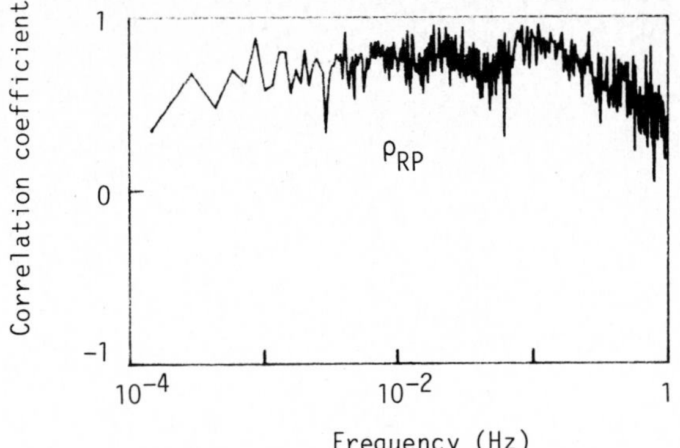

FIGURE 4
Spectrum of the variance-resistance correlation

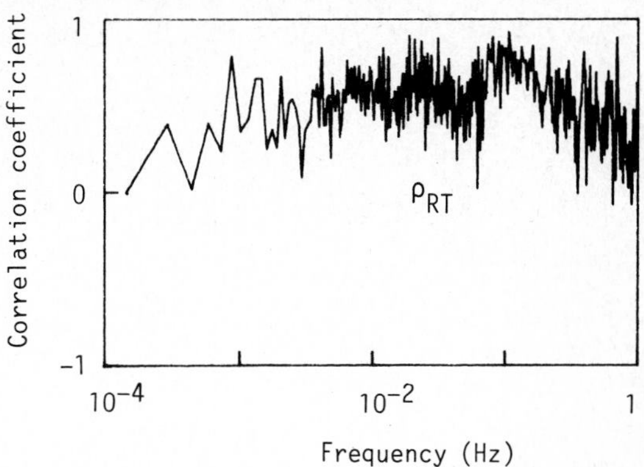

FIGURE 6
Spectrum of the temperature-resistance correlation

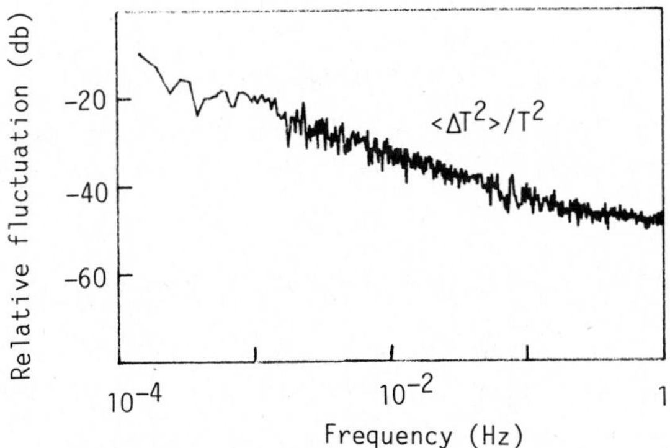

FIGURE 5
Spectrum of the temperature fluctuation

Figure 6 shows the spectrum of the correlation coefficient between the resistance fluctuation and the temperature fluctuation, which is calculated according to eq.(2). It is positive, which coincides with the sign of the temperature coefficient of the resistance. Gradually, it increases towards 1.0 with the increase of frequency, and then starts to decrease in the highest decade.

The effective carrier number is estimated from the relative resistance fluctuation as

$$N=2\times10^{-3}<\Delta R^2>/R^2, \qquad (3)$$

and is of the order of 10^4.

The atomic density of indium is about $4\times10^{22}cm^{-3}$ and there are three valence electron per atom. The carrier number of 10^4 corresponds to a volume of $8\times10^{-20}cm^3$, whose weight is $6\times10^{-19}g$. Heat capacity of this quantity of .1h8 indium is $1.4\times10^{-19}J/K$.

The equilibrium temperature fluctuation in this heat capacitance is

$<\Delta T^2>/T^2=9\times10^{-5}$;

i.e., for T=290K

$<\Delta T^2>=7.6K^2$

within the bandwidth determined by the thermal time constant, which is of the order of 1s.

The temperature fluctuation thus estimated agrees with the value obtained from the noise measurement within a factor of 2.

Table 1 is the summary of the results of the samples with different volume. There are good agreements between the calculated and measured values of the temperature fluctuation for the samples with positive temperature coefficient of resistance. On the other hand, the agreement is poor for the sample with negative temperature coefficient.

The measured values of the temperature coefficient of the resistance shown in Table 1 are about 10 times larger than those calculated

TABLE 1
Summary of the results

Sample	Resistance	Relative res. fluct. at 1Hz	Carrier number	Temp. fluct. Meas.	Temp. fluct. Calc.	Corr. coeff. between temp. and res. fluct.	Temp. coeff. of resistance
A	24.4kΩ	-50db/Hz	200	$40.6K^2$	$382K^2$	1.0	$3.3x10^{-3}$
B	9.8	-60	2000	7.3	38	0.2	5.0
C	10.2	-67	10000	4.3	7.6	0.6	1.0
D	114	-63	4000	0.5	19	-0.5	-5.8

from the ratio of the resistance fluctuation to the temperature fluctuation.

The resistance fluctuation and the temperature fluctuation are closely correlated, and the sign of the correlation coefficient coincides with the sign of the temperature coefficient of the resistance.

Further investigations are necessary on the relation between the temperature coefficient and the correlation coefficients.

4. CONCLUSIONS

The temperature fluctuation in a metal microbridge has a 1/f type spectrum, and its magnitude within the band, which is limited by the reciprocal of the thermal time constant, agrees well with the magnitude of the equilibrium temperature fluctuation.

The 1/f fluctuation is closely correlated with the temperature fluctuation, and possibly originates from the equilibrium temperature fluctuation.

REFERENCE

1. S.HASHIGUCHI: Japanese Jour. Appl.Phys., *22*, (1983) L284.

NOISE IN PHYSICAL SYSTEMS AND 1/f NOISE - 1985
A. D'Amico and P. Mazzetti (editors)

ACCURATE MEASUREMENTS OF OPTIMUM NOISE PARAMETERS OF MICROWAVE TRANSISTORS

Giovanni MARTINES and Mario SANNINO

Dipartimento di Ingegneria Elettrica - Universita` di Palermo - Viale delle Scienze, 90128 Palermo, Italy.

A method for measuring losses of the tuner network used as noise source admittance transformer in transistor noise parameter test-set is presented. Since the method is based on noise figure measurements, tuner losses can be determined on-line while performing measurements for determining transistor noise parameters. As experimental verifications the optimum noise parameters of a GaAs FET in the 4 - 12 GHz frequency range, measured through a computer-assisted measuring system, are reported.

1. INTRODUCTION

The noise behaviour of a transistor depends on the reflection coefficient Γ_s of the input termination.

Modern techniques for the determination of the four noise parameters F_o, minimum noise figure, Γ_o, corresponding optimum value of Γ_s (magnitude and phase) and R_n, noise resistance, require measurements of transistor noise figure $F(\Gamma_s)$ for some (redundant, i.e. more then four) values of Γ_s and a proper procedure for computer-aided data-processing, based on the least-squares method [1]. In selecting Γ_s well-established criteria are to be followed [2,3]. The noise figure measuring system contains as admittance transformer a tuner inserted between the noise source and the device under test (DUT).

The DUT output noise powers are then detected by a receiver and a noise figure meter which measure the noise figure of the whole set-up, given by (Friis formula)

$$(1)\quad F_m(\Gamma_{ns}) = \alpha_{\Gamma s}\left[F(\Gamma_s) + \frac{F_r(S'_{22})-1}{G_a(\Gamma_s)}\right]$$

In (1) Γ_s represents the tuner loss which depends on the setting of the tuner; F, G_a and S'_{22} are the noise figure, the available power gain and the output reflection coefficient of the DUT respectively, and F_r is the receiver noise figure [4,5].

Further improvements of the measuring method and of the computer-aided data processing procedure allow simultaneous determining of noise, gain and scattering parameters of transistors through noise figure measurements [6,7].

From (1) it appears that, in order to determine $F(\Gamma_s)$ from $F_m(\Gamma_s)$, previous measurements of the input-tuner loss $\alpha_{\Gamma s}$ are also needed. As far as this measurements is concerned, a recent paper suggests as unique procedure the (obvious) one of characterizing the tuner by its scattering parameters; for any Γ_s the loss $\alpha_{\Gamma s}$ is then computed as the inverse of the tuner available power gain [8]. Unfortunately, this method requires a) two different experimental procedures (noise figure measurements and S-parameter measurements), and b) setting and resetting accurately the tuner in performing the two measurements in different

----

This work was supported by Ministero della Pubblica Istruzione and Consiglio Nazionale delle Ricerche.

times.

The procedure becomes even inapplicable, when the experimenter is interested in the determination of the transistor optimum noise parameters only, i.e. F_o, $|\Gamma_o|$ and $\angle\Gamma_o$. Actually, in this case the optimum input termination reflection coefficient Γ_{mo} of the whole measuring system is first determined by adjusting the tuner until the minimum reading F_{mo} of the noise figure meter is obtained; then F_o is computed from (1) after that $\alpha_{\Gamma mo}$ and $G_a(\Gamma_{mo})$ are measured, and the value Γ_{mo} is assumed as Γ_o.

Obviously, this procedure is inadequate because it does not take into account in a correct way the influence of the tuner loss and of the receiver noise contribution (both variable while adjusting the tuner) on the determination of the optimum input condition of the transistor.

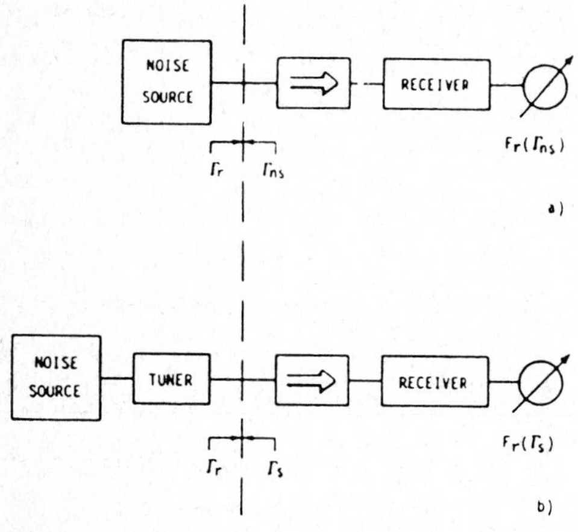

Fig. 1
The two different configurations of the measuring system for determination of tuner loss through noise figure measurements.

All the above drawbacks are fully avoided by the method presented here, which allows measuring tuner losses and DUT available power gain through the same set-up used for noise figure measurements. In other words, F_m, G_a and $\alpha_{\Gamma s}$ are simultaneously measured, for each Γ_s, while adjusting the tuner; thus the DUT noise figure F is computed in real-time from (1). Of course this is possible only if a computer-assisted measuring set-up is available.

2. MEASUREMENTS OF TUNER LOSS THROUGH NOISE FIGURE MEASUREMENTS

Let us first refer to the set-up for measurements of tuner loss only shown in Fig.1. How the method can be applied through a transistor noise parameter test-set is discussed later.

Supposing now, for sake of simplicity, that the noise source is a well matched one, the noise figure for given Γ_s as measured by the meter in Fig. 1,b is given by

$$F_{r\alpha}(0) = \alpha_{\Gamma s} F_r(\Gamma_s) \qquad (2)$$

The receiver noise figure $F_r(\Gamma_s)$ can be computed by first measuring the receiver noise figure $F_r(0)$ in input matched condition by directly connecting the noise source to the isolator, i.e. without the tuner (Fig. 1,a). Taking into account that the receiver is an input-isolated one, and under the simplifying hypothesis $\Gamma_r = 0$, we have [6,7,9]

$$F_r(\Gamma_s) = \frac{F_r(0)}{1 - |\Gamma_s|^2} \qquad (3)$$

From (2) and (3) the tuner loss $\alpha_{\Gamma s}$ is determined. The more general case in which the above assumptions do not hold and the noise source is mismatched is discussed elsewhere [10].

As an example the losses of a tuner as measured by means of the (computer-assisted) set-up of Fig. 1 are reported in Fig. 2.

3. MEASUREMENTS OF OPTIMUM NOISE PARAMETERS

With reference to the simplified (computer-assisted) measuring set-up sketched

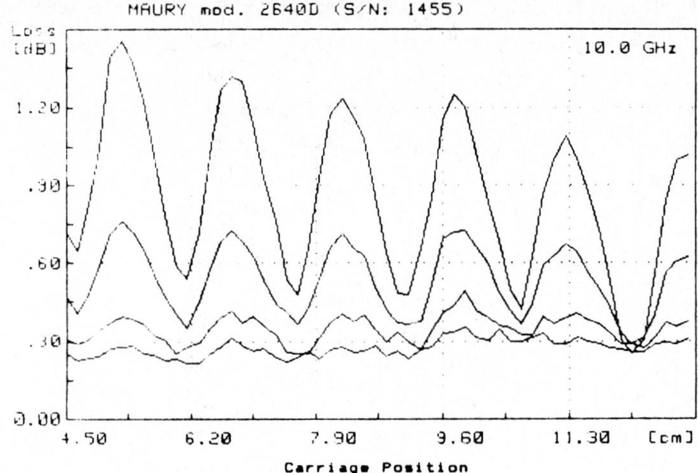

Fig. 2
Loss at 10 GHz of the slide double-screw tuner Maury Microwave 2640D vs. carriage position for some depths of one of the screws (with the other fully extracted).

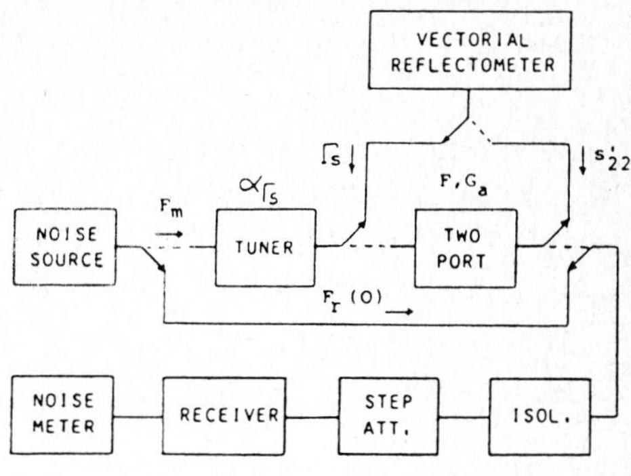

Fig. 3
Simplified block diagram of the measuring system.

in Fig. 3, the measuring steps to carry out, by acting on the switches, are the following

a) for given frequency and transistor bias condition

1) by adjusting the tuner determine the optimum input termination reflection coefficient Γ_{mo} of the whole system; measure Γ_{mo} (and F_{mo});

2) by acting on the tuner realize a value Γ_s of the transistor input termination reflection coefficient in the neighbourhood of Γ_{mo}, and measure it;

3) measure $F_{r\alpha}(0)$ of the tuner-receiver cascade as shown in Fig. 1,b;

4) measure the receiver noise figure $F_r(0)$ as shown in Fig. 1,a);

5) compute $F_r(\Gamma_s)$ through (3) and $\alpha_{\Gamma s}$ through (2);

6) measure the transistor output reflection coefficient S'_{22};

7) measure the noise figure $F_m(\Gamma_s)$ of the tuner-DUT-isolator-attenuator-receiver cascade with the step attenuator set to the minimum value (0 nominal);

8) repeat step 7) for other values of F_r (more than one for redundancy), i.e. for other values of the attenuation (say 3,6,10 dB);

9) compute $G_a(\Gamma_s)$ and $F(\Gamma_s)$ simultaneously through (1) [4,5];

10) repeat steps 2) to 9) for other values of Γ_s until the minimum F_o of $F(\Gamma_s)$ is derived; measure it and the corresponding value Γ_o of Γ_s;

b) repeat a) for other bias conditions of the transistor;

c) repeat a) and b) for other frequency.

Obviously this procedure is also useful when the characterization of a transistor in term of noise, gain and scattering parameters is required. In this case step 10) must be substituted by the following:

10`) repeat steps 2) to 9) for other values of Γs selected through proper criteria 2,3; determine noise, gain and scattering parameters by means of a suitable data processing procedure as already shown elsewhere [6,7].

4. EXPERIMENTAL VERIFICATIONS

As experimental verifications of the method presented, the optimum noise parameters F_o, $|\Gamma_o|$ and $\angle\Gamma_o$ of a GaAs FET NE24483 in the frequency range 4 - 12 GHz for I_{DS} =15%I_{DSS} are reported in Fig. 4.

The computer-assisted version of the measuring system used is shown in Fig. 5.

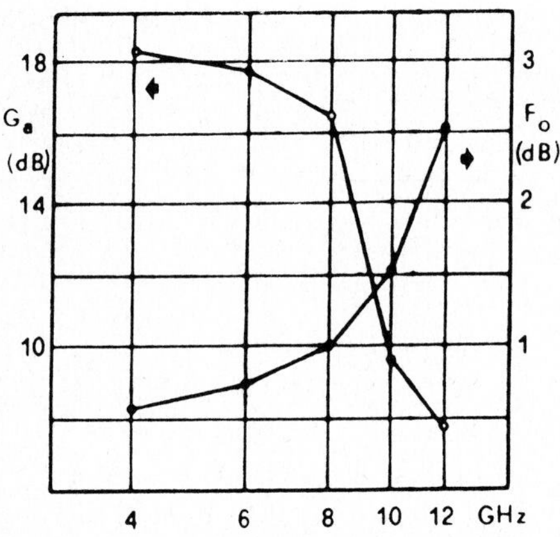

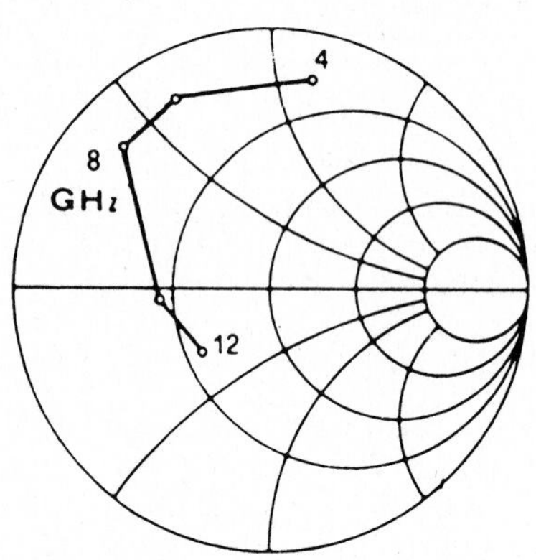

Fig. 4
Optimum noise parameters and associated gain $G_a(\Gamma_o)$ of a transistor NE 24483 (V_{DS} = 3 V; I_{DS} = 15% I_{DSS}) vs. frequency.

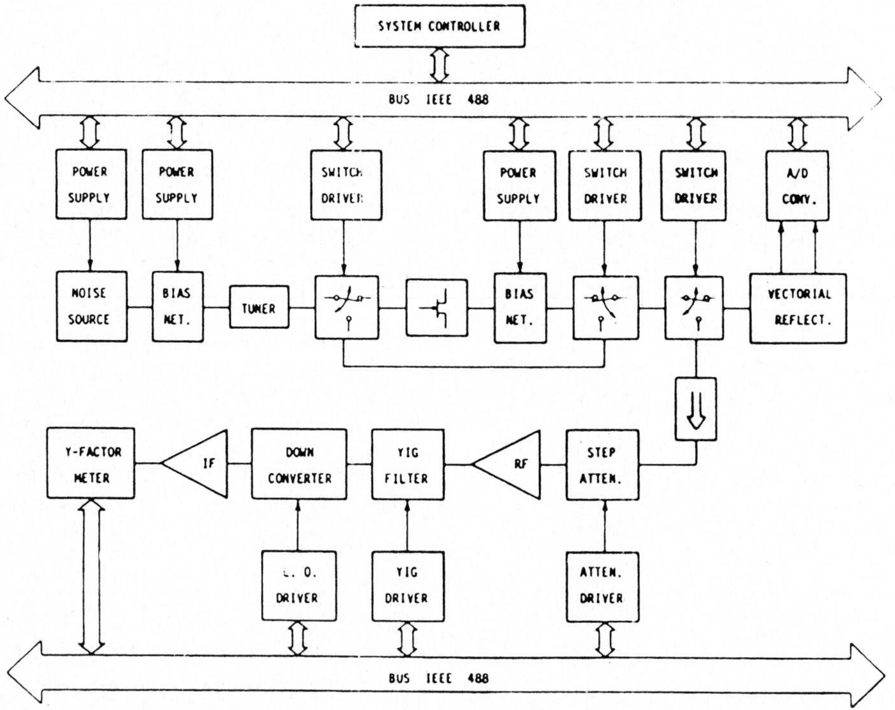

Fig. 5 - Detailed block diagram of the computer-assisted measuring set-up.

REFERENCES

1. R. Q. Lane, Proc. IEEE, vol. 57, pp. 1461-1462, Aug. 1969.
2. G. Caruso and M. Sannino, IEEE Trans. Microwave Theory Tech., vol. MTT-26, pp. 639-642, Sept. 1978.
3. M. Sannino, Proc. IEEE (Letters), vol. 67, pp. 1364-1366, Sept. 1979; also in "Low-noise microwave transistors and amplifiers," edited by H. Fukui, IEEE Press, 1981.
4. M. Sannino, Proc. IEEE (Letters), vol. 68, pp. 1343-1345, Oct. 1980.
5. G. Martines and M. Sannino, IEEE Trans. Microwave Theory Tech., vol. MTT-30, pp. 1255-1259, Aug. 1982.
6. E. Calandra, G. Martines and M. Sannino, IEEE Trans. Microwave Theory Tech., vol. MTT-32, pp. 231-237, March 1984.
7. G. Martines and M. Sannino, IEEE Trans. Instrum. Meas., vol. IM-34, pp. 89-91, March 1985.
8. E. W. Strid, IEEE Trans. Microwave Theory Tech., vol. MTT-29, pp. 247-253, March 1981.
9. M. Sannino, Proc. IEEE (Letters), vol. 70, pp. 100-101, Jan. 1982.
10. G. Martines and M. Sannino, submitted to IEEE Trans. on Microwave Theory Tech.

NOISE IN PHYSICAL SYSTEMS AND 1/f NOISE - 1985
A. D'Amico and P. Mazzetti (editors)

MEASUREMENT OF ELECTRONIC DEVICES NOISE WITH ELIMINATION OF SELF-NOISE OF MEASURING SYSTEM

Ludwik SPIRALSKI

Institute of Electronic Technology, Technical University of Gdańsk, ul.Majakowskiego 11/12, 80-952 Gdańsk, Poland

A method of excess noise measurement for resistors and the two new methods of noise factors measurement for linear circuits with elimination of the self-noise of the measuring system have been described.

The error of a standard noise parameters measurement of low-noise electronic devices strongly depends on the self-noise of the measuring system. The influence of self-noise may be eliminated using the methods developed by the author of this paper.

Well known measurement methods of the excess noise of the low-noise one-port require three steps of the measurement procedure and accounting of corrections[1]. In the system using the calibrated signal or noise generator (Fig. 1), results of measurements are independent of the self-noise of the measuring system[2]. Excess (current) noise of the resistor R_m would be negligible in comparison with the current noise of the element R_x being under test. The measurement is carried out in two steps. In the first one (switch P in the position 1), device R_x under test is biased, and the calibrated signal or noise generator is disconnected. At that time the indication α_1 of power indicator is proportional to the excess noise of R_x and to the total self-noise of the measuring system

$$\alpha_1 = c\left(E^2_{nf} + E^2_{ns}\right) \qquad (1)$$

where c is the constant proportional to the effective gain and to the effective bandwidth of the measuring system. In the second step (switch P in the position 2) the device R_x under test is unbiased and a generator with neglibible output resistance in comparison to R_x is attached. Then the indication α_2 of the power indicator is proportinal to the total self-noise

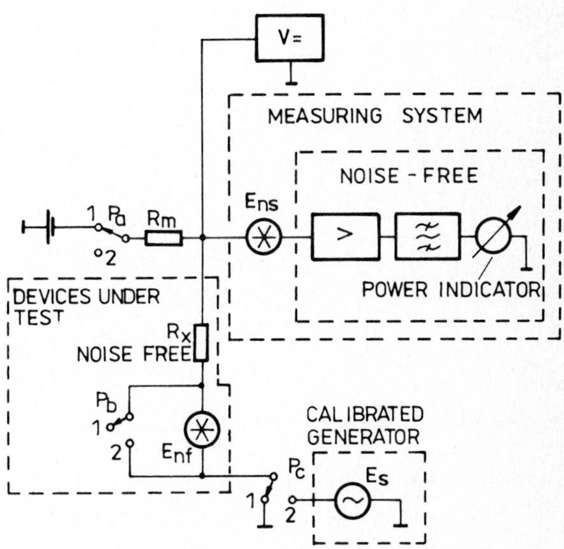

FIGURE 1
Method of measurement of resistor excess noise with elimination of self-noise of measuring system

E_{nf} - rms of the excess noise voltage of the resistor R_x under test,
E_{ns} - rms of the self-noise voltage of the measuring system,
E_s - rms of the voltage of the calibrated signal (noise) generator

of the measuring system and to the power level of the calibrated signal generator

$$\alpha_2 = c\left(E_s^2 + E^2_{ns}\right) \qquad (2)$$

If in both steps of measurement, indication of the power indicator is the same $\alpha_1 = \alpha_2$ then the excess noise level of device under test R_x is accurately equal to the level of the calibrated signal generator $(E_{nf} = E_s)$.

The influence of the self-noise of the measuring system may be eliminated in the case of measurement of the two-port standard noise parameters, e.g. noise figure F. The methods of measurement developed by the author are characterized by the noise figure measurement on the basis of recording the noise power which was measured at least during two measuring steps. In one (two) step of measurement, the noise properties of the measuring system are deteriorated to cause the increase of the noise factor with the value of 0,5 or 1 (Figure 2a and 2b).

In the system shown in Figure 2a we have the 3 dB attenuator following the measured device and in the first step of measurement the noise figure of the system is increased by 0,5 kT_o as compared to its value in the second step of measurement (switch P in the position 1). The power indicator shows the value of measurement given by

$$\alpha_1 = c \cdot kT_o \Delta f \; K_p (F_x K_x + F_p - 0,5) \quad (3)$$

In the second step (switch P in the position 2) the indicated value is

$$\alpha_2 = c \cdot kT_o \Delta f \; K_p \left[\frac{(F_x + F_m) K_x}{2} + F_p - 0,5 \right] \quad (4)$$

where:

c - constant proportional to the amplification of the measuring system,

$k = 1,38 \cdot 10^{-23}$ WAs/K (Boltzmann constant)

T_o = 290K (standard temperature),

Δf - effective bandwidth of the measuring system

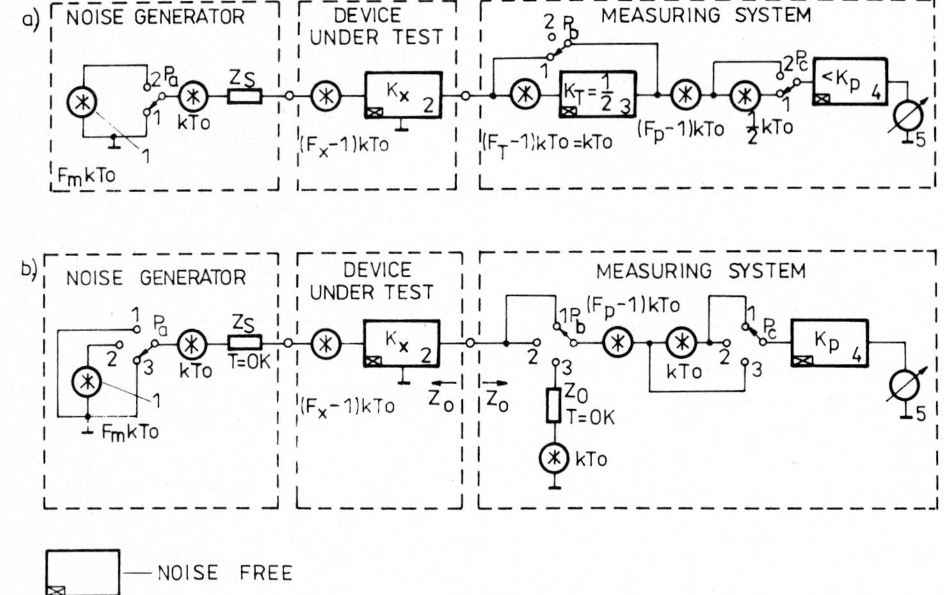

FIGURE 2

The measurement of a noise factor with the full elimination of the self-noise of the measuring system: a) using 3 dB attenuator, b) using the registration of noise power in three phases of measurement; 1 - calibrated noise generator, 2 - measured noisless device, 3 - 3dB attenuator, 4 - amplifier, 5 - power indicator

K_x, F_x (K_p, F_p) - the power amplifications and noise factors of measured device (measuring system), respectively ,

F_m - measured value of noise factor

As it is known from the principle of noise figure measurements the equality $\alpha_1 = \alpha_2$ must be fulfilled, and so from (3) and (4), we have $F_m = F_x$.

The value of the noise factor one may also find using another approach based on the registration of the noise measurement. Using this approach the noise properties of the measuring system are deteriorated appropriately as to have the noise factor increased by 1 kT_o (Figure 2b.) The power inducator shows now the following values of measurements:

- first step (switch P in position 1)

$$\alpha_1 = c \cdot kT_o \Delta f \, K_p (F_x K_x + F_p) \qquad (5)$$

- second step switch P in position 2

$$\alpha_2 = \alpha_1 + c \cdot F_m \, kT_o \Delta f \quad K_x \, F_p \qquad (6)$$

- third step (switch P in position 3)

$$\alpha_3 = c \cdot kT_o \Delta f \, F_p \, K_p \qquad (7)$$

In the third step (measurement value given by (7)) the output of the measured device is disconnected from the measuring system and to its input an impedance Z_o equal to output impedance of measured devices is attached. In this case the value of noise factor of measuring system is decreased by 1 kT_o as compared to its values in both of the proceding cases. Thus, taking into account (5), (6) and (7) we are able to find the noise factor of measured device from the expression

$$F = F_m \frac{\alpha_1 - \alpha_3}{\alpha_2 - \alpha_1} \qquad (8)$$

The laboratory investigations fully confirm the practical usefulness of the presented methods.

REFERENCES

1. American Standard MIL-STD-202B Method 308.

2. Patent PRL 123340 - "Urządzenie do pomiaru szumów nadmiarowych elementów i układów.

3. Patent PRL 95035 - "Sposób eliminacji wpływu szumów własnych wzmacniacza pomiarowego przy pomiarze współczynnika szumów".

4. Patent PRL 117118 - "Sposób pomiaru współczynnika szumów".

NOISE IN PHYSICAL SYSTEMS AND 1/f NOISE - 1985
A. D'Amico and P. Mazzetti (editors)

A NEW METHOD FOR POTENTIOMETER CLICKS EVALUATION

Ludwik SPIRALSKI, Dariusz BARTKIEWICZ

Institute of Electronic Technology, Technical University of Gdańsk, ul. Majakowskiego 11/12
80-952 Gdańsk, Poland

A parameters of clicks for potentiometers and a method of their evaluation have been described.

Deficiences in construction, materials and technology of potentiometers are leading to some undesirable physical phenomena, especially to the noise (Figure 1). A point of importance in the potentiometer construction is a contact

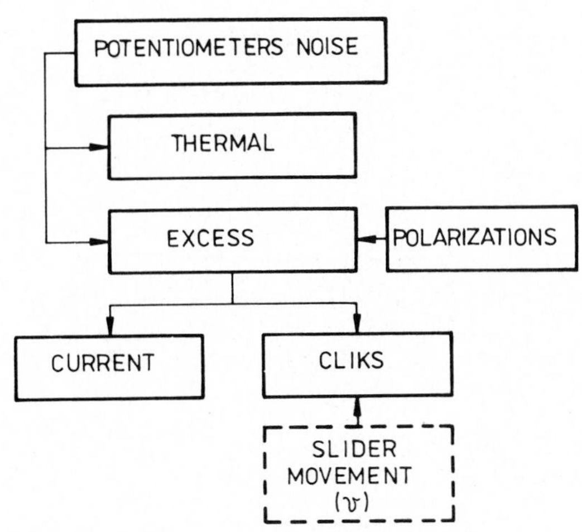

FIGURE 1
Types of noise in potentiometers

formed by two coacting parts: a resistive element and a slider. The surface microtopographic structure of these two parts making a contact is random one, and the intensity of defects occuring in the structure varies during exploitation. These defects are a source of the click noise e_c (t) which is a random process occuring in a surface and is generated when the current flows through the intermediate area existing in the contact place during a slider movement (Figure 2).

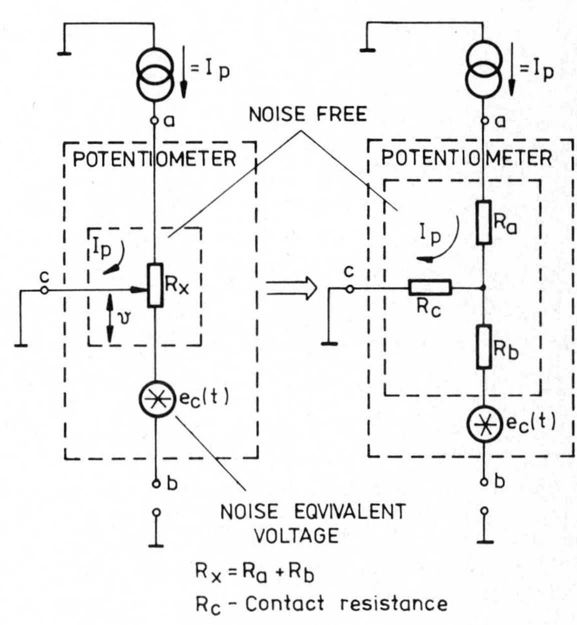

FIGURE 2
Noise equivalent circuit of potentiometer

The click noise depends on the potentiometer type and forms a dominating noise component in these elements.

We have no explicit criterion for the clicks' evaluation. The click detection procedure using the known methods takes into account the quality estimation only, and the results of this procedure depend on the slider movement speed.

The typical method of clicks measurement is based on registration of exceeding a given threshold (level E - Figure 3) at the known speed of slider and current I_p (Figure 2) polarizing the potentiometer R_x under test.

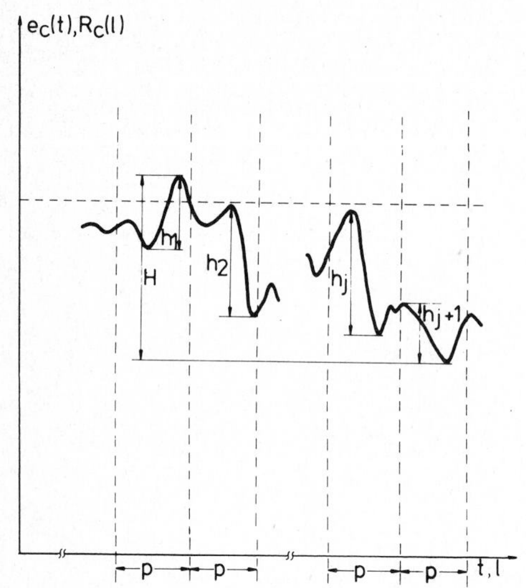

FIGURE 3
Typical waveform of potentiometer clicks
E - treshold level,
H - maximum distance
h_j - distance in j-th section

At the assumed speed of slider the clicks' waveform represents also changes of a contact resistance R_c between the slider and the resistive film in dependence on the length l of the resistive film of the potentiometer (Figure 3). At the current polazization (Figure 2) the intensity of clicks is usually defined by CVR - contact resistance variation (the maximum variation of the contact resistance related to the nominal resistance of the potentiometer). The level of clicks one can also characterize by parameters H and h_j (Figure 3). Those parameters we estimate for the sections of resistance element of the length l. We define them as the difference between the instant values of signals representing clicks of the potentiometer.

In the presented method the result of measurement does not depend on slider movement speed. The new method of clicks measurement with using the linear auxiliary potentiometer R_h polarised by a constant voltage is shown in Figure 4. The sliders of both potentiometers (auxiliary R_h and under test R_x) turn together. Clicks of the potentiometer under test R_x can be recorded on a oscillograph's screen. The results of click measurement obtained by means of the presented method are reproducible.

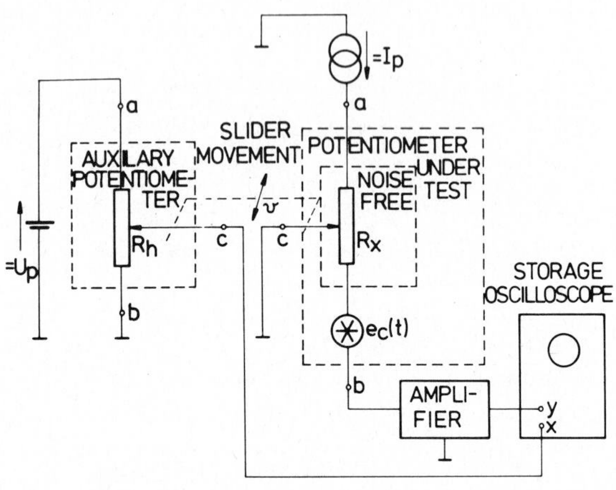

FIGURE 4
A simplified block diagram of oscillographic recording of clicks with elimination of the slider movement speed influence.

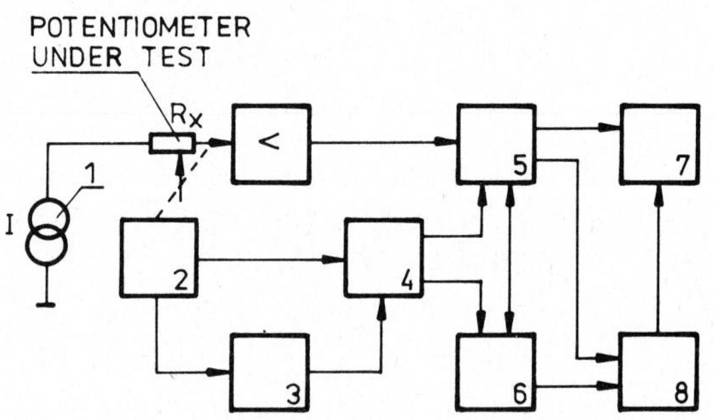

FIGURE 5
System for measuring clicks in potentiometer - block diagram
1 - Current source, 2 - Position-to-electrical signal converter, 3 - Movement direction control unit, 4 - Control unit of the system, 5 - Sample-hold circuit, 6 - Analog-digital converter, 7 - Display of clicks parameters, 8 - Unit processing the results of measurements

An automation of this new procedure of measurement is not difficult (Figure 5) . A current from source flows through a contact between a resistance element and a slider, the last being displaced manually or automatically at a speed chosen arbitrary. The slider position is controlled continuously("on-line") by the converter 2 and by the circuit 3.

Laboratory tests have proved such advantages of presented method as:

- error of the measurement$\leqslant$5%,
- no influence of the slider movement speed on the measurement results,
- possibility of an automation of the measuring process.

MISCELLANEOUS

NOISE IN PHYSICAL SYSTEMS AND 1/f NOISE - 1985
A. D'Amico and P. Mazzetti (editors)

TEMPERATURE DEPENDENCE OF 1/f NOISE IN QUARTZ RESONATORS IN RELATION WITH ACOUSTIC ATTENUATION

Michel PLANAT and J.J. GAGNEPAIN

Laboratoire de Physique et Métrologie des Oscillateurs du C.N.R.S., associé à l'Université de Franche-Comté-Besançon, 32 avenue de l'Observatoire - 25000 Besançon - France

The 1/f noise behaviour of quartz crystal resonators between liquid helium and room temperature is investigated in relation with acoustic attenuation. Large attenuation is observed at 20 K and 50 K The 50 K is due to ionic Na^+ impurity relaxation and the 20 K peak to the interaction of the sound wave with thermal phonons. A strong correlation between 1/f noise level and the acoustic attenuation is found. These results support for quartz crystal the hypothesis of the origin of 1/f noise in phonons fluctuations, and principally in the fluctuations of their relaxation time. The relation between 1/f noise and Q-factor follows a $1/Q^4$ law as it has already been observed at room temperature with resonators of various frequencies. This correlation is not observed on the Na^+ impurity relaxation peak. At very low temperature (below 10 K), the attenuation increases again and 1/f noise level follows approximately the same Q-factor dependence.

1. INTRODUCTION

The 1/f flicker noise still remain widely misunderstood in most domains of the physics. Beyond the fundamental interest to clear the mechanisms of its generation, the subject is of great technical importance in quartz crystal resonators. In fact the short term stability of these resonators used as high quality frequency sources actually is limited by the 1/f noise which correspond in the time domain to a floor.[1]

In previous studies, 1/f noise was measured in various quartz crystal resonators of various frequencies. All the measurements were performed at room temperature[2,3], at the exception of one measurement achieved at low temperature at 4K and at 1K. A strong correlation was observed with the acoustic attenuation (i.e. the inverse unloaded Q-factor) and a theoretical explanation was attempted. The 1/f noise level experimentally follows approximately a $1/Q^4$ law and its origin could be due to fluctuations of the phonon relaxation time.

In this paper experimental work is reported to check this law in the case of a single quartz resonator. The 1/f noise is measured at room temperature and then from liquid helium temperature to about 50K, through the phonon relaxation peak temperature. Results support the previous theoretical hypothesis.

2 - EXPERIMENTAL DEVICE

2.1. Room temperature

The principle of the frequency noise measurement set up is now classical [2]. The resonator of resonance frequency f_0 is inserted in a π transmission network ; the resonance frequency fluctuations induce phase fluctuations of the driving signal, which are detected by means of a balanced phase bridge (Fig. 1).

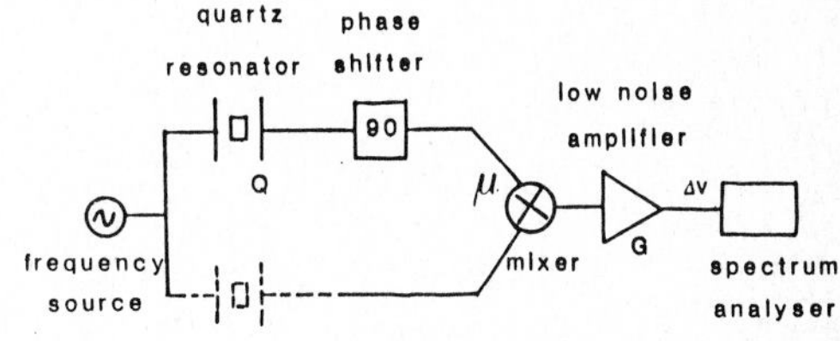

FIGURE 1
Quartz resonator noise measurement set up

The phase bridge of Fig. 1 can be used only when the noise of the resonator to be tested is larger than that of the source. Otherwise two identical resonators must be used, one in each arm of the bridge, to cancell the source frequency fluctuation . In this last case both frequency and Q-factor of the two resonators must be carefully adjusted to obtain the same sensitivity on each arm of the bridge. Q, μ and G are respectively the quality factor of the resonator, the sensibility of the mixer and the gain of the amplifier. Any frequency shift of the resonator Δf induces a voltage shift Δv at the measurement system output following

$$\Delta v = \alpha \, \Delta f \quad \text{where} \quad \alpha = -2Q \, \mu \, G \, / \, f_0 \, . \tag{1}$$

Then the power spectral density of the fractional frequency fluctuations takes the form

$$S_y(f) = \langle \Delta v^2 \rangle \, / \, \alpha^2 \, f_0^2 \tag{2}$$

where $y = \Delta f / f_0$ and $\langle \Delta v^2 \rangle$ is the voltage noise power observed on the spectrum analyser in a 1 Hz bandwidth.

2.2. Low temperature

Low temperature measurements were achieved by immersing the resonator (with a germanium thermometer in contact with its enclosure) inside a liquid helium dewar. The temperature could be shifted from 4.2 K to about 50 K by displacing progressively the whole at an increasing height above the bath. As the frequency temperature curve of the resonator is quite flat in this temperature range, frequency noise measurements were possible with this relatively simple technical arrangement.

3. FLICKER NOISE AT ROOM TEMPERATURE

Flicker noise was measured on various quartz resonators of various frequencies. Either the simple one resonator phase bridge could be used (for resonators of medium quality) or the two resonators phase bridge [2] (for high quality resonators). The results (Fig. 2) show a strong correlation between the levels of the 1/f frequency fluctuations and the unloaded Q-factor of the resonator with approximately the law

$$S_y(1\text{Hz}) \simeq 1/Q^4 \tag{3}$$

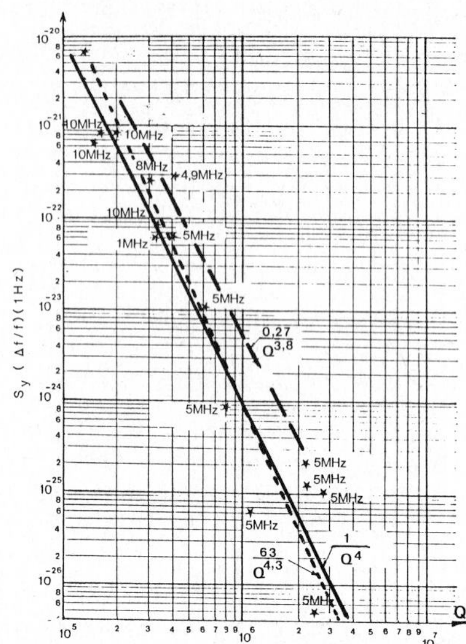

FIGURE 2
Power spectral density of relative frequency fluctuations as a function of unloaded Q-factor for various quartz crystal resonators

4. FLICKER NOISE AT LOW TEMPERATURE

A medium quality 5 MHz resonator was studied in order to use the simpler measurement bridge of Fig. 1. At room temperature we obtained a flicker floor of 10^{-11} for a Q-factor of about 5.10^5. At low temperature the 3 dB bandwidth and the 1/f noise power have been recorded. The 3 dB bandwidth (Fig. 3) shows a clear peak at 50 K which is due to the ionic Na^+ impurity relaxation[5] ; at 20 K the peak looks like a shoulder of the Na^+ peak and is attributed to the interaction of the sound wave with thermal phonons.[6] Then the bandwidth decreases down to the 10 K temperature and increases again beyond 7 K. This phenomenon is not well understood. It can be attributed to the electrodes since an air gap between electrodes and crystal significantly reduce the loss at this temperature (Ref. 5, p. 78). Such behaviour has also been accounted in amorphous glasses[7], it has been established that in addition to phonons, two level low energy excitations exist in glasses. As amorphous silica may be located on the resonator boundary after the crystal cut process both arguments could be connected.

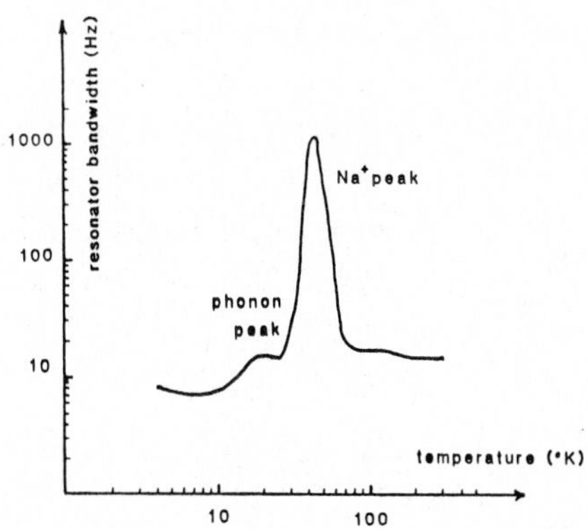

FIGURE 3
3 dB resonator bandwidth as a function of temperature for a medium quality resonator

The corresponding flicker noise power is given on Fig. 4. The curve clearly shows a noise peak ot three orders of magnitude correlated to the phonon relaxation peak. Moreover the noise is also greatly increased down the 7 K temperature. Fig. 4 shows also the noise of the measurement system used in the experiment. The noise of our set up (amplifier, mixer ...) is always several orders of magnitude lower than quartz noise except on the 50 K peak where the sensibility α (see Eqs 1 and 2) is considerably reduced. The quartz noise at this temperature is lower than the set-up noise.

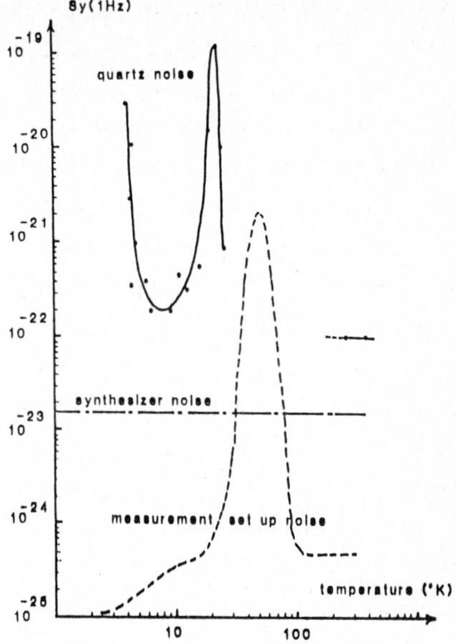

FIGURE 4
Power spectral density of relative frequency fluctuations as a function of temperature

The quartz noise power as a function of the unloaded bandwidth is shown on Fig. 5.

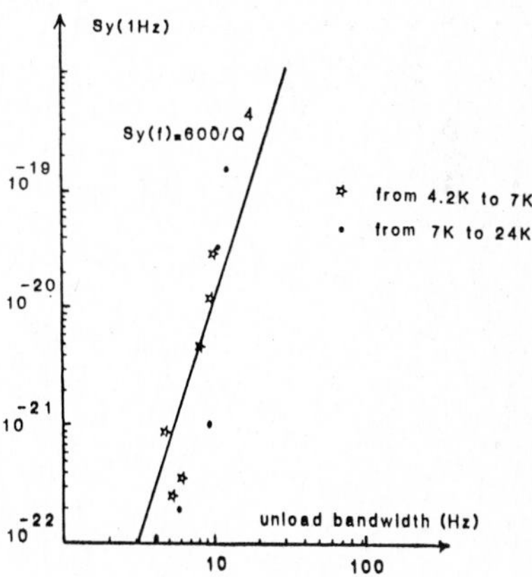

FIGURE 5
Power spectral density of relative frequency fluctuations as a function of unloaded bandwidth for one single resonator excited at various temperatures

The $S_y(1Hz) \simeq 600/Q^4$ dependence is found for all the points distributed along the phonon relaxation peak. Points corresponding to the lower temperature range are represented with stars and the correlation is also satisfactory.

For all points located from 50 K to 300 K the noise measurement was not possible with our actual cryogenic set-up. The crystal has to be inserted in an temperature controlled oven.

The level coefficient 600 obtained for this resonator is higher than the mean coefficient 1 obtained from measurements on resonators of various frequencies. Such drastic shift is not absolutely surprising on account on the number of decades involved in the Q-factor and frequency scales. Nevertheless it can also indicate various mechanisms at the origin of the increasing of flicker noise. This will be verified in future studies.

5. THEORETICAL APPROACH AND CONCLUSION

These measurements support a previous theoretical prediction for the origin of flicker noise of quartz resonators in the fluctuation of the phonon relaxation time τ.[4] From this approach the power spectral density of frequency noise is given by

$$S_y(f) = \frac{1}{Q^4} \left[\frac{E_o}{\Delta E}\right]^2 S_{\Delta\tau/\tau}(f) \tag{5}$$

where E_o is the Young modulus, $\Delta E = CT \langle\gamma^2\rangle$ where C is the specific heat and γ an effective Grüneisen constant.[6]

The 1/f power spectral density is not demonstrated by the theory, but it is in accordance with the Handel's 1/f noise quantum theory.[8]

It has been shown here that the spectral density of relative frequency noise followed the $1/Q^4$ law on the phonon relaxation peak, as expected by the theory. The behaviour at very low temperatures is at first in sight more surprising but not contradictory since the same dependence is obtained. If it is due to the nature of the crystal surface or to very low temperature impurity effects, it can lead also to the fluctuation of the phonon relaxation time. More experimental and fundamental work is necessary to illuminate this question.

REFERENCES

1. J.A. Barnes et al., IEEE Trans. Inst.Meas., 20, 105, 1971.

2. F.L. Walls, A.F. Wainwright, IEEE Trans. Inst. Meas., 24, 15, 1975.

3. J.J. Gagnepain, Proc. 30th Ann. Freq. Cont. Symp., p. 84, 1976.

4. J.J. Gagnepain, J. Uebersfeld, G. Goujon, P. Handel, Proc. 35th Ann. Freq. Cont. Symp., p. 476, 1981.

5. D.B. Fraser, in Physical Acoustics, vol. V, Academic Press, New York (1968).

6. H.E. Bömmel, K. Dransfeld, Phys. Rev., 117, 1245, 1960.

7. S. Hunklinger, W. Arald, in Physical Acoustics, vol. XII, Academic Press, New York (1976).

8. P.H. Handel, Physics Letters, 53A, 438, 1975.

NOISE IN PHYSICAL SYSTEMS AND 1/f NOISE - 1985
A. D'Amico and P. Mazzetti (editors)

FLICKER NOISE IN THICK-FILM (CERMET) RESISTORS: THE EFFECT OF THE FIRING TEMPERATURE

Aldo MASOERO and Anna Maria RIETTO

Istituto Elettrotecnico Nazionale Galileo Ferraris, Torino, Italy

Maria PRUDENZIATI and Bruno MORTEN

Dipartimento di Fisica, via G.Campi 213/A, Modena, Italy

and Gruppo Nazionale di Struttura della Materia, U.R. di Torino e Modena

Thick-film cermet resistors (TFRs) have been prepared at different stages of their microstructure formation by changing the peak temperature T_f of the heating cycle. Measurements have been performed on the power spectral density as a function of frequency (from 0.5 Hz to 20 kHz) measuring temperature (from 95 to 600 K) and firing temperature T_f. Sheet resistance and its temperature coefficient have been also measured in samples with various T_f.

The microstructure of the films deeply affects the level of flicker noise, which cover a range of about three decades by changing T_f and shows a minimum; the microstructural changes has minor influence on the temperature and frequency dependence of the conductance fluctuations. These latter are related with exchanges of charge carriers with localized states having broad distribution of energy, whose position remains unchanged whilst their population and concentration are varied by changing T_f.

1. INTRODUCTION

Thick-film resistors (TFRs) are obtained from screen-printable inks on insulating substrates. The inks contain submicron-sized conducting particles and glassy grains suspended in a viscous fluid which evaporates during the drying process and pyrolyzes during a firing (heating) process. This latter usually involves a well defined cycle, with a specific peak temperature T_f - well above the glass transition temperature - as well as dwell time, and rise and descent rates. The resultant TFR is typically 10-20 µm thick and can be considered as a particular cermet consisting of a glassy matrix where the conductive metal-oxide grains are embedded [1]. Even thought this microstructure is not really stable [2], TFRs can be considered in a "frozen state" since they exhibit chemico-physical and electrical properties stable enought to be exployted for very long times in microcircuits (usually from -40 to 125°C for 10^4-10^5 hours) [3,4] and studied for short times up to at least 600 K without any appreciable change of characteristics. Moreover reproducible samples can be obtained with different ink formulation under a wide range of processing conditions.

These features offer the opportunity for detailed studies of the electrical properties of cermet materials; among others the excess noise was succesfully investigated [5-7]. Recently we have observed that the microstructure and electrical properties of TFRs can be changed in reproducible way by changing the peak firing temperature T_f from near the glass transition temperature up to 1000°C [8]. This observation prompted us to investigate the flicker noise of TFRs prepared in different conditions and then at different stages of their microstructure formation.

2. EXPERIMENTAL

This study is concerned with resistors prepared from a commercial ink (Du Pont 1441). They are essentially made of a matrix of boron lead silicate glass with a transition temperature Tg near 530°C, embedding bismuth-ruthenate grains. Resistors were prepared on 96% alumina and provided with prefired thick-film terminations (DP 9596); all the usual procedures of printing, drying and firing were adered to, except for the value of the peak firing temperature T_f which was chosen as a variable from 600°C (80°C above Tg) up to 1000°C, in steps of

50°C. All the firing dwell times were equal to 60 s and the duration of the whole firing cycle was 2700 s. Sample geometry was $1x1 \cdot 10^{-3}$m. Resistors were studied at room temperature for getting information on the structural and electrical properties. The analysis included measurements of sample thickness, presence of crystalline phases (XRD), surface morphology (SEM), bulk microstructure (SEM and TEM). Measurements of the electrical properties included: sheet resistance at 300 K, temperature dependence of

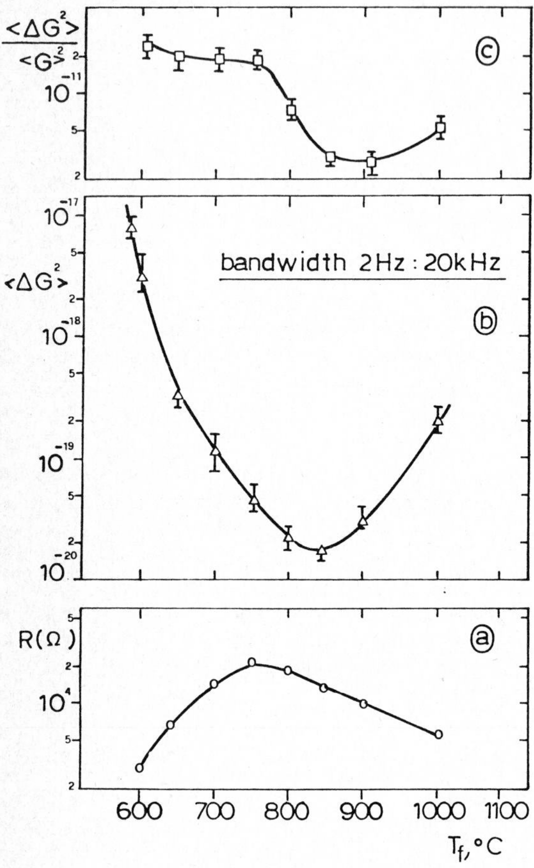

FIGURE 1
Changes of the electrical properties of films as a consequence of a change of firing temperature T_f. Bars cover the scatter of data for different samples prepared in the same conditions.

resistance in the range from 77 K to 625 K, and noise. Experimental procedures for the measurements of electrical properties have been previously described[6]. The experimental results on both resistance and noise of the samples were reproducible even after many cycles of temperature changes over the whole range investigated.

3. EXPERIMENTAL RESULTS

Figure 1 summarizes the main changes in the electrical properties of resistors, i.e. sheet resistance R and noise, expressed in terms of conductance fluctuations $\langle \Delta G^2 \rangle$ and relative conductance fluctuations $\langle \Delta G^2 \rangle / \langle G \rangle^2 = \langle \Delta V^2 \rangle / \langle V \rangle^2$ measured at 300 K. Conductance fluctuations are integrated over the frequency range from 2 Hz to 20 kHz. The resistivity increases about a factor ten by increasing the firing temperature

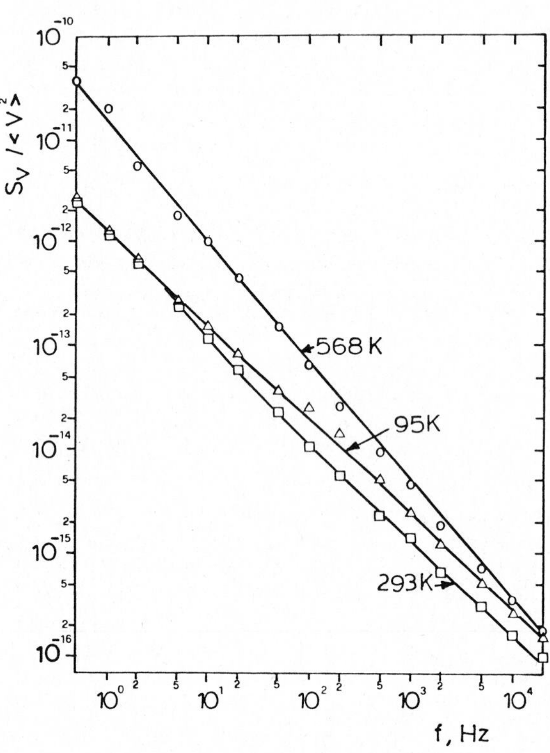

FIGURE 2
Normalized noise power spectral density $S_V/\langle V \rangle^2$ vs. frequency in resistors at different measuring temperatures T. The samples, fired at T_f = 650°C, were $1x10^{-3}$ m wide and $1x10^{-3}$ m long.

T_f from 600 to 750°C and then it decreases going on to T_f=1000°C. Changes of $\langle \Delta G^2 \rangle$ are even larger with values over three decades; here we observe a minimum for samples fired at 850°C.

The plot of $\langle \Delta G^2 \rangle / \langle G \rangle^2$ shows two "critical firing temperatures": T_f=750°C, which marks the transition from samples with higher noise to those of lower noise, and T_f=850°C where the optimum behaviour in terms of relative conductance fluctuations is reached.

Figure 2 exemplifies the features of the power spectral density S_V as a function of the measuring temperature T and frequency f. Here is $S_V/\langle V \rangle^2 = \langle \Delta V^2 \rangle df/ f$ where $\langle \Delta V^2 \rangle \cdot df$ is the mean square voltage noise found in the bandwidth Δf.

Deviations from the pure $S_V$vs. 1/f form are evident in these figures as well as the lower value of S_V (at any f) at 300 K when compared with that of lower and higher temperatures. In fact the relative integral noise $\langle \Delta G^2 \rangle / \langle G \rangle^2$ as a function of the measuring temperature exhibits a complex behaviour (Figure 3) with a broad minimum at a temperature T_{mn} very near to the temperature T_{mr} at which the resistance shows its minimum (see Tab. I)

Table I - T_{mr} and T_{mn} for samples fired at different peak firing temperatures T_f

Peak firing T_f (°C)	T_{mr} (K)	T_{mn} (K)
650	335	300
750	323	300 ± 40
850	293	250 ± 30
1000	273	273 ± 30

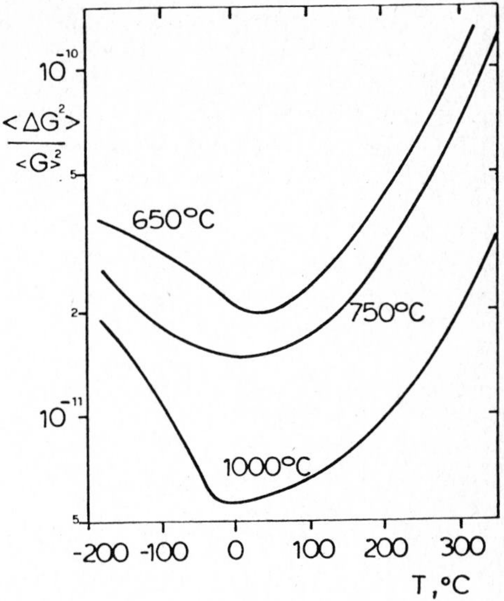

FIGURE 3
Relative integral noise as a function of temperature for samples fired at various peak temperatures T_f.

4. DISCUSSION AND CONCLUSIONS

We have obtained experimental evidences showing that the general features of the flicker noise in TFRs are the same in samples fired at different temperatures,i.e. in different stages of their microstructure formation. However this latter affects deeply the value of sheet resistance and even more the level of the noise.

In order to better understand the origin of this behaviour we have analyzed the power spectral density of the different samples in terms of the island model for flicker noise[7,9]. For this purpose the experimental data $S_V$vs. f were plotted in the form $S_V f$ vs.$\ln(f/f_r)$ where f_r is reference frequency (e.g.see Figure 4). These plots present broad structures and occasionally some peaks whose characteristics (width and frequency of maximum) can be associated with the variance Ψ and the energy spread Φ of trapping stated exchanging charge carriers with the

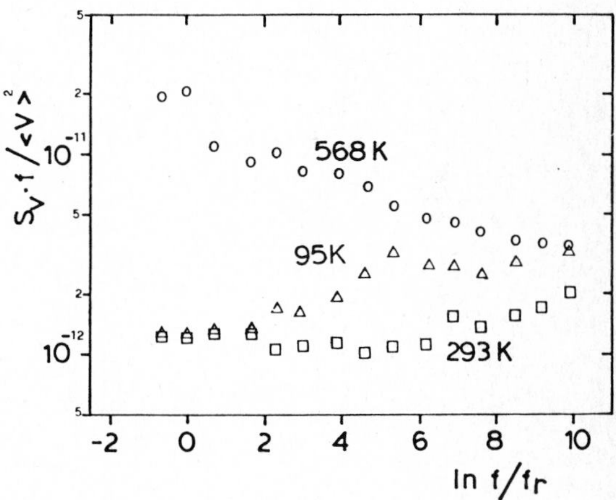

FIGURE 4
Normalized spectral noise power density measured at various temperatures plotted in the form $S_V f$ vs.$\ln(f/f_r)$. The chosen reference frequency is f_r= 1 Hz.

surrounding medium[7]. According to that model the peaks detected at lower temperatures (e.g. the peak at $\ln(f/f_r)=5.2$ at 97 K in Figure 4) corresponds to Φ around 0.2 eV and a relatively narrow Ψ (about 1.5) while the decreasing behaviour of $S_v f$ vs. $\ln(f/f_r)$ for higher temperatures (e.g. 568 K in Figure 4) is expected for high Φ values (larger than 1.5 eV) and large variances (around 8). These figures are common to samples prepared at different peak firing temperatures T_f.

The shapes of these plots are not suitable for a more precise evaluation but they give us two main suggestions: i) charge carriers are trapped and detrapped by a number of relatively shallow (0.2 eV) and deep localized states; ii) the change of the firing temperature affects the number and relative population of these trapping levels, not their energy distribution.

Let us examine now briefly the main results of the characterization of samples in terms of composition and structure. A more detailed presentation of these results will be given elsewhere [8]. Samples fired from 600 to 1000°C have the same thickness, i.e. reached the same stage of densification and sintering of the matrix. When observed with the high resolution allowed for by TEM, the samples appear as cermets whose conductive phase (pyrochlore-grains) are dispersed into the glassy matrix with no physical contact of the grains; the latter are (at any T_f) individual entities separated by glass layers 50- 200 A thick. Size and reciprocal distance as well as presence of agglomerates of smaller grains are scattered inside the same sample in different observed areas and no significant trend can be observed regarding these features by changing T_f. However some meaningful changes in the microstructure of the films prepared at different T_f are related with:

a) Oxidation-reduction reactions concerning the pyrochlore-grains are evidenced by XRD with a reversal of behaviour at 750°C. b) Exchange of Bi from $Bi_2Ru_2O_7$ with Pb of the glass. This exchange reaction is relevant at higher temperatures (from about 800 to 1000°C). c) Phase separation and partial crystallization (SiO_2) of the glassy matrix occurring at higher T_f (850°C) d) Smoothness of the surface of the films. The surface is smoother in samples with T_f from 750 to 850°C; at lower temperatures high viscosity of the glass and less wetting properties gives roughness. At higher temperatures partial crystallization affects the glassy matrix.

On the basis of these observations we argue that there are various possible causes for changes of the electrical properties, and among them excess noise, of cermets at different stages of their formation. The high level of conductance fluctuations in samples fired at T_f lower than 750°C might be related with the presence of different oxides in different states of preparation (see point a). Also a contribution of irregular surfaces on the top of the film can give a contribution to these high level of noise. At higher firing temperatures effects of "doping" of the glass in the intergranular barriers may help the electron transfer between grains [10] and so reduce the "mobility" and its fluctuations while the partial crystallization of the glassy matrix may be in competition and behave as a source of additional scattering of the charge carriers. It is worthnoting that the conventional firing at T_f=850°C results to optimize the noise of the material.

ACKNOWLEDGEMENTS

This work was partially supported by Consiglio Nazionale delle Ricerche and Ministero della Pubblica Istruzione of Italy.

REFERENCES

1. T.V. Nordstrom and H.C.Hills, Int.J.Hybrid Microel. 3 (1980) 4
2. R.W.Vest "Conduction mechanism in thick--film microcircuits" Purdue Final Technical Rep. (1975) ARPA Order 1642, unpublished
3. K.Wilson and N.Sinnadurai, IEEE Trans. on Comp. Hybrid Manufact. Technol. CHMT-5 (1982) 308
4. M.Prudenziati, B.Morten, Int. J. Hybrid Microel. 6 (1983) 96
5. A.Masoero, A.M.Rietto, B.Morten, M.Prudenziati, J.Phys D: Appl.Phys. 16 (1983) 669
6. M.Prudenziati, B.Morten, A.Masoero, Sixth Int.Conf. on Noise in Physical Systems, NBS Special Publ. 614 p. 202-205 (1981)
7. B.Pellegrini, R.Saletti, P.Terreni, M.Prudenziati Phys.Rev. B-27 (1983) 1243
8. M.Prudenziati, B.Morten, C.M.Mari, G.Ruffi, M.Ligabue, in preparation
9. M.Prudenziati, A.Masoero, A.M.Rietto, J. Appl. Phys. 58 (1985) 345
10. M.Prudenziati, Electrocomp. Sci. Technol. 10 (1983) 285

NOISE IN PHYSICAL SYSTEMS AND 1/f NOISE - 1985
A. D'Amico and P. Mazzetti (editors)

THERMAL NOISE AND ZERO-POINT-ENERGY

Werner J. KLEEN

Siemens AG and Technical University Munich*

With reference to 1 during the last years several papers[2,3,4] have been published stating that the zero-point-energy contributes to the available noise energy of resistances, masers and amplifiers. It will be shown that this is not correct.

Planck's law for the average energy density of black radiation at temperature T in vacuum

$$\overline{w}/V = 4\,kT\,\frac{2\pi f^2}{c^3}\,\frac{x}{e^x-1}\,df \qquad (1)$$

(w = energy, V = volume, $x = hf/kT$) is in agreement with all measurements. We have

$$\overline{w}/V = \varepsilon_0\,\overline{E^2}/2 + \mu_0\,\overline{H^2}/2 = \varepsilon_0\,\overline{E^2} \qquad (2)$$

because electrical and magnetic energy are equal. An antenna is introduced into the radiation field. In order to avoid long integrations we choose the simplest one, a Hertz-dipole of length L with the radiation resistance

$$R_r = \frac{2\pi f^2 L^2}{3\varepsilon_0\,c^3}. \qquad (3)$$

The electromagnetic field of the radiation induces an open-circuit voltage v_i and according to eq.(2) we get

$$\overline{v_i^2} = \frac{1}{3}\overline{E^2}\,L^2 = \frac{1}{3}\frac{\overline{w}}{V}\frac{L^2}{\varepsilon_0}. \qquad (4)$$

The factor 1/3 arises from the fact that the dipole picks up energy only from one of three orthogonal directions of polarisation. It follows from eq.(1), (3) and (4)

$$\overline{v_i^2} = 4\,kT\,\frac{2\pi f^2 L^2}{3\varepsilon_0\,c^3}\,\frac{x}{e^x-1}\,df = 4\,kT\,R_r\,\frac{x}{e^x-1}\,df. \qquad (5)$$

This equation is valid not only for a Hertz-dipole but for any form of the antenna as can be shown by more elaborous calculations. A resistance $R = R_r$ equally at temperature T, is matched to the dipole. On account of thermal equilibrium the net power-flow between radiation field and resistance must be zero, i.e. $\overline{v_i^2}/4\,R_r$ must be equal to the available average noise power of the resistance,

$$\overline{P}_{av} = \overline{v^2}/4\,R = kT\,\frac{x}{e^x-1}\,df, \qquad (6)$$

and we obtain the extended Nyquist-equation

$$\overline{v^2} = 4\,kT\,R\,\frac{x}{e^x-1}\,df. \qquad (7)$$

There is no difference whether the resistance is coupled by an antenna to the radiation field of black temperature T or is located in this field, i.e. has the temperature T.

Including zero-point-energy we would have instead of eq.(1)

* Prof. Dr. W. Kleen, Denningerstr. 36, D-8000 München 80

$$w/V = 4\,kT\,\frac{2\pi f^2}{c^3}\left(\frac{x}{2}+\frac{x}{e^x-1}\right)df \qquad (8)$$

The zero-point-energy of the radiation source

$$\overline{w}_0 = 4\,kT\,\frac{2\pi f^2}{c^3}\cdot\frac{x}{2}\,df = \frac{4\pi h f^3}{c^3}\,df \qquad (9)$$

is independent of T. In the range of optical frequencies we have

$$x \gg 1, \qquad x/2 \gg \frac{x}{e^x-1}$$

i.e. $\overline{w}_0$ is predominant. The antenna with the connected resistance is replaced by a detector (photo-diode). From measurements we know that the energy received by the detector approaches zero for $T \to 0$ and does not contain a part independent of the temperature of the radiating source. Therefore we must conclude that zero-point-energy is not contained in the radiation field. Consequently, and on account of the equivalence between received radiation power and noise power of a resistance, as shown before, zero-point-energy is no available noise energy. It is restricted to the interior of resistances and other electronic components where it must exist and was introduced in order to avoid a contradiction to the Heisenberg uncertainty relation at $T \to 0$. It is, however, not radiated, not available, and therefore we believe that an "observation of zero-point-fluctuations" published in 4 must be interpreted differently.

REFERENCES

1. H. B. Callan and T. A. Welton, Phys. Rev. 83 (1951), 34.

2. K. M. van Vliet and A. van der Ziel, Solid-State-Electronics 20 (1977), 931.

3. A. van der Ziel, Solid-State-Electronics 22 (1979), 451; Physica 100B (1980), 97; 101B (1980), 107; 103B (1981), 355; IEEE Journ. of Quantum Electronics QE-17 (1981), 1324.

4. R. H. Koch, D. J. van Harlingen and J. Clarke, Phys. Rev. Lett. 47 (1981), 1216.

NOISE IN PHYSICAL SYSTEMS AND 1/f NOISE - 1985
A. D'Amico and P. Mazzetti (editors)

$\frac{1}{f}$ AND $\frac{1}{\Delta f}$ NOISE IN CARBON-LOADED POLYURETHANE

E.J.P. MAY, S.H. AL-CHARCHAFCHI and H.H. MEHDI

Department of Engineering Science, University of Exeter, Exeter, Devon, EX4 4QF, England.

Electrical conduction in carbon-loaded polyurethane resistors is not well understood at the present time and in any theory of conduction in this material, the ratio of resultant mobility to lattice mobility is likely to be very small. Consequently, according to Hooge and Vandamme,[1] the 1/f noise should be very low. It is shown that 1/f noise is abundantly present when electrical currents are made to flow through this material. This raises doubts regarding the postulation that 'lattice scattering' is the mechanism for 1/f noise.

1. INTRODUCTION

Conductive foam has been used mainly for protecting circuit components from electrostatic effects by shorting out the device leads. The foam is manufactured from carbon-loaded polyurethane. The conduction process is ill understood at the present time but it is suspected that the carbon in the foam has a granular structure. Hooge and Vandamme have shown that $\frac{1}{f}$ noise is very low in materials where scattering is mainly non-lattice and they conclude that $\frac{1}{f}$ noise is due to lattice scattering only. This material, therefore, affords a method of examining their hypothesis.

It will be shown that a sample made from carbon-loaded polyurethane is a very rich source of $\frac{1}{f}$ noise and $\frac{1}{\Delta f}$ noise and these two are related by a factor of four as observed and discussed by Lorteije and Hoppenbrouwers.[2] Also it will be shown that $\frac{1}{f}$ r.f. induced noise is present.[3]

2. EXPERIMENTAL PROCEDURE

2.1. Preparation of the Sample

The conductive foam, known as high density foam, which has a thickness of 5mm, was cut into a rectangular shape with dimensions 5·8cms x 7cms. This piece of foam was sandwiched between two aluminium plates which formed the electrodes. The resistance of the sample was measured and found to be 500 ohms.

2.2. Measurements

The circuit diagrams for the measurements of $\frac{1}{f}$ noise and $\frac{1}{\Delta f}$ noise are shown in Figures 1 and 2 respectively. The measurements were made for various current values.

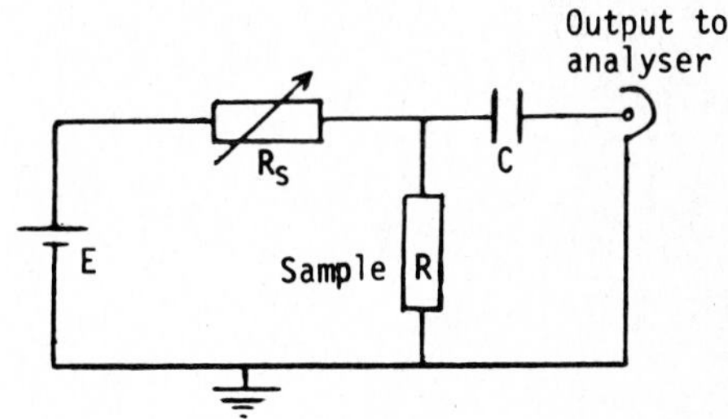

FIGURE 1
Circuit diagram - $\frac{1}{f}$ noise
$R_S >> R$

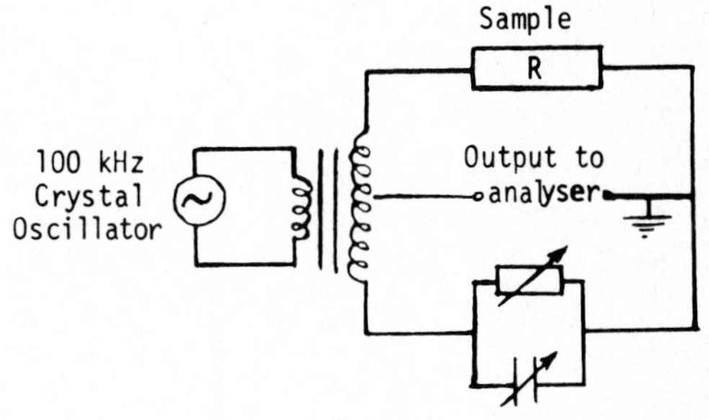

FIGURE 2
Circuit diagram - $\frac{1}{\Delta f}$ noise

The noise power frequency characteristics are shown in Figures 3 and 4. The noise power levels are in dBμ and the bandwidth of noise measurements is 3·16Hz. The frequency of the carrier in the case of $\frac{1}{\Delta f}$ noise was 100kHz. The $\frac{1}{f}$ r.f. induced noise was measured using a circuit similar to that of Figure 2. The results are given in Figure 5, for an r.m.s. current through the sample of 1mA.

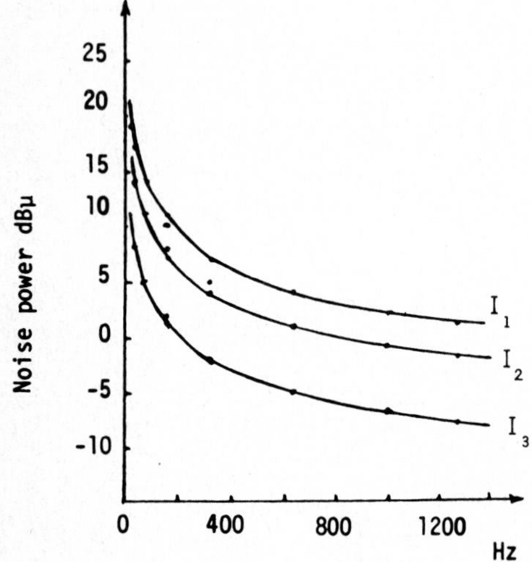

Figure 3
$\frac{1}{f}$ noise in carbon-loaded polyurethane
I_1 = 0·75mA , I_2 = 0·50mA & I_3 = 0·25mA.

3. REMARKS AND CONCLUSIONS

It will be noted from Figure 3 that the $\frac{1}{f}$ noise conforms with the expression given by[4,5],

$$< (\Delta V)^2 > = C\, R^2 I^2 \frac{\delta f}{f}$$

so far as the current, I, is concerned. It is also very close indeed to the ideal $\frac{1}{f}$ law.

Figure 4 reveals the $\frac{1}{\Delta f}$ noise characteristics for the same sample and again the results are very close to the ideal. Furthermore, it will be noted that the amplitudes for the noise powers for the $\frac{1}{f}$ and the $\frac{1}{\Delta f}$ curves are related

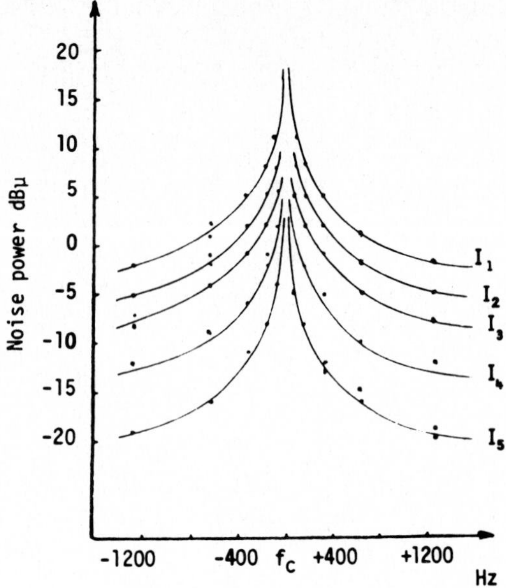

Figure 4
$\frac{1}{\Delta f}$ noise produced by r.f. (100kHz) carrier
I_1 = 1mA , I_2 = 0·75mA, I_3 = 0·50mA,
I_4 = 0·25mA & I_5 = 0·13mA.

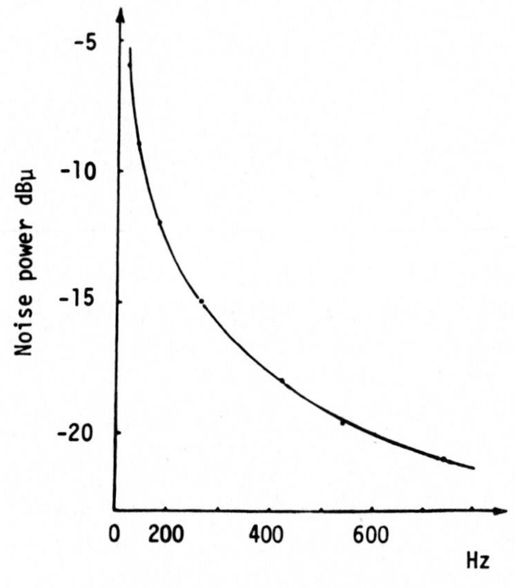

Figure 5
$\frac{1}{f}$ r.f. induced noise

by a factor of four as indicated by Lorteije and Hoppenbrouwers.[2] The curve shown in Figure 5 for $\frac{1}{f}$ r.f. induced noise is also close to the ideal $\frac{1}{f}$ law.

It was found during these experiments that the resistance of 500 ohms was due to the bulk resistance and some surface contact resistance. This observation led to further studies of variation of $\frac{1}{f}$ noise with pressure between the plates. This work when completed will be reported elsewhere. Nevertheless, it is very doubtful if the resistance of the arrangement is due to lattice scattering and one must, therefore, conclude that the mechanism of $\frac{1}{f}$ noise may not lie in the 'lattice scattering-phonon' concept.[1]

REFERENCES

1. F.N. Hooge and L.K.J. Vandamme, Phys. Lett. 66A (1978) 315.

2. J.H.J. Lorteije and A.M.H. Hoppenbrouwers, Philips Res. Repts. 26 (1971) 29.

3. E.J.P. May and W.D. Sellars, Electron. Lett. 11 (1975) 544.

4. F.N. Hooge, Phys. Lett. 29A (1969) 139.

5. F.N. Hooge and A.M.H. Hoppenbrouwers, ibid., 29A (1969) 642.

NOISE IN PHYSICAL SYSTEMS AND 1/f NOISE - 1985
A. D'Amico and P. Mazzetti (editors)

CONTACT-FREE DETECTION OF CURRENT NOISE

Sumihisa HASHIGUCHI, Toru SANADA and Hideki OHKUBO

Faculty of Engineering, Yamanashi University, Kofu, JAPAN

The possibility of a contact-free detection of current noise was theoretically investigated and the necessary conditions for the detection were determined. The detecting system consists of a driver coil, a disk- or toroidal-shaped sample, and a pickup coil. They are magnetically coupled to each other. The fluctuation of the induced current in the disk-shaped sample is detectable for such substance as Bi, whose conductivity is more than 10^6S/m and α/n(Hooge's α normalized by carrier density) is more than 10^{-25}m^3. It is necessary to apply the driving current with the carrier to noise ratio of more than 160db at 1Hz from carrier, and use a low-noise detecting system whose noise level is less than -180dbV/Hz at 1Hz. The threshold of the detection can be improved by increasing the frequency of the driving current. It is also shown that a toroidal sample containing electrolytic solution, the threshold of the detection can be improved by 100db when the effective carrier number is reduced by squeezing the current flow through a perforation provided on a wall put across the current flow.

1. INTRODUCTION

In the conventional methods of current noise detection, the accuracy depends strongly on the quality of the ohmic electrodes which supply the driving current and detect the noise voltage.

In the new method proposed here, the driving current is generated by electromagnetic induction and the fluctuation in the current is detected with a pickup coil coupled to the sample, as shown in figure 1. The voltage induced in the pickup coil contains the carrier component accompanied with the background noise sidebands, and the additional noise sidebands originate from the conductivity fluctuation in the sample. Almost all of the carrier and background noise sidebands are cancelled, and the additional noise sidebands can be obtained by subtracting a fraction of the driving current accordingly.

The magnitudes of the additional noise sidebands were estimated by numerical analysis.

2. CALCULATION PROCEDURES

The sample is divided into thin coaxial rings with the driving and pickup coils.

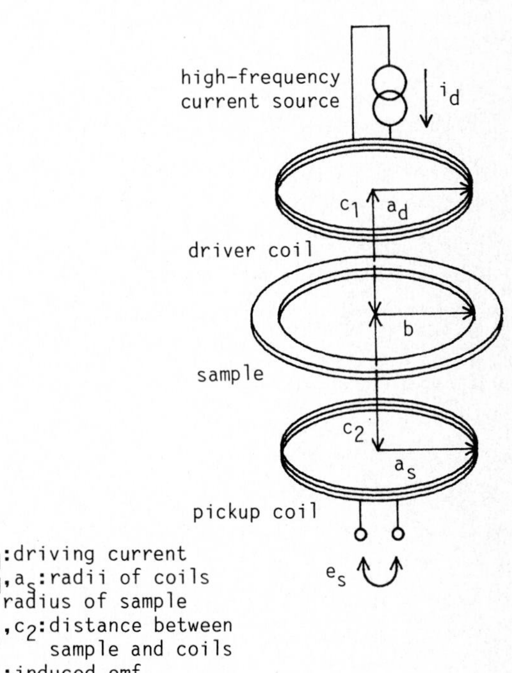

FIGURE 1
A setup for contact-free detection of current noise

This research was granted by Hoso Bunka Kikin, Tokyo, JAPAN.

Each of the following quantities are calculated in the order indicated below.

1) Self inductance of each of the thin rings and two coils. It is a function of the dimensions of the sample and the coils.

2) Mutual inductance between each of the thin rings and each of the two coils. It is also a function of the dimensions.

3) Current i_s each of the thin rings induced by the driving current in the driving coil. It is a function of the inductances mentioned above and the resistance of each ring.

4) The spectral density of the sideband, S_i, at 1Hz from the frequency of the driving current. It is obtained from Hooge's empirical expression of 1/f fluctuation as:

$$S_i/i_s^2=\alpha/\pi ntrdr, \quad (1)$$

where α is Hooge's parameter, and n is the carrier density in the sample, and t, r and dr are respectively the thickness, radius and width of the sample.

5) The electromotive force in the pickup coil induced by each of the rings and the driving coil. Total emf is the vector sum of the contributions of currents in the driving coil and each ring.

6) The spectral density of fluctuation in emf in the pickup coil. It is the sum of the mean square of the contributions of the fluctuating current in each ring.

7) α/n, Hooge's parameter normalized by carrier density. This is adopted as the indicator of the sensitivity of the detection.

The value of the indicator α/n is computed for various values of the geometrical parameters and driving conditions within the practically realizable range.

In the course of calculation, the contributions of the current in each ring to the current in other rings is not taken into account because it is several orders of magnitude smaller than that of the driving coil.

3. RESULTS

Figure 2 shows some of the results of the analysis on the disk-shaped sample. The sample with the diameter of 5cm and the thickness of 100nm is driven by the driving coil of 5cm in diameter set 2mm apart from the sample. The pickup coil which is 5cm in diameter is set 2mm apart from the sample. The driving current is 0.1A.

The ordinate is the ratio of Hooge's parameter α to the carrier density n, and the abscissa is the conductivity σ of the sample. The solid lines show the threshold for the noise detection determined by the frequency f_c of the driving current, and the noise voltage v_{min} of the amplifier connected with the pickup coil. The broken lines show the threshold determined by the noise to carrier ratio N/C of the driving current. Noise in such a sample whose (σ, α/n) plot is above both the solid and broken lines can be detected. It is shown that the higher frequency is preferable. Plots for seven substances are also shown in the figure.

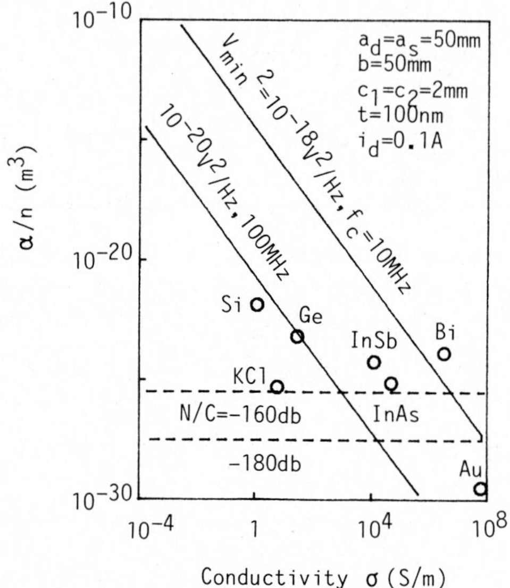

FIGURE 2
Thresholds of contact-free detection of current noise

It can seen that noise in bismuth only is detectable but noise in the other substances are difficult to detect under practically realizable conditions.

As for the toroidal-shaped sample the threshold is almost the same with that of the disk-shaped sample. This is due to the fact that the induced current in the disk-shaped sample flows in a filament-like path, just in the vicinity of the windings of the driving coil.

In order to make the noise level high enough to be detected, it is effective to reduce the carrier number contributing to noise generation.

For electrolytic solutions, it is easy to make samples in a toroidal-shape and to put a wall with a perforation, as shown in fig.3, in order to squeeze the current flow. The toroidal-shaped sample assumed in numerical analysis has the outer radius d of 2.5cm, the width of 1cm and the thickness of 1mm. Inductance of the sample is 1μH. The sample contains 1 normal KCl aqueous solution (concentration of 71.1g/kg), whose conductivity is 11.1S/m and α/n is $8.3\times10^{-27}m^3$, where Hooge's α of 1N KCl is assumed to be 10. The diameter and the length of the perforation are assumed to be 0.01mm. It is also assumed that the inductance of the driver and the pickup coils are 10 μH and 1mH, respectively.

By driving this sample with the driving current and voltage of 2A and 628V, respectively, at 5MHz, the picked-up fluctuation level of -90dbV/Hz at 1Hz is obtained. This is much higher than the noise level of the amplifier and is easily detected, however, the driving requirements are not easy to satisfy.

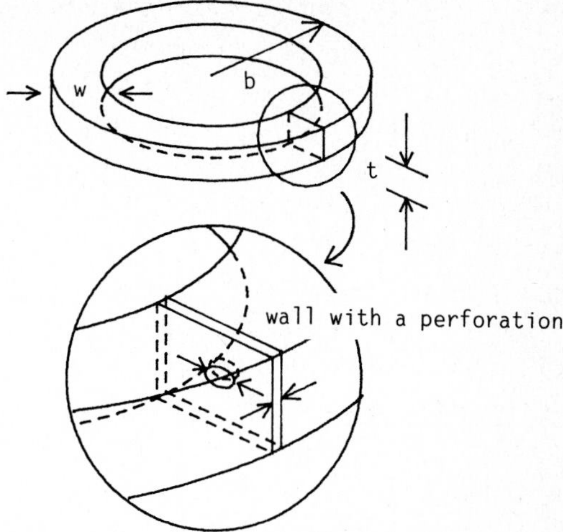

FIGURE 3
Toroidal sample with a current squeeze

If the level of the picked-up fluctuation can be as small as -131dbV/Hz at 1Hz, the driving requirements are relaxed to 0.5A and 31.4V at 1MHz. These requirements can be easily satisfied, and the level of the piched-up fluctuation is 50db higher than that attained with the disk-shaped sample.

4. CONCLUSION

The contact-free detection of current noise is possible for electrolytic solution when the effective carrier number is reduced by squeezing current flow in a toroidol sample with the utilization of a wall having a perforation.

NOISE IN PHYSICAL SYSTEMS AND 1/f NOISE - 1985
A. D'Amico and P. Mazzetti (editors)

DETERMINISTIC NOISE IN EXTRINSIC PHOTOCONDUCTORS

S.W. TEITSWORTH and R.M. WESTERVELT

Division of Applied Sciences and Department of Physics, Harvard University, Cambridge, Massachusetts 02138, U.S.A.

The measured current response of ultrapure Ge far-infrared photoconductors at 4.2 K exhibits behavior characteristic of driven nonlinear oscillators including a period doubling cascade to chaotic oscillation which produces deterministic broadband noise with noise temperature $T_n \sim 10^8$K, frequency locking effects, and intermittency which produces low-frequency noise of the form $1/f^{\alpha}$ with $\alpha \sim 0.5$ to 1.5. This deterministic broadband and low-frequency noise is fundamentally different from noise due to stochastic mechanisms such as Johnson and shot noise. Reduction of the deterministic noise level by greater than 30 dB has been achieved by locking onto a strong sinusoidal drive voltage. These experimental results are in good agreement with numerical solutions of a simple nonlinear rate equation model of extrinsic photoconductors.

Cooled extrinsic photoconductors are among the most sensitive detectors of far-infrared (FIR) radiation, yet they can generate large amounts of noise for reasons which have been poorly understood. Recent experiments on extrinsic Ge,[1,2] GaAs,[3] and InSb[4] photoconductors have shown complex oscillation and broadband noise characteristic of driven nonlinear oscillators.[5] Driven mechanical oscillators, for example, such as the damped pendulum[6] produce broadband and low frequency noise of the form $S(f) \propto 1/f^{\alpha}$ when driven into the nonlinear regime even though they are exactly described by deterministic rate equation models. We show here that analogous phenomena occur in extrinsic FIR photoconductors. This deterministic noise is distinct from noise arising through the well-known stochastic mechanisms such as Johnson noise, shot noise, and generation-recombination noise.[7]

The standard spatially-uniform rate equations describing charge trapping and conduction in extrinsic photoconductors with a single acceptor level are:

$$\frac{dp}{dt} = \gamma a_0 - pra_* + p\kappa_0 \quad (1)$$

$$\varepsilon \frac{dE}{dt} = J_{ext} - pev \quad (2)$$

where p, a_0, and a_* are the concentrations of free holes, and neutral and ionized acceptor levels; E, v, and ε denote the electric field, the drift velocity, and the dielectric constant; and γ is the infrared generation rate. Carrier heating at modest fields $E \sim 1$ V/cm in Ge due to high mobility at low temperature produces strongly nonlinear field dependence in the coefficients r and κ which describe carrier recombination and impurity impact ionization. The reduction in r at high fields increases the carrier lifetime τ_0 and therefore the sensitivity of Ge photoconductors; the abrupt increase in κ at higher fields eventually causes breakdown for $E \sim 3$ V/cm when $\tau_0 \to \infty$.

Equations (1) and (2) constitute a two dimensional dynamical system in the form of a damped, highly nonlinear oscillator. Small signal analysis of this model reveals damped oscillatory transient response, which is commonly observed in actual devices.[7,9] This oscillatory response can be understood as the resonance between the sample capacitance and a ficticious inductance which models changes in dp/dt with

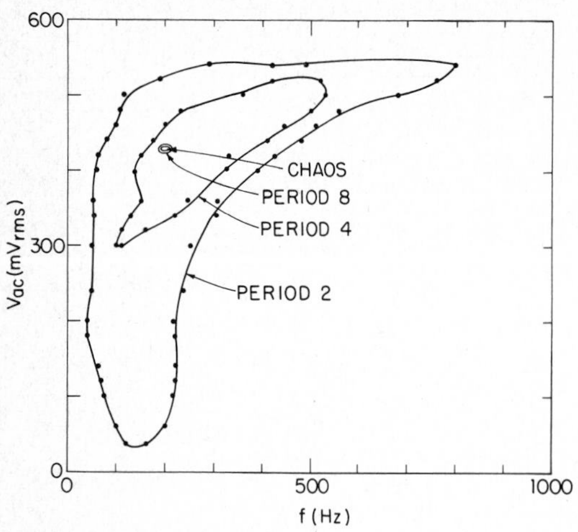

FIGURE 1
Experimental "phase diagram"; the type of subharmonic current response (e.g. period 1, 2, 4, 8, and chaotic) is indicated for applied voltage $V_{dc}+V_{ac}\cos(2\pi ft)$ with $V_{dc} = 0.9$ V.

E through Eq. (1). When driven by a time-dependent current density, $J_{ext} = J_{dc} + J_{ac}\cos(2\pi ft)$, equations (1) and (2) exhibit a variety of characteristic nonlinear phenomena including a period doubling cascade to chaos with broadband noise, frequency locking, and intermittent hopping between modes of oscillation.[8]

The experimental samples were ultrapure p-type Ge crystals fabricated at Lawrence Berkeley Laboratory by E.E. Haller with shallow (~10 meV) acceptor concentration $a \sim 10^{10}$ cm^{-3} to 10^{11} cm^{-3} and B-ion-implanted electrical contacts for holes at low temperature.[1] Results for two sample geometries are reported: 1) rectangular ($8\times8\times4$ mm^3) with electrical contacts on opposite 8×8 mm^3 faces, and 2) rectangular ($0.5\times0.5\times8.0$ mm^3) with an array of five contacts along one 0.5×8.0 mm^2 face. Electrical measurements were made at 4.2 K with the sample enclosed in a cold metal shield at the same temperature. Constant FIR illumination was provided by a blackbody radiator located inside the shield with adjustable temperature ~30 K. A superposition $V(t) = V_{dc} + V_{ac}\cos(2\pi ft)$ of a

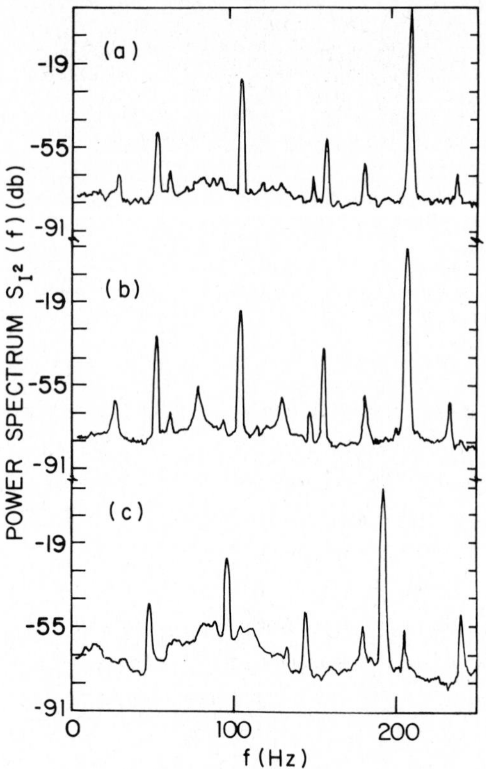

FIGURE 2
Experimental power spectra for the current I(t) showing the transition to chaos for $V_{dc} = 0.9$ V, $V_{ac} = 0.4$ V: (a) f = 209 Hz, period 4; (b) f = 206 Hz, period 8; (c) f = 193 Hz, chaotic period 4.

constant voltage V_{dc} and a sinusoidal voltage with amplitude V_{ac} and frequency f was applied to the crystal and the time-dependent current I(t) was measured. For $V_{ac} = 0$ the measured d.c. I-V characteristics are linear at low average applied fields $E_{dc} = V_{dc}/L$ (L = length of sample), superlinear for $E_{dc} > 0.5$ V/cm, and exhibit breakdown due to impurity impact ionization at $E_{dc} \simeq 3$ V/cm. For moderate FIR illumination the Ge samples show[1] spontaneous oscillation ($V_{ac} = 0$) above a threshold field $E_{th} \sim 0.5$ V/cm to 3 V/cm. Below threshold $E_{dc} < E_{th}$ the samples are stable but exhibit a ringing transient response to steps in applied voltage.

FIGURE 3
Experimental time trace of current I(t) showing intermittent hopping between two modes of oscillation. $E_{dc} = 1.24$ V/cm, length of time record 5 sec.

Figure 1 shows a "phase diagram" in which the qualitative type of current response is plotted vs. V_{ac} and f for the large sample ($8.0 \times 8.0 \times 4.0$ mm^3) with $V_{dc} \cong .9$ volts. For $V_{ac} \lesssim 30$ mVrms the current is simply periodic and no excess noise is observed. However, for increasing V_{ac} and frequencies close to the damped ringing frequency of the sample ~200 Hz the current undergoes a period-doubling transition[5] to period 2 response, then to period 4, then to period 8, and finally to chaotic oscillation in an island in V_{ac} and f as indicated. For higher a.c. drive $V_{ac} \gtrsim 550$ mVrms chaotic response and excess noise is suppressed as shown in Fig. 1 and the system returns to simply periodic behavior. This locking phenomenon can be used to suppress the spontaneous chaotic oscillation[1] by a strong ac drive; we have observed reductions in broadband noise power by >30 dB by this technique. Similar beneficial effects of frequency locking on noise have been observed in solid state microwave oscillators.[10]

Figures 2a to 2c show power spectra of the current response for $V_{ac} = 0.43$ V and three drive frequencies $f = 209$ Hz, $f = 206$ Hz, and $f = 193$ Hz, entering the chaotic island in Fig. 1 horizontally. As shown, the power spectrum develops from period 4 in Fig. 2a, to period 8 in Fig. 2b, to a chaotic period 4 with deterministic broadband noise in Fig. 2c. The corresponding effective noise temperature in Fig. 2c is $\sim 10^7$ K, calculated using the Nyquist formula with the measured dynamic resistance. Note the 80 dB dynamic range in Fig. 2, which demonstrates the low noise level of these devices in the absence of chaos.

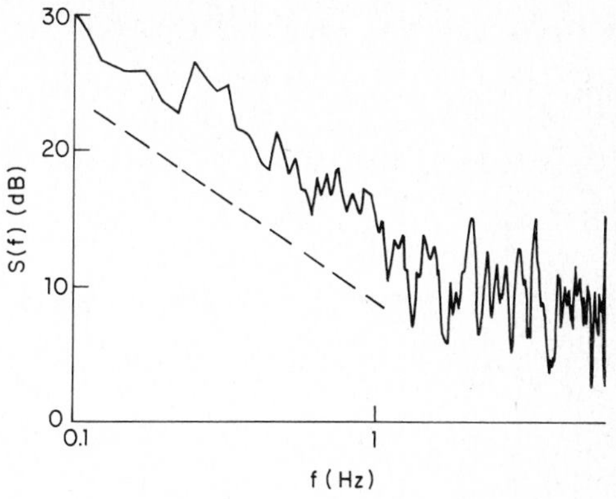

FIGURE 4
Current power spectrum corresponding to Fig. 3 on log-log scale. The dashed line indicates an empirical slope of -1.5.

Nonuniform illumination or multiple electrical contacts can experimentally produce multiple stable modes of spontaneous oscillation for given experimental conditions. Intermittent hopping between different modes of oscillation and accompanying deterministic low-frequency noise with an algebraic power spectrum of the form $1/f^{\alpha}$, $\alpha \sim 0.5$ to 1.5 was observed for the small ($0.5 \times 0.5 \times 8.0$ mm^3) sample. Figure 3 shows the time trace of the current for roughly 450 periods of oscillation with an average electric field of $E_{dc} \simeq 1.3$ V/cm > E_{th} applied across the second and fourth of five contacts. The intermittency shown is produced by hopping between two competing modes of oscillation created by the intervening third contact. The corresponding power spectrum shown in Fig. 4, exhibits low-frequency noise of the form $S(f) \propto 1/f^{\alpha}$ with $\alpha \simeq 1.5$, over a limited range of frequencies. The lowest frequency in Fig. 4 is ~1000 times less than the frequency of oscillation. Intermittency and algebraic power spectra associated with hopping between multiple steady states are well known in numerical studies of nonlinear dynamical systems.[5,6,11]

We have given both experimental and theoretical evidence for the existence of deterministic noise in extrinsic photoconductors. This noise, which may be spectrally flat or in the form of excess low-frequency noise, is a fundamental property of extrinsic photoconductors and is produced by the nonlinearity of the rate equations governing their operation rather than by unwanted defects.

We thank E.E. Haller for providing the Ge samples. This work was supported by the Office of Naval Research under Contract N00014-84-K-0329.

REFERENCES

1. S.W. Teitsworth, R.W. Westervelt, and E.E. Haller, Phys. Rev. Lett. 57, 825 (1983).

2. J. Peinke, A. Mühlbach, R.P. Huebner, and J. Parisi, Phys. Lett. 108A, 407 (1985).

3. E. Aoki and K. Yamamoto, Phys. Lett. 98A, 72 (1983).

4. D.G. Seiler, C.L. Littler, and R.J. Justice, Phys. Lett. 108A, 462 (1985).

5. J. Guckenheimer and P. Holmes, *Nonlinear Oscillations, Dynamical Systems, and Bifurcation of Vector Fields*, (Springer-Verlag, New York, 1983), and references therein.

6. E.G. Gwinn and R.M. Westervelt, Phys. Rev. Lett. 54, 1613 (1985), and references cited therein.

7. P.R. Bratt, in *Semiconductors and Semimetals*, edited by R.K. Willardson and A.C. Beer (Academic, New York, 1977), p. 39.

8. S.W. Teitsworth and R.M. Westervelt, Phys. Rev. Lett. 53, 2587 (1984); and to be published.

9. R.M. Westervelt and S.W. Teitsworth, J. Appl. Phys. 57, 5457 (1985).

10. S.H. Izadpanah, Z. Rav-Noy, S. Mukai, S. Margalit, and Amnon Yariv, Appl. Phys. Lett. 45, 609 (1984).

11. F.T. Arecchi, R. Badii, and A. Politi, Phys. Lett. 103A, 3 (1984).

NOISE IN PHYSICAL SYSTEMS AND 1/f NOISE - 1985
A. D'Amico and P. Mazzetti (editors)

STABILIZATION BY EXTERNAL WHITE NOISE

F.J. DE LA RUBIA

Fisica Fundamental-UNED, Apdo.Correos 50487, 28080 Madrid,Spain.*

The possibility of stabilization by noise is indicated. We show that an unstable deterministic non linear system may become a recurrent diffusion process by perturbing one of the parametersby a white noise,thus avoiding the explosion of the system. The influence of small thermal fluctuations as well as some more general systems are also discussed.

1. INTRODUCTION

In recent years there has been an increasing interest in the study of nonlinear systems with random coefficients or, in other words, in the effect of external noise on dynamical systems.

The most interesting is the fact that this external noise,besides the intuitively obvious effect of perturbing an,otherwise deterministic, behavior by spreading an initially delta-like distribution[1,2], may play a stabilizing role by, for instance,creating new accesible steady solutionsfor the system,not allowed in a purely deterministic description[3-5],changing the critical values for the onset of new regimes[6-8],inducing oscillatory behavior for values of the deterministic parameters that predict a stationary behavior[9], etc. It is worth noting that some of these predictions have been experimentally confirmed[10,11],therefore increasing the relevance of the so called *noise induced transitions*[3].

In this work we generalize previous results[12] and discuss the possibility of stabilizing a broad class of one-dimensional unstable non linear systems by perturbing conveniently the parameters by a white noise.

2. THE GENERAL DETERMINISTIC MODEL

Consider the following general non linear unstable model

$$dx/dt = f(x) = \sum_{r=0}^{n} a_r x^r \qquad (1)$$

where $n \geqslant 2$ and we only assume $a_n > 0$ and $a_0 > 0$. Equation (1) has, for all n ,at most an even number of real and positive steady states, x_i, such that $x_1(s) < x_2(u) < \ldots < x_i(u)$, where s(resp. u) indicates stable (resp. unstable) states. Note that the biggest steady state, x_i, is always unstable and also that the coefficients in (1) may be such that no real and positive steady state exists. Therefore, if the initial condition, $x(0)$, is such that $x(0) > x_i$ (or for all initial conditions if no steady state exists) then $\lim_{t \to \infty} x(t) = \infty$.

As a particular case that will be considered later take n=2. Then if $a_1^2 > 4\, a_2\, a_0$ the system (1) posseses two steady states $(a_1 < 0)$

$$x_{1,2} = [-a_1 \pm (a_1^2 - 4\, a_2\, a_0)^{1/2}] / 2a_2 \qquad (2)$$

where x_1 (x_2) is unstable (stable) and $x_1 > x_2$. On the other hand if $a_1^2 < 4\, a_2\, a_0$ the system has no real and positive steady state and $x(t)$ diverges for all initial conditions.

3. THE STOCHASTIC MODEL

To elucidate whether the unstable system (1) may be stabilized by noise take $a_n(t) = a_n + \sigma\xi(t)$

*This work has been sponsored by the Stiftung Volkswagenwerk.

where $a_n > 0$, $\xi(t)$ is a *white noise* ($\langle\xi(t)\rangle = 0$, $\langle\xi(t)\,\xi(t')\rangle = \delta(t-t')$) and σ is the strength of the noise. It is interesting to remark that the choice of $a_n(t)$ as the stochastic parameter *is not* the only possibility as we will discuss later.

With the assumptions made above, $x(t)$ becomes now a diffusion process[13,14] and obeys Ito's stochastic differential equation

$$dx = f(x)\,dt + g(x)\,dW \qquad (3)$$

where $f(x)$ is given by (1), $g(x) = \sigma x^n$ and W is the Wiener process (Brownian motion). To investigate the process $x(t)$ we must study the nature of the boundaries of the problem, 0 and ∞, as well as its asymptotic behavior. It is well known that most of the qualitative characteristics of the process $x(t)$ can be deduced from the integrability near the boundaries of the function

$$\phi(x) = \exp\{-\int_\beta^x [2 f(z)/ g^2(z)]\, dz\} \qquad (4)$$

where β is an irrelevant arbitrary constant. For the sake of completeness let us summarize some of the rigourous results that can be obtained from $\phi(x)$:

(i) A boundary is *nonattracting* if $\phi(x)$ is not integrable near the boundary[14]. If this is the case the boundary is also *unattainable* and is called *natural*[15].

(ii) If $\phi(x)$ is not integrable near *both* boundaries then the process $x(t)$ is *recurrent*[16,17] and stays finite with probability one at all times[15]. Any other combination leaves the process *transient*[16].

(iii) If a boundary b is natural (in the sense mentioned above) it can also be rigourously proved that $\text{Prob}\{\lim_{t\to\infty} x(t) = b\} = 0$ for all $x(0)$ inside the state space of the process[14].

For the process (3) we easily find

$$\phi(x) = \exp\{(2/\sigma^2)\, x^{1-2n} \sum_{r=0}^{n} (a_r/2n-1-r)\, x^r\} \qquad (5)$$

and near the boundary 0 we have

$$L(0) = \int_0^\beta \phi(x)\,dx \simeq \int_0^\beta \exp\left(\frac{2a_0}{(2n-1)\sigma^2}\, x^{1-2n}\right)dx = \infty$$

since $a_0 > 0$. Near ∞ we get

$$L(\infty) \simeq \int_\beta^\infty \exp\left(\frac{2a_n}{(n-1)\sigma^2}\, x^{1-n}\right) dx = \infty$$

Therefore, for all σ^2, both boundaries are *natural* (it is not difficult to prove the somewhat stronger property that both boundaries are *entrance*[3,14]) and the above properties (i)-(iii) truly apply, in particular $\text{Prob}\{\lim_{t\to\infty} x(t) = \infty\} = 0$ for all $x(0)$, i.e., the process *does not wander out to* ∞.

To further study the behavior of the process we proceed to the Fokker-Planck equation for the probability density associated to (3)

$$\partial P(x,t)/\partial t + \partial J(x,t)/\partial x = 0 \qquad (6)$$

where

$$J(x,t) = f(x)P(x,t) - \frac{1}{2}\frac{\partial}{\partial x}\{g^2(x)P(x,t)\} \qquad (7)$$

is the probability current. Note that since both boundaries are natural the probability current must be zero at both boundaries. On the other hand it can be shown[15] that $P(x,t)$ will converge to the stationary solution $P_s(x)$, such that $J(x) = 0$ for all x, and given by[3,14,15]

$$P_s(x) = N\, x^{-2n} \exp\{-(2/\sigma^2)\, x^{1-2n} \sum_{r=0}^{n} (a_r/2n-1-r)x^r\} \qquad (8)$$

where N is the normalization constant. Note that $P_s(0)=0$ and $P_s(x) = 0\,(x^{-2n})$ for large x. This non exponential decay of the stationary distribution does not depend on the sign of a_n and therefore can also be obtained in some stochastic studies of truly stable deterministic systems[7,8]. This asymptotic behavior also implies that not all the stationary moments exist, to be precise we may calculate $\langle x^q \rangle_s$ only for $q < 2n-1$, a result also found in a different context earlier[18], and also the interesting result that the nonlinearity reinforces the stabilization effect of the noise, i.e., the bigger n the stronger the stabilization.

To clarify the above results we take the simplest non linear model discussed, from a deterministic viewpoint, earlier

$$dx = (ax^2-bx+c)\,dt + \sigma x^2\,dW \qquad (9)$$

where a,b,c > 0. The unique stationary distribution is given by (8) for n=2 and the equation for the extrema of $P_s(x)$, the most probable values, is[12]

$$2\sigma^2x^3 - ax^2 + bx - c = 0 \qquad (10)$$

From (10) we find:

(i) for $\sigma^2 < a^2/6b$ there may be, depending on c, one or three real and positive solutions, leading therefore to the possibility of bistability.

(ii) for $\sigma^2 \geq a^2/6b$ there is only one real and positive solution.

Figure 1 shows the location of the most probable values for different values of σ^2 and c. Note that the stationary distribution exists even for values of the parameters such that the deterministic original problem does not possess steady state.

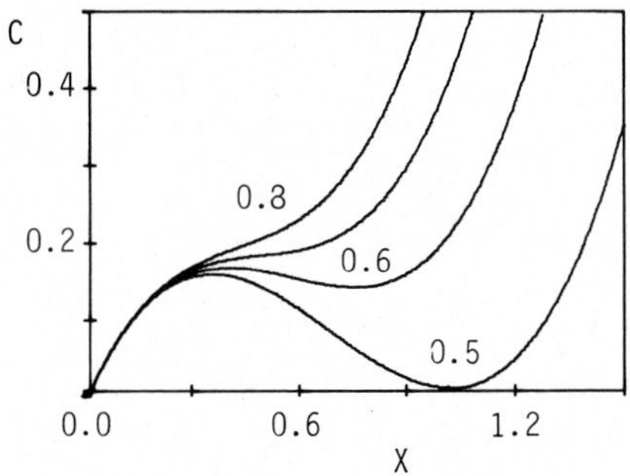

FIGURE 1
Location of the extrema of $P_s(x)$ for a=2, b=1 and various values of c and σ^2.

Figure 2 depicts the shape of the normalized stationary distribution for a=2, b=1, c=0.13 and two values of σ^2. Note that for these values $b^2 < 4ac$ and the deterministic system has no steady state. It is seen that by increasing σ^2, first the bimodal property disappears and then the unique maximum of $P_s(x)$ moves towards smaller values of x. The situation for $b^2 > 4ac$ is qualitatively similar. On the other hand we can also see how the system feels the limit $\sigma^2 \to 0$, where one should recover the unstable character of the deterministic system, since in this case the second maximum becomes more and more important and the process is allowed to reach larger and larger values, in agreement with what one would expect when approaching the limit of zero external fluctuations. It is clear that a similar reasoning applies to the general case of arbitrary n.

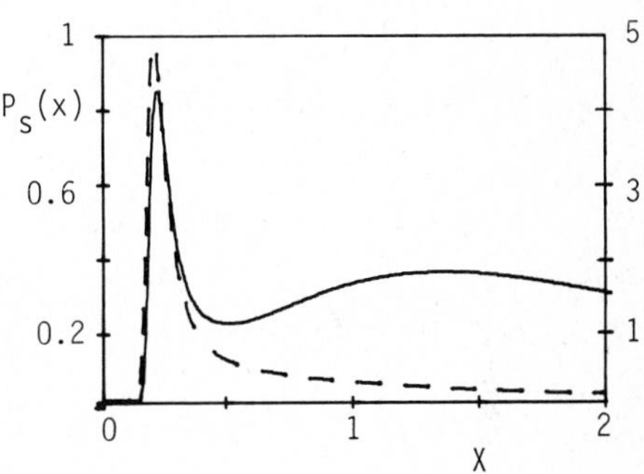

FIGURE 2
Normalized stationary distribution. *Solid line* (left vertical axis): $\sigma^2=0.5$. *Broken line* (right vertical axis): $\sigma^2=0.8$.

4. THERMAL FLUCTUATIONS AND OTHER STOCHASTIC MODELS

In any real experiment a small amount of noise of thermal origin is always present[19,20]. Thus it becomes clear that for the above results to be of any practical use they must be robust against this unavoidable presence of thermal noise. To model this thermal noise we take the standard approach[1,2] of adding a small stochastic term to (3), thus obtaining

$$dx = f(x)\,dt + g(x)\,dW_1 + \alpha^{1/2}\,dW_2 \qquad (11)$$

where W_1 and W_2 are independent Wiener processes and α is a small parameter (usually of the order of the inverse of the system size).

Now, since W_1 and W_2 are independent, (11) represents a two-dimensional process (x,W_2) and the above calculations cannot be directly applied. However to study the qualitative behavior of $x(t)$ we can use an equivalent one-dimensional diffusion process $\bar{x}(t)$ with drift $f(\bar{x})$ and diffusion $g^2(\bar{x}) + \alpha$. It is well known that if $\bar{x}(t)$ has a stationary distribution and Prob$\{\lim_{t\to\infty} \bar{x}(t) = \infty\} = 0$, the same is true for the process $x(t)$, and both stationary distributions coincide Proceeding as before we find that, when $\bar{x}$ becomes very large, $\phi(\bar{x}) \simeq$ constant, and therefore the boundary ∞ is natural and $\bar{x}(t)$ reaches this boundary only with probability zero.

On the other hand since the new noise term enters additively the number and location of the extrema of $P_s(x)$ remains the same, and the only qualitative difference with the thermal noiseless model is the change in the boundary zero that now becomes an accesible boundary. If the process $x(t)$ must stay positive for all time, we should use instantaneous reflection at zero as the appropriate boundary condition.

So far we have interpreted (3) as an Ito equation. It is well known that if (3) is obtained after performing a white noise limit of a colored, i.e., non zero correlation time, noise, one should use the Stratonovich interpretation[1,3,11]. If this is the case, we find that the boundary ∞ is an attracting (regular) boundary and we have Prob$\{\lim_{t\to\infty} x(t) = \infty\} = 1$ for all $x(0)$ and σ^2, i.e., no stabilization is possible. It is worth noting that it is the *spurious drift*[3,13], that must be included when using the Stratonovich interpretation, that makes the stabilization impossible. At this stage it seems clear that the possible relevance of the results presented here is resticted to cases in which the Ito version is the most adequate, as is for instance the case when one takes the continuum limit of a problem that is discrete in nature, as occurs in some biological or ecological applications[13,14]. However, quite recently[6,20] it has been theoretically predicted and experimentally confirmed that in some quite realistic problems neither the Ito nor the Stratonovich interpretation should be used. For, if we have a deterministic system with two widely separated time scales, perturbed by a colored noise with correlation time τ_c, and defining $\lambda = \tau_s/\tau_c$ as the ratio between the short time scale of the deterministic problem and the correlation time of the noise, the evolution of the system on the long time scale could be described by the Ito equation:

$$dx = \left[f(x) + \frac{1/2}{1+\lambda}\, g(x)g'(x) \right] dt + g(x)dW \qquad (12)$$

Note that the pure Ito (Stratonovich) version is recovered when $\lambda = \infty$ (0) . Repeating the above calculations we find that now there exists a *critical value* $\lambda_c = n - 1$ such that if $\lambda < \lambda_c$ the boundary ∞ is regular, i.e., explosion occurs with probability one, whereas if $\lambda \geq \lambda_c$, ∞ becomes natural, i.e., stabilization sets in. This is a quite striking result relating time scales of the problem with the nonlinearity, and that could be checked in an experiment.

5. DISCUSSION

In this work the possibility of stabilizing an unstable system by perturbing one of its parameters has been discussed. We would like to point out that to understand qualitatively the stabilizing effect of the noise one should take into account the peculiarities of the Ito calculus. This can be easily understood by passing to the Stratonovich version where the usual rules of calculus now apply. To do this one should subtract the extra-term $g(x)g'(x)/2$ and it is this term that turns stable the system This opensthe possibility of stabilizing the system by perturbing a different parameter that the one considered here. For, if we perturb $a_s(t)$ then $g(x)\, g'(x) = 0\,(x^{2s-1})$ and therefore if the order of this term is bigger or at least equal than the order of $f(x)$ we can stabilize the system. Then the condition to be fulfilled to obtain the desired result is

$$s \geq (n+1)/2$$

Let us finally indicate that the possibility of stabilizing more general systems perturbed by real noises is an interesting open question. The main problem is now that we have to work with multidimensional processes for wich no complete analytical results are available,although some promising ideas have been advanced quite recently[21].

ACKNOWLEDGEMENTS

I would like to thank Prof.L.Arnold and Dr. W.Kliemann for fruitful conversations.

REFERENCES

1. N.G. Van Kampen, Stochastic Processes in Physics and Chemistry (North-Holland,Amsterdam, 1981).

2. H.Haken, Synergetics:An Introduction (Springer-Verlag,Berlin,1977).

3. W.Horsthemke and R.Lefever, Noise Induced Transitions (Springer-Verlag,Berlin,1984).

4. A.Schenzle and H.Brand,Phys.Rev.,A20(1979) 1628.

5. F.J.de la Rubia and M.G.Velarde,Phys.Lett., A69 (1978) 304.

6. R.Graham and A.Schenzle, Phys.Rev., A26(1982) 1676.

7. F.Moss and G.V.Welland,Phys.Rev.,A25 (1982) 3389.

8. G.V.Welland and F.Moss,Phys.Lett., A89 (1982) 273.

9. F.J.de la Rubia,J.J.Garcia-Sanz and M.G.Velarde, Surface Sci., 143 (1984) 1.

10. S.Kabashima and T.Kawakubo, Phys.Lett., A70 (1979) 375.

11. J.Smythe,F.Moss and P.V.E.McClintock, Phys. Rev.Lett., 51 (1983) 1062.

12. F.J. de la Rubia, Phys.Lett., A110 (1985) 17.

13. L.Arnold, Stochastic Differential Equations (John Wiley,New York,1974).

14. S.Karlin and H.M. Taylor, A Second Course in Stochastic Processes (Academic Press, New York, 1981).

15. I.Gihman and A.Skorohod, Stochastic Differential Equations (Springer-Verlag,Berlin,1972).

16. A.Friedman, Stochastic Differential Equations and Applications,Vol I (Academic Press, New York, 1975).

17. W.Kliemann, Qualitative Theory of Stochastic Dynamical Systems, in: Stochastic Methods in Life Sciences, eds. M.Ianelli and G.Koch, LN in Biomathematics (Springer-Verlag,Berlin, 1981).

18. V.Seshadri, B.J.West and K.Lindenberg,Physica 107A (1981) 219.

19. S.Faetti,C.Festa,L.Fronzoni,P.Grigolini,F. Marchesoni and V.Palleschi, Phys.Lett., A99 (1983) 25.

20. S.Faetti,C.Festa,L.Fronzoni,P.Grigolini and P.Martano, Phys.Rev., A30 (1984) 3252.

21. L.Arnold, Springer Proc. Phys., 1 (1984) 11.

THEORY - 2

NOISE IN PHYSICAL SYSTEMS AND 1/f NOISE - 1985
A. D'Amico and P. Mazzetti (editors)

IS 1/F NOISE FUNDAMENTAL TO THE GENERATION-RECOMBINATION PROCESS?

K.F. KNOTT

University of Salford, Salford M5 4WT, England

Experiments on NPN transistors in which the surface recombination component of base current was greatly increased by electron beam irradiation of the oxide have produced surprising results. For twenty devices of assorted type and manufacture the normalised spectral intensity $S(f).f.I_B^{-2}$. Area of oxide was nearly the same for all devices . The actual result was a mean value of $5.9 \times 10^{-14} cm^{-2}$ with a fractional standard deviation of 0.48. This result is so consistent that one is led to believe that this 1/f noise is a fundamental property of the generation-recombination process.

INTRODUCTION

Experiments on NPN transistors in which an electron beam was used to induce or modify burst noise were described[1] at the 6th International Conference in Washington. At that time the main emphasis was on the correlation of burst noise and specific dislocation sites in the devices. However, several subsidiary results were also observed during those investigations. Firstly, existing burst noise in a high proportion of the devices was unaffected by electron irradiation. Secondly, the large increase in surface recombination current in the base region was accompanied neither by a large increase in 1/f noise nor, in the majority of devices, by the appearance of burst noise. Recently the author has had occasion to re-examine these results in the light of the postulation by Jones[2] that 1/f noise in bipolar transistors is entirely a bulk effect. His results were summarized by the equation,

$$S(f).f.A.\ I_B^{-2} = K^1 \qquad \ldots (1)$$

where $S(f)$ = spectral intensity of base current noise

A = emitter area

K^1 = constant

I_B = non-ideal component of base current

It was not clear from this paper whether the constant K^1 was device dependent. Measurements by the present author on several types of transistor indicated that for devices which did obey the above relationship the constant varied widely between samples.

To further investigate the link between recombination current and 1/f noise it was decided to study a number of devices of different type, manufacture and geometry. The noise and I-V characteristics, were firstly studied. A computer was used to fit the I_B - V_{BE} curves to a simple model which in most cases was found to be of the form,

$$I_{BT} = I_{o_1} \exp\left(\frac{V_{BE}.q}{kT}\right) + I_{o_2} \exp\left(\frac{V_{BE}.q}{mkT}\right) \qquad \ldots (2)$$

where I_{BT} = total base current

and $I_B = I_{O_2} \exp(\frac{V_{BE}\cdot q}{mkT})$ = non-ideal base current component

The devices were then put into scanning electron microscope (SEM) and their base oxide irradiated until the base current increased from 1 nA to 10 nA. The electron energy was chosen such that the oxide was only just being penetrated by the beam. After allowing time for the base current to relax to its final steady state value the noise and I-V characteristics were again measured. Finally, the devices were examined to locate the electrical base-emitter junction and to measure the area of the oxide between it and the edge of the base contact.

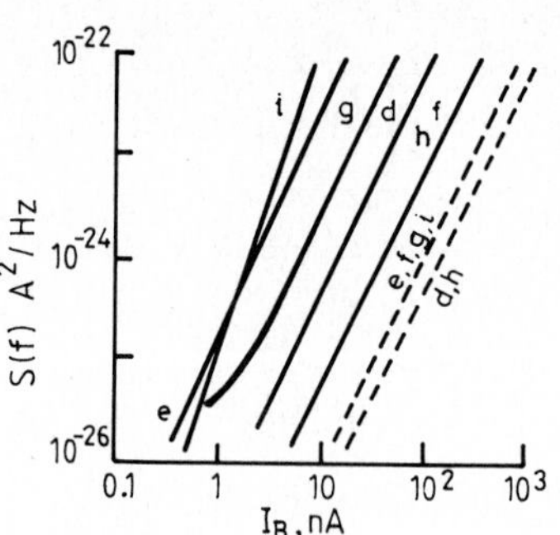

Fig. 1 Excess noise, 2N930
___ against non-ideal component of base current
... against induced surface component

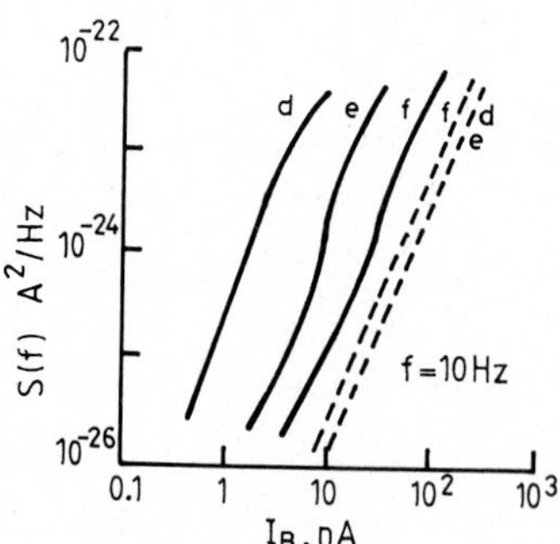

Fig. 2 Excess noise, 2N2484
___ against non-ideal component of base current
... against induced surface component

RESULTS

The devices were divided into two groups, namely, those free of burst noise and those with burst noise which was unaffected by electron irradiation. The excess noise spectra of the first groups were found to be of 1/f form both before and after irradiation. The devices tested were low power types 2N930, 2N2369, 2N2484 and medium power type BFX85. In Fig. 1 the excess noise at 10 HZ before irradiation is shown plotted as a function of the non-ideal component of base current for six samples of burst-noise free 2N930. Also plotted on this figure is the additional excess noise as a function of the induced surface component of base current. The results for three 2N2484 devices having burst noise are shown in Fig. 2.

Similar results were obtained for the rest of the samples. All samples were found to obey an equation of the form of (1) if A was taken to be the area of the oxide above the effective base region and I_B the induced surface current. The index of the law relating induced surface current and V_{BE} was 1.8 $\pm$ 0.1 and the normalised spectral intensity i.e. K^1 had a mean value of 5.9 x 10^{-14} cm^2 with a fractional standard deviation of 0.48. This spread of results is very narrow compared with that one usually encounters in 1/f noise measurements.

DISCUSSION

It is well established that the effect of electron irradiation of the S_iO_2-S_i system is to create trapped positive charge near the interface. In an NPN transistor this leads to depletion of the base surface and eventually to inversion. In the experiments performed here it is thought that the base surface is still depleted but close to inversion in which case the depth of the depleted region will be approaching its limiting value. If this is true the normalising procedure introduced above is equivalent to normalising by the volume of the depleted region. The remarkable uniformity of the results means that either the interface states of all transistors are the same or perhaps that 1/f noise is a fundamental property of the recombination process. The recombination in these experiments is taking place in a well-defined volume.

REFERENCES

1. Knott K F, Proc 6th Int Conf "Noise in Physical Systems" Washington D.C. N.B.S. Spec Publ 614, 97-100(April 1981).

2. Jones B K, Green C T, IEE Colloquium "Low Frequency Noise in Silicon Devices", Digest No. 1984/26 London (Mar 1984).

NOISE IN PHYSICAL SYSTEMS AND 1/f NOISE - 1985
A. D'Amico and P. Mazzetti (editors)

1/f-NOISE GENERATOR

P. GRUBER

LGZ LANDIS & GYR ZUG INC. Central Laboratory, CH-6301 Zug / Switzerland

A noise generator is developed, that approximates a 1/f-noise like behaviour over several frequency decades and that can easily be implemented on a computer for simulation purposes. The approach used here allows to model stationary and nonstationary characteristics.
The model is based on the physical analogon of an infinitely long RC line. Two important interesting cases can be distinguished: the open and the short circuit case, leading to nonstationary and stationary approximations.
The model will be applied in connection with the influence of such noise sources in solid state devices. It can also be used for the development of sophisticated measurement techniques (estimation of slowly varying parameters, measurement of physical quantities, etc).

1. INTRODUCTION

When dealing with solid state devices, 1/f-noise like behaviour of various physical quantities can be observed. One way to study the influence of such noise sources on the performance of solid state devices, is to develop a model with the following features:

i) approximation of 1/f-noise like behaviour over several frequency decades
ii) choice between stationary or nonstationary behaviour
iii) easy implementation on a computer for simulation purpose.

One possibility of such a generator model is described in the paper of Keshner[1]. There the model is found by optimization methods in the sense that for a given maximal relative error over a given frequency range the number of states and the parameters of the model are determined. The resulting model is stationary and is not based on a physical analogon. Its relative error over a given frequency range is nonmonotonic in f. The model derived here, on the other hand is strongly connected to a physical model, requires no optimization, may have a nonstationary component and is characterized by a monotonic dependence of the relative error over several frequency decades.

2. CONTINUOUS 1/f-MODELS

The modelling was attacked from the following point of view: If a white gaussian noise source is given, then what time-invariant linear system described by its impedance function Z(s) generates a 1/f-noise as output?

One answer to this question is: an infinitely long R-C line, as given in Fig. 1, driven by a white current noise source.

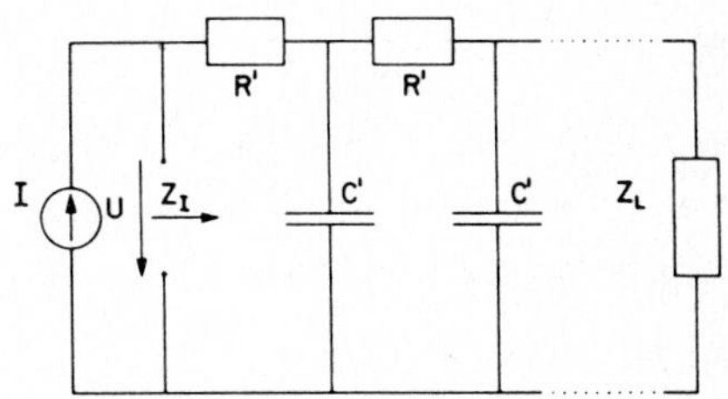

Fig. 1 infinitely long R-C line

The parameters of the line are

R' = resistance per unit length
C' = capacity per unit length
$Z_W(s)$= characteristic impedance $=(R'/sC')^{1/2}$ (1)
ℓ = length of the line
γ = propagation constant
$Z_L(s)$= load impedance
r_2 = reflexion factor

With the above quantities the input impedance $Z_I(s)$ of a R-C line of finite length ℓ can be written as

$$Z_I(s) = \frac{U(s)}{I(s)} = Z_W(s)\frac{e^{\gamma\ell} + r_2 e^{-\gamma\ell}}{e^{\gamma\ell} - r_2 e^{-\gamma\ell}} \quad (3)$$

Four special cases can be distinguished:

1. $\ell \rightarrow \infty \quad Z_I(s) = Z_W(s)$
2. $Z_L(s)=Z_W(s) \quad r_2=0 \quad Z_I(s)=Z_W(s)$
3. open circuit: $Z_L(s); r_2=1$

$$Z_{I_0}(s) = Z_W(s)\frac{\cosh(\ell(sR'C')^{1/2})}{\sinh(\ell(sR'C')^{1/2})} \quad (4)$$

4. short circuit: $Z_L(s) = 0 \quad r_2 = -1$

$$Z_{I_s}(s) = Z_W(s)\frac{\sinh(\ell(sR'C')^{1/2})}{\cosh(\ell(sR'C')^{1/2})} \quad (5)$$

Of interest here are the combinations 1./3. and 1./4. For both combinations $Z_I(s)$ tends towards $Z_W(s)$. If one computes then the quasistationary spectral density $S_{uu}(f)$ of the voltage U(s) given $S_{II}(f)$, one obtains:

$$S_{uu}(f) = Z_W(j2\pi f)\, Z_W(j2\pi f)\, S_{II}(f) \quad (6)$$

With $S_{II}(f) = \sigma^2_I$ and (1), (6) yields

$$S_{uu}(f) = (R'/2\pi fC')\, \sigma^2_I \quad (7)$$

which is the given dependency between S_{uu} and f. The impedance function $Z_I(s)$ however, is not a rational function of s.

3. RATIONAL MODELS

For the derivation of the rational models the following two series for sinhz and coshz are used:

$$\sinh z = z \prod_{n=1}^{\infty}\left(1 + \frac{z^2}{n^2\pi^2}\right) \quad (8)$$

$$\cosh z = \prod_{n=1}^{\infty}\left(1 + \frac{4z^2}{(2n-1)^2\pi^2}\right) \quad (9)$$

Open circuit/short circuit model:

Using (8), (9) and (1) with $z = \ell(sR'C')^{1/2}$ one obtains for (4)

$$Z_{I_0}(s) = \frac{1}{\ell C's}\,\frac{\prod_{n=1}^{\infty}\left(1 + \frac{4\ell^2 sR'C'}{(2n-1)^2\pi^2}\right)}{\prod_{n=1}^{\infty}\left(1 + \frac{\ell^2 sR'C'}{n^2\pi^2}\right)} \quad (10)$$

On the other hand it follows from (9)

$$Z_{I_s}(s) = \ell R'\,\frac{\prod_{n=1}^{\infty}\left(1 + \frac{sR'C'\ell^2}{n^2\pi^2}\right)}{\prod_{n=1}^{\infty}\left(1 + \frac{s4R'C'\ell^2}{(2n-1)^2\pi^2}\right)} \quad (11)$$

Remarks:

- Both (10) and (11) yield rational functions in s for any approximation $n \leq N < \infty$
- The product forms (10) and (11) allow an easy realization of the transfer functions.
- The open circuit model contains an integrating factor, that means for small values of f this model acts like an integrator and thus reflects a nonstationary behaviour (integrated white noise). Its order is N+1.
- $\ell^2 R'C' = \tau$ is the only parameter of the models.
- The short circuit model contains no integrating factor, that means for small values of f this model acts like a pure amplifier and thus is always stationary, Its order is N.
- Apart from that both models have poles and zeroes in the left half plane where the poles and zeroes reverse their meaning.

	poles	zeroes
open circuit	$p_n = -\frac{n^2\pi^2}{\tau}$	$z_n = -\frac{(2n-1)^2\pi^2}{4\tau}$
short circuit	$p_n = -\frac{(2n-1)^2\pi^2}{4\tau}$	$z_n = -\frac{n^2\pi^2}{\tau}$

(12)

- If the order N of the approximation is augmented by one, all the old poles and zeroes remain unchanged.
- By simple frequency transformation the short circuit model can be used for different frequency ranges. These realizations can then be cascaded.
- In the case of the open circuit model the integrator does not allow such frequency shifts of the model.

- Other approximation of coshz and sinhz have the undesired properties of changing poles and zeroes if the approximation order is changed and of complex conjugate poles and zeroes.
- The model developed by Keshner[1] corresponds to the short circuit model with the difference, that the position of the poles and zeroes of the first order sections are optimized. Thus the order N of the approximation can be reduced.

4. DISCRETIZATION AND STATE SPACE FORM

The discretization of the models is carried out by the Euler-backward integration rule, that means s is replaced by $(1-z^{-1})/T$ with T the sampling period.

Open-circuit model:

$$\tilde{Z}_{I_o}(z) = \frac{T}{(\ell C')(1-z^{-1})} \frac{\prod_{n=1}^{\infty}(1+\beta_n(1-z^{-1}))}{\prod_{n=1}^{\infty}(1+\alpha_n(1-z^{-1}))} \qquad (13)$$

$$\beta_n=\left(\frac{\tau}{T}\right)\frac{4}{(2n-1)^2\pi^2} \qquad \alpha_n=\left(\frac{\tau}{T}\right)\frac{1}{n^2\pi^2} \qquad (14)$$

Short-circuit model:

$$\tilde{Z}_{I_s}(z) = \ell R' \frac{\prod_{n=1}^{\infty}(1+\alpha_n(1-z^{-1}))}{\prod_{n=1}^{\infty}(1+\beta_n(1-z^{-1}))} \qquad (15)$$

For computational purposes it is convenient to represent the models in a state space form

$$\begin{aligned} \underline{x}_{k+1} &= A\,\underline{x}_k + \underline{b}\, i_k \\ u_k &= \underline{c}'\underline{x}_k + d\, i_k \end{aligned} \qquad (16)$$

5. COMPUTATION OF THE SPECTRAL DENSITY AND VARIANCE OF THE DISCRETE MODEL

The spectral density is computed with the following relation:

$$S_{uu}(f) = \tilde{Z}_I(z)\, \tilde{Z}_I(z^{-1}) \Big|_{z=e^{j2\pi fT}} \qquad (17)$$

Determination of the variance of the model:
The variance is a function of the time. With the state space form (16) one obtains with $i_K=v_K, v_K \sim (0,\sigma_v^2)$ a white gaussian noise

$$E\{u_k\} = E\{\underline{c}'A^k\underline{x}_0\} \qquad (18)$$

$$\sigma^2_{u_k} = \sigma_v^2\left(\sum_{i=0}^{k-1}(\underline{c}'A^{k-i-1}\underline{b})^2 + d^2\right) \qquad (19)$$

6. EXAMPLES

The following numerical values were chosen:

$\tau = 1$ sec, T=10 sec, $\ell C'=1$, $\ell R'=1$, $\sigma^2_v=1$
N=0, 5, 10, 28

Fig. 2 and Fig. 3 show two spectral densities for N=4 and N=16, Fig. 4 the error curves for N=16. Fig. 5 shows the development of the variance of the output of the noise generators for eight different models. The following remarks can be made:

- The open circuit model of low order has the characteristic that for small and large values of f the spectral density tends towards a 1/f -characteristic (line with a slope the double of the ideal 1/f- model in double logarithmic coordinates).The variances of the open circuit models increase linearly in k for large values of k (N=0 random walk, is linear increasing for all k).
- The short circuit model of low order gets flat for low and large values of f. That is the reason why this model can be put in a cascade.
- The required number of states is as follows (dependent on the relative error):

relative error	number of decades	number of states
5 %	1	4
	2	14
	2 1/2	100

If one compares with the stationary model of Keshner[1], the required number of states for one decade is four instead if one. The advantages however of the models developed here have already been mentioned in Section 3.

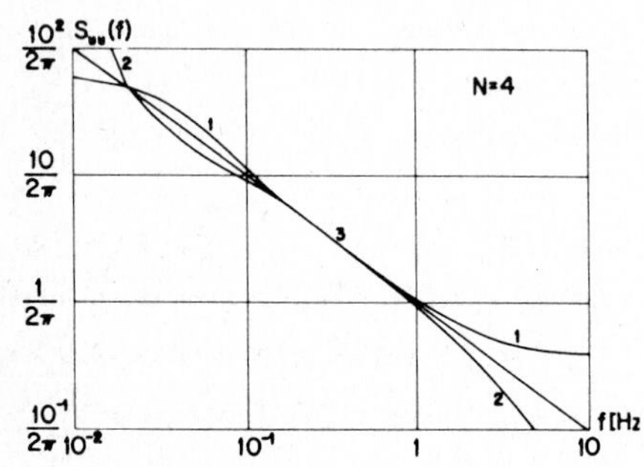

Fig. 2 spectral density: 1) short circuit 2) open circuit 3) ideal

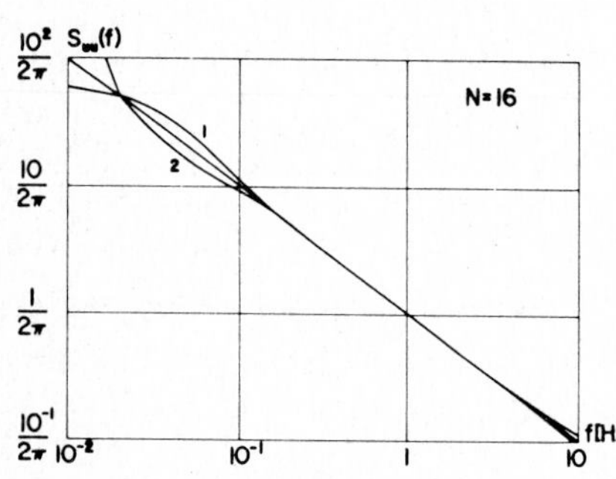

Fig. 3 spectral density: 1) short circuit 2) open circuit

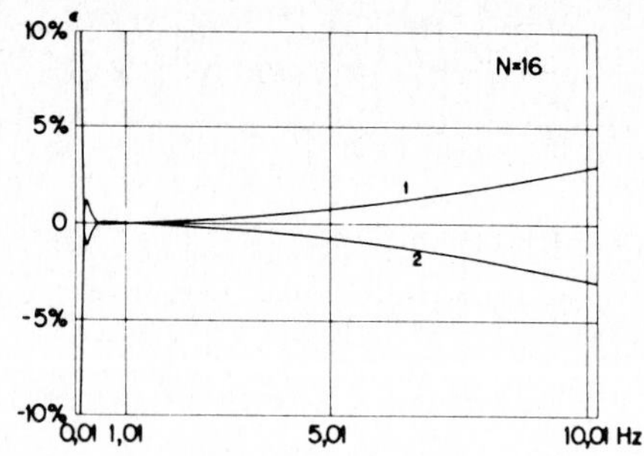

Fig. 4 relative error, N = 16
1) short circuit 2) open circuit

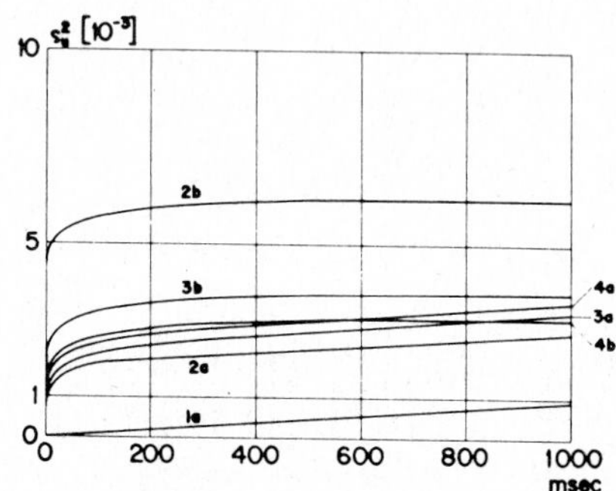

Fig. 5 variance of generator output as function of time
a) open circuit b) short circuit
1) N=0, 2) N=5 3) N=10 4) N=28
(1b out of range)

6. CONCLUSIONS

Simple models have been developed that allow to simulate 1/f-noise like behaviour on a computer. The models have been used in connection with the design of sophisticated measurement techniques (estimation of slowly varying parameters, measurement of physical quantities, etc), where the elimination of such a noise can be of great importance.

REFERENCES

1. M.S. Keshner,
1/f-noise, Proc. IEEE, Vol 70, No. March 1982

2. A. Papoulis,
Probability, Random Variables and Stochastic Processes, McGraw Hill 1965

NOISE IN PHYSICAL SYSTEMS AND 1/f NOISE - 1985
A. D'Amico and P. Mazzetti (editors)

AUTOREGRESSIVE MODEL FOR 1/f NOISE

G. GUTIERREZ, D. CADORIN, R. PADILLA AND M. OCTAVIO

Fundación Instituto de Ingeniería, Apdo. 40200, Caracas 1040-A, Venezuela and
Universidad Central de Venezuela, Caracas, Venezuela

We use a Monte Carlo simulation to generate noise that approaches 1/f over a certain frequency range. A superposition of a small number of noise terms with Lorentzian spectra, for the cases of one rate and two rate kinetics is considered. An autoregressive model is used to calculate the characteristic times. The advantages of this time domain analysis are discussed.

1. INTRODUCTION

One of the problems encountered in the study of 1/f noise is its lack of structural details. Many models have been proposed for 1/f noise [1,2], but it has been very difficult to verify their validity. One promising experimental approach is to study samples that are sufficiently simple and small that individual noise components can be resolved. Recent papers have shown the interest that exists in resolving these individual components and how they combine to produce 1/f noise [3,4,5]. In those papers the noise has been characterized by the sum of a few Lorentzian spectra. We find that for certain cases, an autoregressive (AR) model can be conveniently used to analyze systems in which a small number of individual noise components can be resolved. This type of model has been extensively used in systems control and identification, speech analysis and synthesis and other fields [6,7,8], but little attention has been given to it as a method for analyzing 1/f noise. In this paper we show how this identification method, can, in certain cases, complement the usual spectral analysis, providing physical information in a more direct way.

In section 2, of this paper we indicate how noise with a spectrum that approaches 1/f is generated. In section 3, we describe the AR model used to analyze the generated noise and we derive the relation that was obtained in terms of the parameters of the simulated noise. In the final section we discuss some of the reasons why this time domain analysis can complement the usual spectral analysis increasing the possibilities of obtaining information that may allow us to find the physical source of the noise components.

2. MONTE CARLO SIMULATION

In order to digitally generate Gaussian colored noise with Lorentzian spectrum we consider a Markov chain with two possible states. The characteristic time τ is determined by the probability $P(\tau)$ that the system remains in the same state in the next instant. From the Central Limit Theorem we obtain an approximate Gaussian distribution for the noise components. By considering a Markov process we guarantee that the associated physical process decays exponentially and the individual noise terms give Lorentzian spectra. We use a distribution of characteristic times according to the relation

$$\sigma^2(\tau)p(\tau) = B/\tau^2 \quad \text{(see reference [10])}$$

where $\sigma^2(\tau)$ is the variance, $p(\tau)$ is the probability density and B is a constant.

Figure 1 shows the spectral density vs the frequency, in a log-log plot, for the case of three noise components. With only three components, the resultant noise is 1/f over a range of frequency of approximately one decate. Figure 2 shows the spectral density for the case of six noise components. Notice that the range of freq-uencies for which the resultant noise is ~ 1/f exceeds one decade, for only six components.

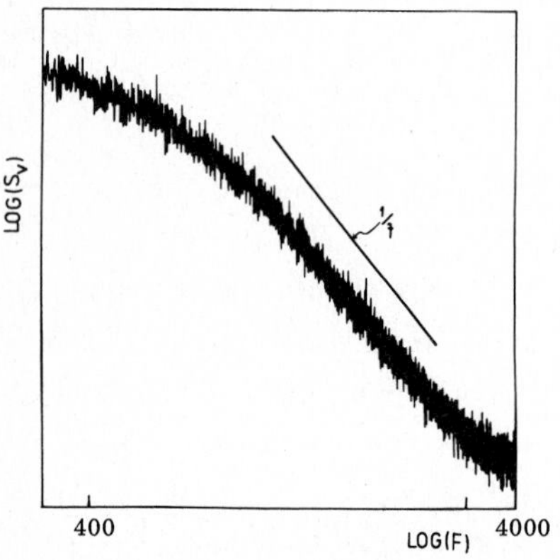

Figure 1: Spectrum for the generated noise. (3 noise components)

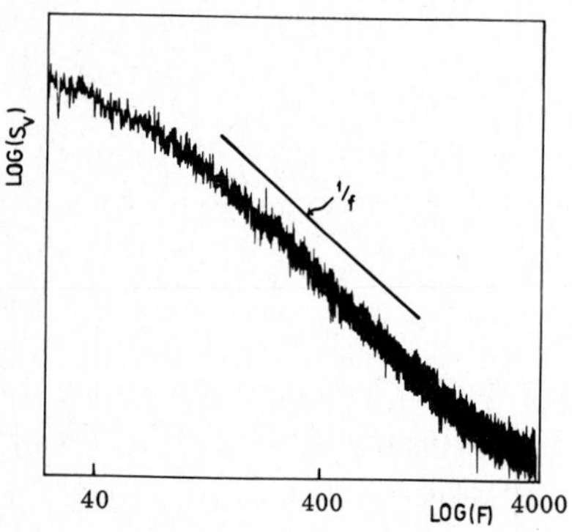

Figure 2: Spectrum for the generated noise. (6 noise components)

We simulated the case of two rate kinetics by introducing the probability $P_1(\tau_1)$ that the process would be switched on and the probability $P_2(\tau_2)$ that the process would be switched off, for each of the noise terms [11]. The above Markov process arises naturally from many physical models. The calculated characteristic time is then an effective time given by

$$\frac{1}{\tau_{eff}} = \frac{1}{\tau_1} + \frac{1}{\tau_2}$$

The spectral densities obtained for the two rate kinetics are similar to the ones shown in Figure 1 and 2. The time series generated with the simulation is analyzed as an autoregressive series. The parameters obtained from the AR model can be shown to be related in a simple manner to the characteristic times used in generating the noise.

3. AUTOREGRESSIVE MODEL

In order to analyze the noise sequence generated by the simulation the model can be represented as follows

$$y(t+k) = a\ y(t) + e(t) \qquad (1)$$

where we can represent y(t+k) as follows

$$y(t+k) = \sum_{i=1}^{n} \{Z_i(t) \prod_{j=1}^{k} u[P_i - x_i(t+j)] + \sum_{j=1}^{k} W_l(t+j)u[x_i(t+j)-P_i] \prod_{l=j+1}^{k} u[P_i - x_i(t+l)]\} \qquad (2)$$

(it can be shown that this expression corresponds to the simulated noise).

Here, $\{Z_i(t),\ i = 1, \ldots, n\}$ are the correlated noises which can be expressed as follows

$$Z_i(t) = Z_i(t-1)u[P_i - x_i(t)] + W_i(t)u[x_i(t) - P_i] \qquad (3)$$

where P_i is the discrete characteristic time for the ith component of the noise, given by the probability that $Z_i(t)$ does not change in the next instant, $x_i(t)$ is the random variable of a uniformly distributed uncorrelated stochastic process, $u(\cdot)$ is the step function defined by

$$u(x) = \begin{cases} 1 & \text{if } x \geq 0 \\ 0 & \text{if } x < 0 \end{cases}$$

and the set $\{W_i(t),\ i= 1, \ldots, n\}$ is a white stochastic sequence with expected value $E\{W_i(t)\}$ equal to zero and variance $\sigma^2_{w_i}$.

The identification of $\underline{a}$ can be accomplished by the minimization of the following functional [6]:

$$J= E_{x,w} \lim_{N\to\infty} \frac{1}{N} \sum_{t=0}^{N} \{y(t+1) - y(t+k|t\} \tag{4}$$

where $\hat{y}(t+k|t)$ is the estimated value of $y(t+k)$ given the information up to the instant t, and

$$\hat{y}(t+k|t)= ay(t)= a \sum_{i=1}^{n} Z_i(t) \tag{5}$$

If we substitute equation (2) and (5) into equation (4), after some calculation we get that

$$J= \sum_{i=1}^{n} \sigma^2_{w_i} [P_i^k + P_i^{-1}(1-P_i^k) + a - 2aP_i^k] \tag{6}$$

If we minimize J with respect to $\underline{a}$, to optimize the value of $\hat{a}$, we obtain the following result

$$\hat{a}_n(k)= \frac{\sum_{i=1}^{n} \sigma^2_{w_i} P_i^k}{\sum_{i=1}^{n} \sigma^2_{w_i}} \tag{7}$$

where k is the degree of decimation in time and n is the number of noise components.

An expression for the $\hat{a}$ parameters can be obtained for the case of a second order system but the expression obtained is more involved.

The importance of equation (7) is that k can be used to identify a particular physical source of a noise component which would otherwise be impossible to measure separately.

If the time constants are sufficiently separated, different values of k can be used to successively filter out the smaller characteristic times so that the different time constants and their relative weights can be calculated. Table 1 shows different combinations of characteristic times calculated in the manner described above. The top line in the table indicates the number of time constants present in the simulated noise, τ_G denotes the characteristic times for the generated noise, and τ_M the correlation times obtained from the AR model.

1		2		3	
τ_G	τ_M	τ_G	τ_M	τ_G	τ_M
.200	.229	.300	.338	.400	.316
.300	.324	.800	.776	.600	.675
.400	.419	.200	.168	.800	.792
.500	.521	.600	.651	.200	.147
.600	.615	.400	.405	.500	.551
.700	.703	.900	.898	.900	.896

Table 1: Characteristic times (arbitrary units).

We found that the error is sufficiently small to allow us to discriminate between characteristic times that are relatively close together, as can be seem from Table 1.

We can summarize the AR algorithm as follows: $\hat{\Theta}(\cdot)= [\hat{a}_1(\cdot) \ \ldots \ \hat{a}_k(\cdot)]'$ is the vector of the parameters, $\phi(t+1)= [y(t)\ y(t-1)\ \ldots\ y(t-k)]'$ is the vector that gives the necessary information to estimate the parameters and $\hat{\varepsilon}(t+1|t)$ is the estimated value of the prediction error given the information in $\phi(t+1)$ and is given by

$$\varepsilon(t+1|t)= y(t+1) - \hat{\Theta}\phi(t+1)$$

Optimum performance of the algorithm should give a white prediction error.

Details of the AR model can be found in references 6 and 9.

4. DISCUSSION

We have simulated a noise that approaches 1/f noise over a frequency range of approximately one decade for only three noise components. We use a system that arises from many physical models. It is interesting to note that in our simulation , only a small number of noise components was required to get a 1/f slope, this is consistent with the results obtained by Rogers and Buhrman for Josephson junctions and Metal-Insulator-Metal tunnel junctions [3,5].

Our calculations give evidence that the AR model can, in some circumstances, be suitable for analyzing 1/f noise. This can be of interest for the particular case in which we can experimentally isolate individual noise components. The components of the measured noise can be conveniently characterized by their time constants and their relative contribution to the resultant noise. If we have a particular physical model that can be expressed in the form of a differential equation or a difference equation, the model can be easily incorporated into a time domain model which can be used to calculate the parameters from the experimental data. This is particularly convenient if the process can be modelled by a low order linear system like the one considered by us. The statistical properties of the time domain analysis are well documental [6,9], so that it can be implemented to complement the spectral methods generally used to analyze 1/f noise.

AKNOWLEDGEMENTS

G. Gutiérrez was supported by a fellowship from Consejo Nacional de Investigaciones Científicas y Tecnológicas (CONICIT), Caracas, Venezuela.

REFERENCES

[1] P. Dutta and P.M. Horn, Rev. of Mod. Phys. 53, 3, p. 497 (1981).

[2] G.N. Bochkov and Yu E. Kuzovlev, Sov. Phys. Usp. 26, 9, p. 829 (1983).

[3] C.T. Rogers and R.A. Buhrman, Phys. Rev. Lett. 53, 13, p. 1272 (1984).

[4] K.S. Ralls, W.J.Skocpol, L.D. Jackel, R.E. Howard, L.A. Fetter, R.W. Epworth and D.M. Tennant, Phys. Rev. Lett. 52, 3, p. 228 (1984).

[5] T. Rogers and R.A. Buhrman, LT-17, Part II, eds. U. Eckern, A. Schmi, W. Weber, H. Wühl (North Holland, Amsterdam, 1984).

[6] L. Ljung, Recursive Identification in Stochastic Systems, The Mathematics of Filtering and Identification and Application, Hazenwinkel, M. and J.C. Willems, Roterdam (1980).

[7] N.S. Jayant and P. Noll, Digital coding of Waveform, Englewood Cliffs, N.J. Prentice Hall (1984).

[8] J. Makhoul, Proc. IEEE, 63, 4, p. 561 (1975).

[9] G.C. Goodwin, R.I. Payne, Mathematics in Science and Engineering, Vol. 136, Academic Press, N.Y. (1971).

[10] D. Halford, Proc. of the IEEE, 56, 3, p. 251 (1968).

[11] J.S. Machlup, J. Appl. Phys. 25, 3, p. 341 (1954).

NOISE IN PHYSICAL SYSTEMS AND 1/f NOISE - 1985
A. D'Amico and P. Mazzetti (editors)

PROPERTIES OF 1/f-NOISE IN HOPPING MODELS

Th.M. NIEUWENHUIZEN and M.H. ERNST

Institute for Theoretical Physics, P.O. Box 80.006, 3508 TA Utrecht, The Netherlands

The basic idea is to associate the phenomenon of 1/f-noise with the singular low frequency behavior of transport properties in stationary, statistically homogeneous, random media. We investigate properties of noise in models where charge particles hop independently on a disordered lattice in the presence of a bias field. Two types of disorder are admitted: temporary traps and deep wells (local scattering), both present in low concentrations and with a wide distribution respectively of release rates and of hopping rates. For suitable distributions of randomness we can describe 1/f-noise. We give microscopic definitions of number fluctuations - related to the traps - and of mobility fluctuations - related to the deep wells. To leading order the current or resistance noise can be decomposed into noise arising from these separate processes. On applying an electric field the 1/f-component of the number fluctuations is unaffected, whereas the 1/f-component of the mobility fluctuations decreases by a factor $(1+E/E_c)^2$.

In the last decades experiments and theory have lead to a better understanding of several aspects of 1/f-noise, such as the equality of current and resistance noise, and the relation with number and mobility fluctuations.

It is our purpose to study the exact status of these relations on the basis of a microscopic model of independent particles, hopping on a lattice with quenched disorder.

From the studies of Lorentz models for the electrical conductivity it is known that weak disorder gives rise to long time tails in the current-current correlation function or to singular low frequency behavior in the spectral density of current fluctuations. If one considers strong disorder, where the variance of the random variable becomes very large, this singularity may become stronger and behave like 1/f.

Also the standard theory of Dutta and Horn[4] requires very strong disorder, described by a distribution $\rho(\tau)$ of relaxation times behaving like $\rho \sim 1/\tau$.

The model to be considered is defined on a hypercubic lattice with M sites in d dimensions. Two states are associated with each site: a conduction state (s=+) and a trapping state (s=-). The characteristic time τ_n for hopping form the well at site n to one of the neighbouring sites (conduction) is a random variable, with the same distribution at each site. The same holds for the random release and trapping rates w_n and $\lambda_n w_n$, respectively from a trapping state into a conduction state and vice versa. This model is a combination of the random jumprate model with random wells[1,2] and the two state model of Machta et al.[3] with random trapping and release rates. The former model shows only mobility fluctuations; the latter one only fluctuations in the number of charge carriers in the conduction band. It is the purpose of this contribution to show that the present model contains both types of fluctuations.

Since the detailed calculations are very similar to those given in [2] and [3], we only sketch the general arguments and omit detailed calculations.

In the presence of a steady bias field along the x-axis the master equation reads

$$\dot{P}_n^+(t) = -\sum_m \Phi_{nm} P_m^+/\tau_m + w_n P_n^- - \lambda_n w_n P_n^+$$
$$\dot{P}_n^-(t) = - w_n P_n^- + \lambda_n w_n P_n^+ \qquad (1)$$
$$\Phi_{n,m} = \sum_{|\rho|=1} \{\delta_{n,m} - \delta_{n,m+\rho}\}\exp(\varepsilon\rho_x)$$

The parameter $\varepsilon = eE\ell/2kT$ measures the gain in electric energy per hop in the direction of the field in units of thermal energy and ℓ is the lattice distance. The stationary distribution is

$$P_n^+(\infty) = \bar{\nu}\tau_n, \quad P_n^-(\infty) = \bar{\nu}\lambda_n\tau_n, \qquad (2)$$

where $\bar{\nu} \equiv 1/\langle\tau(1+\lambda)\rangle$. Hence deeper wells (large τ_n) contain more particles. The fluctuating current in the direction of the field is defined as a spatial average of the single particle currents

$$I(t) = \frac{e}{L}\sum_{j=1}^{N} v_{jx}(t) \qquad (3)$$

where N is the total number of charge carriers and L the length of the sample in the field direction. From (1) and (2) one can deduce that $\langle v_x\rangle = \ell\, \partial_t\langle n_x\rangle_t = 2\bar{\nu}\ell\sinh\varepsilon$. Using (3) one finds that the average current $\langle I\rangle$ satisfies Ohm's law if $\varepsilon \ll 1$, with a resistance $R = kT\ell^2/e^2N\bar{\nu}\ell^2$.

The current correlation function can be related to the velocity auto correlation function $\varphi(t) = \langle v_x(t)\, v_x(o)\rangle - \langle v_x\rangle^2$, defined in our hopping model as

$$\langle v_x(t)\, v_x(o)\rangle = \tfrac{1}{2}\,\partial_t^2\, \langle(x(t) - x(o))^2\rangle$$
$$= \tfrac{1}{2}\ell^2\partial_t^2 \sum_{nn_oss_o} (n_x-n_{ox})^2\langle P(nst;n_os_oo)\rangle.$$

It is calculated by formally solving (1) and taking (2) as initial distribution. The result is

$$S_I(\omega) = 4kTR^{-1} + 4I^2(N\bar{\nu})^{-1}\, \mathbf{Re}\, G(o,i\omega) \qquad (4)$$

where $f = \omega/2\pi$ is the frequency, and **Re** indicates the real part.

The detailed information on the randomness is contained in the Greens function

$$G(q,z) = \langle(zK + \Phi)^{-1}\rangle_{qq} \qquad (5)$$

$$K_{nn'}(z) \equiv \delta_{nn'}\, \tau_n\, \{1+\lambda_n w_n(z+w_n)^{-1}\}$$

where $\langle A\rangle_{qq'}$ denotes the Fourier representation of the matrix $\langle A_{nn'}\rangle$.

The first term in (4) represents the equilibrium Johnson or Nyquist noise; the second is the excess noise S_I^{exc}, the form of which depends on the distribution of disorder, to be specified now. We assume that the τ_n, w_n and λ_n are all independent random variables with site independent distributions. If one considers weak disorder ($\langle\tau^k\rangle,\langle w^{-k}\rangle$ finite (k=1,2,...))only weak (classical) long time tails $\varphi(t) \sim t^{-1-d/2}$ are found, typical for Lorentz models. More interesting is the case of strong disorder, where the distributions of the random variables τ and $1/w$ have power law tails over a large range of τ - and 1/w-values, such that the variances $\langle\tau^2\rangle$ and $\langle w^{-2}\rangle$ become very large. The most relevant case is

(i) a small fraction q of sites contains a trap, described by the λ-distribution $\rho_1(\lambda) = (1-q)\delta(\lambda) + q\delta(\lambda-\lambda_1)$, so that $\langle\lambda\rangle \ll 1$.

(ii) w-distribution $\sim 1/w$ in a large domain (w_-,w_o) with an abundance of slow release rates.

(iii) a small fraction p of the sites contains a deep well, the distribution of which is $\sim \tau^{-2}$ in a large domain (τ_o,τ_+), while the fraction (1-p) of the sites have the host value $\tau_n = \tau_o$.

We calculate G(q,z) in a t-matrix expansion[2], which is essentially an expansion in powers of the concentrations p and q. Only the first two terms are relevant:

$$G(q,z)= g(q,z)- zg^2(q,z)\langle K(1+z\psi K)^{-1}\rangle_{qq}+.. \qquad (6)$$

where

$$g(q,z) = [z\langle K\rangle_{qq} + \Phi(q)]^{-1}$$
$$\langle K\rangle_{qq} = \langle\tau(1+\lambda w/(z+w))\rangle$$

$$\Phi(q)=2\cosh\varepsilon-2\cosh(\varepsilon-iq_x\ell)+2\sum_{\alpha\neq x}(1-\cos q_\alpha\ell)$$

$$\psi(z)=(2\pi)^{-d}\int dq\, g(q,z)$$

Using (i)-(iii) the averages can be calculated and to leading order in $\langle\lambda\rangle$ and p the result reads: $S_I^{exc}(\omega)/I^2=\alpha_H/Nf$, where Hooge's constant is found to be

$$\alpha_H=\langle\lambda\rangle/\log(w_+/w_o)+p, \tag{7}$$

for frequencies in the range

$\max(w_-,1/\tau_+)\ll f\ll\min(w_o,1/\tau_o)$.

For $f\ll\min(w_-,1/\tau_+)$ the excess noise becomes white. Thus, both the traps ($\lambda_n\neq o$) and the deep wells ($\tau_n\neq\tau_o$) can give rise to 1/f-noise. If we would change the powers in (ii) and (iii) into w^{-1+b} and τ^{-2-a}, we could describe behavior close to 1/f-noise:

$S_I^{exc}\sim\langle\lambda\rangle f^{-1+b}+p\,f^{-1+a}$.

Next we consider the relation between current noise in a stationary current and the equilibrium noise (ε=o) in the resistance or conductivity. The latter is equal to $\sigma=n\,e\,\mu$, where n=N/V with V the volume, $\mu=eD/kT$ is the mobility and D the diffusion coefficient. We define the fluctuating diffusion coefficient through

$D=N^{-1}\sum_j\frac{1}{2}\partial_t(r_{j\alpha}(t)-r_{j\alpha}(o))^2$ for fixed α=x,y,.... In the stationary ensemble the outcome does not depend on the direction α because of the isotropy of the jumprates.

$$\langle D\rangle\equiv\tfrac{1}{2}\ell^2\partial_t\sum_{nn_oss_o}(n_\alpha-n_{o\alpha})^2\langle P(nst;n_os_oo)\rangle$$

where s sums over the + and - states. This yields $\langle D\rangle=\bar{\nu}\ell^2$. The conductivity fluctuation for one given particle is

$$\langle\sigma(t)\sigma(t')\rangle=\left(\frac{e^2\ell^2}{2kTV}\right)^2\partial_t\partial_{t'}\sum(n_\alpha-\bar{n}_\alpha)^2\cdot(n'_\alpha-n_{o\alpha})^2\langle P(nst;\overline{nst};n's't';n_os_oo)\rangle$$

where the sum extends over ns, $\overline{ns}$, n's', n_os_o and $t>\bar{t}$, t'>o. After decomposing into conditional probabilities $P(nst|\overline{nst})$ etc. and performing some algebra we find

$$\langle\sigma(t)\sigma(t')\rangle-\langle\sigma\rangle^2=Ne^2V^{-2}\{M^{-1}\sum_{nn'}\langle\mu_n\,P(n+t;n'+t')\mu_{n'}\rangle-\mu^2\} \tag{8}$$

where we have added the contributions of all particles and introduced the local, frozen-in mobility $\mu_n=e\ell^2/kT\tau_n$. From eq. (1) we have

$$\int_{t'}^{\infty}dt\,P(n+t|n'+t')\exp[-z(t-t')]=\tau_n[(zK+\Phi)^{-1}]_{nn'} \tag{9}$$

Thus, using $P(n_os_oo)$ as given in (2), we find from (8)

$$S_R(\omega)/R^2=S_\sigma(\omega)/\sigma^2=\lim_{I\to 0}S_I^{exc}(\omega)/I^2=4(N\bar{\nu})^{-1}\,\mathbf{Re}\,G(o,i\omega)$$

This relation was experimentally observed by Voss and Clarcke[6] and by Beck and Spruit[7]; in our model it is an equality for all frequencies, valid for weak and strong disorder.

We are also able to distinguish between fluctuations in the number of charge carriers $N_c(t)$ and in the (total) mobility $\mu(t)$:

$$N_c(t)=\sum_{j,n}N^+_{j,n}(t)$$

$$\mu(t)=\sum_{j,n}\mu_n N^+_{j,n}(t)/N_c(t)$$

where $N^+_{j,n}(t)=1$ if the j-th charge carrier is in the conduction state (s=+1) at site n at time t and $N^+_{j,n}(t)=o$ otherwise. It is clear that (8) is related to the fluctuation in the product $N_c(t)\mu(t)$, as it should.

The spectral density of number fluctuations can be calculated with the help of (9) and (2) in a manner, similar to (4) and (8):

$$S_{N_c}(\omega)/N_c^2=4(N_c\langle\tau\rangle)^{-1}\mathbf{Re}\langle T(i\omega K(i\omega)+\Phi)^{-1}T\rangle_{oo} \tag{10}$$

where $T_{nn'} \equiv \delta_{nn'}\tau_n$ and N_c is the average number of carriers in the conduction state, and $\langle\ldots\rangle_{oo}$ denotes the (qq) = (oo) element in the Fourier representation. If we assume low concentrations of random sites (p,q<<1) we can forget about the randomness in τ_n and we recover from (10) the contribution to S_I^{exc}/I^2 resulting from the first term in (6).

The expression for the spectral density of the mobility fluctuations is more complicated. If, however p,q<<1 we can set λ_n=o in (8) to leading order in p and put $N_c \simeq N$. We then have the same expression for S_μ/μ^2 as for S_σ/σ^2 (see (8)), keeping in mind that now only τ_n-fluctuations should be taken into account. This corresponds to the second term in (6). As a result we arrive at the decomposition, valid for I → 0,

$$S_I^{exc}/I^2 = S_\sigma/\sigma^2 \simeq S_{N_c}/N_c^2 + S_\mu/\mu^2 \tag{11}$$

This equality is valid solely under the assumption of diluted randomness and holds equally well when (11) does not exhibit 1/f-noise. Result (7) is an example of (11) for a more specific situation. In case the τ_n's are not random ($\tau_n=\tau_o$) the model shows only number fluctuations; if there are no traps (λ_n=o), there are only mobility fluctuations.

These two types of noise can be distinguished, for instance, by applying large fields (hot electrons). In the case described by (i)-(iii) we find for ε>>1 again a noise level of the form (7). Here the strength of $S_{N_c}/N_c^2 \sim \langle\lambda\rangle$ is unchanged[3], but in the contribution from the mobility fluctuations the factor p is replaced by $p(1+E/E_c)^{-2}$, (see[2]). The critical field $E_c = 2kT/pe\ell$ is of the order 10^8V/m in a typical situation ($p \sim 10^{-3}$, $\ell \sim 10^4$Å). This and related type of noise reduction has been observed by Bosman et al.[8] and by Kleinpenning[9], suggesting that in various experimental circumstances 1/f-noise arises from mobility fluctuations and not from number fluctuations[10].

A serious defect of our model is that deep wells accomodate more particles than shallow ones (see (2)). This is the essential reason why in our model the $1/\tau^2$ -distribution of relaxation times occurs, as compared to the $1/\tau$-distribution in the standard theory of Dutta and Horn.

ACKNOWLEDGEMENTS

The authors wish to thank J. Machta for useful discussions. Th.M.N. was supported by the "Stichting voor Fundamenteel Onderzoek der Materie" (F.O.M.), which is sponsored by the "Stichting voor Zuiver Wetenschappelijk Onderzoek" (Z.W.O.)

REFERENCES

1. W. Lehr, J. Machta and M. Nelkin, J. Stat. Phys. 36 (1984) 15.
2. Th.M. Nieuwenhuizen and M.H. Ernst, J. Stat. Phys. to appear (1985).
3. J. Machta, M.Nelkin, Th.M. Nieuwenhuizen and M.H. Ernst, Phys. Rev. B 31 (1985), 7636.
4. P.Dutta and P.M. Horn, Rev. Mod. Phys. 53 (1981) 497.
5. J. Machta, M.H. Ernst, H. van Beijeren and J.R. Dorfman, J. Stat. Phys. 35 (1984) 413.
6. R.F. Voss and J. Clarke, Phys. Rev. B 13 (1976) 556.
7. H.G.E. Beck and W.P. Spruit, J. Appl. Phys. 49 (1978) 3384.
8. G. Bosman, R.J.J. Zijlstra and A. van Rheenen, Phys. Lett. A 78 (1980) 385; ibid. 80 (1980) 57.
9. T.G.M. Kleinpenning, Physica B 113 (1982) 189.
10. F.N. Hooge, T.G.M. Kleinpenning and L.K.J. Vandamme, Rep. Prog. Phys. 44 (1981) 479.

NOISE IN PHYSICAL SYSTEMS AND 1/f NOISE - 1985
A. D'Amico and P. Mazzetti (editors)

IS 1/f NOISE A DOUBLY STOCHASTIC PROCESS?

Ferdinand GRÜNEIS

Consultant to Fraunhofer-Institut für Hydroakustik, Waldparkstr. 41, 8012 Ottobrunn, Federal Republic of Germany

Based on a doubly stochastic process a number fluctuation model has been developed, which is able to generate a $1/f^b$ pattern ($0 \leq b \leq 2$). The specific features of this model regarding power spectrum, limiting frequencies and counting statistics are specified and discussed.

1. INTRODUCTION

Despite considerable effort no generally satisfying interpretation for 1/f noise has been given until now. Rather, it is questionable whether a universal theory is possible at all. In[1] a number fluctuation model has been presented as a further contribution to the discussion about 1/f noise. This model is based on the superposition of a doubly stochastic process (DSP). In the following features of a doubly stochastic Poisson process (DSPP) shall be discussed, which can help to check whether a given 1/f pattern may rely on a DSP.

2. THE NUMBER FLUCTUATION MODEL

As illustrated in fig. 1a a DSP is a random sequence characterized by the fact that the single events occur in a clustered fashion. Depending on the physics underlying a given process the cluster formation can be owing to various mechanisms as Barkhausen bursts[2], the flickering of electron emission from the cathode of a vacuum tube[3], a loitering automobile gathering a cluster of cars behind it[4] or an unsteady running of sandgrains through a hourglass[5]. A stochastic model for a DSP was formulated first with regard to the Barkhausen effect[6]. Later on it was cavitation noise, which provoked a rigorous treatment of stochastic models capable of describing DSPs[7]. The model presented in[1] as well as the subsequent considerations are based

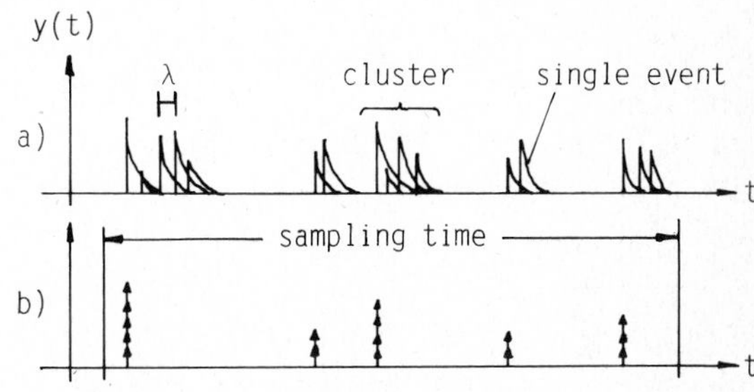

FIGURE 1
Illustration of a DSP
a) Sequence of clusters of single events. The clusters may also overlap depending on the specific process. λ = time interval between events within a cluster.
b) Fluctuation of numbers of single events belonging to the individual clusters shown above.

on formulae derived in[6,7].

To confine the extent of discussion, only the case of a DSPP shall be considered in the following. That means the sequence of clusters is assumed to form a Poisson process; therewith, an overlap of different clusters is a constituent of this model. Moreover, the sequence of single events within the clusters are assumed to form Poisson processes of finite duration. Following[7] and according to[1] the power spectral density (psd) of a DSPP can be written in the form

$$S(f)=2\bar{n}_c\overline{N|X|^2}+2\bar{n}_c|\bar{X}|^2Q(f)+(\bar{n}_c\bar{N}\bar{X}(0))^2\delta(f). \quad (1)$$

Herein X designs the Fourier transform of a single event. X depends on frequency and moreover on parameters as amplitude and duration of the event. The superscribed bars mean the formation of expectation values with regard to these parameters. The function Q is defined by

$$Q(f) = \sum_{m=1}^{N_0} P(N=m) 2\mathrm{Re}(U_\lambda (U_\lambda^m - 1)/(U_\lambda - 1)^2) \tag{2}$$

where P(N=m) is the probability of finding exactly m single events comprised in a cluster and U_λ is the characteristic function based on the probability density of the time intervals λ between successive events within a cluster. As the events are assumed to form Poisson processes, U_λ is given by[7]

$$U_\lambda = 1/(1 - i2\pi f \bar{\lambda}). \tag{3}$$

$\bar{\lambda}$ is the mean value of λ. Finally, $\bar{n}_c$ is the mean rate of cluster occurrence and

$$\bar{N} = \sum_{m=1}^{N_0} m P(N=m) \tag{4}$$

the mean number of events per cluster, whereby m is supposed to range between 1 and some maximum value N_0. Note that

$$\bar{n}_d = \bar{n}_c \bar{N} \tag{5}$$

is the total mean rate of single events.

Obviously, (1) describes simply shot noise provided Q=0. On the other hand, it can be shown that Q≧0 so that Q represents what may be called excess noise. The excess noise described by Q is due to the cluster formation, and may bring about 1/f noise. But the appearance of 1/f pattern depends on the choice of the probability P(N=m) of number of events per cluster. Thus, the model of concern is a number fluctuation model. An appropriate choice for P(N=m) is

$$P(N=m) = m^z / \sum_{\tilde{m}=1}^{N_0} \tilde{m}^z. \tag{6}$$

Numerical computations based on (2), (3), and (6) suggest that Q can be approximated by

$$Q(f) = c/(f\bar{\lambda})^b \tag{7}$$

over a considerable range of frequency, whereby the parameters c and b depend on z and N_0. To study this pattern in context with (1), it is convenient to assume for simplicity $\overline{x^2} = \bar{x}^2$. Though this relates to events of equal shape and size, it is a reasonable assumption anyhow. (1) can then be written in the form

$$S(f) = 2\bar{n}_c \bar{x}^2 R(f) + \bar{y}^2 \delta(f). \tag{8}$$

where

$$R(f) = \bar{N} + Q(f). \tag{9}$$

In addition the general relation

$$\bar{y} = \bar{n}_d \bar{x}(0) \tag{10}$$

has been considered in (8), where $\bar{y}$ has the meaning of the time mean value of the respective process. Fig. 2 shows typical examples of R based on (2), (3), (6), and (9). For f=0, R is maximum and $R(0) = \overline{N^2}$, where $\overline{N^2} = \sum_{m=1}^{N_0} m^2 P(N=m)$. For $f \to \infty$ R approaches asymptotically $R(\infty) = \bar{N}$. Between f=0 and f=∞, the double logarithmic plot of fig. 2 shows almost constant slopes over a considerable range of frequency, corresponding to the exponent b in (7). Provided $R(0)/R(\infty) = \overline{N^2}/\bar{N}$ is sufficiently large, say $> 10^2$, the $1/f^b$ pattern is well pronounced, and the lower and upper limit of this pattern can be estimated at

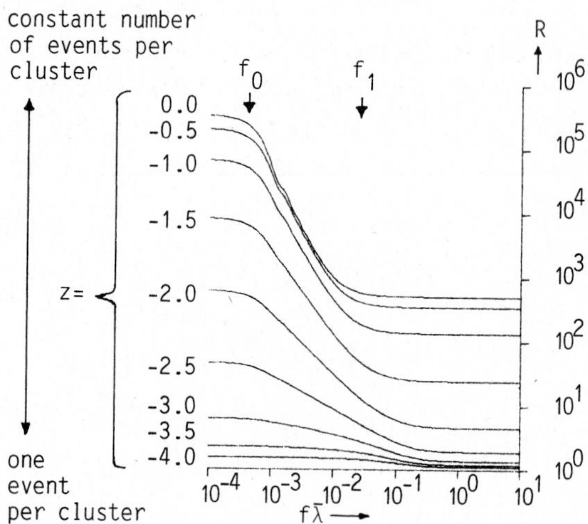

FIGURE 2
Double logarithmic plot of R versus $f\bar{\lambda}$ for several z and $N_0 = 10^3$.

$$f_0\bar{\lambda} \cong (c/\overline{N^2})^{1/b} \quad ; \text{ lower limit} \tag{11}$$

$$f_1\bar{\lambda} \cong (c/\bar{N})^{1/b} \quad ; \text{ upper limit.} \tag{12}$$

Obviously, f_0/f_1 is the larger the larger $\overline{N^2}/\bar{N}$. Combining (7) and (12) yields $Q \approx \bar{N}(f_1/f)^b$ and the psd according to (8) can be schematized by

$$S(f) = 2\bar{n}_c\overline{N^2}\overline{X}^2 + \bar{y}^2\delta(f) \qquad 0 \leq f \leq f_0 \tag{13}$$

$$S(f) = 2\bar{n}_d\overline{X^2}(1+(f_1/f)^b) \qquad f_0 \leq f \leq f_1 \tag{14}$$

$$S(f) = 2\bar{n}_d\overline{X^2} \qquad f_1 \leq f \leq \infty. \tag{15}$$

On closer inspection it is seen that $P(N=1)=1$ for $z=-\infty$ and $P(N=N_0)=1$ for $z=+\infty$. That means degeneration of the DSP to a simple Poisson process for $z=-\infty$ (only one event per cluster) and a tendency to constant number of events per cluster for $z \to +\infty$. Regarding the exponent b numerical calculations suggest

$$\begin{aligned} b &= 0 && \text{for } -\infty \leq z \leq -3 \\ b &= z+3 && \text{for } -3 \leq z \leq -1 \\ b &= 2 && \text{for } -1 \leq z \leq \infty \end{aligned} \tag{16}$$

as a rough but fair approximation. In particular $b=1$ for $z=-2$, that means a pure 1/f pattern is generated for $P(N=m) \sim m^{-2}$. Note, that a $1/f^2$ pattern is approached as an asymptotic behavior for $z \geq -1$. The range $z \leq -3$ is of no practical concern because only negligible excess noise appears compared with shot noise. Moreover, the case $z=-3$ shows a singular behavoir in that rather $Q \sim \log(f\lambda)$ applies instead of (7). For $z \geq -3$

$$c \cong 0.05 + 0.27(z+3)^{-3} \tag{17}$$

may be used as a fair approximation for the parameter c in (7). The expressions (16) and (17) apply the better the larger N_0.

3. A POSSIBLE APPLICATION

In many cases X is flat up to a frequency $f_s=1/2\bar{\tau}_s$ and drops off only beyond this limit, $\bar{\tau}_s$ being the mean duration of the events. Provided $f_1 \geq f_s$, $X(f) \cong X(0)$, and the excess noise $S_e=2\bar{n}_c\overline{X}^2Q(f)$ can then be expressed by

$$S_e = (2\bar{y}^2/\bar{n}_d)(f_1/f)^b \tag{18}$$

considering in addition (5) and (10). For b=1 this resembles Hooge's empirical formula[8] for 1/f noise. This formula relates to processes of electrical nature and $\bar{y}$ is an electric dc. Supposing these processes can be interpreted based on the present model, it was shown in[1] that Hooge's constant can be expressed by

$$\alpha_H = 2f_1\bar{\tau} \tag{19}$$

where $\bar{\tau}$ is the mean transit time needed for the electrons to migrate through a test volume.

(18) may compare with Hooge's formula when the sequence of electrons exhibits a clustered fashion corresponding to the number fluctuation model. But additional clues are required to check whether this model applies. One possibility for this results from (19), which shows α_H not to be a constant but to depend on $\bar{\tau}$ (possibly identical with $\bar{\tau}_s$) and on f_1 and therewith on $\bar{N}$. Another check may be enabled by counting statistics.

4. COUNTING STATISTICS

In this context only the number of single events is of concern, which can be counted in a sampling interval (0,T) of time (see fig. 1b). As a general result, the variance of the overall number of events counted in T is much greater for a DSP compared to a simple Poisson process[1]. Moreover, as shown in fig. 3 the normalized variance of N, which is a measure of the number fluctuations of the events within the clusters, turns out to be extremum for $z=-2$, which is the case generating a pure 1/f shape. Let's consider this in context with the spectrum (see fig. 2). In the extreme cases $z \to -\infty$ (one event per cluster) and $z \to +\infty$ (constant number of events per cluster) one has white noise and a $1/f^2$ pattern, respectively. Between these extreme cases, which show no number fluctuation, 1/f pattern takes in a medium position showing at the same time extreme values of the normalized variance of N.

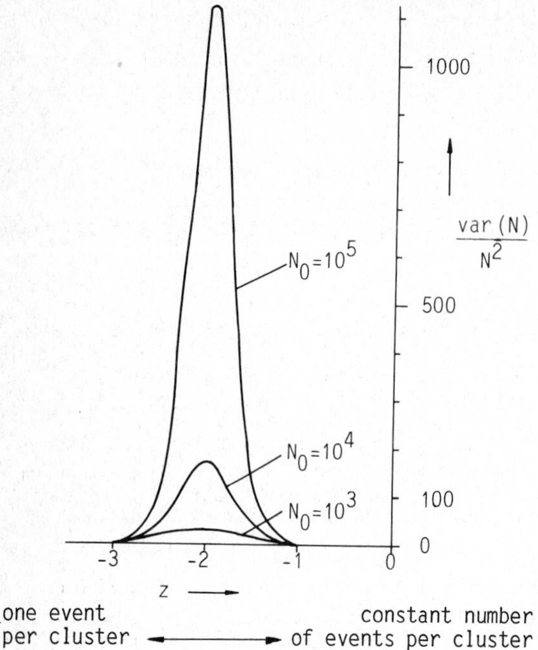

FIGURE 3
Normalized variance of number of events within a cluster.

5. IS THERE A LOWER LIMIT TO 1/f PATTERN?

Imagine, samples of a DSP are taken over a time T; in this case there may be some clusters which do not expire during T. As a consequence the mean duration $N_0\bar{\lambda}$ of these longest clusters will be equal to T. No statistical stationarity is reached within T and the result of averaging depends on T. This model being based on statistical stationarity cannot exactly describe this dependence. Nevertheless, the lower limit to 1/f shape can be estimated to be[1] $f_0=1/2N_0\bar{\lambda}=1/2T$. Considering in addition the fact that spectral analysis is limited at a frequency of about f=1/T, this might explain why it was not successful in some cases[9] to find a lower limit to 1/f pattern by increasing T. The lower limit might have shifted downwards parallel with the lower limit of spectral analysis and T was perhaps too short to include the longest clusters.

6. CONCLUDING REMARKS

A series of stochastic processes show 1/f pattern and can at the same time be suspected to be describable as DSPs. The number fluctuation model discussed above relates 1/f noise with the formation of events as it is typical of DSPs. It is the question how far 1/f pattern is really owing to the doubly stochastic nature.

In some cases cluster formation is quite obvious and it may be possible in some of these cases to directly check the suppositions made. But mostly, this will not be possible, and often even the principal supposition of a DSP is questionable. In these cases indirect checks are required and, if possible, multiple checks of this kind. The above discussed information deduced from the number fluctuation model may provide useful tools in this respect.

ACKNOWLEDGEMENTS

The author expresses his deep appreciation to H.-J. Baiter for his permanent interest in this paper; he has also contributed significantly in making manuscript criticism for which the author is also very grateful.

REFERENCES

1. F. Grüneis, Physica 123A (1984) 149.
2. H. Bittel and I. Westerboer, Ann. d. Physik 7. Folge, Band 4 (1959)203.
3. W. Schottky, Phys.Rev. 28 (1926) 74.
4. T. Musha and H. Higuchi, Jap.J. of Appl. Phys. 15 (1976) 1271.
5. K.L. Schick and A.A. Verveen, Nat. 251 (1974) 599.
6. P.-Y. Arques, J. de Physique 29 (1968) 369.
7. H.-J. Baiter, F. Grüneis and P. Tilmann in ASME, Ann. Meet., Phoenix, AZ (1982) 93.
8. F.N. Hooge, Physics Letters 29A (1969) 139.
9. M.A. Caloyannides, J.Appl.Phys. 45 (1974) 307.

NOISE IN PHYSICAL SYSTEMS AND 1/f NOISE - 1985
A. D'Amico and P. Mazzetti (editors)

FLICKER NOISE IN A CONTINUOUS TIME RANDOM WALK MODEL

J.E. LLEBOT, J. CASAS-VAZQUEZ and J.M. RUBI

Departament de Termologia, Facultat de Ciències, Universitat Autònoma de Barcelona, Bellaterra, (Barcelona) Spain.

In this paper a model based in a continuous random walk mechanism is suggested. A Flicker dependence for the spectral density of the total current I(t) in a small sample is found.

1. INTRODUCTION

There is a continuous interest in the scientific community in both a theoretical description and experimental studies of 1/f noise in different physical systems. Among the great amount of literature on this subject, recently have been published some review papers [1-7] where one can find the more interesting features of the problem.

In the other hand, one of the actives lines of research in solid-state physics has been that of elucidating features of energy transport in disordered structures. Since the lack of a long-range structural order one must drop the lattice picture. The random potential that arises because of the disorder, creates localized states. Carriers then move between localized centers by hopping. A considerable amount of experimental information is available in amorphous semiconductors[8]. Scher and Lax[9] have made important contributions to the problem of hopping transport in amorphous solids using random-walk methodology. They use a generalization of the Montroll-Weiss[10] model of continuous-time random walk on a discrete lattice. (See for a review of this method reference 11). We shall follow their approach.

The purpose of this short contribution is to show how is possible to get a Flicker behaviour of the spectral density in the contesct of the Scher and Lax formalism. Moreover, for those systems where substantial agreement between experiment and theory is obtained, the possible Flicker behaviour of the spectral density would supply additional properties susceptibles to be contrasted.

2. THE MODEL

Let us assume a system where a carrier has a velocity in the x-direction $v_x(t)$ being $\delta v_x(t) = v_x(t)-\bar{v}_x$ the velocity fluctuation. The Fourier transform of the autocorrelation function of $\delta v_x(t)$ is the diffusion constant $D(\omega)$

$$D(\omega) = \int_0^\infty <\delta v_x(t)\ \delta v_x(t+\tau)> e^{-i\omega\tau} d\tau \qquad (1)$$

Taking into account the Wiener-Khinchin theorem, one gets the relation between $S_{\delta v_x}(\omega)$, the spectral intensity and the diffusion constant

$$S_{\delta v_x}(\omega) = 4\ \mathrm{Re}\ (D(\omega)) \qquad (2)$$

where Re stands for the real part.

Then, if one considers a small piece of the sample of cross-sectional area A, length L, with carrier concentration n, carrier charge e, and each carrier gives an independent contribution to the total current I(t), the spectral density for I(t) has the form

$$S_I(\omega) = \frac{4e^2 n \, \mathrm{Re}\,(D)\, A}{L} \qquad (3)$$

From this expression, one deduces that the frequency dependence of $S_I(\omega)$ is given by the behaviour of the real part of the diffussion constant. Following Scher and Lax[9], considering for a hopping conduction mechanism a generalized mobility

$$\mu(\omega) = \frac{e}{kT} D(\omega) \qquad (4)$$

where μ is the mobility and T the temperature, the conductivity is given by

$$\sigma(\omega) = n\ e\ \mu(\omega) \qquad (5)$$

In a hopping model, is inconvenient use relation (1) since it refers to velocities rather than positions. In the quoted reference 9, Scher and Lax rewrite $D(\omega)$ as

$$D(\omega) = -\frac{1}{6}\omega^2 \int_0^\infty e^{-i\omega t} dt \langle |\vec{r}(t)-\vec{r}(o)|^2 \rangle \qquad (6)$$

where $\vec{r}(t)$ is the position of a carrier at time t. Applying a continuous time random walk model where t = 0 coincides with the beginning of a sojourn time and being $\psi(t)$ the probability density for the jump time, they find

$$D(\omega) = \frac{\sigma^2(\omega)}{6} \frac{i\omega\tilde{\psi}(i\omega)}{1-\tilde{\psi}(i\omega)} \qquad (7)$$

where $\tilde{\psi}(i\omega)$ is the Laplace transform of $\psi(t)$, the waiting time destribution function and σ^2 is the variance (assumed finite) associated with a single jump.

The dependence of σ^2 in ω is very low compared with the other terms, changing by a factor of 3 over many decades of ω and can be assumed[9] $\sigma^2(\omega) \equiv \sigma^2$. Therefore, an adequate function $\tilde{\psi}(i\omega)$ would give the frequency dependence of $D(\omega)$ and the corresponding ω dependence of the spectral intensity. We use in this evaluation the simple function used by Tunaley[12]

$$\psi(i\omega) \propto 1 - i\omega\alpha + \beta|\omega|^\nu \exp(+i\pi\nu/2) \qquad (8)$$

where α, β, and ν are parameters restricted by the relations

$$1<\nu<2 \quad ; \quad \beta \geq \alpha^2 \qquad (9)$$

Taking together equations (7) and (8) one gets for the real part of the diffusion constant

$$\mathrm{Re}\ D(\omega) \propto (\alpha - \beta\omega^{\nu-1} \sin \pi\nu/2)$$
$$(\alpha^2 + \beta^2\omega^{2(\nu-1)} - 2\alpha\beta\omega^{\nu-1} \sin \pi\nu/2)^{-1} \qquad (10)$$

Taking the d.c. limit ($\omega \to o$), a familiar result is recovered, provided that

$$D(o) = \frac{\sigma^2}{6\bar{t}} \qquad (11)$$

being $\bar{t}$ the mean waiting time directly related with the parameter α, and the classical Nyquist result is obtained.

Evaluating the limit for a arbitrary small α, together with adequate values for the other two parameters according to the restriction given by (9) drives easily to the following result

$$\mathrm{Re}\ D(\omega) \quad \frac{-\sin\frac{\pi\nu}{2}}{\beta\omega^{\nu-1}} \qquad (12)$$

which shows a Flicker noise behaviour over a range of frequencies of many decades.

As discussed by Machta et al[13] the fundamental difficulty when modeling such systems by means of a waiting time distribution function is the inexistence of a general method for the calculation of the waiting time distribution. As we have already mentioned, in this paper we have used a method slightly different from Tunaley's development which avoids first wainting time density function h(t)[12], but that arrives to a result showing also a Flicker spectra.

ACKNOWLEDGEMENTS

We should acknowledge the partial finantial support of the CAICYT of the Spanish Goverment. One of us (J.E.Ll.) also acknowledges the benefit of a NATO research fellowship.

REFERENCES

1. P. Dutta and P.M. Horn, Rev. Mod. Phys. 53, (1981), 497.

2. D.A. Bell, J. Phys. C: Solid State Phys. 13, (1980), 4425.

3. A. Van der Ziel, Advances in Electronics and Electron Physics, Vol.49, (1979), 225.

4. F.N. Hooge, T.G.M. Kleinpenning and L.K.J. Vandamme, Rep. Prog. Phys. 44, (1981), 479.

5. R.F. Voss , Proc. 33rd Annual Frequency Control, Symp. Atlantic City, N.J. (1979), 40-46

6. L.K.J. Vandamme, in Noise in physical systems and 1/f noise (M. Savelli, G. Lecoy and J.P. Nougier eds.) North-Holland, Amsterdam, 1983.

7. J.E. Llebot in Recent Developments in Nonequilibrium Thermodynamics (J. Casas-Vázquez, D. Jou and G. Lebon eds.) Lecture Notes in Physics nº 199, Springer, Berlin, 1984.

8. J. Mort and J. Knights, Nature 290, (1981) 659.

9. H. Sher and M. Lax, Phys. Rev. B 7 (1973), 4491.

10. E.W. Montroll and G.H. Weiss, J. Math. Phys. 6 (1965), 167.

11. G.H. Weiss and R.J. Rubin in Advances in Chemical Physics, Vol. 52 (I. Prigogine and S. Rice eds.) John Wiley, 1983.

12. J.K.E. Tunaley, J. Stat. Phys. 15 (1976), 157.

13. J. Machta, M. Nelkin, Th.M. Nieuwenchuizen and M.H. Ernst, Phys. Rev. B 31, (1985), 7636.

NOISE IN PHYSICAL SYSTEMS AND 1/f NOISE - 1985
A. D'Amico and P. Mazzetti (editors)
© Elsevier Science Publishers B.V., 1986

ANALYSIS OF 1/F NOISE POWER FLUCTUATIONS BASED ON STRUCTURE FUNCTIONS

Lech Z. HASSE

Institute of Electronic Technology, Technical University of Gdańsk, ul.Majakowskiego 11/12, 80-952 Gdańsk, Poland

The method of noise power instability investigation based on structure functions has been presented. In particular it is shown how this approach can be used to predict reliability of electronic devices by dealing with polynominal drifts of noise power.

The prediction of reliability of electronic devices can be realized by measurement of their noise in the very low frequency range during long-term work, in order to determine changes of noise properties in time. This analysis may be provided by the investigation of noise power fluctuations. The past studies of the variance of noise were carried out also for many other purposes, mostly for the investigation of stationarity of 1/f noise. In this paper the application of structure functions to the description of time-domain power fluctuations is proposed.

The main reasons for using structure functions seem to be the following:

- the mathematical difficulties due to the fact that the total variance of 1/f noise diverges can be avoided,
- 1/f noise may have a mean value as a function of time; an application of structure functions enables detrending measured noise (it is possible even to determine the order of the drift) ,
- a comprehensive structure function representation theory can be developed, which, among other things,makes it possible to calculate the power spectral density of noise fluctuations.

The measured time-dependent energy L of noise may be expressed generally as follows:

$$L(t) = P_o \cdot t + \sum_{l=2}^{R} \frac{a_{l-1}}{l!} t^l + [N(t) - N(0)] + L_o \quad (1)$$

where P_o is the constant average power, the a_l's describe a set of time-independent random variables modelling the lth-order long-term energy drifts, L_o is the initial value of energy (for $t = 0$) , and $[N(t) - N(0)]$ is a second-order stationary process used to characterize the short-term power fluctuations, which may be treated as a "pure" noise. Its first increment $\Delta N(t; T_M) \triangleq N(t + T_M) - N(t)$ is stationary; hence the Qth increment, defined recursively by

$$\Delta^Q N(t) \triangleq \Delta^{Q-1}[\Delta N(t)] = \sum_{l=0}^{Q} (-1)^l \binom{Q}{l} N[t + (Q - l) T_M] \quad (2)$$

where T_M is the duration of individual power measurement, is also stationary.

The Qth structure function of the energy increment is given by

$$D_L^{(Q)}(T_M) \triangleq E\{[\Delta^Q L(t; T_M)]^2\} = E\{[\Delta^{Q-1}[\Delta L(t; T_M)]]^2\} \quad (3a)$$

and its estimator was calculated as follows:

$$\tilde{D}_{\tilde{L}}^{(Q)}(T_M) = \frac{1}{K-Q} \sum_{p=1}^{K-Q} \left\{\Delta^{Q-1}[\Delta\tilde{L}(t_p;\, T_m)]\right\}^2 \tag{3b}$$

for Q = 1,2,3,4 and K = 100 (K is the number of time sections) .

Considering the energy L defined by (1) we can obtain generally

$$\Delta^R L\,(t;\, T_M) = a_R\, R_M^R + \Delta^R N\,(t;\, T_M) \tag{4a}$$

and for $Q > R$ it reduces to

$$\Delta^Q L\,(t,\, T_M) = \Delta^Q\, N\,(t;\, T_M) \tag{4b}$$

Thus, if a_R and $N(t)$ are uncorrelated, then

$$D_L^{(R)}(T_M) = E\left\{[\Delta^R L\,(t;\, T_M)]^2\right\} =$$

$$= T_M^R\, E\,[a_R^2] + E\left\{[\Delta^R N\,(t;\, T_M)]^2\right\} \tag{5}$$

and for $Q > R$ the problem of finding $E\left\{[\Delta^Q L\,(t;\, T_M)]^2\right\}$ reduces to one of finding the structure function $D_N^{(R)}(T_M) = E\left\{[\Delta^Q N\,(t;\, T_M)]^2\right\}$ of the stationary part of noise energy only. For the noise energy /power/ drifts of Rth-order, the Qth increment of energy is stationary for $Q > R$.

Therefore through the investigation of structure functions of energy increments it is possible to evaluate a drift of power and to state, in particular, whether the increase of power with time occurs. The degree of its monotonic increase may even be ascertained, which is very useful for the prediction of reliability of electronic devices. Using the Mellin transform, the spectral density of power fluctuations can also be determined.

In the experimental part the measurements of noise power of 100 selected samples of BC 413 transistors and 100 samples of 741 type operational amplifiers were made in the frequency bandwidth $\Delta f_1 = 1$ Hz and $\Delta f_2 = 10$ Hz using a specially built average power meter. The instrument enables one to choose the sampling frequency and averaging time, and records each short-term measurement (for $T_M = 10$, 100 or 1000 s) . It is also possible to obtain practically continuous long-term checking of noise power by the meter.

The usefulness of the analysis for the investigation of noise power stationarity of semiconductor devices was confirmed. The more complex utility of structure finctions could perhaps give more information than that contained in increments of noise power.

DEVICES - 2

NOISE IN PHYSICAL SYSTEMS AND 1/f NOISE - 1985
A. D'Amico and P. Mazzetti (editors)

1/f NOISE IN THE LIGHT OUTPUT OF LASER DIODES

L.K.J. VANDAMME and J.R. DE BOER

Department of Electrical Engineering, Eindhoven University of Technology, Eindhoven, The Netherlands

We have investigated the lower-frequency fluctuations in the optical output $S_P(f)$ of the spontaneous and stimulated emissions from 1 Hz to 10 Hz. Under spontaneous emission conditions the dependence $S_P \propto P^{3/2}$ is observed with P, the average light power output. For laser diodes with slight scatter losses, a second region is found with $S_P \propto P^{5/2}$. Gain-guided lasers show $S_P \propto P^4$ in the super-radiation region. A model is proposed in which the 1/f fluctuations in the optical output are due to fluctuations in the absorption coefficient and hence in the emission. The model explains the experimentally observed power dependence $S_P \propto P^m$ with m = 3/2, 5/2, 4.

1. INTRODUCTION

Brophy[1] was the first to mention 1/f spectra from intensity fluctuations in the optical emission of luminescence and laser diodes. The fluctuations were investigated well below the lasing threshold in a frequency range between 10 Hz and 10 kHz. A correlation was found between the optical emission noise and the diode-current noise. Tenchio[2] extended these observations above the lasing threshold in the range of 2 Hz to 3×10^5 Hz. Dandridge and Taylor[3] observed a correlation between 1/f intensity fluctuations and frequency fluctuations (FM noise) in the optical emission.

The 1/f noise in the frequency fluctuations increases the spectral width[4]. The spectral properties of the light emission are very important for some applications. These facts[3,4] make 1/f noise in laser diodes an important subject.

In this paper we present experimental results and a model of the 1/f noise in the optical output S_P.

2. LIGHT POWER OUTPUT VS CARRIER CONCENTRATION

First a straightforward single-mode treatment will be presented in terms of photon density S. Then an extension will be made in terms of optical output power P stemming from all modes. The electron density n' is assumed to be confined within the active layer. The electron and photon densities are considered uniform within the active region. In spite of its simplicity, the analytical expression yield insight into the macroscopic and 1/f noise behaviour of the laser diode.

The rate equations from the optically active region are given below:

(i) spontaneous emission rate $r_{sp}(m^{-3}s^{-1})$

$$r_{sp} = A'n'p' \tag{1}$$

(ii) stimulated emission rate r_{st}

$$r_{st} = B'n'p'S \tag{2}$$

(iii) band-band absorption rate $-r_{bb}$

$$-r_{bb} = B'(N_c - n')(N_v - p')S \tag{3}$$

(iv) intraband absorption rate and scattering losses

$$-r_{cl} = (\varphi'n' + \beta')S \tag{4}$$

where $S(m^{-3})$ is the photon density, $A'(m^3/s)$ and $B'(m^6/s)$ are the transition coefficients for spontaneous and stimulated emission. N_c and N_v are the density of states in the conduction and the valence bands, respectively. The free electron and hole concentrations are denoted by n' and p'. The intraband absorption transition is given by φ'. The scatter losses in the optical cavity are determined by β'. In the steady state there is no net increase in photon density, hence $dS/dt = r_{sp} + r_{st} + r_{bb} + r_{cl} = 0$. With eqs. (1) - (4) this results in

$$A'n'p' + B'(n'N_v + p'N_c - N_cN_v)S - (\varphi'n' + \beta')S = 0 \qquad (5)$$

From eq. (5) we obtain for the photon density in the cavity

$$S = \frac{A'n'p'}{B'(N_cN_v - n'N_v - p'N_c)+(\varphi'n' + \beta')} \qquad (6)$$

The optically active region has unintentional dope. The electron-and-hole concentrations in that region results from injection out of the n-type and p-type cladding layers.

Assuming symmetrical injection, $n' = p'$. We now define $n = n'/\sqrt{N_cN_v}$ with n the normalized concentration. The optical output P (Watt) is proportional to the photon density. Equation (6) for the photon density can be rewritten in the optical output as

$$P = \frac{An^2}{B(1 - an) + \varphi n + \beta} \qquad (7)$$

with $a = \sqrt{N_c/N_v} + \sqrt{N_v/N_c} \approx 4.5$ for GaAs, $\varphi(m^{-1})$ the intraband absorption coefficient, $\beta(m^{-1})$ the losses and $B(m^{-1})$ the band-band absorption coefficient.

In order to extend the single-mode treatment of eq. (6) to a multimode treatment A becomes proportional to the line width $\Delta\lambda$ and a mode density proportional to ν^2. A rough estimate for the spontaneous emission-line width is 20 nm. The optical output P is proportional to $S.h\nu.\Phi.v$ with $h\nu \approx E_{gap}$, Φ the area of the facet of the optically active region and v the group velocity of the light in the cavity. The parameter A is calculated from

$$A = \frac{8\pi\nu^2\Phi\Delta\nu h\nu r}{v^2} B = \eta B \ (Wm^{-1}) \qquad (8)$$

with r a dimensionless optical cavity parameter. For $n = (1 + \beta/B)/(a - \varphi/B)$, the denominator in eq. (7) becomes zero. Equation (7) becomes invalid. In the laser action region both rate equations for photons and electrons must be used. For slightly lower concentrations at $n = 1/a$, population inversion, is reached and the material becomes transparent. However, at that point the gain is not high enough to compensate the mirror losses and the laser still operates as a super-radiator below threshold current. In general it holds that $\beta \ll B$ and $\varphi \ll aB$.

3. RELATION FOR THE 1/f NOISE IN THE LIGHT OUTPUT

We assume that the 1/f fluctuations in the optical output are due to fluctuations in the coefficients A and B. Due to the Einstein rate equations and Planck's distribution function, coefficient B' appears in both eqs. (2) and (3) and $A = \eta B$, as can be seen from eq. (8). Hence $S_B/B^2 = S_A/A^2$.

All electrons in the optically active region act independently of each other. Hence the noise is considered as a bulk effect. The noise power S_B in the absorption coefficient is proposed as being of the same form as Hooge's empirical relation[5]

$$\frac{S_A}{A^2} = \frac{S_B}{B^2} = \frac{\varepsilon}{n'\Omega f} \qquad (9)$$

where Ω is the volume of the optically active region, B denotes the average value of the band-band absorption coefficient in an optically active region, with $n'\Omega$ electrons and holes. ε is a dimensionless parameter.

In this model a change ΔB brings about a ΔA and hence $\Delta P = (dP/dA)\Delta A$. The sensitivity coefficient dP/dA it follows from eq. (7) and $A = \eta B$ that

$$\frac{dP}{dA} = \frac{P^2(\varphi n + \beta)}{(nA)^2} \qquad (10)$$

With eqs. (9), (10) the noise spectral density $S_P = (dP/dA)^2 S_A$ becomes

$$S_P = \frac{\varepsilon(\varphi n + \beta)^2 P^4}{A^2 n^5 \sqrt{N_cN_v}\,\Omega f} \qquad (11)$$

For $n \ll 1/a$ we find from eq. (11) that $S_P \propto P^4/n^5$ and $P = An^2/B$ from eq. (7). This results in

$$S_P = \frac{\varepsilon\beta^2 A^{1/2} P^{3/2}}{B^{5/2}\sqrt{N_cN_v}\,\Omega f} \propto P^{3/2} \qquad (12)$$

The dependence $S_P \propto P^{3/2}$ is in agreement with

experimental results at low P-levels as presented in fig. 1.

When $\beta/\varphi < n << 1/a$ the power P remains proportional to n^2. For S_P the proportionality becomes P^4/n^3, so that from eqs. (7) and (11) it follows that

$$S_P = \frac{\varepsilon\varphi^2 P^{5/2}}{A^{1/2}B^{3/2}\sqrt{N_c N_v}\Omega f} \propto P^{5/2} \qquad (13)$$

The dependence $P^{5/2}$ has been observed in Sharp diodes (fig. 1). The region with the $P^{5/2}$ proportionality disappears when $a\beta/\varphi > 1/2$. This means that the ratio of scatter losses to intraband absorption becomes large. Then the interval for n becomes too small.

When stimulated emission becomes important for $n \rightarrow 1/a$ then P continues to increase because the denominator in eq. (7) reduces to zero. Then from eqs. (7) and (11) we find

$$S_P \approx \frac{\varepsilon(\varphi + a\beta)^2 a^3 P^4}{A^2\sqrt{N_c N_v}\Omega f} \propto P^4 \qquad (14)$$

This proportionality has been observed in gain-guided laser diodes in the region of stimulated emission below threshold (super-radiation range).

4. EXPERIMENTAL RESULTS

Figure 1 shows the 1/f noise in the optical emission at f = 1 Hz below and above the lasing threshold. The Philips CQP 10 is a gain-guided diode and the Sharp, VSIS, is an index-guided laser diode. The noise has been observed after a life test for a full week.

5. PARAMETER VALUES

Before calculating the noise parameter ε from the experimental results in fig. (1) and eq. (12), the numerical values of the parameters must be estimated. Table I presents estimated values, from literature or experiment for both index and gain-guided lasers.

From a diagram in ref. [6] showing absorption coefficient due to intraband absorption and scatter losses we approached $\beta + \varphi n$ as

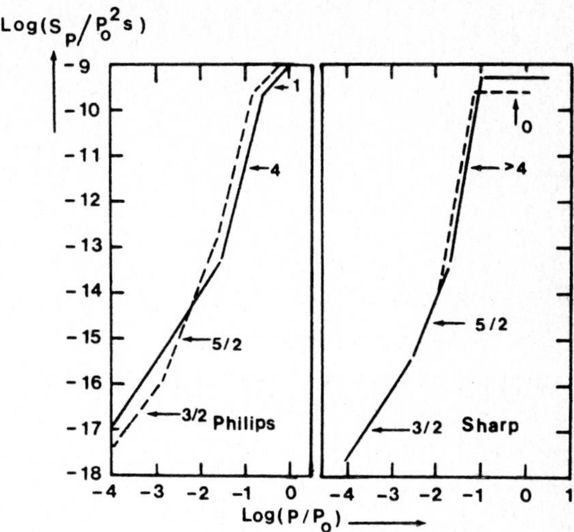

FIGURE 1

Spectral density of the optical output S_P at 1 Hz versus average optical output P. The full lines represent the experimental observations at T = 287 K, the dotted lines at T = 197 K. The scales are made dimensionless by introducing P_0. P_0 is chosen at 3 mW, corresponding to a photodetector diode current of 1 mA.

$$\beta + \varphi n = 10 + 10^{-17}\, n' \ (\mathrm{cm}^{-1}) \qquad (15)$$

which means that $\varphi = 10^{-17}\sqrt{N_c N_v} \approx 20\ \mathrm{cm}^{-1}$ and $\beta = 10\ \mathrm{cm}^{-1}$.

The parameter B is the band-band absorption coefficient of the order of $10^6\ \mathrm{m}^{-1}$ for GaAs[6]. The emission parameter A is calculated from eq. (8) where $h\nu \approx 1.524$ eV, the band gap of the optical active region for the Philips laser, and $h\nu = 1.611$ eV for the Sharp laser. This corresponds to an emitted-light frequency ν of 3.7×10^{14} Hz and 3.9×10^{14} Hz. The refractive index $\bar{n} \approx 3.63$ for both lasers. Hence the group velocity $v = c/\bar{n}$ is obtained. The velocity of light in vacuum is denoted by c. The effective bandwidth on spontaneous-emission conditions, is according to theory[6], of the order of $\Delta\nu_{sp} \approx kT/h$. A spontaneous emission line width of $\Delta\lambda = 20$ nm has been experimentally observed, corresponding to a $\Delta\nu_{sp} = 9 \times 10^{12}$ Hz. In the spontaneous emission region $r = 1 - R$, with R

TABLE I PARAMETER VALUES

	Gain-guided Philips CQL10	Index-guided Sharp VSIS
facet area Φ (μm^2)	0.2 x 5	0.08 x 4
cavity length ℓ(μm)	250	250
volume cavity Ω(cm^3)	2.5×10^{-10}	3.2×10^{-11}
$\sqrt{N_c N_v}$ (cm^{-3})	2×10^{18}	2.2×10^{18}
intraband absorption φ(m^{-1})	2×10^3	$\varphi \geq 2.2 \times 10^3$
scatter losses β(m^{-1})	10^3	$\beta < 10^3$
emission coefficient A(W m^{-1})	750	280
interband (band-band) absorption B(m^{-1})	10^6	10^6
wavelength λ(nm) in vacuum	820	770

the reflection coefficient at the facets $r \approx 0.68$. The noise parameter ε as calculated from experimental results eq. (12) and the above parameters is $\varepsilon = 10^4$ for the Philips (no. 381) and $\varepsilon = 10^3$ for the Sharp (no. 6).

6. DISCUSSION AND CONCLUSIONS

The fact that the proportionalities $S_P \propto P^m$ with m = 3/2, 5/2, 4 have not been hitherto observed, is due to the normal procedure of anlysing S_P as a function of the diode current I_ℓ. Sometimes one finds[2] log S_P/P versus lin I_ℓ and at other times[3] S_P/P^2 versus lin I_ℓ has been presented. If one used an empirical relation in which the relative noise in the absorption coefficient is normalized, not on the number of electrons in the optically active region, but on the number of photons in absorption and emission at every moment, a fraction of n' in eq. (9) would be used. The creation time of a photon is of the order of $\tau_{cr} = 1/\nu \simeq 3 \times 10^{-15}$ s. The electron lifetime is τ_e circa 10^{-8} s. Hence the fraction of electrons involved in a photon interaction (creation or absorption) is

$$\theta = \frac{\tau_{cr}}{\tau_e} = 3 \times 10^{-7} \tag{16}$$

Proposing an empirical relation for the absorption coefficient normalized on the number of electron-photon interactions, results in

$$\frac{S_B}{B^2} = \frac{\alpha}{\theta n'\Omega f} = \frac{\varepsilon}{n'\Omega f} \tag{17}$$

with $\alpha = \varepsilon \times \theta$. From our experimental results it follows that $\alpha \approx 3 \times 10^{-3}$ (Philips) and $\alpha \approx 3 \times 10^{-4}$ (Sharp).

Such α-values also appear in the 1/f conductivity fluctuations[5].

The model presented explains experimentally found power dependences $S_P \propto P^m$ with m = 3/2, 5/2, and 4.

In the laser action regions eq. (7) becomes invalid because both rate equations for photons and electrons must be used. Due to the stimulated emission, electrons are mainly dependent on one other and eq. (9) must be adapted.

ACKNOWLEDGEMENT

This work has been sponsored by the "FOM".

REFERENCES

1. J.J. Brophy, J. Appl. Phys. 38, 2465, (1967)
2. G. Tencio, Electron. Lett. 13, 614, (1977)
3. A. Dandridge and H.F. Taylor, IEEE QE-18, 1738 (1982)
4. M. Ohtsu and S. Kotajima, Jpn. J. Appl. Phys. 6, 760 (1984)
5. F.N. Hooge, T.G.M. Kleinpenning and L.K.J. Vandamme, Rep. Prog. Phys. 44, 479 - 532 (1981)
6. H.C. Casey, jr. and M.B. Panish, "Heterostructure Lasers" Part A p. 174, Academic Press, New York, 1978

NOISE IN PHYSICAL SYSTEMS AND 1/f NOISE - 1985
A. D'Amico and P. Mazzetti (editors)

CORRELATION BETWEEN 1/f FLUCTUATIONS IN LASER DIODES

J.R. DE BOER and L.K.J. VANDAMME

Department of Electrical Engineering, Eindhoven University of Technology, Eindhoven, The Netherlands

The coherence function between the fluctuations in the optical power outputs P and Q from both facets of laser diodes has been investigated from 1 Hz up to 10 Hz. The coherence function $\gamma^2_{PQ}(f)$ is defined as $\gamma^2_{PQ}(f) = S^2_{PQ}(f)/S_P(f) \cdot S_Q(f)$. In the frequency range where P and Q show pure 1/f noise, a frequency independent coherence γ^2_{PQ} has been observed between P and Q.
A model is proposed that explains the experimentally observed γ^2_{PQ} versus P. Due to the optical cavity there is always some correlation between ΔP and ΔQ. The model explains the decreasing γ^2_{PQ} with increasing P in the LED region by increasing intraband absorption. Due to the increasing stimulated emission in the super-radiation region, γ^2_{PQ} increases even to $\gamma^2_{PQ} = 1$ when threshold current is reached and laser action starts.
The stimulated emission is assumed to be correlated over the whole optically active region. The experimentally observed $\gamma^2_{PQ}(P)$ is in good agreement with the calculated values.

1. INTRODUCTION

The coherence function between the noises in the optical power outputs P and Q from both facets has been investigated from 1 Hz up to 10 kHz. The coherence function $\gamma^2_{PQ}(f)$ is defined as

$$\gamma^2_{PQ}(f) = \frac{\overline{\Delta P \cdot \Delta Q}^2}{\overline{\Delta P^2} \cdot \overline{\Delta Q^2}} \qquad (1)$$

where ΔP and ΔQ are the deviations from the mean optical outputs observed through a band-pass filter with central frequency f. Our experimental results always showed a frequency-independent plateau value γ^2_{PQ} in the frequency range where both quantities P and Q showed pure 1/f noise.

Figure 1 represents these plateau values γ^2_{PQ} as function of the optical power output for a gain-guided Philips laser diode, where $P_0 = 3$ mW. The dip in γ^2_{PQ} in figure 1 occurs below threshold at the transition between the spontaneous emission region (LED) and the super-radiation region at P_2.

2. MODEL

We consider the fluctuations Δp and Δq in the emitted light that arrives at the facets for the first time. (see figure 2) The coherence function (inside the optical cavity) between Δp and Δq is given by

$$F^2 = \frac{\overline{\Delta p \cdot \Delta q}^2}{\overline{\Delta p^2} \cdot \overline{\Delta q^2}} \qquad (2)$$

Further on, an expression will be derived for F in the case of spontaneous and stimulated emission. Due to the optical cavity, there is a relation between the observed coherence function

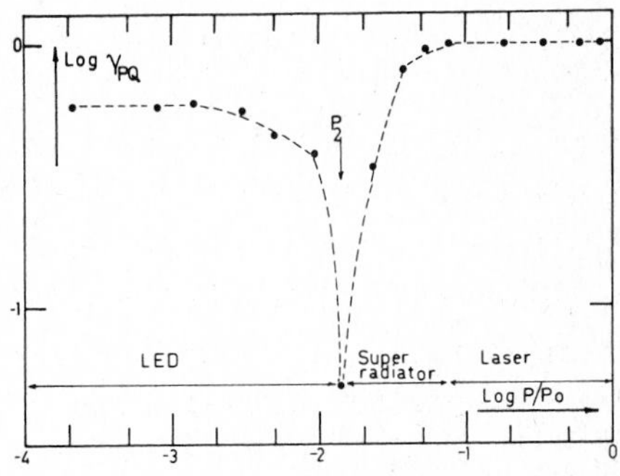

FIGURE 1
Experimentally observed coherence function at low frequencies as a function of the normalized optical output

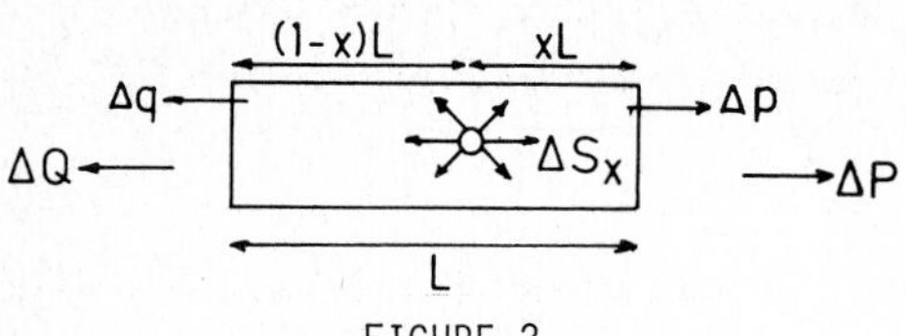

FIGURE 2
Optical cavity of length L

γ_{PQ}^2 outside the optical cavity and the coherence function F^2 inside it. Considering the low frequency fluctuations of at most 10 kHz and a travelling time of 3×10^{-12} s for a wave crossing the cavity from one facet to the other, a quasi-static treatment is justified. With R the reflection coefficient for the facets and α the absorption coefficient for fluctuations, the part of Δp that finally leaves the cavity at the right-hand side in figure 2 becomes

$$\Delta p_r = \theta \cdot \Delta p \tag{3}$$

$$\text{with } \theta = (1 - R)\left\{1 + (Re^{-\alpha L})^2 + (Re^{-\alpha L})^4 + \ldots\right\} \tag{4}$$

The part of Δp that finally leaves the cavity at the left-hand side is

$$\Delta p_\ell = \theta \cdot Re^{-\alpha L} \cdot \Delta p \tag{5}$$

Likewise, the parts of Δq which leave the cavity at both sides become

$$\left.\begin{aligned} \Delta q_r &= \theta \cdot Re^{-\alpha L} \cdot \Delta q \\ \Delta q_\ell &= \theta \cdot \Delta q \end{aligned}\right\} \tag{6}$$

From eqs. (3-6) we obtain for ΔP and ΔQ

$$\left.\begin{aligned} \Delta P &= \theta(\Delta p + Re^{-\alpha L} \cdot \Delta q) \\ \Delta Q &= \theta(Re^{-\alpha L} \cdot \Delta p + \Delta q) \end{aligned}\right\} \tag{7}$$

We assume the cavity to be statistically homogeneous and put $\overline{\Delta p^2} = \overline{\Delta q^2}$. With equations (1), (2) and (7) the relation between the coherence functions outside and inside the cavity becomes

$$\gamma_{PQ}^2 = \left\{\frac{F + 2Re^{-\alpha L} + F(Re^{-\alpha L})^2}{1 + 2FRe^{-\alpha L} + (Re^{-\alpha L})^2}\right\}^2 \tag{8}$$

From eq. (8) it follows that, in the absence of internal coherence (F = 0), there is still an outside coherence $\gamma_{PQ}^2 \neq 0$ due to the reflection at the mirrors. For a complete internal coherence (F = 1) under full-stimulated emission γ_{PQ}^2 becomes 1, as can be seen from eq. (8).

3. COHERENCE FUNCTION IN THE CASE OF SPONTANEOUS EMISSION

Figure 2 shows a local noise source ΔS_x at a distance xL from the right-hand facet. We assume uncorrelated noise sources radiating in every direction and distributed homogeneously over the channel. In other words,

$$\overline{\Delta S_x \cdot \Delta S_{x'}} = \overline{\Delta S^2} \cdot \delta(x - x') \tag{9}$$

The contribution of ΔS_x to Δp and Δq now becomes

$$\left.\begin{aligned} \Delta p_x &= e^{-\alpha x L} \cdot \Delta S_x \\ \Delta q_x &= e^{-\alpha(1 - x)L} \cdot \Delta S_x \end{aligned}\right\} \tag{10}$$

Considering the sum of all local contributions, Δp and Δq become

$$\left.\begin{aligned} \Delta p &= \int_0^1 e^{-\alpha x L} \cdot \Delta S_x dx \\ \Delta q &= \int_0^1 e^{-\alpha(1 - x)L} \cdot \Delta S_x d_x \end{aligned}\right\} \tag{11}$$

and

$$\left.\begin{aligned} \overline{\Delta p^2} &= \overline{\Delta q^2} = \left(\frac{1 - e^{-2\alpha L}}{2\alpha L}\right)\overline{\Delta S^2} \\ \overline{\Delta p \cdot \Delta q} &= e^{-\alpha L}\,\overline{\Delta S^2} \end{aligned}\right\} \tag{12}$$

This yields the coherence function inside the cavity:

$$F_{sp}^2 = \left(\frac{2\alpha L\, e^{-\alpha L}}{1 - e^{-2\alpha L}}\right)^2 \tag{13}$$

With equation (8) and (13), γ_{PQ}^2 is calculated as a function of the absorption coefficient α. The result with R = 0·32 is shown in figure 3.

Considering the γ_{PQ}^2 dependence on α in fig.3 and the observed trend in γ_{PQ}^2 in figure 1, the experimental results could be explained in the LED region by an increasing absorption coefficient with increasing light output P. There are two possibilities to support this suggestion. (i) Increasing optical output in the LED region

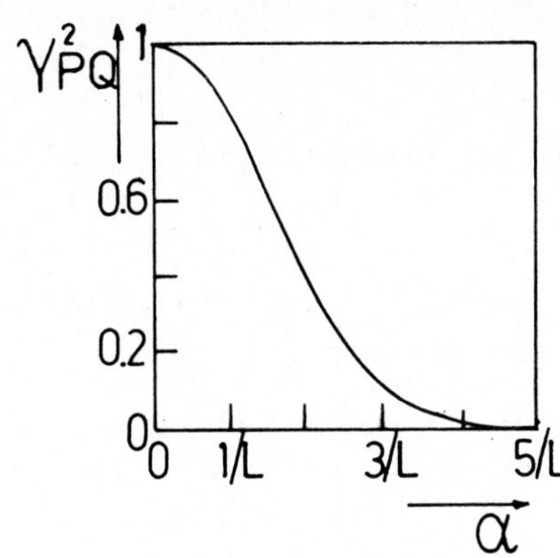

FIGURE 3
Reduction in γ_{PQ}^2 with increasing absorption α in the case of spontaneous emission

results in a slight decrease in the average wavelength and thus an enhanced absorption coefficient. (ii) Increasing optical output goes hand in hand with an increase in free carrier concentration, n', and therefore an increasing absorption coefficient.

Since we observed a shift of at most 4 nm in the mean wavelength we have chosen for the second option. The three most important absorption mechanisms in the laser diode are the band-band absorption, the free carrier absorption and the scatter losses. Generally, the band-band absorption makes the largest contribution. However, band-band absorption decreases at increasing carrier concentration and therefore this phenomenon cannot be used to explain the decrease of γ_{PQ}^2. It is not the number of the absorbed photons that is important for the coherence function at low frequencies, but the way in which the photons are absorbed.

With free carrier absorption the electrons are excited over an energy distance hν in the conduction band. This is proportional to the electron concentration for the conduction band absorption. The electron (or hole) which has absorbed a photon can easily drift away over the cladding layer outside the optically active region. The carrier confinement due to band gap differentials at the interfaces is of the order of 0.4 eV.

The scatter losses are independent of the carrier concentration. We will use the free carrier absorption and the scatter losses as the decorrelating absorption mechanism α for the fluctuations to calculate the coherence function γ_{PQ}^2. Then α becomes

$$\alpha = \varphi n' + \beta \tag{14}$$

with n' the free carrier concentration.

4. COHERENCE FUNCTION AT STIMULATED EMISSION

As we can see from experimental results in figure 1, the coherence function increases in the super-radiation region where stimulated emission becomes important. The addition of the effects of the uncorrelated 1/f noise sources as in eq. (11) is not allowed, because of possible amplification of ΔS_x over the whole cavity. Hence the sources cannot be treated as independent. The parts in Δp and Δq which are caused by stimulated emission are fully correlated. For that part it is correct to say that

$$F_{st} = 1 \tag{15}$$

Above threshold, only stimulated emission is important and, using eq. (8) the coherence function γ_{PQ}^2 outside the cavity becomes unity.

5. RATIO OF STIMULATED AND SPONTANEOUS FLUCTUATIONS IN THE CASE OF SUPER-RADIATION

When Δp_{sp}, Δq_{sp} are caused by spontaneous emission and Δp_{st}, Δq_{st} are caused by stimulated emission, then the total fluctuations are given by

$$\left.\begin{aligned} \Delta p &= \Delta p_{sp} + \Delta p_{st} \\ \Delta q &= \Delta q_{sp} + \Delta q_{st} \end{aligned}\right\} \tag{16}$$

The fluctuations caused by stimulated emission are assumed to be independent of those caused by spontaneous emission, which means $\overline{\Delta p_{sp} \cdot \Delta p_{st}} = \overline{\Delta q_{sp} \cdot \Delta q_{st}} = 0$. Then, with eqs. (2) and (16), F becomes

$$F = F_{sp}(1 - Z) + F_{st} \cdot Z \tag{17}$$

with Z the ratio between the stimulated fluctuation and the total fluctuation.

$$Z = \frac{\overline{\Delta p^2_{st}}}{\overline{\Delta p^2_{sp}} + \overline{\Delta p^2_{st}}} = \frac{\overline{\Delta q^2_{st}}}{\overline{\Delta q^2_{sp}} + \overline{\Delta q^2_{st}}} \tag{18}$$

The factors F_{sp} and F_{st} are the square root of the coherence functions as described in eq. (2). For the spontaneous emission region we observed that[1] the fluctuations in the optical output are proportional to the average optical output as shown below

$$\overline{\Delta P^2_{sp}} \sim P^{3/2} \tag{19}$$

For the stimulated emission region a stronger dependence has been observed[1] to wit

$$\overline{\Delta P^2_{st}} \sim P^m \tag{20}$$

with m = 4 for a Philips gain-guided laser diode and m ≈ 6 for a Sharp index-guided one. From equations (18), (19) and (20) it follows for Z that

$$Z = \frac{1}{\left(\frac{P_2}{P}\right)^{m - 3/2} + 1} \tag{21}$$

with P_2 the light output where the fluctuations in optical output change from LED to the super-radiation region. This is the point where the dependence on P changes from $P^{3/2}$ to P^m. Fig. 4 shows Z as a function of the light output.
A strong increase in Z is observed from fig. 4 entering the super-radiation and laser regions. In the spontaneous emission region it follows that[1]

$$P \sim n^2 \tag{22}$$

with eqs. (8), (13), (14), (15), (17), (21), (22) and R = 0·32, which is the value for our GaAs laser diodes, γ^2_{PQ} has been calculated. The dotted lines in figure 5 represent the calculations, the full lines the experimental results. The calculations are carried out with an experimentally observed value for P_2. The fitting parameters φ and β in eq. (14) are in the range of experimentally observed values.[2]

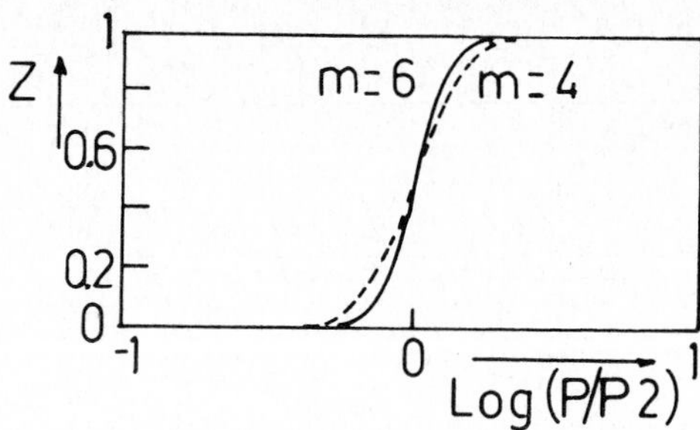

FIGURE 4
The ratio Z between correlated fluctuations from stimulated emission and the total fluctuations versus light output

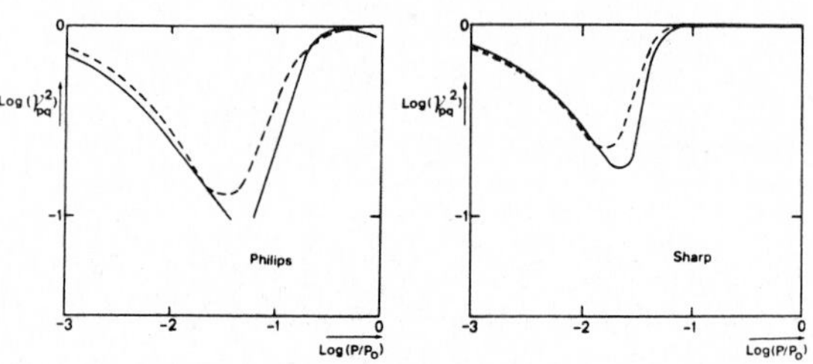

FIGURE 5
Comparison between experimental and calculated results

6. CONCLUSION

Assuming spatially uncorrelated noise sources under conditions of spontaneous emission and spatially correlated sources with stimulated emission, the model explains the experimental dependence of γ^2_{PQ} on P. The reduction in γ^2_{PQ} with increasing P in the LED region is caused by scatter and intraband absorption losses. The transition from LED to the super-radiation region goes hand in hand with an increasing fraction of spatially correlated stimulated light. This explains the increase of γ^2_{PQ} with P in the super-radiation region.

REFERENCE

1. L.K.J. Vandamme and J.R. de Boer, 1/f noise in the light output of laser diodes, Rome noise conference 1985

2. M.B. Panish and H.C. Casey Jr., Heterostructure Laser Part A, p. 179, Academic Press, New York, 1978

NOISE IN PHYSICAL SYSTEMS AND 1/f NOISE - 1985
A. D'Amico and P. Mazzetti (editors)

ON 1/f MOBILITY FLUCTUATIONS IN BIPOLAR TRANSISTORS (I): THEORY

T.G.M. KLEINPENNING

Eindhoven University of Technology, Eindhoven, The Netherlands

Expressions are presented for the 1/f current noise in bipolar transistors. These expressions are based on mobility-fluctuation 1/f noise. The current dependences of the 1/f noise intensities are found to be $S_{I_C} \sim I_C$ and $S_{I_B} \sim I_B{}^n$, where I_C is the collector current and I_B the base current. The exponent n is in the range of 0.5 - 2. An expression for the base current noise, induced by intrinsic base resistance fluctuations, is derived. The current dependence of the induced noise is $S_{I_B} \sim I_B{}^4$.

1. INTRODUCTION

Up to the present a number of papers have been published on 1/f noise in bipolar transistors.

In an extensive review paper on 1/f noise Van der Ziel has discussed 1/f noise in transistors[1]. The source of the 1/f noise in a transistor was generally accepted as being located in the emitter-base region. Mostly, 1/f noise was assumed to be a surface effect caused by carrier trapping in surface states. Experimental investigations prove, however, that 1/f noise is due to fluctuations in free-carrier mobility or diffusivity and not to free-carrier density fluctuations by trapping[2]. 1/f noise in p-n junctions can also be interpreted very well in terms of mobility fluctuations[3,4]. In a recent paper, Kilmer et al.[5] showed that 1/f noise in transistors can be interpreted in terms of mobility fluctuations.

The purpose of the present paper is to present a number of expressions for the mobility 1/f noise in currents in bipolar transistors.

2. 1/f CURRENT NOISE FORMULAE

Let us consider a pnp transistor with $V_{EB} >> kT/q$ and $V_{CB} < 0$, in which the following currents are dominant (see figure 1).

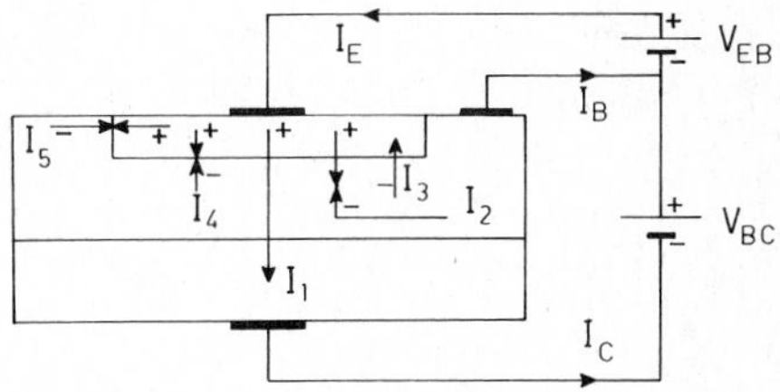

FIGURE 1
A pnp transistor

I_1 = emitter-collector current ($\sim \exp(qV_{EB}/kT)$)
I_2 = current due to holes injected into the base and recombining there ($\sim \exp(qV_{EB}/kT)$)
I_3 = current due to electrons injected from the base into the emitter ($\sim \exp(qV_{EB}/kT)$)
I_4 = current due to carriers recombining in the bulk space-charge region between emitter and base ($\sim \exp(qV_{EB}/2kT)$)
I_5 = current due to carriers recombining in the surface space-charge region ($\sim \exp(qV_{EB}/2kT)$)

The emitter current is $I_E = I_1 + I_2 + I_3 + I_4 + I_5$, the base current $I_B = I_2 + I_3 + I_4 + I_5$, and the collector current $I_C = I_1$. These currents show shot noise and 1/f noise due to mobility fluctuations.

For the noise in current I_i we can write

$$S_{I_i}(f) = 2qI_i(1 + f_{c_i}/f) \quad (1)$$

with f_{c_i} the corner frequency, i.e. the frequency at which the 1/f noise equals the shot noise.

If current I_1 is determined by diffusion in the base, then the 1/f noise is equal to the noise in a short p^+-n diode, so that[3,4]

$$f_{c_1} = \frac{\alpha_p D_p}{2W_B^2} \ln\left[\frac{p(0)}{p(W_B)}\right] = \frac{\alpha_p D_p}{2W_B^2} \ln\left[\frac{s_p + D_p/W_B}{s_p \exp(-qV_{EB}/kT) + D_p/W_B}\right] \quad (2)$$

Here α_p is the Hooge parameter, D_p the hole diffusivity in the base, W_B the base width, p(0) and $p(W_B)$ the hole density at the emitter and collector sides of the base, and s_p the recombination velocity of holes at the end of the base. Equation (2) can be applied, provided the base width W_B is greater than that of the base-collector junction W_j. A more accurate analysis gives the 1/f noise in the current I_1 as[3,6]

$$S_{I_1}(f) = \frac{2qI_1 f_{c1}}{f} = \frac{I_1^2}{fA\ell^2}\int_0^\ell \frac{\alpha_p(x)}{p(x)}dx \quad (3)$$

Here we assume that the effective device length $\ell = W_B + W_j$ extends over the region where the hole density depends on I_1. In the b-c junction the electric field strength is often so high that the holes have the saturation velocity v_s. In that case we have $I_1 = qpv_sA$, which implies that mobility fluctuations in the b-c junction have no influence on I_1, so that eq. (3) leads to eq. (2) with W_B^2 replaced by ℓ^2. Note that the emitter-to-collector transit time τ is also determined by the device length ℓ,

$$\tau = W_B^2/2D_p + W_j/v_s.$$

Mobility fluctuations lead to fluctuations in the current I_2; the corner frequency f_{c_2} (see appendix) is found to be

$$f_{c_2} = (\alpha_p/4\tau_p)\ln[p(0)/p(W_B)] \quad (4)$$

where τ_p is the hole lifetime in the base. For electron injection from base into emitter (I_3), we have a result similar to that given by eq. (2)

$$f_{c_3} = \frac{\alpha_n D_n}{2W_E^2} \ln\left[\frac{n(0)}{n(-W_E)}\right] \quad (5)$$

where $n(-W_E)$ is the electron density at the emitter contact.

For the 1/f noise in the recombination currents (I_4, I_5) we find[3,4,7]

$$f_{c_4} = \alpha/3\tau_{jb} \text{ and } f_{c_5} = \alpha/3\tau_{js} \quad (6)$$

where τ_{jb} and τ_{js} are the carrier lifetimes in the bulk and in the surface space-charge region, respectively. Here it should be noted that prevailing ideas that the 1/f noise in recombination currents in the space-charge region can only be interpreted in terms of fluctuations in the recombination velocity or in the recombination cross-section are incorrect. Mobility fluctuations have been recently shown to induce fluctuations in the recombination current[3,7].

In figure 2 we show the bias dependences of the currents I_B and I_C and their 1/f noise intensities. This figure leads to four possibilities for the current dependence of S_{I_B}:

(i) $V_{EB} < V_S, V_I$: $S_{I_B} \sim I_B \sim I_C^{\frac{1}{2}}$

(ii) $V_S < V_{EB} < V_I$: $S_{I_B} \sim I_B^2 \sim I_C$

(iii) $V_I < V_{EB} < V_S$: $S_{I_B} \sim I_B^{\frac{1}{2}} \sim I_C^{\frac{1}{2}}$

(iv) $V_{EB} > V_S, V_I$: $S_{I_B} \sim I_B \sim I_C$

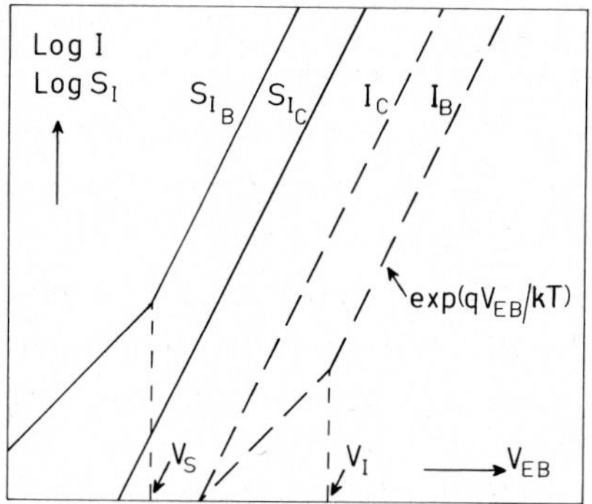

FIGURE 2
Diagrammatic representation of calculated I-V and S_I-V characteristics

3. BASE RESISTANCE FLUCTUATIONS

As put forward by Stoisiek and Wolf[8], an additional 1/f noise source in transistors can be due to fluctuations in the intrinsic base resistance r_b modulating the base current. However, their qualitative elaboration of this idea is incorrect. Here we shall recalculate the relation between the induced current noise in the base S_{I_B} and the resistivity fluctuations in the base for a circular pnp transistor with emitter radius R (see figure 3). The density of the base current flowing vertically through the junction at

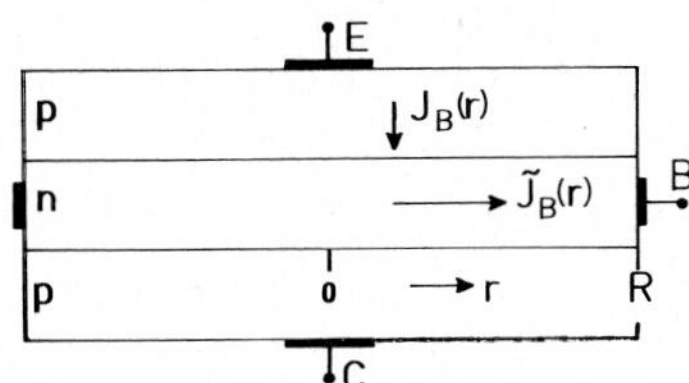

FIGURE 3
Scheme of emitter-base junction

position r is

$$J_B(r) = J_B(R)\, e^{-qV(r)/\eta kT} \tag{7}$$

where η is the ideality factor $1 \leq \eta \leq 2$. The voltage V(r) is given by

$$V(r) = \int_r^R E(r')dr' = \int_r^R \rho\tilde{J}_B(r')dr' \tag{8}$$

where E(r) is the electric field strength, $\tilde{J}_B(r)$ the current density in lateral direction in the neutral base, and V(R) = 0. The resistivity of the base is ρ. The relation between $\tilde{J}_B(r)$ and $J_B(r)$ is given by

$$\tilde{J}_B(r) = \frac{1}{2\pi r W_B}\int_0^r 2\pi r' J_B(r')dr' \tag{9}$$

At low base currents we have $V(0) \ll \eta kT/q$ and hence

$$J_B(r) \simeq J_B(R) \simeq I_B/\pi R^2 \tag{10}$$

and

$$\tilde{J}_B(r) \simeq (r/2W_B)J_B(R) \tag{11}$$

According to eqs. (8,9) the inequality $V(0) \ll \eta kT/q$ is equivalent to

$$\int_0^R \rho\tilde{J}_B(r)dr = \frac{\rho I_B}{4\pi W_B} = 2r_b I_B \ll \eta kT/q$$

where $r_b = \rho/(8\pi W_B)$.

Let us consider the situation where V_{EB} is constant, $V(0) \ll \eta kT/q$, and the resistivity ρ shows mobility fluctuations according to[2]

$$S_\rho(x,x',f) = \frac{\alpha\rho^2}{fnA}\,\delta(x-x') \tag{12}$$

These ρ-fluctuations lead to fluctuations in V(r) and thus in $J_B(r)$ and $\tilde{J}_B(r)$. Using eqs. (7) and (10) the fluctuations in I_B are given by

$$\delta I_B = \int_0^R 2\pi r\delta J_B(r)dr = \frac{-2qI_B}{\eta kTR^2}\int_0^R r\delta V(r)dr \tag{13}$$

From eqs. (8-11) we obtain

$$\delta V(r) = \int_r^R [\delta\rho(r')\tilde{J}_B(r') + \rho\delta\tilde{J}_B(r')]dr' =$$

$$\frac{I_B}{2\pi W_B R^2}\int_r^R r'\delta\rho(r')dr' +$$

$$+ \frac{\rho}{W_B}\int_r^R \frac{1}{r'}\left[\int_0^{r'} r''\delta J_B(r'')dr''\right]dr' \tag{14}$$

Substituting eq. (14) into eq. (13) yields

$$-\delta I_B = \frac{4I_B\chi}{\rho R^4}\int_0^R r\left[\int_r^R r'\delta\rho(r')dr'\right]dr +$$

$$+ \frac{8\pi\chi}{R^2}\int_0^R r\left\{\int_r^R \frac{1}{r'}\left[\int_0^{r'} r''\delta J_B(r'')dr''\right]dr'\right\}dr \tag{15}$$

with $\chi = q\rho I_B/(4\pi\eta kTW_B) \ll 1$.

Partial integration of eq. (15) results in

$$-\delta I_B = \frac{2I_B\chi}{\rho R^4}\int_0^R r^3\delta\rho(r)dr +$$

$$+ \frac{2\pi\chi}{R^2}\int_0^R (R^2 - r^2)r\delta J_B(r)dr \tag{16}$$

The last term in eq. (16) can be neglected because its contribution is of the order of

$$\frac{2\pi\chi}{R^2}\int_0^R R^2 r\delta J_B(r)dr = \chi\delta I_B \ll \delta I_B.$$

Using eq. (12) with $A = 2\pi rW_B$ and x = r and eq. (16) we find

$$\frac{S_{I_B}}{I_B^2} = \frac{(q\rho I_B)^2\,\alpha}{(4\pi\eta kTR)^2 3\pi W_B^3 nf} = \frac{4\alpha}{3fN}\left[\frac{r_b}{r_\pi}\right]^2 \qquad (17)$$

Here $N = \pi R^2 W_B n$ is the number of electrons in the base and $r_\pi = \eta kT/qI_B$ the input resistance. According to the model, 1/f fluctuations in r_b lead to 1/f fluctuations in I_B with $S_{I_B} \sim I_B^4$. However, the observed 1/f noise in I_B is orders of magnitude greater than predicted by eq. (17) and it is nearly always proportional to I_B^γ with $1 \leq \gamma \leq 2$.

Here it can be remarked that fluctuations in r_b also lead to fluctuations in the collector current I_C. Instead of eqs. (7) and (13) we have

$$J_C(r) = J_C(R)\, e^{-qV(r)/kT}$$

and

$$\delta I_C = \int_0^R 2\pi r \delta J_C(r)dr = \frac{-2qI_C}{kTR^2}\int_0^R r\delta V(r)dr$$

Thus

$$\frac{\delta I_C}{I_C} = \frac{\eta \delta I_B}{I_B} \text{ and } S_{I_C} = \left(\frac{\eta I_C}{I_B}\right)^2 S_{I_B}$$

APPENDIX

Here we shall derive eq. (4). Thus we have to show that I_2 fluctuates due to mobility fluctuations in the base and that $S_{I_2} = 2qI_2 f_{c_2}/f$. The recombination current I_2 fluctuates if the hole density in the base fluctuates like

$$\delta I_2 = \frac{qA}{\tau_p}\int_0^{W_B} \delta p(x)dx = \frac{q\delta P}{\tau_p}$$

and

$$S_{I_2} = (q/\tau_p)^2 S_P \qquad (18)$$

where τ_p is the hole lifetime. The fluctuations in p(x) can be derived from the relations

$$I_1(x) = -qD_p A dp(x)/dx$$

and

$$-\delta I_1(x) = \{I_1(x)/D_p\}\delta D_p(x) + qD_p A d\delta p(x)/dx \qquad (19)$$

Since the junctions are a.c. short-circuited we have the boundary conditions[9] $\delta p(0) = \delta p(W_B) = 0$. The current $I_1(x)$ is almost independent of the position x, hence $\delta I_1(x)$ is almost independent of x. From eq. (19) we find consequently that

$$\delta I_1 = -(I_1/W_B)\int_0^{W_B} \{\delta D_p(x)/D_p\}dx \qquad (20)$$

and

$$\delta p(x) = \frac{-1}{qD_p A}\left[x\delta I_1 + I_1\int_0^x \frac{\delta D_p(y)}{D_p}dy\right] \qquad (21)$$

Equations (20) and (21) lead to

$$\delta P = A\int_0^{W_B} \delta p(x)dx = \frac{I_1}{qD_p^2}\int_0^{W_B}\left(x - \frac{W_B}{2}\right)\delta D_p(x)dx$$

Using the cross-correlation spectral density in the diffusivity[3]

$$S_{D_p}(x,x') = [\alpha_p D_p^2/fAp(x)]\delta(x-x')$$

and

$$p(x) = p(0) - [p(0)-p(W_B)]\frac{x}{W_B} = p(0)\left[1 - \frac{ax}{W_B}\right]$$

we obtain

$$S_P = \frac{\alpha_p I_1^2}{fAq^2 D_p^2}\int_0^{W_B}\frac{(x - W_B/2)^2}{p(x)}dx =$$

$$\frac{\alpha_p I_1^2 W_B^3}{fAq^2 D_p^2 p(0)}\left[\frac{1}{2a} - \frac{1}{a^2} - \frac{(1-a/2)^2 \ln(1-a)}{a^3}\right]$$

Since $a = 1 - p(W_B)/p(0) \simeq 1$, the term between brackets can be approximated by $-(1/4)\ln(1-a)$. With eq. (18) we now find $S_{I_2} = 2qI_2 f_{c2}/f$, where f_{c2} is given by eq. (4) and the current I_2 by $qP/\tau_p \approx qAW_B p(0)/2\tau_p$. The current I_1 is equal to $qAD_p p(0)/W_B$.

REFERENCES

1. A. van der Ziel, Adv. in Electronics and Electron Physics 49 (1979) 225.
2. F.N. Hooge, T.G.M. Kleinpenning and L.K.J. Vandamme, Rep. Prog. Phys. 44 (1981) 479.
3. T.G.M. Kleinpenning, Physica 98B (1980) 289; 121B (1983) 81.
4. T.G.M. Kleinpenning, J. Vac. Sci. Technol. A3 (1985) 176.
5. J. Kilmer, A. van der Ziel and G. Bosman, Solid-State Electronics 26 (1983) 71.
6. A. van der Ziel and C.M. van Vliet, Phys, Stat. Sol. (a) 72 (1982) K53.
7. T.G.M. Kleinpenning, Physica B, accepted for publication.
8. M. Stoisiek and D. Wolf, IEEE-ED 27 (1980) 1753.
9. A. van der Ziel, Solid-State Electronics 25 (1982) 141.

NOISE IN PHYSICAL SYSTEMS AND 1/f NOISE - 1985
A. D'Amico and P. Mazzetti (editors)

ON 1/f MOBILITY FLUCTUATIONS IN BIPOLAR TRANSISTORS (II): EXPERIMENT

T.G.M. KLEINPENNING

Eindhoven University of Technology, Eindhoven, The Netherlands

1/f Noise experiments are presented for planar silicon bipolar transistors. The experimental results can be explained in terms of fluctuations in the mobility of free charge carriers. Experimental results from literature on 1/f noise in transistors are also discussed here in terms of mobility fluctuations. The agreement between experiment and theory is often found to be satisfactory.

1. INTRODUCTION

The purpose of this paper is to describe the 1/f noise, observed in bipolar transistors, in terms of mobility fluctuations of the free charge carriers. We shall present experimental results which will be compared with the 1/f noise formulae based on mobility fluctuations presented in ref.[1]. We shall also compare experimental results from literature with these 1/f noise formulae.

2. CIRCUIT FOR NOISE MEASUREMENT

Usually the noise of a bipolar transistor is measured with the transistor in a common-emitter or in a common-collector configuration (see figure 1)

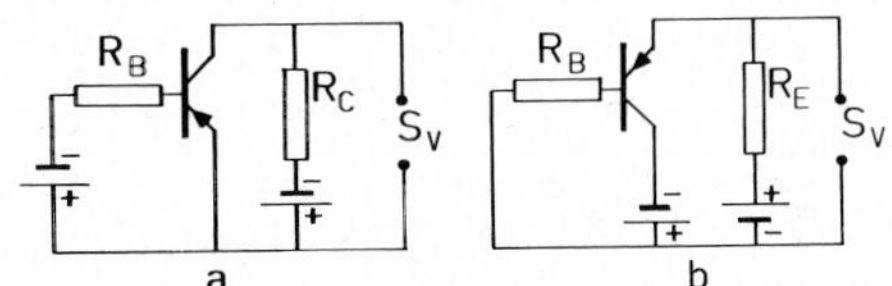

FIGURE 1
Circuits of a pnp transistor in a common-emitter (a) and in a common-collector configuration (b)

In a common-emitter configuration the voltage noise across the resistor R_C is given by[2,3]

$$S_V(f)=\left[\frac{g_m r_\pi R_B R_C}{r_\pi + R_B}\right]^2 \left[S_{I_B}(f)+\frac{4kT}{R_B}\right]+R_C^2\left[S_{I_C}(f)+\frac{4kT}{R_C}\right] \quad (1)$$

where $g_m = dI_C/dV_{EB}$ is the transconductance, $r_\pi = dV_{EB}/dI_B$ the input resistance, R_B the sum of external and internal base resistances, $\beta = g_m r_\pi$ the current amplification factor, and S_{I_B} and S_{I_C} the sum of shot noise and 1/f noise in the currents I_B and I_C. The noise in I_B and in I_C can be determined by measuring $S_V(f)$ as a function of R_B.

In the common-collector configuration the voltage noise across the resistor R_E is given by[4,5]

$$S_V(f) = \left[r_e - \alpha_T R_B\right]^2 S_{I_B}(f) + \left[r_e + (1 - \alpha_T)R_B\right]^2 S_{I_C}(f) + 4kTR_B \quad (2)$$

provided $R_E \gg r_e$, with $r_e = dV_{EB}/dI_E$ and $\alpha_T = \beta/(\beta + 1)$. The 1/f noise in I_B and in I_C can be determined by measuring $S_V(f)$ at $R_B \ll r_e$ and at $R_B = r_e/\alpha_T$.

3. EXPERIMENTAL RESULTS

Noise investigations were made on several types of bipolar transistors; pnp planar silicon transistors and commercial npn micro-wave transistors made by Hewlett and Packard (types 35824A, 35826E, HXTR2101, HXTR2102, HXTR3101, HXTR3102). The noise was measured using the common-collector and/or the common-emitter configuration. Both configurations lead to the same experimental results for S_{I_B} and S_{I_C}. A constant 9 V collector-base bias was used. The measurements were done at room temperature.

3.a. pnp transistor

Transistors with an emitter area A = 80 x 210 μm^2, a base thickness W_B = 10 μm and an emitter thickness W_E = 3 μm were investigated. The dopings in the emitter, base and collector are $N_E = 5 \times 10^{18}$ cm^{-3}, $N_B = 10^{16}$ cm^{-3}, and $N_C = 3 \times 10^{15}$ cm^{-3}, respectively. The collector and base current versus V_{EB} at V_{BC} = 9 V are plotted in fig. 2. The currents did not change appreciably when V_{BC} was varied between 0 and 27 V. The collector current obeys the relation

$$I_C = I_{CO}\exp\left(\frac{qV_{EB}}{kT}\right), \; I_{CO} = \frac{qAD_p n_i^2}{N_B W_B} \tag{3}$$

Experimentally we found $I_{CO} = 3.2 \times 10^{-15}$ A and theoretically $I_{CO} = 4 \times 10^{-15}$ A with D_p = 10 cm^2/s and $n_i = 10^{10}$ cm^{-3}. The base currents have an ideality factor $\eta \approx 2$ at low currents and $\eta = 1$ at higher currents.

The cut-off frequency $f_T \approx D_p/\pi W_B^2$ was experimentally found to be 1.5 MHz at I_C = 0.8 mA. Noise spectra were measured in the frequency range of 10 Hz to 100 kHz in the c-c configuration. At low frequencies the spectra show 1/f noise and at high frequencies white noise. The 1/f noise can be minimised by adjusting R_B, see figure 3. From this plot the internal base resistance is found to be r_b = 20 Ω. The 1/f noise density in I_B versus V_{EB} is plotted in figure 2. The 1/f noise in I_C was hardly detectable. From figure 3 we find $S_{I_C} \leq 5 \times 10^{-24}$ A^2s at I_C = 165 μA. Since $S_{I_B} \sim \exp(qV_{EB}/2kT)$ we conclude that the 1/f noise in I_B is a result of fluctuations in the recombination current in the space charge[1]. So we have

$$S_{I_B} = 2\alpha q I_B''/3f\tau_j \tag{4}$$

$$I_B'' = (qWAn_i/\tau_j)\exp(qV_{EB}/2kT) \tag{5}$$

Here $W = \pi kT/2qE_m \approx 0.015$ μm is the effective width of the space-charge region and E_m the maximum field strength[6]. From eqs. (4,5) we find experimentally for transistor A:

$\alpha/\tau_j \approx 2 \times 10^6$ s^{-1}, $A/\tau_j \approx 5 \times 10^4$ cm^2/s and $\alpha/A \approx 40$ cm^{-2}, and for transistor B: $\alpha/\tau_j \approx 8 \times 10^5$, $A/\tau_j \approx 7 \times 10^3$ and $\alpha/A \approx 10^2$.

If the recombination current occurs in the bulk space-charge region, then $A = 1.7 \times 10^{-4}$ cm^2 and thus $\alpha \approx 10^{-2}$ and $\tau_j \approx 10^{-8}$ s. If the recombination current occurs at the surface, then $A = \ell d$ with ℓ the perimeter of the emitter and d a characteristic depth ($\sim$ 0.1 μm). In this case we find $A \approx 10^{-7}$ cm^2, $\alpha \approx 10^{-5}$ and $\tau_j \approx 10^{-11}$ s.

From $S_{I_C} \leq 5 \times 10^{-24}$ A^2s at I_C = 165 μA and f = 1 kHz, using eqs. (2,3) in ref.[1] we find the Hooge parameter for the holes in the base to be at most $\alpha_p = 4 \times 10^{-6}$.

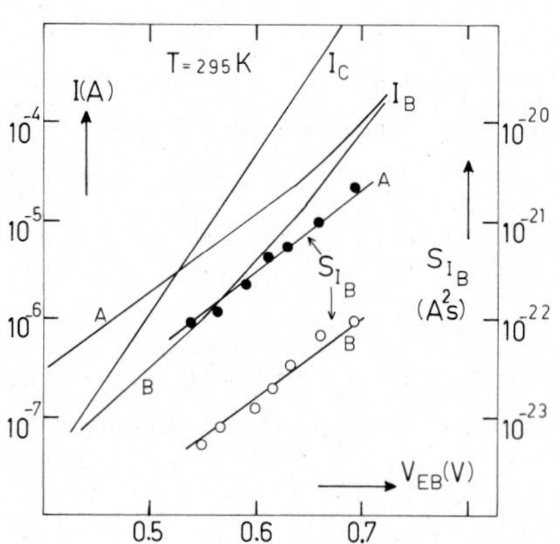

FIGURE 2
I_C, I_B and S_{I_B} at 1 kHz versus V_{EB} at V_{BC} = 9 V of pnp transistors A and B

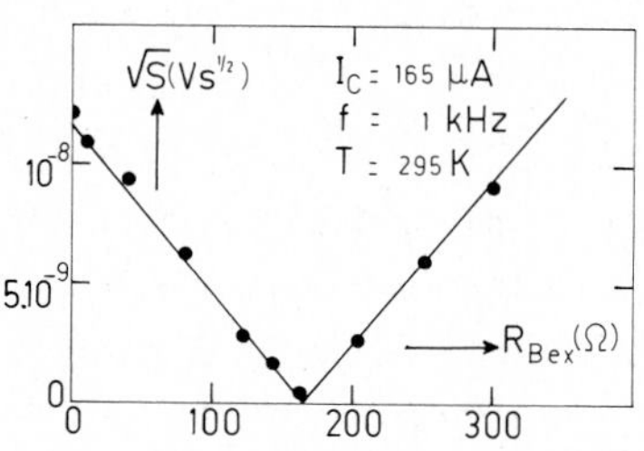

FIGURE 3
1/f Voltage noise versus external base resistance for transistor A in c-c configuration.

3.b. npn microwave transistors

The emitter area of the microwave transistors is about 2×10^{-5} cm^2. The I_C - V_{BE} characteristics at V_{CB} = 9 V of the transistors are almost the same (figure 4). The current I_C is nearly independent of V_{CB} for $0 < V_{CB} < 10$ V. Some I_B - V_{BE} plots are also shown in figure 4. The base currents have an ideality factor $\eta = 2$ at low currents and $\eta = 1$ at higher currents. The collector current obeys eq. (3) with $I_{CO} \approx 3 \times 10^{-16}$A. Using the typical parameters for a microwave transistor $D_n = 10$ cm^2/s, $N_B = 4 \times 10^{17}$ cm^{-3} and $W_B = 0.2$ µm, we calculate I_{CO} to be 4×10^{-16}A.

The noise spectra show 1/f noise at low frequencies and white noise at high frequencies. In the c-c configuration the 1/f noise can be strongly reduced by adjusting $R_B = r_e/\alpha_T$. Thus according to eq. (2) for the 1/f noise we have $S_{I_B} >> S_{I_C}$. For some transistors we have plotted S_{I_B} and S_{I_C} versus V_{BE} in figure 4.

In microwave transistors the base width W_B and the emitter width W_E are thin (< 1 µm), therefore the part of the base current with $\eta = 1$ is due to holes injected from the base into the emitter. Consequently we have[1]

$$S_{I_B} = 2qI_B'f_c'/f + 2qI_B''f_c''/f \tag{6}$$

with I_B' the current due to holes injected into the emitter ($\eta = 1$) and I_B'' the current due to recombination in the space-charge region between emitter and base ($\eta = 2$). The current I_B' is given by

$$I_B' = (qAD_pn_i^2/N_EW_E)\exp(qV_{BE}/kT) \tag{7}$$

the current I_B'' by eq. (5) and the corner frequencies f_c' and f_c'' given by eqs. (5,6) in ref.[1]. Since S_{I_B} was found to be proportional to $\exp(qV_{BE}/kT)$, we conclude that the 1/f noise is dominated by fluctuations in I_B', thus

$$S_{I_B} = 2qI_B' \frac{\alpha_p D_p \ln[p(0)/p(-W_E)]}{2W_E^2 f} \tag{8}$$

Using $D_p = 1.5$ cm^2/s, and $W_E = 0.5$ µm, we obtain α_p values in the range of 10^{-5} to 10^{-3}.

In transistor HXTR3102 we have measured $S_{I_C}/I_C \approx 2 \times 10^{-19}$ A^{2}s at f = 1 kHz. With the help of eqs. (2,3) in ref.[1] this results in $\alpha_n \approx 2 \times 10^{-6}$ for electrons in the base. Here we have taken $W_B = 0.2$ µm, $W_j \approx (2\varepsilon V_{CB}/qN_C)^{\frac{1}{2}}$ = 3.5 µm and $D_n = 10$ cm^2/s.

It should be noted that, for highly doped emitters and bases, the α-values calculated above have to be corrected according to[7]

$$\alpha_{observed} = (\mu/\mu_\ell)^2\alpha_{true} = (D/D_\ell)^2\alpha_{true}$$

with μ the mobility and μ_ℓ the lattice mobility.

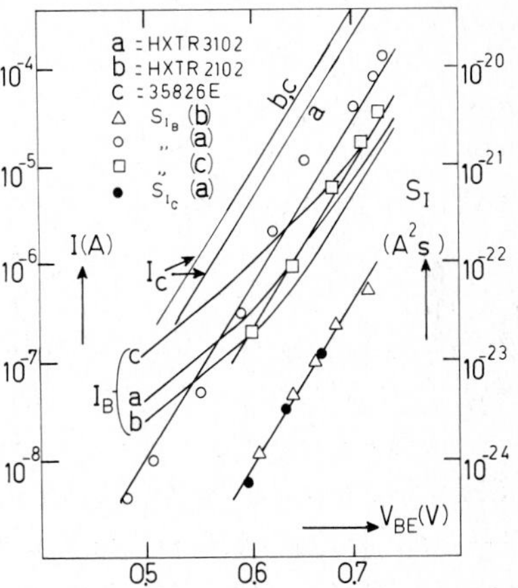

FIGURE 4

I_C, I_B, S_{I_B} and S_{I_C} at 1 kHz versus V_{BE} at V_{CB} = 9 V for several npn transistors

4. COMPARISON WITH LITERATURE

Here we review some 1/f noise data of transistors reported in the literature. First of all, it can be concluded both from our experiments and those of other investigators, that the 1/f noise in I_B is dominant[2,4,8-12].

Conti[10] has measured the 1/f noise in gated transistors. He found $S_{I_B} \sim I_C \sim I_B \sim \exp(qV_{EB}/kT)$, which is in agreement with the mobility fluctuation model[1]. Plumb and Chenette[4] found for a typical transistor $S_{I_B} \sim I_B \sim I_E^{1/2}$, a result similar to our pnp transistors. For low-noise transistors Gibbons[3] found $S_{I_B} \sim I_E \sim \exp(qV_{EB}/kT)$, a result similar to our npn microwave transistors. Higuchi and Ochi[13] and Green and Jones[12] found that the 1/f noise in their transistors can be described by the relation

$$S_{I_B} = KI_B''^2/Af \quad (9)$$

where I_B'' is the nonideal base current ($n = 2$) and K a constant in the range of 10^{-14}-10^{-12} cm^2. We found a similar result for the npn transistors with $K \approx 10^{-13} - 10^{-11}$ cm^2. According to the mobility fluctuation model the constant K is found with help of the eqs. (5,7,8) to be

$$K = \alpha_p D_p^2 \tau_j^2 \ln[n(0)/n(-W_E)]/W_E^3 W^2 N_E \quad (10)$$

For $\alpha_p = 10^{-3}$, $D_p = 1.5$ cm^2/s, $\tau_j = 10^{-8}$ s, $n(0)/n(-W_E) = 100$, $W_E = 0.5$ μm, $W = 0.01$ μm and $N_E = 10^{20}$ cm^{-3}, equation (10) gives $K \approx 10^{-13}$ cm^2. Here τ_j is valued using eq. (5).

5. CONCLUSIONS

The 1/f noise in bipolar transistors can be satisfactorily described in terms of mobility fluctuations with α-values in the range of 10^{-6} to 10^{-3}. The 1/f noise in the base current is nearly always dominant. The mobility fluctuation model predicts that, at a fixed collector current, the 1/f noise in the base current drops with decreasing base current. For example, if the lifetime in the emitter-base space-charge region increases, then the base current and its 1/f noise decreases, whereas the current gain increases. Thus the 1/f noise in high-gain transistors has to be lower then that in low-gain units, which is also confirmed experimentally[13-15]. Investigators often present the observed 1/f noise in the base current versus base current with the result $S_{I_B} = I_B^{\gamma}$ with $1 < \gamma < 2$. We prefer S_{I_B} vs V_{EB} plots (see figs. 2,4) instead of S_{I_B} vs I_B plots (see figure 5), as the last ones are less illustrative.

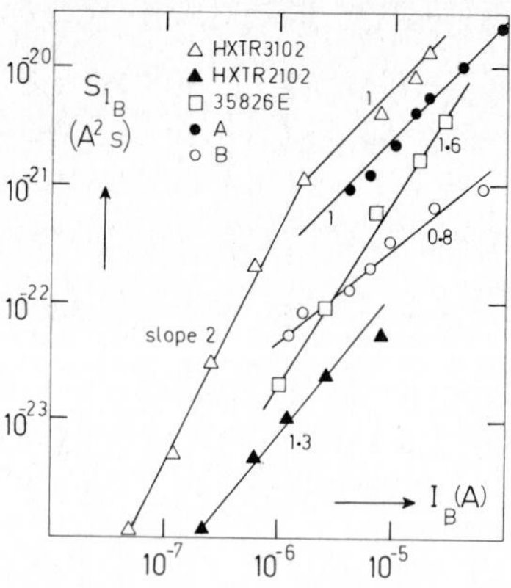

FIGURE 5
Relationship between 1/f noise intensity in S_{I_B} at 1 kHz and base current of the transistors from figures 2 and 4.

REFERENCES

1. T.G.M. Kleinpenning, On 1/f mobility fluctuations in bipolar transistors (I): theory, 8th Int. Conf. on Noise in Physical Systems, Rome, 1985
2. J. Kilmer, A. van der Ziel and G. Bosman, Solid-State Electronics 26 (1983) 71.
3. J.F. Gibbons, IRE Trans. El. Dev. ED-9 (1962) 308.
4. J.L. Plumb and E.R. Chenette, IEEE-ED 10 (1963) 304.
5. A. van der Ziel, Solid-State Electronics 25 (1982) 141.
6. A. van der Ziel, Solid State Physical Electronics (Prentice-Hall, Englewood Cliffs, N.J., 1968).
7. F.N. Hooge, T.G.M. Kleinpenning and L.K.J. Vandamme, Rep. Prog. Phys. 44 (1981) 479.
8. M. Stoisiek and D. Wolf, IEEE-ED 27 (1980) 1753.
9. M.B. Das, IEEE-ED 22 (1975) 1092.
10. M. Conti, Solid-State Electronics 13 (1970) 1461.
11. K.F. Knott, Solid-State Electronics 16 (1973) 1429.
12. C.T. Green and B.K. Jones, J. Phys. D.: Appl. Phys. 18 (1985) 77.
13. H. Higuchi and S. Ochi, Proc. Symp. on 1/f Noise, Tokyo, 1977, p. 140.
14. S.T. Hsu, IEEE-ED 18 (1971) 425.
15. M. Stoisiek and D. Wolf, Solid-State Electronics 23 (1980) 1147.

NOISE IN PHYSICAL SYSTEMS AND 1/f NOISE - 1985
A. D'Amico and P. Mazzetti (editors)

LOCATION OF 1/f NOISE SOURCES IN BIPOLAR TRANSISTORS

X.N. ZHANG and A. VAN DER ZIEL, EE Department, Univ. of Minnesota, Minneapolis, MN 55455 USA

H. MORKOC, EE Department, University of Illinois, Champaign-Urbana, IL USA

The location of the 1/f noise sources in 2N2955 Ge transistors and double heterojunction bipolar transistors (DHBJTs) was studied by a new method. This method can generally be utilized to locate any low frequency noise source in bipolar transistors.

1. THE MEASURING METHOD

The measuring circuit used in this investigation is shown in Fig. 1. All measurements were done at 20 Hz and at room temperature. We also measured the spectra themselves to be sure that they were of the 1/f type.

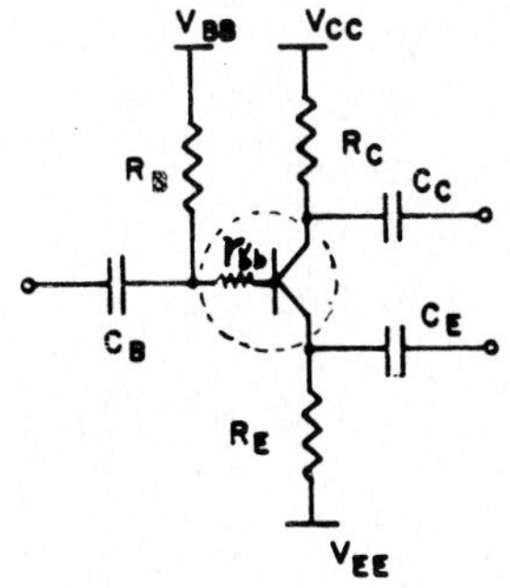

Fig. I The Bias - Circuit

A resistance R_e was inserted in the emitter lead to provide the possibility of negative feedback; the emitter bias V_{EE} was so adjusted that the emitter voltage was near ground potential. Without altering the bias conditions we can measure six different current spectra as follows:

(1) $S_{IC}(f)$; collector noise spectrum without feedback, with base and emitter h.f. grounded.

(2) $S_{IC}'(f)$; collector noise spectrum with feedback (C_E open) and base h.f. grounded.

(3) $S_{IC}''(f)$; collector noise spectrum without feedback, with emitter h.f. grounded and C_B open.

(4) $S_{IC}'''(f)$; collector noise spectrum with feedback (C_E open) and C_B open.

(5) $S_{IB}(f)$; base noise spectrum without emitter feedback, collector and emitter h.f. grounded.

(6) $S_{IB}'(f)$; base noise spectrum with emitter feedback, collector h.f. grounded and C_E open.

With R_e omitted we could only measure three configurations (2), (5) and (3) and that would not be sufficient for locating the noise sources. In principle one could also measure $S_{IE}(f)$, but since the emitter has the low internal resistance $R_{eo}=1/g_{me}$, the voltage gain would be very small and the device noise might drown in the amplifier noise. Emitter feedback via R_e accomplishes the same purpose more easily.

Since $[S_I(f)\Delta f]^{1/2}$ is a current generator we must calculate $S_I(f)$ by h.f. short-circuiting the terminals in question. For example, for calculating $S_{IC}'(f)$ one must h.f. short-circuit the collector and evaluate all contributions to the collector noise current i_c.

2. NOISE IN 2N955 Ge p^+-n-p TRANSISTORS

$S_{IB}(20)$ and $S_{IB}'(20)$ versus I_B at constant V_{CE} were found to be identical. This is not surprising, since the feedback hardly affects the base noise because the base-emitter conductance $g_{be}=1/R_{BE}=g_m/\beta_F$ is so small.

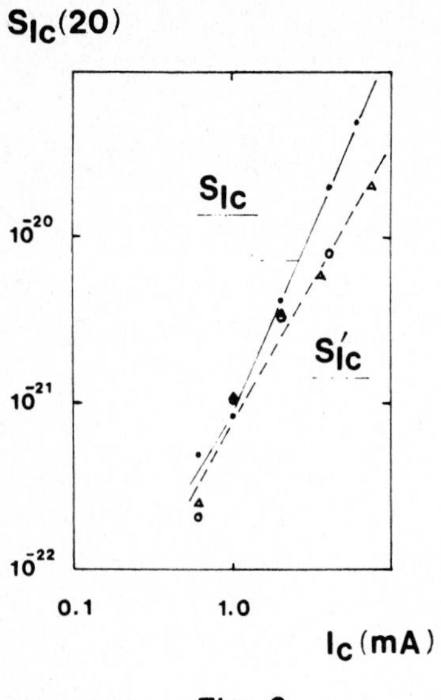

Fig. 2

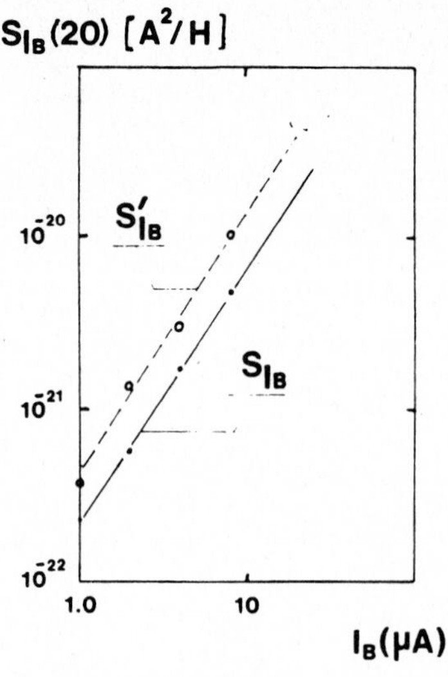

Fig. 3

Figure 2 shows $S_{IC}(20)$ and $S_{IC}'(20)$ versus I_C at constant V_{CE}. We note that $S_{IC}'(20)$ is somewhat smaller than $S_{IC}(20)$ because of the feedback. Also shown is $S_{IB}(f)$, and it is seen that $S_{IC}'(f)=S_{IB}(f)$. This indicates that the 1/f noise current generator i_p is located between base and collector and that it is caused by fluctuations in the recombination of hole-electron pairs in the base region and in the collector space charge region. Since an injected hole either recombines or is collected, this gives rise to partition 1/f noise.

Why is $S_{IC}(f)$ somewhat larger than $S_{IC}'(f)$? There are two reasons. In the first place $S_{IC}(f)$ may have a small noise component i_{ce} flowing between emitter and collector; it is strongly reduced by the feedback. In the second place i_p flows through $r_{b'b}$ and this gives a current $g_m r_{b'b}\, i_p$ in the collector. Hence

$$S_{IC}(20) = S_{IB}(20)[1+g_{mc}r_{b'b}]^2 > S_{IB}(20) \quad (1)$$

where $g_{mc}=qI_C/kT$, and

$$S_{IC}'(20) = S_{IB}(20)[1+g_{m'}r_{b'b}]^2 \simeq S_{IB}(20) \quad (1a)$$

because $g_{m'}=g_{mc}/(1+g_{me}R_e)\simeq 0$, where $g_{me}=qI_E/kT$.

3. NOISE IN DOUBLE HETEROJUNCTION BIPOLAR TRANSISTORS

Figure 3 shows $S_{IB}(f)$ versus I_B at constant V_{CE}; this noise is quite small. Figure 4 shows $S_{IC}(f)$ versus I_C at constant V_{CE}, this noise is much larger. The two current scales compare as follows: I_B=3.0 μA corresponds to I_C=100μA so that β_F=33, whereas the a.c. current amplification factor β_f is somewhat larger. At this point $S_{IB}(20)=1.2\times10^{-21}$ Amp^2/Hz and $S_{IC}(20) =1.9\times10^{-18}$ Amp^2/Hz so that $S_{IC}(20)/S_{IB}(20)=1600$ which is somewhat larger than β_f^2. The main noise source is therefore the collector noise.

Figure 4 also shows $S_{IC}'(20)$ versus I_C; it is much smaller than $S_{IC}(20)$ because of the emitter feedback. Obviously

$$S_{IC}'(20) = S_{IC}(20)/(1+g_{me}R_e)^2 \quad (2)$$

This indicates that the current generator $[S_{IC}(f)\Delta f]^{1/2}$ is connected between collector and emitter. We also note that the slope of $S_{IC}(20)$ versus I_C is much larger than the slope of $S_{IB}(20)$ indicating that $S_{IB}(f)$ and $S_{IC}(f)$ come from different 1/f noise mechanisms.

Finally, Figure 4 shows $S_{IC}''(20)$ and

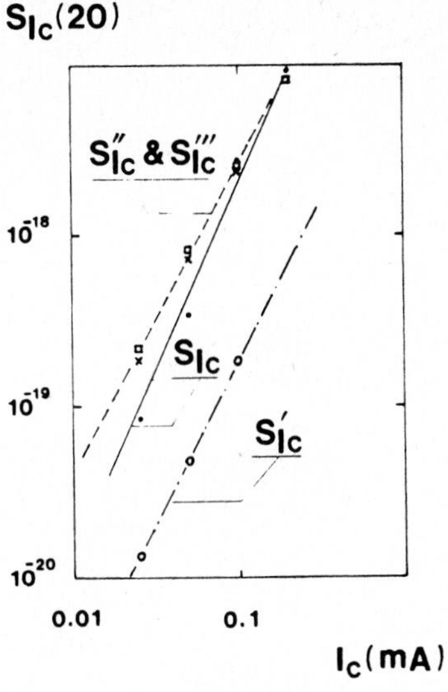

Fig. 4

$S_{IC}'''(20)$ versus I_C. These spectral intensities are equal and are somewhat above $S_{IC}(20)$. Since

$$S_{IC}''(20) = S_{IC}(20)+\beta_f^2 S_{IB}(20) \quad (3)$$

because the noise sources are independent, the last term in (3) is comparable to the first term. Since $S_{IB}(20)$ versus I_B has a smaller slope than $S_{IC}(20)$, the ratio $S_{IC}''(f)/S_{IC}(f)$ decreases somewhat with increasing current.

To understand why $S_{IC}''(20)$ and $S_{IC}'''(20)$ are nearly equal we turn to Fig. 1. If i_{be} is the base noise and i_{ce} the collector noise, then

$$i_c''' = \beta_f i_{be} + i_{ce} g_{mc} R_e/(1+g_{me}R_e) \simeq \beta_f i_{be} + i_{ce} = i_c'' \quad (4)$$

as had to be proved.

We note from Figure 3 that $I_B'(20)$ is larger than $I_B(20)$. This comes from feedback of the collector noise i_{ce}. According to Figure 1

$$i_b' = i_{be} + \frac{i_{ce}R_e/R_{BE}}{1+g_{me}R_e} \simeq i_{be} + i_{ce}/\beta_f$$

so that

$$S_{IB}'(20) = S_{IB}(20)+S_{IC}(20)/\beta_f^2 > S_{IB}(20) \quad (5)$$

since $\beta_f^2 S_{IB}(20)$ and $S_{IC}(20)$ are comparable.

There are two possible locations of the base noise current generator. It can be located between base and emitter or between base and collector. Since here $S_{IB}(f) \ll S_{IC}(f)$ we cannot discriminate between these two possibilities.

4. BURST NOISE

We also applied the above method to low-frequency burst noise and found that this noise source was located between the emitter and the outside of $r_{b'b}$, as found earlier by another method.[1]

ACKNOWLEDGMENT

This work was supported by an NSF grant.

REFERENCES

1. R. C. Jaeger, A. J. Brodersen and E. R. Chenette, Record of the 1968 Region III IEEE Convention, 58 (1968) see also A. van der Ziel, Noise, Sources, Characterization, Measurements, Prentice Hall, Inc. (1970), Section 6.5.

NOISE IN PHYSICAL SYSTEMS AND 1/f NOISE - 1985
A. D'Amico and P. Mazzetti (editors)

1/f NOISE MODIFICATIONS IN ELECTRICALLY DEGRADED SHORT-CHANNEL MOSFETs

Z.H. FANG*, S. CRISTOLOVEANU, A. CHOVET

Laboratoire de Physique des Composants à Semiconducteurs (U.A. CNRS 840)
Institut National Polytechnique de Grenoble
ENSERG, 23 rue des Martyrs 38031 GRENOBLE CEDEX - FRANCE

The 1/f noise of short-channel MOSFETs is measured before and after an electrical stress. The relative spectral density of drain current fluctuations S_I/I_D^2 is shown to increase after aging, due to a stress-induced generation of traps in the oxide. The noise experiments from weak to strong inversion also show that traps are mainly generated in a very narrow region near the drain and have an inhomogeneous energetical distribution.

1. INTRODUCTION

It has been recently confirmed [1,2] that low-frequency 1/f noise in MOSFETs, even when made with optimized technology, still essentially comes from fluctuations in the number of carriers in the inversion channel, due to trapping effects through interface states and/or traps in the oxide.

Therefore, 1/f noise measurements can be an interesting tool to investigate the interface degradation caused by the injection of hot electrons into the gate oxide. Such a mechanism is responsible for MOSFET aging, which is an essential problem to be solved for small-size device applicability.

It was demonstrated by REIMBOLD et al [1,2] that an elegant way to analyse the 1/f noise in MOS transistors is to describe it in the form of the relative spectral density of drain current fluctuations S_I/I_D^2, according to the relation :

$$\frac{S_I(f)}{I_D^2} = \frac{K}{f}\,\frac{N_t}{(q\,N_{ss} + C_D + C_{ox} + C_n)^2} \qquad (1),$$

which is valid (at least for low drain bias) from weak to strong inversion provided that the trap density N_t ($m^{-3}eV^{-1}$) in the oxide at the Fermi level is uniform along the whole channel. K is a constant inversely proportional to the device area and $C_{ss} = qN_{ss}$ (N_{ss} : $m^{-2}eV^{-1}$), C_D, C_{ox} and C_n are capacitances (per unit area) associated with surface states, depletion region, oxide and inversion channel; $C_n = \alpha(q/kT)Q_n$, with $\alpha = 1$ in weak inversion and $\alpha = 0.5$ in strong inversion.

We shall use this model to explain the influence of the aging induced by electrical stress on short-channel (0.4 to 2.5 μm) n-type MOSFETs made at the LETI Laboratories (Grenoble). Then, from 1/f noise measurements, information will be obtained on the defect creation, localisation and energy distribution.

Throughout this work, 1/f noise was observed below 100 Hz and at room temperature, and measured using a digital spectrum analyser system, while $I_D(V_D, V_G)$ curves were automatically plotted by a computer-controlled pA-meter.

---

* On leave from the Dept. of Physics, Wuhan University, P.R.China.

2. WEAK INVERSION REGION

2.1. Aging effects

Typical noise spectra are shown in Fig. 1 for a short-channel MOSFET (effective channel length $L \simeq 0.5\ \mu m$) before and after a strong electrical stress ($V_D = 5V$, $V_G = 10V$ during 15 hours) in the ohmic region (a) or in the saturation region (b).

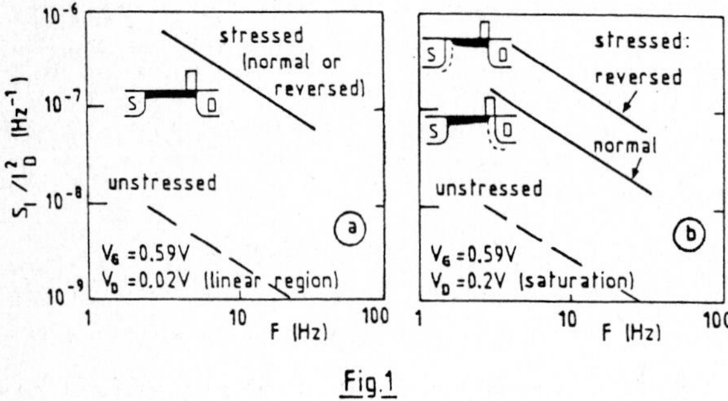

Fig.1

It appears that S_I/I_D^2 is greater after the electrical stress, due to an increase in the mean trap density N_t. More precisely, for ***weak inversion*** *(where $C_n << C_{ox} + C_D + C_{ss}$), the extension of eq. (1) for a two-part model of the channel gives, at any drain bias :*

$$\frac{S_I(f)}{I_D^2} = \frac{K}{f}\,\frac{1}{L}\left[\frac{\Delta L\, N_{td}}{(qN_{ssd} + C_D + C_{ox})^2} + \frac{(L-\Delta L)N_{to}}{(qN_{sso} + C_D + C_{ox})^2}\right] \qquad (2),$$

because N_t and N_{ss} take the values N_{td} and N_{ssd} over a distance ΔL, and N_{to} and N_{sso} elsewhere.

Aging effects observed through 1/f noise reveal, therefore, an important increase in slow states N_t (traps inside the oxide).

2.2. Trap localisation and density

Noise spectra were also measured in the reversed operating mode of the MOSFET (i.e. source and drain connections inter-changed). Before the electrical stress, no difference was observed in the 1/f noise between normal (N) and reversed (R) modes, for both ohmic and saturation regions, which is a clear indication of the uniformity of N_t (and N_{ss}) along the channel.

After stress, the 1/f noise level increases, but no difference is observed between N and R modes for the linear $I_D(V_D)$ region because the two configurations are equivalent (Fig. 1a). For the saturation region, however, the noise is lower in the normal mode (Fig. 1b) and could even be identical to its initial level before stress.

This is due to the fact that the degradation (trap creation) during the electrical stress occurs near the drain contact and this is proved in Figs. 2 and 3 where S_I/I_D^2 is plotted against drain voltage V_D at $f = f_0 = 10$ Hz ; this localisation could also be deduced from static $I_D(V_D, V_G)$ curves [3, 4] although relevant parameters (threshold voltage, carrier mobility, etc.) are much less sensitive than noise.

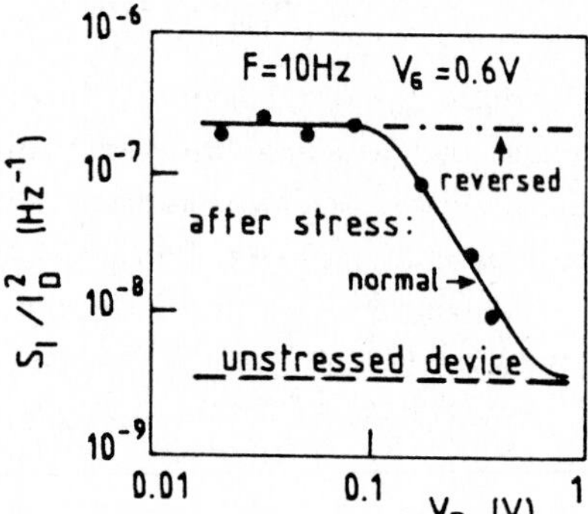

Fig. 2 : Same MOSFET as in Fig. 1 : $L \simeq 0.5\ \mu m$, stressed 15 hours at $V_D = 5V$, $V_G = 10V$.

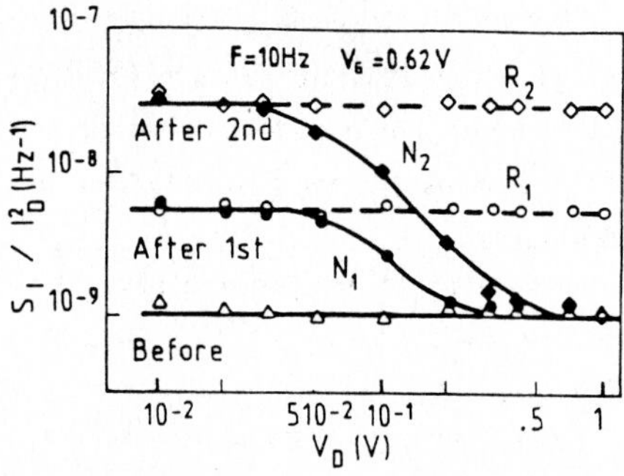

Fig. 3 : MOSFET with $L \simeq 2.4\ \mu m$. Each stress : $V_G = V_D = 9V$, 2 hours.

In weak inversion, for a homogeneous (unstressed) device, S_I/I_D^2 *is a constant* K_0 *independent of* V_D, *in agreement with eq. (1) (or eq. (2) for* $N_{td} = N_{t0}$) :

$$K_0 \simeq \frac{K}{f_0} \frac{N_{t0}}{(C_D + C_{ox})^2} \quad \text{if } C_{ss} \ll C_D + C_{ox}$$

After the aging, S_I/I_D^2 *remains a constant* $K_d > K_0$ *for the R mode while, for the N mode, the plateau value* K_d *also observed in the ohmic region is followed by a decrease to the initial value* K_0 *(which is reached for a* V_D *bias corresponding to the saturation region).*

This result is explained by the progressive screening of the damaged region (ΔL*) as the "pinch-off point" moves away from the drain (N mode) : from eq. (2), we have*

$$K_d \simeq \frac{K}{f_0} \frac{\Delta L}{L} \frac{N_{td}}{(qN_{ssd} + C_D + C_{ox})^2} \qquad (3)$$

$$\text{if } N_{td} \gg \frac{L}{\Delta L} N_{t0} .$$

Since the stress induced traps N_{td} *become inactive when the surface beneath them is depleted (i.e. when the drain depletion region is increased by* ΔL*), we get for a sufficient drain bias, from eq. (2) :*

$$\frac{S_I}{I_D^2}(f_0) \simeq \frac{K}{f_0} \frac{N_{t0}}{(C_D + C_{ox})^2} = K_0 .$$

ΔL *can then be estimated : from weak inversion equations [5] and Fig. 2, we have found that (for strong aging)* $\Delta L \simeq 50$ *nm ; it therefore follows from Fig. 2 (or Fig. 3) that, after the electrical stresses applied to our MOSFETs, the trap density is increased several hundred times.*

From Fig. 3, it also appears that after successive aging :

- the length ΔL *increases (a higher drain voltage is required in N mode to obtain* $S_I/I_D^2 \simeq K_0$*).*

- the density of interface states (N_{ss}*) is either little affected by the aging process (*$C_{ssd} < C_{ox} + C_D$*), or it increases much more slowly than* N_t *(see eq. (3)).*

In addition, our latest experiments seem to show that after strong enough aging, the trap transverse distribution is no longer uniform in the oxide but is greater close to the interface ; this leads to a G.R.- type noise observed at higher frequencies (beyond 100 Hz the noise can slowly decreases as $f^{-0.3}$ *over nearly two decades of frequency, while the* V_D *influence in N and R modes remains significant [6]).*

3. FROM WEAK TO STRONG INVERSION

While in weak inversion C_n *is negligible in eq. (1), it prevails in strong inversion ; for a uniform (low* V_D*) and homogeneous channel the equation must be written :*

$$\frac{S_I}{I_D^2} \propto \frac{1}{f} \frac{N_t}{Q_n^2} \simeq \frac{1}{f} \frac{N_t}{C_{ox}^2 (V_G - V_T)^2} \qquad (4)$$

Eq. (4) is also approximately valid for a stressed device, as when passing from eq. (1) to eq. (3), if N_t *is replaced by* $(\Delta L/L)N_{td}$ *and* V_T *by* V_{Td}, *since the threshold voltage can also be affected by the aging process.*

3.1. 1/f noise in strong inversion

Curves S_I/I_D^2 *vs.* V_D, *similar to Fig. 3 but now relative to strong inversion, are plotted in Fig. 4 for the same transistor operated in the N or R mode before and after the same two aging steps.*

Very similar qualitative observations can be made : in the ohmic region, S_I/I_D^2 *is a constant, as suggested either by eq. (4) or its equivalent for a damaged channel. Before aging,* S_I/I_D^2 *first increases with* V_D *when leaving the ohmic region because* $\int Q_n^{-2}(y)dy$ *is proportional to* $(V_G - V_T - V_D/2)^{-2}$ *; following this,* S_I/I_D^2 *becomes approximately constant in the saturation region.*

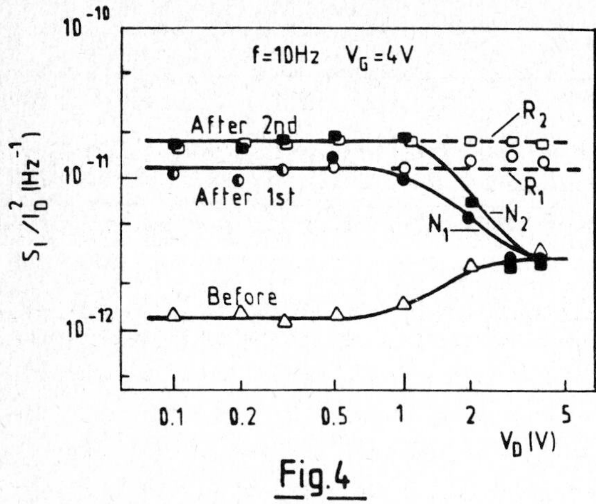

Fig.4

After aging, and only in the N mode, S_I/I_D^2 decreases with V_D to the same saturation value observed before stress : this is the "basic noise" related to the part of the channel, unaffected by trap creation, between the source and the true pinch-off point of the channel.

3.2. Estimation of the trap energetical distribution

In Fig.5, S_I/I_D^2 at 10 Hz is plotted as a function of V_G, at a fixed, low V_D.

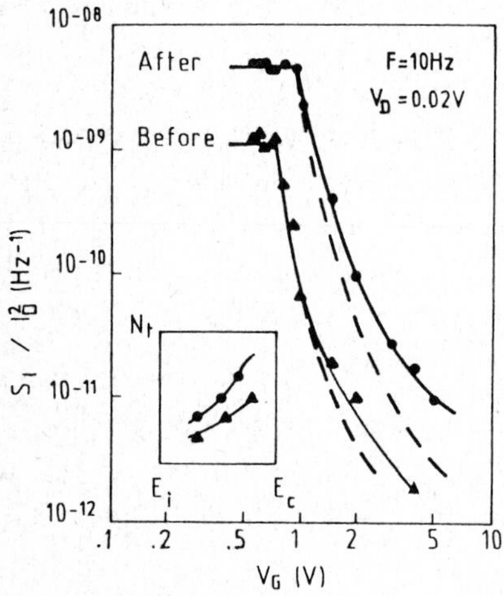

Fig.5 : Same MOSFET as in Fig.3 ; dotted lines : theory (eq.(4))

The usual plateaux [1,2] are observed in weak inversion, followed by steep decrease.

For both the stressed and unstressed MOSFETs, the experimental decrease of S_I/I_D^2 with V_G is slightly slower than expected from eq.(4), due to an increasing value of N_{to} (or N_{td}) with energy.

Moreover, once the modification of V_T has been taken into account, it appears that $N_{td}(E)$ is roughly proportional to $N_{to}(E)$, since the curves are homothetic. This has also recently been confirmed by charge pumping measurements [7] and means that the aging process creates more traps near the conduction band.

4. CONCLUSION

The relation between 1/f noise in MOSFETs and the aging process clearly appears from the analysis of the drain current relative spectral density. This quantity was shown to increase with the electrical stress, in the ohmic region, due to a significant generation of slow states (traps) in the gate oxide, near the drain. Noise measurements, therefore, appear to be a very sensitive tool for the characterisation of such aging processes.

REFERENCES

[1] G.REIMBOLD, P.GENTIL, A.CHOVET, Proc. 7th Int. Conf.on Noise, Montpellier, 1983, North-Holland, pp.295-298.

[2] G.REIMBOLD, IEEE Trans. Electron. Dev. ED-31 (1984), pp. 1190-1198.

[3] S. CRISTOLOVEANU, B. CABON-TILL, K.N.KANG, P.GENTIL, Rev. Phys. Appl., 19, (1984) pp. 933-939.

[4] F.C.HSU, S.TAM, IEEE Electron. Dev. Lett. EDL-5 (1984), pp. 50-52.

[5] R.J.Van OVERSTRAETEN, G.J.DECLERCK, P.A.MULS, IEEE Trans. Electron. Dev., ED-22 (1975), pp. 282-288.

[6] Z.H.FANG, to be published.

[7] Z.H.FANG, H.HADDARA, S.CRISTOLOVE-ANU, G.GHIBAUDO, A.CHOVET, ESSDERC, Aachen (1985) ; and H.HADDARA, S.CRIS-TOLOVEANU, to be published.

NOISE IN PHYSICAL SYSTEMS AND 1/f NOISE - 1985
A. D'Amico and P. Mazzetti (editors)

FLICKER NOISE DUE TO MINORITY CARRIER TRAPPING IN THE BULK OF BIPOLAR TRANSISTORS

G. BLASQUEZ and D. SAUVAGE*

C.N.R.S. - L.A.A.S., 7, avenue du Colonel Roche, 31077 TOULOUSE CEDEX, France

The expression of the current fluctuations due to minority carriers trapping within short diodes is established. Experiments carried out in pnp integrated bipolar transistors are consistent with this model.

1. INTRODUCTION

Experimental studies of 1/f noise in pnp integrated bipolar transistors demonstrated that bulk current noise exists in the neighbourrhood of the emitter base junction [1]. KLEINPENNING [2], KILMER, Van DER ZIEL and BOSMAN [3], interpreted this noise as mobility fluctuations. An alternative based on minority carriers trapping is given in the following.

2. NOISE MODEL

Let us consider an abrupt short diode of the p^+n type as shown in Fig. 1. At low level biasing conditions, the current flowing through the diode is a hole current and the hole distribution p(x) in the n region is practically linear. REY and LETURCQ [4] expressed this current as follows :

$$I_E = \frac{q^2 A_E n_i^2 D_p}{Q_B + Q_{BS}} \exp \frac{V_{EB}}{U_T} \qquad (1)$$

where q, n_i, D_p, V_{EB}, U_T have their usual meanings, A_E is the (emitter) junction area, $Q_B = q\int_0^{W_B} N_D\, dx$ and $Q_{SB} = q\int_0^{W_B} p(x)\, dx$.

The traps situated within the n region (base) modulate the local density p(x) of the injected holes. Consequently, I_E fluctuates and generates current noise. In the simplest case, these traps are of the SHOCKLEY-READ - HALL type with a concentration N_t, a capture cross section c_p and energy level E_t. Fundamentally, trapping noise is of the Lorentzian type with a time constant τ_t equal to :

$$1/\tau_t = c_p\,(p+p_1) \qquad (2)$$

where $p_1 = n_i \,\mathrm{ex}\left[(E_i - E_t)/(kT) \right]$, E_i is the intrinsic level (midgap).

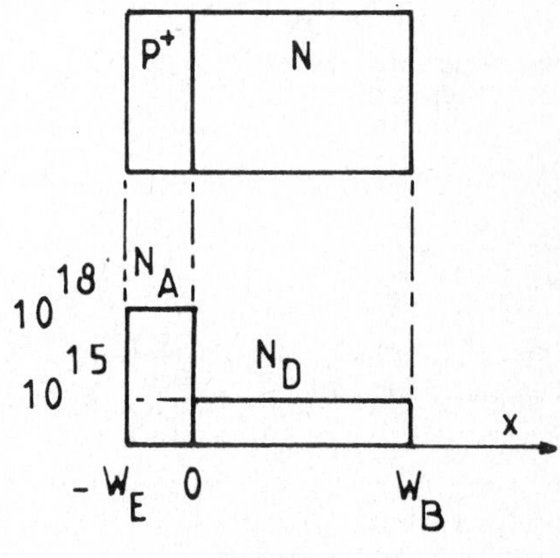

FIGURE 1

Illustration of a short diode corresponding to the emitter base junction of integrated pnp transistors.

*CIT-ALCATEL, Centre de Villarceaux, 91000 LA VILLE DES BOIS (France).

Actually, since p(x) is a decreasing function of x, τ_t is rather distributed than constant. By noting P the total number of holes withing the base, $P = A_E \int_0^{W_B} p(x)\,dx$, the spectral density of P has to be written :

$$S_p = A_E \int_0^{W_B} 4\,N_t f_t (1 - f_t)\,\frac{\tau_t}{1 + \omega^2 \tau_t^2}\,dx \qquad (3)$$

where $f_t = p_1/(p + p_1)$

Taking into account that :

$$p(x) = p(o)\left[\,1 - x/W_B\,\right] \qquad (4)$$

the calculation of $S_p(f)$ leads to

$$S_p(f) = \frac{A_E\,W_D\,N_t\,P_1}{p(o)}\,\frac{1}{f} \qquad (5)$$

for $f_{min} < f < f_{max}$ where $f_{min} = c_p\,p_1/(2\pi)$ and $f_{max} = c_p\,p(o)/(2\pi)$.

Expression (5) shows that noise generated by this mechanism can be of the 1/f type.

Now let us derive the expression of the current noise. Let δP be a fluctuation of P and δI_E the corresponding fluctuation of I_E. At low level biasing conditions ($Q_B >> Q_{SB}$) (1) gives :

$$\delta I_E = -\,I_E\;\delta Q_{SB}/Q_B \qquad (6)$$

$$S_{I_E}(f) = I_E^2\;S_{Q_{SB}}/Q_B^2 \qquad (7)$$

where $S_{I_E}(f)$ and $S_{Q_{SB}}$ are respectively the power spectral density of δI_E and δQ_{SB}. Furthermore, at low biasing conditions :

$$p(o) = (n_i^2 \exp\frac{V_{EB}}{U_T})\,/\,N_D$$

From (7), (5) and (1), it is easily shown that:

$$S_{I_E}(f) = \frac{qD_p\,N_t\,p_1}{(N_D\,W_B)^2}\,\frac{I_E}{f} \qquad (8)$$

If there are m independent trap levels a similar calculation leads to :

$$S_{I_E}(f) = \frac{q\,D_p}{(N_D\,W_B)^2}\,\frac{I_E}{f}\sum_{i=1}^{m} N_{ti}\,p_{1i} \qquad (9)$$

3. VALIDITY AND CONSISTENCY OF THE NOISE MODEL

In order to make clear the frequency bandwith in which this noise can be observed, the case of the EB junction of pnp integrated bipolar transistors is now considered. Typically $N_D \approx 10^{15}\,cm^{-3}$ and $W_B \approx 10\ \mu m$. Assuming $2\pi\,10^{-12} < c_p < 2\pi\,10^{-9}\ cm^{-3}\,s^{-1}$, $10^7 < p_1 < 10^{10}\ cm^{-3}$, at $V_{EB} \approx 480$ mV, the conditions associated with (5) give $1 < f_{max}/f_{min} < 10^9$, $10^{-5} < f_{min} < 10$ Hz. These values are spread out and consistent with the frequency bandwidth in which 1/f noise is usually observed.

Now, let us estimate the magnitude of this noise. If $10^{13} < N_t < 10^{14}\ cm^{-3}$, $10^7 < p_1 < 10^{10}\ cm^{-3}$, $D_p \approx 10\ cm^2\,s^{-1}$, equation (8) yields for f = 16 Hz and I_E = 100 μA, $10^{-26} < S_{I_E}(f) < 10^{-22}\ A^2/Hz$.

Then let us compare this result with some experimental date. Figures 2 and 3 give respectively the variations of the r.m.s. value of 1/f noise as a function of the current and of the

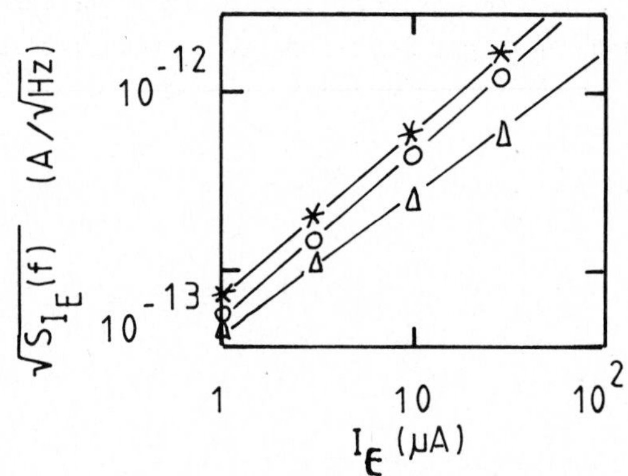

FIGURE 2

1/f input current noise vs emitter current in integrated pnp transistors. Measurement conditions : V_{CE} = 4V, T = 20°C, f = 10 Hz.

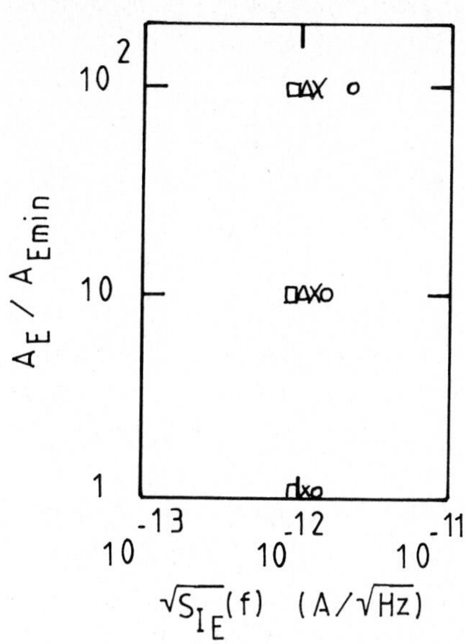

FIGURE 3

1/f input current noise vs emitter area.
V_{CE} = 4 V, T = 20°C, f = 10 Hz, I_E = 30 μA.

emitter area. On the basis of the three examples presented (Fig. 2) it can be seen that the r.m.s. value of the noise is proportional to the square root of the current. This is in good agreement with (8). In addition, for I_E = 100 μA and f = 10 Hz, $S_{I_E}(f) \approx 2.10^{-24}$ A^2/Hz. The theoretical estimation given in the preceding sentences is consistent with this experimental result.

Finally, from the measurements carried out on twelve transistors (see Fig. 3), the 1/f appears to be independent of the emitter area. Once again the deduction from the model developped in Section 2 are confirmed experimentally.

4. CONCLUSION

The preceding considerations and observations lead us to suggest that the combination of a trapping process with a distribution of carriers can produce 1/f noise in the bulk of short diodes and bipolar transistors.

REFERENCES

1. G. Blasquez and D. Sauvage in Noise in Physical systems and 1/f noise, Ed. Savelli, Lecoy and Nougier, North-Holland (1983) 275.

2. T.G.M. Kleinpenning, Physica 98B (1980) 289.

3. J. Kilmer, A. Van Der Ziel and G. Bosman, in Noise in Physical systems and 1/f noise, Ed. Savelli, Lecoy and Nougier, North Holland (1983) 271.

4. G. Rey and P. Leturcq, Physique des composants actifs à semiconducteurs, Dunod, Paris (1978).

VARIOUS PHYSICAL SYSTEMS - 2

NOISE IN PHYSICAL SYSTEMS AND 1/f NOISE - 1985
A. D'Amico and P. Mazzetti (editors)

DEPENDENCE OF THE 1/f NOISE PARAMETER α_ℓ ON VOLUME AND TEMPERATURE

R.H.M. CLEVERS

Eindhoven University of Technology, Eindhoven, The Netherlands

Experimental results are presented concerning Hooge's 1/f noise parameter α_ℓ in silicon. It is shown that the α_ℓ value does not depend on the effective noise volume. An α_ℓ value of the order of 10^{-6} was found for volumes between 10^{-16} m^3 and 10^{-11} m^3. The temperature dependence of α_ℓ between 77 K and 300 K is measured. We observed a weak dependence of α_ℓ on temperature.

1. INTRODUCTION

The 1/f noise in the conductance of homogeneous semiconductors can be described with Hooge's empirical relation

$$S_V/V^2 = \alpha/fN = (\mu/\mu_\ell)^2\, \alpha_\ell/fN \qquad (1)$$

in which μ is the mobility, μ_ℓ the lattice mobility, f the frequency, N the number of carriers and α_ℓ and α are dimensionless parameters. Experiments have indicated mobility fluctuations rather than number fluctuations as the source of 1/f noise. Furthermore, it has been shown that 1/f noise is caused by fluctuations in the mobility due to lattice scattering. This leads to the correction factor $(\mu/\mu_\ell)^2$ in (1).[1]

Values for α_ℓ between 10^{-3} and 10^{-8} have been observed.[2,3] This broad range of α_ℓ values is not yet understood. Bisschop[2] suggested that the parameter α_ℓ depends on the effective noise volume. His collection of α_ℓ values didn't give a decisive answer.

The temperature dependence of α_ℓ is still ambigious. From 300 K to 77 K in Si and Ge Bisschop[2] found an α_ℓ value which decreased by a factor of 10 to 100. Palenskis and Shoblitzkas[4] found a constant α_ℓ between 400 K and 77 K in n- and p-Si.

2. EXPERIMENTS

For our experiments we used point contacts on an n-Si and a p-Si substrate. The radii of the point contacts ranged from 100 μm to 1 μm. On the same chip of 3·5 x 3·5 x 0·4 mm^3 there is a large ohmic contact of 1·2 x 1·2 mm^2 for measuring bulk properties. For the p-type substrate the p^+ contacts were made by means of a boron implant and for the n-type substrate the n^+ contacts by a phosphorus diffusion. The depth of implantation or diffusion is 1 μm.

The resistivity ρ of the wafer is determined by measuring the resistance of the various contacts. The resistance of the point contacts is given by[5].

$$R = \frac{\rho r_0}{2\pi a} = \frac{\rho}{2\pi a}\left\{\frac{\arctan(d/\varepsilon a) - \arctan(b/\varepsilon a)}{\varepsilon}\right\} \qquad (2)$$

with a the radius of the point contact, b the implantation or diffusion depth, d the thickness of the wafer and ε equals $(1-(b/a)^2)^{\frac{1}{2}}$.

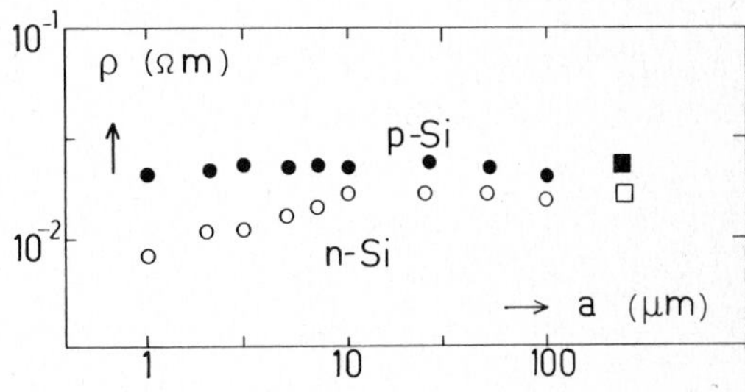

FIGURE 1
The resistivity obtained from point contacts (o, ●) and bulk contact (□, ■) as functions of mask dimensions

Figure 1 shows ρ as computed for each contact with a the radius expected from mask dimensions. For the smallest contacts on the n-Si substrate, because of 1 μm underdiffusion, the actual radii are larger than the expected radii. When this is accounted for, the resistivity of the smallest point contact becomes 0·016 Ωm. For the implanted contacts there is no difference between actual and expected radii. At 300 K the resistivities are 0·022 Ωm and 0·016 Ωm for the p- and n-Si substrates, respectively.

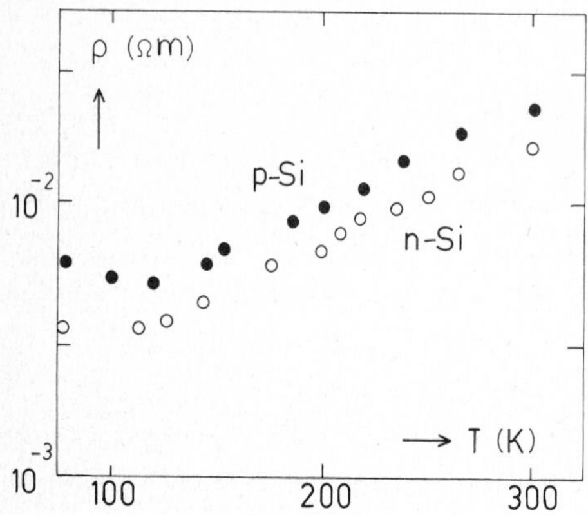

FIGURE 2
Temperature dependence of the resistivity of p- and n-Si

Figure 2 gives the measured temperature dependence of the resistivity. With these and other data from literature[6,7,8] the mobility and the carrier concentration are computed. The temperature dependence of the carrier concentrations is given in figure 3 for each. Figure 4 gives the temperature dependence of the mobilities and lattice mobilities μ_ℓ taken from[6,7,8]. At 300 K the carrier concentrations are $6{\cdot}3 \times 10^{21}$ m^{-3} and $3{\cdot}0 \times 10^{21}$ m^{-3} and the mobilities $0{\cdot}045$ m^2/V s and $0{\cdot}131$ m^2/V s for p- and n-Si, respectively.

The noise spectra were obtained by a real time spectrum analyser. The measurements were carried out between 2 Hz and 16 kHz. The coefficient $\gamma(= - d\ell n(Sv)/d\ell n(f))$ is found to be 1·0.

3. RESULTS AND DISCUSSION

In the case of an inhomogeneous current pattern, as with point contacts, the total number of

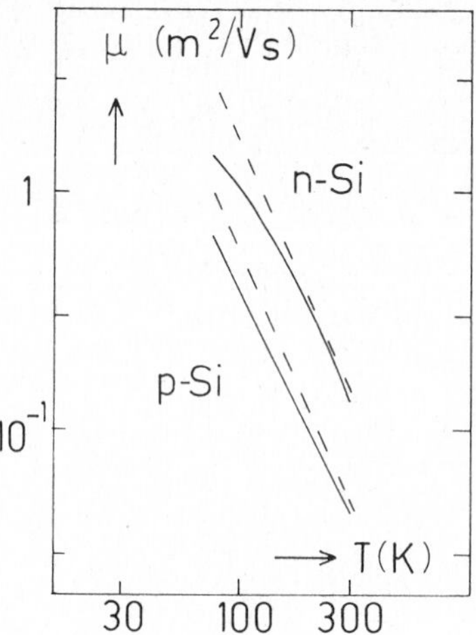

FIGURE 3
Temperature dependence of computed mobility (solid line) and lattice mobility (dashed line) of p- and n-Si

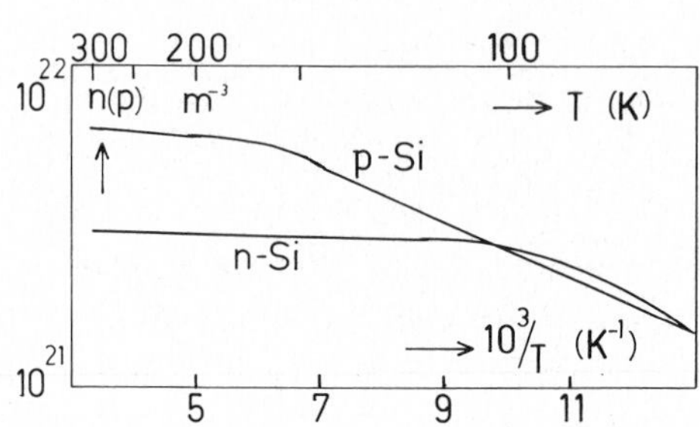

FIGURE 4
Temperature dependence of the carrier concentration (computed) of p- and n-Si

carriers N in (1) must be replaced by $N_{eff} = n\Omega$, with Ω the effective noise volume[1]. For point contacts with resistance R, the effective noise volume[5] is given by:

$$\Omega = (5\rho^3/4\pi^2R^3)(r_o{}^5/s_o) \qquad (3)$$

The factors r_o, s_o and $s_o/r_o{}^5$ are given in figure 5.

With this effective noise volume and with equation (1), the α values are computed.

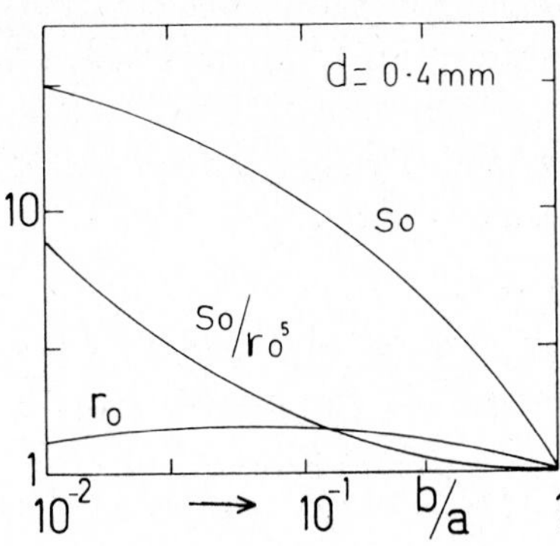

FIGURE 5
The factors r_0, s_0, s_0/r_0^5 as functions of b/a

Figure 6 shows α as a function of the effective noise volume. We find an α value of the order of 10^{-6} at 300 K and 77 K which is independent of this volume over five decades. For the point contacts with the largest radii, due to the large ohmic contact on the other side of the chip, a change of the current pattern occurs. This will reduce the effective noise volume and the α value for at most a factor of two. We have not taken this into account.

Figure 7 shows the measured temperature dependence of α and α_ℓ of the point contact having a radius of 5 μm. The temperature dependence of α and α_ℓ of the other point contacts was the same. The α and α_ℓ values for n-Si show a small, but significant temperature dependence.

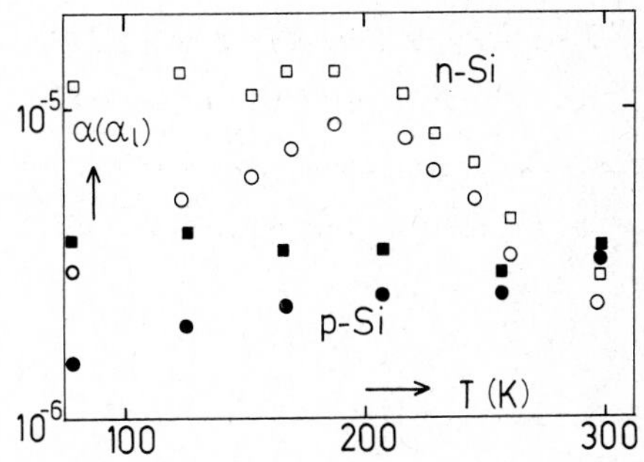

FIGURE 7
Temperature dependence of α and α_ℓ of p- and n-Si

Figure 8 compares the temperature dependence of α_ℓ as measured by Bisschop[2] and Palenskis and Shoblitzkas[4] with our results.

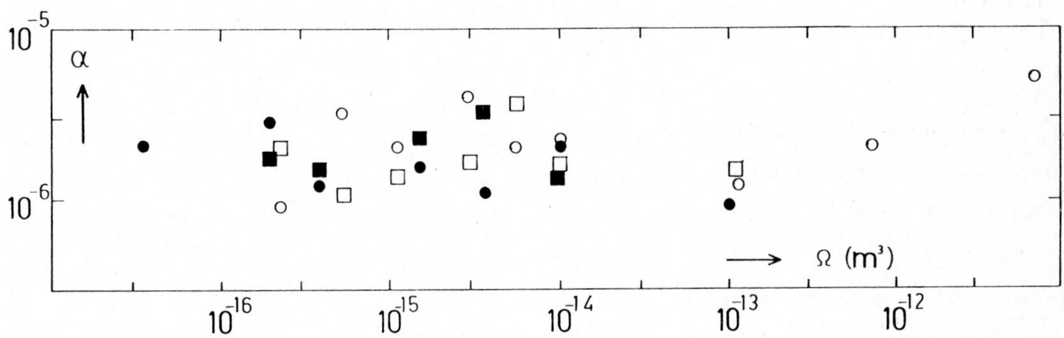

FIGURE 6
α as function of the effective noise volume for p-Si at 300 K (●) and 77 K (■) and for n-Si at 300 K (o) and 77 K (□)

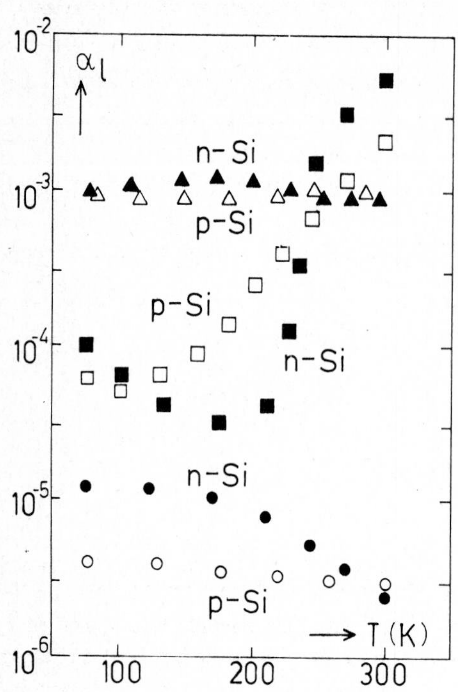

FIGURE 8
Temperature dependence of α_ℓ taken from literature (■ and □ from[2], ▲ and △ from[4]) and our results (o and •)

4. CONCLUSIONS

We have shown that the 1/f noise density in Si is inversely proportional to the effective noise volume over five decades. These results confirm Hooge's empirical relation[1]. Thus, 1/f noise is a volume effect in this case also, where an α_ℓ value of 10^{-6} is observed. It can be concluded that α_ℓ is independent of the effective noise volume.

5. ACKNOWLEDGEMENTS

The author thanks T.G.M. Kleinpenning and L.K.J. Vandamme for stimulating discussions and the Eindhovense Fabricage Faciliteit voor ICs (founded in 1984) for the preparation of the samples using modern MOS technology.

REFERENCES

1. F.N. Hooge, T.G.M. Kleinpenning and L.K.J. Vandamme, Rep. Prog. Phys. 44 (1981) 479.

2. J. Bisschop, Ph.D. thesis (University of Technology Eindhoven, 1983).

3. G.S. Bhatti and B.K. Jones, J. Phys. D., 17 (1984) 2407.

4. V. Palenskis and Z. Shoblitzkas Solid State Commun. 43 (1982) 761.

5. G.W.M. Coppus and L.K.J. Vandamme, Appl. Phys. 20 (1979) 119.

6. S.S. Li and W.R. Thurber, Solid-State Electron. 20 (1977) 609.

7. S.S. Li, Solid-State Electron. 21 (1978) 1109.

8. N.D. Arora, J.R. Hauser and D.J. Roulston, IEEE ED-29 (1982) 292.

NOISE IN PHYSICAL SYSTEMS AND 1/f NOISE - 1985
A. D'Amico and P. Mazzetti (editors)

THE EFFECTS OF 500 KEV ELECTRON IRRADIATION AND SUBSEQUENT ANNEALING ON 1/F NOISE IN COPPER FILMS

Jonathan PELZ and John CLARKE

Department of Physics, University of California, Berkeley, Berkeley CA 94720, and
Materials and Molecular Research Division, Lawrence Berkeley Laboratory, Berkeley CA 94720

Polycrystalline copper films were maintained at 90K on the cold stage of an electron microscope and irradiated with 500keV electrons to induce defects. With an electron dose of about $5 \times 10^{20} cm^{-2}$, the spectral density of the noise voltage across the films increased by an order of magnitude while the electrical resistivity increased by at most 10%. The films were annealed at progressively higher temperatures; after each annealing process the 1/f noise and resistivity were remeasured at 90K. Both the 1/f noise and resistivity were reduced, but at the lower annealing temperatures the fractional reduction in the added noise was substantially more than in the added resistivity. These results suggest that a large fraction of the added noise may be generated by a small mobile fraction of the added defects that are more readily annealed than the majority of the defects. After a room temperature annealing process, both the noise and resistivity returned nearly to their initial values. The temperature dependence of the noise after irradiation and partial annealing was consistant with the Dutta-Dimon-Horn thermal activation model.

The origin of 1/f noise in thin metal films has been a puzzle for many years[1,2]. Dutta, Dimon and Horn[3] (DDH) proposed a general model that explained many features of the noise but did not identify the particular microscopic process responsible. Recently there has been growing evidence that crystal defects are involved in the noise[4,5,6,7], and theoretcial models have been proposed explaining the noise in terms of defect motion[8,9].

It is well known that bombardment by electrons with kinetic energy > 400 keV will create defects within pure bulk copper, mostly in the form of Frenkel Pairs (FP), i.e. vacancy-interstitial pairs[10]. These defects are mobile at room temperature and anneal via recombination; thus, an enhanced FP population is retained only at lower temperatres. In this paper, we show that 500 keV electron irradiation of Cu films maintained at 90K increases both the resistance and the level of 1/f noise. Subsequent annealing at progressively higher temperatures reduces both the resistance and the noise very nearly to their initial values. The temperature dependence of the noise depends strongly on the history of irradiation and annealing, and is consistant with the DDH thermal activation model[3].

In our experiments, 90μm × 4μm × 100nm polycrystalline Cu films were mounted on a custom-built cold stage of an Hitachi HU-650 electron microscope, and the noise (over a 0.1Hz-25Hz frequency band) and resistance were measured <u>in-situ</u>. The sample preparation and experimental apparatus have been described elsewhere[11]. All irradiations were performed with the samples held at 90K. The samples were considered fully annealed or "unirradiated" after an extended (>12hr) room temperature annealing process, since both the resistance and noise returned nearly to their initial values. All data reported in this paper are from a single Cu sample; similar results were obtained from two other Cu samples.

For pure, bulk-like Cu at 90K it is generally believed that a radiation-induced vacancy is frozen in place, while an interstitial migrates freely until it recombines with a vacancy (removing a FP), is trapped at another defect (such as a grain boundary, surface, or impurity), or clusters with other interstitials[10,12]. The defect concentration thus builds up in the form of frozen vacancies and trapped or clustered interstitials, which can be monitored by the change in sample resistivity[10]: $\Delta\rho \approx 3 \times 10^{-6}$ Ωcm/at.% FP. This type of trapping dynamics, in the form of the "unsaturable trap model", has been used to explain the observation that $\Delta\rho$ scales as $\phi^{1/2}$ in bulk materials irradiated near 90K, where ϕ is the electron dose[12].

In our first series of measurements, the sample was cooled to 90K inside the microscope, and the resistance and noise were measured. We then irradited the sample with a known dose of electrons, and remeasured the resistance and noise. For three separate irradiation sequences, each starting with the sample in the unirradiated state, we found that $\Delta\rho$ scales approximately as $\phi^{1/2}$ for 3.5nΩcm < $\Delta\rho$ < 100nΩcm and 10^{18}cm^{-2} < ϕ < 5×10^{20}cm^{-2}. This scaling is similar to the behavior seen in bulk copper; however the magnitude of $\Delta\rho$ we measure for a given dose ϕ is significantly larger than expected in bulk copper. We also measure significant "subthreshold damage" for incident electron energy < 400keV. We note here that our samples (with small crystallites, probable oxidation at surfaces and grain boundaries, and the presence of a substrate) are quite different from the freely suspended bulk-like materials used in previous studies. We suspect that these different sample conditions are responsible for the anomalous behavior.

It is convenient to characterize the measured 1/f noise in terms of the parameters m and α, where the frequency exponent m is the fitted slope of a log-log plot of the noise spectral density, and

$$\alpha = f_o S_V(f_o) N/\overline{V}^2 \approx f_o S_R(f_o) N/R^2.$$

Here, $S_V(f)$ and $S_R(f)$ are the spectral densities of the sample voltage and resistance fluctuations, respectively, $\overline{V}$ is twice the rms voltage across half the sample, R is the sample resistance, N=2.9×10^{12} (±20%) is the estimated number of atoms in the sample, and f_o = 1Hz. Before each irradiation sequence, the initial value of α at 90K was within 10% of 5.5×10^{-4}, and m was within 3% of 0.98. In Fig. 1, we plot $\Delta\alpha$ vs. $\Delta\rho$ (along the dashed line) for the three irradiation sequences described in the previous paragraph. We see that α increases by an order of magnitude. Simultaneously, m increases by about 10%, with most of the increase occurring after the first irradiation (not shown). The data fall approximately on the dashed line $\Delta\alpha \propto \Delta\rho^{0.6}$. Assuming $\Delta\rho$ to be proportional to the added defect concentration n_d, these data indicate that $\Delta\alpha$ scales as $n_d^{0.6}$. We note here that n_d is a measure of the total number of added defects, including many which are essentially frozen at 90K. Existing defect-noise models[8,9] however, relate 1/f noise only to mobile defects which change position or orientation in the same frequency range as the observed noise. Thus the observed scaling law does not test directly the dependence of the noise magnitude on mobile defect concentration predicted by these models.

In our second series of measurements, we

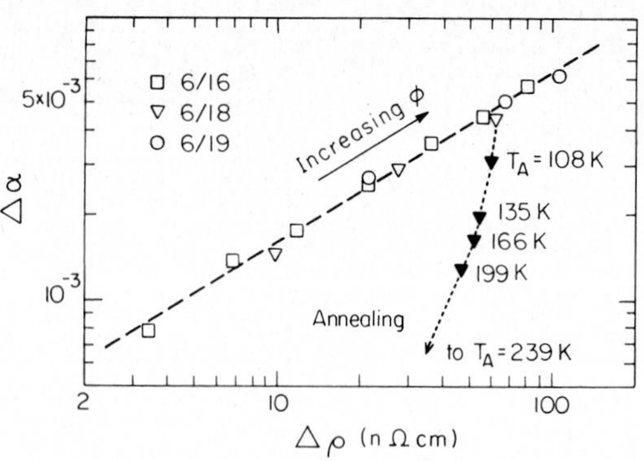

Fig. 1. Change in 1/f noise magnitude $\Delta\alpha$ vs. change in sample resistivity $\Delta\rho$. The dashed line $\Delta\alpha \propto \Delta\rho^{0.6}$ is drawn for comparison. Points along this line correspond to increasing electron dose ϕ, while points along the dotted line correspond to annealing at successively higher temperatures. The data point for T_A = 239K (not shown) is $\Delta\rho \approx 11.6$nΩcm, $\Delta\alpha \approx 7 \times 10^{-5}$.

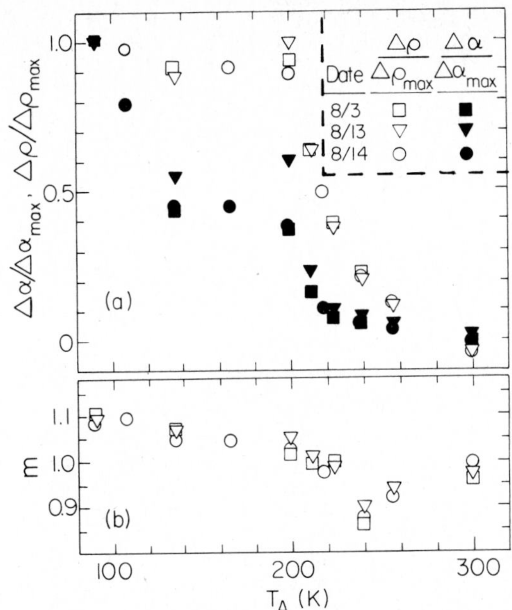

Fig. 2. Annealing behavior of the irradiated Cu film with $\Delta\rho_{max} \approx 90$nΩcm and $\Delta\alpha_{max} \approx 6 \times 10^{-3}$ prior to annealing. (a) Recovery of the 1/f noise magnitude ($\Delta\alpha/\Delta\alpha_{max}$) and resistivity ($\Delta\rho/\Delta\rho_{max}$) vs. annealing temperature T_A; (b) the frequency exponent m vs. T_A.

heated the sample (after irradiation) to a temperature T_A for five minutes, then cooled it to 90K to remeasure the noise and resistance. The sequence is repeated for higher T_A. The dependence of $\Delta\alpha$ on $\Delta\rho$ after annealing is shown by the dotted line in Fig. 1. The annealing reduces the noise much more rapidly than the resistance, producing hysteresis in the plot of $\Delta\alpha$ vs. $\Delta\rho$. This behavior is also illustrated in Fig. 2(a), which shows annealing behavior in the form of recovery curves. Most of the resistivity recovery occurs in the range 200K < T_A < 250K, and is similar to the "stage III recovery" well documented for irradiated bulk copper[10] although occurring at somewhat lower temperatures. Recent studies indicate that this recovery step in bulk copper is connected to the free migration of a monovacancy[13]. The noise magnitude, $\Delta\alpha$, recovers partially over the range 200K < T_A < 300K in which the resistivity recovers, but also exhibits a strong recovery at temperatures below 135K that is not readily apparent in the resistivity curve. We note here that a small fraction (presumed to be mobile) of the added defects may be responsible for much of the added noise. The observed difference in the recovery of $\Delta\alpha$ and $\Delta\rho$ is readily explained if one assumes that these "noisy" defects anneal at lower temperatures than the bulk of the added defects.

The frequency exponent m of the noise also changes during the annealing experiments. The annealing behavior, shown in Fig. 2(b), shows a striking dip at T_A = 240K. We note that this annealing temperature falls within the "stage III" recovery of $\Delta\alpha$ and $\Delta\rho$.

In a third set of measurements, we first irradiated the sample to $\Delta\rho \approx 85$nΩcm and annealed it at a temperature T_A before measuring the noise as a function of temperature for T < T_A. In Fig. 3 the noise magnitude NS_R(1Hz) is plotted as a function of T, for T_A = 201K, T_A = 239K, and for the sample in the "unirradiated" (i.e. fully annealed) state. We see that the temperature dependence of the noise magnitude is a strong function of the annealing state of the sample. The three curves in Fig. 3 are quadratic fits to the three sets of data points. In Fig. 4, the frequency exponent m is plotted for the same data series shown in Fig. 3. Shown also as curves are the predictions for the temperature dependence of m, where we have used the fitted curves from Fig. 3 and the DDH model[3]:

$$m(\omega,T) = 1 - [1/\ln(\omega\tau_o)][\partial \ln S_R(\omega,T)/\partial \ln T - 1].$$

Here, $\omega/2\pi$ = 1Hz and τ_o is taken to be 10^{-14}s. The curves precict the general trends of the data points rather well, indicating that our results are consistent with the DDH model.

In summary, we have shown that the 1/f noise in polycrystalline Cu films increases in a systematic way with 500 keV electron bombardment, and thus have demonstrated a direct connection

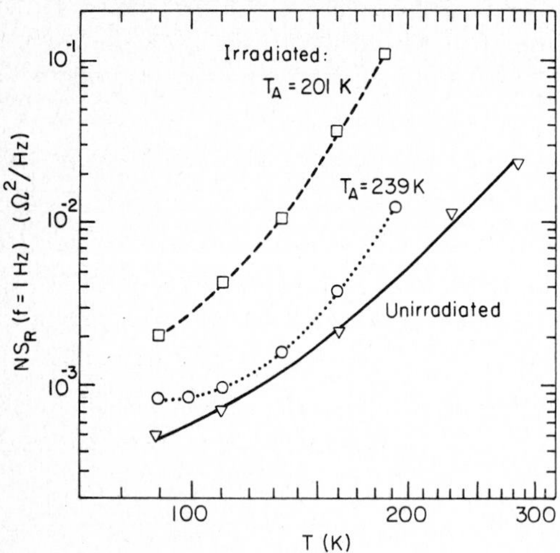

Fig. 3. Dependence of NS_R(1Hz) on temperature T for the sample in the unirradated state, and for $\Delta\rho_{max} \approx 85 n\Omega cm$ followed by a 5 min anneal at T_A. Curves are quadratic fits to the data points.

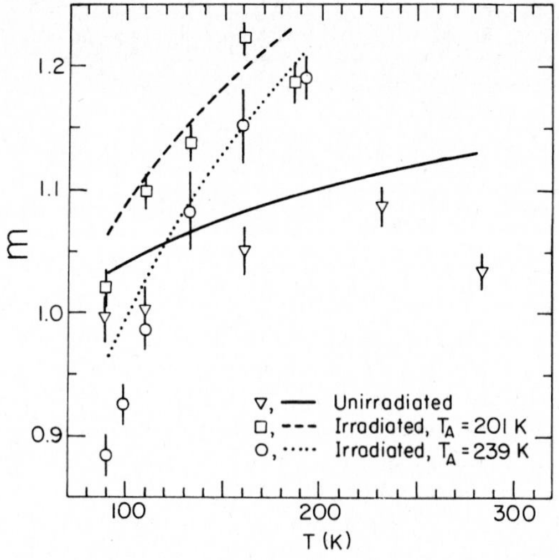

Fig. 4. Dependence of the frequency exponent m on T for the same data series as Fig. 3. Curves are predictions of the DDH model.

between 1/f noise and defects in metals. The difference in the recovery of $\Delta\alpha$ and $\Delta\rho$ obtained after successive annealing steps suggests that a large fraction of the added noise is generated by a subpopulation of "mobile" defects that are more readily annealed than the majority of added defects. The temperature dependence of the noise magnitude and frequency exponent m after irradiation is consistent with the Dutta-Dimon-Horn model, indicating that thermally activated kinetics govern the added 1/f noise.

We gratefully acknowledge the use of the National Center for Electron Microscopy of the Lawrence Berkeley Laboratory. We thank D.W. Ackland, R. Gronsky, E.E. Haller, F.N.H. Robinson, D.Turnbull, E.R. Weber, and D.J. Jurica for valuable discussions and assistance. This work was supported by the Director, Office of Energy Research, Office of Basic Energy Sciences, Materials Sciences Division of the U.S. Department of Energy under Contract No. DE-AC03-76SF00098.

1. F. N. Hooge and A. M. F. Hoppenbrouwers, Physica 45, 386 (1969)
2. P. Dutta and P. M. Horn, Rev. Mod. Phys. 53, 497 (1981)
3. P. Dutta, P. Dimon, and P.M. Horn, Phys. Rev. Lett. 43, 646 (1979).
4. J. W. Eberhard and P. M. Horn, Phys. Rev. B 18, 6681 (1978).
5. D. M. Fleetwood and N. Giordano, Phys. Rev. B 28, 3625 (1983).
6. G. P. Zhigal'skiy, YU. YE. Sokov, and N. G. Tomson, Radiotekh. and Elektron. 24, 410 (1979) [Radio Eng. and Electron. Phys. 24, 137 (1979)].
7. D. M. Fleetwood and N. Giordano, Phys. Rev. B 31, 1157 (1985).
8. F. N. H. Robinson, Phys. Lett. 97A, 162 (1983).
9. Sh. M. Kogan and K. E. Nagaev, Fiz. Tverd. Tela. 24, 3381 (1982) [Sov. Phys. Solid State 24, 1921 (1982)].
10. J. W. Corbett, Electron Radiation Damage in Semiconductors and Metals (Academic Press, NY, 1966), Chaps. 24 and 25.
11. Jonathan Pelz and John Clarke, Phys. Rev. Lett. 55, 738 (1985).
12. R. M. Walker, Radiation Damage in Solids, D. S. Billington, ed. (Academic Press, NY, 1962), p.594.
13. Th. Wichert, Point Defects and Defect Interactions in Metals, J. Takamuru, M. Doyama, M. Kiritani, eds. (North-Holland, Amsterdam, 1982), p.19; C. Minier and M. Minier, Ibid, p.27.

NOISE IN PHYSICAL SYSTEMS AND 1/f NOISE - 1985
A. D'Amico and P. Mazzetti (editors)

SHOT AND FLICKER NOISE OF P-N JUNCTION IN THE GENERATION-RECOMBINATION, DIFFUSION AND HIGH-INJECTION REGIONS

Bruno PELLEGRINI

Istituto di Elettronica e Telecomunicazioni, Università di Pisa, Via Diotisalvi 2, 56100 Pisa, Italy

The capacitance, the current, the shot and flicker noise spectra of p-n junction diodes are measured versus the voltage, and the correlations among them are shown. The great current interval explored, extending from $5x10^{-9}$ to $5x10^{-2}$ A, allows one to find that the values of the shot and $1/f^{\gamma}$ noises and, in particular, the frequency exponent γ change appreciably in the generation-recombination, diffusion and high-injection regions which are located by means of the measures both of the capacitance and of the current versus the voltage. Such results can be justify by means of the island model.

1. INTRODUCTION

Most of the experiments and of the models dealing with the low frequency noises, and in particular with the $1/f^{\gamma}$ noise, concerns the unipolar electron devices (UED), whereas a smaller number regards the bipolar ones (BED). Such a difference is due to the charge carrier type, to the electric field, to the conduction and noise mechanisms, and to their temperature dependence, which are greatly different in the two cases.

In fact, while in UED we have an alone type of carriers and a their electric drift in neutral zone, in BED we have two types of carriers subjected to several processes. They may be stored, generated and destroied as charge carriers, their transport mechanisms may be, comtemporely or separately, the electrical drift and the diffusion both in neutral zones and in depleted regions characterised by high electric field too; moreover, finally, the various phenomena and quantities are very sensitive to the temperature.

Such a multiplicity and complexity of processes and conditions coexisting in BED have as a consequence that for them the flicker noise spectrum has not the simple and general form $S_A = \alpha A^2/Nf^{\gamma}$, holding true for about all the UED, where α is a constant, N is the sample carrier number, f is the frequency, γ is an exponent near to one and A is the quantity, voltage, current, resistence, mobility and so on, of whom one measures the fluctuations.

In BDE A reduces itself to the alone current, or voltage, the dependence of S_A on the varions quantities is not so well defined and general, it is different for the various BED and, in particular, the exponent γ varies in a range much greater than that of the UED.

Such uniquess lack of experimental results and also the difficulty in obtaining them explain the smaller number of experiments and of model researches for BED than for UED, in spite of their scientific and technical importance.

On other hand, viceversa, the deep differences between BED and UED, the multiplicity of conditions in which the BED may work and which may be properly screened, and the problem difficulty itself, induce to study the BED noises and, especially, their flicker noise, since just from such experimental and teoretical situations,very different from the ones of UED, new physical data and insight about the origin of the $1/f^{\gamma}$ noise itself may derive.

With this aim we have measured the capacitance, the current, the shot and flicker noise spectra of p-n jniunction diodes 1N4001 in order to research correlations among them in the generation-recombination (GRR), diffusion (DR) and high-injection (HJR) regions which we have explored by performing the measures from $5x10^{-9}$ to $5x10^{-2}$ A.

2. CAPACITANCE AND CURRENT

When the measures of the jnction capacitance C versus the voltage v, performed at 1 MHz in the interval between -40 and 0.25 V, are plotted as C^{-3} vs v, they give a straight line, shown in Fig.1, which intercepts the v axis at v_{in}=.32 V. Such linear dependence of C^{-3} on v proves that the junction is linearly graded[1].

By assuming that the junction is of type n^+-p, i.e. with $N_D \gg N_A = Ax$, N_D and N_A being the donor and acceptor concentrations, respectively,from the solution of the Poisson equation we obtain[1-3]

$$1/C^3 = (3/S^3 Aq\epsilon^2)(v_{in}-v) \qquad , \qquad (1)$$

and

$$v' = v_{in} + kT/q \qquad , \qquad (2)$$

where $S=1.26x10^{-2}cm^2$ is the diode section, q is the electron charge, ϵ is the dielectric constant, v' is diffusion potential[1-3], k is the Boltzamann constant and T is the abslute temperature.

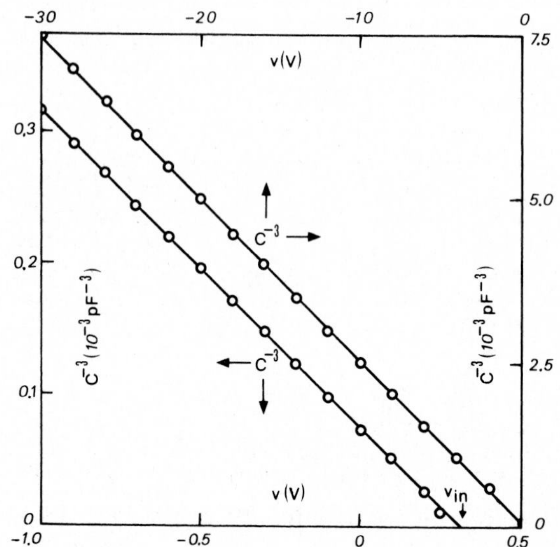

Fig.1-Plots of C^{-3} vs v referred to two different axis systems.

From the slope of the C^{-3}-v plot we obtain $A=3.5x10^{16}cm^{-4}$ and then, from the expression $W=[3\epsilon(v_{in}-v)/qA]^{1/3}$ of the depletion layer width, for v=0 we have $W=W_0=5.6\ \mu m$ to which it correspnds the acceptor concentration $N_A(W_0)=AW_0=1.96x10^{13}cm^{-3}$ and a Fermi level $F_p(W_0)=F_p$ equal to 0.33eV (see insert of Fig.2). Therefore from (2) it follows that the depletion layer in the p region disappears for a bias voltage v=v'=0.345 V, whereas the density of the minority electrons injected into p region equals the hole concentration for $v \geq v'' \simeq (E_G-2F_p)q^{-1}=0.46$ V, $E_G=1.12$ eV being the silicon energy gap[1].

In conclusion for 0<v < v' the generation-recombination process in the p depletion region (GRR) prevails; for $v' < v < v''$ the depletion region desappears and the electrons, which remain minority carriers, diffuse into the neutral p region (DR) and, finally, for $v > v''$, the electron density tends to become greater

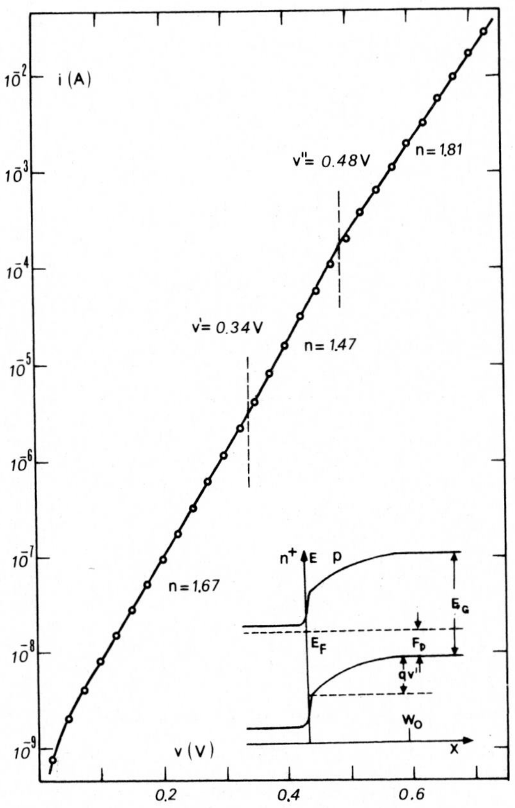

Fig.2-i-v characteristics. The insert represents the energy bands of a n^+-p junction.

than the hole one and the high-injection regime (HJR)[1], characterized by both drift and diffusion curents, is established.

In such three regions, located by means of capacitance measures, the i-v characteristics should have different shapes.

Such an expectation is confirmed by the i-v characteristics, reported in Fig.2, which shows three values, 1.67, 1.47 and 1.81, of the ideality factor n in the voltage intervals $v<0.34$ V, $0.34 < v < 0.48$ V and $v > 0.48$ V, respectively, which agree well with the three regions GRR,DR and HJR, respectively, above found by means of C^{-3}-v characteristics, and with the p-n junction theory[1].

3. SHOT AND FLICKER NOISES

The voltage noise spectra have been measured, on the frequency band comprised between 0.1 and 2.5×10^5 Hz, for mamy values of v in the interval 80, 720 mV and for several 1N4001 diodes.

At 1 Hz, where normally the flicker noise prevails on the shot one, we have observed a great spread of the noise values from sample to sample. In same cases, according also to the bias current, the flicker noise did not emerge from the shot one, or from the background noise of the preamplifier, whereas for other diodes it was much greater.

For one of these more "noisy" diodes, to which the C^{-3}-v and i-v characteristics of Fig.1 and 2 refer, we give in Fig.3 the power spectral density S_s vs v of the shot noise evaluated at high frequency where the spectrum itself is frequency indipendent.

In order to evaluate the "ideality" of the shot noise, in Fig.3 we report also the current "fraction" $\eta \equiv S_s/2qir^2$ which contributes to the shot noise itself, $r=kT/nqi$ being the diode differential resistance evaluated, for each v value, by means of the data of i-v characteristics. From the plot we deduce that η changes abruptly for v=v', owing to the r discontinuity at v' (Fig. 2), and it has values substantially different from 1 and between themselves in GRR and DR.

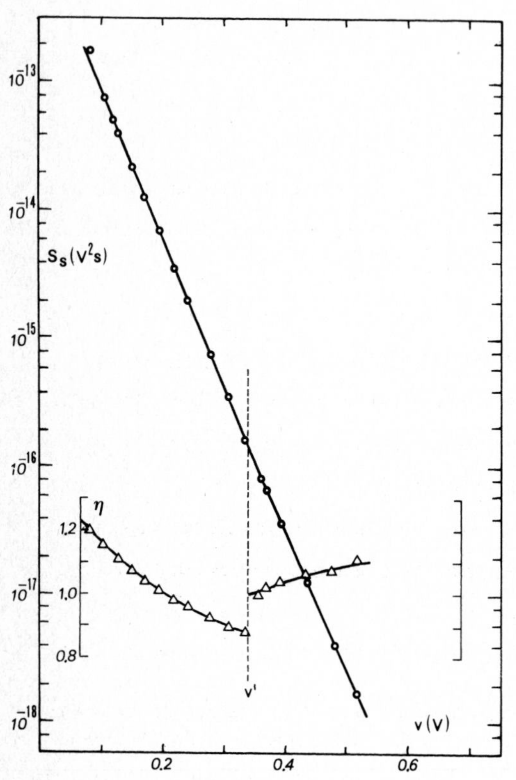

Fig.3-Power spectral density S_S of the shot noise and current "fraction" η which contribute to it.

In HJR, i.e. for high current, S_S is not measurable because it becomes smaller than the background noise.

In Fig.4 we report the spectrum $S_F \propto 1/f^{\gamma}$ and its frequency exponent γ vs v at 1 Hz, where the flicker noise prevails on the shot one. The exponent greatly changes, from 0.65 to 1, in GRR, it is equal to 1 in DR and it slowly increases in HJR, whereas $S_F(1)$ is nearly constant in HJR and it increases, when v decreases, in the other two zones.

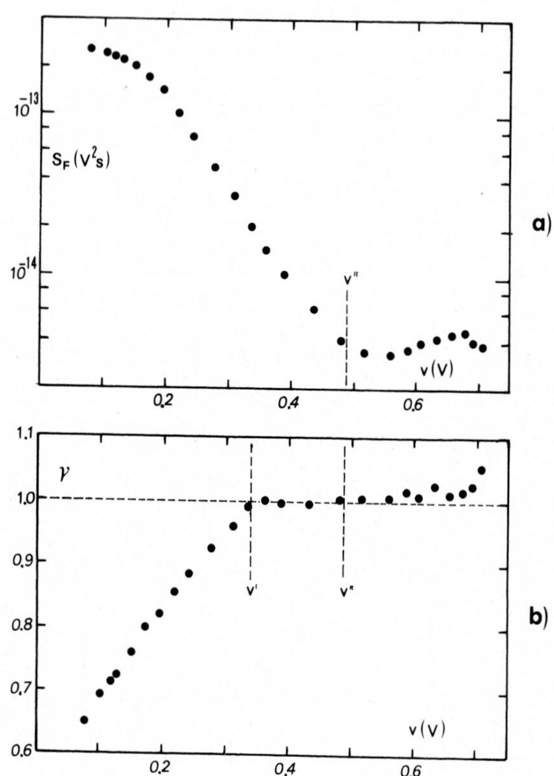

Fig.4-Power spectral density S_F of the flicker noise (a) and of its frequency exponent at 1Hz (b).

From another point of view, by indicating a with $S_i=S_v/r^2=Ai^{\beta}/f^{\gamma}$ the current noise spectrum, from Fig. 4a we have also $\beta(v)\simeq 2$ in HJR and $1<\beta<2$ in the other two regions; in particular $\beta \simeq 1.43$ for $0.18 < v < 0.43$ V.

Analogans results have been obtained for other "noisy" diodes.

3. CONCLUSIONS

Shot and flicker noise measures of n^+-p junctions have been performed on so very large voltage (or current) range that all the three GRR, DR and HJR regions are analysed contemporarely on the same sample. We seen that this has been carried out for the first time(4).

The new ewperimental results which have been reached are the close correlations among the C^{-3}-v, i-v, S_s-v, S_F-v and γ-v characteristics. A noticeable result is that the shot and flicker noises and in particular γ strongly depend on the carrier transport mechanisms.

It does not appear easy to explain such a result by means of mobility fluctuation model(4), expecially in GRR where γ is very different from 1 and the generation-recombination in the depletion layer is the prevaling conduction mechanism across the p-n junction.

It seems possible, instead, to justify the experimental results by means of the island model(5) by taking into account in it also the effects of the charge fluctuations of each defect on the fluctuations of the others and on the generation-recombination current through them.

In fact such a coupling effect can not be neglected for the islands in depletion layer in GRR because each of them is not screened by free carriers.

Such a reciprocal island interaction should justify the dependence of S_F and, especially, the one of γ on v in GRR which, in fact, disapears in DR and HJR when the depletion layer width becomes null, that is when its defects are screened by the free carriers and so they become independent each other and modulate separately the drift and diffusion currents according to the island approach(5).

However the problem of traslating such ideas into an analytical form apt to calculate the experimental results, i.e. of extending the island model from the unipolar to the bipolar devices in their various and complex working conditions remains an open matter.

ACKNOWLEDGEMENTS

The work has been supported by the CNR, through CSMDR and CCTE, and by the MPI of Italy.

REFERENCES

1- S.M.Sze, "Physics of semiconductor devices", (Wiley, New York, 1969), pp. 92 and 96

2- B.Pellegrini, Solid-St.Elect.13, 1175 (1970)

3- B.Pellegrini, Solid-St.Elect.18, 887 (1975)

4- T.G.M.Kleinpeming, Physica 98B, 289 (1980)

5- B.Pellegrini, Phys.Rev. B 24, 7071 (1981)

NOISE IN PHYSICAL SYSTEMS AND 1/f NOISE - 1985
A. D'Amico and P. Mazzetti (editors)

$1/f^{\nu}$ NOISE GENERATORS

B. PELLEGRINI, R. SALETTI*, B. NERI and P. TERRENI

Istituto di Elettronica e Telecomunicazioni and *CSMDR-CNR, Università di Pisa, Via Diotisalvi 2, 56100 Pisa, Italy

A preceding numerical simulation has shown that the sum of a few Lorentzian spectra may be sufficient to yield $1/f^{\nu}$ spectra over frequency bands of many decades. Such a result is here utilized to build two standard $1/f$ and $1/f^{0.9}$ noise generators over the 5×10^{-2}, 10^{5} Hz band. The goal is achieved in a simple way by connecting at the input of an ultralow noise, very high input resistance amplifier purposely designed, a proper number of resistor-capacitance parallels which generate just the Lorentzian spectrum sum required by the numerical simulation. The proposed design method may be extended to synthetize other types of coloured noise standard generators.

1. INTRODUCTION

Several flicker noise models(1-6) show that a sum of N Lorentzian spectra

$$S=\frac{2\sigma}{\pi}\sum_{h=1}^{N}\frac{\varphi_h}{\varphi_h^2+f^2}, \qquad (1)$$

with a proper distribution of the pole-frequencies φ_h can give a $1/f^{\nu}$ spectrum, σ being the variance of the considered quantity and f the frequency.

To this end, in order to obtain S in a closed form, the various models change the discrete variable φ_h into a continuous one φ, the sum (1) into the integral

$$S'=\frac{2\sigma}{\pi}\int_0^{\infty}\frac{\varphi}{\varphi^2+f^2}D_{\varphi}\,d\varphi \qquad (2)$$

and they prove or assume that the pole distribution D_{φ} is of the type $D_{\varphi}\propto 1/\varphi$ for $0<\varphi<\infty$. In this way the classical spectrum $S'\propto 1/f$ is obtained.

More in general, if it is $D_{\varphi}\propto 1/\varphi^{\nu}$, the equation (2), for $0<\nu<2$, gives(7)

$$S'\propto 1/f^{\nu} \qquad (3)$$

The problem to be solved is to find the shape and the minimum value of D_{φ}, for each φ, which allows one to substitute the sum (1) with the integral (2), with a given maximum error in the considered frequency band f_1, f_2.

To this end, owing to the power form of S', it is convenient to use a logarithmic transform by defining $Y=\lg(\varphi/f_r)$, $\Theta'=10\lg S'$, where f_r is an arbitrary reference frequency; in such a way Θ' becomes a straight line in the plane Θ', Y more simple to compare with $\Theta=10\lg S$(7).

By computing, with the help of a computer, the sum (1) for various shapes and values of the distribution $D_{\varphi}(\varphi)$ and/or of the corresponding one $D_y(Y)$, it has been found that several types of them are able to give $1/f^{\nu}$ noise(7), and that, in particular, a constant distribution D_y=1.4 p/d (poles per decade) in the band $10^{-2} f_1/f_r$, $10^2 f_2/f_r$ is sufficient to yield a 1/f noise with a maximum error less than 1%; in this way, for istance, one obtains that N=20 poles uniformely distributed over 14 decades are sufficient to generate 1/f noise over 10 decades!

The aim of this work is to design and to build $1/f^{\nu}$ noise standard generators by generating and summing an appropriate number of voltage Lorentzian spectra characterized by proper allocations of their poles. Each spectrum may be easily obtained by using the thermal noise generator of a resistor R_h connected in parallel to a capacitance C_h=C, and their sum may be achieved by putting, as shown in Fig. 1, N of such R_h-C_h groups in series at the input of an amplifier characterized by proper input impedance, current and voltage noise generators, $Z_i=R_i/(1+j\omega C_i R_i)$, i_n and e_n, respectively. It is worth noting that such a design approach of $1/f^{\nu}$ generators is quite different from any other preceding one[8], based on the filtering of a white noise with networks having proper trasfer functions.

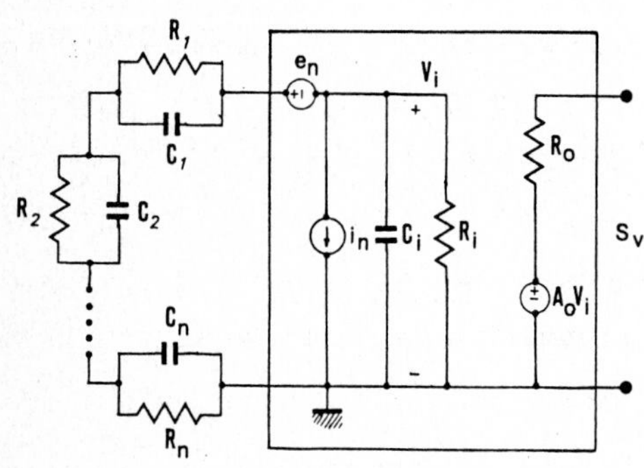

Fig. 1: $1/f^{\nu}$ noise generator.

2. $1/f^{\nu}$ GENERATOR DESIGN

A. Amplifier

The power spectral density of the voltage noise generated by the N parallels R_h-C has just the form (1), where now it is σ=kT/C and $\varphi_h=1/2\pi CR_h$. Therefore if e_n, i_n and $1/Z_i$ were null, the amplifier output spectrum would be

$$S_v=|A_0|^2 S \quad , \qquad (4)$$

where A_0 is the voltage gain, independent of the frequency.

In the real case the second member has to be multiplied for the attenuation

$$\alpha^2=\left|1+\frac{1+j2\pi fR_iC_i}{2\pi R_iC}\sum_{h=1}^{N}\frac{1}{\varphi_h+jf}\right|^{-2} \qquad (5)$$

which becomes equal to 1 for f=0 and f much greater than the maximum φ_h, if it is, respectively,

$$R_i \gg \sum_{h=1}^{N} R_h=R_T \quad , \quad C_i \ll C/N \quad . \qquad (6)$$

Moreover in order to obtain (4), for each f in the generator band, it has to be

$$S \gg S_e+S_i\left|\sum_{h=1}^{N} R_h/(1+jf/\varphi_h)\right|^2 \quad , \qquad (7)$$

where S_e and S_i are the spectra of e_n and i_n, respectively; (7) holds true on the assumption that (6) is verified. In particular (4) and (7) are verified if

$$S \gg S_e+S_iR_T^2 \quad . \qquad (8)$$

In order to satisfy (6) and (8) we have used

an ultralow noise and very high input resistance FET preamplifier, built by ourselves[9], which is characterized by C_i=45pF, $R_i \gg 10G\Omega$ and $i_n \ll 5fA/\sqrt{Hz}$ at any frequency, whereas S_e is plotted in Fig. 2 together with the generated spectra $S_v/|A_0|^2 \propto 1/f^\gamma$.

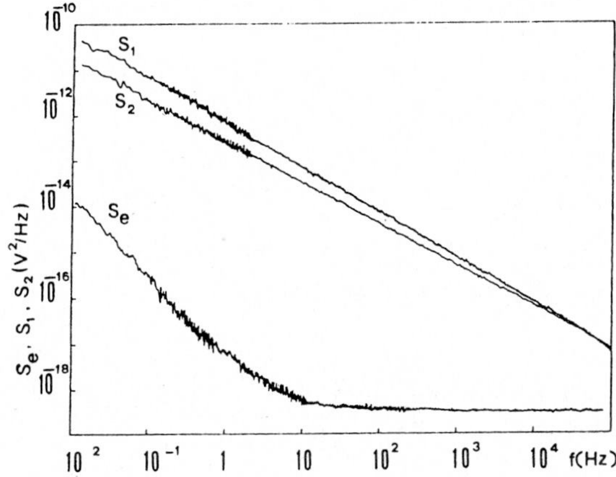

Fig. 2 : Spectral densities of the input voltage noise generator and of the generated noises $S_{1,2}=S_v/|A_0|^2 \propto 1/f^\gamma$; $S_1 \propto 1/f$ and $S_2 \propto 1/f^{0.9}$

B. 1/f noise generator

The problem is now to find the allocation of the poles φ_h in order to obtain the spectrum $1/f^\gamma$ with given exponent γ in a defined band.

We have solved this problem in two cases. In the first one we consider a constant pole density D_y=1.4 p/d in the frequency band 10^{-3}, 10^5 Hz; it leads to γ=1 and to N=11 poles characterized by the parameters C=32nF, $R_1=5\times10^8\ \Omega$ and $R_{h+1}=R_h/5$, chosen in such a way to satisfy the relationships (6) and (8), as the experimental spectra S_1 of Fig. 2 confirms.

The error $\eta=\Theta-\Theta'$ and the relative one $\epsilon=(S-S')/S=10^{\eta/10}-1$ between the spectrum S given by (1) and the one S' given by (3), in which now γ=1, have been computed vs f. The maximum value ϵ_M of $|\epsilon|$ is less than 5% and it is not experimentally detectable in the 5×10^{-2}, 10^5 Hz band.

In order to confirm the validity of the analytical model we have built a second generator with D_y=0.7 p/d by short-circuiting, alternatively, one half of the R_h-C parallels in the previous generator.

The calculated errors η and ϵ are then plotted in Fig. 3 (line), for both the cases D_y=1.4 and 0.7 p/d, together with the measured values (dots) for the second one. The good agreement confirms the success in obtaining the $1/f^\gamma$ generators.

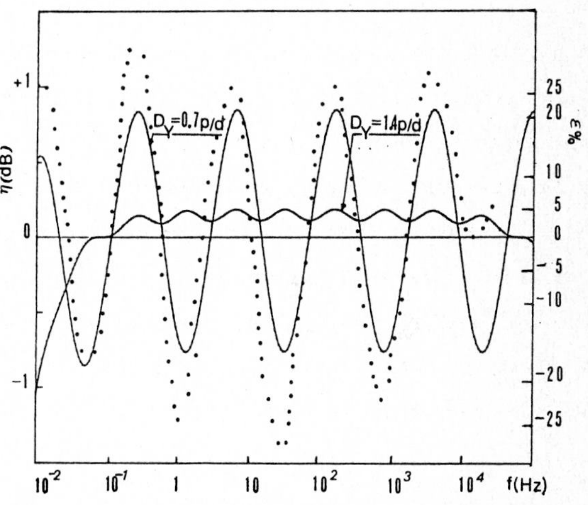

Fig. 3: Relative errors between S given by (1) and $S' \propto 1/f$; dots indicate the experimental values for D_y=0.7 p/d.

C. $1/f^{0.9}$ noise generator

A third $1/f^\gamma$ noise generator, with γ=0.9 in the frequency comprised again between 5×10^{-2} and 10^5 Hz, has been built by means of 16 R_h-C parallels whose parameters C=100nF and $R_h=5\times10^8$,

39×10^6, 5.1×10^6, 0.94×10^6, 221×10^3, 62×10^3, 20×10^3, 7300, 2900, 1250, 560, 270, 138, 73, 40, 23 Ω have been chosen, as confirmed by the experimental plot S_2 of Fig. 2, according to (6), (8) and the pole distribution $D_y \propto \exp(0.23Y)$, to which it corresponds $D_\varphi \propto 1/\varphi^{0.9}$. In this case the maximum analytical error ε_M is less than 5%(7). This result is confirmed by its measured value.

In the same way we can build other $1/f^\gamma$ noise generators with any value of γ between 0 and 2. However when $|\gamma-1|$ tends to 1, the pole number, for a given band, greatly increases.

3. CONCLUSIONS

The numerical simulation(7) allows one to find that the sum of a surprising low number of Lorentzian spectra is sufficient to produce $1/f^\gamma$ noises with negligible errors in wide frequency bands.

Such a result has been utilized to design $1/f^\gamma$ noise generators, in given frequency bands, by summing N Lorentzian noise spectra generated by as many resistor-capacitance parallels connected in series at the input of an ultralow noise, very high input resistance amplifier purposely designed.

Two 1/f and $1/f^{0.9}$ noise generators have been built on a band greater than six decades, from 5×10^{-2} to 10^5 Hz, by choosing the pole frequency of the R-C groups according to their logarithmic distribution which is uniform in the first case and exponential in the second one.

Since the generated output noise spectrum is $S_v=\beta|A_0|^2kT/Cf^\gamma$, where β is a well defined numerical parameter(7), and since A_0 and C are reliable quantities, the proposed systems may be considered as standard flicker noise generators.

Finally the new tecnique proposed to obtain $1/f^\gamma$ noise spectra suggests an its extention to the synthesis of reliable noise generators with various and given shape of the power spectral densities.

ACKNOWLEDGEMENTS

The work has been supported by the CNR, through CSMDR and CCTE, and by the MPI of Italy.

REFERENCES

|1| J.Bernamont, Proc.Phys.Soc. 49, (1937).

|2| A.L.McWorter,"1/f noise and related surfaces effects in germanium", R.L.E.295 and Lincoln Lab.Tech., Rep. 80,M.I.T., (1955).

|3| A. Van der Ziel, "Noise in measurements", (Wiley, New York 1976), p. 77.

|4| P.Dutta, P.M.Horn, Rev.of Mod.Phys. 53, 497, (1981).

|5| B.Pellegrini, Phys.Rev. B 24, 7071, (1981).

|6| B.Pellegrini, Phys.Rev. B 26, 1791, (1982).

|7| B.Pellegrini, R.Saletti, B.Neri, "Minimum number of Lorentzian spectra sufficient to yield $1/f^\gamma$ spectrum" to be published.

|8| M.S.Keshner, Proc. of the IEEE 70, 212, (1982).

|9| B.Neri, R.Saletti, P.Terreni, Proc. of 85th AEI Annual Meet., Rep. 47, Riva del Garda, Oct.1984.

NOISE IN PHYSICAL SYSTEMS AND 1/f NOISE - 1985
A. D'Amico and P. Mazzetti (editors)

SOURCES OF 1/f NOISE IN RuO_2-BASED THICK RESISTIVE FILMS

Andrzej KUSY and Andrzej SZPYTMA

Department of Electrical Engineering, Technical University of Rzeszów, P.O. Box 85, 35-959 Rzeszów, Poland

A model of 1/f noise in thick resistive films (TRFs) has been presented. Two types of 1/f noise sources have been considered: (i) one is located in the conducting neck between two adjacent and sintered conducting particles, (ii) the other deals with the tunneling process between two adjacent but isolated conducting particles. Relative power spectral density (RPSD) of source (i) has been estimated on the basis of Hooge semiempirical formula while for the source (ii) using Kleinpenning model of 1/f noise in tunnel diodes. A network of parallel chains constructed of the two types of 1/f noise sources has been considered and equations have been derived for RPSD and resistance of the network. The experimental results on RPSD and resistance both versus volume fraction of conducting component v for two groups of TRFs with RuO_2 particle size d=10 nm and d=300 nm have been used as the input data in the calculation with the help of the model. As the result a number of barriers N_B versus v for both groups of TRFs has been given. On that basis the observed differences between RPSD within the two groups of films and between them have been explained.

1. INTRODUCTION

Transport of electric charge in thick resistive films (TRFs) occures via chains of conducting particles formed in glass matrix during firing process[1,2]. According to Pike and Seager[1] the particles in the chains are separated by very thin layers of glass forming metal-insulator-metal (MIM) units.

To develope a model of 1/f noise in TRFs we introduce a modification to this type of conduction mechanism. We assume on the basis of conducting particles sintering taking place during firing of the films[3] that not all but some of adjacent conducting particles are separated by thin glass layers and some are sintered. Next we estimate relative power spectral density (RPSD) of 1/f noise in sintered contacts and in MIM units. On that basis the RPSD of 1/f noise is calculated and compared with the one measured for RuO_2 based TRFs.

2. THEORY

RPSD of 1/f noise from homogeneous conducting sample can be described by[4]

$$\frac{S_{U_C}}{U_C^2} = \frac{\alpha}{V_{ef}n}\frac{1}{f} \tag{1}$$

where U_C is constant voltage applied to the sample, S_{U_C} its power spectral density (PSD) of voltage fluctuations, α the Hooge parameter, V_{ef} and n sample effective volume and free charge carriers concentration, f frequency. For our sample V_{ef} can be found as

$$V_{ef} = 20\pi a^3 \tag{2}$$

where a is the radius of neck cross section area.

Resistance of sintered contact can be described by Holm equation[5]

$$R_C = \rho/(\pi a) \tag{3}$$

where[3]

$$\rho = \rho_b(1 + \frac{3}{16}\frac{\lambda}{a}) \tag{4}$$

ρ_b is bulk resistivity and λ electron mean free path of the conducting component.

To estimate the RPSD of 1/f noise generated from tunneling process in MIM units we utilize Kleinpenning theory of 1/f noise in tunnel diodes[6]. According to his model, 1/f noise in devices of this type arises in result of the fluctuation of barrier height U_0 caused by Nyquist noise of the insulator. Assuming that the loss tangent of the insulator, tan δ, is independent of frequency and that the fluctuations of electric field within the barrier are spatially independent, Kleinpenning has shown that[6]

$$\frac{S_{U_B}}{U_B^2} = \frac{mqkTs^2 \tan\delta}{3\pi U_0 \hbar^2 C(s,d)} \frac{1}{f} \tag{5}$$

where U_B is a constant voltage on the barrier, S_{U_B} PSD of U_B fluctuations, -q and m are the electron charge and effective mass, k Boltzmann constant, T absolute temperature, $\hbar$ Planck constant, C(s,d) the copacitance of the system of two RuO_2 particles having diameter d and separated by glass layer of thickness s. The capacitance C(s,d) can be estimated in the following way[7]

$$C(s,d) = \pi\varepsilon_0\varepsilon_r d \sinh\beta \sum_{i=1}^{\infty} [\mathrm{csch}\,(2i-1)\beta + \mathrm{csch}\,2i\beta] \tag{6}$$

where csch $\beta = (d + s)/d$, ε_0 is electrical permitivity of vacuum and ε_r is relative electrical permitivity of glass.

The series resistance of MIM structure 1/f noise source, R_B can be calculated from $(U_B \ll U_0)$[6]

$$R_B = \frac{4\pi^2\hbar^2 s}{q^2 A (2mqU_0)^{1/2}} \exp\,(8mqs^2 U_0/\hbar^2)^{1/2} \tag{7}$$

where A is tunneling area.

We consider now a special network constructed of two types of the described above 1/f noise generators together with their series resistances R_C and R_B. We assume that giant cluster responsible for conduction and noise phenomena observed between two opposite electrodes to the film can be substituted by a system of M parallel chains. Finally assuming that all M chains contain equal numbers of identical noise sources and resistances of particular type it can be shown that 1/f noise RPSD and resistance of the network are given by

$$\frac{S_U}{U^2} = N_C^{-1} \left(1 + \frac{N_B R_B}{N_C R_C}\right)^{-2} \frac{S_{U_C}}{U_C^2} + N_B^{-1} \left(1 + \frac{N_C R_C}{N_B R_B}\right)^{-2} \frac{S_{U_B}}{U_B^2} \tag{8}$$

and

$$R = (N_C R_C + N_B R_B)/M^2 \tag{9}$$

where N_C and N_B are numbers of contacts and barriers (MIMs) in the network.

To estimate the number M we determine its lower and upper bounds M_l and M_u. The lower bound is found from the condition that R_B in equation (9) must be positive which is fulfilled when $M > M_l$ where

$$M_l = (N_C R_C/R)^{1/2} \tag{10}$$

The upper bound M_u is calculated from the formula

$$M_u = N_C(d+s)/L \tag{11}$$

where L is the length of the resistive film. We estimate M by geometric mean $M = (M_l M_u)^{1/2}$.

3. EXPERIMENTAL RESULTS

The resistance and the RPSD of 1/f noise obtained for two groups of RuO_2-based TRFs versus volume fraction of conducting component has been presented in Figures 1 and 2. Particle size of RuO_2 powder used to prepare the films was d=10 nm for one group and d=300 nm for the other one as estimated from BET surface area measurements. The glass powder used for preparation of all the films has the composition: 65% PbO, 25% SiO_2, 10% B_2O_3 (% by weight), average particle size 1.6 μm and relative electrical permitivity ε_r=8.3. All the films were fired in air on alumina substrates at peak temperature equal to 800°C. The films are

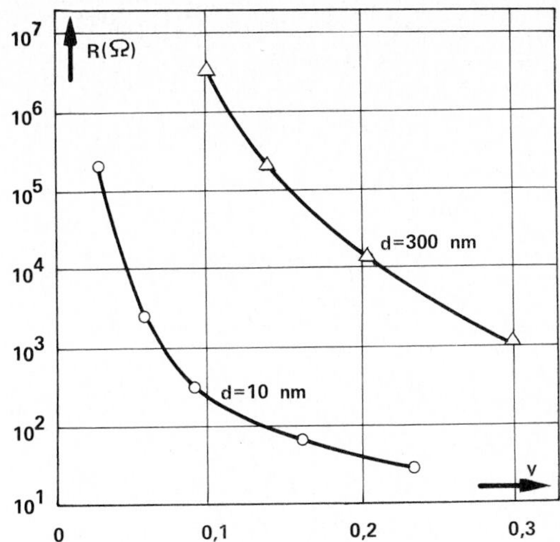

FIGURE 1
Resistance of TRFs made of 10 nm and 300 nm RuO_2 particles versus volume fraction of RuO_2

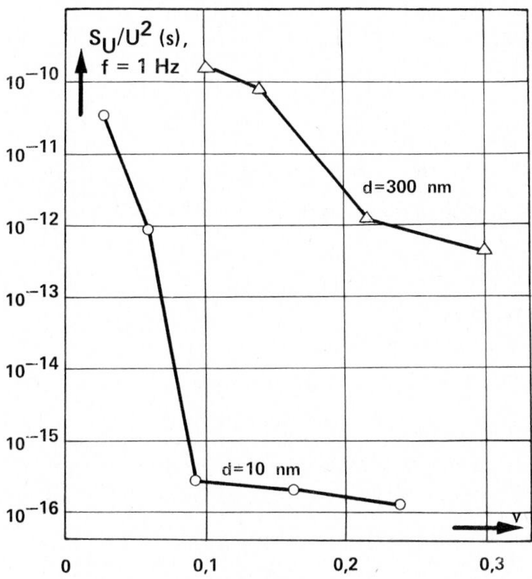

FIGURE 2
1/f noise RPSD at f=1 Hz of TRFs made of 10 nm and 300 nm RuO_2 particles versus volume fraction of RuO_2

squares 5x5 mm^2 with thickness in the range from 10 μm to 23 μm.

4. CALCULATION RESULTS AND CONCLUSIONS

Using equations (1) - (11) and the experimental data given in Figures 1 and 2 the relations N_B versus volume fraction of conducting component v have been estimated. In the calculations the following data have been used: $n=6.4\cdot10^{28}\ m^{-3}$ [8], $\lambda=2.2$ nm [3], $\rho_b=3.5\cdot10^{-7}\ \Omega m$ [8] $\alpha=10^{-3}$ [4], T=293 K, $m=m_0$ (i.e. the rest mass of an electron) $\tan\delta = 0.1$, $U_0=1$ V, $\varepsilon_r=8.3$, L=5 mm, a=1.6 nm and a=8.5 nm for d=10 nm and d=300 nm respectively[3]. Tunneling area A has been estimated as $A = \pi a^2$. The values of N_C have been estimated assuming that $N_C=N/3$, where N, number of conducting particles in the giant cluster of the film has been found from the volume of the film, volume fraction of RuO_2 and RuO_2 particle size. The procedure of calculation of N_B was the iteration type one and was done with the help of microcomputer.

The final results of calculations have been presented in Figure 3.

On the basis of the carried out researches the following conclusions can be put forward:

(i) TRFs made of RuO_2 having mean diameter equal to 10 nm indicate 1/f noise RPSD 4-6 orders of magnitude smaller than TRFs made of RuO_2 with mean diameter 300 nm, which is in qualitative agreement with the paper by Vandamme[9].

(ii) This difference in RPSDs can be explained by much greater (up to 6 orders of magnitude) number N_B of tunneling barriers in small RuO_2 particle size TRFs than in large RuO_2 particle size ones.

(iii) 1/f noise RPSD decreases with v increasing; it can be explained by increasing the number of tunneling barriers in the film.

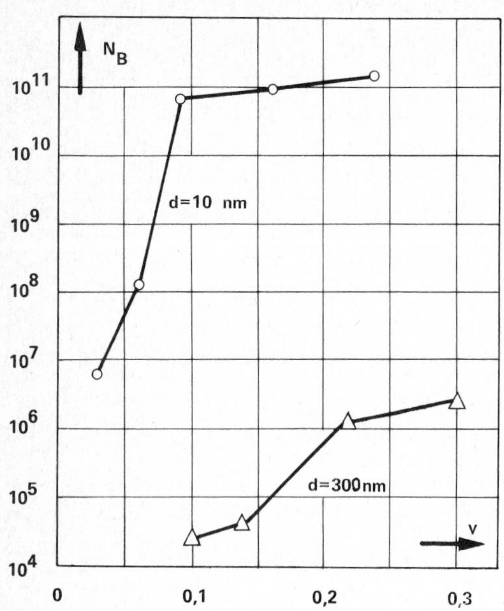

FIGURE 3
Calculated number of MIM barriers N_B versus volume fraction of RuO_2 v in 10 nm and 300 nm RuO_2 particle size films

REFERENCES

1. G.E. Pike, C.H. Seager, J. Appl. Phys. 48 (1977) 5152.

2. A. Kusy, Thin Solid Films, 43 (1977) 243.

3. R.W. Vest, Conduction mechanisms in thick film microcircuits, Final Tech. Rep., Purdue University Research Foundation, Advanced Research Projects Agency order 1642, December 1975.

4. F.N. Hooge, A.M.H. Hoppenbrouwers, Physics Letters 29 (1969) 642.

5. R. Holm, Electric contacts. Theory and applications (Springer, Berlin 1981) p. 16.

6. T.G.M. Kleinpenning, Solid State Electronics 21 (1978) 927 and 25 (1982) 78.

7. W.R. Smythe, C. Yeh, Electricity and magnetism, American Institute of Physics Handbook, ed. D.E. Gray (Mc Graw-Hill, 1972) p. 5-12.

8. D.B. Rogers, R.D. Shannon, A.W. Sleight, J.L. Gillson, Inorganic Chemistry 8 (1969) 841.

9. L.K.J. Vandamme, Electrocomponent Science and Technology 4 (1977) 171.

NOISE IN PHYSICAL SYSTEMS AND 1/f NOISE - 1985
A. D'Amico and P. Mazzetti (editors)

LATTICE VIBRATIONS IN SILICON BY 1/f NOISE SPECTROSCOPY

Mihai MIHAILA

R and D Center for Semiconductors(CCSITS), Str. Erou Iancu Nicolae 32B, Bucharest 72996, Romania

In bipolar transistor structures the mobility fluctuation 1/f noise parameter for holes injected from base into the emitter shows singularities attributed to phonon-carrier interactions. It results that the noise parameter is a variable with spectroscopic character. Phononic energies in phosphorus-doped silicon were determined for the first time by 1/f noise spectroscopy.

1. INTRODUCTION

We found recently that in some bipolar transistors the mobility fluctuation 1/f noise parameter for holes(α_p) injected from base into the emitter shows resonances due to phonon-carrier interaction[1]. We further investigated the dependence of the 1/f noise parameter α_p on the base current. From Kleinpenning's[2] and Van der Ziel's[3] theories, the following simplified formula was used to *estimate* α_p values:

$$\alpha_p = \beta S_{I_B}/I_B , \qquad (1)$$

where S_{I_B} is the noise spectral density of the base current(I_B). The cofficient β depends on some injection and electrical characteristics of the devices. For the devices used in this experiment $\beta \sim 6.25\times10^{10}A^{-1}\cdot Hz$ at a measuring frequency f=10Hz. All measurements were performed *at room temperature*.

2. RESULTS AND DISCUSSION

As can be seen in Fig.1, for a given transistor structure the current dependence of α_p shows a large number of shoulders and maxima. To give a feeling for the mechanisms involved in the generation of such structures, the qV values for each singularity were determined from a Shokley-type formula:

$$I_B=I_R\cdot\exp(\Delta E_g/kT)\cdot[\exp(qV/kT)-1] , \qquad (2)$$

were $I_R=6.6\times10^{-11}$.A is the reverse current of the emitter-base junction, and ΔE_g=0.12eV stands for the band-gap narrowing due to heavy doping effects in emitter. Table I shows that our qV values are of the same order of magnitude as the phononic energies in silicon. The qV values were succesively compared with the phononic energies in silicon determined by different methods such as inelastic neutron scattering[4], tunneling spectroscopy[5], magnetophonon resonance[6], point-contact spectroscopy[7], photoconductivity[8,9], and infrared absorbtion[10].

TABLE I: Phonon energies(meV) in Si:P

Sing. No.	qV from 1/f noise	Other methods	Ref.	Phonon type
1	11.4	11.3	6	g(TA)
2	15.6	?	-	-
3	17.8	17.9	7	g(LA)
4	18.7	18.8	6	g
5	27.4	26.9	9	-
6	30.6	31.5	7	Acoustic at K with Σ_3 symmetry
7	38	38.3	7	Σ_1 acoustic(atK)
8	42.43	42.5	7	Σ_3 optical(atK)

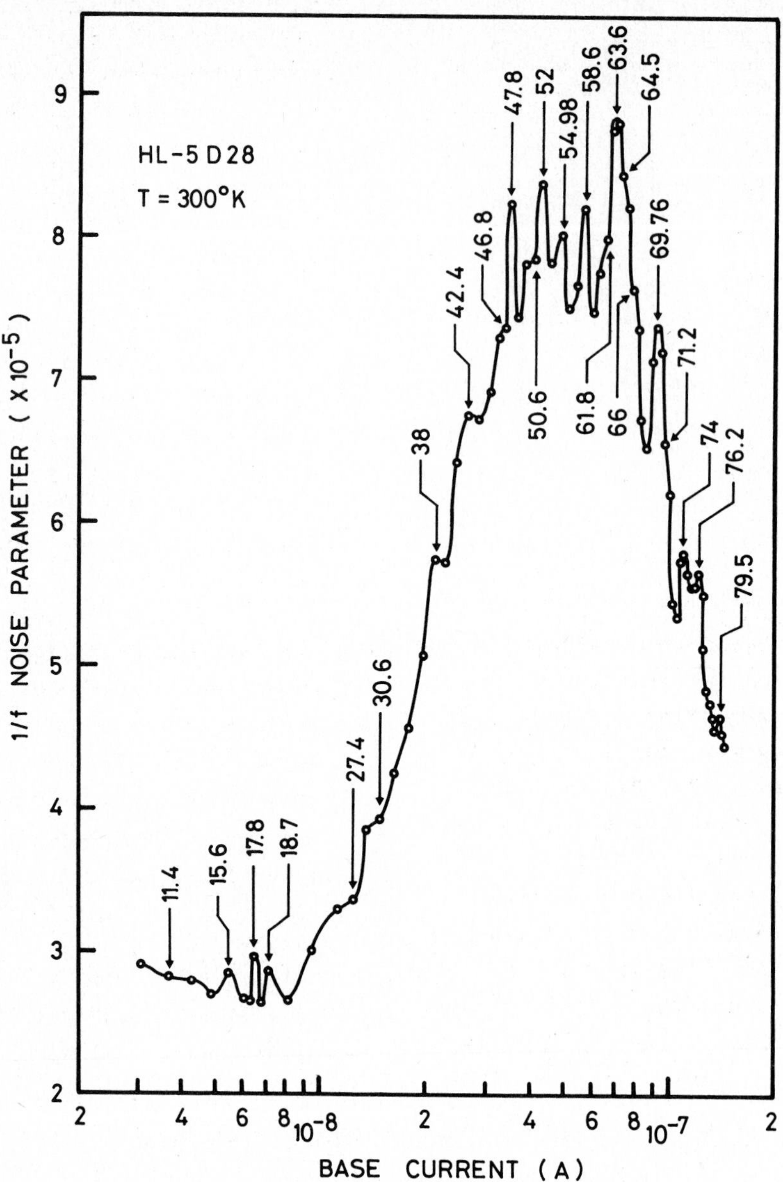

FIGURE 1

Dependence of the 1/f noise parameter(α_p) for holes injected from base into the emitter on the base current. The observed singularities(shoulders or maxima) along with the corresponding energies(meV) are indicated by arrows.

TABLE I (continued)

Sing. No.	qV from 1/f noise	Other methods	Ref.	Phonon type
9	46.8	46.4	6	f
10	47.8	47.4	9	f
11	50.6	50.4	6	f
12	52	51.8	7	f
13	54.98	?	-	kT_{Debye} in silicon?
14	58.6	58.7	6	f
15	61.8	62	-	1 + 11
16	63.6	63.4	-	1 + 12
17	64.5	64.8	-	Raman?
18	66	66.4	-	1 + 13
		65.5	7	g(TO) or TO intravalley
19	69.76	69.8	-	3 + 12
20	71.2	71.2	8	-
21	74	74.2	-	2 + 14
22	76.2	76.4	-	3 + 14
		76.9	10	boron-local mode
23	79.5	79.4	-	5 + 12
		79.9	10	boron-local mode

The qV values determined from 1/f noise measurements are particularly in very good agreement with Pepper's[6] and Eaves et.al.'s[6] results obtained , respectively, by point-contact spectroscopy and magnetophonon resonance in phosphorus-doped silicon(Si:P), which is our case also(n^{++} emitter).
In the TABLE I, the phonon type is the one indicated by the cited authors. Here *g* and *f* phonons are intervalley phonons between ellipsoids whose principal axes are parallel and perpendicular, respectively. Note with Pepper[7] that the new scattering modes introduced by impurities are well resolved by 1/f noise spectroscopy. Structures corresponding to three-phonon combinations and quantum(?) oscillations were observed in the 1/f noise parameter of other devices. Of particular interest is the shape of the curve in Fig.1 which shows similar recemblances to the phonon density of states(Debye spectrum). However, this question requires further investigations mainly because the use of Eq.1 to determine α_p is strictly valid when the recombination factor n=1.

3. CONCLUSIONS

We conclude that *the mobility fluctuation 1/f noise parameter is a variable with spectroscopic character.* Based on this observation, phononic structures were observed in the 1/f noise parameter of copper, silver, and silicon[11]. It seems that the possible new method, tentatively called *1/f noise spectroscopy*, might be useful in search for the microscopic origin of 1/f noise in condensed matter, which remained unknown for over half a century now.

REFERENCES

1. M. Mihaila, Phys. Lett. 104A(1984)157.

2. T.G.M. Kleinpenning, Physica 98B(1980) 289.

3. A. van der Ziel, Solid-St. Electron. 25(1980)141.

4. B.N. Brockhouse, Phys. Rev. Lett. 2 (1959)256.

5. A.G. Chynoweth, R.A. Logan, D.E. Thomas Phys. Rev. 125(1962)877.

6. L. Eaves *et.al.*, J. Phys. C:Solid St. Phys. 8(1975)1034.

7. M. Pepper, J. Phys. C: Solid St. Phys. 13(1980)L709.

8. G.M. Guichar *et.al.*, Phys. Rev. 5B (1972)3013.

9. A. Onton, Phys. Rev. Lett. 22(1969) 288.

10 M. Balkanski and W. Nazarewicz, J. Phys. Chem. Solids 27(1965)671.

11 M. Mihaila, Phys. Lett.107A(1985)465.

NOISE IN PHYSICAL SYSTEMS AND 1/f NOISE - 1985
A. D'Amico and P. Mazzetti (editors)

PHONONIC STRUCTURES IN THE 1/f NOISE PARAMETER OF THE GOLD FILMS

Mihai MIHAILA

R and D Center for Semiconductors(CCSITS), Str. Erou Iancu Nicolae 32B, Bucharest 72996, Romania

Phononic structures were identified in the 1/f noise parameter of the gold films with different structures. This sustains the idea that regardless of the film structure the phonon-carrier interaction is the common source of the 1/f noise in gold films.

1. INTRODUCTION

Different physical mechanisms such as equilibrium temperature fluctuations[1], activated random fluctuations[2] or vacancy diffusion[3] were taken into consideration as possible sources of 1/f noise in continuous metal films. Enhanced 1/f noise levels observed in discontinuous metal films are usually explained in terms of some tunneling mechanisms[4-7].

As to the gold films, either continuous or discontinuous, their low-frequency noise properties are well studied although a still persisting question is whether 1/f noise is a bulk or a surface effect. Hooge and Hoppenbrowers[8] conjectured that 1/f noise in continuous gold films is a bulk phenomenon because their results were well described by the Hooge's formula:

$$S_V/V^2 = \gamma/N.f^\alpha \quad , \qquad (1)$$

where γ is the Hooge's constant(2×10^{-3}) and the remaining terms have their usual meaning. Bulk hypothesis is strongly sustained by the Eberhard and Horn's results[4] as well as by Fleetwood and Giordano's ones[9], although not all of their findings can be adequately described by Eq.1. It seems that this situation can be avoided by taking into consideration that the 1/f noise parameter is not a constant but a variable with spectroscopic character. Starting from this observation, phononic structures were observed in the 1/f noise parameter of copper, silver and silicon[10]. We report in this work the observation of the phononic structures in the 1/f noise parameter of gold films.

2. RESULTS AND DISCUSSION

The temperature dependence of the 1/f noiise parameter($\gamma=S_VNf^\alpha/V^2$) for an 800Ä continuous gold film was determined from Eberhard and Horn's results(Fig.7, Ref.4) by taking into account the temperature dependence of the exponent α(Fig.4,Ref.4). Figure 1 shows that the dependence γ vs. temperature exhibits seven singularities at temperatures corresponding to the L[001] lattice vibration mode or to two-phonon or even three-phonon combinations in gold[11]. TABLE I shows that the phononic energies ($k_B.T$) determined from Eberhard and Horn's 1/f noise measurements in gold films are in good agreement with the data from inelastic neutron scattering[11].

Celasco *et.al.*[6,12] determined that

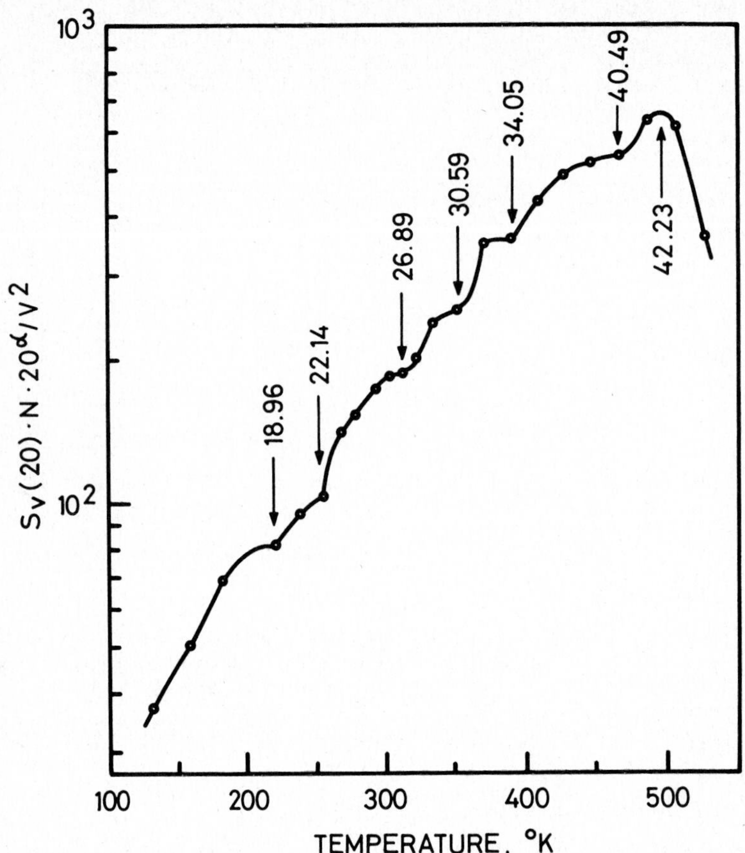

FIGURE 1
Dependence of the 1/f noise parameter $\gamma=S_v(20).N.20^{\alpha}/V^2$ on vs. temperature for an 800Å gold film on sapphire substrate. The points were determined from Eberhard and Horn's results(Ref.3). Phononic structures(along with the corresponding energies) are indicated by arrows.

the temperature dependence of the 1/f noise in discontinuous gold films exhibits broad peaks with maximaplaced between 350-400 K(Ref.6,Fig.6) or 400-470 K(Ref.12,Fig.1). As indicated in the TABLE I, these temperature ranges would correspond to the phonon combinations proposed for peaks 3-5 and 5-7, respectively. Moreover, Celasco *et.al's* results(Ref.12,Fig.1) features some possible singularities at about 80K , 135K, and 260K(a change in slope) which would correspond, respectively, to $T[\frac{1}{2}\frac{1}{2}\frac{1}{2}]$, T[001], and two 11.38meV a combination. It seems now reasonable to suppose that a phonon-assisted mechanism (most probable, tunneling) is also responsible for the 1/f noise generation in discontinuous gold films.

Fleetwood and Giordano[13] have found that 1/f noise in an AuPd film is affected by annealing, the noise level decreasing substantially after some annealing steps. Usually the defect

TABLE I: Phononic energies(meV) in gold

Sing. No.	T(K)	K_BT	Ref.11	Assign.#
1	220	18.96	19	L[001]
2	257	22.14	2×11.38	T[001],L[110], T_1[110]
3	312	26.89	19+7.7	L[$\frac{1}{2}\frac{1}{2}\frac{1}{2}$]+T[$\frac{1}{2}\frac{1}{2}\frac{1}{2}$]
4	355	30.59	30.47	L[100]+T[100] T_1[110]+T_2[110]
5	395	34	3×11.38	T[001],L[110], T_1[110]
6	470	40.49	40.5	L[$\frac{3}{4}\frac{3}{4}0$]+T_1[$\frac{3}{4}\frac{3}{4}0$] +T_2[$\frac{3}{4}\frac{3}{4}0$]
7	490	42.85	41.85	L[110]+T_1[110] +T_2[110]

#Some of these combinations might be forbidden by the selection rules.

annealing is accompanied by phenomena of vacancy-interstitial annihilation. If a phonon model of vacancy-interstitial pair annihilation is assumed[14] it results that the phonon density of states is considerably modified by annealing. However, the characteristic energies of the phonon spectrum are presumable insensitive to disorder[15] and, consequently, to annealing. Without going into details, one observes that the 1/f noise parameter in an AuPd film(Ref.13,Fig.1, curve C) shows singularities corresponding to the phononic energies in gold(see TABLE I) at the following temperatures:90K(7.7meV), 140K(12meV), 220K(18.96meV), and 260K(22.4meV). The second and the third singularity occur more clearly as maxima in the distribution of the activation energies(Ref.13,Fig.3, curve C). The Fleetwood and Giordano's results point on the phonon-carrier interaction as a possible source of 1/f noise in gold alloys.

One notes that some structures observed in the 1/f noise parameter of gold fils by Kilmer *et.al.*[16] seem to correspond to T[$\frac{1}{2}\frac{1}{2}\frac{1}{2}$] mode phonons in gold.

3. CONCLUSIONS

The 1/f noise existing data for gold films were reinterpreted in the light of the fact that the 1/f noise parameter has a spectroscopic character. Phononic structures were observed in the 1/f noise parameter of the gold films, either continuous or discontinuous, as well as in an AuPd alloy. It seems reasonable now to consider that the phonon-carrier interaction is the bulk, common source of the 1/f noise in all these systems.

REFERENCES

1. R.F. Voss and J. Clarke, Phys. Rev. Lett. 36(1976)42.
2. P. Dutta, P. Dimon, and P.M. Horn, Phys. Rev. Lett. 43(1979)646.
3. J.W. Eberhard and P.M. Horn, Phys.Rev. 18B(1978)6681.
4. J.L. Williams and R.K. Burdett, J. Phys. C2(1969)298.
5. J.L. Williams and I.L Stone J. Phys. C5(1972)2105.
6. M. Celasco, A. Masoero, P. Mazzetti, A. Stepanescu,Phys. Rev.17B(1978)2553; Phys. Rev. 17B(1978)2564.
7. A.Ludviksson ,R. Kree, A. Schmid, Phys. Rev. Lett. 52(1984)950.
8. F.N. Hooge and A.M.F. Hoppenbrowers, Physica 45(1969)386.
9. D.M. Fleetwood and N. Giordano, Phys. Rev. 27B(1983)667.
10. M. Mihaila, Phys. Lett.107A(1985)465.
11. J.W. Lynn,H.G. Smith and R.M. Nicklow, Phys. Rev. 8B(1973)3493.
12. M. Celasco, F. Fiorillo,and A. Masoero Phys. Rev 19B(1979)1304.
13. D.M. Fleetwood and N. Giordano, Phys. Rev. 31B(1985)1157.
14. D.C. Mattis,Phys.Rev.Lett.53(1984)568.
15. S.J. Poon and T. Geballe, Phys. Rev. 18B(1978)233.
16. J. Kilmer, C.M. van Vliet, G. Bosman, A. van der Ziel, and P. Handel, Phys. Stat. Sol.(b) 121(1984)429.

NOISE IN PHYSICAL SYSTEMS AND 1/f NOISE - 1985
A. D'Amico and P. Mazzetti (editors)

1/f NOISE IN THE MICRO-HERTZ RANGE

A.H. DE KUIJPER and T.G.M. KLEINPENNING

Eindhoven University of Technology, Eindhoven, The Netherlands

In this paper we present investigations on the variance of a 1/f noise source. A theoretical relation is derived between the variance σ^2 and the observation time T. The 1/f character of the noise source in the micro-Hertz range is checked by measuring the variance as a function of T, with $T < 3 \times 10^5$ s. At $f = 3.3 \times 10^{-6}$ Hz the spectrum of the noise source was still 1/f.

1. INTRODUCTION

The frequency range in which a noise signal can describe 1/f behaviour has long been the subject of research.

The lowest frequency at which 1/f noise can be observed is especially interesting because, (a) it can provide insight into the origin of 1/f noise and (b) the physical limitation is that any stationary noise signal must have a finite variance.

Long-term instability of sample and measuring equipment sets a limit to the observation time. The best results (in terms of low frequencies) were obtained by Caloyannides[1] who presented measurements as low as 5×10^{-7} Hz. This is one cycle in 3 weeks !

Caloyannides obtained his results by taking data samples from ten noise sources. The time data were converted into the power spectrum by a mathematical operation, including a doubtful removal of the mean.

In the present paper the problem is treated somewhat differently.

Consider a stochastic signal x(t) with $\lim_{T\to\infty} < x(t) >_T = 0$. Here T is the observation time. The variance of this signal

$$\sigma^2 = \lim_{T\to\infty} < x^2(t) >_T = \int_0^\infty S_x(f)df \qquad (1)$$

where $S_x(f)$ is the power spectral density function. When x(t) is a 1/f noise signal with $S_x(f) = C/f$ (C is a constant) we are confronted with the problem that σ^2 is divergent.[2]

A possible solution of this problem is the existence of a low-frequency cut-off point f_ℓ and a high-frequency cut-off point f_h.

Applying $S_x(f) = C/f$ to equation (1) we obtain

$$\sigma^2 = C\ell n(f_\infty/f_0) \qquad (2)$$

where $f_\infty = \min(f_h, f_e)$ with f_e the equipment bandwidth and $f_0 = \max(f_\ell, 1/T)$ and with T the observation time. For $T < 1/f_\ell$, that is, the observation time shorter than the (hypothetical) inverse low-frequency cut-off point, equation (2) becomes

$$\sigma^2 = C\ell n(Tf_\infty) \qquad (3)$$

Here we see that the variance grows logarithmically with the observation time T.

We have investigated the relation between σ^2 and T of a 1/f noise signal up to $T = 3 \times 10^5$ s in order to search for the cut-off frequency f_ℓ.

Brophy[3] has found logarithmic growth of the variance of a 1/f noise signal up to $T = 5 \times 10^3$ s. We extended the measurements of Brophy by almost a factor of 100 up to $T = 3 \times 10^5$ s. Contrary to Brophy, we took into account the contribution of the higher frequency components of x(t) to the variance σ^2. The measurement technique is described in section II.

In section III we derive an expression for the variance.

Finally, in section IV we shall present and discuss the experimental results.

2. MEASUREMENT TECHNIQUE

As already mentioned, the fundamental problem in measuring very low frequencies (long observation times) is the need for stable sample and measuring equipment during the observation time. Sample and equipment should be insensitive to fluctuations in environmental variables such as temperature and supply voltage.

As 1/f noise source we used a noisy carbon sheet resistor. The current through the sample was supplied by a battery. During the measurement, less than 1 o/oo of the capacity of the battery was used. The sample was placed in a special temperature bath kept constant within 10^{-5} K. The measurement set-up is given in figure 1.

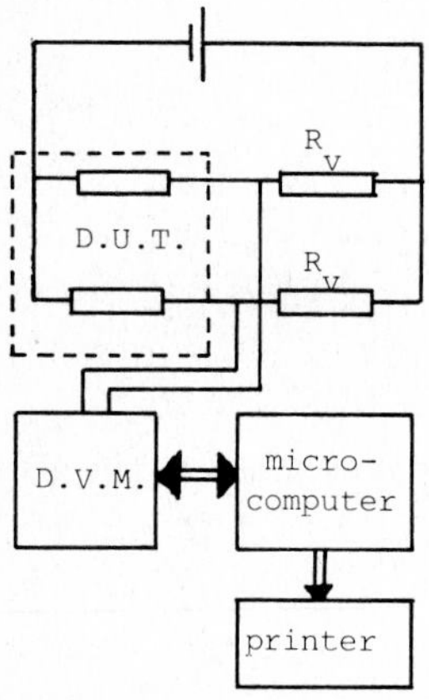

FIGURE 1
Measurement set-up

Samples of v(t) were taken with a stable and accurate digital voltmeter (HP 3456 A) using the D.C. 100 mV range. (The 90-day error in this range is less than $\pm$ 4.6 μV, which is far below the level of the fluctuations of v(t)).

The variance of $x(t_i)$ has to be currently updated in order to decide whether the experiment should be continued or stopped.

The variance after k + 1 samples is

$$\sigma_{k+1}^2 = \frac{1}{k} \sum_{i=1}^{k+1} \left(x_i^2 - M_{k+1}^2\right) \tag{4}$$

with M_{k+1} as the mean.

Using $$\sigma_k^2 = \frac{1}{k-1} \sum_{i=1}^{k} \left(x_i^2 - M_k^2\right) \tag{5}$$

it follows from (4) and (5) that

$$\sigma_{k+1}^2 = \frac{1}{k}\left[(k-1)\sigma_k^2 + kM_k^2 - (k+1)M_{k+1}^2 + x_{k+1}^2\right] \tag{6}$$

With the help of the relation

$$M_{k+1} = \frac{kM_k + x_{k+1}}{k+1} \tag{7}$$

equation (6) leads to

$$\sigma_{k+1}^2 = \frac{1}{k}\left[(k-1)\sigma_k^2 + \frac{k}{k+1}(M_x - x_{k+1})^2\right] \tag{8}$$

Thus, using the previous results of σ_k^2, M_k and the last time sample x_{k+1}, the values of σ_{k+1}^2 and M_{k+1} can be calculated.

3. CALCULATION OF THE VARIANCE

In this section we shall derive an expression for the variance of a 1/f noise signal measured with the equipment described in figure 1.

The digital voltmeter has an analog-digital converter which uses the multislope technique (a special form of the dual slope technique). This means that the analog signal v(t) is integrated during a time τ, the integration time of the voltmeter.

The noise signal is averaged over τ seconds, so that the time signal from which we derive the variance is

$$v_a(t) = \frac{1}{\tau} \int_{t-\tau}^{t} v(t')dt' \tag{9}$$

Writing

$$v(t') = \int_{-\infty}^{+\infty} V(f)\exp(j2\pi ft)df \tag{10}$$

we obtain

$$v_a(t) = \int_{-\infty}^{+\infty} \frac{V(f)}{j2\pi f\tau}[1-\exp(-j2\pi f\tau)]\exp(j2\pi ft)df \tag{11}$$

where V(f) is the amplitude spectrum of v(t).

Rewriting (11) into

$$v_a(t) = \int_{-\infty}^{+\infty} V_a(f)\exp(j2\pi ft)df \tag{12}$$

with $V_a(f)$ the amplitude spectrum of $v_a(t)$, we find

$$V_a(f) = \frac{V(f)}{j2\pi f\tau}[1 - \exp(-j2\pi f\tau)] \tag{13}$$

For $\tau \to 0$ the Fourier transforms of $v(t)$ and $v_a(t)$ are equal.

From Parseval's theorem we have

$$\sigma^2(T) = \frac{1}{T}\int_0^T |v_a(t)|^2 dt = \frac{1}{T}\int_{1/T}^{\infty} |V_a(f)|^2 df \tag{14}$$

From eqs. (13) and (14) we find the observed variance to be

$$\sigma^2(T) = \frac{1}{T}\int_{1/T}^{+\infty} \frac{V(f)V^*(f)}{(2\pi f\tau)^2} |1-\exp(-j2\pi f\tau)|^2 df$$

$$= \int_{1/T}^{\infty} V(f).V^*(f) . \frac{\sin^2(\pi f\tau)}{(\pi f\tau)^2} df \tag{15}$$

In the case of 1/f noise with spectrum

$$S_v(f) = C/f^\beta \tag{16}$$

where $\beta = 1$, equation (15) leads to

$$\sigma^2(T) = \int_{1/T}^{\infty} \frac{C}{f} . \frac{\sin^2(\pi f\tau)}{(\pi f\tau)^2} df$$

$$\approx C\ \ell n(0.4\ T/\tau) \tag{17}$$

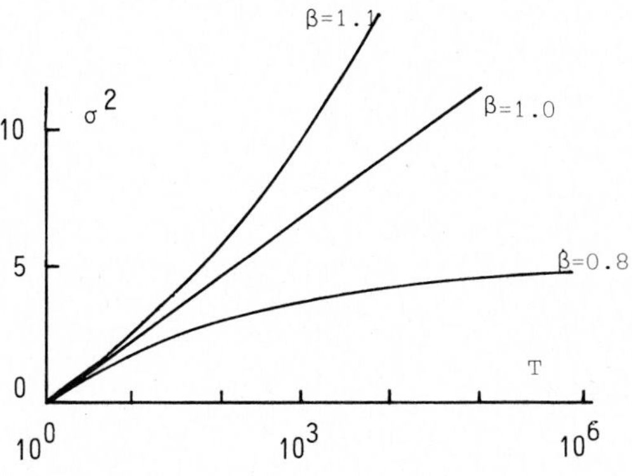

FIGURE 2

σ^2 versus T according to

$\sigma^2 = \int_{1/T}^{1} \frac{1}{f^\beta} df$ for β = 0.8, 1.0 and 1.1.

Note that $\int_{1/T}^{0.4/\tau} \frac{C}{f} df = C\ \ell n(0.4\ T/\tau)$, so that the effective bandwidth $f_e = 0.4/\tau$.

From eq. (17) we see that the variance grows logarithmically with the observation time T in the case $\beta = 1$.

In figure 2 we have plotted σ^2 versus T for β = 0.8, 1.0 and 1.1. Note that the σ^2 versus T behaviour is very sensitive to the value of β.

4. EXPERIMENTAL RESULTS

Two noisy carbon-sheet resistors were inserted in a Wheatstone bridge providing a relatively high noise level at $\langle v(t)\rangle \approx 0$.

We first measured the spectrum of the sample from 1 - 10 kHz. The result is given in figure 3.

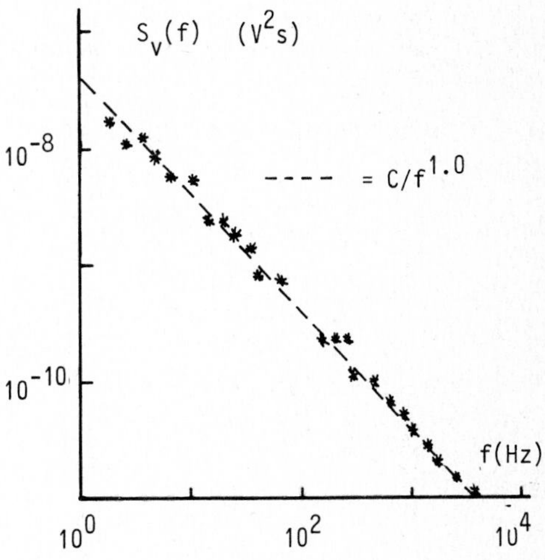

FIGURE 3

Spectrum of sample I.

$\beta = 1.0$ $C = 6 \times 10^{-8}\ V^2$.

As τ can be varied with the voltmeter it gave us the opportunity to check the dependence between $\sigma^2(T)$ and τ (see eq. 17) at a fixed value of the observation time T. The result for T = 50 s is presented in figure 4. We find that there is good agreement between theory and experimental result.

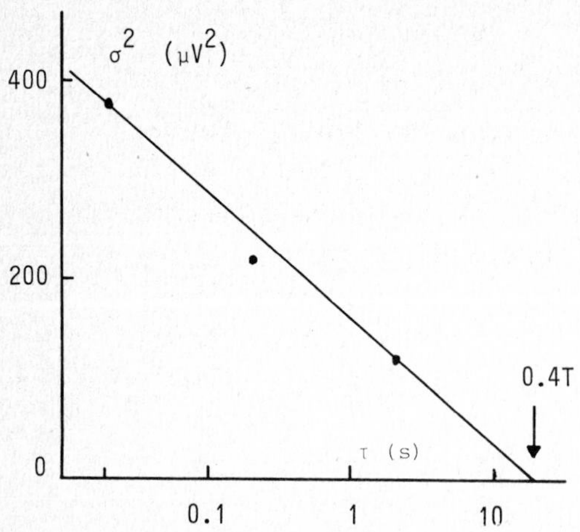

FIGURE 4
Variance σ^2 as a function of the integration time τ. The value of T is 50 s. The solid line is proportional to $\ell n(0.4\ T/\tau)$. Experimental data are expressed by dots. Sample II.

To obtain $\sigma^2(T)$ from experiments, we collect a set of samples $x(t_i)$ with the digital voltmeter. The variance was determined using eq. 8. The interval time was 2.5 s and the number of samples about 10^5.

The variance $\sigma^2(T)$ in figure 5 is the geometric mean of ten measurements.

Figure 5 also gives the graphical representation of eq. (17) for $\beta = 1$ and $\beta = 1.05$.

From figure 5 we conclude that:

- the experimentally observed variance $\sigma^2(T)$ agrees with the predicted variance given in eq. (17),
- $S_V(f)$ is still almost 1/f at $T = 3 \times 10^5$ s. ($f = 3.3 \times 10^{-6}$ Hz).

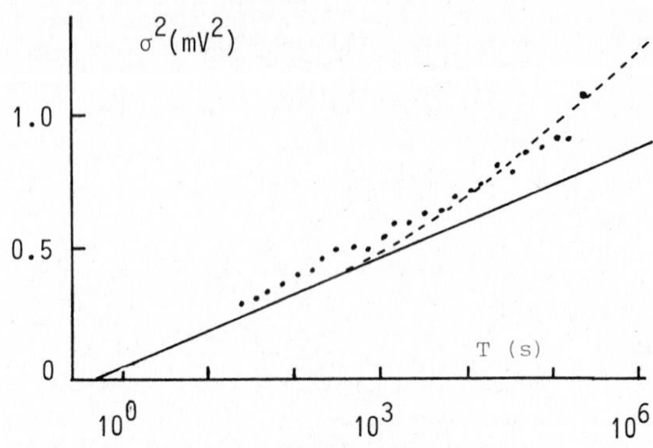

FIGURE 5
Variance σ^2 as a function of the observation time T. The value of τ = 200 ms. The solid line is based on eq. (17) with $C = 6 \times 10^{-8}\ V^2$. The dashed line is based on eqs. (15,16) with $\beta = 1.05$. Experimental data are expressed by dots. Sample I.

REFERENCES

1. M.A. Caloyannides, J. of Appl. Phys. 45 (1974) 307 - 316.

2. M.S. Keshner, Proc. of the IEEE, 70 (1982) 212 - 218.

3. J.J. Brophy, J. of Appl. Phys. 41 (1970) 2913 - 2916.

NOISE IN PHYSICAL SYSTEMS AND 1/f NOISE - 1985
A. D'Amico and P. Mazzetti (editors)

1/f NOISE GENERATOR

András AMBRÓZY and László B. KISS

Technical University Budapest, H-1521 Budapest
József Attila University, H-6720 Szeged, Dóm tér 9.

Noise measurements generally need some kind of precision noise sources for comparison. Several types of white noise sources have been known for some time. No similar sources exist which generate 1/f noise.
The frequency range of interest is between $10^{-5}...10^{5}$ Hz. Our target is to build a simple, inexpensive 1/f source for that range.
Random pulses of $v = 1(t)\ A/\sqrt{t}$ are generated by a random telegraph signal passed through a nonlinear network which realizes the differential equation of $dv/dt= -v^3/2A^2$.
Analyzed are the ratio of upper and lower corner frequencies and the amplitude density function of the resulting pulse train.
With a slight effort almost Gaussian distribution can be obtained.

1. INTRODUCTION

Noise measurements generally need some kind of precision sources for comparison. Several types of white noise sources have been known for some time, e.g. saturated vacuum diode, avalanche diode, generator of pseudorandom noise.No similar devices or equipment exist which generate 1/f noise.

The frequency range of interest is between $10^{-5}...10^{5}$ Hz since above the upper limit the 1/f component is buried in the white noise.Our target is to build a simple, inexpensive unit for that range

Two possibilities exist:

i/ The output of a precision white noise source is fed to the input of a special filter having a transmission proportional to $1/\sqrt{f}$ (not realizable with lumped components) [1].

ii/ Randomizing time functions of known spectrum to obtain 1/f noise [2].

The advantage of i/ is the Gaussian output, its disadvantages are the expensive realization and no possibility of integration.

As for randomized time functions, there are two direct ways:

a/ "McWorther generator" can be obtained if random telegraph waves with an average time constant τ_i are summed with weights $1/\tau_i$. Advantages are the Gaussian output and easy integration, disadvantage the complexity which probably exceeds that of the filter method.

b/ The "Campbell 1/f generator" (Fig.1) produces random pulse train of time function $v = A/\sqrt{t}$. Its benefit that in a very simple form it gives 1/f spectrum and easily integrable.

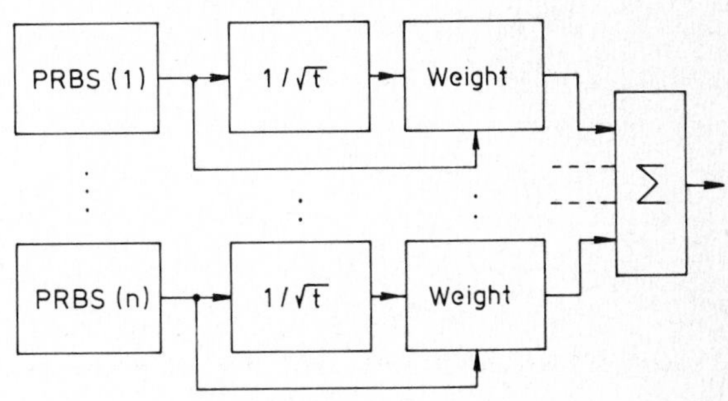

FIGURE 1
1/f "Campbell" generator

Almost Gaussian output can be obtained by weighting although precision Gaussian output requirement leeds to more complicated circuitry.

2. THEORETICAL CONSIDERATIONS

According to the Campbell-theorem the superposition of transients v(t) gives

$$V(t) = \sum_{-\infty}^{\infty} Av(t-t_i)(-1)^i \qquad (1)$$

whose power spectrum is

$$S(f) = \nu A^2 [V_F(f)]^2 \qquad (2)$$

where t_i are the random starting points of transients belonging to a Poisson-process with parameter ν, V_F the Fourier-transform of v(t).

If

$$v(t-t_i) = \frac{A}{\sqrt{t-t_i}} 1(t-t_i) \qquad (3)$$

where 1(t) is the unit step function then Eq. (2) leads to 1/f power spectrum:

$$S(f) = \nu A^2/f \qquad (4)$$

Of course, the transients given in Eq.(3) are limited twofold: if the time interval of a transient is limited ($T=t_{imax}-t_i$) then $v_{min} = A/\sqrt{T}$ and if the rise time of the pulse is τ then $v_{max} = A/\sqrt{\tau}$ (Fig.2.).

It can be easily shown that

$$f_\ell \approx \frac{1}{T}; \quad f_u \approx \frac{1}{\tau} \qquad (5)$$

or

$$\frac{f_u}{f_\ell} = \left(\frac{v_{max}}{v_{min}}\right)^2 \qquad (6)$$

According to the central limit theorem the output distribution of the Campbell-generator approximates the Gaussian one if the transients strongly overlap. There is, however, a practical problem: one elementary transient $1(t-t_i)(t-t_i)^{-1/2}$ has an amplitude-density function proportional to v^{-3} therefore in the case of high v-s (i.e. shortly after commencement of transients) the probability of interference of waves with similar amplitudes is very low even with great number of individual generators. So in practical cases (when number of generators < 10) the amplitude density function of the output is almost Gaussian at low levels but proportional to v^{-3} at high levels.

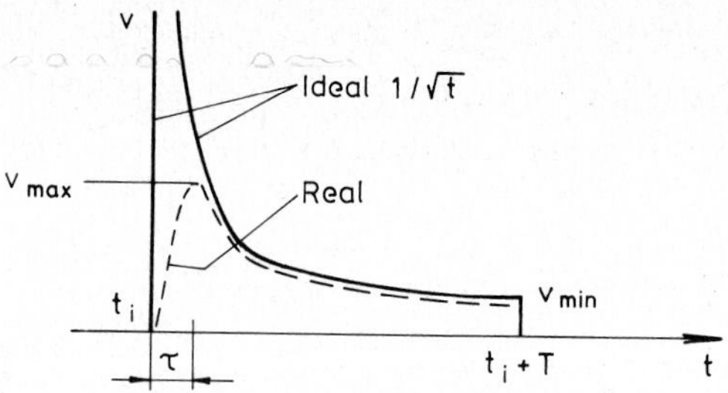

FIGURE 2
Ideal and real transient

To accelerate the Gaussian convergence there are two possibilities:

i/ An all-pass filter smooths the peaks and so makes easier the interference of waves with same magnitude.

ii/ Weighting the amplitudes of different transients very rapid convergence can be obtained.

3. ELECTRONIC IMPLEMENTATION

The transient of $v = 1(t)A/\sqrt{t}$ can be generated by the solution of the differential equation

$$\frac{dv}{dt} = -\frac{1}{2A^2} v^3 \qquad (7)$$

In integrated form:

$$v = -\frac{1}{2A^2}\int_0^t v^3 dt \qquad (8)$$

The electronic implementation of Eq.(8) can be seen in Fig. 3. The voltage v(t) is raised to the third power by two four-quadrant analog multipliers. The transients are initialized by a pseudorandom binary pulse train coupled to the input of the integrator via C′.

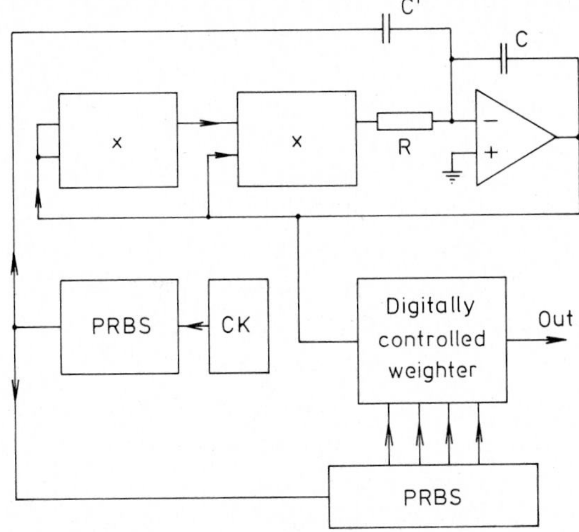

FIGURE 3
Transient generator and weighter

In Fig.3. a possible setup of weighting is also given. The weighter is allowed to change its state synchronized to the transients to avoid the distortion of the 1/f spectrum.

According to the above description one can get a 1/f spectrum only in a region $f_\ell < f < f_u$ but changing the coupling capacitor C′ and C and the clock frequency of the binary driver signal, the complete spectrum can be shifted upward and downward while f_u/f_ℓ remains constant.

4. EXPERIMENTS

The power spectrum and the amplitude density function was examined in separate experiments. For spectrum evaluation the setup of Fig. 3. was used without weighting network. The realized circuit was built with commercial components, the multipliers were the old 795 type whose nonlinearity is in the order of 2-4% even at optimal adjustment. Any nonlinearity, offset, etc. distorts the 1/f spectrum.

Stable 1/f spectrum could be obtained within 3 decades, with proper adjustment within 4 decades. Investigating Eq.(6) it can be stated that better quality components will result in even wider spectrum.

The amplitude density function was studied by numerical experiments. Monte-Carlo method was used both for unweighted and weighted transients. (Weighting was done by multiplication with discrete levels which were gained from the inverse Gaussian density function.)

The convergence to Gaussian of the unweighted transients is very slow at high amplitudes. With weighting, even one transient generator can supply almost Gaussian signal, although not at high amplitudes. With 7 discrete levels and 2 transient generators the deviation from pure Gaussian is ±3% in the range of 2 ... 98%. With 9 levels the same precision can be obtained even with one transient generator.

REFERENCES

1. M.S.Keshner, Proc. IEEE 70 (1982) 212.
2. D.Halford, Proc. IEEE 56 (1968) 251.

NOISE IN PHYSICAL SYSTEMS AND 1/f NOISE - 1985
A. D'Amico and P. Mazzetti (editors)

LOW FREQUENCY NOISE IN A SYSTEM COMPOSED OF TWO ELECTRODES WITH A BUTANE FLAME BETWEEN THEM

E.J.P. MAY and H.H. MEHDI

Department of Engineering Science, University of Exeter, Exeter, Devon, EX4 4QF, England.

It is found that the flames produced by butane gas are conductive. Measurements have shown that a current flowing through a flame between two electrodes exhibits a type of 1/f noise.

1. INTRODUCTION

Electrical current can be made to flow through certain parts of a gas-flame. It is found that these currents have strong components of low-frequency. A preliminary investigation of this noise was undertaken and will be described below.

2. MEASURING CIRCUIT DETAILS

Two metal electrodes spaced about 1cm apart and having an area of $1cm^2$ were placed in a butane flame. Connections were made to these electrodes from a d.c. source through a relatively low value resistor. The voltage across this resistor was used to measure current fluctuation. The circuit is shown in Figure 1.

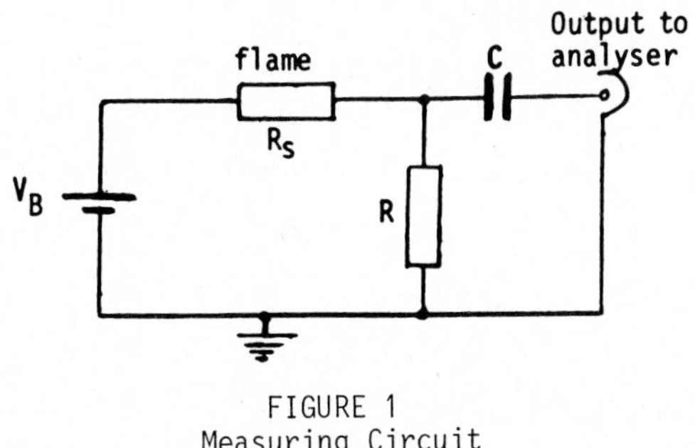

FIGURE 1
Measuring Circuit
$R_S >> R$

The output of the circuit goes to a wave analyser (bandwidth 3·16Hz). The butane flame was adjusted so that the voltage across the resistor R at the output was 0·5V when the battery voltage was 40V. The output of the frequency analyser was observed to be 'bistable'. The meter needle seemed to dwell for short

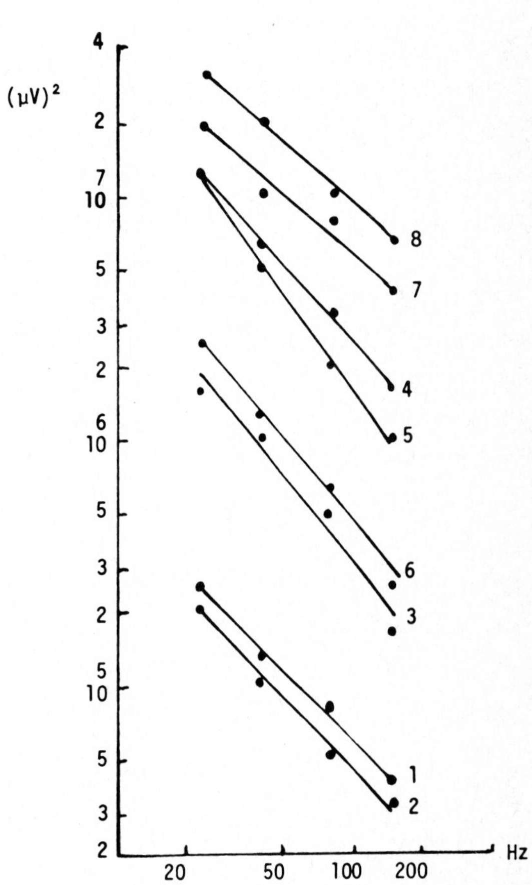

FIGURE 2
$R = 3{\cdot}3k\Omega$ $C = 80\mu F$

1	V_B = 40V	V_R = 0.5V
2	V_B = 20V	V_R = 0.5V
3	V_B = 50V	V_R = 0·65V
4	V_B = 50V	V_R = 1·8V
5	V_B = 50V	V_R = 2·5V
6	V_B = 83V	V_R = 0·65V
7	V_B = 83V	V_R = 1·8V
8	V_B = 83V	V_R = 2·5V

periods of time at the low end. The high end was not so well defined particularly at the low frequency end (10 - 40Hz) of the spectrum. In fact, the reading at the high end is due to convection and movement of the flame. The assumption was made that a low noise existed on which a high level intermittent noise was superimposed. The more stable low level readings were taken for various values of voltage and flame size. These are shown in Figure 2.

3. CONCLUSIONS

The results shown in Figure 2 indicate a low frequency spectrum which is quite close to $\frac{1}{f}$ noise. It must be confessed, however, that these readings are somewhat subjective because one had to distinguish between two sets of conditions. The impression one got was that one observed information in a time division system where the divisions were somewhat random. Nevertheless, two workers were able to agree independently on the readings and this is the reason why these results are being presented.

In accepting these readings, one is making the assumption that $\frac{1}{f}$ noise is present in the electrode flame arrangement as a constant component in the presence of an occasional very strong component, very often prompted by a draught. It may well be that the noise is produced in the plasma of the flame, but there is the possibility that it is produced at the interface between the plasma and the electrodes. Further work is being undertaken to locate the source of noise.

NOISE IN PHYSICAL SYSTEMS AND 1/f NOISE - 1985
A. D'Amico and P. Mazzetti (editors)

1/F NOISE IN MOS INVERSION LAYERS

S.A. HAYAT and B.K. JONES

Department of Physics, University of Lancaster, Lancaster LA1 4YB, United Kingdom.

Measurements are reported of 1/f noise in MOS channels at very low V_{DS} between 77K and 360K. Detailed characterisation of the channels was also made. Both enhancement and depletion devices showed a noise with two components. One appears to be a number fluctuation independent of the channel carrier number while the other is proportional to the carrier number. From the temperature and interface state density dependence the latter component is ascribed to a number fluctuation.

1. INTRODUCTION

In electronic systems excess noise, above thermal and shot noise, consists of a resistance fluctuation. It can be divided into two types : generation-recombination noise with a Lorentz spectrum and 1/f noise with a $1/f^{\beta}$ spectral density with $\beta \approx 1$. The behaviour of the generation-recombination noise is generally thought to be well understood. There is however no general consensus about the physical origin of the 1/f noise and whether it has the same origin in all systems, whether it is a bulk or surface effect or whether it is a mobility or number fluctuation.

The Lancaster Noise Group has been carrying out a series of detailed experiments on excess noise in silicon since it is a very pure and crystalline material with well known properties. The specimens are commercial devices with known geometry. An understanding has been made of the noise in junctions[1,2,3]. However, the simplest system in which to study the effect is the uniform cross-section silicon resistor. This is experimentally the channel of a field-effect transistor with very small source-drain voltage (V_{DS}). A detailed study of JFET channels[4,5] showed a complex behaviour because of the non-uniform properties across the channel but it suggested a number fluctuation at the boundaries. In that geometry the boundaries of the resistor are defined by depletion regions.

The aim of the present study is to investigate the mechanism of the 1/f noise in MOST channels in which one boundary is a silicon-silicon oxide interface while the other is a depletion region. The channel is two dimensional so that all the carriers experience both boundaries.

2. EXPERIMENT

A limited set of the specimens and results only are reported here. The specimens are BSV81 depletion mode and IT 1750 enhancement mode n-channel MOST. The measurements were made over the range 77K to 360K. The noise was measured in strong inversion at very small V_{DS} with the criterion that the noisiness, $\overline{v_n^2}/V_{DS}^2$, was taken to the limit as V_{DS} went to zero. The intensity spectrum was fitted to the sum of a white component and a $1/f^{\beta}$ excess component The magnitude and exponent, β, were recorded.

In order to understand the properties of the specimens extra measurements were made. The device resistance $R_{DS} \equiv V_{DS}/I_D$ against V_G enabled the threshold voltage, V_T, and parasitic series resistance, R_{SER}, to be determined. The effect of this series resistance is not easy to distinguish from the mobility degradation due to the gate field but there is good evidence to assume that the latter effect is small[6]. From this the average carrier mobility, μ, in the channel could be derived, $\mu = L/WC'_{OX}R_{DS}(V_G-V_T)$, where C'_{OX} is the oxide capacitance per unit area. A mutual conductance measurement gave the field-effect mobility, μ_{FE}. With an assumed series resistance μ and μ_{FE} were constant for $(V_G-V_T) \gtrsim 2V$. A charge-pumping technique[7] gave the surface state density D_{it}.

3. RESULTS

The measurements showed that the excess noise followed a $1/f^{\beta}$ relationship with β close to unity. The magnitude of $\overline{v_n^{\,2}}/V_{DS}^2$ varied greatly with (V_G-V_T) and T. The method of analysis used was to calculate quantities which should be constant over the range of experimental variables if a particular model were correct.

One major model for 1/f noise is that of mobility fluctuations. In its simplest form it is assumed that for all systems

$$S_v/V_{DS}^{\,2} \equiv \overline{v_n^{\,2}}/\Delta f V_{DS}^2 = \alpha/Nf \qquad (1)$$

where N is the total number of free carriers in the specimen and S_v is the spectral density. The quantity α is taken as 2×10^{-3} from the results of many systems. In this model the experimental function

$$H \equiv Nf\, S_v/V_{DS}^{\,2} = C_{OX}(V_G-V_T)\, f\, S_v/eV_{DS}^2 \qquad (2)$$

should be equal to $\alpha = 2 \times 10^{-3}$ and be independent of (V_G-V_T) and T. The results are shown in figure 1. It can be seen that H is very dependent on (V_G-V_T) and hence N since $N = C_{OX}(V_G-V_T)/e \propto (V_G-V_T)$.

In more sophisticated models of the mobility fluctuation the noise is assumed to depend only on the lattice scattering so that the function

$$H' = H(\mu_\ell/\mu)^2 \qquad (3)$$

should be equal to $\alpha = 2 \times 10^{-3}$. Here μ_ℓ is the mobility due to lattice or phonon scattering only. Figure 2 shows H and H' as a function of temperature for the two specimens. The data is taken at very strong inversion at $(V_G-V_T) = 6V$.

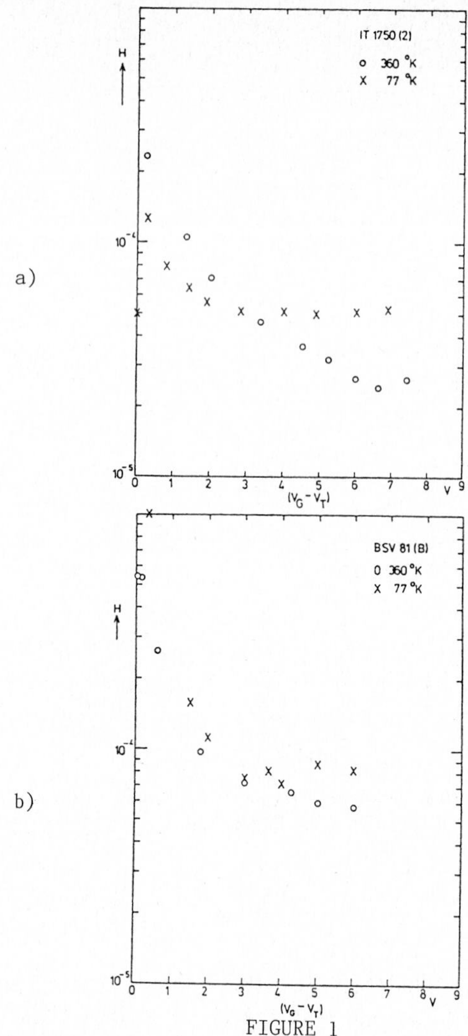

FIGURE 1
The mobility fluctuation parameter H as a function of (V_G-V_T) which is proportional to the number of free carriers N. (a) IT 1750(2), (b) BSV81(B).

The value of (μ_ℓ/μ) was calculated assuming that the lattice scattering has a $T^{-3/2}$ dependence and fitting this to the high-T part of the μ data. In this case $(\mu_\ell/\mu) \sim 1$ above 220K.

The other main model assumes that the noise is caused by a carrier number fluctuation due to oxide and interface states. The number fluctuation per unit area of boundary (one side only) is given by the function of the experimental variables -

$$S'_n \equiv \frac{\overline{n_n^2}}{\Delta f} = \frac{S_v f}{V_{DS}^2}\frac{N^2}{LW} = \frac{S_v f}{V_{DS}^2}\frac{L^3}{We^2}\frac{1}{R_{DS}^2\mu_{FE}^2} \qquad (4)$$

where L and W are the channel length and width.

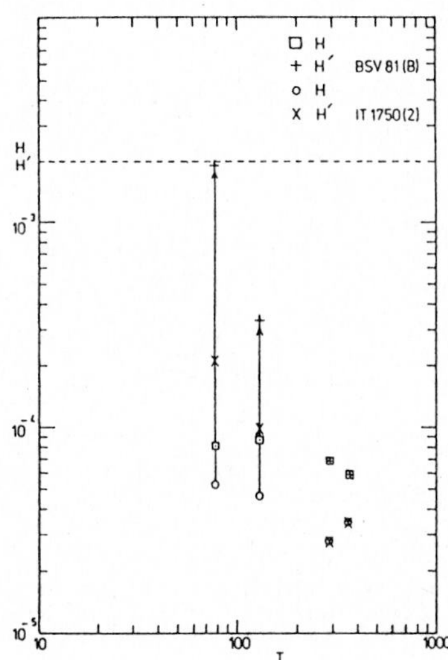

FIGURE 2
The mobility fluctuation parameters H and H' as a function of temperature. The data is taken at (V_G-V_T) = 6V IT 1750(2) BSV81(B)

a) 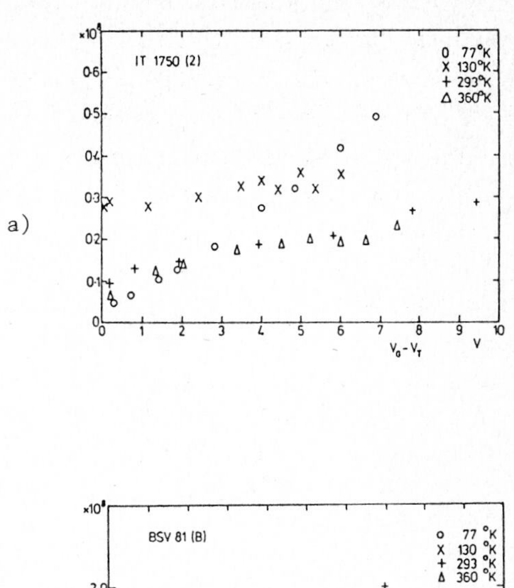

b) 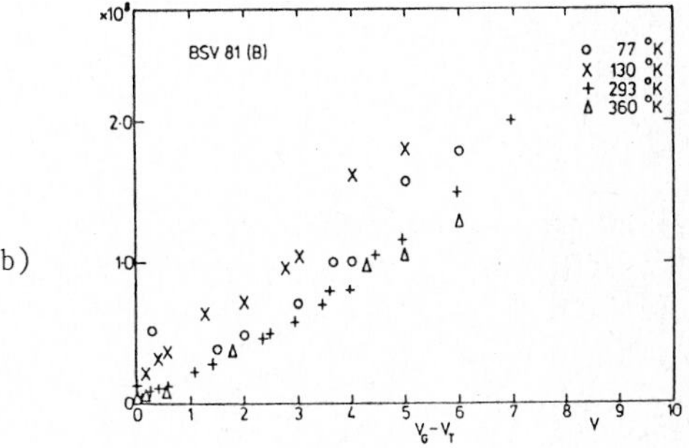

FIGURE 3
The number fluctuation per unit area S'_n as a function of (V_G-V_T) which is proportional to the number of free carriers N. (a) IT 1750(2), (b) BSV81(B)

This quantity is shown in figure 3. The data apparently can be expressed as a constant plus a component proportional to (V_G-V_T) or the number of carriers per unit area $(N' = C'_{OX}(V_G-V_T)/e)$. We take this latter component and calculate its slope divided by the interface state density D_{it} and C'_{OX}/e, shown in figure 4. This quantity represents the number fluctuation per unit area per carrier per surface state.

4. DISCUSSION

The simple interpretation of the mobility fluctuation model shows that the magnitude of H is smaller than predicted and that there is a large (V_G-V_T) dependence. The function H is temperature dependent. The lattice scattering correction increases H' to nearer the value $\alpha = 2 \times 10^{-3}$ but the temperature dependence is very large. This model has been expanded considerably for inversion layers but involves a fitting procedure which is beyond the present analysis[8].

The number fluctuation interpretation produces a much smaller variation with (V_G-V_T) and T. The magnitude of the effect, the number fluctuation per unit area, is about 10^7 compared with 10^3 for JFET channels[5].

There are two components in the number fluctuation; one proportional to the number of carriers and the other a constant. The first is reasonable since each carrier in the two-dimensional channel can fluctuate independently and has an equal probability of interacting with the silicon oxide interface. We find that this component is proportional to the interface trap density as found elsewhere[9].

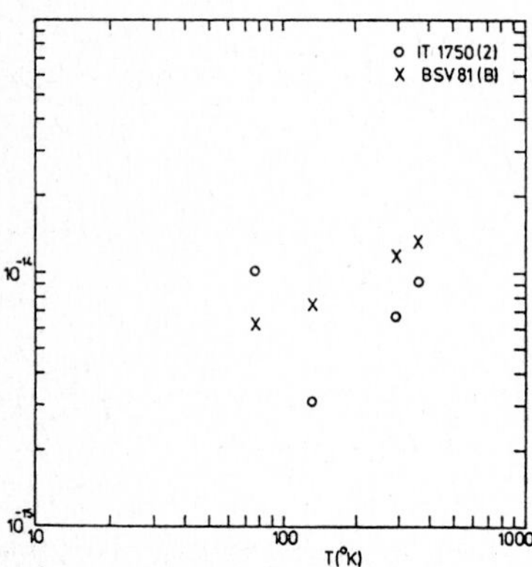

FIGURE 4
The quantity $(e/D_{it}C'_{OX})(dS'_n/d(V_G-V_T))$ against temperature for IT 1750(2) and BSV81(B)

This number fluctuation per unit area per carrier per surface state appears to be temperature independent.

The contribution which is independent of the carrier number is not large and could be due to the fluctuation of the parasitic series resistance. We conclude that these results suggest a number fluctuation model for the 1/f noise.

It should be noted that we have shown an experimental relationship of the form

$$(S_v/V_{DS}{}^2)\ f \approx A/N^2 + B/N \tag{5}$$

where the first term corresponds to a constant number fluctuation (see equation (4)) and the second has a characteristic of independent fluctuations of each carrier. If the first term is ascribed to the parasitic series resistance then either a number fluctuation or a mobility fluctuation model may be appropriate for the data due to the channel itself. We differentiate between these two models by the dependence on the surface state density and temperature.

REFERENCES

1. C.T. Green and B.K. Jones J.Phys.D. 18 77-91 (1985).

2. C.T. Green and B.K. Jones J.Phys.D. (1985) in press.

3. B.K. Jones and J.D.Stocker J.Phys.D. 18 93-102 (1985)

4. G.S. Bhatti and B.K. Jones J.Phys.D. 17 2407-22 (1984).

5. G.S. Bhatti Ph.D. Thesis (1983), University of Lancaster.

6. B.K. Jones and P.C. Russell J.Phys.D. (1985) in press.

7. G. Groeseneken, H.E. Maes, N. Beltran and R.F. de Keersmaecker IEEE Trans. Electron Devices ED31 42-53 (1984).

8. L.K.J. Vandamme Solid State Electron 23 317-23 (1980).

9. H.E. Maes and S.H. Usmani J.Appl.Phys. 54 1937-49 (1983).

NOISE IN PHYSICAL SYSTEMS AND 1/f NOISE - 1985
A. D'Amico and P. Mazzetti (editors)

$\frac{1}{f}$ NOISE IN FERRITES

E.J.P. MAY and H.M. ANIAGYEI

Department of Engineering Science, University of Exeter, Exeter, Devon, EX4 4QF, England.

This work is mainly concerned with the effect of the ambient atmosphere, particularly the humidity on the 1/f noise produced by ferrite resistors - it is shown that this effect which is quite pronounced is due to the fact that ferrites are porous. At least part of the change in noise power is due to change in resistance.

1. INTRODUCTION

Early investigations[1] on $\frac{1}{f}$ noise using graphite paper revealed that the noise depended on the humidity of the atmosphere. It was decided to investigate this effect on a more stable and well defined porous material. Ferrites were selected for this purpose and in order to make comparisons with non-porous materials, a conductive diamond type IIB was chosen. It also became apparent that the resistance of one of the ferrite samples changed with humidity and a study of this was incorporated in this investigation. This afforded an excellent opportunity to examine the Hooge formula.[2,3]

$$\left\langle\left(\frac{\Delta V}{V}\right)^2\right\rangle = \frac{\alpha}{N}\ \frac{\Delta f}{f} \qquad \ldots\ldots\ (1)$$

particularly with regard to the constant α and N, the total number of carriers in the sample.

2. EXPERIMENTAL PROCEDURE

2.1. Preparation of Samples

Two types of ferrites were tested, both manganese-zinc manufactured by Mullard U.K. Ltd. One was ferrox cube grade A4 which has a resistivity of 200 ohm-mm. This was in the form of an I-piece (FX 1106), and is used for transformer cores. The other was a grade A1 with a resistivity of 800 ohm-mm. This was in the shape of a bead and is normally used for decoupling and anti-parasitic purposes. Several methods of connecting electrodes to these samples and the diamond sample were tested and all gave the same results - which were repeatable.

The resistance of the A4 grade sample was measured at room temperature in a normal laboratory atmosphere and found to be 800 ohms (I = 50 μA). The grade A1 sample had a resistance of 6 k ohm (I = 20 μA) under the same conditions.

2.2. Measurements

The noise measurements on the ferrites were made using the circuit shown in Figure 1 because the resistances of the ferrite samples were low.

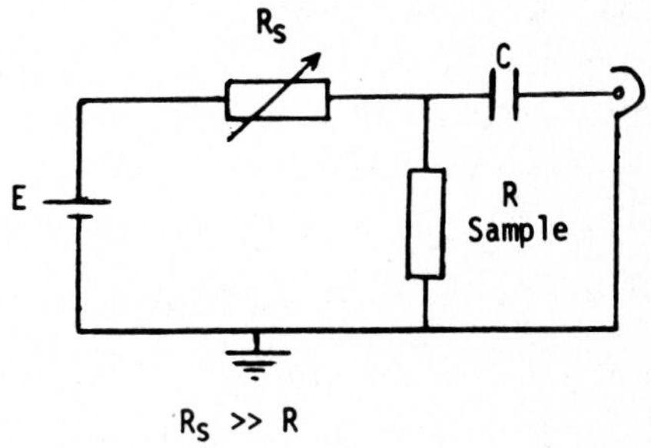

FIGURE 1

Diamond samples, which were of high resistance, were measured using the circuit shown in Figure 2. The bandwidth for all noise measurements was 10Hz and all levels were measured in dBμ.

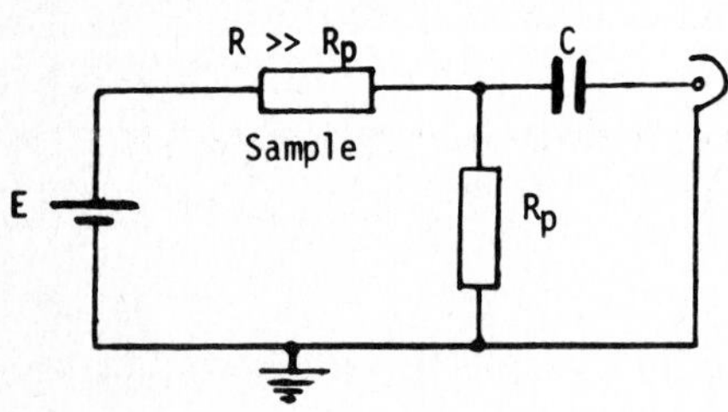

FIGURE 2

The first experiment involved the measurement of $\frac{1}{f}$ noise power at different frequencies for all the samples as shown in Figures 3 and 4.

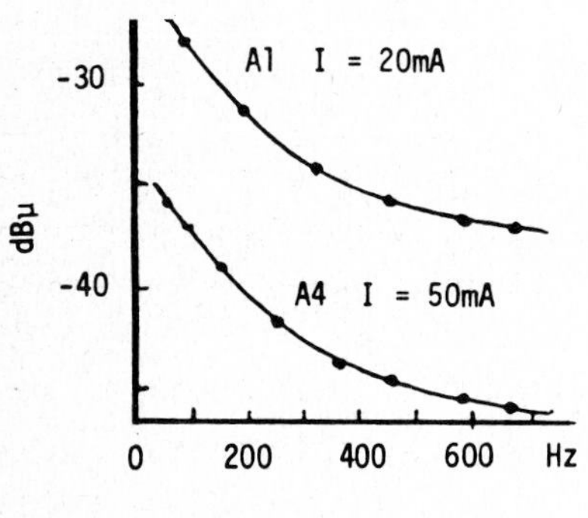

FIGURE 3

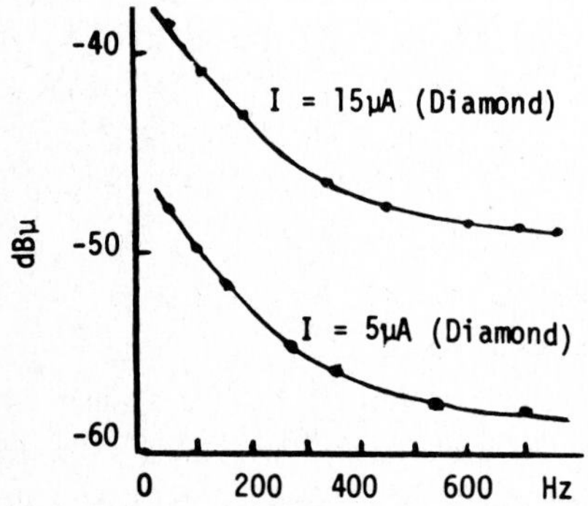

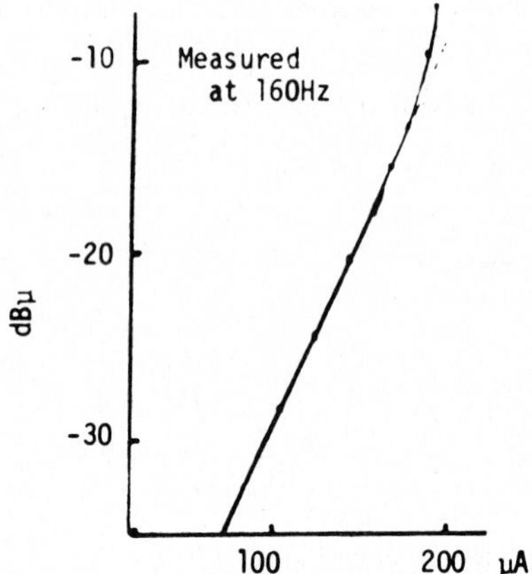

FIGURE 5
Noise power vs current A4 grade

In the second experiment the frequency was kept constant at 160Hz and the $\frac{1}{f}$ noise power was measured versus current for the grade A4 ferrite as shown in Figure 5. Figure 6 shows the corresponding curve for the A1 grade ferrite.

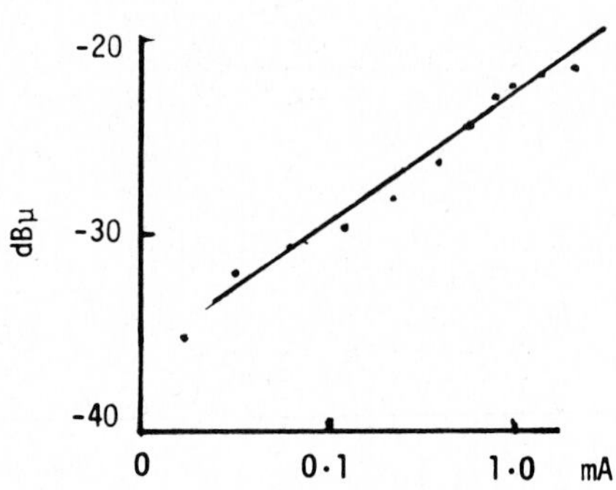

FIGURE 6
Noise power vs current A1 grade

FIGURE 4

All the noise measurements described so far were carried out in a normal laboratory atmosphere (with a humidity of about 54%). In order to investigate the effect of change in ambient the sample was enclosed in a vessel and noise measurements were carried out using pure dry oxygen, argon, nitrogen and dried atmospheric air. There was no appreciable difference between these. The effect of humidity was then examined using the A1 grade sample. Argon was introduced into the chamber and the humidity measured was 4%. The noise power frequency spectrum was measured and plotted in Figure 7. Atmospheric air was then introduced into the chamber and the experiment was repeated. The humidity of the air in the chamber was found to be 54%. These results were also plotted in Figure 7. It will be observed that the noise level increased with % humidity (10·5 dB at the low frequency end).

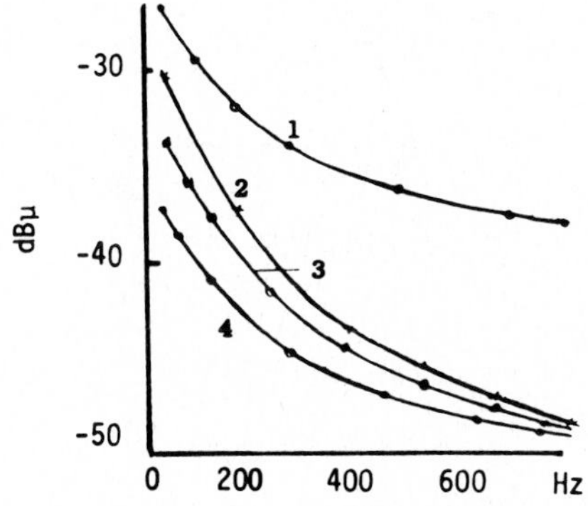

FIGURE 7
Noise power vs frequency
1. 54% humidity
2. Water
3. Glycerine
4. 4% humidity argon

It was decided to examine the noise-frequency characteristics of the ferrite sample when totally immersed in distilled water and glycerine. These results are all shown in Figure 7.

During the measurement of noise power versus current for the A1 ferrite, it was observed that the resistance varied considerably with current. It was, therefore, decided to measure the resistance for different currents in the grade A1 ferrite sample and the resistance versus current is plotted in Figure 8 for several ambient conditions.

All the effects described above using distilled water or humid atmosphere were found to be completely reversible, although in some cases it took a considerable time for the samples to dry out. In the case of the ferrite immersed in water, 12-24 hours were needed before the samples returned to their original resistance values. It was found that changes in the ambient had no effect on the diamond sample.

3. DISCUSSION AND CONCLUSIONS

It is clear from Figures 3 and 4 that ferrite and diamond (type IIB) resistors generate $\frac{1}{f}$ noise. In the case of grade A4 ferrite the noise power versus frequency curve, given in Figure 5, is as one would expect from the Hooge expression

$$< (\Delta V)^2 > = \frac{\alpha}{N} I^2R^2 \frac{\Delta f}{f} \qquad \text{...... (2)}$$

This is because the resistance R does not change with current. In the case of the noise power versus current characteristic for the ferrite grade A1 given in Figure 6, this is not the case. If one considers the room atmosphere case (i.e. 54% humidity) one observes that the increase in noise power is not 20dB for a current change from 0·1 to 1mA but it is only 9dB. The resistance as seen in Figure 8 drops from 4·5k ohm at 0·1mA to 1·9k ohm at 1mA. From the Hooge expression it is seen that this causes a drop in noise power level of 7·5dB making the expected noise power gain 12·5dB. In addition, when the value of the resistance changes from 4·5k ohm to 1·9k ohm one expects an increase in the value of N by a factor of 2·4. This assumes that all the changes in the resistance are due to changes in N. This increase in the value of N would according to equation (2) account for a further 3·7dB loss. Hence the expected noise power change amounts to 8·8dB. This is very close to the 9dB obtained from the actual measurement given in Figure 6.

An other interesting observation is that the noise power level changes with humidity as seen from Figure 7. This change amounts to about -11dB for a change in humidity from 54% to 4%. A prediction from Hooge's expression can be arrived at by considering the change in resistance and the accompanying change in N due to the change in humidity. The change in resistance is given in Figure 8 and for a current of 20μA, the resistance changes from 5·9k ohm to 2·1k ohm. This would give rise to a drop in noise power of 9dB according to the Hooge expression. The odd 2dB discrepancy can be accounted for by the corresponding change in N which brought about the change in resistance.

A main observation is that the changes in noise produced by changing the humidity or immersing the ferrites in distilled water or glycerine took a considerable time to settle - as long as 24 hours, but all changes were reversible. One concludes that the changes were in the bulk of the porous material. It should be noted that no changes occurred in the case of the conducting diamond.

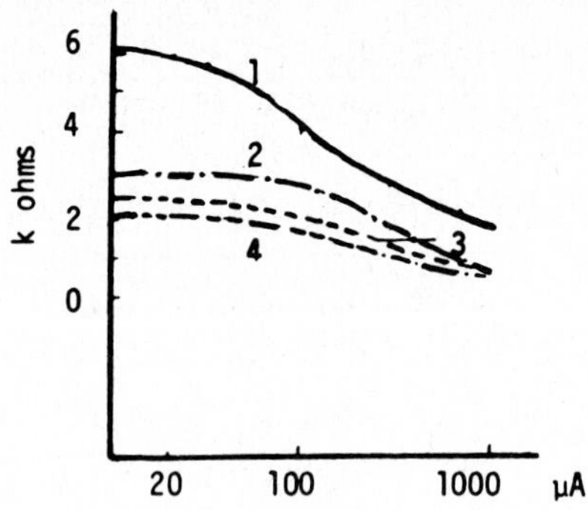

FIGURE 8
Resistance vs current A1 grade for:
1. 54% humidity
2. Water
3. Glycerine
4. 4% humidity argon

REFERENCES

1. H.M. Aniagyei, M.Sc. thesis,(1974) University of Exeter, England.

2. F.N. Hooge, Phys. Lett., 29A (1969) 139.

3. F N. Hooge and A.M.H. Hoppenbrowers, ibid., 29A (1969) 642.

NOISE IN PHYSICAL SYSTEMS AND 1/f NOISE - 1985
A. D'Amico and P. Mazzetti (editors)

TEMPERATURE DEPENDENCE OF 1/f NOISE IN THIN NiCr FILMS

A. BELU-Marian, D. URSUTIU*, S. DUMITRU*, R. MANAILA and A. DEVENYI

Institute of Physics and Technology of Materials Bucharest, P.O.B. MG-7, Romania
*Department of Physics, University of Braşov, Romania

It is shown that the voltage noise in NiCr thin films with different structure and conduction mechanisms seems to correlate with the phonon-carrier interaction.

1. INTRODUCTION

The electrical properties and structure of NiCr films depend on many factors such as film composition, reactive gas partial pressure during deposition and heat treatment after deposition. The purpose of this paper was to find out the experimental correlations between the temperature dependence of the electrical resistivity and that of the voltage noise for NiCr thin films showing an activated or a metallic conduction mechanism, the structure of which was investigated by X-ray diffraction.

2. EXPERIMENTAL

NiCr thin films have been reactively sputtered in Ar with different partial pressures of ethanol, using a d.c. diode system and cathodes with the Ni:Cr ratio 50:50, onto ultrasonically cleaned silica substrates held at $T < 100^oC$. The base pressure before admission of ethanol and argon was less than $1x10^{-3}$ Pa. After deposition the samples were annealed in vacuum at 275^oC.

The Ni:Cr content of the films, as determined by X-ray fluorescence, ranged between 63:37 at % and 70/30 at %. The film thickness, measured using the multiple beam interference technique, was about 100 nm for the samples intended for electrical and voltage noise measurements. Thicker samples were prepared for the X-ray investigation.

The diffraction patterns of the films were taken with monochromatized CuKα radiation. The small thickness of the films and/or preferred orientation phenomena rendered difficult a precise recording of the whole diffraction pattern. Therefore, we measured the profiles of the most intense lines (111) and (200) by a step-scanning procedure, with a counter step of 0.02^o (2θ).

The temperature dependence of the electrical resistivity $\rho(T)$ as well as of the voltage noise were measured on the same samples in a vacuum better than $1x10^{-3}$ Pa with conventional methods, in the temperature range between -150^oC and the annealing temperature.

The noise spectral density $Sv(V^2/Hz)$ versus frequency f was taken in the 8-8000 Hz frequency range. Current densities of $(1-7)x10^3$ A/cm^2 were used and the thermal noise measured in the absence of current was subtracted from the total noise power.

3. RESULTS AND COMMENTS

3.1. Structure

The films consists of crystallites of a γ - NiCr phase with numerous stacking faults along the [111] direction and of a disordered, quasi-amorphous phase (Fig. 1). The latter can be attributed to the relatively thick intercrystallite boundary regions, where a

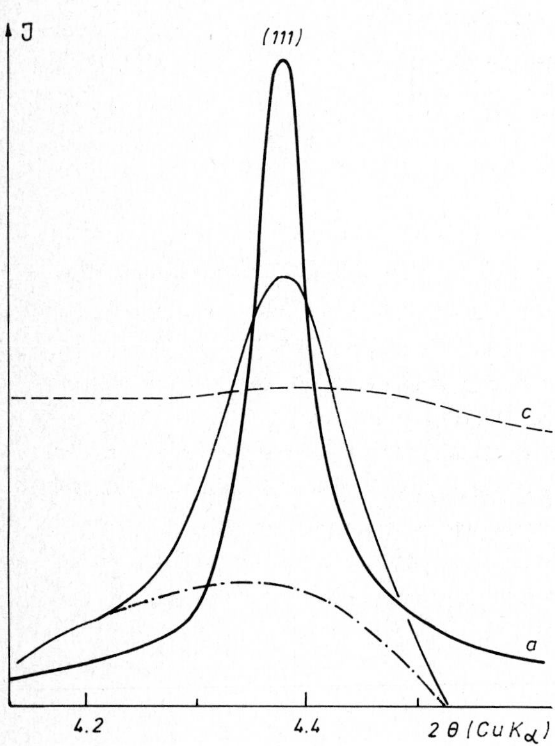

FIGURE 1
Profile of the [111] diffraction maximum for d.c.-sputtered films in ethanol partial pressure of: a:0, b: $4x10^{-3}$ Pa, c: $12x10^{-3}$ Pa.

high concentration of O_2, C and may be excess Cr is expected. The higher the volume concentration of the disordered phase, the higher was found to be the resistivity of the sample.

3.1. Electrical measurements

Figure 2 shows the temperature dependence of the resistivity measured for two samples with very different structures and resistivity at room temperature (ρ_{RT}) values. It can be seen that there is a significant difference in the variation of the resistivity with temperature. Sample 1 with higher ρ, consisting in the disordered phase, exhibits an activated conduction mechanism due to the intercrystallite phase. The $\rho(T)$ dependence is given by $\rho \sim \exp.(A/T^n)$ where n = 1/2. A dependence of that form is found in a wide range of granular metals and disordered systems and can be derived under different assumptions[1].

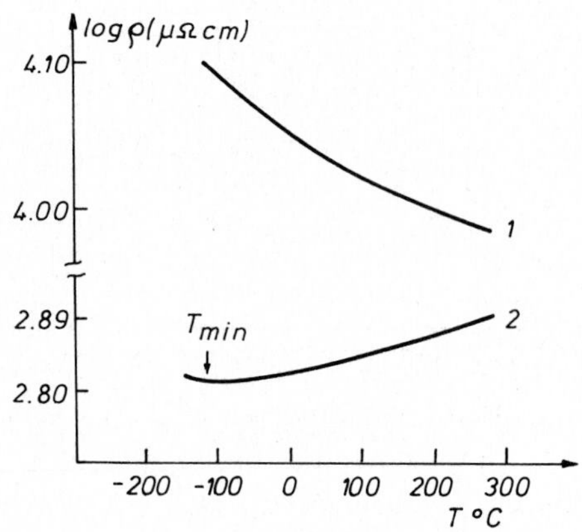

FIGURE 2
Resistivity vs.T measured of two NiCr samples with different structure and ρ_{RT} values.

Sample 2 showing crystallites as well as disordered phase has a minimum in the $\rho(T)$ dependence at T_{min}. This fact suggests the coexistence of two temperature dependend conduction mechanisms, related to the crystallite and inter-crystallite phases, of which the one of higher resistivity prevails at a given T. At $T > T_{min}$ the metallic contribution due to the crystallites prevails in $\rho(T)$ behaviour.

3.3. Voltage noise measurements

A linear dependence was found for the normalized noise spectral density S_V/V^2 (where V is the d.c. voltage across the sample) versus log f at all the measured temperatures, for both samples, despite their showing strikingly different $\rho(T)$ dependences (Fig. 2). This allows us to use the well known phenomenological expression : $S_V/V^2 = C/f^m$ [2]. We should remark the absence of any kinks or bent in the S_V/V^2 vs. f spectrum at all temperatures and m could be determined with a maximum error $\pm$ 0.04. Figure 3 shows m as function of T for both samples. In spite of the experimental points being widely scattered it is evident that m is temperature dependent, increasing with

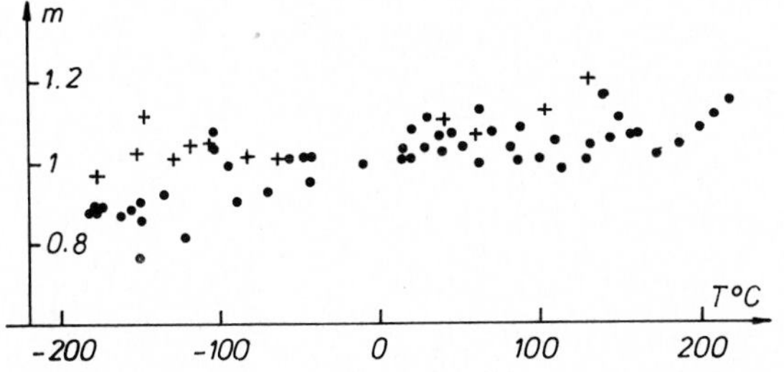

FIGURE 3
Frequency exponent m vs.T • for sample 1, + for sample 2.

temperature and taking values between 0.80 (at lower T) and 1.20 (at higher T). These values of m are still well within the range reported for 1/f noise[3]. A similar temperature dependence of m was found in other materials too [4,5]. The temperature dependence of the 1/f noise parameter C of NiCr samples 1 and 2 is presented in Fig. 4. It can be noticed that there is no direct link between the temperature dependence of the electrical resistivity and that of C (Figs. 2 and 4). The temperature behaviour of C for both samples exhibits maxima and minima. Though, it seems that this structure can be correlated with the fundamental lattice vibration modes[6].

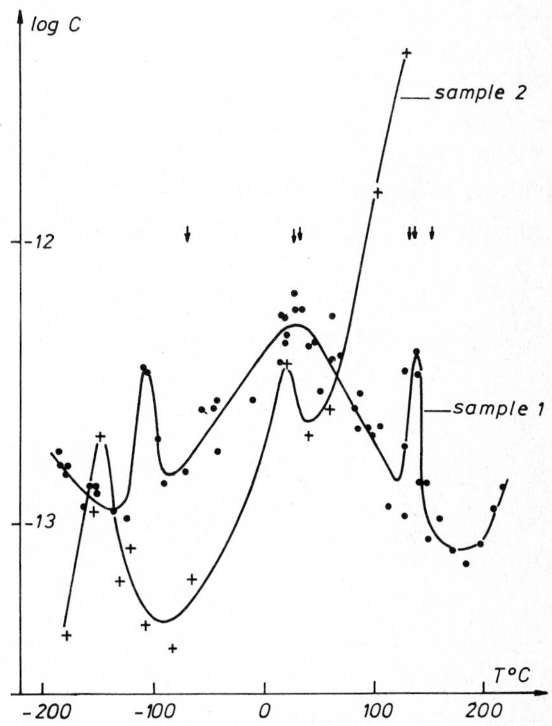

FIGURE 4
Noise parameter C vs.T • for sample 1, + for sample 2. The arrows indicate the temperatures corresponding to the energies of the lattice vibration modes in Ni[7].

The arrows in Fig.4 indicate the temperatures corresponding to the energies of the lattice vibration modes in Ni, obtained by inelastic neutron scattering. We took the lattice vibration modes of Ni because the crystalline lattice of the samples belongs to the same γ-type. Figure 4 shows that there is a good agreement between the temperatures corresponding to the energies of the lattice vibration modes in Ni and maxima occuring at 30^{0}C in samples 1, 2 and 140^{0}C for sample 2. The maximum observed at T = -110^{0} C for sample 1 and T = -150^{0}C for sample 2 cannot be correlated in this way, unless we assume that the lattice vibration mode in Ni, corresponding to T = -70^{0}C, is moved to lower energies in NiCr.

4. CONCLUSIONS

The voltage noise in NiCr thin films with very different structure and conduction mechanism has the same physical origin; it can be described by the empirical expression $S_V/V^2 = C/f^m$.

There is not a common physical origin for the temperature dependence of resistivity and noise.

The common source of 1/f noise in NiCr samples seems to be the phonon-carrier interaction.

REFERENCES

1. B.Abeles, Ping Sheng, M.D.Coutts, Y.Arie, Adv.Phys. 24 (1975) 407.
2. F.N.Hooge, T.G.M.Kleinpenning and L.K.J.Vandamme, Rep.Prog.Phys. 44 (1981) 479.
3. P.Duta, P.M.Horn, Rev.Mod.Phys. 53 (1981) 497.
4. Maria Prudenziati and Bruno Morten Proc.of the Sixth Int.Conf.on Noise in Phys.Systems 1981, Washington.
5. D.M.Fleetwood and N.Giordano Phys.Rev.B 31 (1985) 1157.
6. M.Mihăila, Phys.Lett. 107A (1985) 564.
7. R.J.Birgeneau, J.Cordes, G.Dolling and A.D.B.Woods, Phys.Rev. 136A (1964) 1359.

QUANTUM NOISE - 2

NOISE IN PHYSICAL SYSTEMS AND 1/f NOISE - 1985
A. D'Amico and P. Mazzetti (editors)

COHERENT STATES QUANTUM 1/f NOISE AND THE QUANTUM 1/f EFFECT

Peter H. HANDEL

Physics Department, University of Missouri, St. Louis, MO 63121, USA

An electrically charged particle includes the bare particle and its field. The field has been shown in the last two decades to be in a coherent state, which is not an eigenstate of the Hamiltonian. Consequently, the physical particle is not described by an energy eigenstate, and is therefore not in a stationary state. In this paper we show that the fluctuations arising from this non-stationarity have a 1/f spectral density and affect the ordered, collective, or translational motion of the current carriers. This "coherent" quantum 1/f noise should be present along with the familiar quantum 1/f effect of elementary cross sections and process rates introduced ten years ago, just as the magnetic energy of a biased semiconductor sample coexists with the kinetic energy of the individual, randomly moving, current carriers. The amplitude of the quantum 1/f effect is always the difference of the coherent quantum 1/f noise amplitudes in the "out" and "in" states of the process under consideration and dominates in small samples, while large samples should exhibit the larger coherent quantum 1/f noise.

1. INTRODUCTION

A physical, electrically charged, particle should be described in terms of coherent states of the electromagnetic field, rather than in terms of an eigenstate of the Hamiltonian. This is the conclusion obtained from calculations [1] of the infrared radiative corrections to any process performed both in Fock space (where the energy eigenstates are taken as the basis, and the particle is considered to have a well defined energy) and in the basis of coherent states. Indeed, all infrared divergences drop out already in the calculation of the matrix element of the process considered, as it should be according to the postulates of quantum mechanics, whereas in the Fock space calculation they drop out only a posteriori, in the calculation of the corresponding cross section, or process rate. From a more fundamental mathematical point of view, both the description of charged particles in terms of coherent states of the field, and the undetermined energy, are the consequence of the infinite range of the Coulomb potential [2]. Both the amplitude and the phase of the physical particle's electromagnetic field are well defined, but the energy, i.e. the number of photons associated with this field, is not well defined. The indefinite energy is required by Heisenberg's uncertainty relations, because the coherent states are eigenstates of the annihilation operators, and these do not commute with the Hamiltonian.

A state which is not an eigenstate of the Hamiltonian is nonstationary. This means that we should expect fluctuations in addition to the (Poissonian) shot noise to be present. What kind of fluctuations are these? This question was answered in a previous paper [3]. The additional fluctuations were identified there as 1/f noise with a spectral density of $2\alpha/\pi f$ arising from each electron independently, where $\alpha = 1/137$ is the fine structure constant.

We will briefly derive this result again in the present paper, but we will stress the connection between the coherent quantum 1/f noise and the usual quantum 1/f effect.

2. COHERENT QUANTUM 1/f NOISE

The coherent quantum 1/f noise will be derived again in the three steps: first we consider just a single mode of the electromagnetic field in a coherent state and calculate the autocorrelation function of the fluctuations which arise from its nonstationarity. Then we calculate the amplitude with which this mode is represented in the field of an electron. Finally, we take the product of the autocorrelation functions calculated for all modes with the amplitudes found in the previous step.

Let a mode of the electromagnetic field be characterized by the wave vector q, the angular frequency $\omega = cq$ and the polarization λ. Denoting the variables q and λ simply by q in the labels of the states, we write the coherent state [1-3] of amplitude $|z_q|$ and phase arg z_q in the form

$$|z_q\rangle = \exp[-(1/2)|z_q|^2 \exp[z_q a_q^+]\ |0\rangle \qquad (1)$$

$$= \exp[-(1/2)|z_q|^2] \sum_{n=0}^{\infty} (z_q{}^n)/n!\ |n\rangle.$$

Let us use a representation of the energy eigenstates in terms of Hermite polynomials $H_n(x)$

$$|n\rangle = (2^n n!\sqrt{\pi})^{-1/2} \exp[-x^2/2]\ H_n(x)\ e^{in\ \omega t}. \qquad (2)$$

This yields for the coherent state $|z_q\rangle$ the representation

$$\Psi_q(x) = \exp[-(1/2)|z_q|^2]\exp[-x^2/2]$$

$$\sum_{n=0}^{\infty} \{[z_q e^{i\omega t}]^n/[n!(2^n \sqrt{\omega})]^{1/2}\} H_n(x)$$

$$= \exp[-(1/2)|z_q|^2]\exp[-x^2/2]$$

$$\exp[-z^2 e^{-2i\omega t} + 2xze^{i\omega t}]. \qquad (3)$$

In the last form the generating function of the Hermite polynomials was used [3]. The corresponding autocorrelation function of the probability density function, obtained by averaging over the time t or the phase of z_q, is, for $|z_q|<<1$,

$$Pq(\tau,x) = |\Psi_q|^2_t |\Psi q|^2_{t+\tau} > \qquad (4)$$

$$= \{1 + 8x^2|z_q|^2[1 + \cos\omega\tau] - 2|z_q|^2\} \exp[-x^2/2].$$

Integrating over x from $-\infty$ to ∞, we find the autocorrelation function

$$A^1(\tau) = (2\pi)^{-1/2}\{1 + 2|z_q|^2\cos\omega\tau\} \ . \qquad (5)$$

This result shows that the probability contains a constant background with small superposed oscillations of frequency ω. Physically, the small oscillations in the total probability describe a particle which has been emitted, or created, with a slightly oscillating rate, and which is more likely to be found in a measurement at a certain time that at other times in the same place. Note that for $z_q = 0$ the coherent state becomes the ground state of the oscillator which is also an energy eigenstate, and therefore stationary and free of oscillations.

We now determine the amplitude z_q with which the field mode q is represented in the physical electron. One way to do this [3] is to let a bare particle dress itself through its interaction with the electromagnetic field, i.e. by performing first order perturbation theory with the interaction Hamiltonian

$$H' = A_\mu j^\mu = -(e/c)\vec{v}\cdot\vec{A} + e\phi \ , \qquad (6)$$

where A is the vector potential and ϕ the scalar electric potential. Another way is to Fourier expand the electric potential $e/4\pi r$ of a charged particle in a box of volume V. In both ways we obtain [3]

$$|z_q|^2 = (e/q)^2(hcqV)^{-1} \ . \qquad (7)$$

Consider now all modes of the electromagnetic field; we obtain from the single - mode result of Eq. (5)

$$A(\tau) = C\ \Pi_q\{1 + 2|z_q|^2\cos\omega_q\tau\}$$

$$= C\{1 + \Sigma_q 2|z_q|2\cos\omega_q\tau\}$$

$$= C\{1 + 2V/2^3\pi^3)\int d^3q|z_q|^2\cos\omega_q\tau\} \qquad (8)$$

Here we have again used the smallness of z_q and we have introduced a constant C. Using Eq. (7) we obtain

$$A(\tau) = C\{1 + 2(V/2^3\pi^3)(4\pi/V)(e^2/2hc)\int(dq/q)\cos\omega_q\tau\}$$

$$= C\{1 + 2(\alpha/\pi)\int\cos(\omega\tau)d\omega/\omega \ . \qquad (9)$$

Here $\alpha = e^2/4\pi hc$ is the fine structure constant 1/137. The first term in curly brackets is unity and represents the constant background, or the d.c. part. The autocorrelation function for the relative, or fractional density fluctuations, or for current density fluctuations in the beam of charged particles is obtained therefore by dividing the second term in curly brackets by the first term. The constant C drops out when the fractional fluctuations are considered. According to the Wiener-Khintchine theorem, the coefficient of $\cos\omega\tau$ is the spectral density of the fluctuations, $S_{|\Psi|^2}$, or S_j for the current density $J = e(k/m)|\Psi|^2$

$$\frac{S_{|\Psi|^2}}{<|\Psi|^2>^2} = \frac{S_j}{<j>^2} = 2\frac{\alpha}{\pi}\frac{1}{fN} = 4.6\cdot10^{-3}f^{-1}N^{-1}. \qquad (10)$$

Here we have included the total number N of charged particles which are observed simultaneously in the denominator, because the noise contributions from each particle are independent.

This result is related to the well known quantum 1/f effect [4-9]. If a beam of charged particles is scattered, passes from one medium into another medium (e.g. at contacts), is emitted, or is involved in any kind of transitions, the amplitudes z_q which describe its field will change. Then, even if the initial state was prepared to have a well-determined energy, the final state will have an indefinite energy, with an uncertainty determined by the difference between the new and old z_q amplitudes, Δz_q. This, however, is just the bremsstrahlung amplitude Δz_q. We thus regain the familiar quantum 1/f effect, according to which the small energy looses from bremsstahlung of infraquanta yield a final state of indefinite energy, and therefore lead to fluctuations of the process rate, or cross section, of the process in which the electrons have participated, and which has occasioned the bremsstrahlung in the first place. The calculation of piezoelectric 1/f noise [10] which deals with phonons as infraquanta, was phrased in terms of the coherent field amplitudes z_q for the first time, although it is concerned only with the usual quantum 1/f effect. It has α substituted by the piezoelectric coupling constant g.

3. CONNECTION WITH THE USUAL QUANTUM 1/f EFFECT

The assumptions included in the derivation of the above coherent quantum 1/f noise result are:

1 - The "bare particle" does not have compensating energy fluctuations which could cancel the fluctuations present in the field. The latter are due to the interaction with distant charges, and have nothing to do with the bare particle. Therefore, this assumption is quite reasonable.

2 - The experimental conditions do not alter the physical definition of the charged particle as a bare particle dressed by a coherent state field. This second assumption depends on the experimental conditions.

One way to understand this second assumption is based on the spatial extent of the beam of particles or of the physical sample containing charged particles, and is specifically based on the number of particles per unit length of the sample. According to this model, the coherent state in a conductor or semiconductor sample is the result of the experimental efforts directed towards establishing a steady and constant current, and is therefore the state defined by the collective motion, i.e. by the drift of the current carriers. It is expressed in the Hamiltonian by the magnetic energy E_m, per unit length, of the current carried by the sample. In very small samples or electronic devices, this magnetic energy

$$E_m = \int (B^2/8\pi) d^3x = [nevS/c]^2 \ln(R/r) \qquad (11)$$

is much smaller than the total kinetic energy E_k of the drift motion of the individual carriers

$$E_k = N'mv^2/2 = nSmv^2/2 = E_m/s. \qquad (12)$$

Here we have introduced the magnetic field B, the carrier concentration n, the cross sec-electric circuit, and the "coherent ratio"

$$s = E_m/E_k = 2ne^2S/mc^2 \ln(R/r) = 2e^2N'/mc^2, \qquad (13)$$

where N' = nS is the number of carriers per unit length of the sample and the natural logarithm ln(R/r) has been approximated by one in the last form. We expect the observed spectral density of the mobility fluctuations to be given by a relation of the form

$$(1/\mu^2)S_\mu(f) = (1/1+s)[2\alpha A/fN] + (s/1+s)[2\alpha/\pi fN] \qquad (14)$$

which can be interpreted as an expression of the effective Hooge constant if the number N of carriers in the (homogeneous) sample is brought to the numerator of the left hand side. Eq. (14) needs to be tested experimentally. In this equation $A = 2(\Delta v/c)^2/3\pi$ is the usual nonrelavistic expression of the infrared exponent, present in the familiar form of the quantum 1/f effect [4-9]. This equation does not include the quantum 1/f noise in the surface and bulk recombination cross sections, in the surface and bulk trapping centers, in tunneling and injection processes, in emission or in transitions between two solids.

Note that the coherence ratio s introduced here equals the unity for the critical value N' = N" = $2\cdot 10^{12}$/cm., e.g. for a cross section S = $2\cdot 10^{-4}$cm^2 of the sample when n = 10^{16}. For larger samples with N' >> N"/A only the second term remains in Eq. (14). We hope that an expression silimar to Eq. (14) will allow us to extend the present good agreement between theory and experiment to the case of large semiconductor samples [11-14].

ACKNOWLEDGEMENT

This work was supported by the Air Force Office of Scientific Research and by the U.S. Army NVEOL.

REFERENCES

[1] Chung, V., Infrared divergence in quantum electrodynamics, Phys.Rev. 140B (1965) 1110-1122.
[2] Jauch, J.M. and Rohrlich, F., The Theory of Photons and Electrons (Springer, Heidelberg, 1976).
[3] Handel, P.H., Any particle represented by a coherent state exhibits 1/f noise in Noise in Physical Systems and 1/f Noise, M. Savelli, G. Lecoy, and J.P. Nougier (Eds.), Elsevier Science Publ. B.V., 1983.
[4] Handel, P.H., 1/f Noise - an "infrared" phenomenon, Phys.Rev.Lett. 34 (1975) 1492-1495.
[5] Handel, P.H., Quantum approach to 1/f noise, Phys.Rev.A22 (1980) 745-757.
[6] Sherif, T.S. and Handel, P.H., Unified treatment of diffraction and 1/f noise, Phys.Rev.A26 (1982) 596-602.
[7] Handel, P.H. and Sherif, T.S., Direct calculation of the Schroedinger field which generates quantum 1/f noise in Noise in Physical Systems and 1/f Noise, M. Savelli, G. Lecoy, and J.P. Nougier (Eds.), Elsevier Science Publ. B.V., 1983.
[8] Handel, P.H. and Wolf, D., Characteristic functional of quantum 1/f noise, Phys.Rev.A26 (1982) 3727-3730.
[9] Van Vliet, K.M., Handel, P.H., and Van der Ziel, A., Superstatistical emission noise, Physica 108A (1981) 511-513.
[10] Handel, P.H., and Musha, T., Quantum 1/f noise from piezoelectric coupling, in Noise in Physical Systems and 1/f Noise, M. Savelli, G. Lecoy, and J.P. Nougier (Eds.), Elsevier Science Publ. B.V., 1983.
[11] van der Ziel, A., Handel, P.H., Zhu, X.C., and Duh, K.H., A theory of the Hooge parameters of solid state devices, IEEE Transactions ED-32, 667-671 (1985).
[12] Kilmer, J., van Vliet, C.M., Bosman, G., van der Ziel, A. and Handel, P.H., Evidence of electromagnetic quantum 1/f noise found in gold films, Phys. Status Solidi 121B, 429-432 (1984).
[13] van der Ziel, A. and Handel, P.H., 1/f Noise in n^+p diodes, IEEE Transactions, ED-September or October 1985.
[14] van der Ziel, A., Invited paper in this volume.

NOISE IN PHYSICAL SYSTEMS AND 1/f NOISE - 1985
A. D'Amico and P. Mazzetti (editors)

1/F NOISE IN ALPHA-PARTICLE RADIOACTIVE DECAY OF $_{94}Pu^{239}$, $_{95}Am^{241}$, AND $_{96}Cm^{244}$

G.S. KOUSIK, C.M. VAN VLIET, G. BOSMAN, W.H. ELLIS and E.E. CARROLL

Electrical Engineering and Nuclear Science Departments, U. of Florida, Gainesville, FL 32611, USA

Data obtained from extensive measurements on counting techniques for alpha particle radioactive decay from a source containing $_{94}Pu^{239}$, $_{95}Am^{241}$ and $_{96}Cm^{244}$ are presented. These data show that the statistics are non-Poissonian for large counting times (of the order of 1000 minutes) contrary to the popular belief that alpha decay is an example of Poissonian statistics. Measurements of the Allan variance indicated the presence of a slow Lorentzian flicker noise and of 1/f noise, and the magnitude of the noise for large counting times is considerably larger than that predicted by Poissonian statistics.

1. INTRODUCTION

Heavy nuclei undergoing spontaneous radioactive decay emit monoenergetic alpha particles. This emission process is a charged particle scattering event and must be accompanied by soft photon emission. The quantum theory of 1/f noise[1] applies to this situation, and one must expect a 1/f-type fluctuation in the particle current, i.e., in the counting statistics. In addition there may be Lorentzian flicker noise. MacDonald's theorem is the main link between counting statistics and particle current noise. For 1/f noise this theorem is not applicable because the integral diverges.

A useful concept to study counting statistics is the "Allan variance," and the Allan variance theorem that relates the Allan variance of the total number of counts to the noise spectral density of the counting rate[2,3]. We assume that the spectral density of the counting rate m, $S_m(\omega)$ contains shot noise, 1/f noise and Lorentzian noise, i.e.,

$$S_m(\omega) = 2m_0 + \frac{2\pi C m_0^2}{|\omega|} + 4B\frac{\gamma}{\gamma^2+\omega^2}, \qquad (1)$$

where γ is the reciprocal of a time constant. Then the Allan variance, using the Allan variance theorem

$$\sigma^{A2}_{M_T} = \frac{4}{\pi}\int_0^\infty \frac{S_m(\omega)}{\omega^2}\sin^4(\omega T/2)d\omega, \qquad (2)$$

is found to be

$$\sigma^{A2}_{M_T} = m_0T+2Cm_0^2T^2\ell n2+\frac{B}{\gamma^2}(4e^{-\gamma T}-e^{-2\gamma T}+2\gamma T-3), \qquad (3)$$

where $M_T = mT$ is the count in an interval T. Defining the relative Allan variance as $\sigma^{A2}_{M_T}/\langle M_T\rangle^2$, we have

$$R(T) = \frac{\sigma^{A2}_{M_T}}{\langle M_T\rangle^2} = \frac{1}{m_0T} + 2C\ell n2 + \frac{B}{m_0^2T^2\gamma^2}(4e^{-\gamma T}- e^{-2\gamma T} + 2\gamma T - 3). \qquad (4)$$

For short-time intervals the term $1/m_0T$ (due to Poissonian statistics) dominates and R(T) is proportional to 1/T. When T is long enough and there is no Lorentzian noise, then R(T) approaches $2C\ell n2$, a constant called the "flicker floor." In the case when Lorentzian noise dominates (T large enough so that shot noise is negligible), we find for long-time intervals that the relative Allan variance does not show a flicker floor but increases beyond the flicker floor. Expanding the third term of eq. (4) to three orders in γT, we find

$$R(T) \simeq \frac{2B\gamma T}{3m_0^2} \quad (\gamma T << 1), \qquad (5)$$

indicating that R(T) is proportional to T. If T becomes very large, i.e., $\gamma T >> 1$, then

$$R(T) \simeq \frac{2B}{m_0^2 \gamma T} \qquad (6)$$

indicating a decrease as 1/T. Of course, if one could measure for very long time intervals, one would expect to see the flicker floor due to 1/f noise. The value of the flicker floor given by the quantum theory of 1/f noise requires a detailed quantum mechanical description of the emission process.

2. EXPERIMENTAL METHOD

The block diagram of the counting system is shown in Fig. 1. The source of alpha particles consists of $_{94}Pu^{239}$, $_{95}Am^{241}$, and $_{96}Cm^{244}$, emitting alpha particles of three different energies. The essential data for this source are given in Table 1. The present setup is similar to that used by Gong[3] except for the silicon detector being reverse biased at 90 volts using batteries.

Table 1

Source	Alpha Particle Energy (MeV)
$_{94}Pu^{239}$	5.155
$_{95}Am^{241}$	5.486
$_{96}Cm^{244}$	5.805

A typical full energy spectrum is shown in Fig. 2. The three main peaks correspond to the alpha particles emitted from the three sources. sources. The total number of counts M_T is taken under a fixed portion of the full spectrum for each peak. The systematic range used in this experiment was from peak channel - FWHM × 6 to peak channel + FWHM × 2, where FWHM is the full width at half maximum.

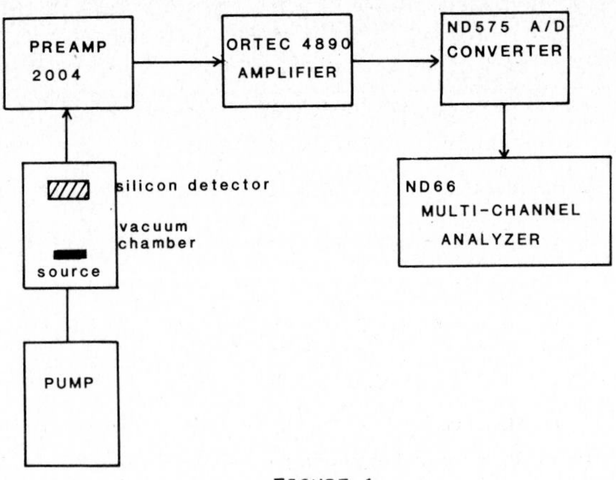

FIGURE 1
Experimental setup.

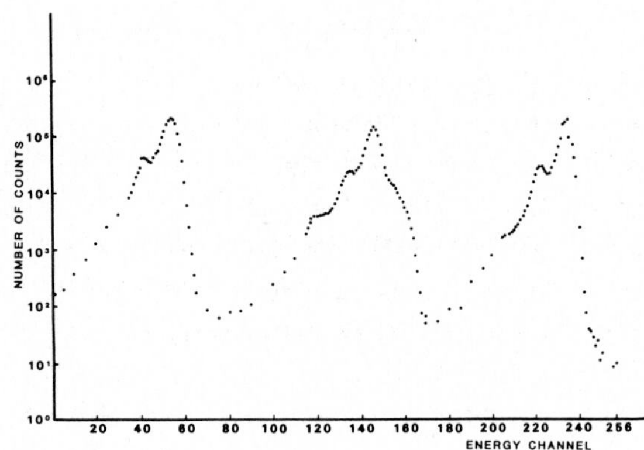

FIGURE 2
Energy spectrum

The total count M_T in each region can be read directly from the memory of the ND66 (multichannel analyzer) and the Allan variance calculated using

$$\sigma^{A2}_{M_T} = \frac{1}{2(N-1)} \sum_{i=1}^{N-1} [M_T^{(i)} - M_T^{(i+1)}]^2, \qquad (7)$$

and

$$\langle M_T \rangle = \frac{1}{N} \sum_{i=1}^{N} M_T^{(i)}, \qquad (8)$$

where N is the number of intervals. Part of the data was found by the "add-up" technique described in Gong[3], where, for example, $M_{4000}^{(1)}$ was taken equal to $M^{(1)}_{2000} + M^{(2)}_{2000}$ and $M^{(2)}_{4000}$ equal to $M^{(3)}_{2000} + M^{(4)}_{2000}$, etc. This is useful for long-time intervals and has been shown to give very little error[3]. This method was also used for $M^{(i)}_{400}$.

3. DISCUSSION OF RESULTS

The relative Allan variance itself fluctuates with the number of counting intervals used. It converges to the correct value only if N, the number of counting intervals, is sufficiently large; we found that N should be 50-100.

Results are shown in Fig. 3, where $\langle R_n(T)\rangle$ is plotted against 1/T, and Figs. 4 and 5, where $\langle \sigma^{A2}_{M_T}\rangle$ is plotted against T. As seen from the relative Allan variance and Allan variance, all three sources show Poissonian statistics, i.e. shot noise, for short time intervals (up to about 200 minutes). The $_{94}Pu^{239}$ source shows Poissonian statistics for the interval lengths up to about 1000 minutes. For counting intervals longer than 1000 minutes, $_{94}Pu^{239}$ source shows excess noise, i.e., in excess of the shot noise levels. The relative Allan variance (Figure 3) shows a flicker floor, indicative of the presence of 1/f noise. The Allan variance (Fig. 4) indicates the presence of 1/f noise, since it is proportional to T^2 for long time intervals.

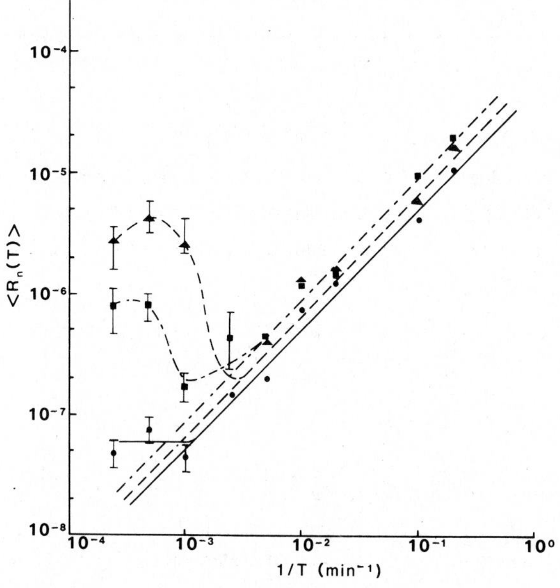

FIGURE 3
Relative Allan variance vs. 1/T $_{94}Pu^{239}$; $_{95}Am^{241}$; $_{96}Cm^{244}$.

The level of the flicker floor found for the $_{94}Pu^{239}$ source is the same order of magnitude as that found for $_{95}Am^{241}$ by J. Gong[3], but slightly smaller. On the other hand, the $_{95}Am^{241}$ and $_{96}Cm^{244}$ sources show a linear dependence on T (for $\gamma T \ll 1$) for the relative Allan variance (Fig. 3) and T^3 dependence for the Allan variance (Fig. 5), indicative of Lorentzian flicker noise.

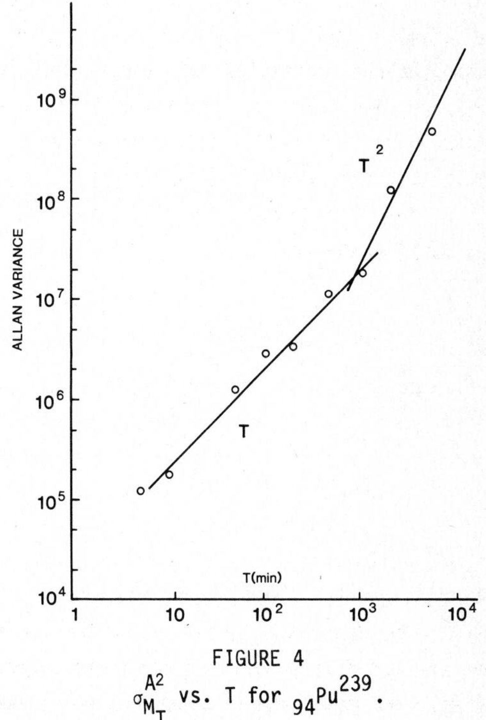

FIGURE 4
$\sigma^{A2}_{M_T}$ vs. T for $_{94}Pu^{239}$.

The Lorentzian flicker noise dominates $\langle R_n(T)\rangle$ of the $_{95}Am^{241}$ and $_{96}Cm^{244}$ sources for large T. We can estimate γ for each source from the maximum $\langle R_n(T)\rangle$ of each source. Differentiating the third term of (4) with respect to T and setting the result equal to zero, we have

$$e^{-2\gamma T'}(1+\gamma T')-2e^{-\gamma T'}(2+\gamma T')-\gamma T'+3 = 0; \tag{9}$$

T' is where $<R_n(T)>$ has its maximum value.

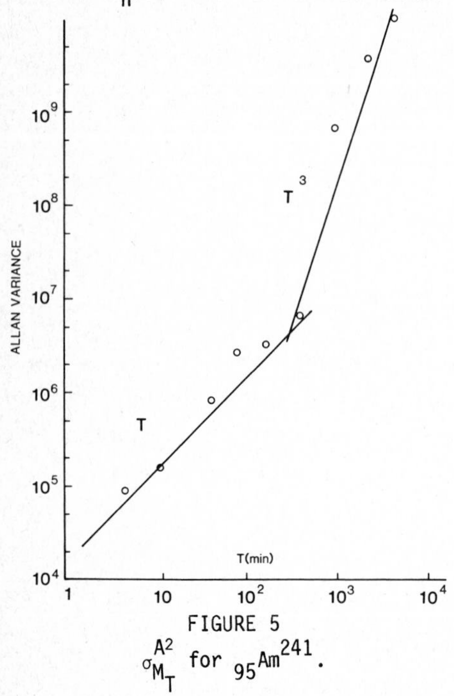

FIGURE 5

$\sigma^{A2}_{M_T}$ for $_{95}Am^{241}$.

The value of $\gamma T' \approx 1.85$ satisfies this equation. Therefore, from Fig. 3 we can estimate γ (or the time constant $1/\gamma$ of the Lorentzian flicker noise) for $_{95}Am^{241}$ to be 9.25×10^{-4} min^{-1} (for T' ~ 2000 minutes) or the time constant ($1/\gamma$) of about 18 hours. Similarly, for $_{96}Cm^{244}$ source T' is approximately 3000 minutes, and therefore γ is about 6×10^{-4} or $1/\gamma$ is approximately 27 hours. The origin of the Lorentzian flicker noise in these two sources and the long time constants found above is as yet unknown.

The flicker floor for $_{94}Pu^{239}$ is about 0.5×10^{-7}, whereas for another americium source $_{95}Am^{241}$, Gong[3] found a flicker floor of 10^{-7}. In Gong's thesis[3] it was indicated that the Am flicker floor is of order of magnitude as expected from the quantum 1/f noise theory. The Pu flicker floor should be lower, since its energy is lower (see Table 1). This is found to be the case.

The pulse height produced by the detector when an alpha particle is detected depends on the bias voltage. If the bias voltage is unstable, then alpha particles with the same energy coming at different times will be recorded in different energy channels. The Eveready batteries of 90 volts each were used and the spectrum was found to be stable. Also, since a fixed portion of the full spectrum is counted, the particles with energies around the boundaries falling inside or outside the boundaries could contribute to the fluctuations. The boundaries were so chosen as to cover the entire peak in each of the three cases. This ensures that the alpha particles with energies near the peaks were always counted, and those near the boundaries were so few in number that they didn't contribute significantly to the counts.

In view of the fact that the above-mentioned contributions to fluctuations in the counts (other than processes within the sources) were insignificant, the data from the counting experiment indicate the presence of excess noise in alpha-particle counting statistics for long time intervals. This could be of either the Lorentzian flicker or 1/f noise types.

References

1. P.H. Handel, Phys. Rev. Lett. 34(24) (1975) 1492.

2. C.M. Van Vliet and P.H. Handel, Physica 113A (1982) 261.

3. J. Gong, Noise associated with electron statistics in avalanche photodiodes and emission statistics of α-particles, Ph.D. dissertation, U. of Florida, Gainesville, FL (1983).

NOISE IN PHYSICAL SYSTEMS AND 1/f NOISE - 1985
A. D'Amico and P. Mazzetti (editors)

QUANTUM 1/F NOISE IN SOLID STATE SCATTERING, RECOMBINATION, TRAPPING AND INJECTION PROCESSES

Peter H. HANDEL

Physics Department, University of Missouri, St. Louis, MO 63121, USA

C.M. VAN VLIET

University of Montreal, Canada, and University of Florida-Gainesville, USA

The unified description of quantum 1/f noise in all elementary process rates and cross sections of electric transport processes in solids presented in 1982 is further developed in the present paper. 1/f fluctuations in the mobility, diffusion constant, surface and bulk recombination speed, tunneling and trapping rate are all derived from the same fundamental formula in terms of the fine structure constant.

1. INTRODUCTION

Ten years ago, in December 1984, quantum 1/f noise was first recognized as a fundamental property of physical cross sections and process rates.[1] Therefore, it will also be present in kinetic processes such as electric transport in semiconductors and metals. The purpose of this contribution is to estimate quantum 1/f noise in electric transport phenomena.

The spectral density of fractional cross section fluctuations of electric electromagnetic type is

$$\sigma^{-2}S_{\sigma}(f) = (4\alpha/3\pi fN)(\Delta\vec{v}/c)^2 , \qquad (1)$$

where $\Delta\vec{v}$ is the velocity change of the carriers in the process whose cross section we consider, N is the effective number of carriers testing the cross section simultaneously, α the fine structure constant, and c the speed of light.

Therefore[2] we expect to see the highest fractional 1/f noise in umklapp scattering which has the largest velocity change. The second largest fractional fluctuations should be registered by inter-valley scattering in indirect gap semiconductors, followed by normal phonon scattering and optical phonon scattering. Finally, the lowest 1/f noise is associated with the (Conwell - Weiskopf) scattering on ionized impurities, which is low-angle scattering par excelence. These remarks apply only to the elementary processes mentioned; to find the magnitudes of the actual noise contribution in a sample we need to multiply also with a factor which indicates to what extent the scattering process considered is represented in the reciprocal mobility of the solid at hand.

From

$$1/\mu = \sum_i 1/\mu_i \tag{2}$$

we obtain

$$\mu^{-2}S_\mu(f) = \sum_i (\mu/\mu_i)^2 \mu_i^{-2} \; S_{\mu_i}(f) \; . \tag{3}$$

2. SCATTERING PROCESSES

For umklapp scattering

$$|\vec{\Delta v}| = h/am^* \; , \tag{4}$$

where h is Plack's constant, a the lattice constant, and m^* the effective mass of the carriers. This yields

$$\vec{\mu}_u S_{\mu_u}(f) = (4\alpha/3\pi fN) \; [h/m^*ac]^2 \tag{5}$$

as a first approximation for umklapp scattering.

For inter-valley scattering

$$|\vec{\Delta v}| = h/2am^* \; , \tag{6}$$

which yields as a first approximation

$$\mu_i^{-2}S_{\mu_i}(f) = (4\alpha/3\pi fN) \; [h/2m^*ac]^2 \tag{7}$$

Considering an average momentum exchange which corresponds to the thermal energy, we obtain for normal phonon scattering

$$\overline{|\vec{\Delta v}|} = 6kT/m^*)^{1/2} \tag{8}$$

at temperature T, and similarly

$$\mu_n^{-2}S_{\mu_n}(f) = (4\alpha/3\pi fN) \; [6kT/m^*c^2]. \tag{9}$$

Let f(T) be the mobility ratio applicable for umklapp processes, f' the ratio for normal phonon scattering, and g(T) the similar ratio prescribed by Eq. (3) for inter-valley scattering. Then, if we neglect the quantum 1/f noise from scattering on impurities and defects, we obtain

$$\mu^{-2}S_\mu(f) = (4\alpha/3\pi fN)\{(h/am^*c)^2f^2 + (h/2am^*c)^2g^2 + 6kT/m^*c^2)(1 - f^2 - g)^2\} \; . \tag{10}$$

To get a rough approximation, we write for metals $f = \exp(-T_D/8T)$ and for semiconductors $f = \exp(-T_D/4T)$, where T_D is the Debye temperature. The direct contribution of electronic umklapp processes to the resistivity of semiconductors may be nebligible, but for 1/f noise it may not be negligible, as our calculations show, because of the additional $(\vec{\Delta v})^2$ factor in the quantum 1/f noise formula, and because of the association of n - processes with umklapp processes (indirect electron-phonon umklapp processes).

The mobility fluctuations derived above induce similar fluctuations in the diffusion constant $D = \mu kT/e$. The above expressions involve averages of the velocities of the carriers over the first Brillouin zone in semiconductors, or over the part below the Fermi surface in metals and degenerate semiconductors. They also involve integrals over the phonon wave vectors, and will be given in detail in the paper with Van Vliet and Kousik in this volume, in terms of the scattering fluctuation spectrum

$$S_w(k_1k_2,k'k'') = 2\alpha(A_{12}/f)\langle w_{k_1}k'\rangle\langle w_{k_2}k''\rangle \quad (11)$$

with

$$2\,A_{12} = (4\alpha/3\pi)|\vec{\Delta v}_1||\vec{\Delta v}_2|/c^2 \quad (12)$$

where the velocity changes $\vec{\Delta v}_1$ and $\vec{\Delta v}_2$ are given by q_1/m^* and q_2/m^* in terms of phonon wave vectors.

3. RECOMBINATION, TRAPPING AND INJECTION.

For recombination and generation processes, $|\vec{\Delta v}| = |\vec{v}-0| = (3kT/m^*)^{1/2}$, because the final velocity of the carriers is zero. In abrupt junctions at a bias V we have

$$|\vec{\Delta v}| = [(Vdiff-V+U)+3kT]^{1/2}(e/m)^{1/2}\ , \quad (13)$$

i.e. we take the acceleration in the built-in junction potential V_{diff} and in the surface potential jump U into account for injection processes. The corresponding spectral density of fractional cross section fluctuations is

$$\sigma^{-2}S_\sigma(f) = 4\alpha/3\pi Nf)(3kT/m^*c^2) \quad (14).$$

Similar relations apply to traps. In both cases U is applicable only for surface recombination, and for surface traps. Finally, injection-extraction noise includes a contribution which describes the generation of shallow electron-hole pairs in the Fermi surface of the electrodes contacting the semiconductor. Most of the 1/f noise contributions mentioned above have been verified experimentally by A. van der Ziel and his students in various ways.

ACKNOWLEDGEMENT

This work was supported by the Air Force Office of Scientific Research and by the U.S. Army NVEOL.

REFERENCES

1. Handel, P.H., "1/f Noise - an 'Infrared' Phenomenon", Phys. Rev. Lett. 34, (1975) 1492.

2. Handel, P.H., "Umklapp Scattering - the Strongest Source of 1/f Noise" in the Final Technical Report, ONR Contract N-0014-69-0405, July 1982, p.16.

3. van der Ziel, A., Handel, P.H., Zhu, X.C. and Duh, K.H., A theory of the Hooge parameters of solid state devices, IEEE Transactions ED-32, 667-671 (1985).

NOISE IN PHYSICAL SYSTEMS AND 1/f NOISE - 1985
A. D'Amico and P. Mazzetti (editors)

GRAVIDYNAMIC QUANTUM 1/f NOISE

Peter H. HANDEL

Physics Department, University of Missouri, St. Louis, MO 63121, USA

Gravitons of negligible energy and momentum subtly help shape the distribution of matter in space and time with a 1/f spectrum at low frequencies.

A brief calculation in a recent paper[1] has shown that in a charged particles scattering experiment the outgoing beam carries a current which exhibits 1/f noise at low frequencies. Therefore all cross sections and process rates will present this quantum 1/f effect with a spectral density given by the infrared exponent divided by f. The purpose of the present paper is to generalize this result to the non-electodynamic case and to use it as a probe for the existence of infraquanta, i.e. of very low frequency photons, gravitons, or other massless particles or elementary excitations, present in condensed matter. After a review of the previously published results, I will state the general quantum 1/f principle, consider the case of gravitons, and suggest some experiments.

To review the previous result, consider the action of a force of any nature on an incident electron beam $\exp(ip_0r/h)$. In the case of elastic scattering from a fixed potential (e.g. Coulomb scattering from a fixed force center) we are used to represent the outgoing wave in the form a $\exp(ip\cdot r/h)$, with $|p| = |p_0|$. This implies a constant outgoing current $|a|^2p/m$. However, a bremsstrahlung amplitude appears whenever charged particles are accelerated. Therefore, the outgoing wave includes an incoherent mixture of components characterized by a slight energy loss hf, i.e. components of slightly lower frequencies and wave numbers. These components will interfere with the non-bremsstrahlung part. The boundary between the former and the latter is defined by the finite low frequency threshold of the measuring apparatus, given by the inverse of the duration T of the experiment. As a result of this self-interference process, quantum current fluctuations will be exhibited by the outgoing current. The beat frequency of the interference process equals the frequency f of the bremsstrahlung photon involved. Consequently, the spectrum of the current fluctuations is the same as the bremsstrahlung photon number spectrum, i.e. proportional to 1/f.

Infrared divergences occur in many interactions in which massless particles can be emitted. The general quantum 1/f principle states that any current (cross section) with infrared divergent coupling to a system of massless infraquanta will exhibit macroscopic quantum fluctuations with a 1/f spectral density at sufficiently low frequencies. The massless particles may be photons, gravitons, tranversal phonons or other elementary excitations in solids, gluons for colored currents, etc.

The case of gravitons is particularly interesting, since it may open the way to quantum-gravitational experiments. Although the gravitational interaction is much weaker than the electromagnetic interaction, sizable 1/f noise may occur due to the fact that large masses are much easier to assemble than large, uncompensated electrical charges. Thus, the fluctuations may become observable precisely because of the weakness of gravitational interactions.

Although the calculation for gravitons is similar to that for photons, I will present it in some detail in order to exhibit the correspondence explicitly, and in order to include some new notations introduced in a more detailed treatment.[2-4] Very helpful for this purpose is the synthetical treatment of infrared quantum-gravitational divergences by Weinberg[5] whose notations will be used here for the benefit of the reader.

Let $\Gamma_{\beta\alpha}$ be the rate of an arbitrary process, defined as the absolute square of the S matrix element for calculated by ignoring all radiative corrections. If one includes contributions from virtual infrared gravitons of momenta $<|q|<\Lambda$, one obtains the rate

$$\Gamma_{\beta\alpha} = (\lambda/\Lambda)^B \Gamma^{o}_{\beta\alpha} \qquad (1)$$

where $B>0$ is an infrared exponent which replaces here the exponent αA introduced in quantum electrodynamics. The ultraviolet cuttoff $\Lambda >>> \lambda$ is fixed, and is chosen about one order of magnitude below the characteristic energy of the process $\alpha\to\beta$, i.e. below the kinetic energy of the colliding particles in a nonrelatavistic collision process. This way, $\Gamma_{\beta\alpha}$ includes practically no radiative corrections and, on the other hand, Λ is small enough, so that the soft photon approximations used in the factorization of the infrared divergences can be considered justified. The infrared cutoff λ is taken very small (smaller than any limit that would affect the non-infrared lines), in order to display the infrared divergences. λ will drop out of the calculation, finally.

In the limit of vanishing λ, the rate given by Eq. (1) is zero. Physically this means that there is no perfectly elastic rate. To retain a result of physical significance, we have to include the infrared terms describing the emission of real gravitons. Indeed, these processes will be included experimentally, if we measure the rate with energy loss below the detection threshold of the experiment $\varepsilon_o(\gg\lambda)$ left arbitrary here (although $<\Lambda$) for generality. The result does not allow factorization and exponentiation of the infrared divergences at the S matrix level (as for virtual gravitons), but does allow it at the level of the rate of the process, which turns out to be, with energy loss below a given ε_1,

$$\Gamma^{g}_{\beta\alpha}(<\varepsilon_1) = (\varepsilon_1/\lambda)^B\, b(B)\Gamma_{\beta\alpha} \simeq (\varepsilon_1/\Lambda)^B \Gamma^{o}_{\beta\alpha} \qquad (2)$$

Here we have used Eq. (1) to obtain the final form. The function b(B) which corresponds to F(A), is given by $b(x) = 1 - \pi^2x^2/12 + \ldots \simeq 1$. In fact, only the $\varepsilon_1{}^B$ dependence and the independence on λ is important for our purpose which is to illustrate the 1/f principle on the quantum-gravidynamical (QGD) case.

The fact that real graviton divergences exponentiate with the same B as in Eq. (1) has resulted in the elimination of all divergences in the final form of Eq. (2). For non-relativistic elastic two-body scattering the expression of B simplifies to

$$B = (8G/5\pi\hbar c^5)\mu^2 v^4 \sin^2\Theta_c \qquad (3)$$

where μ is the reduced mass, $v = |\vec{v}_1-\vec{v}_2|$ is the relative velocity, and Θ_c is the scattering angle in the center of mass system. The gravitational constant $G = 6.6732\ 10^{-8}$dyn cm^{2}g^{-2} replaces here the fine structure constant α of quantum electrodynamics (QED) according to the scheme $\alpha \to 3G\mu^2v^2/5\hbar c^3$ and $\Theta_c/2 \to \Theta_c$. The largest effects are expected from $\pi/2$ scattering. In Eq. (2) we write

$$(\varepsilon_1/\Lambda)^B = (\varepsilon_o/\Lambda)^B[1 + B\int_{\varepsilon_o}^{\varepsilon_1} (\varepsilon/\varepsilon_o)^B\, d\varepsilon/\varepsilon] \qquad (4)$$

If ε_o/h is the graviton detection threshold, i.e. the lowest 1/f noise frequency measured (not less than the reciprocal noise measurement time T), the first term in the square brackets yields the rate of elastic scattering , while the second gives the bremsstrahlung scattering rate up to the energy loss ε_1. The corresponding scattering matrix elements are obtained by taking the square root and restoring the phases, say ϕ and $\phi + \gamma(\varepsilon)$. For elastic scattering we obtain from Eqs. (2) and from Eq. (4) the amplitude $(\varepsilon_o/\Lambda)^{B/2}[b(B)\Gamma_{\beta\alpha}]^{1/2}\exp(i\phi)$.

For bremsstrahlung in the energy interval $d\varepsilon$ we obtain the same amplitude, multiplied by $\rho_\varepsilon \exp[i\gamma(\varepsilon)d\varepsilon/\varepsilon]$, where

$$\rho_\varepsilon = \sqrt{B}(\varepsilon/\varepsilon_o)^{B/2}, \qquad (5)$$

and $\gamma(\varepsilon)$ is a random phase which reflects the uncertainty of the moment of graviton emission (and removal from the system). Consequently, if the incoming beam is described by a plane wave $\exp[i(p\cdot r-Et)]$, with $\hbar = c = 1$, the scattered wave will be a stochastic mixture

$$\Psi = ae^{-iEt}\{e^{i\vec{p}\vec{r}} + \int_{\varepsilon_o}^{\varepsilon_1}\rho_\varepsilon e^{i\vec{p}\vec{r}+i\varepsilon t}\, a^+_\varepsilon d\varepsilon/\sqrt{\varepsilon}\} \qquad (6)$$

where

$$a \equiv (\varepsilon_o/\Lambda)^{B/2}[b(B)\Gamma^{o}_{\beta\alpha}]^{1/2}e^{i\phi} \qquad (7)$$

and $e^{i\gamma(\varepsilon)}$ has been included in the definition of a^+_ε. Here a^+_ε are graviton creation operators of indefinite direction angles and polarization, introduced in a similar way[1,2] as in QED (see [2], Eqs. 8-11) with the commutation relations

$$[a_\varepsilon, a^+_{\varepsilon'}] = \delta(\varepsilon - \varepsilon');\ [a_\varepsilon, a_{\varepsilon'}] = [a^+_\varepsilon, a^+_{\varepsilon'}] = 0 \qquad (8)$$

Details on gravitational radiation, its helicity, and polarization states can be found in the literature.[6] Note that p^- differs from p only due to the radiative energy loss ε, if the process is otherwise elastic and described in the center of mass system. Consequently,

$$\vec{p}^- = \vec{p} - (\partial E/\partial\vec{p})^{-1} = \vec{p} - \hat{p}\varepsilon/v;\ \hat{p} = \vec{p}/|\vec{p}|, \qquad (9)$$

and Eq. (6) can be rewritten in the form

$$\Psi = ae^{i(\vec{p}.\vec{r} - Et)}[1 + \int_{\varepsilon_o}^{\varepsilon_1}\rho_\varepsilon{}^{i\varepsilon\Theta}a^+_\varepsilon\, d\varepsilon/\varepsilon], \qquad (10)$$

where $\Theta = t - \vec{p}\cdot\vec{r}/v$, and v is the speed of the beam.

The nonrelativistic current density operator corresponding to Eq. (10) is

$$\vec{j} = -i/2m[\Psi^+\nabla\Psi - (\nabla\Psi)^+\Psi]$$

$$= \vec{p}|a|^2/m\ \{1 + \int_{\varepsilon_o}^{\varepsilon_1}\rho_\varepsilon(e^{i\varepsilon\Theta}a^+_\varepsilon + e^{-i\varepsilon\Theta}a_\varepsilon)d\varepsilon/\varepsilon^{1/2}$$

$$+ \int_{\varepsilon_0}^{\varepsilon_1} d\varepsilon/\sqrt{\varepsilon'}\rho_\varepsilon\rho_\varepsilon' e^{i(\varepsilon' - \varepsilon)\Theta} a_\varepsilon \, a^+_{\varepsilon'}\} \quad (11)$$

Its expectation value over the vacuum state of the Hilbert space of detectable gravitons ($\varepsilon > \varepsilon_0$) is

$$\langle\vec{j}\rangle = \vec{p}(|a|^2/m)[1+\int_{\varepsilon_0}^{\varepsilon_1}\rho_\varepsilon d\varepsilon/\varepsilon] = p(|a|^2/m)(\varepsilon_1/\varepsilon_0)^B \quad (12)$$

Substituting Eq. (7) we obtain a result which agrees with what is to be expected from Eq. (2), and does not depend on ε_0.

Defining the current fluctuations $\delta j = j - \langle j\rangle$, we construct the autocorrelation function in time and space

$$A(\eta) = 1/2[\delta j^+(\Theta + \eta)\delta j(\Theta) + c.c.]$$

$$= 2|a|^4 \vec{p}^2/m^2\{\int_{\varepsilon_0}^{\varepsilon_1}\rho_\varepsilon^2 \cos\varepsilon\eta d\varepsilon/\varepsilon + \int_{\varepsilon_0}^{\varepsilon_1} d\varepsilon/\varepsilon \int_{\varepsilon_0}^{\varepsilon_1} d\varepsilon'/\varepsilon'$$

$$\rho_\varepsilon\rho_\varepsilon' \cos(\varepsilon + \varepsilon')\eta \quad (13)$$

where $\eta = \tau - \hat{p}\cdot\vec{s}/v$ is a shift due to a time dalay τ and a spatial displacement s. The second term in the curly brackets is the square of the first term, if we look at it before adding the complex conjugate denoted by c.c. Therefore, it can be neglected, as long as B is small. Taking s = 0 and applying the Wiener-Khintchine theorem, we obtain the spectral density of relative current noise

$$\langle j\rangle^{-2}\langle(\delta j)^2\rangle_f = 2\rho_f^2 \, f^{-1}[1 + \int_{f_0}^{f_1} df'/f']^{-2}$$

$$= 2B(ff_0/f_1^2)^B \, f^{-1} \quad (14)$$

where the units have been restored and $f = \delta\varepsilon/h$. If we fix τ instead of s in Eq. (13), we obtain the same spectrum in wave numbers along the scattered beam. The factor $(ff_0/f_1^2)^B$ is negligible in most cases, since Eq. (3) shows that B<<1 usually. The same result given by Eq. (14) is obtained if the incoming beam is a mixture of wave packets.[7] To include the whole current, and to get the complete spectrum we may formally set $\varepsilon_1 = hf_1 = \Lambda$, although only the low frequency part will not be covered by thermal and shot noise effects. Eqs. (13) and (14) can be interpreted best as fluctuations of the scattering cross section, or process rate $\Gamma_{\beta\alpha}$, in time.

A tacit assumption in this derivation was that the energy of the emitted gravitons has been lost and is no longer present in the scattered state. The emitted gravitons are not part of the system we are considering in Eq. (6), although we include the operators a_ε which describe "missing gravitons" of negative energy and which include the randum phases. Note that these operators could be replaced simply by exp $[i\gamma(\varepsilon)]$, and the expectation value over the vacuum state of the Hilbert space of detectable gravitons by a phase average.

Consider the experiment depicted by Fig. 1, in which a large laminar column of water is partially deflected at a angle by a fixed target 1. In an effort to preserve the laminar character of the column, superficial friction and turbulence are reduced by using a huge, elastic, freely moving conveyer belt as the target. The scattered part of the column impinges on a small platform (2) which is critically damped with a variable time constant $T'_1 = 1/f'_1$, of the order of seconds or larter. Fluctuations of the pressure exerted on the platform go through a multichannel Fourier analyzer after which they are amplified, squared, and averaged over several days. The power spectrum is compared to Eq. (14) and the value of B is compared with the value predicted by Eq. (3) for various values of Θ_c, v, and μ. The value of μ is approximated by the mass M of fluid which is scattered together, i.e. as a whole. For example, consider a column of 1m in diameter, coming in with a speed of 10 m/s. The mass M accelerated by the conveyer belt at a time is about 10^5g. According to Eq. (3), B = 1.075 $10^9(v/c)^4(M/1g)^2\sin^2\Theta$. This yields B = 1.235 10^{-11} and according to Eq. (14), $\langle j\rangle^{-2} \cdot \langle(j)^2\rangle_f = 2.47/f \; 10^{-11}$. The electrical 1/f noise of the transducer can be of the order of $10^{-15}/f$, i.e. well below the effect to be measured.

A practical experimental opportunity is offered by a hydroelectrical plant or by natural waterfalls. For example, the Taum Sauk pump-storage hydro plant at Lesterville, Missouri in the Ozarks, has a 5.8 $10^6 m^3$ reservoir 270 m above the larger lower reservoir. The pumping (or generation) period is 9.5 hrs. During this time the water may flow at a speed of 70 m/s through the 2.43 m^2 cross section of the 2100 m long shaft and tunnel. With M = 10^5g we obtain B = 3.18 10^{-8} for an experiment similar to Fig. 1. Both in the generation and in the pumping period, the frequency of the reversible turbines should exhibit 1/f noise of gravitational origin, although M and B may have different values. This low frequency flicker effect should be present in the a.c. frequency generated. To verify the calculation of matter 1/f noise experimentally, all we have to do is to record and Fourier analyze these turbine frequency fluctuations. The spectral density of turbine frequency fluctuations under constant load should be $(3/f)\ 10^{-8}$ if 100 kg of water are scattered as a whole, and 100 times less, if that coherent mass is only 10 kg. In a different experiment, gravel, or a stream of rocks, could replace the water in Fig. 1. Then M would be the characteristic rock mass which is easier to define. Various forms of matter 1/f noise may have been observed already[8] in the universe, or in ocean currents. Note that instead of measuring current fluctuations, temporal or spatial fluctuations of the density $|\Psi|^2$ can be measured and analyzed.[9] In general, I expect a 1/f spectrum at frequencies which are so low, that no other causes for a nonuniform distribution of matter are left.

At any finite temperature a very large number kT/hf of gravitons will be present in the low frequency modes. By generalizing the theory of infrared radiative corrections and bremsstrahlung to include the thermal radiation background, I have recently proven[10] that the 1/f noise is the same as in the absence of the radiation background which only contributes a small additional white noise term. Furthermore, the derivation of matter 1/f noise presented in this paper can also be performed semiclasically[3], without the use of a and a^+ operators; see also[12]. Finally, it is important to mention that the obtained 1/f noise can be interpreted as a fundamental variation of the corresponding process rate, or cross section. Indeed, the flow rate varies as if there would be a fluctuation in the cross section of the duct, or of the opening in the fluid reservoir where the flow originates, or of the viscous resistance of the pipe.

If the measurement of matter 1/f noise succeeds at sufficiently low frequencies, against the turbulence background, it will provide an extremely sensitive gravitational wave detector. Equating the gravitational zero point energy density per unit frequency interval $4\pi hf^3/c^3$ to the value $\phi/(c\Delta f)$ generated by an energy flux incident in a frequency interval Δf, we obtain the c.w. detection limit $\phi_0 = 10^{-46}$erg sec^{3}cm$^{-2}\Delta f f^3$. For $f = 10^{-2}$Hz this lowest flux ϕ_0 detectable through current noise spectroscopy corresponds to a c.w. sensitivity of $\phi_0/\Delta f = 10^{-52}$erg/cm^2 = 10^{-57}GPU/sec. Such flux levels can be generated in the laboratory. By definition, 1 GPU = 10^5erg/cm^{2}Hz is the time - integrated flux unit. Note: this paper has first been submitted for publication to Phys. Rev. Letters on December 12, 1975 as manuscript #LM700, but has not yet been accepted for publication. The references have been updated.

ACKNOWLEDGEMENT

This work was supported by the Air Force Office of Scientific Research.

REFERENCES

1. Handel, P. H., Phys. Rev. Letters 34, 1492 (1975).
2. Handel, P. H., Proc. XVII NATO Advanced Study Inst. "Linear and Nonlinear Electron Transport in Solids" Antwerpen, J. T. Devreese and V. van Doren, Eds. (Plenum Press, New York, 1976), p. 515.
3. Handel, P. H., Phys. Rev. A 22, 745 (1980).
4. Sherif, T. S., and Handel, P. H., Phys. Rev. A 26, 596 (1982).
5. Weingerg, S., Phys. Rev. 140B, 516 (1965).
6. See, e.g., S. Weinberg: Gravitation and Cosmology, Wiley New York, 1972.
7. Handel, P. H., and Eftimiu, C., Proc. of the Symposium on 1/f Fluctuations, T. Musha Ed., Tokyo Inst. of Technology Press (1977), p. 183. [8]. A recent review is in W. H. Press, Comments on Astrophysics 7, 103 (1978).
9. Schick, K. L., and Verveen, A. A., Nature 251, 559 (1974).
10. Handel, P. H., Proc. II. Int. Symp. on 1/f Noise, Orlando, FL, 1980, K. M. van Vliet Editor, p. 96.
11. Handel, P. H., and Wolf, D., Phys. Rev. A26, 3727 (1982).
12. Handel, P. H., and Wolf, D., in "Noise in Physical Systems" p. 169, Springer Series in Electrophysics, Vol. 2, Springer Verlag Berlin, Heidelberg, New York 1978.

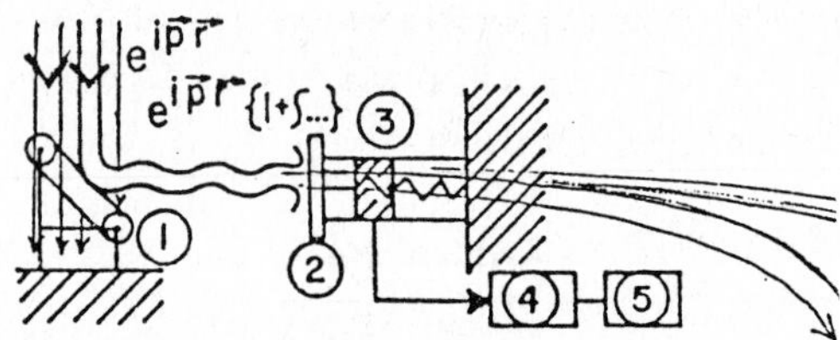

FIGURE 1: Matter 1/f noise experiment: 1 scatterer, 2 platform, 3 pressure transducer, 4 electric spectral analyzer, 5 quadratic meter. For low frequencies digital Fourier analysis is recommended.

NOISE IN PHYSICAL SYSTEMS AND 1/f NOISE - 1985
A. D'Amico and P. Mazzetti (editors)

1/f NOISE IN n^+-p JUNCTIONS CALCULATED WITH QUANTUM 1/f THEORY

A. VAN DER ZIEL, EE Department, U. of Minnesota, Minneapolis, MN 55455, USA

P. H. HANDEL, Physics Department, U. of Missouri, St. Louis , MO 63121 USA

Two new quantum 1/f noise effects for n^+-p junctions are calculated, (a) generation-recombination 1/f noise in the space charge region of junctions, (b) injection-extraction 1/f noise in carriers crossing the junction. Both effects come about by 1/f fluctuations in the cross sections and process rates and by the acceleration or deceleration of carriers crossing the space charge region. The effects will be masked by Umklapp quantum 1/f noise unless the latter is absent or $|V_{dif}-V|$ is large.

1. GENERATION-RECOMBINATION 1/f NOISE[1,2]

The starting point of the discussion is the Stockley-Hall-Read theory of the diode. According to this theory the recombination rate R and the generation rate G have a difference

$$R-G = \frac{pn-n_i^2}{(n+n_1)\tau_{po}+(p+p_1)\tau_{no}} \quad (1)$$

where τ_{no} and τ_{po} are the Shockley-Hall-Read lifetimes for electrons and holes and n_1 and p_1 are hole and electron densities when the Fermi level lies at the trap level. If the trap level lies at the intrinsic level $n_1=p_1=n_i$; we assume the latter to be the case. If A is the crosssectional area of the junction, the current I_{gr} is

$$I_{gr}=\int_0^w e(R-G)Adx=e\int_0^w \frac{(pn-n_i^2)}{(n+n_i)\tau_{po}+(p+n_i)\tau_{no}}Adx \quad (2)$$

where w is the width of the space charge region.

Writing $\tau_{no}=1/C_n$ and $\tau_{po}=1/C_p$, where C_n and C_p are the generation rates for a single electron and a single hole, respectively, we have

$$\delta\tau_{no}/\tau_{no} = -\delta C_n/C_n;\ \delta\tau_{po}/\tau_{po} = -\delta C_p/C_p \quad (3)$$

so that

$$\delta[R(x)-G(x)] = [R(x)-G(x)].$$

$$\frac{(n+n_i)\tau_{po}\delta C_p/C_p+(p+n_i)\tau_{no}\delta C_n/C_n}{(n+n_i)\tau_{po}+(p+n_i)\tau_{no}} \quad (4)$$

Hence the noise is

$$\delta I_{gr} = e\int_0^w \delta[R(x)-G(x)]A\delta x \quad (4a)$$

and the spectrum is

$$S_I(f) = e^2\int_0^w\int_0^w [R(x)-G(x)][R(x')-G(x')]A^2.\ F(x,x',f)dxdx' \quad (5)$$

where, since δC_p and δC_n are independent,

$$F(x,x',f) = \frac{(n+n_i)^2\tau_{po}^2 S_{Cp}(x,x',f)/C_p^2}{[(n+n_i)\tau_{po} + (p+p_i)\tau_{no}]^2} + \frac{(p+n_i)^2\tau_{no}^2 S_{Cn}(x,x',f)/C_n^2}{[(n+n_i)\tau_{po} + (p+n_i)\tau_{no}]^2} \quad (5a)$$

We next observe that, in analogy with a Hooge-type formula[3]

$$\frac{S_{Cp}(x,x',f)}{C_p^2} = \frac{\alpha_{Hp}/f}{|R(x')-G(x')|(\tau_{no}+\tau_{po})A}\delta(x'-x) \quad (6)$$

$$\frac{S_{Cn}(x,x',f)}{C_n^2} = \frac{\alpha_{Hn}/f}{|R(x')-G(x')|(\tau_{no}+\tau_{po})A}\delta(x'-x) \quad (7)$$

The factor $(\tau_{no}+\tau_{po})$ enters in because $S_{cp}(x,x',f)/C_p^2$ and $S_{cn}(x,x',f)/C_n^2$ must be independent of τ_{po} and τ_{no} when $p \approx n \approx n_i \exp(eV/2kT)$, where V is the applied voltage. This yields, if we integrate over the delta function,

$$S_I(f) = e^2 \int_0^w \frac{|R(x)-G(x)|A}{(\tau_{po}+\tau_{no})} \cdot \frac{(n+n_i)^2\tau_{po}^2\alpha_{Hp}/f + (p+n_i)^2\tau_{no}^2\alpha_{Hn}/f}{[(n+n_i)\tau_{po} + (p+n_i)\tau_{no}]^2} dx \quad (8)$$

We now observe that the second factor in (8) is practically a constant as long as p and n are comparable. We may then substitute $n+n_i \approx p+n_i$ in that factor, bring it outside the integral sign and call it α_H. Then

$$S_I(f) = \frac{e\alpha_H}{f(\tau_{no}+\tau_{po})} \cdot e\int |R(x)-G(x)|Adx = \frac{\alpha_H e|I_{gr}|}{f(\tau_{no}+\tau_{po})} \quad (9)$$

where

$$\alpha_H = \left(\frac{\tau_{po}}{\tau_{po}+\tau_{no}}\right)^2\alpha_{Hp} + \left(\frac{\tau_{no}}{\tau_{po}+\tau_{no}}\right)^2\alpha_{Hn} \quad (9a)$$

Now $\tau_{po}=(V_pA_pN_t)^{-1}$ and $\tau_{no}=(V_nA_nN_t)^{-1}$, where N_t is the density of the recombination centers, A_p and A_n are cross sections and V_p and V_n are carrier speeds. Since V_p is inversely proportional to $(m^*_p)^{1/2}$ and V_n is inversely proportional to $(m^*_n)^{1/2}$, where the m^* represent the effective masses, τ_{no} and τ_{po} are proportional to $(m^*_n)^{1/2}$ and $(m^*_p)^{1/2}$, respectively, so that

$$\alpha_H = \frac{m^*_n\,\alpha_{Hn} + m^*_p\alpha_{Hp}}{[(m^*_n)^{1/2}+(m^*_p)^{1/2}]} \quad (9b)$$

We now show that

$$\alpha_{Hn} = \frac{4\alpha}{3\pi}\left[\frac{2ea/V_{dif}-V) +3kT}{m^*_nc^2}\right] \quad (9c)$$

$$\alpha_{Hp} = \frac{4\alpha}{3\pi}\left[\frac{2e(1-a)(V_{dif}-V) +3kT}{m^*_pc^2}\right] \quad (9d)$$

so that

$$\alpha_H = \frac{4\alpha}{3\pi}\frac{[2e(V_{dif}-V) +6kT]}{[(m^*_n)^{1/2}+(m^*_p)^{1/2}]^2} \quad (10)$$

The proof is simple. In the first place C_p and C_n fluctuate because A_p and A_n fluctuate. This yields the Hooge parameters α'_{Hp} and α'_{Hn},

$$\alpha'_{Hp} = \frac{4\alpha}{3\pi}\frac{\overline{\Delta v_p^2}}{c^2} = \frac{4\alpha}{3\pi}\left(\frac{3kT}{m^*_pc^2}\right)$$

$$\alpha'_{Hn} = \frac{4\alpha}{3\pi}\frac{\overline{\Delta v_n^2}}{c^2} = \frac{4\alpha}{3\pi}\left(\frac{3kT}{m^*_nc^2}\right)$$

since $1/2m^*_n\overline{\Delta v_n^2} = 1/2m^*_p\overline{\Delta v_p^2} = 3/2kT$ is the energy gained by the particles in the emission process. In the second place the electrons and holes in the space charge region are decelerated or accelerated and this produces 1/f noise in the same manner as the cathode quantum 1/f noise in vacuum diodes and triodes. Most of the generation and recombination takes place in a narrow strip of the space charge region of width $w' \ll w$. Let that strip have a potential difference a $(V_{dif}-V)$ with respect to the n-region. Electrons excited from the trap level are accelerated and they gain an energy $E_n = ea(V_{dif}-V) = 1/2m^*_n\Delta v_n^2$ and holes excited from the trap level are accelerated and gain an energy $E_p = e(1-a)(V_{dif}-V) = 1/2m^*_p\Delta v_p^2$. This produces Hooge parameters α''_{Hp} and α''_{Hn} such that

$$\alpha''_{Hp} = \frac{4\alpha}{3\pi}\left(\frac{\overline{\Delta v_p^2}}{c^2}\right) = \frac{4\alpha}{3\pi}\left[\frac{2e(1-a)(V_{dif}-V)}{m^*_pc^2}\right]$$

$$\alpha''_{Hn} = \frac{4\alpha}{3\pi}\left(\frac{\overline{\Delta v_n^2}}{c^2}\right) = \frac{4\alpha}{3\pi}\left[\frac{2ea(V_{dif}-V)}{m^*_pc^2}\right]$$

Since $\alpha_{Hp} = \alpha'_{Hp}+\alpha''_{Hp}$ and $\alpha_{Hn} = \alpha'_{Hn}+\alpha''_{Hn}$, we obtain Eqs (9c) and (9d) as had to be proved.

2. A GENERAL QUANTUM 1/f NOISE THEOREM[4]

The following general theorem can be derived

(a) Let a single elementary event be described by a parameter u_i and let there be N such events. If the total process is described by a parameter u such that

$$u = \sum_{i=1}^{N} u_i \tag{11}$$

then

$$\bar{u} = N\bar{u}_i \quad ; \quad S_u(f) = NS_{ui}(f) \tag{11a}$$

(b) If $S_{ui}(f)/(\bar{u}_i)^2 = \alpha_H/f$

$$\text{then } S_u(f)/(\bar{u})^2 = \alpha_H/fN \tag{11b}$$

(c) Let u and u_i represent currents and let the electrons in the system have a time constant τ. An "elementary event" then means that there is on the average one electron in the system so that $u_i = e/\tau$. Then

$$\bar{u} = Ne/\tau \ ; \ S_u(f) = (\bar{u})^2\alpha_H/fN = \alpha_H e\bar{u}/f\tau \tag{11c}$$

This applies to g-r 1/f noise. Here $u = |I_{gr}|$, $\tau = \tau_{no} + \tau_{po}$ is the time constant associated with a single hole-electron pair. Consequently

$$S_I(f) = \alpha_H e|I_{gr}|/[f/\tau_{po}+\tau_{no})]$$

which corresponds to (9).

3. INJECTION-EXTRACTION 1/f NOISE ACROSS n+-p JUNCTIONS

The above theorem also holds for the injection-extraction noise mechanism. Here electrons are injected across the space charge region, are there decelerated and this produces 1/f noise in the usual manner. Equation (11c) applies in that case. The time constant τ for short p-regions is equal to the diffusion time $\tau_d = w^2/2D_n$ of the carriers through the p-region whereas for long p-regions τ is equal to the carrier life time τ_n. For short p-regions

$$S_I(f) = \alpha_{Hn}\, eI/f\tau_d \tag{12}$$

where in analogy with the previous case (section 1)

$$\alpha_{Hn} = \frac{4\alpha}{3\pi}\left[\frac{2e(V_{dif}-V) + 3kT}{m^*_n c^2}\right] \tag{12a}$$

Here the term 3kT comes from the recombination at the contact to the p-region. The general problem of p-regions of arbitrary length will be discussed elsewhere.

In $Hg_{1-x}Cd_xTe$ m^*_n is very small and much smaller than $[(m^*_n)^{1/2}+(m^*_p)^{1/2}]^2$ so that α_{Hn} in (12a) is much larger than α_H in (10). Hence g.r. 1/f noise will be much smaller than injection-extraction 1/f noise[1,2]. Both can be smaller than Umklapp 1/f noise so that they will only be observable if the latter is absent, or if $|V_{dif}-V|$ is large.

There seems to be no Umklapp 1/f noise in the collector current of a bipolar transistor.[5] Zhang and van der Ziel[6] were also unable, however, to observe injection-extraction 1/f noise in the collector current, even at large collector back bias. We refer to their paper for details.

REFERENCES

1. W. A. Radford and C. E. Jones, Proc. U.S. Workshop on HgCdTe, J. Vac. Sci. Techn., A-3, 183 (1985).

2. A. van der Ziel and P. H. Handel, IEEE Trans. El. Dev., ED-32 (Oct. 1985).

3. F. N. Hooge, Phys. Lett., A-29, 139 (1969); Physica, 83B, 14 (1976).

4. A. van der Ziel, accepted by Physica B.

5. A. van der Ziel, P. H. Handel, X. C. Zhu and K. H. Duh, IEEE Trans. El. Dev., ED-32, 667 (1985).

6. X. N. Zhang and A. van der Ziel, this conference.

NOISE IN PHYSICAL SYSTEMS AND 1/f NOISE - 1985
A. D'Amico and P. Mazzetti (editors)

TEST FOR THE PRESENCE OF INJECTION-EXTRACTION AND OF UMKLAPP QUANTUM 1/f NOISE IN THE COLLECTOR OF SILICON TRANSISTORS

X. N. ZHANG and A. VAN DER ZIEL

Electrical Engineering Department, Univ. of Minnesota, Minneapolis, MN 55455 USA

It was found that injection-extraction quantum 1/f noise and Umklapp quantum 1/f noise could not be observed in the collector current of p^+-n-p (GE82-185) and n^+-p-n (NEC57807) silicon bipolar transistors. Possible reasons for these results are discussed.

1. INTRODUCTION

An investigation on the existence of the injection-extraction and the Umklapp quantum 1/f noise processes in the collector current of bipolar transistors was carried out for GE82-185 (p^+-n-p) and NEC57807 (n^+-p-n) silicon transistors. It was earlier observed[1,2] that Umklapp 1/f noise seemed to be absent in the collector current of these devices. We have here attempted to lower the upper limit of the Hooge parameters α_H found earlier.

It is believed[3] that the injection-extraction 1/f noise can sometimes be smaller than the Umklapp 1/f noise. It thus seemed that the above devices would be ideal for testing the presence of injection-extraction 1/f noise in the collector current of these transistors, since Umklapp 1/f noise would not mask the injection-extraction 1/f noise.

The experiments were performed at room temperature in the common emitter bias configuration, with the base terminal h.f. grounded. The frequency range was 20Hz-100KHz.

2. UMKLAPP 1/f NOISE IN THE COLLECTOR

The measurement of the spectrum $S_{IC}(f)$ of the above transistors gave no indication of 1/f noise. Only the shot noise of the collector current I_C and the amplified thermal noise of the series base resistance $r_{b'b}$ were observed, so that

$$S_{IC}(f) = 2qI_C + 4kTr_{b'b}g_m^2 \qquad (1)$$

We therefore assumed that at 20Hz the 1/f part of $S_{IC}(f)$ was at most 1/5 of the measured value of $S_{IC}(f)$. Denoting this part by $S_{IC}'(f)$ we find:

$S_{IC}'(f) \leq 1.40\times10^{-22}A^2/Hz$ for the GE82-185 at $I_C = 1.0$ mA ($r_{b'b} = 15\Omega$)

$S_{IC}'(f) \leq 1.40\times10^{-22}A^2/Hz$ for the NEC57807 at $I_C = 1.0$ mA ($r_{b'b} = 15\Omega$)

According to van der Ziel[4] for p^+-n-p transistors

$$S_{IC}'(f) = \frac{\alpha_{Hp}}{f}\frac{qI_CD_p}{W_B^2}\ln\left[\frac{p(0)}{p(W_B)}\right] \qquad (2)$$

and for n^+-p-n transistors

$$S_{IC}'(f) = \frac{\alpha_{Hn}}{f}\frac{qI_CD_n}{W_B^2}\ln\left[\frac{n(0)}{n(W_B)}\right] \qquad (2a)$$

Using the same parameters as Zhu and van der Ziel GE82-185: D_p=12.5 cm^2/sec, $p(0)/p(W_B)$=133, W_B=1.63 μmeter. NEC57807: D_n=35cm^2/sec, $n(0)/n(W_B)$=11.7, W_B=0.373 μmeter, we find $\alpha_{Hp} \leq 7.6\times10^{-9}$($p^+$-n-p); $\alpha_{Hn} \leq 2.8\times10^{-10}$($n^+$-p-n) which are about a factor 5 lower than Zhu's result.[1] This is due to the lower 1/f noise in our units.

Van der Ziel et al[2] find for Umklapp 1/f noise $\alpha_{Hn}=2.2\times10^{-8}$ and $\alpha_{Hp}=3.9\times10^{-7}$, thus indicating complete absence of Umklapp noise in the collector. The value of α_{Hn} is even less than the value 0.94×10^{-9} expected for elastic collision 1/f noise. A more accurate calculation by Kousik[5] shows a value of α_{Hn} caused by phonon collisions that is about a factor 3 larger (3×10^{-9}). It thus seems that even normal collision 1/f noise is absent in the collector current of NEC57807 n^+-p-n transistor.

Why is there no collision nor Umklapp 1/f noise in the collector current? As the theory indicates these noise processes are generated in the base. The minority carrier density in the base has a constant gradient and the current flow is practically independent of the collector-emitter voltage V_{CE}. It might be that this will make the collision 1/f noise processes, both those of the phonon type and those of the Umklapp type, inactive. This needs further investigation.

3. INJECTION-EXTRACTION 1/f NOISE IN THE COLLECTOR

We also tried to measure injection-extraction 1/f noise in the collector of GE82-185 p^+-n-p and NEC57807 transistors and could not find any at f=20Hz, even for large collector back bias.

According to the theory[2]

$$S_{IC}(f) = \alpha_{Hni} \frac{2qI_C D_n}{f W_B^2} \tag{3}$$

and, as the previous case, the negative outcome of the experiment yields $S_{IC}(20) \leq 1.40\times10^{-22} A^2/Hz$ at I_C=1.0mA. Applying this to the NEC57807 n^+-p-n transistors with W_B=0.373μmeter and D_n=35cm²/sec yields

$$\alpha_{Hni} \leq 3.4\times10^{-10} \tag{3a}$$

Neglecting a small term, the injection-extraction 1/f noise theory yields[2]

$$\alpha_{Hni} = \frac{4\alpha}{3\pi}\left[\frac{2q|V_{dif}-V|}{m_n^* c^2}\right] \tag{3b}$$

where α=1/137 is the fine structure constant. Taking $|V_{dif}-V|$=2.0V, $m_n^*=m$ (n-type silicon) yields

$$\alpha_{Hn} = 2.4\times10^{-8} \tag{3c}$$

This is about a factor 70 larger than our estimate (3a). We therefore conclude that the effect is absent.

Why was injection-extraction 1/f noise not observed? There are two possible explanations: (1) The collector junction is back-biased and does not control the current flow at all; this may make the injection-extraction noise process inactive.
(2) Equation (3b) comes from Handel's relation

$$\alpha_{Hn} = \frac{4\alpha}{3\pi}\left(\frac{\overline{\Delta V^2}}{c^2}\right) \tag{4}$$

where ΔV is the change in velocity in the injection-extraction process. Neglecting collisions yields $\frac{1}{2} m_n^* \overline{\Delta V^2} = q|V_{dif}-V|$ and thus gives (3b). But the carriers passing through the space charge region probably suffer many collisions. This may make $\overline{\Delta V^2}$, and hence α_{Hn}, much smaller.

At present we prefer the first explanation but cannot rule out the second.

4. CONCLUSION

Neither collision quantum 1/f noise, either the normal collision one or the Umklapp one, nor the injection-extraction 1/f noise seems to be present in the observed noise in the collector current of bipolar transistors.

ACKNOWLEDGEMENT

This work was partly supported by an NSF grant.

REFERENCES

1. X. C. Zhu and A. van der Ziel, IEEE Trans. El. Dev., ED-32, 658 (1985).

2. A. van der Ziel, P. H. Handel, X. C. Zhu and K. H. Duh, IEEE Trans. El. Dev., ED-32, 667 (1985).

3. A. van der Ziel and P. H. Handel, IEEE Trans. El. Dev., ED-32, (Oct. 1985).

4. A. van der Ziel, Solid State Electron, 25, 141 (1982).

5. G. Kousik, Ph.D. Thesis, University of Florida (1985).

NOISE IN PHYSICAL SYSTEMS AND 1/f NOISE - 1985
A. D'Amico and P. Mazzetti (editors)

QUANTUM 1/f NOISE IN SQUIDS

Peter H. HANDEL

Physics Department, University of Missouri-St. Louis, MO 63121, USA

The fundamental quantum 1/f fluctuations of the cross sections and transition rates which determine the normal resistance are evaluated for the case of a Josephson junction. Considering the velocity change in the quantum 1/f formula equal to twice the Fermi velocity and the concentration of carriers in the barrier 10^{19} cm^{-3}, a spectral density of fractional fluctuations in the normal resistance of the barrier of 4 $10^{-14}/f$ is obtained for a junction with a volume of the barrier 10^{-12} cm^3. These fluctuations are inversely proportional to the barrier volume and result in voltage fluctuations both directly and through the dependence of the critical current on the normal resistance, in good agreement with the experimental data.

Any cross section or process rate defined for electrically charged particles must fluctuate in time with a 1/f spectral density according to quantum electrodynamics, as a consequence of infrared - divergent coupling to low - frequency photons.[1] This fundamental effect discovered 10 years ago, leads to quantum 1/f noise observed in many systems with a small number of carriers, and is also present in the cross sections and process rates which determine the resistance and tunneling rate in Josephson junctions, providing a lower limit of the observed 1/f noise.

The present paper gives a brief and physical explanation and derivation of the quantum 1/f effect, followed by a discussion of the application to Josephson junctions and SQUIDS.

Consider a scattering experiment, e.g. Coulomb scattering of electrons on a fixed charge and focus on the scattered current which reaches the detector. Part of the Schroedinger field of this outgoing beam has lost energy hf due to bremsstrahlung in the scattering process. This part interferes with the main, nonbremsstrahlung part yielding beats at any frequency f in the outgoing DeBroglie waves. These beats are observed as fluctuations of the scattered current and interpreted as cross section fluctuations.

The bremsstrahlung amplitude is known to be $(2\alpha/3\pi f)^{1/2}v/c$, where v is the change in the velocity vector of the particles during the scattering process, and α is the fine structure constant $e^2/hc = 1/137$. The beat current is proportional to this amplitude, and the spectral density of the fractional current j (or cross section σ) fluctuations is therefore

$$j^{-2}S_j(f) = \sigma^{-2}S_\sigma(f) = 4\alpha v^2/3\pi Nfc^2 \qquad (1)$$

which duplicates the number spectrum of the emitted photons. Here N is the number of the particles which are simultaneously measuring the cross section.

In a Josephson junction the normal resistance R_n of the barrier is proportional to a scattering cross section or transition rate experienced by the electron in quasiparticle tunneling and by Cooper pairs below the critical current I_c. Therefore

$$R_n^{-2}S_{Rn}(f) = (4\alpha/3\pi)(v^2/c^2Nf)$$
$$= (8\alpha/3\pi)(v_F^2/c^2Nf) = 4\ 10^{-14}/f\Omega \qquad (2)$$

where we have approximated v^2 with $2v_F^2$, v_F being the Fermi velocity, and the number of carriers N simultaneously present in the barrier of volume Ω (in μ^3) by 10^7, for barriers wider than 10^{-7}cm.

Assuming a linear relationship between the critical current I_c and $G_n = R_n^{-1}$., we obtain similar to Rogers and Buhrman[2]

$$S_V(F) = (4/f)\ 10^{-12}(T/3K)[R_s g(V)/(R_s + R_j)]^2$$
$$[I_c R_n(I^2/I_c^2 - 1)^{-1/2} + g(V)V]^2\Omega^{-1}, \qquad (3)$$

where $R_j(V)$ is the junction resistance, R_s the shunt resistance, and $g(V) = R_n/R_j$.

The noise caused in a SQUID by the source considered above can be obtained as the sum of the noise contributions from the two junctions.

In addition to this noise present in each junction, SQUIDS' may also allow us to see coherent quantum 1/f noise[3]

$$I^{-2}S_I(f) = 4.6\ 10^{-3}/fN_{eff}, \qquad (4)$$

where N_{eff} is an effective number of carriers which define the coherent current state in the vicinity of the two Josephson junctions. This noise is caused by the coherent character of the field of each current carrier, which leads to uncertainty in its energy and thereby generates an additional form of quantum 1/f noise in the current. In practice the current will fluctuate, therby exhibiting a departure from a perfectly coherent field state, i.e. 1/f amplitude and phase fluctuations.

The above quantum 1/f results of Eqs. (2) and (3) are in good quantitative agreement with the experimental data[2], but the coherent quantum 1/f result (4) needs further investigation before a meaningful comparison with the experiment can be performed.

ACKNOWLEDGMENT

This work was supported by the Air Force Office of Scientific Research

REFERENCES

1. P.H. Handel, Phys. Rev. Lett. 34, 1492, 1495 (1975); Phys. Rev. A22, 745 (1980); T.S. Sherif and P.H. Handel, A26, 596 (1982); P.H. Handel and D. Wolf, A26, 3727 (1982); K.M. Van Vliet, P.H. Handel and A. van der Ziel, Physica 108A, 511 (1981); A. Widom et al, Phys. Rev. B26, 1475 (1982); B27, 3412 (1983); A. van der Ziel and P.H. Handel, Physica 125B, 286 (1984), IEEE Trans. ED-32, 667 (1985).

2. C.T. Rogers and R.A. Buhrman, IEEE Trans MAG-19, 453 (1983).

3. P.H. Handel in "Noise in Physical Systems and 1/f Noise" Eds., M. Savelli, G. Lecoy, and J.P. Nougier (North Holland Pub. Co., Amsterdam, 1983,) pp. 97-100. See also the other papers on quantum 1/f noise in the same volume, by A. Widom et al, by P.H. Handel et al, by J. Kilmer et al, and by J. Gong et al.

NOISE IN PHYSICAL SYSTEMS AND 1/f NOISE - 1985
A. D'Amico and P. Mazzetti (editors)

EXPERIMENTS ON HIGH-FREQUENCY AND 1/f NOISE IN LONG n^+-p $Hg_{1-x}Cd_xTe$ DIODES

MS. X. L. WU and A. VAN DER ZIEL

EE Department, U. of Minnesota, Minneapolis, MN 55455, USA

Thermal noise and shot noise at 50KHz was measured in long n^+-p $Hg_{1-x}Cd_xTe$ diodes for x=0.30, and compared with the theoretical diode model. Low-frequency noise spectra were measured at 300°K, 193°K and 77°K and found to be of the 1/f type. In diodes only the excess minority carriers generate 1/f noise. At 300°K the 1/f noise came from the series resistance R_s of the diode except near saturation. Both for the diode 1/f noise and the 1/f noise generated in R_s Handel's quantum limit is not reached.

1. HIGH-FREQUENCY NOISE IN $Hg_{1-x}Cd_xTe$

The h.f. noise of n^+-p $Hg_{1-x}Cd_xTe$ photodiodes with x=0.30 was measured at 50KHz at 300°K, 193°K and 77°K. The diodes were provided by Dr. G. R. Williams, Rockwell International, Thousand Oaks, CA and came from batch 5384. The 1/f noise at 50KHz was always neglible; so that the only remaining noise sources were shot noise of the diode current and thermal noise of the series resistance R_s. Hence, the mean square value of the noise emf e_{tot} of the diode can be written as

$$\overline{e_{tot}^2} = \overline{i_d^2} R_d^2 + 4kTR_s\Delta f \tag{1}$$

where $\overline{i_d^2}$ represents the diode shot noise and R_d the diode conductance,

$$\overline{i_d^2} = 2q(I+2I_o)\Delta f \ ; \ R_d = kT/[q(I+I_o)], \tag{1a}$$

I is the current, and I_o the saturation current of the diode. The short circuit noise current may thus be written as

$$S_{Itot}(f) = \frac{\overline{e_{tot}^2}/\Delta f}{(R_s+R_d)^2} = 2q(I+2I_o)\left(\frac{R_d}{R_s+R_d}\right)^2 + \frac{4kTR_s}{(R_s+R_d)^2} \tag{2}$$

Good agreement was obtained at 300°K with I_o=121 μA and R_s=215 ohm.

2. LOW-FREQUENCY NOISE IN $Hg_{1-x}Cd_xTe$

We measured low frequency noise spectra at 300°K, 193°K and 77°K as a function of bias and found that in all cases the spectrum was of the 1/f type.

Next we investigated whether <u>all</u> the minority carriers contributed to the 1/f noise or only the <u>excess</u> minority carriers. According to van der Ziel and Anderson[1] the diode spectrum can be written in the form

$$S_I(f) = \alpha_{Hn}(qI/f\tau_n)f(\gamma) \tag{3}$$

for a long diode. Here I is the current, τ_n the electron life time and $\gamma=[\exp(qV/kT)-1]$, where V is the applied bias. Finally

$$f(\gamma) = \left[\frac{1}{3} - \frac{1}{2\gamma} + \frac{1}{\gamma^2} - \frac{1}{\gamma^3}\left(\frac{qV}{kT}\right)\right] \tag{3a}$$

when <u>all</u> minority carriers contribute to the 1/f noise, and

$$f(\gamma) = \pm 1/3 \tag{3b}$$

(plus sign for forward bias, minus sign for back bias) when only the <u>excess</u> minority carriers contribute. Note that I and f(γ) are both negative for back bias so that $S_I(f)$ is always positive.

We also note that |f(γ)| has a zero for V=0(γ=0), approaches 1/3 for large forward bias (γ>>1), and approaches |1.83-qV/kT| for large

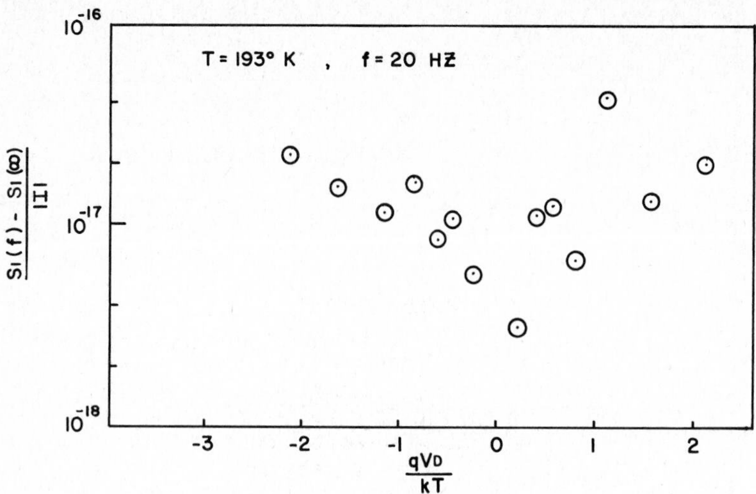

Fig.1 - The Ratio of 1/f NOISE and Current vs. Normalized Voltage Curve

back bias (γ=1) if all minority carriers contributed to the 1/f noise. If only the excess minority carriers contribute to the 1/f noise then $|f(\gamma)|=1/3$ throughout. This allows discrimination between the two cases.

Figure 1 shows $[S_I(20)-S_I(40KHz)]/I$ versus qV/kT at T=193°K. One notes that there is no zero at qV/kT=0, nor is there the large increase in noise at back bias that Eq. (3a) requires. Rather one sees, except for some spread due to errors in the measurements, that the expression is practically independent of bias. This indicates that only the excess minority carriers contribute to the 1/f noise of diodes. The same was found to be true at 77°K.

At T=300°K, however, $[S_I(20)-S_I(50KHz)]/|I|$ gave a clean zero at V=0. But $[S_I(20)-S_I(50KHz)/|I|^2$ showed a constant value as a function of bias, except near saturation (Fig. 2). This would be expected if the 1/f noise of the device were due to current 1/f noise generated in the series resistance R_s, it varies as I^2 and should thus predominate if $R_d^2 << (R_s+R_d)^2$.

For according to Hooge[2]

$$[S_I(f)]_{R_s} = \frac{\alpha_{Hn}}{fN_o} I^2 \tag{4}$$

where it is assumed that the p-region is intrinsic, with $N_o = n_oV_{eff}$ being the total number of electrons in the p-region; here n_o is the electron density and V_{eff} the effective volume of the series resistance. The holes give a negligible contribution to the 1/f noise.[3]

If we calculate α_{Hn} from (4) the main inaccuracy in α_{Hn} is due to the inaccuracy in V_{eff}. If the resistor is rectangular, V_{eff} = AL, where A is the cross-sectional area; in all other cases a suitable value for V_{eff} must be chosen. If we calculate α_{Hn} from (3) the inaccuracy in α_{Hn} is due to the inaccuracy in τ_n.

3. QUANTUM 1/f NOISE APPROACH

According to Handel's quantum 1/f noise theory[4] the Hooge parameter can be expressed as

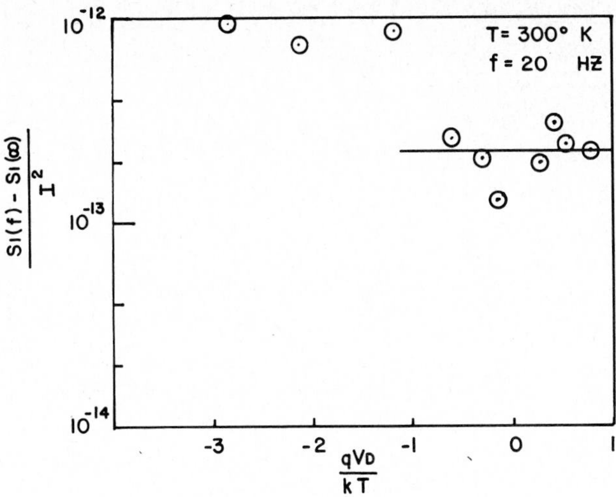

Fig. 2 - The Ratio of 1/f Noise and Current Square vs. Normalized Voltage Curve.

$$\alpha_H = \frac{4\alpha}{3\pi}\left(\frac{\overline{\Delta V^2}}{c^2}\right) \qquad (5)$$

α=1/(137), C the velocity of light and ΔV the change in carrier velocity in the collision processes. Van der Ziel et al[5] applied this to the case of Umklapp collisions, added a term $\exp(-\theta_D/2T)$ where θ_D is the Debye temperature, and obtained

$$\alpha_H = \frac{4\alpha}{3\pi}\left(\frac{h}{m^*ac}\right)^2 \exp\left(\frac{-\theta_D}{2T}\right) \qquad (5a)$$

This expression gave very good agreement with the experimental data for p-channel MOSFETs and n-channel JFETs,[5] in the latter case the temperature dependence of α_{Hn} was also very well described by (5a).[6]

In order to apply this to $Hg_{1-x}Cd_xTe$ diodes with x=0.30 it must be taken into account that m_n^* is extremely small, that $m_p=0.55m_o$ and that $\theta_D \simeq 250$°K. This yields[3] theoretically $\alpha_{Hp}=13\times10^{-8}$ at 300°K, and $\alpha_{Hn}=2.6\times10^{-5}$, 5.0×10^{-5} and 4.8×10^{-5} at 77°K, 193°K and 300°K, respectively, so that $\alpha_{Hp} \ll \alpha_{Hn}$.

Equations (3)-(3b) yield

$$\frac{S_I(f)}{|I|} = \alpha_{Hn}\frac{q}{3f\tau_n} \qquad (6)$$

so that the measurement of $S_I(f)/|I|$ at 20 Hz and near saturation yields α_{Hn}/τ_n. As shown elsewhere, $\tau_n(T)$ has a maximum at intermediate temperature and can be considerably lower at much higher and much lower temperatures. Assuming $\tau_n(300) = 10^{-8}$ sec, $\tau_n(193) = 5\times10^{-8}$ sec and $\tau_n(77) = 0.5\times10^{-8}$ sec yields $(\alpha_{Hn})_{exp}$. Evaluating (α_{Hn}) theory for x = 0.30 from (5a) yields the results

T	$(\alpha_{Hn})_{exp}$	$(\alpha_{Hn})_{theory}$
300°K	3.4×10^{-4}	4.8×10^{-5}
193°K	3.4×10^{-4}	5.0×10^{-5}
77°K	8×10^{-4}	2.6×10^{-5}

Since $(\alpha_{Hn})_{exp}$ is about one order of magnitude larger than (α_{Hn})theory, we conclude that the Umklapp 1/f noise is masked by a Hooge-type 1/f noise.

We now turn to the 1/f noise of the series resistance R_s at T = 300°K. We measured $S_I(f)/I^2 = 2.2\times10^{-13}$/Hz at f = 20 Hz and forward bias. Now A = 150μx10μ = $1.5\times10^{-5}cm^2$ and L ≃ 0.2 cm. Finally[3,7] $n_i = 3.44\times10^{15}/cm^3$

for $x = 0.30$, so with $n_a = 2x10^{15}/cm^3$ we have $n_o = 2.58x10^{15}/cm^3$ so that $(\alpha_{Hn})_{exp} = 3.4x10^{-2}$, whereas Handel's coherent state quantum 1/f noise yields[7] $(\alpha_{Hn}) = 4.6x10^{-3}$. We thus see that $(\alpha_{Hn})_{exp}$ is about 7 times larger than $(\alpha_{Hn})_{theory}$ so that the quantum 1/f noise limit hsa not been reached here either.

4. CONCLUSION

The quantum 1/f noise limit has not yet been reached in the observed 1/f noise of $Hg_{1-x}Cd_xTe$ diodes with x=0.30.

ACKNOWLEDGEMENT

This work was partly supported by an NSF grant and by the Night Vision Electro-Optics Laboratory through the Army Research Office under contract DAAG29-81-D-0010. We are indebted to Dr. G. M. Williams, Rockwell International, Thousand Oaks, CA for providing the samples.

REFERENCES

1. J. B. Anderson and A. van der Ziel, to be published.
2. F. N. Hooge, Phys. Lett., A29, 139 (1969); Physica, 125B, 14 (1976).
3. X. C. Zhu, X. L. Wu, A. van der Ziel and E. G. Kelso, IEEE Trans. El. Dev., ED-32 (July 1985).
4. P. H. Handel, Phys. Rev. Lett, 34, 1492 (1975); Phys. Rev., 22, 745 (1980).
5. A. van der Ziel, P. H. Handel, X. C. Zhu and K. H. Duh, IEEE Trans. El. Dev., ED-32, 667 (1985).
6. A. Pawlikiewicz and A. van der Ziel, to be published.
7. A. van der Ziel, Invited Paper, this conference.

APPLICATIONS AND MEASUREMENT TECHNIQUES - 2

NOISE IN PHYSICAL SYSTEMS AND 1/f NOISE - 1985
A. D'Amico and P. Mazzetti (editors)

OPTIMIZATION OF SPLIT-DRAIN MAGFET GEOMETRIES WITH RESPECT TO 1/f NOISE

H. P. BALTES, A. NATHAN, D. R. BRIGLIO, and Lj. RISTIĆ

Department of Electrical Engineering, University of Alberta, Edmonton, Alberta, Canada T6G 2E1

The sensitivity and signal-to-noise ratio of dual- and triple-drain magnetic-field sensitive MOSFET configurations are computed. The minimum detectable magnetic induction is estimated under flicker noise.

1. INTRODUCTION

A magnetic field sensor (MFS) is an entrance transducer that converts a magnetic field into an electronic signal. Integrated semiconductor, notably silicon, MFS are manufactured using integrated circuit technology[1]. Examples are integrated bulk Hall plates, vertical or lateral magnetotransistors, magnetodiodes, and magnetic-field sensitive MOSFET structures (also referred to as MAGFET). In low-frequency application, flicker noise is the fundamental limitation of MFS performance with respect to the minimum detectable magnetic induction B_{min}. This holds in particular for MAGFET structures, such as the MOSFET with Hall probe contacts or with dual- or triple-drains[2]. Flicker noise in bulk Hall devices[3,4] and magnetodiodes[5] has been investigated only recently.

2. THEORY

In this paper, we estimate B_{min} for various split-drain MAGFET structures. To this end we combine our previous numerical modeling results[6] with Leventhal's model[7] for 1/f noise in silicon inversion layers. We assume that the 1/f noise is not affected by the presence of the magnetic field. We present the potential and current distributions in the MAGFET channel, the magnetic-field sensitivity, and the signal-to-noise ratio (SNR) associated with the various device configurations. The SNR is computed using the mean square drain noise current per unit bandwidth

$$\langle i_{DF}^2\rangle/\Delta f = q^2\mu_s^2 V_D^2 W_{eff} N_s/L^3 f \qquad (1)$$

adapted from[7]. Here, q denotes the elementary charge, μ_s the (electric-field dependent) surface mobility of the electrons, V_D the applied drain voltage, W_{eff} an appropriately defined effective drain width, N_s the (process dependent) density of surface states, and L the channel length. For N_s we assume 10^{10} cm^{-2}.

Two device configurations are considered viz., the dual- and triple-drain (abbreviated as "DD" and "TD" respectively) NMOS sensors (see Fig. 1). We recall that the carriers (electrons) in the channel are deflected by the Lorentz force due to a magnetic induction, B perpendicular to the chip surface. This leads to an increase of one drain current, say I_{D1} at the expense of the other drain current, I_{D2}. For simplicity, we assume that the noise currents in the two drains D1, D2 are independent, i.e.,

$$\langle i_{DF}^2\rangle = \langle i_{D1F}^2\rangle + \langle i_{D2F}^2\rangle \qquad (2)$$

or

$$W_{eff} = W_{D1} + W_{D2}, \qquad (3)$$

e.g., $W_{eff} = W - d$ for the dual-drain device (see Fig. 1(a)). This means we assume the split-drain MAGFET to behave as two parallel noise durrent sources.

The sensitivity of the MAGFET devices is defined in terms of the current imbalance in the two drains i.e.,

$$\delta = (I_{D1} - I_{D2})/(I_{D1} + I_{D2}). \quad (4)$$

For a unit bandwidth,

$$\begin{aligned} SNR &= (I_{D1} - I_{D2})/\langle i_{DF}^2 \rangle^{1/2} \\ &= (I_{D1} + I_{D2})\, |\delta| / \langle i_{DF}^2 \rangle^{1/2}. \end{aligned} \quad (5)$$

3. NUMERICAL RESULTS

A systematic comparison of results for a variety of aspect ratios (W/L) shows the following. For a given channel area L · W, the square channel geometry, L = W, provides a broad optimum with respect to δ/B. Equation (1) indicates that L > W is preferable with respect to SNR, but technology prevents the choice of too small W, since d $\gtrsim$ 5 μm. Tables 1 and 2 compare the sensitivity δ and SNR for different device configurations and dimensions. It turns out that δ is proportional to B for B $\leq$ 1T, and is somewhat superior for the triple-drain device (TD) over the dual-drain device (DD). A dual-drain NMOS sensor with W = L = 25 μm, and drain separation, d = 5 μm operating in the linear region shows a minimum detectable field, B_{min} = 4.6 μT. Here, the minimum detection limit for the magnetic induction, B_{min} is determined by taking SNR = 1. Increasing the device dimensions to W = L = 100 μm and d = 20 μm, yields B_{min} = 1.1 μT. An increase in SNR is found in the triple-drain MOSFET. A minimum detectable field, B_{min} = 0.7 μT is obtained for the W = L = 100 μm, the drain regions W_{D1}, W_{D2} being 20 μm wide and d = 20 μm (see Fig. 1(b)).

TABLE 1

a: W = L = 25 μm, $W_{D1} = W_{D2}$ = 10 μm, d = 5 μm, B_{min} = 4.6 μT.
b: W = L = 100 μm, $W_{D1} = W_{D2}$ = 40 μm, d = 20 μm, B_{min} = 1.1 μT.

B (T)	$\delta_{a,b}$ (%)	SNR_a ($\times 10^5$ $H_z^{-1/2}$)	SNR_b ($\times 10^5$ $H_z^{-1/2}$)
±0.25	±3	0.55	2.2
±0.50	±6	1.10	4.3
±0.75	±9	1.60	6.5
±1.00	±12	2.20	8.9

TABLE 2

a: W = L = 25 μm, $W_{D1} = W_{D2}$ = 5 μm, d = 5 μm, B_{min} = 2.9 μT.
b: W = L = 100 μm, $W_{D1} = W_{D2}$ = 20 μm, d = 20 μm, B_{min} = 0.7 μT.

B (T)	$\delta_{a,b}$ (%)	SNR ($\times 10^5$ $H_z^{-1/2}$)	SNR ($\times 10^5$ $H_z^{-1/2}$)
±0.25	±5	0.84	3.4
±0.50	±10	1.70	6.9
±0.75	±15	2.50	10.3
±1.00	±20	3.40	13.9

4. MEASUREMENTS

In our numerical calculations (Tables 1 and 2), we have assumed a Hall mobility μ_n^* = 1700 cm^2/Vs. The relative current imbalance, δ was measured for both the DD and TD MAGFET's under the same operating conditions as described in Fig. 1 (see Table 3). As predicted by theory, the TD exhibits relative current imbalance superior to the DD. We note a systematic difference between experimental and numerical results for both devices. This could be explained by the difference in the actual value of Hall mobility in the MAGFET channel and the value assumed in the numerical calculations.

TABLE 3

The measured relative current imbalance, δ for the dual- and triple-drain MAGFET's under the same operating conditions as described in Fig. 1.

DD: $W = L = 100\ \mu m$, $W_{D1} = W_{D2} = 40\ \mu m$, $d = 20\ \mu m$, $V_T = 1$ V.

TD: $W = L = 100\ \mu m$, $W_{D1} = W_{D2} = 20\ \mu m$, $d = 20\ \mu m$, $V_T = 1$ V.

B(T)		-1.0	-0.75	-0.5	-0.25	0.0	0.25	0.5	0.75	1.0
δ (%)	DD	-3.4	-2.60	-1.7	-0.90	0.0	0.80	1.7	2.50	3.3
	TD	-5.1	-3.90	-2.5	-1.20	0.0	1.20	2.5	3.70	4.9

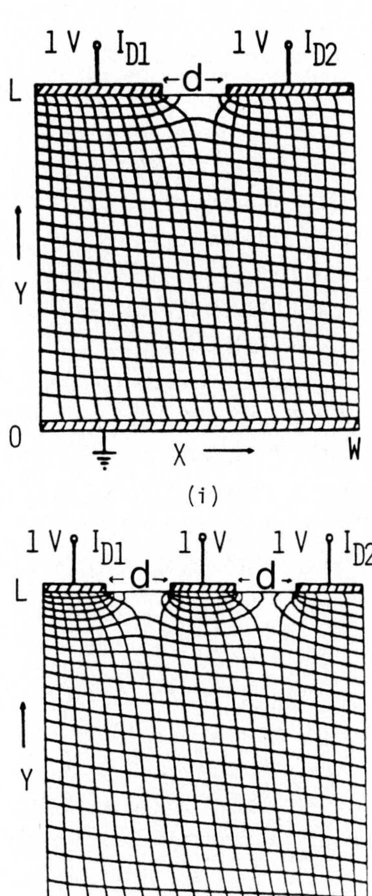

FIGURE 1
Equipotential and current lines for the
(i) DD MAGFET
(ii) TD MAGFET
$W = L = 100\ \mu m$, $d = 20\ \mu m$, $B = 1$ T, and $V_T = 1$ V.

5. CONCLUSION

In view of the multi-drain structure of the MAGFET, partition noise[8] may also be involved, but is not considered here. Noise measurements on a number of MAGFET samples is currently under progress.

ACKNOWLEDGEMENTS

This work has been supported by grants from the Natural Sciences and Engineering Research Council of Canada (NSERC). We would like to thank Phillip Haswell for providing technical assistance with the measurements.

REFERENCES

1. H. P. Baltes, Magnetic sensors, in: Trends in Physics 1984, Proc. 6th Gen. Conf. Europ. Phys. Soc., Prague, Aug. 27-31, 1984, J. Janta and J. Pantoflíček, Eds., Europ. Phys. Soc. 1985, pp. 606-611.

2. R. S. Popović and H. P. Baltes, IEEE J. Solid-State Circuits, SC-18 (1983) 426-428.

3. T. G. M. Kleinpenning, Sensors and Actuators, 4 (1983) 3-9.

4. R. S. Popović, Proc. 12th Yugoslav Conference on Microelectronics, MIEL '84, Nis, Yugoslavia, I (1984) 229-307.

5. A. Chovet, S. Christoloveanu, A. Mohaghegh, and A. Dandache, Sensors and Actuators, 4 (1983) 147-153.

6. A. Nathan, A. M. J. Huiser, and H. P. Baltes, IEEE Trans. Electron Devices, ED-32 (1985) 1212-1219.

7. E. A. Leventhal, Sol. St. Electronics, 11 (1979) 621-627.

8. A. Ambrózy, Electronic Noise (McGraw Hill, New York, 1982).

NOISE IN PHYSICAL SYSTEMS AND 1/f NOISE - 1985
A. D'Amico and P. Mazzetti (editors)

TIME MEASUREMENT BY NOISE: REDUCTION OF FLICKER-FLOOR

László B. KISS and András AMBRÓZY

József Attila University, H-6720 Szeged, Dóm tér 9., Hungary
Technical University Budapest, H-1521 Budapest, Hungary

The possibility of time measurement by a stable Gaussian process has been investigated. This principle enlarges the set of physical processes which can be used for time measurement and in some special cases it is more accurate than other ones.

1. INTRODUCTION

The possibility of time measurement by a stable Gaussian process has been investigated (Fig.1.). This principle enlarges the set of physical processes which can be used for time measurement and in some special cases it is more accurate than the other ones.

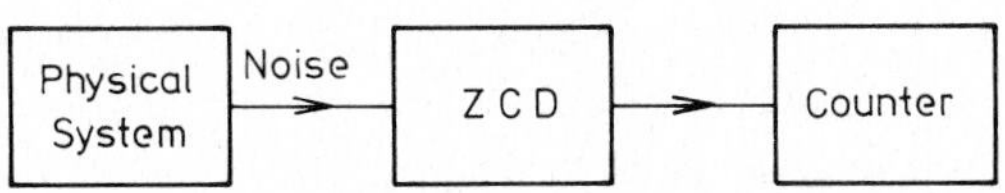

FIGURE 1
Noise-clock. ZCD: Zero Crossing Detector

2. GENERAL CONSIDERATIONS

The mathematical background is well known long ago: the expected frequency ν of zero crossings of a Gaussian process with power spectrum S(f) is given by [1]:

$$\nu = 2\left[\frac{\int_0^\infty f^2 S(f)df}{\int_0^\infty S(f)df}\right]^{1/2} \qquad (1)$$

The accuracy of a noise-clock is governed by laws different from the ones governing the accuracy of a classical clock.

This can be illustrated by a simple example: the zero crossing detector ZCD senses the zeros of the current flowing through the coil of a resonant circuit fed by white noise current (Fig.2.). The following paradox is known for a long time: the shape of the current timefunction $i^*(t)$ strongly depends on the series resistance r, but the expected frequency of zero crossings

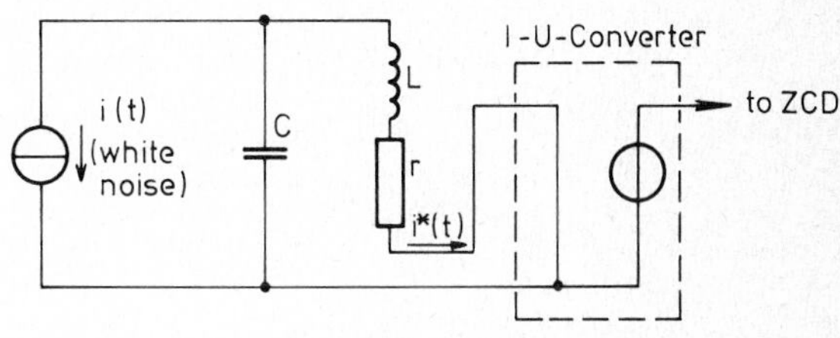

FIGURE 2
Noise-clock with parallel LC Circuit

$$\nu = \pi^{-1}(LC)^{-1} = 2f_0$$

is independent on r [1]. The same resonant circuit in a sine wave oscillator would give:

$$\nu_{osc} = 2f_0\left[1-\frac{r^2C}{L}\right]^{1/2} = 2f_0\left[1-\frac{1}{Q^2}\right]^{1/2} \qquad (2)$$

wich depends on r. It can be easily shown on the basis of Eq.2 that in the case of fluctuating r the frequency of oscillator is fluctuating and the corresponding fractional frequency spectrum [2] (in the high Q-factor limit) is:

$$S_y(f) = \frac{1}{16\cdot Q^4} S_r(f) = \frac{1}{16\cdot Q^4} S_Q(f) \qquad (3)$$

where $S_r(f)$ and $S_Q(f)$ are spectra of quantities

$\delta r/r$ and $\delta Q/Q$ respectively.

It is obvious that in the case of 1/f fluctuation of r the oscillator has a flicker floor [2] while the noise-clock has not. We will investigate this property of resonators on a more general way in Sec.4.

Consider the excess inaccuracy of the noise-clock originating from the nonzero bandwidth Δf around the centerfrequency f_0 of the clock-driving-noise. The driving-noise has a correlation-time: $\tau_{corr} \approx (\Delta f)^{-1}$.

If the average clock-frequency deviation is $\Delta\nu_{corr}$ during τ_{corr}, then:

$$\frac{\Delta\nu_{corr}}{\nu} < \frac{\Delta f}{2f_0},$$

and the measured time deviation ΔT during τ_{corr} is:

$$\Delta T_{corr} = \tau_{corr}\frac{\Delta\nu_{corr}}{\nu} < \tau_{corr}\frac{\Delta f}{2f_0} = \frac{1}{2f_0}$$

For longer measuring time $T = n.\tau_{corr}$ ($n \gg 1$) to calculate the uncertainty ΔT we summarize the n quasi-independent random variables $\Delta T^{(i)}_{corr}$ and we get:

$\Delta T \approx \sqrt{n}.\Delta T_{corr} < \dfrac{\sqrt{n}}{2f_0}$. The relative inaccuracy during time T is:

$$\frac{\Delta T}{T} < \frac{\sqrt{n}}{2f_0 T} = \frac{1}{\sqrt{\Delta f T}}\frac{1}{2Q} = \frac{1}{2\sqrt{f_0 T Q}}$$

It is obvious from Eq.4 that for an accurate noise-clock high center frequency f_0 and Q-factor are necessary (e.g. Q-multiplying technique), therefore we will use the high Q-factor approximation in our calculations.

3. TWO KINDS OF NATURAL NOISE SOURCES

On the basis of Eq.1 we investigated the case of Lorentzian spectra (e.g. natural broadening and collisional broadening of excited gas radiation), and the case of Gaussian spectra (e.g. Doppler-broadening of excited Maxwell-Boltzmann gas radiation). For Lorentzian spectra $S_L(f)=[(f-f_0)^2+\Delta f^2]^{-1}$ Eq.1 is divergent [1] however in practical cases the noise-clock has several (not too stable) cut-off frequencies $f_c > f_0$ (i.e. cut-off frequencies of ZCD), thus we calculated the integrals to an upper limit f_c. In the high f_c limit ($f_c \gtrsim 10 f_0$):

$$\nu_L = 2f_0\left[1 + \frac{\Delta f}{f_0^2}(Af_c + Bf_0) - \frac{\Delta f^2}{2f_0^2}\right] \qquad (5)$$

where A and B are quasi constants in the order of 1. ν_L weakly depends on f_c consequently Lorentzian noise is usable in a noise-clock. In the case of a Gaussian spectrum $S_G(f) = \exp\left[-\dfrac{(f-f_0)^2}{2\Delta f^2}\right]$ the clock-frequency by Eq.1. is:

$$\nu_G \cong 2f_0\left(1 + \frac{\Delta f^2}{2f_0^2}\right) \qquad (6)$$

4. NOISE-CLOCKS WITH RESONATORS

To make a narrowband driving-noise we can use (active or passive) resonator in the noise-clock. Most of the resonators in the case of white driving-noise and high Q-factor have the following output noise spectrum:

$$S_f(f) = \frac{k}{1+Q^2\left(\frac{f_0}{f} - \frac{f}{f_0}\right)^2}$$

Let us make a new driving-noise by integration of the filtered noise (Fig.3.):

$$S_i(f) = \frac{1}{f^2}\frac{k}{1+Q^2\left(\frac{f_0}{f} - \frac{f}{f_0}\right)^2}$$

Then the noise-clock-frequency given by Eq.1 is:

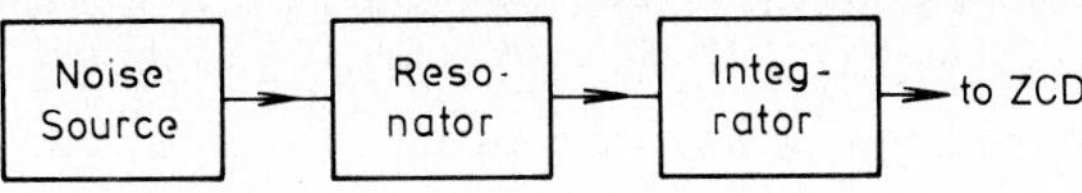

FIGURE 3
Noise-clock with resonator and integrator in it

$\nu_i = 2f_0$, thus it is independent on the Q-factor (like the noise-clock with parallel LC-circuit mentioned in Sec.2.). The Q-factor is connected with the losses of the resonator, and since the losses are dissipative, thus usually the Q-factor is modulated by 1/f noise (e.g. quantum-1/f fluctuation of scattering cross section [3]). Therefore the noise-clock descibed above has a reduced flicker floor: the fluctuation of the Q-factor will not appear in the clock-frequency. Moreover if even a systematic drift of the Q-factor exists (e.g. increase of any frictions) it will not appear in the clock-frequency.

The question of resonators mentioned above can be illustrated by the example of a quartz-resonator. It can be easily shown, that the relative frequency shift of the (series-) phase-resonance of the equivalent circuit (Fig.4.) is:

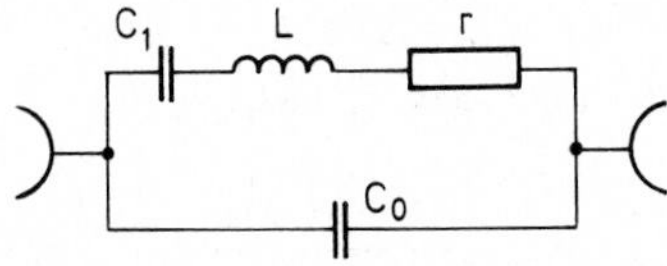

FIGURE 4
Equivalent circuit of a quartz-resonator

$$\frac{\delta f}{f_0} = \frac{1}{2Q^2}\,\frac{C_0}{C_1},$$

thus in the case of fluctuating Q-factor the corresponding fractional frequency spectrum of the quartz-oscillator is:

$$S_y^*(f) = \frac{1}{16Q^4}\,S_Q(f)\left(\frac{C_0}{C_1}\right)^2 = \frac{1}{16Q^4}\,S_\tau(f)\left(\frac{C_0}{C_1}\right)^2$$

Where $S_\tau(f)$ is the spectra of quantity $\delta\tau/\tau$ (τ is the relaxation time of the resonator). In Ref. [4] the fluctuation of the center frequency of quartz-resonators, caused by fluctuating τ, is investigated, and the fractional frequency spectrum is given as

$$S_y^1(f) = \frac{1}{Q^4}\left(\frac{E_0}{\Delta E}\right)^2 S_\tau(f)$$

where $(E_0/\Delta E)^2 \approx 10^5$. Whereas in a classical clock both $S_y^*(f)$ and $S_y^1(f)$ exist, in a noise-clock $S_y^1(f)$ will only remain.

The reduction of the flicker floor in a noise-clock for a 5 MHz, 5th overtone resonator [5]:

$$\frac{S_y^1(f)}{S_y^*(f)} = \frac{1}{256}$$

5. SUMMARY

The possibility of time measurement by noise was described. Usually the short term stability of a noise-clock is lower, the long term stability may be higher than the same properties of the corresponding classical clock.

ACKNOWLEDGEMENT

One of the authors (L.B.K.) is very grateful to the Organizing Committee for the financial support.

REFERENCES

1. S.O.Rice, Bell.Syst.Techn.Journ 23 (1944) 282, 24 (1945) 46.

2. J.J.Gagnepain, Noise in Physical Systems and 1/f Noise (Ed.M.Savelli & al.) North Holland (1983) 309.

3. P.H.Handel,Phys.Rev. 22A (1980) 745.

4. J.J.Gagnepain & al., Relation between 1/f noise and Q factor in quartz resonators at room and low temperatures,first theoretical interpretation, in: Proc.35th Annual Frequency Control Symposium,Philadelphia (1981)

5. J.J.Gagnepain, Fundamental noise studies of quartz crystal resonators, in: Proc. 30th Annual Frequency Control Symposium, Atlantic City (1976)

NOISE IN PHYSICAL SYSTEMS AND 1/f NOISE - 1985
A. D'Amico and P. Mazzetti (editors)

DATA ACQUISITION SYSTEM FOR NOISE CHARACTERIZATION IN THE VERY LOW FREQUENCY RANGE

Lech Z. HASSE, Alicja KONCZAKOWSKA, Ludwik SPIRALSKI

Institute of Electronic Technology, Technical University of Gdańsk, ul.Majakowskiego 11/12, 80-952 Gdańsk, Poland

A system for digital noise measurements in the very low-frequency range has been presented. The data acquisition is carried out by a new original method of collecting the optimum number of data. The system makes it possible to perform comprehensive noise investigations of electronic devices using the method outlined in the paper.

1. THE ADVANTAGES OF 1/F NOISE INVESTIGATIONS

Detailed investigations of electronic devices noise properties in the very low frequency range consist in measuring of the statistical characteristics of noise. In this way it is possible to construct mathematical models of noise and to draw conclusions from physical implications of the obtained models. The correlation between 1/f noise and long-term drift of electronic device and its reliability results from high sensitivity of 1/f noise to the conditions on the structure surface. The investigations of 1/f noise as a function of voltage bias, current level, temperature etc. make it possible to examine the degree of surface contamination, to conclude about the phenomena in the vicinity of contacts, and material defects. They also allow us to localize noise sources and give a convenient, non-destructive method of the evaluation of structure fabrication technology, providing the most general information on both, the state of the device surface and bulk region. Therefore, these investigations provide a possibility of full selection of the devices as regards their reliability in the case of special apjlications.

2. THE METHOD OF COLLECTING THE OPTIMUM NUMBER OF DATA

The measuring system constructed on the basis of the special method of data acquisition enables to do a number of such investigations as were mentioned above.

The design of analog antialiasing filters preventing spectrum superposition is difficult or even practically impossible in the very low frequency range in view of required filter time constants. Therefore, usually a very large number of samples from every noise waveform is recorded to estimate a characteristics in the correspondingly low frequency range. All those data have to be fed to a computer where, after preliminary data processing, the values of individual parameters or the characteristics are calculated. This procedure occupies and creates an unnecessary load to computer memory and the data have to be recorded previously on a large amount of carrier (tape, for example). These drawbacks are eliminated in the system which contains, among others, the following modules (Figure 1):

- analog filter with cut off frequency f_{max},
- A/D converter,
- digital low pass filters with cut off frequencies f_i,
- data reduction system which supplies only every kth sample to the record system,
- recorder.

To reduce farther the maximum frequency contained in the analyzed signal, additional tracks composed of digital low pass filters and data reduction modules can be attached to the data reduction output, depending on the needs. This operation may be repeated many times. So, for example, the data for the power spectral density estimation are recorded separately for every decade of frequency.

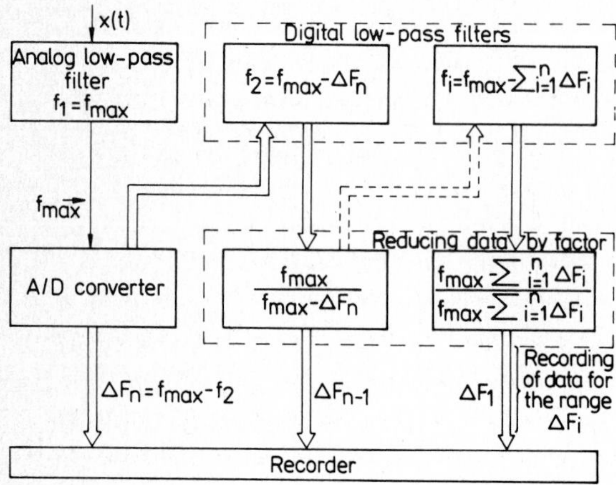

FIGURE 1
A method of recording the optimum number of data using a single analog filter and simultaneous multiple low pass digital filtration and data reduction.

The main advantages resulting from the application of this type of system consist in elimination of analog filters in the range of very low frequencies, thus enabling the operation of analog waveform discretization to be performed at a single fixed sampling frequency, providing simultaneously the collection of the minimum number of samples required for the assumed accuracy of estimation.

3. THE SYSTEM OF COLLECTING THE OPTIMUM NUMBER OF DATA

The method of data collection outlined in the previous paragraph has served as a basis for the construction of a system for data recording in the frequency range from 10^{-4} Hz to 10 Hz. The general functional block diagram of the system is presented in Figure 2.

The solid-state devices under test are placed in a thermostat with two-stage temperature stabilization to obtain a temperature stability at a level of 10^{-4} K. Measurement signals collected at the output of amplifiers attached to the devices under test are transmitted to preamplifiers and analog antialias filters and next to DAS 1128 (Analog Devices) circuit which acts as a commutator of measurement channels and A/D converter. The maximum number of measured devices equals to 15. The results of measurements from individual channels are transmitted to minicomputer subsystem which directs them for recording either directly or after filtering and reducing the data. The minicomputer subsystem consists of: processor module, program memory module, buffer memory module, real-

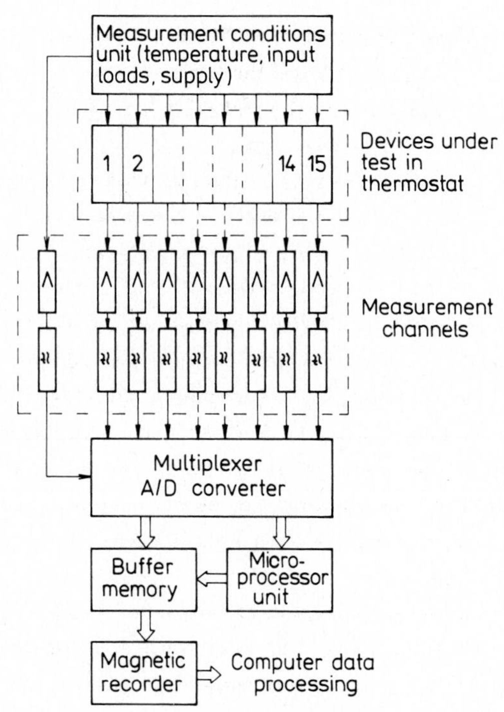

FIGURE 2
Functional scheme of the system of data collecting in the range of low and very low frequencies.

time clock module, module of fast multiplication of 16-bit data and the modules of control desk service and recorder. The digital part of the system is controlled at the operation desk.

All arithmetic operations forming the algorithm of digital filtration have to be performed in the intervals between the successive data introduced at the filter input.

The digital filtration in the system is realized by means of Butherworth's filters, within a parallel structure.

In time domain the digital filtering consists in performing the following arithmetic operations for each channel:

$$y_i[n] = A_{oi}\ x[n] + A_{1i}\ x[n-1] +$$

$$- B_{1i}\ y_i[n-1] - B_{2i}\ y_i[n-2],$$

$$y[n] = \sum_{i=1}^{M} y_i[n], \qquad n = 1,2....,N$$

where: M - the number of filter branches,

$A_{j,i}$ $B_{j,i}$ - filter coefficients,

$x[n]$ - input signal of filters at a moment $n\Delta t$,

$y_i[n]$ - output signal of filter in the ith branch at a moment $n\Delta t$,

$y[n]$ - output signal of filter at a moment $n\Delta t$.

The time of single-operation execution (one step of calculation at a time moment $n\Delta t$) for a specified number of branches M is obviously not constant and in the constructed system with application in hardware of an additional multiplying circuit amounts to 45 ms, 65 ms, 85 ms and 105 ms for M = 2, M = 3, M = 4 and M = 5, respectively.

The system makes it possible to record data (the same number N of data for the operation modes in the frequency range from 10^{-4} to 10 Hz) to estimate the power spectral density of noise.

Before starting the recording of noise waveforms, the parameters characterizing the operation mode of the system should be selected and set at appropriate values at the operation desk. These parameters are, among others: number of recording channels, number of digital filter branches M, number of recorded samples N, digital filter coefficients, operation mode.

The 16 channels of this system work parallely and therefore the record on a magnetic tape is in a multiplexed form of data from each channel, after in real-time digital filtering and decimation. For the goals of correlation and spectral analysis the recorded data should be demultiplexed and storaged in computer memory.

4. APPLICATION OF THE MEASURING SYSTEM

Using appropriate test circuits determining the measurement parameters it is possible to measure all standard noise parameters, such as noise figure, equivalent input noise voltage and current, noise temperature, equivalent noise resistance and noise index. The digitally recorded waveform allow us also to construct more complex mathematical description used subsequently for the determination of statistical characteristic estimators such as power spectral density, autocorrelation and cross-correlation functions or probability density function.

Most of these characteristics can be determine using various methods of their estimation depending on the specific subject of investigations and the required final form of the results. It is also possible to conclude about stationarity of the estimated parameters and characteristics.

The system makes it also possible to detect changes of temperature and of the supply voltage of the measured devices. This is essential in long-time measurements since it allows us to investigate the effect of temperature and supply conditions on the measurement results.

Concluding, the above described system is suitable for complex investigations of the noise properties of electronic devices in the range of very low frequencies.

NOISE IN PHYSICAL SYSTEMS AND 1/f NOISE - 1985
A. D'Amico and P. Mazzetti (editors)

MEASUREMENT OF THE ELECTRON PAIR TRANSFER FREQUENCY IN A SUPERCONDUCTING WEAK LINK DEVICE

R.J. PRANCE, H. PRANCE, T.D. CLARK, J.E. MUTTON, T.P. SPILLER and R. NEST

School of Mathematical and Physical Sciences, University of Sussex, Brighton, Sussex BN1 9QH, U.K.

1. INTRODUCTION

The concept of quantum mechanical electron pair transfer through a superconducting weak link was first introduced by Josephson in 1962 (1). Although in Josephson's work weak links were described in terms of quantum mechanical pair transfer processes (considered as a barrier penetration problem) it has recently become apparent (2,3,4,5) that when appropriate experimental techniques are adopted circuits comprising a weak link enclosed by a thick ring, either superconducting or normal, can be observed to act in a fully quantum mechanical way. This requires that, for a weak link capacitance C, the magnetic flux Φ threading the ring and $C\dot{\Phi}$, the effective charge at the weak link, be treated as canonically conjugate variables. Thus

$$[\Phi, C\dot{\Phi}] = i\hbar \tag{1}$$

with the uncertainty relation

$$\Delta\Phi.\Delta(C\dot{\Phi}) \gtrsim \hbar \tag{2}$$

From (2) it can beinferred that weak link ring circuits should exhibit two limits of behaviour. These are

(i) where the flux threading the ring is rather well defined about integer ($n = 0, \pm 1, \pm 2,..$ etc.) values of the superconducting flux quantum $\Phi_0 = h/2e$, i.e. when the charge at the weak link is very uncertain, $\Delta Q >> q$, with $Q = C\dot{\Phi}$ and $q = 2e$ is the pair charge.

(ii) where the charge at the weak link is rather well defined (localised) about integer ($N = 0, \pm 1, \pm 2, ...$ etc.) values of q, i.e. when the flux threading the ring is very uncertain, $\Delta\Phi >> \Phi_0$.

Case (i) corresponds to the limit in which SQUID magnetometer behaviour can be observed and has been discussed by us at length in a previous publication (2). We shall refer to this as the flux mode regime. Case (ii) corresponds to the canonically conjugate limit of behaviour which we shall call the charge mode regime. In this limit it is convenient to describe the weak link, capacitance C, in terms of a set of polarisation states $\pm Nq$ ($N = 0, \pm 1, \pm 2, ...$ etc.). The energy W_N stored on this capacitor is due to this polarisation charge and the displacement charge ($Q_x = CV_x$, $V_x = -\dot{\Phi}_x$) created by any time rate of change of externally generated flux Φ_x cutting across the weak link. Thus

$$W_N = (Nq + Q_x)^2/_{2C} = (Nq + CV_x)^2/_{2C} \tag{3}$$

We can write down a Schrödinger equation for the weak link of the form

$$W_N A_N = E(Q_x)A_N \tag{4}$$

where A_N is the amplitude for the weak link to be in the Nq^{th} polarisation state. If we then allow weak quantum mechanical coupling between nearest neighbour polarisation states, as Josephson did (1), then we can rewrite (4) to incorporate these charge state transition process as

$$W_N A_N - \frac{\hbar\nu}{2}(A_{N+1} + A_{N-1}) = E(Q_x)A_N \tag{5}$$

where $\nu/4\pi$ is the frequency for the quantum mechanical transfer of one pair through the weak link.

Equation (5) yields the ground state energy of the weak link operating in the charge mode limit, as shown in figure 1a for a small value of $\nu(\hbar\nu = 0.05\ q^2/C)$.

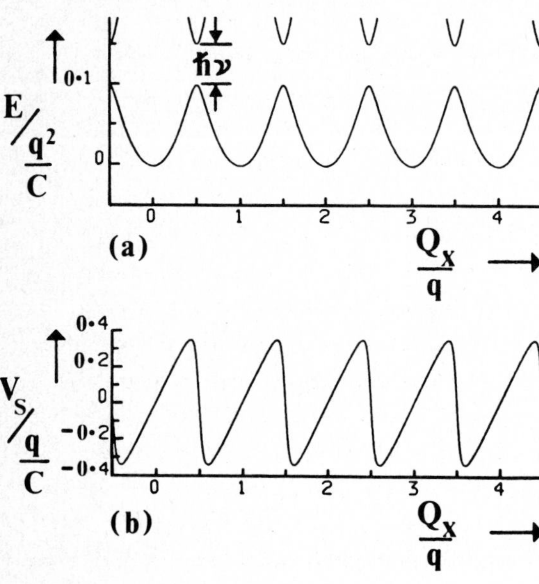

FIGURE 1

By analogy with electron energy bands in condensed matter physics, we shall, for convenience, refer to this as the ground state charge space (Q_x) energy band of the weak link. The screening voltage (4) $V_X = dE(Q_x)/dQ_x$ for this ground state band is shown in figure 1b.

2. CHARGE BAND DYNAMICS

Recently (3) we have demonstrated that it is possible to investigate charge band behaviour in superconducting weak links using radio frequency (rf) SQUID techniques. With the weak link enclosed by a ring, either superconducting or normal, to allow inductive coupling to a parallel capacitor inductor tank circuit, we have shown (5) that the cycle averaged rf voltage V_{OUT} developed across the tank circuit at zero phase (in phase) with respect to the input rf drive current I_{IN} to the tank circuit is given by the quantum mechanical expression

$$V_{OUT} = Q\omega_R L_t[I_{IN}\sin^2\delta - \frac{\mu}{\Lambda_r\omega_r\pi}\int_{\pi}^{\pi} d\zeta \sin\zeta \langle V_S(Q_X^{tot}, \zeta)\rangle] \qquad (6)$$

with the weak link operating in a single charge band - here the ground state band - with screening voltage $\langle V_S(Q_X^{tot},\zeta)\rangle$. In (6) Q is the quality factor of the unloaded tank circuit, $\omega_{R/2\pi}$ is the resonant frequency of the tank circuit - weak link ring combination, L_t, Λ and M_{ts} are, respectively, the tank circuit coil inductance, the ring inductance and the mutual inductance between the tank circuit coil and the ring, $\mu = M_{ts/L_t}$ with coupling coefficient between the ring and tank circuit coil $K_{ts} = [M_{ts}{}^2/L_t\Lambda]^{\frac{1}{2}}$, δ is the phase of the unrenormalised tank circuit impedance at frequency ω_r. The renormalised ring inductance and resonant frequency due to the ring-tank circuit interaction are, respectively $\Lambda_r = \Lambda(1-K_{ts}{}^2)$ and $\omega_r = \omega_R(1-K_{ts}{}^2)^{-\frac{1}{2}}$ and the argument of the screening voltage $\langle V_S\rangle$ is $Q_X^{tot} = -C[\dot{\Phi}_X + \mu\dot{\Phi}_{xrf}]$ where $\mu\Phi_{xrf}$ is the flux felt by the ring due to the rf current in the tank circuit.

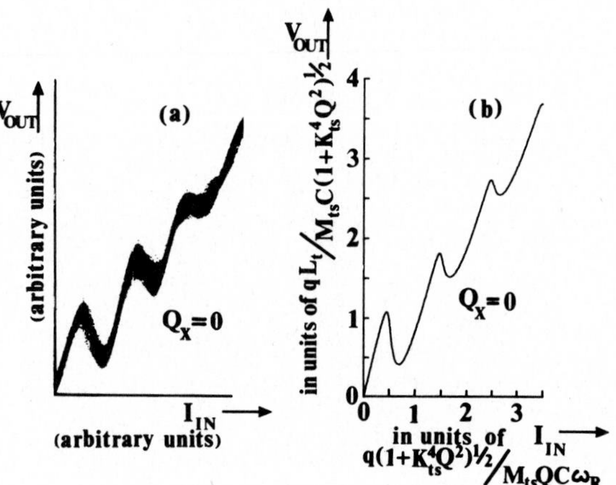

FIGURE 2

In figure 2a we show an experimental in phase rf-biased ($\omega_{R/2\pi}$ = 27 MHz) V_{OUT} versus I_{IN} characteristic plotted at a temperature T = 1.5K with $Q_x = 0$ for a very small cross-section niobium point contact weak link (comparable in section with the superconducting penetration depth) enclosed within a standard two hole Zimmerman type SQUID block (2) machined from copper. The rf voltage and current scales are approximately one hundred times larger than required to observe flux mode SQUID magnetometer behaviour, as required in the charge mode limit (3,4). In figure 2b we show the theoretical in phase, $Q_x = 0$, V_{OUT} versus I_{IN} characteristic, computed from (6), for the SQUID parameters pertaining to figure 1a, namely $Q = 40$, $K_{ts}^2 Q = 1.58$, $\Lambda = 5 \times 10^{-10}$H, $L_t = 2 \times 10^{-7}$H and the weak link operating in the ground state charge band with $\hbar\nu = 0.05q^2/C$ (figure 1a).

3.CHARGE BAND SUSCEPTIBILITIES

In figure 2a we have, in effect, monitored the first derivative of the ground state charge band energy as a function of Q_x[i.e. $\langle V_s \rangle = dE(Q_x)/dQ_x$]. As we have shown in earlier experiments on external flux (Φ_x) space energy bands in flux mode weak links enclosed by a superconducting ring the susceptibility of the weak link ring, operating in a particular flux band, can be monitored by the method of double derivative noise spectroscopy (2). All that is required is that the weak link ring be coupled to a tank circuit and the mean square noise voltage across the tank circuit $\langle|\Delta V_t|^2\rangle$ recorded as a function of slowly varying (quasistatic) flux Φ_x. For a tank circuit resonant frequency of $\omega_{R/2\pi}$ the mean square voltage for the weak link ring operating in a particular flux band (i.e. κ = 1, 2, 3, ... etc.) is given by

$$\langle|\Delta V_t|^2\rangle \simeq \omega_R^2 k_B T L_t [1 - K_{ts}\Lambda d^2E_\kappa(\Phi_x)/d\Phi_x^2] \quad (7)$$

where $\chi_\kappa^m(\Phi_x) = \Lambda\, d^2E_\kappa(\Phi_x)/d\Phi_x^2$ is the in-band magnetic susceptibility of the superconducting weak link ring.

We have found that we can use the same passive (i.e. no rf current applied to the tank circuit) technique to plot the in band electric susceptibility

$$\chi_\kappa^e\,(Q_x) = -\,C.d\langle V_x(Q_x)\rangle/dQ_x = -\,C.d^2E(Q_x)/dQ_x^2 \quad (8)$$

of a superconducting weak link operating in the charge mode limit. The displacement charge Q_x can be generated by a Faraday Law voltage coil (3) or, where the weak link is enclosed (shunted) by a copper ring, simply by passing a DC current I_x through the normal shunt. Since the shunt has finite resistance this current creates a voltage V_x across the weak link. In figure 3 we show a plot of $\langle|\Delta V_t|^2\rangle$ versus V_x for the hybrid copper-niobium SQUID ring of figure 2a. The resonant frequency of the coupled ring-tank circuit system is still 27 MHz at T = 1.5K. The total shunt resistance of the two hole copper SQUID block is approximately $10^{-4}\Omega$ and the bias current I_x(i.e. V_x in figure 3) is arranged to increase linearly with time, from negative to positive. The $V_x = 0$ origin in figure 3 is at the midpoint between the two susceptibility peaks. We note that the second derivative, d^2E/dQ_x^2 versus Q_x, of the ground state charge band in figure 1a consists of a set of cusp-like spikes, spaced by q along the Q_x axis, with a half width determined by the matrix element $\hbar\nu$. The cusp-like spikes are evident in figure 3. The linearly rising background between these spikes is caused by breakthrough of the rather large I_x drive (± 1 A from the origin to the susceptibility spikes) into the rf detection electronics. There is also some truncation in the amplitude of the spikes due to the filtering required to produce an adequate signal to noise. For comparison, we show inset in figure 3 a small section of a magnetic susceptibility pattern [equation (7)] for a superconducting flux mode weak link ring operating in the ground state flux space band (2) recorded using the noise spectroscopy technique. A best theoretical fit for this pattern is found for a quantum mechanical flux transfer matrix element

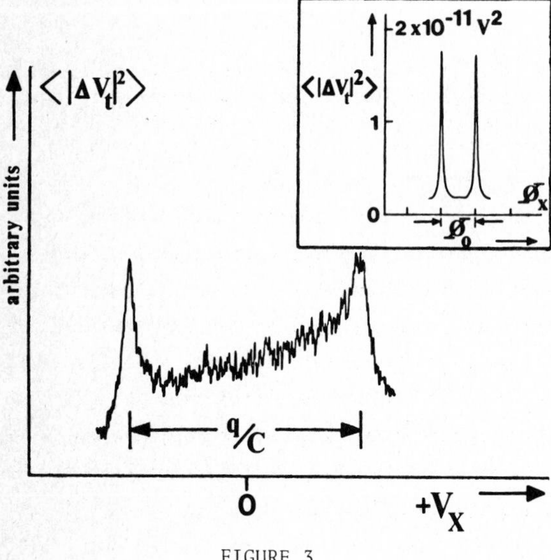

FIGURE 3

$\hbar\Omega = 0.05\ \Phi_0^2/\Lambda$, where $\Omega/4\pi$ is the frequency at which Φ_0 flux bundles cross the weak link. From figure 3 the ratio of the spike half width to the inter-spike spacing is 0.063 which corresponds very closely to the value of this ratio for the second derivative of the ground state energy band of figure 1a (not shown). We therefore conclude that for this particular point contact weak link, at T = 1.5K, the quantum mechanical electron pair transfer frequency $\nu/4\pi$ is very close to $0.05q^2/hC$ where, using a Faraday Law voltage source to split the triangular features in figure 2a, we estimate $C \simeq 1 \times 10^{-16}$ F(3). This sets $\nu/4\pi = 4 \times 10^{10}$ Hz.

REFERENCES

1. B. D. Josephson, Phys. Letts. 1, 251 (1962);

2. R. J. Prance, J. E. Mutton, H. Prance, T. D. Clark, A. Widom and G. Megaloudis, Helv. Phys. Acta 56, 789 (1983).

3. R. J. Prance, T. D. Clark, J. E. Mutton, H. Prance, T. P. Spiller and R. Nest, Phys. Letts. 107A, 133 (1985).

4. H. Prance, R. J. Prance, J. E. Mutton, T. P. Spiller, T. D. Clark and R. Nest, "Quantum Duality in SQUID rings", published in "Festkörperprobleme-Advances in Solid State Physics XXV, 337 (1985).

5. T. P. Spiller, T. D. Clark, R. J. Prance, H. Prance, J. E. Mutton and R. Nest, "Quantum Mechanical derivation of the V_{OUT}-I_{IN} characteristic for an AC-biased SQUID Magnetometer" submitted to Physics Letters.

NOISE IN PHYSICAL SYSTEMS AND 1/f NOISE - 1985
A. D'Amico and P. Mazzetti (editors)

THE METHOD AND SYSTEM FOR SEMICONDUCTOR NOISE EVALUATION

Alicja KONCZAKOWSKA

Institute of Electronic Technology, Technical University of Gdańsk, ul.Majakowskiego 11/12
80-952 Gdańsk, Poland

In this paper the method and system for data acquisition and analysis of semiconductor noise in very low frequency ranges is presented. The analog and digital lowpass filters as antialiasing filters were applied. The digital filtration during the sampling interval has been realized. The power spectral density function for the acquisited data evaluation were used.

1. INTRODUCTION

Contemporary trends in the reliability research of the semiconductor devices yield that the measurement and analysis their noise in very low frequency ranges are very useful. Commonly used parameter for noise evaluation in frequency domain is the power spectral density function (PSDF) of the noise measurement signal. The shape and level values of PSDF make possible the classification of the semiconductor devices and identification some of their malfunctions and defects [2,3].

One of the most effective method of the PSDF estimation is the digital method (algorithm of fast Fourier transform - FFT).

The objective of this paper is to show the system for data acquisition in sequence frequency decades below 1 Hz and the transistor noise PSDF estimation in these frequency ranges.

2. BASIC PRINCIPLE

Decreasing of maximal frequency of the analysis range can be obtained by lowpass filtration. The construction of analog lowpass filters in very low frequency ranges is very difficult to realize.

In this situation the application in the data acquisition system only one analog lowpass filter with cut off frequency f_{max}, a/d converter with sampling frequency f_s (where $f_s \geqslant 2f_{max}$), digital lowpass filter with cut off frequency $0{,}1f_{max}$ and decimation unit are adventaged. The decimation of filtrated data consists of keeping only every 10-th datum. The digital filtration, decimation and data transmission must be realized in realtime.

The transfer characteristic $H(z)$ of the realized parallel form of the digital filter is expressed in z-domain as follows:

$$H(z)=\sum_{i=1}^{M}\frac{A_{oi}+A_{1i}\cdot z^{-1}}{1+B_{1i}\cdot z^{-1}+B_{2i}\cdot z^{-2}} \qquad (1)$$

where: M - filter order,

A_{oi}, A_{1i}, B_{1i}, B_{2i} - filter coefficients of i-th branch,

$z^{-1}=\exp(-j2\pi f/f_s)$ - unit delay operator.

This M th-order filter is defined by the following input - output relation:

$$y_i(n\Delta t) = A_{oi}x(n\Delta t) + A_{1i}x\left[(n-1)\Delta t\right] + - B_{1i}y_i\left[(n-1)\Delta t\right] - B_{2i}y_i\left[(n-2)\Delta t\right] \qquad (2)$$

$$i = 1,2,\ldots,M$$

$$y(n\Delta t) = \sum_{i=1}^{M} y_i(n\Delta t) \qquad (3)$$

$$n = 0,1,\ldots,10(N-1)$$

where: $x(n\Delta t)$ - filter input data,

$y(n\Delta t)$ - filter output data,

$y_i(n\Delta t)$ - i th-branch output data,

$\Delta t = 1/f_s$ - sampling interval.

After decimation the output sequence has the form $y(10n\Delta t)$ n = 0,1,...,N-1. The time for the digital filtration and decimation depends on the filter order M, and it must be lower than sampling interval, because all digital procedures are provided during Δt.

In presented system the noise signals of semiconductor devices are collected and next analysed in decades: 10^{-1}- 1 Hz, 10^{-2}-10^{-1} Hz and 10^{-3}-10^{-2} Hz. In this system for each decade suitable analog or analog and digital lowpass filters were applied. The digital filtration and decimation have been realized by means minicomputer.

After acquisition of the required number of data for each decade the PSDF was estimated based on the same procedure FFT[1]. The segment averaging procedure (for Q_s time slices) is used to reduce the normalized standard error ε_r.

standard error of spectra estimate ε_r = 0,1.

The spacing Δf of spectra estimate is equal to the lower frequency limit of the analysed decade. For example in decade 10^{-1} - 1 Hz frequency spacing is

$$\Delta f = \frac{1}{N \cdot \Delta t} = \frac{3,2 f_{max}}{N} = 0,1 \text{ Hz.}$$

The presented main principles of the data acquisition and analysis are the base for the design and realization of the system.

3.THE SYSTEM AND SOFTWARE DESCRIPTION

The system for the data acquisition in frequency range 10^{-3} - 1 Hz is consisted of:

- test head,
- two analog lowpass filters with cut off frequency f_{max} = 1 Hz or 10^{-1} Hz, are used exchangeable,
- a/d converter with sampling frequency f_s=3,2 Hz or 3,2 10^{-1}Hz,
- translator consists of: data translator unit

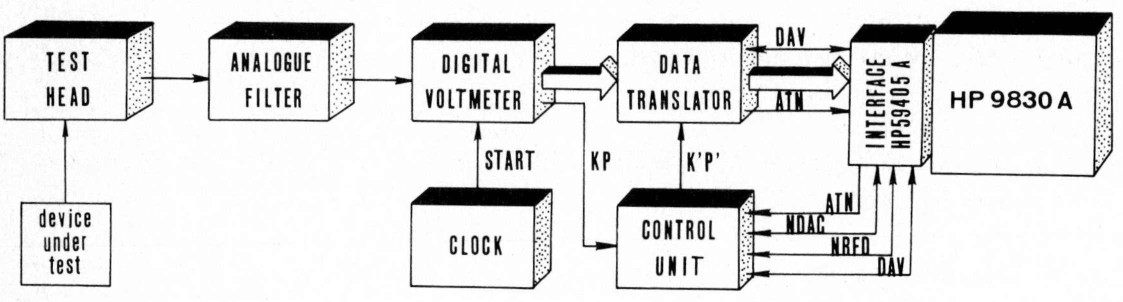

FIGURE 1

Block diagram of the data acquisition and analysis system

In the presented system the sampling frequency was chosen such as $f_s = 3,2 f_{max}$, where f_{max} is cut off frequency of the used analog lowpass filter.

In each decade the number N of data for PSDF estimation is equal to 32, the number of time slices is Q_s = 100, so it gives the normalized

and control unit,

- HP Interface Bus 95405A,
- HP Calculator 9830A.

Block diagram of the system is shown in Figure 1.

The test head consists of: the biasing circuit of the transistor under test and preamplifier. The transistor works in common emitter mode. The input stage of preamplifier is constituted of a set of FET to obtain the lowest noise.

The analog filters are a Butterworth's type with cut off frequency f_{max} = 1Hz or 10^{-1}Hz and slope 48 dB/octave.

As the a/d converter has been used the digital voltmeter V 628, which is controlled by quartz stabilized clock with the sampling frequency f_s. The signal START from the clock with frequency f_s is the control signal for the voltmeter output data in parallel mode and BCD code. The control signal KP from the a/d converter indicates the end of the convertion procedure.

The data in BCD code are translated to the ASCII code in the data translator unit. The signal K'P' from the data translator unit depends on signal KP and indicates the data multiplexion process.

The HP Interface Bus is a standard unit.

The HP 9830A Calculator for the data filtration, decimation, acquisition and transmission control has been applicated.

The acquisition and pre- processing of the transmitted data in the HP 9830A are realized based on the software procedures:

1. data transmission and acquisition,
2. data transmission, filtration, decimation and acquisition.

The flow chart of the procedure N° 2 is presented in Figure 2.

The data are recorded in the calculator memory in array mode $A(Q_s, 32)$. The number Q_s means number of time slices, which are necessary for power spectral estimation.

Q_s, z and coefficient of the digital filter are the input parameters. The digital filtration is realized based on equation (2) for all input data, but calculation of the results from equation (3) is realized only for decimated data. The values of z rules of a "print mode":
z = 1 the results (data) are printed in real-

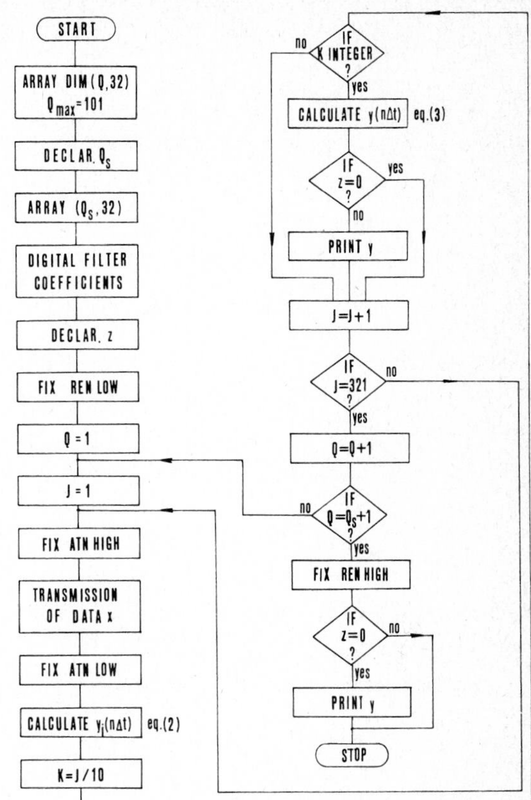

FIGURE 2
The flow chart of the data transmission, filtration, decimation and acquisition procedures

time, z = 0 results are printed in the last step.

The time t_f for every filtration step (eqs 2, 3) depends on the order M of of digital filter and for example for M = 3 and M = 5 is respectively equal to 280 ms and 460 ms. The transmission time t_t for a datum is equal to 10 ms and the integration time t_i of voltmeter V 628 t_i= 20 ms. The operation time $t_o = t_i + t_t + t_f$ should be smaller than the sampling interval Δt.

The maximal number of the data possible to record by this program is equal to 3200.

The next step in processing of the acquisited data in HP is the PSDF estimation via FFT procedure. Every raw spectra estimator is calculated within 32 data and final smooth spectra estimate is computed by Q_s segment averaging.

The leakage is limited by chosen taper data window.

4. EXPERIMENTAL RESULTS

This presented system has been tested by means of the sine signal. Also the experimental evaluation for the noise measurement signal of the transistors BC 107 have been done. For example in Figure 3 the PSDF of npn transistor BC 107 is presented. Multi level burst noise was generated by this transistor, and this phenomena was additionally observed on the osciloscope screen. The appearance of multi level burst noise causes in effect the shape of PSDF with local maximum as it is shown in Figure 3.

The results obtained in experimental tests yield good agreement to the theoretical and practical patterns. So, this system is useful in the spectral analysis of semiconductor noise for the reliability goals.

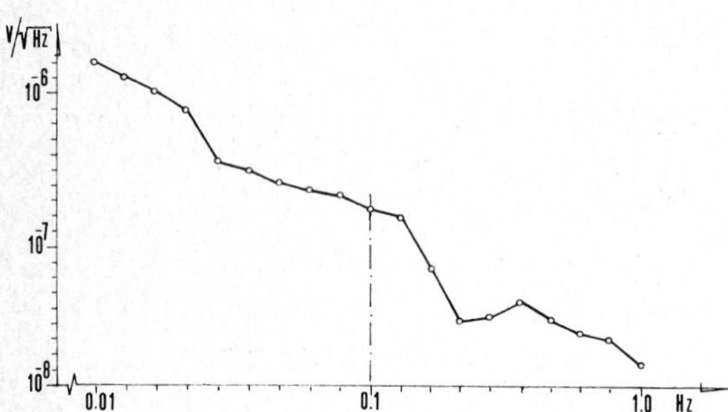

FIGURE 3
The power spectral density function of npn transistor BC 107

REFERENCES

1. J.S. Bendat and A.G. Piersol, Random Data, (Wiley-Interscience, 1971).

2. M. Mihaila and K. Amberiadis and A. Van der Ziel, Solid-State Electronics 27,7 (1984), pp. 675-676.

3. D. Sauvage and M. Gueguen, 4th International Conference on Reliability and Maintainability, France, (1984).

LIST OF PARTICIPANTS

R. ALABEDRA,
C.E.M., C.N.R.S. - UA 391,
Université des Sciences et Techniques
du Languedoc,
34060 Montpellier Cedex,
France.

A. AMBROZY,
Technical University,
Budapest,
Hungary.

H.P. BALTES,
Department of Electrical Engineering,
University of Alberta,
Edmonton,
Alberta,
Canada T6G 2E1.

A. BEARZOTTI,
I.E.S.S./CNR,
Via Cineto Romano 42,
00156 Roma,
Italy.

T. BELATOUI,
Equipe de Microoptoélectronique de Montpellier
(EM^2) U.S.T.L.,
34060 Montpellier Cedex,
France.

G. BERTOTTI,
Istituto Elettrotecnico Nazionale,
Corso Massimo d'Azeglio 42,
10125 Torino,
Italy.

S. BHATTACHARYA,
Exxon Research and Engineering,
Clinton Township Rt. 22E,
Annandale,
NJ 08801,
U.S.A.

G. BLASQUEZ,
CNRS - L.A.A.S.,
7 Avenue du Colonel Roche,
31077 Toulouse Cedex,
France.

M.F. BOCKO,
Department of Electrical Engineering,
Hopeman Hall,
University of Rochester,
Rochester,
NY 14627,
U.S.A.

F. BORDONI,
I.F.S.I.,
Frascati,
Roma,
Italy.

G. BOSMAN,
Department of Electrical Engineering,
University of Florida,
Gainesville,
FL 32611,
U.S.A.

J.J. BROPHY,
University of Utah,
Salt Lake City,
UT 84112,
U.S.A.

P. CARELLI,
I.E.S.S./CNR,
Via Cineto Romano 42,
00156 Roma,
Italy.

A. CHOVET,
Laboratoire de Physique des Composant
à Semiconducteurs - U.S., CNRS 840,
Inst. National Polytechnique de Grenoble Enserg,
23 rue des Martyrs,
38031 Grenoble Cedex,
France.

L.O. CHUA,
University of California,
College of Engineering,
Department of Electrical Engineering,
Berkeley,
California,
U.S.A.

J. CLARKE,
Materials and Molecular Research Division,
Lawrence Berkeley Laboratory,
Berkeley,
CA 94720,
U.S.A.

R.H.M. CLEVERS,
Department of Electrical Engineering,
University of Technology,
Eindhoven,
The Netherlands.

J. CLUZEL,
C.E.N.G./L.E.T.I./L.I.R. 85x,
38041 Grenoble,
France.

N. COUGH,
GEC Research Laboratories,
Hirst Research Center,
East Lane,
Wembley,
HA9 4PP Middlesex,
England.

R. CRISTIANO,
Istituto di Cibernetica del C.N.R.,
Via Toiano 6,
80072 Arcofelice,
Napoli,
Italy.

A. D'AMICO
I.E.S.S./C.N.R.,
Via Cineto Romano 42,
00156 Roma,
Italy.

J.R. DE BOER,
Department of Electrical Engineering,
University of Technology,
Eindhoven,
The Netherlands.

J. DE GOEDE,
Department of Physiology and
Physiological Physics,
State University of Leiden,
Leiden,
The Netherlands.

A.H. DE KUYPER,
Eindhoven University of Technology,
Department of Electrical Engineering,
P.O. Box 513,
5600 MB Eindhoven,
The Netherlands.

F.J. DE LA RUBIA,
Fisica Fundamental - UNED,
Apdo Correos 50, 487,
28080 Madrid,
Spain.

S. DEMOLDER,
Laboratory of Electronics,
Ghent State University,
Sint Pietersniewstraat 41,
B - 9000 Gent,
Belgique.

S. DE PANFILIS,
I.F.S.I.,
Frascati,
Roma,
Italy

M. DE MURCIA,
Centre d'Electronique de Montpellier,
Université des Sciences et Techniques
du Languedoc,
34060 Montpellier Cedex,
France.

R. DE WAMES,
Rockwell International,
1049 Camino Dos Rios, 5,
P.O. Box 1085,
Thousand Oaks,
CA 91360,
U.S.A.

P. DUTTA,
Department of Physics,
Northwestern University,
2145 Sheridan Road,
Evanston,
IL 60201,
U.S.A.

M.H. ERNST,
Institut for Theoretical Physics,
P.O. Box 80.006,
5308 TA Utrecht,
The Netherlands.

F. FIORILLO,
Istituto Elettrotecnico Nazionale
Galileo Ferraris,
Corso Massimo d'Azeglio 42,
10125 Torino,
Italy.

J.J. GAGNEPAIN,
Laboratoire de Physique et Mètrologie des
Oscillateurs du C.N.R.S.,
Université de Franche-Comté-Besancon,
32 Avenue de l'Observatoire,
25000 Besancon,
France.

D. GASQUET,
Centre d'Electronique de Montpellier,
Université des Sciences et Techniques
du Languedoc,
34060 Montpellier Cedex,
France.

Z.D. GENCHEV,
Institute of Electronics,
Bulgarian Academy of Sciences, 1784,
Sofia,
Bulgaria.

M. GENTILI,
I.E.S.S./CNR
Via Cineto Romano 42,
00156 Roma,
Italy.

J.P. GILLHAM,
A.G.I.O. Royal Signal and Radar Establishment,
Malvern,
WR14 3PS Worcestershire,
England.

P. GRUBER,
LGZ Landis and GYR Zug Inc.,
Central Laboratory,
CH - 6301 Zug,
Switzerland.

F. GRÜNEIS,
Consultant to Fraunhofer Institut für
Hydroakustik,
Waldparkstrasse 41,
8012 Ottobrunn,
F.R.G.

G. GUTIERREZ,
Fundacion Instituto de Ingenieria,
Apdo 40200,
Caracas 1040 - A,
Venezuela.

S.A. HAYAT,
Department of Physics,
University of Lancaster,
Lancaster LA1 4YB,
U.K.

P.H. HANDEL,
Physics Department,
University of Missouri,
St Louis,
MO 63121,
U.S.A.

S. HASHIGUSCHI,
Faculty of Engineering,
Yamanashi University,
Kofu,
Japan.

L.Z. HASSE,
Institute of Electronic Technology,
Technical University of Gdansk,
Gdansk,
Poland.

C. HEIDEN,
Institut für Angewandte Physik der
Justus-Liebig-Universität Giessen,
D - Giessen 6300,
F.R.G.

E.A. HENDRIKS,
Laboratory for Experimental Physics,
University of Utrecht,
Princetonplein 5,
3584 CC Utrecht,
The Netherlands.

F. HUET,
Physique des Liquides et Electrochimie,
Tour 22,
4 Place Jessieu,
75230 Paris Cedex 05,
France.

R.P. JINDAL,
AT&T Bell Laboratories,
Murray Hill,
NJ 07974,
U.S.A.

B.K. JONES,
Department of Physics,
University of Lancaster,
Lancaster LA1 4YB,
England.

A. JOULLIE,
Equipe de Microoptoélectronique de Montpellier,
U.S.T.L.,
34060 Montpellier Cedex,
France.

H.F.F. JOS,
Fysisch Laboratorium Rijksuniversiteit Utrecht,
Princetonplein 5,
3584 CC Utrecht,
The Netherlands.

R. JUNG,
Institut für Theoretische Physik,
Universität Stuttgart,
D - Stuttgart 80,
Germany.

K. KANDIAH,
AERE,
Building 347.3,
Harwell, Didcot,
OX11 ORA Oxfordshire,
England.

T. KAWAKUBO,
Department of Applied Physics,
Tokyo Institute of Technology,
Oh-Okayama Meguro,
Tokyo,
Japan.

L.B. KISS,
Jozsef Attila University,
Szeged,
Hungary.

W. KLEEN,
Denningerstrasse 36,
8 Munchen 80,
Germany.

T.G. KLEINPENNING,
Eindhoven University of Technology,
Eindhoven,
The Netherlands.

K.F. KNOTT,
University of Salford,
Salford,
M5 4WT,
England.

R. KOCH,
IBM Research Center,
Yorktown Heights,
NY 10598,
U.S.A.

A. KONCZAKOWSKA,
Institute of Electronic Technology,
Technical University of Gdansk,
Gdansk,
Poland.

A. KUSY,
Department of Electrical Engineering,
Technical University of Rzeszów,
Rzeszów,
Poland.

G. LECOY,
C.E.M., C.N.R.S. - UA 391,
Université des Science et Techniques
du Languedoc,
Montpellier Cedex,
France.

J.E. LLEBOT,
Department de Termologia,
Facultat de Ciences,
Universita Autonoma de Barcelona,
Bellaterra,
Spain.

D. LOVISOLO,
Dipartmento de Biologia Animale,
Laboratorio di Fisiologia Generale
Università di Torino,
10100 Torino,
Italy.

A. LUCCHESINI
I.E.S.S./CNR,
Via Cineto Romano 42,
00156 Roma,
Italy.

P. LUGLI,
Dipartimento di Fisica e Centro
Interuniversitario di Struttura della
Materia dell'Università di Modena,
Via Campi, 213/A,
41100 Modena,
Italy.

C. MACCONE,
Dipartmento di Matematica,
Politecnico di Torino,
Corso Duca degli Abruzzi, 24,
10129 Torino,
Italy.

A.J. MADENACH,
Max-Planck Institute,
Heisenbergstrasse 1,
D - Stuttgart 80,
F.R.G.

E.J.P. MAY,
Department of Engineering Science,
University of Exeter,
Exeter,
Devon EW4 4QF,
England.

P. MAZZETTI,
I.E.N.G.F.,
Corso Massimo d'Azeglio 42,
10125 Torino,
Italy.

S. MICCIANCIO,
Via Resuttana Colli, 352,
90146 Palermo,
Italy.

T. MIMAKI,
University of Electro Communications,
Chofu,
Tokyo 182,
Japan.

I. MODENA,
I.E.S.S./CNR,
Via Cineto Romano 42,
00156 Roma,
Italy.

F. MOSS,
University of Missouri at St. Louis,
Saint Louis,
MO 63121,
U.S.A.

T. MUNAKATA,
Institut für Angewandte Physik der Universität,
D - 6000 Frankfurt a.M.,
F.R.G.

J.L. MUNOZ-COBO,
Catedra de Fisica Universidad Politecnica de Valencia,
Campus Canino de Vera,
Valencia,
Spain.

T. MUSHA,
Tokyo Institute of Technology,
Nagatsuta,
Midoriku,
Yokohana 227,
Japan,

NASUASHIBI,
Thorn EMI,
Central Research Laboratories,
Dawney Road,
Hayes,
Middlesex UB3 IHH,
England.

B. NERI,
Istituto di Elettronica e Telecomunicazioni,
Università di Pisa,
Via Diotisalvi 2,
56100 Pisa,
Italy.

J.P. NOUGIER,
Centre d'Electronique de Montpellier,
Université des Sciences et Techniques du Languedoc,
34060 Montpellier Cedex,
France.

B. ORSAL,
C.E.M., C.N.R.S. - UA 391,
Université des Sciences et Techniques du Languedoc,
34060 Montpellier Cedex,
France.

G.V. PALLOTTINO,
Istituto Plasma Spazio, CNR,
Via Galilei,
00044 Frascati,
Italy.

G. PADOVINI,
Istituto di Fisica Politecnico,
Piazza L. da Vinci, 32,
20133 Milano,
Italy.

B. PELLEGRINI,
Istituto di Elettronica e Telecomunicazioni,
Università di Pisa,
Via Diotisalvi 2,
56100 Pisa,
Italy.

J. PELZ,
Department of Physics,
University of California,
Berkeley,
CA 94720,
U.S.A.

J.M. PERANSIN,
CEM/USTL,
Place E. Bataillon,
34060 Montpellier,
France.

G. PETROCCO,
I.E.S.S./CNR,
Via Cineto Romano 42,
00156 Roma,
Italy.

C. PICKUP,
CSIRO Division of Applied Physics,
P.O. Box 218,
Lindfield,
NSW 2070,
Australia.

M. PLANAT,
Laboratoire de Physique et Metrologie des Oscillateurs,
32 Avenue de l'Observatoire,
25000 Besancon,
France.

H. PRANCE,
School of Mathematical and Physical Sciences,
University of Sussex,
Brighton,
BN1 9QH,
England.

M. PRUDENZIATI,
Dipartimento di Fisica,
Via G. Campi, 214/A,
41100 Modena,
Italy.

L. REGGIANI,
Dipartimento di Fisica e Centro Interuniversitario di Struttura della Materia,
Via Campi, 213/A,
41100 Modena,
Italy.

R. SALETTI
Istituto di Elettronica e Telecomunicazioni,
Università di Pisa,
Via Diotisalvi 2,
56100 Pisa,
Italy.

M. SALVATI,
I.E.S.S./CNR,
Via Cineto Romano 42,
00156 Roma,
Italy.

J.F. SAUTERAU,
LAAS - CNRS,
7 Avenue Colonel Roche,
31400 Toulouse,
France.

M. SAVELLI,
C.E.M., C.N.R.S. - UA 391,
Université des Sciences et Techniques du Languedoc,
34060 Montpellier Cedex,
France.

A. SCAVENAC,
C.N.E.T.,
196 Rue de Paris,
92220 Bagneux,
France.

D.T. SMITH,
Clarendon Laboratory,
University of Oxford,
Oxford OX1 3PH,
England.

L. SPIRALSKI,
Institute of Electronic Technology,
Technical University of Gdansk,
Gdansk,
Poland.

A. STEPANESCU,
I.E.N.G.F.,
Corso Massimo d'Azeglio 42,
10125 Torino,
Italy.

L. STORM,
Westfälische Wilhelms-Universität,
Institut für Angewandte Physik,
Corrensstrasse 2/4,
4400 Münster/Westfalen,
B.R.D.

R. TETZLAFF,
Institute für Angewandte Physik der Universität,
Robert Mayer Strasse, 2 - 4,
6000 Frankfurt a.M.,
B.R.D.

A. TICHY-RACS,
Research Institute for Technical Physics HAS,
1325 Budapest,
Hungary.

P. TRIBOLET,
C.E.N.G./L.E.T.I./L.I.R. 85x,
38041 Grenoble,
France.

M. TRIPPE,
Department of Electrical Engineering,
University of Florida,
Gainesville,
FL 32611,
U.S.A.

M.J. UREN,
Royal Signals and Radar Establishment,
Great Malvern,
Worcestershire,
England.

M. VADACCHINO,
Dipartimento di Fisica,
Politecnico di Torino,
10100 Torino,
Italy.

M. VALENZA,
C.E.M., C.N.R.S. - UA 391,
Université des Sciences et Techniques du Languedoc,
34060 Montpellier Cedex,
France.

L.K.J. VANDAMME,
Department of Electrical Engineering,
University of Technology,
Eindhoven,
The Netherlands.

R.J. VAN DEN BERG,
Department of Physiology and Physiological Physics,
State University of Leiden,
Leiden,
The Netherlands.

A. VAN DER ZIEL,
Department of Electrical Engineering,
University of Florida,
Gainesville,
FL 32611,
U.S.A.

A.D. VAN RHEENEN,
Department of Electrical Engineering,
University of Florida,
Gainesville,
FL 32611,
U.S.A.

C.M. VAN VLIET,
E.E. and Nuclear Science Departments,
University of Florida,
Gainesville,
FL 32611,
U.S.A.

S. VITALE,
Dipartimento di Fisica,
Università di Trento,
38050 Povo,
Italy,

U. WEISS,
Institut fur Theoretische Physik,
Universitat Stuttgart,
D - 7000 Stuttgart 80,
Germany.

R.M. WESTERVELT,
Division of Applied Physics,
Harvard University,
Cambridge,
MA 02138,
U.S.A.

C.F. WHITESIDE,
Department of Electrical Engineering,
University of Florida,
Gainesville,
FL 32611,
U.S.A.

D. WOLF,
Institut für Angewandte Physik der Universität,
D - 6000 Frankfurt a.M.,
F.R.G.

R.J.J. ZIJLSTRA,
Fysisch Laboratorium Rijksuniversiteit Utrecht,
Princetonplein 5,
3584 CC Utrecht,
The Netherlands.

AUTHOR INDEX

SUBJECT INDEX